PROCEEDINGS OF THE SECOND INTERNATIONAL CONFERENCE ON
PRESS-IN ENGINEERING 2021, KOCHI, JAPAN

T0314466

Organized by
International Press-in Association

Subsidized by
THE MAEDA ENGINEERING FOUNDATION
THE KAJIMA FOUNDATION

Supported by
Asian Civil Engineering Coordinating Council
GIKEN LTD.
GIKEN SEKO CO., LTD.
International Society for Soil Mechanics and Geotechnical Engineering
Japan Construction Machinery and Construction Association
Japan Press-in Association
Japan Society of Civil Engineers
Japanese Technical Association for Steel Pipe Piles and Sheet Piles
Kami City
Kochi City
Konan City
Kochi Construction Industry Association
Kochi Construction Managing Engineers Association
Kochi Industrial Association
Kochi Institute of Invention and Innovation
Kochi Prefecture
Kochi Survey and Planning Association
Kochi University of Technology
National Institute of Technology, Kochi College
Shikoku Railway Company
The Kochi Shimbun
THE BANK OF KOCHI, LTD.
The Institution of Professional Engineers, Japan, Shikoku Regional Headquarters
The Japan Civil Engineering Consultants Association, Shikoku Branch
The Japanese Geotechnical Society
The Shikoku Bank, Ltd.
West Nippon Expressway Company Limited, Shikoku Branch

Proceedings of the Second International Conference on Press-in Engineering 2021, Kochi, Japan

Editors

Tatsunori Matsumoto
Kanazawa University, Japan

Katsutoshi Ueno
Tokushima University, Japan

Koichi Isobe
Hokkaido University, Japan

Hidetoshi Nishioka
Chuo University, Japan

Koji Watanabe
Aichi Institute of Technology, Japan

CRC Press
Taylor & Francis Group
Boca Raton London New York

CRC Press is an imprint of the
Taylor & Francis Group, an **informa** business

A BALKEMA BOOK

CRC Press/Balkema is an imprint of the Taylor & Francis Group, an informa business

© 2021 Copyright: International Press-in Association

Typeset by Integra Software Services Pvt. Ltd., Pondicherry, India

Library of Congress Cataloging-in-Publication Data

Applied for

Published for: International Press-in Association
          Address: 5F, Sanwa Konan Bldg, 2-4-3 Konan, Minato-ku,
          Tokyo 108-0075, Japan
          Tel: +81-3-5461-1191
          URL: https://www.press-in.org

ISBN: 978-1-032-10414-0 (Hbk)
ISBN: 978-1-032-10416-4 (Pbk)
ISBN: 978-1-003-21522-6 (eBook)
DOI: 10.1201/9781003215226

Proceedings of the Second International Conference on Press-in Engineering 2021, Kochi, Japan – Matsumoto et al (eds)
© 2021 Taylor & Francis Group, London, ISBN 978-1-032-10414-0

Table of contents

Session B: Infrastructure development

Session C: Disaster prevention and mitigation

Session D: Case histories

Proceedings of the Second International Conference on
Press-in Engineering 2021, Kochi, Japan – Matsumoto et al (eds)
© 2021 Taylor & Francis Group, London, ISBN 978-1-032-10414-0

Preface

This proceedings contains all the papers presented in the Second International Conference on Press-in Engineering (ICPE2021) held in Kochi, Japan, with on-line style during the period of June 19 to 20, 2021.

The conference was organized by International Press-in Association, IPA. IPA is an international organization of Press-in Engineering founded in 2007, with challenging aims of fusion among various engineering disciplines, such as geotechnical engineering, environmental engineering, mechanical engineering, monitoring engineering, data and information processing and other engineering disciplines.

Having experienced more than ten years' various activities, IPA held the First International Conference on Press-in Engineering in Kochi in 2018, collecting case histories from various parts of the world and presenting the State-of-the-Art Report in this unique discipline. IPA has kept a strong commitment to hold the second conference again in Kochi in 2021, the place where a piling machine called 'Silent Piler' was born in 1977. Since then, the family of Silent Piler has been playing a key role of Press-in Engineering up to the present time.

The main theme in ICPE2021 is "Evolution and Social Contribution of Press-in Engineering for Infrastructure Development, and Disaster Prevention and Mitigation".

The conference theme covers

1. Disaster prevention and mitigation, such as countermeasure against Tsunami, erosion control, slope protections and others
2. Renovation or development of infrastructure, such as road, railway, adjacent construction, rural area developments and others
3. Improvement of productivity and environmental issues, such as development of devices, construction managements, logistics, noise, vibration and others
4. Methods or case studies of project evaluation
5. Performance assessment of structures, such as pressed-in piles/sheet piles, and retaining wall/deep foundations constructed by pressed-in piles/sheet piles and others
6. Education
7. Others

During the two-day conference, there were two keynote lectures and three State-of-the-Art reports. All the materials presented in the above lectures are also included in the proceedings.

The conference would not have been possible without supports from various organizations and individuals. On behalf of the Organizing Committee, I would like to express my deepest gratitude to these organizations and individuals. The Organizing Committee has received supports from 27 organizations, locally and internationally, including International Society for Soil Mechanics and Geotechnical Engineering, Asian Civil Engineering Coordinating Council, Japan Society of Civil Engineers, The Japanese Geotechnical Society, The Institution of Professional Engineers, Japan (Shikoku Regional Headquarters), Kochi Construction Managing Engineers Association, Japanese Technical Association for Steel Pile Piles and Sheet Piles, Japan Press-in Association, Japan Construction Machinery and Construction Association,

The Japan Civil Engineering Consultants Association, Kochi Construction Industry Association, Kochi Industrial Association, Kochi Survey and Planning Association, Kochi Institute of Invention and Innovation, Kochi Prefecture, Kochi City, Kami City, Konan City, Kochi University of Technology, National Institute of Technology (Kochi College), Shikoku Railway Company, West Nippon Expressway Company Limited (Shikoku branch), The Shikoku Bank, Ltd., THE BANK OF KOCHI, LTD., The Kochi Shimbun, GIKEN LTD. and GIKEN SEKO CO., LTD. Funding has been committed from two organizations, which are, THE KAJIMA FOUNDATION and THE MAEDA ENGINEERING FOUNDATION.

My special thanks go to Mr. S. Hamada, the Governor of Kochi Prefecture, Prof. M. Isobe, the President of Kochi University of Technology, Prof. Chun Fai Leung, the President of IPA, Prof. Malcolm Bolton, the founding President of IPA, and Mr. A. Kitamura, the Executive Chairman of Giken Ltd.

As the Chair of the Organizing Committee, I would like to record my thanks to all the members of the International Advisory Board and the Organizing Committee for their dedicated efforts to make this event a great success. It is our sincere hope that this proceedings remains as a valuable source of reference on the Press-in Engineering for many years to come.

J. Matsumoto

Tatsunori Matsumoto
Chair of the Organizing Committee
Vice President of International Press-in Association

Proceedings of the Second International Conference on
Press-in Engineering 2021, Kochi, Japan – Matsumoto et al (eds)
© 2021 Taylor & Francis Group, London, ISBN 978-1-032-10414-0

International advisory board

Proceedings of the Second International Conference on
Press-in Engineering 2021, Kochi, Japan – Matsumoto et al (eds)
© 2021 Taylor & Francis Group, London, ISBN 978-1-032-10414-0

Organizing committee

CHAIR
Tatsunori Matsumoto (Japan)

VICE CHAIR
Jiro Takemura (Japan)

SECRETARY GENERAL
Hisanori Yaegashi (Japan)

DEPUTY SECRETARY GENERAL
Yukihiro Ishihara (Japan)

MEMBERS
Katsutoshi Ueno (Japan)
Kojiro Okabayashi (Japan)
Masahiro Ouchi (Japan)
Tetsuo Morita (Japan)

Yoshihiko Tani (Japan)
Hideki Hosoda (Japan)
Takao Kishida (Japan)
Yoshihisa Fujisaki (Japan)

Scientific working group

CHAIR
Katsutoshi Ueno (Japan)

CO-CHAIR
Kojiro Okabayashi (Japan)

MEMBERS
Yukihiro Ishihara (Japan)
Koichi Isobe (Japan)
Hidetoshi Nishioka (Japan)
Masahiro Ouchi (Japan)
Koji Watanabe (Japan)

Xi Xiong (Japan)
Tsunenobu Nozaki (Japan)
Nanase Ogawa (Japan)
Yumi Kocho (Japan)

Proceedings of the Second International Conference on
Press-in Engineering 2021, Kochi, Japan – Matsumoto et al (eds)
© 2021 Taylor & Francis Group, London, ISBN 978-1-032-10414-0

Administrative working group

CHAIR
Yoshihisa Fujisaki (Japan)

CO-CHAIR
Katsuhiko Tanouchi (Japan)

MEMBERS
Akihito Kawaguchi (Japan)
Riki Fukuhara (Japan)
Kimihiko Hayashi (Japan)
Tsunenobu Nozaki (Japan)

IPA Secretariat

SECRETARY GENERAL
Hisanori Yaegashi (Japan)

SECRETARY
Tsunenobu Nozaki (Japan)
Yuki Hirose (Japan)
Naoki Suzuki (Japan)
Hongjuan He (Japan)

Proceedings of the Second International Conference on
Press-in Engineering 2021, Kochi, Japan – Matsumoto et al (eds)
© 2021 Taylor & Francis Group, London, ISBN 978-1-032-10414-0

Review policy and list of reviewers

Originally 81 abstracts were submitted, and 60 technical papers for sessions A - E were accepted for publication in this proceedings, based on a careful review in terms of quality, reliability, originality and readability, basically by three members of a panel consisting of the following experts:

M. Bouassida (Tunisia)
M. Doubrovsky (Ukraine)
K. Fujiwara (Japan)
M.M. Futai (Brazil)
K. Gavin (Netherlands)
S.K. Haigh
(United Kingdom)
J. Hamada (Japan)
M. Hamada (Japan)
K. Horikoshi (Japan)
Y. Ishihara (Japan)
K. Isobe (Japan)
K. Itoh (Japan)

K. Kasama (Japan)
Y. Kikuchi (Japan)
P. Kitiyodom (Thailand)
M. Koda (Japan)
O. Kusakabe (Japan)
C.F. Leung (Singapore)
T. Matsumoto (Japan)
A. McNamara
(United Kingdom)
R. Motamed (USA)
S. Moriyasu (Japan)
K. Nakai (Japan)
H. Nishioka (Japan)

T. Nozaki (Japan)
K. Sawada (Japan)
Y. Sawamura (Japan)
E. Serino (Japan)
H. Suzuki (Japan)
S. Taenaka (Japan)
J. Takemura (Japan)
V.A. Tuan (Vietnam)
K. Ueno (Japan)
K. Watanabe (Japan)
X. Xiong (Japan)
N.A. Yusoff (Malaysia)

We appreciate all the reviewer's contributions through their constructive comments.

Invited lectures

Proceedings of the Second International Conference on
Press-in Engineering 2021, Kochi, Japan – Matsumoto et al (eds)
© 2021 Taylor & Francis Group, London, ISBN 978-1-032-10414-0

Design considerations for piles jacked or driven into strong soil or weak rock

(IPA 15th Anniversary Special Keynote Lecture)

M.F. Randolph
The University of Western Australia

ABSTRACT: Piles are often used in weathered soil profiles, perhaps needing to be embedded down to a hard, less weathered zone. The weathered material may vary from a strong soil to weak rock, often comprising a mix of partially weathered blocks embedded in a soil matrix. These are difficult materials for estimation of design parameters but may also cause problems during the installation of jacked or driven piles. The paper first reviews some of the design approaches for estimating engineering parameters such as shaft friction and end-bearing in these types of soil and weak rock. It then discusses the potential for pile tip damage during the installation process, presenting preliminary results from current doctoral research to assess conditions for pile tip damage more accurately.

1 INTRODUCTION

Although cast-in-situ piles such as rock sockets are the most common type of pile used in weak to moderate strength rock, jacked or driven piles also have a place, particularly in variably weathered profiles. There are, however, significant design challenges, ranging from characterization of the strength of weak rock and the associated design parameters to the risks of premature refusal on less weathered layer or of damage to the pile tip as it is penetrated through the layers.

Open-ended pipe piles are the most frequent pile type for coastal developments such as wharves and, increasingly, the large diameter monopiles used to support offshore wind turbines.

This paper reviews design approaches for jacked and driven piles in strong soil and soft rock, including characterization of the mass strength of the soil and correlations proposed in the literature for key engineering parameters such as shaft friction and end-bearing resistance. This material draws on the discussion in Randolph (2019).

In the second half of the paper, application of steel pipe piles for offshore wind turbines is considered, in particular the risk of tip damage as such relatively thin-walled piles are penetrated through the soil and rock profile. Current doctoral research by Juliano Nietiedt at the University of Western Australia has been addressing this problem with the objective of developing quantitative design guidelines for potential tip damage and the propagation of a mild dent into severe deformation.

2 CHARACTERIZATION OF ROCK

2.1 *Penetrometer correlations for weak rock*

In weak rock, it is possible to use conventional cone penetrometer tests (CPTs) in addition to taking samples and testing them in the laboratory. This provides a transition from strong soil to weak rock, allowing correlations to be developed as has been undertaken for calcareous sediments such as calcarenite (or weak limestone) and chalk.

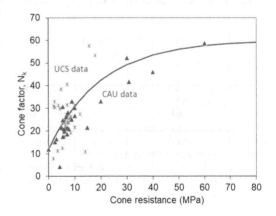

Figure 1. Correlation of cone factor with cone resistance.

Correlations relate the cone resistance q_{CPT} and rock shear strength ($s_u = q_{ucs}/2$) via a cone factor N_k,

DOI: 10.1201/9781003215226-63

just as for clays, where q_{ucs} is the unconfined compression strength or equivalent failure stress applied under triaxial conditions. Figure 1 shows one such correlation, where a reasonable fit to the data is obtained using an exponential increase in N_k with increasing cone resistance, according to

$$N_k = N_{k,\max} - \left(N_{k,\max} - N_{k,\min}\right)e^{-0.005 q_{CPT}/p_a} \quad (1)$$

taking $N_{k,\min}$ as 12 and $N_{k,\max}$ as 60 (with p_a standing for atmospheric pressure of 0.1 MPa).

The increase of N_k with cone resistance arises partly because of an increase in the ratio of rock mass strength to unconfined compression strength as the quality of the rock increases, as discussed below, but also because of increasingly partially drained conditions during penetration of the cone.

2.2 Mass rock properties for design

Characterizing the strength of and how it responds during failure (i.e. whether exhibiting brittle fracture or merely compressing into a rubble) is extremely challenging. Different failure modes will depend on the degree of confinement, and even on the progress of a particular load application, such as the response in bearing beneath the base of a pile.

The unconfined compression strength (UCS) test is one of the most common forms of laboratory strength measurement for rock, with the deviatoric strength at failure referred to here as q_{ucs}. The UCS may be 'extended' to a non-linear failure envelope using models such as that developed by Hoek and Brown (Hoek & Brown 1997, Hoek et al. 2002).

The Hoek-Brown failure envelope is expressed as

$$\sigma'_1 = \sigma'_3 + \sigma_{ci}\left(m_b \frac{\sigma'_3}{\sigma_{ci}} + s\right)^a \quad (2)$$

where σ_{ci} is the UCS for intact (undamaged) rock. In high quality rock this becomes equivalent to q_{ucs}, assuming that the laboratory unconfined strength test is carried out on intact, undamaged, rock. In general, however, σ_{ci} exceeds q_{ucs} due to the presence of internal weaknesses in the sample.

Hoek et al. (2002) correlated the various parameters in Eq. (2) to the geological strength index (GSI) according to

$$m_b = m_i \exp\left(\frac{GSI - 100}{28 - 14D}\right)$$

$$s = \exp\left(\frac{GSI - 100}{9 - 3D}\right) \quad (3)$$

$$a = 0.5 + \left[e^{-GSI/15} - e^{-100/15}\right]/6 \sim 0.5$$

where D is a damage factor (0 undamaged to 1 completely fractured) due to blasting or other form of excavation, and m_i is the material constant for the rock in question.

The geological strength index itself is a rather subjective parameter based on the joint spacing (RQD) and surface quality of the joint, varying between about 30 (below which the rock is essentially rubble) and 100.

Essentially Eq. (3) provides adjustment factors to account for the properties of the jointed rock mass relative to intact blocks of rock. For pile construction, with relatively low damage and often in softer, e.g. carbonaceous, rocks, D may be assumed to be close to zero, and m_i in the range 3 to 10.

From the Hoek-Brown relationship in Eq. (2), the equivalent unconfined strength q_{ucs} and (bilateral) tensile strength q_t for the rock mass may be deduced by setting σ'_3 to zero to obtain $q_{ucs} = \sigma_{ci}s^a$ and setting $\sigma'_1 = \sigma'_3 = -q_t$ to obtain $q_t = s\sigma_{ci}/m_b$. Hence the ratio of tensile to compressive strength is about $s^{0.5}/mb$. This ratio is often estimated as 0.1 for soft rock, although lower ratios are obtained for different combinations of s and m_b.

For pile design, where the rock is confined, it may be appropriate to consider what Hoek and Brown referred to as a 'global' rock mass strength σ_{cm}, allowing for confinement by stresses in the range from zero to $\sigma_{ci}/4$, expressed as

$$q_{cm} = \sigma_{ci} \frac{m_b + 4s - a(m_b - 8s)}{2(1+a)(2+a)(m_b/4 + s)^{1-a}} \quad (4)$$

The other important design parameter for rock is its modulus, generally expressed in terms of the Young's modulus. Liang et al. (2009) proposed a correlation (based on Bieniawski 1978), with the rock mass modulus E_m related to that of intact rock (E_i as measured in an unconfined compression test on an assumed undamaged sample) and the GSI by

$$E_m = \frac{e^{GSI/21.7}}{100} E_i \quad (5)$$

These various relationships are illustrated in Figure 2 for the case of zero damage ($D = 0$). It may be seen that the unconfined compressive strength ratio decreases exponentially, by a factor of 10 for each reduction in GSI by ~40, whereas the global mass strength ratio shows rather higher values for low and intermediate values of GSI.

Pile design parameters for soft rock are often correlated against the q_{ucs} so it is relevant to consider the ratio q_{ucs}/q_{cm}, which is the ratio of the unconfined compression strength measured in the laboratory to the in situ rock mass strength. That ratio is plotted in Figure 3, showing that the true mass strength of rock increases to double the unconfined compression strength for GSI of 50%.

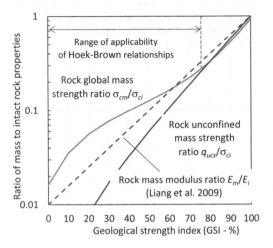

Figure 2. Ratios of mass to intact rock properties.

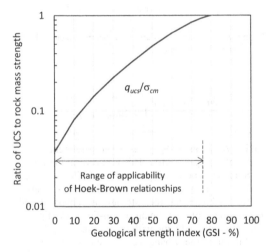

Figure 3. Ratios of mass to intact rock properties.

3 AXIAL PILE DESIGN PARAMETERS

3.1 *Construction effects*

Although the focus here is on jacked or driven piles, it is worth considering how design parameters might differ from the more common (in rock) construction technique of cast-in-situ rock sockets or grouted piles.

For the base resistance, mobilization of end-bearing capacity will require smaller displacements for jacked or driven piles (particularly the former) compared with a cast-in-situ pile. Figure 4 shows a bi-directional load cell developed by the Bolivian company Incotec, which can be lowered on the reinforcing cage prior to casting the pile. After curing of the pile the cell can be expanded to pre-load the pile base before it is finally grouted solid.

In principle, the reverse is likely to be true for the shaft resistance, with cast-in-situ piles offering higher shaft resistance due to interlocking effects, and hence dilation induced enhancement of the lateral effective stress as the pile is loaded, for cast-in-situ piles.

Figure 4. Expanded base pile construction.

Figure 5. Underreaming tool developed for Pluto project (https://www.lddrill.com/our-projects/pluto-jacket-installation/).

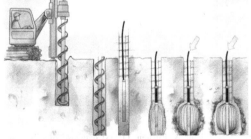

Figure 6. Expanded base pile construction.

5

As is discussed further below, in very weak rock (or strong soils) or very strong rock, the degree of interlocking for cast-in-situ piles may be small, requiring additional construction techniques in order to maximize shaft friction. Examples are shown in (a) Figure 5, where an underreaming tool was used to create grooves in the pile shaft for grouted insert piles constructed in high-quality limestone; and (b) Figure 6, which shows expandable 'lanterns' developed by Incotec and the construction process to expand them following casting of a pile, thus creating an elliptical expansion and enhanced load transfer between pile and soil. The concept is similar to the technology used for expanded base piles, patented in 1982 (Massarsch 2019).

Although the expanded base and shaft technology has generally been used in uncemented silts and sandy deposits, it would prove equally effective in ensuring robust load transfer for piles socketed into weak rock.

3.2 Base resistance

An extensive review of design approaches for the end-bearing resistance of piles in rock was given by Zhang & Einstein (1998). They documented a database of field measurements on piles from 0.3 m to 1.9 m diameter, for reported q_{ucs} values spanning from 0.6 MPa to 55 MPa. They proposed a non-linear fit to the data, expressed in non-dimensional form as

$$q_b = 15\sqrt{q_{ucs}p_a} \text{ or } \frac{q_b}{q_{ucs}} = \frac{15}{\sqrt{q_{ucs}/p_a}} \quad (6)$$

where p_a is atmospheric pressure (0.1 MPa).

The above expression is shown with the database in Figure 7, together with what may be regarded as an upper bound design approach of $q_b/q_{ucs} = 5$.

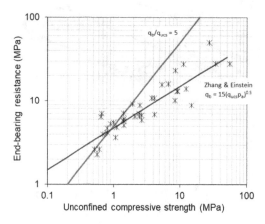

Figure 7. Zhang & Einstein (1998) database of pile base resistance in rock.

As indicated by (6), the Zhang & Einstein approach leads to very low 'bearing factors' in strong rock, which fall below unity for q_{ucs} values exceeding ~20 MPa. This does not seem reasonable physically, and may reflect vagaries in the UCS data, or limited base displacements in some cases.

An alternative hyperbolic variation of bearing factor $N_b = q_b/q_{ucs}$ may be expressed as

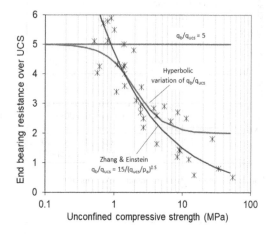

Figure 8. Hyperbolic variation of bearing factor shown with Zhang & Einstein (1998) database.

$$N_b = N_{b,\min} + \frac{(N_{b,\max} - N_{b,\min})}{1 + (q_{ucs}/q_{ucs,ref})^2} \quad (7)$$

taking $N_{b,\min}$ as 2, $N_{b,\max}$ as 5 and a reference UCS value of $q_{ucs,ref} = 2$ MPa. This is shown in Figure 8.

Since the end-bearing response of piles is closely linked to cavity expansion, the reduction in bearing factor for increasing rock strength is consistent with the stiffness of rock being a non-linear multiple of rock strength, reducing with increasing strength.

Although the Zhang & Einstein database was developed for cast-in-situ piles, it is also applicable to jacked or driven piles. Assessment of the rock quality and strength is critical. As a guide, Stevens et al. (1982) comment that refusal of driven piles will generally occur in rock with UCS exceeding 5 MPa. Also, if piles are to be driven a significant distance into the rock, consideration should be given to potential tip damage and premature refusal. Careful monitoring of hammer energy and blowcount is needed, ideally with stress-wave monitoring as well, in order to minimize risk of over-stressing the pile tip (Wiltsie et al. 1985). Also, as discussed later, the potential for hard inclusions such as boulders or unweathered clumps of rock to dent the tip of open-ended pile needs to be assessed.

3.3 Shaft resistance

The shaft friction of cast-in-situ piles relies heavily on interlocking between the pile and surrounding rock (Seidel & Haberfield 1995). The resulting magnitude of shaft friction is a function of the height and roughness angle of the rock asperities created by drilling, which tend to reach a maximum in rock of intermediate strength (q_{ucs} of ~2 MPa) but reduce significantly for high strength rock (Figure 9a).

This leads to a relationship for the ratio of shaft friction to the rock shear strength expressed as

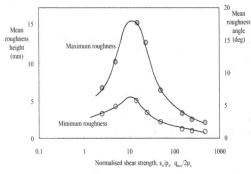

(a)   Assumption regarding borehole roughness

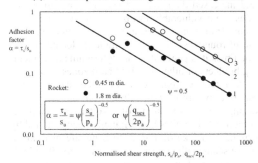

(b)   Shaft friction ratios as function of strength

Figure 9. Predictions of rock-socket shaft friction from ROCKET (Seidel & Haberfield 1995).

$$\frac{\tau_s}{(q_{ucs}/2)} = \frac{\psi}{\sqrt{q_{ucs}/2p_a}} \text{ or } \tau_s = \psi\sqrt{\frac{q_{ucs}}{2}p_a} \qquad (8)$$

illustrated in Figure 9b. As evident from this figure, there is also an effect of the pile diameter (assuming that the rock asperity heights are independent of diameter), with reducing effect of dilation as the diameter increases.

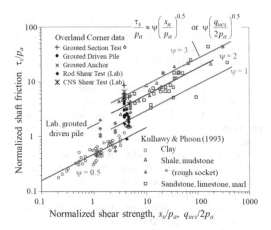

Figure 10. Variation of shaft friction with rock shear strength (extended from Kulhawy & Phoon 1993).

The power law variation of τ_s/q_{ucs} with rock strength is consistent with the data assembled by Kulhawy & Phoon (1993) shown in Figure 10, supplemented by data from field tests (Randolph et al. 1996). The bulk of the data for rock follow the trend line for $\psi = 2$, which is equivalent to $\tau_s = 0.45q_{ucs}^{0.5}$, using units of MPa.

In weak rock, where cone penetration testing is possible, the shaft friction for cast-in-situ piles may also be correlated with the cone resistance. Consistent with the trend for τ_s/q_{ucs}, the ratio of shaft friction to cone resistance reduces with increasing q_{cone}, with a trend of (Randolph et al. 1996)

$$\frac{\tau_s}{q_{cone}} \approx 0.02 + 0.2e^{-0.04q_{cone}/p_a} \qquad (9)$$

In most cases the lower limit of τ_s of about 2% of the cone resistance will apply.

The mineralogy of a given rock may be expected to have a much greater effect on the shaft friction for jacked or driven piles than for cast-in-situ piles. For example, in carbonate material the process of pile driving creates an annulus of completely destructured material adjacent to the pile, which leads to very low shaft friction for piles driven into limestone and also, prior to set up, in chalk. Typical design values of shaft friction for weak limestone are in the range 5 to 15 kPa unless site specific test data are available that can justify higher values. Indeed, higher values of shaft friction are appropriate for some of the carbonate soils in the Middle East, particularly where calcium carbonate content falls below 70 % (Thomas et al. 2010).

In chalk, however, although shaft friction during pile installation is similarly low, with destructured 'puttied' chalk adjacent to the pile, relatively strong setup is observed, potentially by a factor of five or more (Dürhkop et al. 2015; Ciavaglia et al. 2017; Buckley et al. 2018). During installation, shaft friction is typically around 15 to 25 kPa apart from near the pile tip where values as high as 200 kPa may be reached. The long-term shaft friction after set up may reach 150 kPa or more, but for design purposes is often limited to 100 kPa (Augustesen et al. 2015).

For non-carbonate rocks such as mudstones, much higher values of shaft may be achieved, although there are rather limited data from full scale tests. Design approaches vary from treating the rock as a strong clay, following American Petroleum Institute design guidelines (API 2011), or using empirical correlations based on test data. For the former, and assuming strength ratios (s_u/σ'_{vo}) exceeding unity, the shaft friction would vary as

$$\tau_s = 0.5\sigma'^{0.25}_{v0}(q_{ucs}/2)^{0.75} \qquad (10)$$

Terente et al. (2017) showed that this can be very conservative, based on shaft friction values deduced from dynamic load tests, which were well in excess of 500 kPa at depths of 15-25 m below seabed in rock with q_{ucs} values of 1-2 MPa. The shaft friction values also tended to increase with depth, suggesting that an effective stress approach based on estimated in situ horizontal effective stresses might be more appropriate.

In summary, shaft friction for piles jacked or driven into rock varies significantly with the mineralogy of the rock and the extent to which (a) it is destructured and undergoes compaction during pile installation, and (b) physiochemical processes that may lead to time-dependent recovery of structure and strength. Data from full-scale testing, including monitoring of piles during installation, is needed in order to reduce the level of conservatism necessary in its absence.

4 TRENDS IN OFFSHORE WIND INDUSTRY AND PILE TIP DAMAGE

4.1 Introduction

The offshore wind industry has relied heavily on the use of large diameter so-called 'monopiles' to support offshore wind turbines. To date, developments have been concentrated in Europe, in the heavily glaciated dense sands and overconsolidated clays such as found in the North Sea. As the industry now expands into North America and Asia, more varied and less competent sediments are encountered, which have increased design challenges.

Typical monopile diameters have increased gradually over the last two decades, in response to the increase in generating power and size of the wind turbines (see Figure 11). Diameters have now reached 8 to 10 m, with diameter to wall thickness (D/t) ratios generally exceeding 80 and sometimes greater than 110.

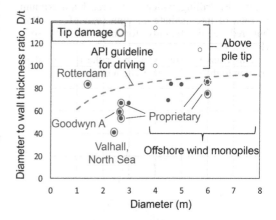

Figure 11. Trends for diameter and D/t in offshore wind industry and incidents of pile tip damage.

The combination of more challenging sediments and increased diameter and D/t ratios has made the piles more vulnerable to tip damage. This may originate from minor fabrication imperfections such as out of roundness, or from asymmetric loading of the pile tip as it is penetrated through sediments containing boulders or other localized hard zones.

A rather extreme case of such damage was reported by Broos et al. (2017) from piles extracted during the expansion of Rotterdam Harbour (Figure 12. The piles had originally been installed by driving at a rake of 1 in 5, to penetrate medium to dense sands with cone resistance of 25 to 40 MPa. It is possible that the raking angle (around 11 degrees) contributed to the distortion, since the pile tip would have encountered any stronger stratum at one edge.

Figure 12. Pile tip damage from Rotterdam harbour.

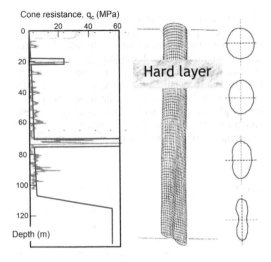

Figure 13. Soil strength profile and measured growth in pile distortion at Goodwyn.

Although initial damage to a pile tip may be relatively minor, once the pile tip cross-section is no longer circular, either because of an originally elliptical shape or a minor dent, the soil can act as a dye, with the pile 'extruding' through the soil following the current shape of the tip. This type of failure occurred for Woodside's Goodwyn platform on the North-West shelf of Australia as a result of driving 2.65 m diameter piles through a 5 m thick layer of strong calcarenite (cone resistance estimated as ~80 MPa), as shown in Figure 13 (Erbrich et al. 2010).

The process of extrusion buckling is very challenging numerically because of the extreme geometric and material non-linearities. Barbour & Erbrich (1995) developed an ABAQUS-based analysis procedure called BASIL to address this, using a system of pile-soil interaction springs attached at nodes to the 3-dimensional model of the pile, avoiding the need to model the soil continuum explicitly. An example outcome of a BASIL analysis is shown in Figure 14.

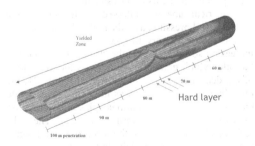

Figure 14. Example pattern of plastic strains for a pile pushed 25 m beyond the hard layer.

Doctoral research being undertaken at UWA by Juliano Nietiedt has resulted in quantitative guidelines to estimate the potential for (a) inelastic denting of a pile tip as a result of boulder impact, and (b) propagation and growth of a small dent by extrusion buckling. The work combines data from centrifuge model tests (Nietiedt et al. 2020) and numerical modelling. The results presented below are preliminary findings that are in the process of being submitted as journal publications (Nietiedt et al. in prep a,b,c).

4.2 Pile tip damage from boulder impact

Pile tip damage can be initiated by contact with a sloping hard layer or a localized zone of strong material such as a boulder or unweathered rock, as illustrated in Figure 15. Large deformation three dimensional (3D) finite element analyses were undertaken in which an embedded was forced downward under an imposed vertical displacement, meanwhile being allowed to rotate freely. The ellipsoidal boulder was characterized by the long and short axis dimensions, $a_{boulder}$ (or a_b) and $b_{boulder}$ (or b_b) and the contact point by an eccentricity e_b.

Figure 16 Shows typical vertical and horizontal reaction curves for the worst value of eccentricity $e_b/a_b = 0.3$. The reactions are normalized by the cone resistance established by separate numerical analyses of cone penetration and also by the maximum cross-sectional area of the boulder A_{pi}.

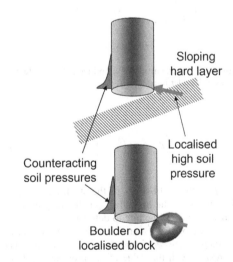

Figure 15. Initiation of damage by localized contact with hard material.

The maximum horizontal reaction force can be compared with the horizontal force F_p required to cause plastic deformation of the pile tip. This force may be expressed as

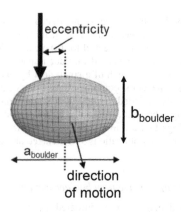

Figure 16. Simplified modelling of boulder impact.

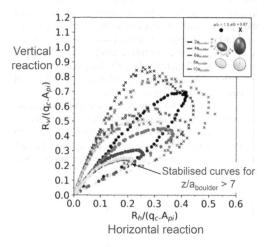

Figure 17. Reaction curves for simplified 3D finite element analyses of boulder impact.

$$\frac{F_p}{f_y t^2} \approx 0.25 \left(\frac{s_{load}}{t}\right)^{0.4} \qquad (11)$$

where f_y is the yield stress of the steel, t is the pile wall thickness at the tip and s_{load} is the length of the line load applied to the pile to deform it.

The value of $F_p/f_y t^2$ varies between 0.5 and 1.2 as s_{load}/t ranges between 5 and 40. The coefficient may be compared with the value of 1.2 proposed by HSE (2001).

4.3 Conditions for extrusion buckling

Conditions for extrusion buckling may be assessed by considering the forces imposed on the pile as it advances into the soil. Consider first an idealized slightly ellipsoidal deformation of the pile tip, leading to a trumpet shape on one vertical plane through the

pile axis, and a slight inward taper on the orthogonal plane (see Figure 18). As the pile advances, external lateral pressure will build up on the outside of the tapered profile and, to a lesser degree, internal lateral stresses will increase within the trumpet profile. The key geometric detail is the angle that the inwardly tapered wall makes with the pile axis. The greater that angle, the larger will be the buildup of lateral soil pressure – essentially a feedback process that encourages further growth of the pile deformations.

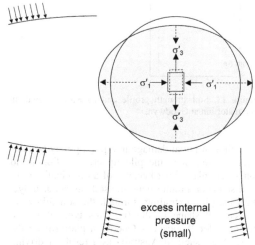

Figure 18. Feedback loop of stress changes due to elliptical deformation of the pile tip.

Laboratory testing of thin-walled piles, with D/t of 50 and 100, allowed the shape of asymmetric dents at the pile tip to be parameterized. Figure 19 shows how a dent may be quantified by the maximum departure δ from the original circular shape, the chord width w and the length s_{dent} along the length of the pile. The figure also indicates the expected gradual buildup of external soil pressure as the pile is penetrated. This starts from zero at the pile tip as the pile cuts into fresh soil.

Numerical analysis allowed evaluation of the resulting soil pressure, as indicated in Figure 20.

It is then necessary to quantify the effect of the increasing soil pressure on the structural response of the pile, which gradually weakens as the magnitude of the dent increases. As indicated in Figure 21, the reloading stiffness reduces as the dent grows, while the force F to cause additional plastic deformation increases but trends towards a plateau. The subscripted F_p and δ_p refer to the force and dent magnitudes that first cause plasticity in the pile.

Eventually, the results of boulder impact and conditions for extrusion buckling may be combined to allow estimation of the overall pile response. If the pile tip impacts a sufficiently large boulder at only

10

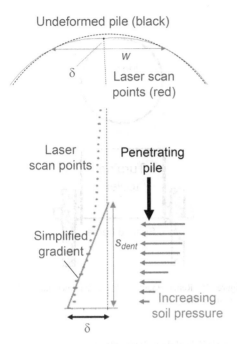

Undeformed pile (black)

Laser scan points (red)

δ

w

Laser scan points

Penetrating pile

Simplified gradient

S_{dent}

Increasing soil pressure

δ

Figure 19. Dent shape and indicative soil pressures.

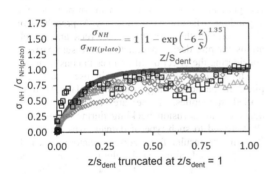

$$\frac{\sigma_{NH}}{\sigma_{NH(plato)}} = 1\left[1 - \exp\left(-6\frac{z}{s}\right)^{1.35}\right]$$

z/s_{dent}

z/s_{dent} truncated at $z/s_{dent} = 1$

Figure 20. Evolution of external soil stress along the length of the pile.

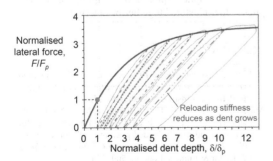

Normalised lateral force, F/F_p

Reloading stiffness reduces as dent grows

Normalised dent depth, δ/δ_p

Figure 21. Evolving structural response of the pile.

Large dent followed by tip crumpling

0.34 D_{pile}

0.32 D_{pile}

Small dent followed by extrusion buckling

1.1 D_{pile}

0.83 D_{pile}

$\delta/D_{pile} = 18\%$

$\delta/D_{pile} = 16\%$

Figure 22. Comparison of pile shapes from centrifuge model tests (left) and 3D finite element analyses.

a small eccentricity, the impact force is likely to exceed significantly the force to cause plasticity (assuming that the boulder is embedded in reasonably strong soil). In that case a large dent will eventuate, as evidenced from both physical and numerical modelling (upper part of Figure 22).

Alternatively, if the pile tip hits a boulder near the edge, or the surrounding soil is less strong, only a small (or no) dent will eventuate. However, that dent may grow through extrusion buckling as the pile is penetrated further (lower part of Figure 22).

Although the predictive framework is still being developed (Nietiedt et al., in prep. a,b,c), preliminary comparisons with experimental and numerical results shows reasonably good agreement, as shown in Figure 23.

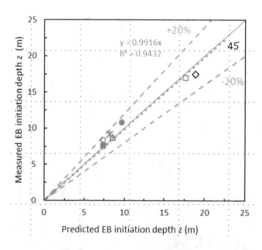

Figure 23. Comparison of predicted and measured depths to initiate extrusion buckling.

4.4 *Alternative approaches for OWF foundations*

Alternative foundation types are being explored by the wind industry for conditions that are not conducive to large diameter monopiles. The most common alternative is the use of a jacket structure to support the wind turbine, with the structure itself supported on smaller diameter driven piles or suction caissons.

Recently, one of the major offshore installers (Heerema) publicized alternative 'silent' foundation approaches including screw piles and hydraulically jacked piles using a similar approach to Giken's Silent Piler. The Heerema technique is illustrated in Figure 24.

Figure 24. Heerema's silent piling technique using small group of jacked piles.

This development raises the issue of other substitute arrangements for large diameter monopiles, such as that illustrated in Figure 25. The monopile is replaced by a ring of smaller piles, grouted (after installation) into a transfer template that links the turbine tower to the pile group. Such an arrangement, which requires rather less steel than the monopile for a given rotational stiffness, lends itself to push-in technology.

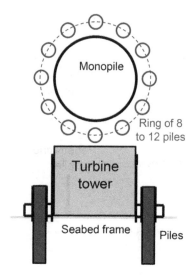

Figure 25. Replacement of large diameter monopile by ring of smaller piles.

5 CONCLUDING REMARKS

This paper has explored design challenges for jacked and driven piles in strong soil and weak rock, both for general application and with respect to the expanding offshore wind industry. Approaches for characterizing the strength of soft rock and correlations for pile design parameters were summarized, considering the effect of differences in construction.

The offshore industry routinely uses steel pipe piles, in particular large diameter relatively thin-walled monopiles. These are vulnerable to tip damage and extrusion buckling during installation. Background to such type of damage was discussed. Preliminary results from current doctoral research were presented, aimed at developing a quantitative framework to predict the potential for tip damage from either boulder damage or extrusion buckling of an out-of-cylindrical thin-walled pile.

Alternative foundation concepts that might lend themselves to 'silent piling' were discussed briefly, with a monopile replaced by a group of smaller piles, either clustered together or in a ring.

REFERENCES

API 2011. *Recommended Practice 2GEO: Geotechnical and Foundation Design Considerations, 1st*Ed. American Petroleum Institute, Washington, DC.

Augustesen, A.H., Leth, C.T., Østergaard, M.U., Møller, M., Dührkop, J. & Barbosa, P. 2015. Design methodology for cyclically and axially loaded piles in chalk for Wikinger OWF. *Proc. Int. Symp. Frontiers in Offshore Geotechnics III*. Taylor & Francis, London, 509–514.

Barbour, R. & Erbrich, C.T. 1995. Analysis of soil skirt interaction during installation of bucket foundations using ABAQUS. *Proc. Of ABAQUS Users Conference*, Paris.

Bieniawski, Z.T. 1978. Determining rock mass deformability: Experience from case histories. *Int. J. Rock Mech. Min. Sci. Geomech. Abstr.*, 15:237–248.

Broos, E., Sibbes, R. & de Gijt J. 2017. Widening a harbor basin, demolition of a deep see quay wall in Rotterdam. *Proc. of COME2017: Decommissioning of Offshore Geotechnical Structures*, Hamburg, Germany, 251–262.

Buckley, R.M., Jardine, R.J., Kontoe, S., Parker, D. & Schroeder, F.C. (2018). Ageing and cyclic behaviour of axially loaded piles driven in chalk. *Géotechnique*, 68 (2),146–161.

Ciavaglia, F., Carey, J. & Diambra, A. 2017. Time-dependent uplift capacity of driven piles in low-to-medium-density chalk. *Géotechnique Letters*, 7(1): 1–7.

Dührkop, J., Augustensen, A.H. & Barbosa, P. 2015. Cyclic pile load tests combined with laboratory test results to design offshore wind turbine foundations in chalk. *Proc. Int. Symp. Frontiers in Offshore Geotechnics III*. Taylor & Francis, London, 533–538.

Erbrich, C.T., Barbosa-Cruz, E. & Barbour, R. 2010. Soil-pile interaction during extrusion of an initially deformed pile. *Proc. of 2nd International Symposium on Offshore Geotechnics*, Perth, 489–494, Taylor & Francis, London, Australia.

Hoek, E. & Brown, E.T. 1997. Practical estimates of rock mass strength. *Int. J. Rock Mech. Min. Sci.* 34(8): 1165–86.

Hoek, E., Carranza-Torres, C.T. & Corkum, B. 2002. Hoek-Brown failure criterion - 2002 edition. *Proc. 5th North American Rock Mech. Symp.*, Toronto, Canada.

HSE 2001. *A Study of Pile Fatigue During Driving and In-Service and of Pile Tip Integrity*. Offshore Technology Report 2001/018, Health and Safety Executive, UK, published by Her Majesty's Stationery Office, Norwich.

Kulhawy, F.H. & Phoon, K.K. 1993. Drilled shaft side resistance in clay soil to rock. *Geotech. Spec. Pub. No. 38, Design and Performance of Deep Foundations*, ASCE, New York, 172–183.

Liang, R., Yang, K. & Nusairat, J. 2009. P-y criterion for rock mass. *J. Geotech. & Geoenviron. Eng.*, ASCE, 135 (1): 26–36.

Massarsch, K.R. 2019. The evolution of the expander body concept and future applications. Keynote Lecture, *Proc. 4th Int. Congress of Deep Foundations of Bolivia*. [https://www.cfpb4.com/]

Nietiedt, J.A., Randolph, M.F., Gaudin, C., Doherty, J.P., Kallehave, D., Gengenbach, J. & Shonberg, A. 2020. Physical modelling of pile tip damage arising from impact driving. *Proc. 4th Int. Symp. on Frontiers in Offshore* Geotechnics, *ISFOG2020*, Austin, Deep Foundations Institute, Hawthorne, USA, 787–797.

Nietiedt, J.A., Randolph, M.F., Gaudin, C. & Doherty, J.P. In prep. a. Numerical assessment of tip damage during pile installation in boulder rich soils. In preparation.

Nietiedt J.A., Randolph M.F., Gaudin C. Doherty J.P. In prep. b. Centrifuge model tests investigating initiation and propagation of pile tip damage during driving. In preparation.

Nietiedt J.A. Randolph M.F., Gaudin C. & Doherty J.P. In prep. c. Calculation approach for initiation of extrusion buckling of steel pipe piles. In preparation.

Randolph, M.F. 2019. Considerations in the design of piles in soft rock. Keynote paper. *Proc. Int. Conf. on Geotechnics for Sustainable Infrastructure Development*, Hanoi.

Randolph, M.F., Joer, H.A., Khorshid, M.S. & Hyden, A.M. 1996. Field and laboratory data from pile load tests in calcareous soil. *Proc. 28th Annual Offshore Tech. Conf.*, Houston, OTC 7992, 1: 327–336.

Seidel, J. & Haberfield, C.M. 1995. The axial capacity of pile sockets in rocks and hard soils, *Ground Engineering*, 28(2): 33–38.

Stevens, R.S., Wiltsie, E.A. & Turton, T.H. 1982. Evaluating pile drivability for hard clay, very dense sand, and rock. *Proc. 14th Annual Offshore Tech. Conf.*, Houston, OTC 4205:465–469.

Terente, V., Torres, I., Irvine & Jaeck, C. 2017. Driven pile design method for weak rock. *Proc. 8th Int. Conf. Offshore Site Investigation and Geotech.*, Society for Underwater Technology, London, 1: 652–657.

Thomas, J., van den Berg, M., Chow, F. & Maas, N. 2010. Behaviour of driven tubular steel piles in calcarenite for a marine jetty in Fujairah, United Arab Emirates. *Proc. Int. Symp. on Frontiers in Offshore Geotechnics*, Perth, Taylor & Francis, London, 549–554.

Wiltsie, E.A., Stevens, R.F. & Vines, W.R. 1985. Pile installation acceptance in strong soils. *Proc. 2nd Int. Conf. on Application of Stress Wave Theory on Piles*, Balkema, Stockholm, 72–78.

Zhang, L. & Einstein, H.H. 1998. End bearing capacity of drilled shafts in rock, *J. Geot. & Geoenviron. Eng.*, ASCE, 124(7): 574–584.

Proceedings of the Second International Conference on Press-in Engineering 2021, Kochi, Japan – Matsumoto et al (eds)
© 2021 Taylor & Francis Group, London, ISBN 978-1-032-10414-0

Research and development for infrastructure maintenance, renovation, and management

Y. Fujino
Institute of Advanced Sciences, Yokohama National University, Yokohama, Japan
Josai University, Saitama, Japan

D.M. Siringoringo
Institute of Advanced Sciences, Yokohama National University, Yokohama, Japan

ABSTRACT: The collapse of Sasago Tunnel, located on the Chuo Expressway about 80 kilometers west of Tokyo in 2012, has led to questions about the current quality and safety of infrastructure, and immediately brought public attention to the issue of infrastructure degradation in Japan. Japanese government decided to invest on the research and development for efficient management of infrastructure through implementation of science and advanced technologies. The new research and development (R & D) program named "Infrastructure maintenance, renovation and management" is started in 2014 under the Council of Science, Technology and Innovation (CSTI)'s Strategic Innovation Program (SIP). The 5-year program covers various subjects of infrastructure maintenance with the key technologies in condition assessment using advanced technologies such as non-destructive testing, monitoring and robotics, long-term performance prediction of infrastructure, development of durable high-quality of material for repair and replacement, and data management of bridges and other infrastructure using advanced information and communication technologies (ICT). In this paper, outline and outcomes of this program are explained.

1 INTRODUCTION

Economic sustainability, security and well-being of a nation depend heavily on the reliable functioning of infrastructures typically such as roads, highways, and bridges. Civil and transportation infrastructures in Japan need to cope with not only the vast territorial challenging landscape and large population, but also natural disasters, such as earthquakes, tsunamis, and typhoons. They have suffered severe damage from natural disasters, so that the design specifications have been regularly revised to provide more resilient infrastructure (Fujino and Siringoringo 2008; Fujino et al. 2016; Fujino, 2018). In addition, large amount of infrastructure stocks built during the peak of economic periods have now started to enter the last periods of their service. Significant portion of civil and transportation infrastructures such as bridges and tunnels are now deteriorating, and they require extensive efforts for maintenance. As a tool for infrastructure condition assessment, structural health monitoring (SHM) is becoming more and more used in civil engineering infrastructures and they have been implemented for infrastructure assessment in many parts of the world, including Japan (Boller et al. 2009).

On December 2nd, 2012, suspended roof for air ventilation in Sasago Tunnel of the Chuo Expressway near Tokyo collapsed onto traffic causing nine deaths. More casualties were avoided even though the collapse roof was over 130 m length because the traffic volume was low at the time of accident. This is the first maintenance-related accident with involvement of human loss. The accident has attracted public attention to the issue of infrastructure degradation and led to strong doubts about current quality and safety of infrastructures. Realizing this condition, Japanese government decided to invest on research and development for efficient infrastructure management. A new research and development program named "Infrastructure maintenance, renovation and management" was launched in 2013 under the Japanese Council of Science, Technology and Innovation (CSTI)'s Cross-ministerial Strategic Innovation Promotion Program (SIP). The 5-year program covers various subjects with key technologies in condition assessment, non-destructive testing, monitoring, and robotics; long-term performance prediction, development of high-quality durable material for repair and replacement, and infrastructure management using advanced information and communication technologies (ICT). The program consists of over 60 research

DOI: 10.1201/9781003215226-1

projects involving universities, research institutes and industries. This initiative is expected to prevent further accidents and setting an example for efficient infrastructure maintenance by reducing the burden of maintenance works and cost (SIP, 2014a).

Five research and development fields were set up under this program with the budget of about 30 million US$ allocated to all themes every year since 2014. The five research and development fields programs are as follows (Figure 1).

1. Inspection, diagnosis, and monitoring technologies. To develop efficient and effective inspection or monitoring technologies that can effectively gather data showing the infrastructures deterioration.
2. Structural materials, deterioration mechanisms, repair and strengthening technologies. To develop simulation technologies for modelling deterioration mechanisms of structural materials or predict deterioration process of structures.
3. Information and communication technologies. To develop data management technologies and systems that can maximize the use of information on infrastructure maintenance, renovation, and repair.
4. Robotics technologies. To develop maintenance or repair robots that can implement efficient and effective inspection/diagnosis. To develop the robots that allow investigation and/or work needed at dangerous disaster sites.
5. Asset management technologies. To develop technologies or systems that can maximize the use of information and facilitate efficient and effective infrastructure asset management.

The research and development for asset management technologies includes development of systematic management technologies aiming at minimization of lifecycle costs, development of asset management technologies applicable to local governments and nationwide, and establishment of basis for overseas expansion by organizing technical exchange platform with overseas infrastructure owners and individuals with relevant knowledge and experience. In the following sections, examples of research and development projects in the five sectors are given. More detailed explanations on the projects are given in (Fujino & Siringoringo 2020).

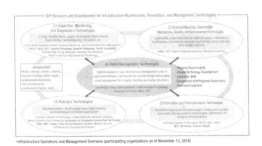

Figure 1. Five fields of R&D in SIP-Infrastructure.

2 RESEARCH AND DEVELOPMENT PROGRAMS FOR INSPECTION, DIAGNOSIS AND MONITORING TECHNOLOGIES

2.1 Tunnel inspection technology using rapidly scannable non-contact radar

Generally, tunnel inspections consist of close-range visual examination, but the conventional inspection technology has been criticized for inadequate safety and objectivity, high oversight risks, and the inability to properly evaluate the deformation progress because: inspection records are in the form of hand drawings and judgments are subjectively made by inspectors. To solve these issues, a mobile tunnel inspection vehicle called MIMM-R is developed under SIP program (Figure 2). In the system, the inspection vehicle is equipped with MMS (Mobile Mapping System; co-developed by Mitsubishi Electric Corporation), MIS (Mobile Imaging System; co-developed by Keisoku-kensa.Co.,Ltd), and MRS (co-developed by Walnut Ltd.). MIMM-R does not require traffic control but allows the high-precision and objective detection of irregularities while traveling at high-speed, typically 50-70 km/h and has been used at practical level. (Yasuda et al., 2014, 2016).

Details of specifications and sensors mounted on the inspection vehicle are provided on Table 1. The purpose of this new inspection technology is to detect inner defects in lining concrete using a rapidly scannable non-contact radar system. Ultimately, a synthetic diagnosis system to assess the soundness of tunnel comprehensively is expected. The inspection system also provides database compilation of various defects including inner defects by a 3D visualizing technology.

The inspection system employs a principle of subsurface indirect radar survey. The radar utilizes physical characteristics of electromagnetic (EM) waves that reflect at boundaries between different

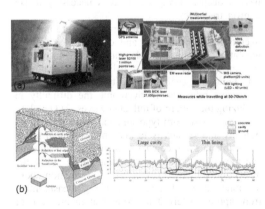

Figure 2. (a) mobile tunnel inspection vehicle MIMM-R and its specification, (b) Basic principle of EM wave reflection for tunnel inspection, and example of tunnel inspection by EM wave reflection-based principle (Yasuda et al. 2014).

15

Table 1. Detail specifications and sensors mounted on the inspection vehicle MIMM-R (Yasuda et al., 2016).

High-precision topographic survey	Road data and road framework data (based on the 3D point cloud data)
Laser Tunnel Surveys Deformation mode analysis	The high-precision laser device (1 million-point data per second), highly dense point cloud data, identification of lining shapes and deformations
Tunnel Image surveys Soundness assessment	Cracks as small as 0.2 mm (while travelling at 50 to 70 km/h). Identify the progressive of unsoundness conditions, the causes of deformation.
Radar Tunnel Surveys Cavity Evaluation	Non-contact radar system to detect lining thickness and back cavities (while travelling at 50 km/h) The system aims to quickly detect hazard locations with thin lining and cavities, as shown in Figure 2.

substances. When electromagnetic (EM) wave released from an antenna into the ground, reflected wave will have different electrical properties on the boundaries of layers consisting of different substances. Typical cross-section of tunnel is shown in Figure 2 which has three layers consisting of different substances, namely, concrete lining, cavities, and ground layers, respectively. As shown in the figure, the reflected EM waves from each layer boundary are received by the antenna. Some of the energy is absorbed when an EM wave propagates through a medium with different damping capabilities and this results in the reduced amplitude of reflected waves. Because of this, the general characteristics of the reflected waves coming from deep locations lack necessary strength and thus are undetectable on radar records. This damping effect is closely related with frequency, the greater the frequency the greater the damping effect, and vice versa. Therefore, to have a deeper survey depth, it is necessary to set a lower frequency.

Key innovations and developments were made to overcome the challenges and difficulties in implementation of high-speed non-contact inspection system (Yasuda et al., 2016): 1) Developing a technique to evaluate polarity and coefficient of EM wave reflection to overcome the difficulty in analyzing the pattern of reflected waves, 2) Development of a new horn-type antenna with high directionality and sensitivity to keep measurement target accessible in a non-contact manner. The new antenna can move up and down, slide and rotate above the inspection vehicle to gather more information about the inner defect (Figure 2) allowing measurement from approximately 3 m. Note that in the case of contact antennas, the diffusion characteristics of EM waves create difficulties in keeping an adequate distance with measurement target, and 3) Performance improvement of controller and sampler to allow the non-contact radar system operates on inspection

vehicle moving at normal speed 50 km/h or faster. To handle the extreme speed of EM waves, a sampler is used to divide a single trace, obtain the divided pieces, and then reconstruct the same single trace shape. High speed data collection was made possible by enhancing the sampling speed and increasing the speed of analog/digital converter. This is necessary to obtain data in the same volume as the contact type while travelling at a speed of 50 km/h or faster.

2.2 Ultrasensitive magnetic nondestructive testing for evaluation of steel infrastructure

A non-destructive testing (NDT) method using highly sensitive magnetic measurements for evaluation of deeper and extended defects on steel infrastructure such as bridges was developed in the SIP-sponsored study (Tomioka et al., 2017). The system is named an extremely low-frequency eddy current testing (ELECT) and the basic instruments are schematically shown in Figure 3. The system consists of oscillator, AC power supply, magnetic field applying coil, compensation coil, anisotropic magnetic resistive (AMR) sensor probe, amplifier, and PC. The measurement system operates in the following procedure. A compensation coil circled around AMR sensor probe will generate magnetic field in the opposite direction to the applied magnetic field of the AMR sensor. The AMR sensor and modulation coil are driven by the power supply with the current applied 0.15A. A magnetic field in the z-axis direction in the figure is measured by the AMR sensor. For thickness estimation, spectrum analysis of the magnetic field (SAM) is applied by measuring multiple frequency magnetic field vectors. Using SAM analysis, measurement of the thickness changes and

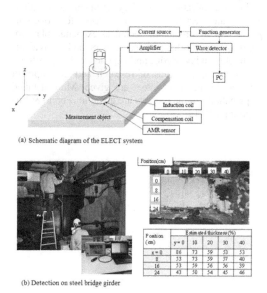

(a) Schematic diagram of the ELECT system

(b) Detection on steel bridge girder

Figure 3. Bridge corrosion investigation using ELECT.

16

imaging of the steel plate are possible. In addition, a serious problem with magnetic noise caused by the magnetization fluctuation of iron steel is resolved using ELECT-SAM.

Two defects can be detected by this system, namely, the thickness reduction due to corrosion and inner steel crack. To detect the depth reduction of steel plate caused by corrosion of thick steel plate, extremely low-frequency eddy current testing (ELECT) with an applied magnetic field ranging from 1 Hz to 1 kHz was developed. The steel thickness is estimated based on the magnetic spectroscopy, which is traced using the obtained multi-frequency magnetic vector signals as shown in Figure 3. As a result, steel plate thinner than 20 mm can be measured within 0.1 mm resolution. Moreover, the shape of the back-side corrosion is determined by scanning with a magnetic probe. Compared with ultrasonic testing, ELECT has the advantage of noncontact, which enables detection even on rough, corroded, or coated surfaces. For inner crack in steel, unsaturated AC magnetic flux leakage (USAC-MFL) testing using an MR gradiometer was developed. Conventional MFL methods need a strong applied magnetic field to ensure that the magnetization characteristic of the steel becomes saturated so that the magnetic field leakage is measurable. They need a strong power source, and therefore they are not suitable for field testing. To solve this problem, a gradiometer with two highly sensitive MR sensors to detect weak magnetic field intensity and the change in the phase-shifted signal caused by the crack were applied. A sharp signal change is detected just above not only the surface but also the inner crack without being influenced by the variations in steel remnant magnetization. As the frequency of the applied magnetic field is decreased, a deeper inner crack is observed, and then an inner crack at a depth of 10 mm can be detected. USAC-MFL is helpful for covering undetectable regions of the sur-face and subsurface (the dead zone) of ultrasonic testing.

The two detection procedures were tested experimentally and then applied for steel bridge inspection. Based on the above fundamental experimental results, ELECT-SAM was applied to the real corroded parts of the girder of a bridge, where it was not easy to apply ultrasonic testing. The applied current was 0.15A and the frequencies of the applied magnetic field were between 1 Hz to 20 Hz to obtain the differential vector. Figure 3(b) shows the detection results on a severely corroded steel bridge plate. Measurement was taken at 20 different points (5 points at the intervals of 10 cm in the x direction and 4 points at the intervals of 8 cm in the y direction). The reduction in thickness was successfully estimated even at the surface of the corroded part where it was difficult to apply the ultrasonic testing. More detailed explanations on the measurement system and practical application are provided by Tomioka et al. (2017) and Tsukada et al. (2016).

2.3 Concrete slab condition assessment using a vehicle equipped with Ground Penetrating Radar (GPR)

Bridge slabs are important parts of bridges while the evaluation of their structural conditions requires significant manpower and time because dense hammering tests must be conducted; the need on efficient and reliable assessment of typical damage on RC bridge slab is high. Presently, a GPR system mounted on a moving vehicle are utilized for high-speed assessment (Figure 4). Even at a speed of 80 km/h, the radar signals reflected from the slab can be captured; the obtained signals are visually examined by inspectors to assess the slab condition. However, the signal is not so sensitive to the damage because the wavelength of the existing GPR system is much larger than the damage scale; the accurate detection is difficult, and the accuracy depends on inspectors. Image checking by inspectors is also time consuming. Thus, recently an algorithm to automatically detect damage from the GPR signals has been proposed (Mizutani et. al. 2017).

The algorithm first calculates the cross-correlation between signal from non-damaged area called 'reference' and that from target area. If the target area is not damaged, the waveform of the GPR signal has high similarity to the reference, resulting in a large cross-correlation value. On the other hand, the cross-correlation at damaged area becomes small. By applying a certain threshold to the cross-correlation, the damage and non-damaged areas are distinguished. The hammering test and GPR assessment are in good agreement.

3 R&D PROGRAMS FOR STRUCTURAL MATERIALS, DETERIORATION MECHANISMS, REPAIR AND STRENGTHENING TECHNOLOGIES

3.1 Remaining fatigue life assessment of damaged RC decks using data assimilation of multiscale model and site inspection

Reinforced concrete (RC) bridge decks constructed in urban areas sustain heavy traffic load during its lifetime. The increase of traffic loads and intrusive

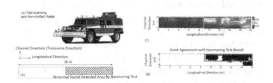

Figure 4. Concrete slab condition assessment using a vehicle equipped with GPR. (a) GPR measurement vehicle. Comparison between hammer test results and GPR signals after processing: (b) measurement bridge slab and its hammering test result, (c) Maximum of cross-correlation function, (d) after applying threshold (Mizutani et al. 2017).

environment condition have increased the risk of bridge deck deterioration and even failure. Damages on RC bridge deck is not uncommon in Japan, especially on the old bridge decks because they were designed for thinner slab decks to reduce the top-heavy mass required in seismic resistance design. Typical damages on the RC bridge deck such as lattice cracks developing over the bottom face of the decks. Repair works of the damage on the bridge deck is not easy technically, and it may create extended problem financially to the highway or road network such as traffic closure or detour. Therefore, maintenance of reinforced concrete bridge decks has been a primary concern for roadway or highway operators.

Estimating the remaining fatigue life of RC bridge decks subjected to traveling wheel-type loads is an important aspect of maintenance. Operators needs a good model to estimate the remaining life of bridge deck so that the necessary corrective actions such as repair and replacement can be taken timely and properly. Given that periodical inspection data of real bridges is available for the whole life, the series of data could be useful to verify its reliability of remaining life assessment. However, past ambient states and traffics to individual bridge have been rarely recorded and the initial quality of constructed concrete on which the scientific discussion can be based on is generally unknown or unrecorded in practice. One of the ways to verify the life assessment method is to apply for statistical analysis of big data of maintenance (Yamazaki and Ishida, 2015). The other way is to follow the mechanical and chemical states of the target.

The SIP-sponsored program proposed a method to estimate the remaining fatigue life is proposed using data assimilation procedure, i.e., coupled life-span simulation with inspection data at site. Multiscale analysis with hygro-mechanistic models is employed as the platform of data assimilation on which the visual inspection of cracking on the members' surfaces and the acoustic emission (AE) tomography are numerically integrated. For verification, the wheel running load experiments of slabs (Figure 5) were conducted with continuous data acquisition of both crack patterns and the acoustic emission data over the life till failure (Tanaka et al., 2017b). Visually inspected cracks are converted to space-averaged strains with Bernoulli-Euler theory. Imbalance in deformation and forces are compensated by numerical predictor-corrector cycles and non-inspected internal cracks are reproduced. During the loading, non-destructive tests (NDT) were conducted to detect the degree of structural concrete deterioration at periodical interval of times. The crack patterns observed on the lower surface, flexural cracks were induced at the very beginning of loads. Number of cracks gradually developed with increase in the wheel load passages. Radial cracks were formed by 100000 cycles. Then, both number and average crack width increased by 200000 cycles

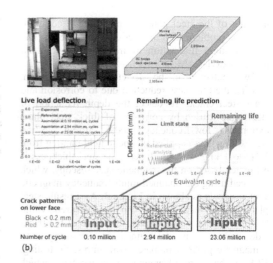

Figure 5. (a) wheel running test machine and dimensions of RC deck specimen and position of wheel load, (b) Data assimilation with crack patterns in wheel running test (Tanaka et al 2017).

(2.94 million equivalent cycles). At 250 thousand cycles (23.06 million equivalent cycles), lattice-type cracks were clearly formed, and the large crack width was observed (Figure 5).

Data assimilation method to combine the visual crack inspection data at site and the multiscale modelling was developed for remaining fatigue lifetime to failure, and its applicability was experimentally verified with moving load slab experiments. The numerical predictor-corrector method is used as a search-engine of most plausible solution of internal damages for existing slabs, where the visible cracks are set to be flexural ones as a start-up of data assimilation. As the first set-up for the subsequent predictor-corrector cycles, the elastic wave velocity field is converted to the fictitious isotropic stiffness, based on which the most possible cracking is searched. The proposed assimilation method successfully reproduces most probable internal cracks over the volume of analysis domains, and with this approach the remaining life of the deck slabs inspected can be successfully estimated (Tanaka et al. 2017 a,b).

3.2 Performance assessment of Chromium bearing steel in concrete under salt-intrusive coastal environment

In the SIP-program research, Nishimura (2018) studied the performance of Cr bearing steel in concrete on the bridge located in a coastal environment in Miyako-island in Okinawa prefecture. The nano structure of the rust of Cr bearing steel in concrete was investigated in detail using TEM – EELS (Electron Energy Loss Spectroscopy). The objective was to find relationship between the high corrosion resistance and the rust formation was discussed for the Cr bearing steel in concrete exposed under coastal environment. The exposure

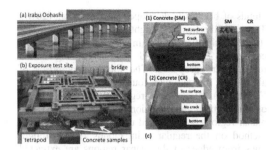

Figure 6. Photos for the (a) Irabu Oohashi and (b) Exposure test site. (c) Photos for the concrete samples and embedded (1) SM (carbon steel) and CR (Cr steel) after the exposure test (Nishimura, 2018).

test was conducted for 2 years at the exposure test site near Irabu Oohashi (bridge) in Miyako-island as shown in Figure 6. As the test samples were exposed on the tetrapod located 4 m from the sea, the sea water was splashed directly on the samples. The chemical composition of Cr bearing reinforcing steel (CR) was 7% Cr - 2% Si–Fe in mass%, additionally, carbon steel (SM) was used for the comparison. After the exposure test for 2 years, the steels and concrete blocks were compared. The carbon steel (SM) has a considerable corrosion on the surface, and the concrete block had large cracks. This behavior is explained that the expansion of the corrosion product (rust) causes the cracks of the concrete block. In contrast, CR has little corrosion on the surface, and there is no crack on the concrete. Thus, it is possible to demonstrate that CR has significantly higher corrosion resistance than that of SM in concrete at the exposure site.

From the exposure test, it was demonstrated that Cr bearing steel (CR) has significantly higher corrosion resistance than SM in concrete. SM had a considerable corrosion on the surface, and the concrete block had large cracks. However, Cr had little corrosion on the surface, and there was no crack on the concrete. Based on the TEM-EELS measurements, the chemical shift of Cr–L3 and Cr–L2 were recognized in inner rust, which corresponded to Cr (II) and Cr (III) oxide state. In inner rust of CR, Cr and Si were enriched in nano iron oxides, which was believed to increase the corrosion resistance of CR in concrete. The importance of this study is to confirm that CR has high corrosion-resistant performance in concrete, thus is effective to be implemented in the bridges located on the coastal environment.

4 R & D PROGRAMS FOR INFORMATION AND COMMUNICATION TECHNOLOGIES

4.1 *Automated recognition and 3D CAD modelling of standardized steel bridge members in a laser scan*

Management of bridge inspection records based on 3D models facilitates not only the intuitive understanding of damage distribution on the structure by the inspectors, but also ensures smooth communication among stakeholders involved in the maintenance. However, a major obstacle for the 3D management model is that 3D models of the existing bridges, such as computer-aided design (CAD) models, are not provided in most cases. Laser scanning is a promising method for obtaining reliable 3D measurements of large-scale structures, such as bridges. Therefore, several researchers have studied automated 3D as-built modelling of bridge structures based on laser-scanned point clouds. The proportion of steel bridges in Japan is quite high; therefore, these bridges strongly require an automated as built 3D modelling technique. In particular, the superstructures of steel bridges generally include a lot of standardized steel members, such as L-shaped or H-shaped angle steels. Therefore, for standardized steel members, there is a strong need for automated 3D modelling method based on laser-scanned point clouds captured from bridge superstructures. The primary algorithm for the 3D modelling method is to recognize the steel members using a variety of standardized cross-sectional shapes in laser-scanned point clouds.

In a SIP-sponsored research (Kanai et al., 2016), a fully automatic method to recognize standardized steel members from point clouds captured by a single laser scan is developed. The recognition process is illustrated in Figure 7. The process mainly consists of three stages: (1) the creation of a cross-sectional database, (2) the segmentation of primary planar region groups from the one-scan point clouds, and the extraction of feature dimensions of the region groups, and (3) the estimation of the type and size of the standardized steel members. The first stage is the pre-process phase; the second and third stages are executed in the actual recognition phase.

In the first stage, the cross-sectional database was built by registering the following geometric attributes characterizing the visually feasible cross-sectional shape of a standardized steel member. Using the attributes, a broad range of steel bridge member types, including L-shaped, CT-shaped, H-shaped, and U-shaped members, whose size variations were standardized according to ISO 657-1:198. In the second stage, the extraction of primary planar regions, segmentation of the planar region groups, and extraction of feature dimensions of the groups were carried out using the region growing method. Finally, in the last step of the recognition, geometric attributes of each planar region group were cross-checked with those registered in the collating table in the cross-sectional database to identify the type and size of the standardized steel member that fitted to the laser-scanned point cloud of the planar region group. Detailed information on the complete algorithms is described in (Kanai et al., 2016).

The proposed automated recognition system was tested on a short-span steel beam bridge in Yamanote, Sapporo, Japan. The bridge was scanned from the dry

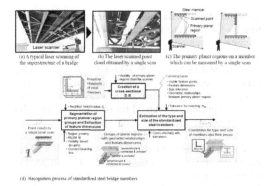

(a) A typical laser scanning of the superstructure of a bridge | (b) The laser scanned point cloud obtained by a single scan | (c) The primary planer regions on a member which can be measured by a single scan

(d) Recognition process of standardized steel bridge members

Figure 7. Laser scanning of steel bridge members and a scheme for automatic recognition process of standardized steel bridge members (Kanai et al. 2016).

riverbed by using a terrestrial laser scanner (FARO Focus 3D). As shown in Figure 7, the point cloud captured from the partial structure between two consecutive support bracings by a single laser scan was used as the input for recognition. The point cloud included 4,037,897 points, and the distance between the adjacent sample points ranged from 3 to 8 mm.

Three types and sizes of standardized steel members were used in the structure: three L-shaped, two L-shaped, and four CT-shaped steel members. The actual condition and sizes were manually examined in advance by the steel rule according to the registered cross-sectional database of steel members. The results show that the developed automatic recognition system can successfully recognize all L-shaped and CT-shaped steel members.

4.2 *Efficient registration of laser scanned point clouds of bridges using linear features*

The use of Terrestrial Laser Scanning (TLS) for bridge survey and inspection is quite common. Large number and precise 3D point clouds can be acquired with TLS measurement. The large numbers of point clouds from laser scanning and as-built 3D models from the point clouds of the structures can be utilized to support efficient maintenance, for examples, regular monitoring, detecting geometric changes of the structure, inspection planning and inspection data management. In the laser scanning of bridges, multiple scanning by TLS at different positions are required for acquiring point clouds without lack of points caused by obstruction. The origin of the coordinate system for the point clouds of each scan is the position of the TLS system. Therefore, an efficient, and accurate registration is required. Efficient registration of point clouds from terrestrial laser scanners enables us to move from scanning to point cloud applications immediately.

In the SIP-sponsored research, a new efficient rough registration method of laser-scanned point

clouds of bridges is developed by Date et al. (2018). The method relies on straight-line edges as linear features, which often appear in many bridges. Efficient edge-line extraction and line-based registration methods are proposed. The method comprises of algorithms divided into four steps as explained in Figure 8(a). At first, the regular point clouds based on the azimuth and elevation angles are created, and planar regions are extracted using the region growing method on the regular point clouds. Then, straight lines from edges of the planar regions are extracted as linear features. Next, vertical, and horizontal line clusters are created according to the direction of the lines. To align the position and orientation of two-point clouds, two corresponding nonparallel line pairs from line clusters are used. Finally, in the registration process the well-known RANSAC approach with a hash table of line pairs is used. In this process, the hash table is used for finding candidates of corresponding line pairs efficiently. Sampled points on the line pairs are used to align the line pairs, and occupied voxels and down-sampled point clouds are used for efficient consensus calculation. The flowchart of proposed algorithms for pairwise rough registration is shown in Figure 8(a) (Date et al., 2018).

Performance of the methods was evaluated using laser-scanned point clouds of three data sets from different types of bridges: a small steel bridge, a middle-size concrete bridge, and a high-pier concrete bridge as shown in Figure 8(b). In the experiments, successful rates of the rough registration were 100%. It was shown that the registration based on linear features was effective for registering the laser-scanned point clouds of the bridges. The

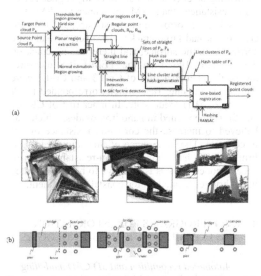

Figure 8. In the inspection of bridges using Terrestrial Laser Scanning (TLS), dense and precise 3D point clouds (a) Proposed algorithm for pairwise rough registration, (b) Precise results of application and scanning point registration results (Date et al., 2018).

observed registration errors were in the order of tens centimeters and they were modified by precise registration with processing time for rough registration of 19-point clouds was about 1 min.

5 R&D PROGRAMS FOR ROBOTIC TECHNOLOGIES

5.1 Unmanned Aerial Vehicle (UAV) for bridge inspection

A full-scale maintenance that includes the implementation of routine close visual inspection is recommended for all bridges every 5 years. However, the number of personnel for bridge inspectors in Japan is insufficient to cover all domestic bridges and will not increase significantly in the future because of safety concern about inspectors, decreasing birthrate, and an aging population. In addition, there are issues that manual inspection requires special access to bridge components through temporary scaffolds, special cranes or overhang buggy that are not always available, expensive and take time to maneuver, as well as ladder trucks that often block traffic. For these conditions unmanned aerial vehicle with its operationability, and accessibility is seen as a promising tool. UAVs have also become attractive in the field of disaster robotics, where it is expected to be used for exploration of disaster sites. However, during exploration, the UAV may encounter complex spaces such as a narrow space in a partially damaged building, between steel frame of bridges, connections of bridge deck and pylon, and piping of factories. Thus, an ordinary UAV without proper protection and a mechanism to maintain stability during a collision with obstacles cannot be used in this type of environment due to the high risk of falling.

Figure 9 shows PRSS-UAV and its main components. Spherical shell that can rotate freely covers the entire body of the UAV and protects it from colliding with obstacles without compromising the UAV's attitude and flight stability. Additionally, the system allows for flexible movement that is possible when the system contacts any surface because its spherical shell can function as a wheel. This mechanism enables the UAV to operate in a complex environmental condition without the need for complex sensing and avoidance strategies. In this program the PRSS-UAV is developed for close visual

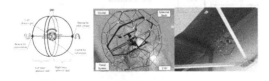

Figure 9. Unmanned Aerial Vehicle (UAV) with a passive rotating spherical shell (PRSS), (a) Basic configuration, (b) Photos of the components, (c) Operating for with a close space during steel bridge inspection.

bridge inspection. There are two research issues to solve based on these requirements: (a) the hardware (mechanism) and (b) software (image processing).

The system of visual inspection requires a camera to capture images of cracks, corrosions, or any other problem in all critical parts of the bridge. A good image resolution is necessary for better analysis of the bridge condition. Further, at the visual inspection, bridge damages may appear from different angles (at the side, top, or edges), thus this requires a camera that can adjust its position with the addition of yaw rotation/movement, that enables a full overhead view for the application. The system utilizes the lightweight and high-resolution GoPro Hero camera, which is good for indoor and outdoor use. The system also provides a real-time wireless transmission of video from the camera using DJI Lightbridge. The raw data acquired from the motion capture system were used for postprocessing, extracting useful data for analysis. The recorded average flight duration of several test flights was 5.7 min. The average speed of the PRSS UAV was 0.65 m/s.

The overall weight of PRSS-UAV is equivalent to 2,583 g, which is about 92.25% of the maximum take-off weight of the UAV. Likewise, the power rating of the system was more than doubled using a 7,800 mAH battery at 11.1 V. The system also provides easy maintenance by allowing the PRSS-UAV assembly to be split into two subassemblies. Detailed specification of PRSS-UAV system is given in Table 2. In this configuration, it is possible to easily remove and insert the UAV and other components inside the spherical shell if needed. The optimal size, weight, and configuration of the system was also designed for ease of transport and deployment (Salaan et al., 2018).

The second type of UAV for bridge inspection developed in the SIP program is two-wheeled multicopter with three-dimensional modeling technology (Hada et al., 2017). The system consists of multicopter as an inspection support robot, automatic geotagging technology and 3D model-based bridge maintenance database. For the inspection robot, a UAV two-wheeled multicopter shown in Figure 10 is developed. The two-wheeled multicopter system can crawl along a surface vertically and get into small space in the bridge. It has two main advantages compared to conventional UAV system. 1) Wind-resistance capability against crosswind by using contact friction between wheels and structure surface. This allows measurement along tall bridge pylon. 2) It allows movement along the surface of structure with constant distance between onboard camera and the structure surface that make it possible to capture close-up image of structure surface.

In most cases, the UAV used for inspection need to operate under the bridge girder where GPS signal is unreliable or even unavailable. To realize a position estimation of the UAV, a method of position estimation using automatic geotagging technology based on the structure from motion (SFM) technique using a 360-degree spherical camera is developed. The position estimation is used for geotagging, which is

Table 2. Specification of Passive Rotating Spherical Shell Unmanned Aerial Vehicle (PRSS-UAV) (Salaan et al., 2018).

	Initial quad	New quad	Unit
Frame diameter	450	450	mm
Propeller diameter	10	11	in
Overall diameter	704	730	mm
Overall weight	1	1.45	kg
Max take-off weight	2	2.8	kg
Battery@11.1 V	3300	7800	mAH

Figure 10. (a) model of two-wheeled multicopter UAV system and the ability of the system to operate in difficult conditions: (b) crawling on tall bridge pylon subjected to crosswind, (c) crawling into close-space between bridge pylon and girder.

a process of adding information to photos of video. The geotagging software can show both the close-up of deterioration images and their locations. The bridge inspection system consists of two main works, namely the field work and office work. The field work consists of capturing the images of inspected bridge components, automatic geotagging of inspection images, and employing three-dimensional laser scanning that can be used for semiautomatic 3D CAD model generation. Based on this information, the 3D model-based bridge maintenance database is constructed in the office work. Bridge conditions, possible damage or defects can be located. The system is tailored to accommodate supervisor's usage such as scheduling for bridge inspection, planning for bridge repair and prediction of bridge deterioration. This system is developed within collaboration of research groups from Fujitsu Limited, Hokkaido University, The University of Tokyo, Nagoya Institute of Technology and Docon Co. Ltd.

6 R & D PROGRAMS FOR ASSET MANAGEMENT TECHNOLOGIES

6.1 *Bridge Information Modelling (BIM) based on IFC for supporting maintenance management of existing bridges.*

One of significant information technologies in the Architecture, Engineering, and Construction (AEC) industry is Building Information Modelling (BIM). Maintenance of infrastructures that consist of

inspection and repair process in the life cycle can also take advantage of the advanced development in BIM for more effective and efficient asset management.

In the SIP program a research on bridge information model that extends the Industry Foundation Classes (IFC) international standard is proposed (Tanaka et al., 2016). This information model satisfies the information requirements for inspection, evaluation, and maintenance processes. The bridge information model consists of interactive web system that combines the history data related to inspection, evaluation, and repair of the bridge is recorded in the 3D bridge model. A web content providing system is constructed to show the model in the outside field of maintenance. The web content system includes the product data, data extractor, data converter and webserver. The web content providing system is based on the three.js library (WebGL) that makes it available for the outside field maintenance purpose. The design data from CAD are exported as IFC-data and stored in the product data. Measured data such as photo image of degradation and as is bridge model are also stored in the product model. The current bridge conditions are obtained from latest measurement using advanced technologies such as laser scan and radar, and they are stored using IFC Engine library.

The system input past inspection reports as a PDF format. For paper reports, scans and scanning software are employed to obtain PDF file. From PDF data, text data, degradation photo image data and degradation sketch data are extracted by software. From degradation sketch, position of degradation is extracted using image recognition software. Character recognition techniques are utilized to identify type of degradation and photographic image. Later, these data are integrated into bridge information model.

The information system has function of viewing not only 3D shape model but also 3D model with photo texture. By extracting information from the past and associating it with 3D model, it is easier for bridge operator to create a repair plan. Figure 11

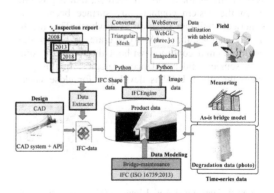

Figure 11. Structure of developed web content for supporting maintenance management of existing bridges (Tanaka et al., 2016).

shows the structure of developed web content for the system. More detailed information of the system is provided by (Tanaka et al., 2016).

7 CONCLUSIONS

This paper describes the new research and development (R & D) program named "Infrastructure maintenance, renovation and management". It is started in 2014 under the Japan Council of Science, Technology and Innovation (CSTI)'s Strategic In-novation Program (SIP) and covers various subjects of infrastructure maintenance with key technologies in condition assessment using non-destructive testing, monitoring, and robotics; long-term performance prediction of infrastructure, development of durable high-quality of material for repair and replacement, and management of large number of bridges and other infrastructure data using advanced information and communication technologies (ICT). The program consists of about 60 research projects involving universities, government research institutes and industries. The common goal is to have all the out-comes finally implemented in infrastructure maintenance, renovation, and management. The outcomes of SIP-Infrastructure would be implemented inside and outside Japan.

REFERENCES

Boller, C., Chang, F. K., & Fujino, Y. (Eds.). (2009). *Encyclopedia of structural health monitoring*. Wiley.

Date, H., Yokoyama, T., Kanai, S., Hada, Y., Nakao, M., & Sugawara, T. (2018). Efficient Registration of Laser-Scanned Point Clouds of Bridges Using Linear Features. *International Journal of Automation Technology*, *12*(3), 328–338.

Fujino, Y. (2018). Vibration-based monitoring for performance evaluation of flexible civil structures in Japan. *Proceedings of the Japan Academy, Series B, 94*(2), 98–128.

Fujino, Y. and Siringoringo, D.M., (2020). Recent research and development programs for infrastructures maintenance, renovation, and management in Japan. Structure and Infrastructure Engineering, 16(1), pp.3–25.

Fujino, Y., Siringoringo, D. M., & Abe, M. (2016). Japan's experience on long-span bridges monitoring. *Structural Monitoring and Maintenance, 3*(3), 233–257.

Fujino, Y. and Siringoringo, D.M., (2008). Structural health monitoring of bridges in Japan: An overview of the current trend. In *Fourth International Conference on FRP Composites in Civil Engineering (CICE2008)* (pp. 22–24).

Hada, Y., Nakao, M., Yamada, M., Kobayashi, H., Sawasaki, N., Yokoji, K., … & Yamashita, A. (2017). Development of a Bridge Inspection Support System using Two-Wheeled Multicopter and 3D Modeling Technology. *Journal of Disaster Research, 12*(3), 593–606.

Kanai, S., Hashikawa, M., & Date, H., (2016), Automated recognition and 3D CAD modeling of standardized steel bridge members in a laser scan, *Proc. of 16th Int. Conf. on Computing in Civil and Building Engineering (ICCCBE2016)*,July 6–8, 2016, Osaka, Japan.

Mizutani,T, Nakamura,N., Yamaguchi,T, Tarumi, M., Ando, Y. and Hara,I, (2017), Bridge slab damage detection by signal processing of UHF-band ground penetrating radar data, Journal of Disaster Research, vol. 12, no.3, pp. 415–421, Aug. 2017.

Nishimura, T. (2018). Exposure Test Performance for Chromium Bearing Steel in Concrete under Coastal Environment. *ISIJ International, 58*(5), 936–942.

Salaan, C. J. O., Okada, Y., Mizutani, S., Ishii, T., Koura, K., Ohno, K., & Tadokoro, S. (2018). Close visual bridge inspection using a UAV with a passive rotating spherical shell. *Journal of Field Robotics, 35* (6), 850–867.

SIP (2014a) (Cross-ministerial Strategic Innovation Program), Pioneering the Future: Japanese Science, Technology, and Innovation, Council for Science, Technology and Innovation, Cabinet Office, Government of Japan, 2014 (http://www8.cao.go.jp/cstp/gaiyo/sip/index.html, access Feb 18th, 2018)

SIP (2014b) (Cross-ministerial Strategic Innovation Program), Pioneering the Future: Japanese Science, Technology, and Innovation, Council for Science, Technology and Innovation, Cabinet Office, Government of Japan, 2014.

Tanaka, F., Hori, M., Onosato, M., Date, H., Kanai, S., (2016), Bridge Information Model Based on IFC Standards and Web Content Providing System for Supporting an Inspection Process, *Proc. of 16th Int. Conf. on Computing in Civil and Building Engineering (ICCCBE2016)*, July 6–8, 2016, Osaka, Japan

Tanaka, Y., Maekawa, K., Takahashi, Y., and Ishida, T. (2017a), Life assessment of bridge decks by data assimilation and survival analysis, Proceedings of the 2nd ACF symposium 2017.

Tanaka, Y., Maekawa, K., and Takahashi, Y. (2017b), Remaining fatigue life assessment of damaged RC decks-Data assimilation of multi -scale model and site inspection-, Journal of Advanced Concrete Technology, Vol.15, pp.328–345.

Tomioka, T., Goda, T., Sakai, K., Kiwa, T., Tsukada, K., (2017), Imaging of internal corrosion of steel structures using an extremely low-frequency eddy-current testing method. *Proc. of 15th Asia Pacific Conference for Non-Destructive Testing (APCNDT2017)*, Singapore.

Tsukada K., Haga Y., Morita K., Nannan S., Sakai K., Kiwa T, and Cheng W, (2016),Detection of Inner Corrosion of Steel Construction Using Magnetic Resistance Sensor and Magnetic Spectroscopy Analysis, IEEE Trans.Magn., 52, 6201504

Yamaguchi,T, Mizutani T,Tarumi M, and Su,D (2018). Sensitive Damage Detection of Reinforced Concrete Bridge Slab by Time-Variant Deconvolution of SHF-Band Radar Signal, IEEE Transactions on Geoscience and Remote Sensing, 2018 (under peer review).

Yamazaki, T. and Ishida, T. 2015. *Application of Survival Analysis to Deteriorated Concrete Bridges in East Japan*. J. JSCE, Ser.F4, Vol. 71, No.4, 2015, pp. I_11-I_22.

Yasuda, T., Yamamoto, H., & Kitazawa, R. (2014) Cavity Detection behind Tunnel Lining using Non-contact Radar at 50km/h, *Construction Machinery and Construction*, 66(12), 51–56.

Yasuda, T., Yamamoto, H., Shigeta, Y., Ishii, H., & Kitazawa, R. (2016), Tunnel Inspection Technology using Rapidly Scannable Non-Contact Radar, *Proc. of EASEC-15*, January 6–8, 2016, Ho Chi Minh City, Vietnam.

23

State-of-the-art report

Proceedings of the Second International Conference on
Press-in Engineering 2021, Kochi, Japan – Matsumoto et al (eds)
© *2021 Taylor & Francis Group, London, ISBN 978-1-032-10414-0*

State of the art report on application of cantilever type steel tubular pile wall embedded to stiff grounds

J. Takemura

Tokyo Institute of Technology, Japan

ABSTRACT: IPA-TC1 was set up in 2017 to figure out the issues for further application of cantilever type steel tubular pile wall embedded to stiff grounds and establish a rational design procedure of embedded cantilever steel tubular pile wall as the final goal. This state of the art report overviews the research activities done by TC1, i.e. case study, physical modeling, numerical analyses, parametrical study by design models, and gives considerations and remaining challenges for the rational design of this type of wall.

1 INTRODUCTION

1.1 *Motivation*

Various types of earth retaining structures or walls have been employed in the history of civil engineering, such as masonry walls, RC concrete walls, sheet pile wall, mechanically stabilized earth wall. Similar to the other civil engineering structures, these retaining walls should satisfy the required performance under the design conditions. Several conditions should be considered in the selection and design of the retaining structures, such as design loads (actions) and site environment. Among the several options of the earth retaining systems, embedded retaining walls are common one for both temporary and permanent structures and various types of embedded retaining walls are used are used depending on the site conditions (Gaba et al. 2017). Among the embedded retaining walls, cantilever wall is the simplest wall, of which stability is relied on only the embedment soil or rock against the load from the retained side. With the simple retaining mechanism and a relatively large wall deflection, this type of wall has been mostly used for temporary work or for the permanent wall with small height.

However, thanks to the innovative pile installation method, like rotary cutting press-in method (e.g. Gyropress), the applicability of steel tubular pile wall (STPW) has increased significantly for various structures (road, harbor, railway) and objectives, not only ordinary retaining structures (Miyanohara et al. 2018, Suzuki & Kimura 2021), but also restoration, rehabilitation and reconstruction of disaster areas (Takada 2016, He 2018). The installation abilities of large steel tubular pile (STP) into very stiff grounds with low noise and vibration and without damage of

pile end are all critical advantages of the rotary cutting press-in method (Table 1). The damage of pile is a main concern in the pile penetration in stiff layer (Randolph 2021). Figure 1 shows a typical example of STPW application for road widening project using a narrow steep slope reinforced by ground anchors with several requirements from road traffic and residential sides (Kitamura & Kitamura 2019).

The combination of large diameter and high rigidity STP, and the stiff embedment ground enables the application of the cantilever embedded walls with large retained height (Figure 2). Figure 3 shows the relationship between the calculated wall deflection and wall retained height for two ground conditions (relatively dense sand with SPT N-value=50 and soft rock with unconfined compression strength q_u = 1.5MPa). The calculations were made assuming the ordinary static load based on Cantilever Steel Sheet Piles Retaining Wall - Design Manual (JTASP-PACTC 2007) with the conditions shown in Figure 4. In Figure 3, allowable displacements of requirement 1 are also shown on the stability of embedment soil (δ_{gs}) and serviceability of the facility on the retained soil (δ_T). From the calculated results, it can be confirmed that by the combination of large diameter STP and stiff embedment ground, the wall top displacement caused by the wall bending deflection and the wall rotation in the embedment can be controlled below the required displacement. However, the current design method of embedded cantilever wall has been developed for the relatively flexible steel sheet pile wall into soft grounds for small retained height (e.g. less than 4m, JTASPPACTC 2007). Therefore, simple application of the current design method to the cantilever type STPW embedded in stiff grounds may require

DOI: 10.1201/9781003215226-64

Table 1. Advantages and concerns of rotary press-in cantilever large diameter steel tubular pile wall.

Advantages

- Applicability to severe construction site environment, such as small working space, steep slope, hard ground, noise- vibration restriction, remote operation at the failure risk slope;
- Construction accuracy and safety;
- No traffic interruption, short construction time;
- Continuous recording of pile installation process.

Concerns

- Technology advanced without rational design method and currently adopted design methods not considering the specific features of the wall, namely, very large stiffness piles in stiff ground;
- Few records on the critical performance of the wall;
- Relatively expensive compared to the other common retaining walls; beside the facility cost, material, welding and transportation could be reduced by the rational design and construction practice.

Figure 1. Application of STPW (H=7.7-12.4m, l_p =20.5-24.0m) for road widening projects (Kitamura & Kitamura 2019).

excessive embedment depth, or increase a risk of failure caused by the unexpected performance of the wall.

Several concerns can be pointed out in the commonly used current design methods, such as,

1) The minimum embedment depth (d_0) requirement using characteristic value (β), such as $d_0 \geqq 3/\beta$ (Figure 4) should be verified, as it is based on long flexible wall behavior in an infinite uniform elastic media, which conflicts the rigid nature of large diameter steel tubular piles;

2) As the pile diameter (Φ) increases, the relative embedment depth (d_e/Φ) and wall thickness and diameter ratio (t/Φ) tend to decrease. Furthermore, near the surface of stiff ground, especially rock, the stress concentration could occur on the front side of

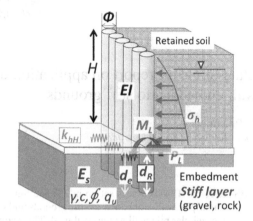

Figure 2. Cantilever STPW embedded in stiff ground and conditions used in the design.

tubular pile. These particular conditions could enhance local and 3D behavior, which is not considered in the conventional 2D analytical/numerical models;

3) High confinement or fixity of the piles by the stiff ground could generate the large resilience of pile, which affects the wall-soil interaction, the wall pressure from the retained soil, and the residual wall displacement after the temporal loading event, e.g. earthquake.

1.2 Objectives of TC1

To answer the above-mentioned concerns and establish a rational design procedure, Technical Committee TC1 "Application of Cantilever Type Steel Tubular Pile Wall embedded to Stiff Grounds" was set up in IPA. Four working groups were created in TC1 with several tasks as shown below. Some findings related to the task were presented in the report.

1.2.1 Tasks of Working Group
WG1 on design method:

- Task 1: to investigate design methods presently used, and identify the issues such as embedment depth, soil characteristics, seismic design.
- Task 2: to analyze the design procedure of existing large diameter tubular steel pipe walls.
- Task 3: to propose new rational design method of large diameter tubular steel pipe wall including seismic design.

WG2 on centrifuge model test:

- Task 1: to clarify mechanical behavior of large diameter tubular steel pipe wall subjected to static load in stiff ground.
- Task 2: to analyze influence of critical conditions such as embedment depth, ground stiffness and strength on the behavior of wall.

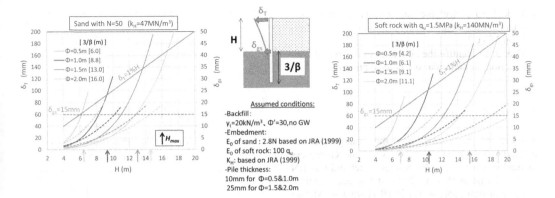

Figure 3. Wall displacement estimated simplified method (JTASPPACTC 2007; IPA 2014 & 2021).

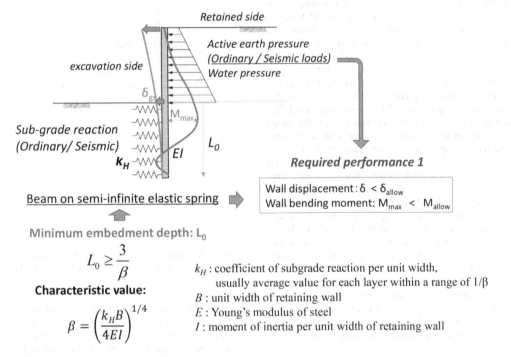

Figure 4. Simplified method, static verification method with a linear elastic subgrade reaction (beam on elastic base) model, which is commonly used performance verification for permanent and variable situations (ordinary static/seismic load), (JTASPPACTC 2007; IPA 2014 & 2021).

- Task 3: to discuss difference between the behavior of actual structures and that predicted by the simplified design model.
- Task 4: to simulate the deformation and failure behavior against earthquake load.

WG3 on numerical analyses:

- Task 1: to verify and calibrate 3D FEM method by centrifuge modeling.

- Task 2: to analyze the and local behavior of wall and ground, which cannot be observed in the centrifuge model tests.
- Task 3: to analyze the influence of parameters on behavior of large diameter tubular steel pipe wall using simple flame analysis and 2D FEM.

WG4 on case study of construction:

- Task 1: to collect construction cases with design details as much as possible.

- Task 2: to collect the data observed during and after construction, if available, with the collaboration of TC2.
- Task 3: to identify the concerns in the actual construction, in particular on the cost and time.

2 CASE STUDIES

Number of applications of the rotary press-in method is shown in Figure 5. Since the Gyropress method was developed by Nippon Steel and GIKEN LTD and first applied in 2004, it has been applied more than 400 projects till 2019 (Hirata & Matsui 2016; IPA 2019; Suzuki & Kimura 2021). Figure 6 shows some summary of the case records, giving the number of projects in terms of pile diameter, pile length with the joint number and maximum converted SPT N-values with the site ground type. As for the pile diameter, 1m piles have been most commonly used about 40%, but the large diameter piles over 1.0m have been used more than 30% with 2.0m maximum. The project shown in Figure 1 is an example of the 2.0m pile wall. While for the pile length, the most frequently used length is about 18m and very long piles over 30m were constructed. The number of welding joint depends on the site condition, such as upper space clearance, and the required pile length. The most of piles have been embedded in the ground with maximum N-value over 50. About 20 % of recorded cases, the walls were constructed in gravel or rock ground with the converted N-value of 300 or higher. The application of large diameter pile in the stiff grounds for large height retaining walls is increasing trend especially for the site with severe conditions, such as spatial restriction, low noise and vibration requirements, and short construction period.

Among the records of which design procedure were confirmed, the minimum embedment depth $(d_o \geqq 3/\beta)$ were adopted especially for the road construction projects. Although the number of applications has increased significantly, very limited records

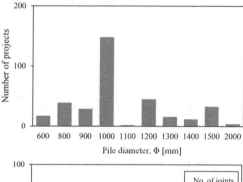

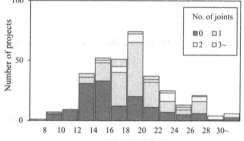

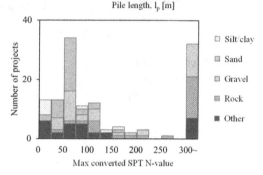

Figure 6. Summary of case records of STPW installed by rotary cutting press-in method (Suzuki & Kimura 2021).

are available on the monitored behavior of the wall during and after the construction. The data related to serviceability limit state (SLS) and the ultimate limit state (ULS), and the wall performance from SLS and ULS under various actions, e.g. static and seismic loadings with detailed site conditions are critically important to rationalize the design procedure.

Case studies of foundation in stiff grounds, e.g. soft rocks, with detail field measurements are rather limited, which are mostly on the end bearing capacity of piles. Nanazawa et al. (2015) conducted an intensive study on the end bearing capacities of piles installed in rock ground, which covering various codes and design methods, analyses of field loading tests on 94 sites. At the most of the sites only standard penetration test were conducted and due to the capacity limitation of the machine, the majority of piles were installed in the rock ground of equivalent N-values less than 200. Beside N-values, the other ground properties are limited, such as unconfined

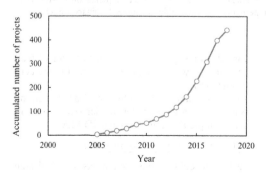

Figure 5. Accumulated number of the application of the rotary press-in method (Suzuki & Kimura 2021).

30

compression (UCC) test data available for 10% of the cases, and very few on the rock quality (RQD, classification). As for the driven steel pile into soft rock, several researches have been done. Randolph (2019) delivered a keynote lecture on various aspects of the design of piles embedded in soft rock, covering rock properties of pile design, the effect of construction methods and techniques for optimising pile performance, axial load transfer parameters, effects of strain softening and cyclic loading, and analysis approaches for laterally loaded piles in rock. On the contrary to the piles driven or socked in the soft rock, very limited researches have been made for the retaining wall embedded in rocks, especially field tests, including the monitoring of actual structures.

3 BEHAVIOR OF LARGE DIAMETER STPWS IN STIFF GROUND AND CRITICAL CONDITIONS

Over the past decades, extensive investigations based on the physical modelling of excavations on a cantilever and propped walls embedded in clays and sandy soils have contributed to the development of design codes and the calibration of numerical models (Padfield & Mair 1984, Bolton & Powrie 1987 & 1988, Richards & Powrie 1998, Day 1999). Also, the observations from numerous case histories (Long 2001, Ou et al. 1993) related to the real field applications in a vast range of soil conditions and the failure of earth retaining structures (D'Andrea & Day 1998, Whittle & Davies 2006) oftentimes revised the codes for the safe and economic design of retaining structures in sand and clays. However, the available literatures based on physical models or the real field experiments to illustrate the behaviour of self-standing walls embedded in soft rocks are extremely rare (e.g. Richards et al. 2004). Perhaps it might be attributed to the difficulties in the installation and creating the failure of such large retaining structures in the real field or even in centrifuge models with required dimensions.

To fill the gap or limitation of field records and investigate the critical behavior of large diameter STPW embedded in stiff grounds, several series of centrifuge model tests and numerical analyses have been carried out to discuss the effects of critical conditions, such as embedment depth, and ground conditions, 3D effect of tubular pile wall, and static and dynamic loading.

3.1 Centrifuge model study

Three series of centrifuge model tests were conducted, 1) Simulation of excavation and loading using cantilever plate wall embedded in soft rock (Kunasegaram et al. 2018, Kunasegaram & Takemura 2019), 2) Horizontal loading tests to single STP pile and STP wall socketed in a soft rock with and without overlying sand, and 3) Dynamic loading tests to

STPW in a soft rock. The main parameter studied is the embedment depth to the stiff layer (d_R). For the uniform soft rock ground, the wall embedment depth (d_e) is equal to d_R, while for the two-layers ground de is the sum of the top layer depth and d_r (Figure 2). To generalize the embedment or socket depth, normalized embedment depth ($d_e\beta$ or $d_R\beta$) were estimated. The d_e and d_R adopted in the centrifuge tests were all far below the minimum embedment depth ($d_o = 3/\beta$).

3.1.1 Series I model: 2D retaining wall
Model test setup developed for this series is shown in Figure 7. This setup can model the high stiffness embedded cantilever wall behavior with large retaining height from the serviceability limit state (SLS) to the ultimate limit state (ULS) in a geotechnical centrifuge. The former performance corresponding an excavation process (ordinary loading) can be simulated by draining the water from the closed rubber box placed in front of the wall and the latter extreme loading process is created by feeding the drained water to backfill sand contained in a rubber box behind the wall. For the model which did not exhibit large displacement after rising the water level, the centrifugal acceleration was increased stepwise with a 5g increment up to 95g to observe the large movement of the wall. In this series, several centrifuge model tests were conducted in plane strain (2D) condition using aluminum plate walls with per width flexural rigidity (EI) equivalent to STPW with $\Phi = 2.5$ m & t = 25 mm and 1.0 m & t = 10 mm in a prototype scale under 50g centrifugal acceleration. Artificial soft rocks and sand were used as wall embedment media (Kunasegaram et al. 2018, Kunasegaram & Takemura 2019).

Observed wall top displacements and rotations of 12m high rigid walls ($\Phi_{eq} = 2.5$ m) with $d_e = 2.5$ m

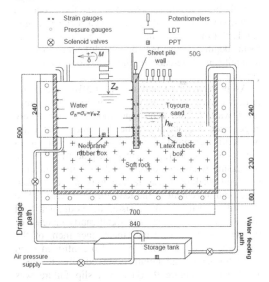

Figure 7. Centrifuge model test setup on rigid plate wall: Series I (Kunasegaram & Takemura 2019).

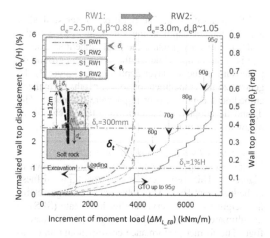

Figure 8. Effect of embedment depth observed in Series I centrifuge tests (Kunasegaram & Takemura 2019).

$(d_e\beta = 0.86)$ and 3.0 m $(d_e\beta = 1.05)$ are compared in Figure 8. As a unified loading index in the two processes, the moment load applied at the excavation bottom is used in the horizontal axis. Though the $d_e\beta$s of the model wall are much smaller than the minimum requirement $(d_0\beta = 3)$, the wall displacements by the excavation were well controlled below the target allowable displacement $(\delta_t = 120\text{mm}, 1.0\%$ wall height (H)). Taking wall top displacement $\delta_t = 300\text{mm}$, which is an allowable displacement as required performance 2 against level 2 earthquake (JTASPPACTC 2007), as a reference of ULS, the safety margin at SLS to the requirement failure are about 25% (from 3100 to 4000 kNm/m) for $d_e = 2.5\text{m}$ and about 40% (from 4000 to 5650 kNm/m) for $d_e = 3.0\text{m}$ respectively. These margins seem not large enough, but it should be pointed out that these required performances are introduced for small retained height, e.g. H < 4 m. Though the margins for the two embedment depths might not be so different, there is a significant difference in the behaviour over $\delta_t = 300\text{mm}$.

As pointed by Li and Lehane (2010), creep has critical effects on the behaviour of embedded cantilever wall. In the centrifuge model the relatively large creep displacements, the disp. increment without load increment, were observed as shown in Figure 8. However, all creep displacements were decreasing with time, except of $d_e = 2.5\text{m}$ wall after the final loading. Clear failure took place without additional increment of the load for $d_e = 2.5\text{m}$, while the wall with $d_e = 3.0\text{m}$ resisted the additional load more than 7000 kNm/m. The significantly increase of the wall stability by a half meter increment of the embedment for this case can be also confirmed from the deformation and failure observed after the tests as shown in Figure 9. Backward slip failure was confirmed for the wall with $d_e = 2.5\text{m}$ wall, but for

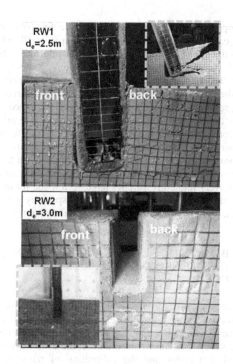

Figure 9. Observed deformation and failure of the cantilever walls (Kunasegaram & Takemura 2019).

$d_e = 3.0\text{m}$, the embedment portion was securely fixed by the soft rock, preventing the catastrophic failure.

3.1.2 Series II model: Parametric study on rock sock depth and ground conditions

To investigate the effect of embedment depth (d_e) under clear loading conditions, lateral resistances of the tubular steel pipe wall socketed into soft rock were investigated by centrifuge model tests in 50g centrifugal acceleration. They are two simplified models from the targeted structures and conditions, namely wall model and single pile model, as shown in Figure 10. Two types of model ground were prepared for the two models, single layer of soft rock, and soft rock with overlying sand. Lateral loading tests were performed for $\Phi = 2\text{m}$ single STP (SP) and STP wall (RPW) made socketed in the two model grounds with different wall/pile embedment depth, d_e, or rock socket depth, d_R (Figure 11). Considering the loading conditions and the displacement behavior of the embedded cantilever wall (Figure 10a), lateral load, P_L, were imposed to pile/wall top by one-way alternate manner as depicted in Figure 12. The load – displacement curves are compared in Figure 13. Details of the tests are given in Kunasegaram et al. (2019), but the test conditions, e.g. ground conditions and socket depth with normalized depth $[d_R, d_R\beta]$ are shown in Figure 11. From Series II tests, several findings are derived.

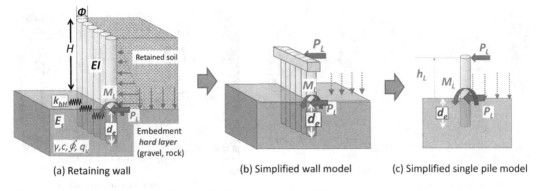

(a) Retaining wall   (b) Simplified wall model   (c) Simplified single pile model

Figure 10. Target structure and simplified models (Kunasegaram et al. 2019).

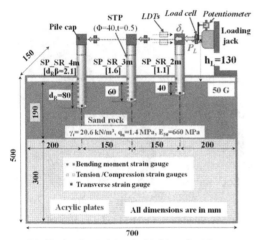

(a) Single pile model embedded in soft rock.

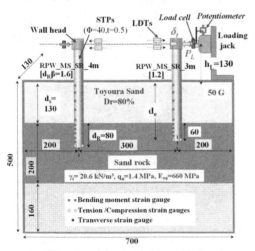

(b) STP wall model embedded in sand/soft rock.

Figure 11. Centrifuge model test setups for Series II (Kunasegaram et al. 2019).

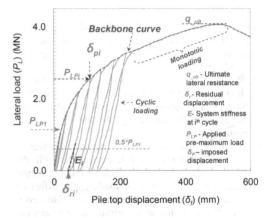

Figure 12. Typical load - displacement curves, and definitions of load and displacement parameters (Kunasegaram et al. 2019).

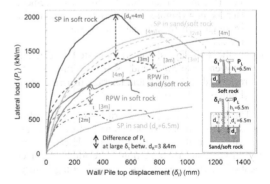

Figure 13. Observed deformation and failure of the cantilever walls (Kunasegaram et al. 2019).

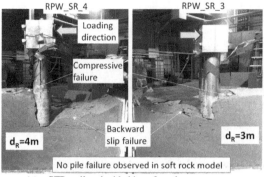
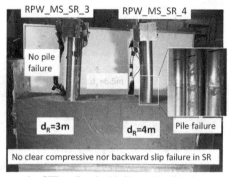

(a) STP wall embedded in soft rock.　　　　(b) STP wall embedded in sand/soft rock.

Figure 14. Observed failures of the STP wall model (Kunasegaram et al. 2019).

1) Lateral resistances of wall and single pile increase with d_R, but the trend of the increase depends on imposed displacement and, ground condition (Figure 13).
2) The observed failures of soft and pile/walls are shown for STP wall models and single pile models in Figures 14 & 15 respectively. These failures are controlled by d_R and d_e and the ground types.
3) Optimum socket depth, over which the effect of d_R is insignificant, is much smaller than $3/\beta$. As shallow rock part mainly resists the horizontal load in the early stage of loading, the effect of socket depth may not be so apparent. However, once the rock initiates yielding at the shallow depth, the influence of socket depth becomes eminent.
4) The single piles have higher lateral resistance per unit width than the walls both for the initial sub-grade reaction and the ultimate resistance.
5) The effects of socket depth and the difference between the pile and the wall on the lateral resist-ance (Figure 13) and the residual displacement after loading (Figure 16) are more significant for the single soft rock layer than the sand/soft rock layers.
6) In the two layers, the complicated interaction between pile/wall and soil determines the residual displacement and lateral stiffness of wall (Figures 17 & 18).

The above findings are all critically concerned to the issues for the rational design procedures, such as, critical embedment depth, non-linearity of p-y curves, and accumulation of residual displace-ment of the wall subjected to various loading history.

3.1.3 Series III model: Dynamic loading on STP retaining wall embedded in soft rock

A 12m high cantilever walls embedded in the soft rock with backfill sand, which is similar to the

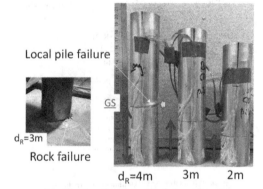

(a) Single piles embedded in soft rock.

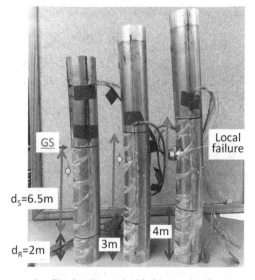

(b) Single piles embedded in sand/soft rock.

Figure 15. Observed failure of STP (Kunasegaram et al. 2019).

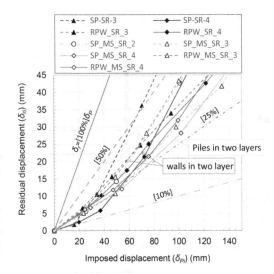

Figure 16. Residual displacement against imposed displacement (Kunasegaram et al. 2019).

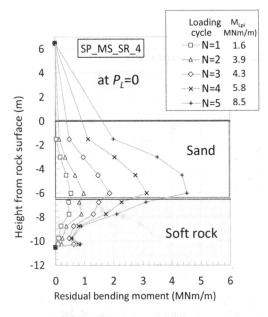

Figure 17. Residual bending moment of the STP wall embedded in sand/rock (Kunasegaram et al. 2019).

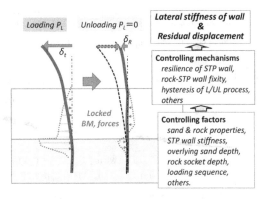

Figure 18. Hypothetical mechanism of residual displacement after unloading and lateral subgrade reaction in reloading (Kunasegaram et al. 2019).

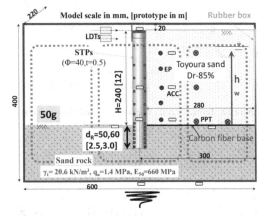

Figure 19. Centrifuge model test setup of STP wall for dynamic loading: Series I (Shafi et al. 2019).

model in Series I, were made using the same STP wall in Series II as illustrated in Figure 19. Several dynamic loadings were applied by sinusoidal input acceleration to the models with different embedment depths of d_R = 2.5 m and 3.0 m (Shafi et al. 2021). In the loading sequence, water was fed in the backfill to rise the water level (Figure 20). Typical observed wall top displacement and earth pressured behind the wall are

shown in Figure 21. In the early stage of cyclic loading, the amplitudes from the trends and the residual wall displacement are relatively large compared to the later stage of loading where steady cyclic behavior is observed for the all measurements. This typical behavior could be confirmed for the wall with d_R = 3.0 m ($d_R\beta$ = 1.2), but it is the case only for dry shaking for the wall with d_R=2.5m ($d_R\beta$ = 10) as shown in Figure 20. The accumulated wall displacements observed in the entire loading processes were plotted against the cumulative Arias intensity for the walls with d_R = 2.5m and 3.0 m in Figure 22. Similar to Series I, the dynamic stability of the wall is significantly increased by a half meter increase of the embedment into the soft rock. It was also found that high confinement or fixity of the piles by the stiff ground could generate the large resilience of pile, which resulted in the increase of wall earth pressure after shaking (Figure 23). This pressure should be considered as an action to

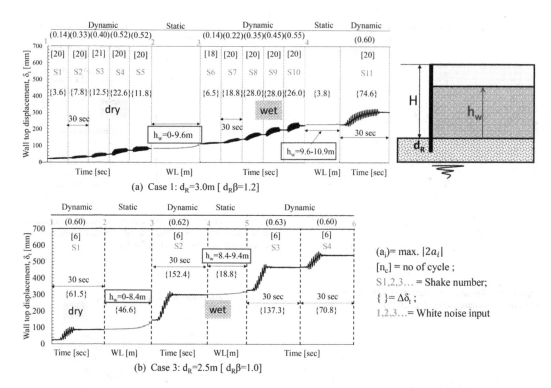

Figure 20. Dynamic and static loading conditions and observed wall top displacement (Shafi et al. 2019).

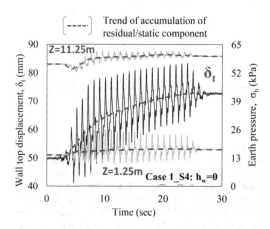

Figure 21. Observed wall top displacement and earth pressure during shaking (Shafi et al. 2021).

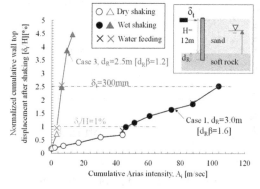

Figure 22. Effect of embedment depth under dynamic loading Series III centrifuge tests (Shafi et al. 2021).

examine the structural safety of the pile and the wall residual displacement after the earthquake. It should be not that no clear ground failure, as observed in Series II (Figure 14), was observed for the wall with $d_R = 3.0$ m, which had been loaded under very critical conditions (high water table and large intensity of dynamic loading) and $d_R = 2.5$ m, which was displaced nearly 5% of wall height at wall top (Figure 24). The vertical overburden stress on the

rock surface behind the all could prevent the backward slip failure.

3.2 Numerical studies: Issues remained

Though the centrifuge model tests could provide valuable results on the mechanical behavior of the cantilever STPW embedded in stiff grounds, there are still many remaining concerns which might affect the wall behavior or be critical in the rational and safe design procedure, such as,

36

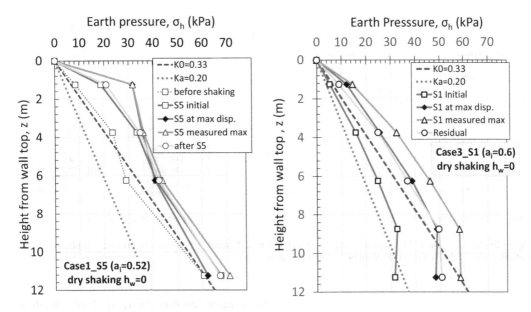

Figure 23. Change of earth pressure distribution on the wall (Shafi et al. 2021).

Figure 24. STP wall and embedded soft rock after dynamic loading: Series III centrifuge tests (Shafi et al. 2021).

- plugging of rock in the socket part,
- local 3D effect of deformation of thin wall tubular pile.

In particular, the pile structural failure with local buckling (Figures 14 & 15) could be a common ULS for the wall relatively large d_e, even $d_e\beta$ are well below the minimum requirement ($d_e\beta > 3.0$). Using 3D FEM, Ishihama et al. (2019) investigated the plugging effect, the local stresses at the pile tip, and pile deformation at the pile tip and the rock surface, and found that the plugging effect and 3D effects could not significantly affect the overall behavior of laterally loaded STPW in soft rock.

TC1_WG3 further conducted 3D and 2D FEM analyses for the wall model embedded in the soft

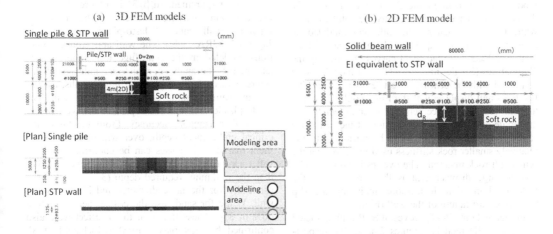

Figure 25. (a) 3D FEM models of single STP and STP wall (Centrifuge test Series II) and (b) 2D FEM model solid beam wall with EI equivalent to that of the STP wall.

37

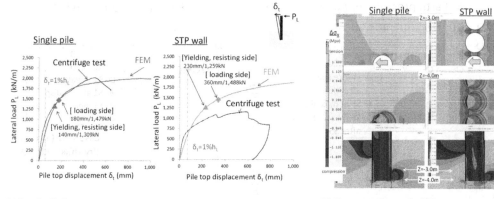

(a) Load - displacements

(b) Contours of $\Delta\sigma_h$ at δ_t=200mm

Figure 26. Results of 3D FEM analyses of single STP and STP wall (Centrifuge test Series II) (a) load – displacement with centrifuge test results, and (b) contours of horizontal stress increment.

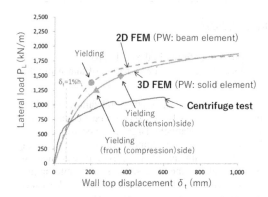

Figure 27. Comparison of 3D and 2D FEM of STP wall embedded in soft rock (d_R=4.0m, Centrifuge test Series II).

with d_R = 4.0 m (Centrifuge test Series II), of which models are given in Figure 25. In the 3D analysis, the actual tubular pile was modeled by solid element, while in the 2D analysis, the pile was modeled by a beam element with equivalent EI of the actual STP wall. The load – displacement curves obtained by the 3D analyses (Single pile and STP wall models) were compared with the centrifuge test result in Figure 26. For the small displacement the FEM could well capture the load – displacement behavior especially for the single pile. The under-estimation of the resistance at the small displacement could be attributed to the relatively smaller rock stiffness used in FEM than the actual soft rock material. The over-estimation of STP wall at large displacement is the limitation of the FEM used in strain localization including the slip type backward failure of the wall (Figure 14).

In Figure 27 the P_L- δ_t curves of STP wall predicted by 3D and 2D models (Figures 25a & 25b respectively) are compared. At the relatively small displacement less than 50mm, which corresponds to allowable value or SLS, no significant difference can be seen in the 3D and 2D models. Over this SLS, the resistance of 2D model becomes larger than that of the 3D model. This can be attributed to local yielding of the ground and pile due to stress concentration at the front toe near the rock surface and back toe near the pile bottom (Figure 26b). As mentioned above, the analyses overestimate the ultimate resistance, however, from Figure 27 it can be said that the 2D model can be applied without specific consideration of 3D effects of STP wall for the displacement prediction till the SLS.

4 ANALYTICAL STUDIES: SUBGRADE REACTION (P-Y) METHOD

The top wall displacements analyzed by subgrade reaction method using bi-linear p-y relation are plotted against to the relative embedment depth to the minimum requirement ($d_e\beta/3$) in Figure 28. This is the common analytical model for the design of retaining wall called "elasto-plastic analysis" in Japan (e.g. JRA 1999). The wall and ground conditions modeled in the centrifuge test Series I were assumed in the analyses, which are indicated in the figure with the centrifuge experiment results, assuming an ordinal loading condition of h_w = 0 m. The elasto-plastic analysis could predict the wall deflection with reasonable accuracy. From the figure the critical embedment depth, over which the displacement markedly increases can be confirmed. For the soft rock cases, the critical depths are much smaller than the minimum required depth ($3/\beta$), 20 to 30%, especially for the large diameter and high retaining wall. While for sand case the critical depth is about 60% of $3/\beta$. These trends of the d_e effect were also confirmed by rigid-plastic FEM (Mochizuki et al. 2019&2021).

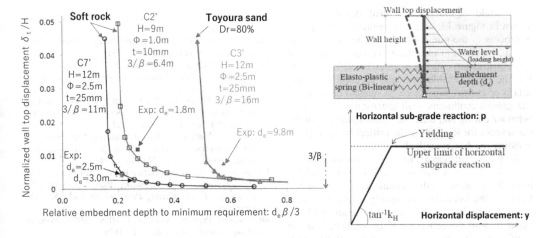

Figure 28. Relationship between wall top displacement and relative embedment depth to the minimum requirement $(d_e\beta/3)$ calculated by subgrade reaction method using bi-linear p-y curves for the conditions of the walls in Series I centrifuge model at the end of excavation process, h_w=0m (Ishihama et al. 2019).

Developments of plastic region in the embedment against the water rise, h_w, obtained by the elasto-plastic analysis are depicted in Figure 29. The "plastic" corresponds to the subgrade reaction reaching to the upper limit as showing in Figure 28.

Detailed discussion about the design method for the embedment length of STP wall pressed-in stiff ground was made by Sanagawa (2021).

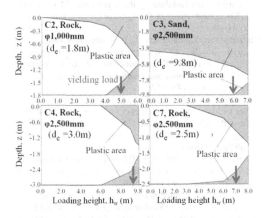

Figure 29. Development of plastic region in the embedded depth with the loading by increasing the water level in the backfill: Series I centrifuge model (Ishihama et al. 2019).

5 CONCLUDING REMARKS: CONSIDERATION OF RATIONAL DESIGN

Current design practice for the embedded cantilever wall using the minimum embedment depth $(d_o = 3/\beta)$ adopted in the simple design method (JTASPPACTC 2007, IPA 2014&2021) is based on the assumption of infinite beam on the uniform linear elastic subgrade. But the subgrade reaction of the pile and wall are very non-linear and change with depth (API 2002, Liang et al. 2009, Reese 1997). As its simplicity, this simple method could be beneficial for the small height flexible sheet pile wall. However, in the application of STP wall in stiff grounds this minimum embedment depth requirement tends to be over conservative and not economical, and even inconsistent with limit states design concept, especially for the high retaining height and large diameter STP wall. Structural pile failure becomes dominant ultimate limit state over a certain embedment depth (d_e). This can be considered as a critical depth, over which no significant contribution of the d_e increase can be expected for ULS, and this critical depth might be much less than the minimum required embedment depth. The common practice using minimum embedment depth in the simple method should be considered as an option, not requirement and the required performances of the limit states, e.g. SLS and ULS should be examined by reasonable methods, considering the non-linearity of soil – structure interaction.

Combination of elasto-plastic analysis and limit equilibrium method should be a common design practice for the large diameter STP wall in stiff ground. However, in the application of the non-linear subgrade reaction method, there are several issues remained for the further research. For examples, modeling of p-y relation, including the evaluation of the parameters used in the p-y curves. The estimation of modulus and strength of soil/rock is key issues for the very stiff ground with very large SPT-N values. The variability of ground conditions are also major uncertainties in the design and construction of the wall. Especially as the depth of stiff

39

layer could significantly control deformation as shown in Figure 13. The data recorded in the press-in process of the pile could contribute to the reduction of the effect of uncertainty and economical and safe construction (Suzuki et al. 2021a&b). To establish the more reliable methods, accumulating field data is of critical importance, as reliable database on site ground conditions, wall specifications, and wall behavior during and after the construction should be the sources for identifying the critical issues and for updating the design method.

As for the earthquake loading, two issues can be pointed out, one is dynamic earth pressure and the other the residual earth pressure after the loading. Even for the large flexural rigidity of the wall with secured fixity by the stiff ground, a relatively large dynamic wall deflection could be generated in the cantilever wall, which might cause the seismic earth pressure different from the one on the rigid wall and increase the residual earth pressure. They are closely related to the wall stress and residual displacement (see Figures 17&18,Figure 24). To have answers on these issues, the further researches should be performed by the physical and numerical studies and field study of the wall after the strong earthquake.

The outcome derived from the TC1 activity will be summarized and reported as a form of final report soon.

ACKNOWLEDGEMENT

All contents in this SOA report are outcomes of TC1 activity. The author must express his deep appreciation to the efforts and contribution of all committee members, especially leader and Ex-leaders of Working group, Dr. T. Sanagawa (WG1), Mr. Y. Ishihama & Mr. K. Toda (WG3) and Mr. N. Suzuki (WG4), and Dr. K. Sawada (General Secretary). The enthusiasm and hardworking of the student members (Dr. V. Kunasegaram and Mr. SM Shafi (Tokyo Tech), Mr. K. Mochizuki (Hokkaido Univ) must be acknowledged. The author would also like to express his gratitude to Prof. Osamu Kusakabe, the immediate past President of IPA, for his suggestion to initiate this TC.

REFERENCES

American Petroleum Institute (API). 2002. Recommended Practice for Planning, Designing and Constructing Fixed Offshore Platforms—Working Stress Design: 226p.

Bolton, M.D. & Powrie, W. 1987. The collapse of diaphragm walls retaining clay. Géotechnique, 37(3): 335–353.

Bolton, M.D. & Powrie, W. 1988. Behaviour of diaphragm walls in clay prior to collapse. Géotechnique, 38(2): 167–189.

D'Andrea, R. & Day, R.W. 1998. Discussion and closure: design and construction of cantilevered retaining walls.

Practice Periodical on Structural Design and Construction 3(2): 87–88.

Day, R.A. 1999. Net pressure analysis of cantilever sheet pile walls. Géotechnique, 49(2): 231–245.

Gaba, A.R., Simson, B., Powrie, W. Beadman, D.R. & D. Selemetas. 2017. Guidance on embedment retaining wall design, CIRIA: 441p.

He, H. 2018. Gyropress (Rotary Cutting Press-in) method in Disaster Recovery Project (Kagoshima Prefecture, Japan), IPA Newsletter, 3(1): 11–15.

Hirata, H. & Matsui, N. 2016. Expanding Applications of the Gyro-press Method, Nippon Steel & Sumitomo Metal Technical Report, No.113: 42–48.

International Press-in Association (IPA). 2014. A handbook of Steel Tubular Pile Retaining Wall Constructed by Gyropress (Rotary Cutting Press-in) Method: 152p. (in Japanese).

International Press-in Association (IPA). 2019. Press-in Piling Case History (Volume 1): 168p.

International Press-in Association (IPA). 2021. Press-in retaining structures: a handbook (2nd) edition..

Ishihama, Y., Takemura, J. & Kunasegaram, V. 2019. Analytical evaluation of deformation behavior of cantilever type retaining wall using large diameter tubular piles into stiff ground, Proceedings of the 4th International Conference on Geotechnics for Sustainable Infrastructure Development, Hanoi: 91–98.

Japan Road Association (JRA). 1999. Road Earthwork, Guideline for the Construction of Temporary Structures, (in Japanese).

Japanese Technical Association for Steel Pipe Piles and Advanced Construction Technology Centre (JTASP-PACTC) 2007. Cantilever Steel Sheet Piles Retaining Wall - Design Manual: 138p. (in Japanese).

Kitamura, M. & Kitamura, S. 2019. Cantilevered Road Retaining Wall Constructed of 2,000mm Diameter Steel Tubular Piles Installed by the Gyro Press Method with GRB system. Press-in Piling Case History, IPA, Volume 1: 41–48.

Kunasegaram, V., Takemura, J., Ishihama, Y. & Ishihara, Y. 2018. Stability of Self-Standing High Stiffness-Steel Pipe Sheet Pile Walls embedded in Soft Rock. Proceedings of 1st International Conference on Press-in Engineering, Kochi: 143–152.

Kunasegaram, V. & Takemura, J. 2019. Deflection and failure of high stiffness cantilever retaining wall embedded in soft rock. International Journal of Physical Modelling in Ge-otechnics, http://dx.doi.org/10.1680/jphmg.19.00008

Kunasegaram, V., Shafi, SM, Takemura, J. & Ishihama, Y. 2019. Centrifuge model study on cantilever steel tubular pile wall embedded in soft rock, Proceedings of the 4th International Conference on Geotechnics for Sustainable Infrastructure Development, Hanoi: 1045–1052.

Li, A.Z. & Lehane, B.M. 2010. Embedded cantilever retaining walls in sand. Géotechnique, 60(11): 813–823.

Liang, R. Yang, K. & Nusairat, J.M. 2009. p-y Criterion for Rock Mass. J. Geotech. Geoenviron. Eng., 135(1): 26–36.

Long M. 2001. Database for retaining wall and ground movements due to deep excavations. J. Geotech. Geoenviron. Eng., 127(3): 203–224.

Miyanohara, T., Kurosawa, T., Harata, N., Kitamura, K., Suzuki, N. & Kajino, K. 2018. Overview of the Self-standing and High Stiffness Tubular Pile Walls in Japan, Proceedings of the First International Conference on Press-in Engineering, Kochi: 167–174.

Mochizuki, K., Isobe, K., Takemura, J. & Ishimhama, Y. 2019. Numerical simulation for centrifuge model tests on the stability of self-standing steel pipe pile retaining wall by Rigid Plastic FEM. *Proceedings of the 4th International Conference on Geotechnics for Sustainable Infrastructure Development, Hanoi*: 481–488.

Mochizuki, K., Isobe, K., Takemura, J. & Toda, K. 2021. Numerical simulation for centrifuge model tests on cantilever type tubular pile retaining wall by rigid plastic FEM. *Proceedings of 2nd International Conference on Press-in Engineering, Kochi*. (to be published).

Nanazawa, T., Kouno, T. & Tanabe, M. 2015. Studies on the end bearing capacities of pile in rock grounds. *Technical note of PERI, Japan*: 80p (in Japanese).

Ou, C. Y., Hsieh, P.G. & Chiou, D.C. 1993. Characteristics of ground surface settlement during excavation. *Canadian Geotechnical Journal*, 30(5): 758–767.

Padfield C.J. & Mair R.J. (eds) 1984. The Design of Propped Cantilever Walls Embedded in Stiff Clays. Construction Industry Research and Information Association (CIRIA), Westminster, London. CIRIA Rep: 104p.

Randolph, M. 2019. Considerations in the design of piles in soft rock -Keynote Paper. *Proceedings of the International Conference on Geotechnics for Sustainable Infrastructure Development. Hanoi*: 1297–1312.

Randolph, M. 2021. Design Considerations in the Tip Resistance of Piles Jacked or Driven into Strong Soil or Weak Rock. Keynote lecture 2, *Proceedings of 2nd International Conference on Press-in Engineering, Kochi*. (to be published).

Reese, L. C. 1997. Analysis of laterally loaded piles in weak rock. *J. Geotech. Geoenviron. Eng.*, 123(11): 1010–1017.

Richards, D.J. & Powrie, W. 1998. Centrifuge model tests on doubly propped embedded retaining walls in overconsolidated kaolin clay. *Géotechnique*, 48(6): 833–846.

Richards, D. J., Clayton, C.R.I., Powrie, W. & Hayward, T. 2004. Geotechnical analysis of a retaining wall in weak rock.

Sanagawa, T. 2021. Discussion about design method for embedded length of self-standing steel tubular pile walls pressed into stiff ground. *Proceedings of 2nd International Conference on Press-in Engineering, Kochi*. (to be published).

Shafi, S, M., Takemura, J., Kunasegaram, V., Ishihama, Y., Toda, K. & Ishihara, Y. 2021. Dynamic behavior of Cantilever Tubular Steel Pile Retaining Wall Socketed in Soft Rock. *Proceedings of 2nd International Conference on Press-in Engineering, Kochi*. (to be published).

Suzuki, N. & Kimura, Y. 2021. Summary of case histories of retaining wall installed by rotary cutting press-in method. *Proceedings of 2nd International Conference on Press-in Engineering, Kochi*. (to be published).

Suzuki, N., Nagai, K. & Sanagawa, T. 2021a. Reliability analysis on cantilever retaining walls embedded into stiff ground (Part 1: contribution of major uncertainties in the elasto-plastic subgrade reaction method), *Proceedings of 2nd International Conference on Press-in Engineering, Kochi* (to be published).

Suzuki, N., Ishihara, Y. & Nagai, K. 2021b. Reliability analysis on cantilever retaining walls embedded into stiff ground (Part 2: construction management with piling data), *Proceedings of 2nd International Conference on Press-in Engineering, Kochi*. (to be published).

Takada, H. 2016. Restoration and Reconstruction of Kyu-Kitakami River, *IPA Newsletter*, 1(1): 3–5.

Whittle, A. J. & Davies, R. V. 2006. Nicoll highway collapse: evaluation of geotechnical factors affecting design of excavation support system. *In International Conference on Deep Excavations, Singapore*.

Proceedings of the Second International Conference on
Press-in Engineering 2021, Kochi, Japan – Matsumoto et al (eds)
© *2021 Taylor & Francis Group, London, ISBN 978-1-032-10414-0*

State of the art report on the use of press-in piling data for estimating subsurface information

Y. Ishihara
Giken, Kochi, Japan

O. Kusakabe
International Press-in Association, Tokyo, Japan

ABSTRACT: The construction and design of structures with piles are usually conducted based on limited number of results of subsurface investigations, which will be conducted typically at the intervals of several tens or one hundred meters. On the other hand, there are local variations in the actual ground, which could exist in a smaller area than the spatial intervals of the subsurface investigations. As a result, the prior information and the actual condition of the ground can be different, which deteriorates the rationality of the construction and design. The press-in piling data obtained for every single pile are expected to provide an effective solution, by being applied to the automatic machine operation or to the estimation of the subsurface information. This paper introduces the methods of estimating the subsurface information from press-in piling data, most of which were organized into a technical material in Japanese under the activity of IPA-TC2.

1 INTRODUCTION

1.1 *Outline of the Press-in Method*

The Press-in Method is one of the piling methods. As it installs piles using a static jacking force, it generates less noise and vibration. It gains a reaction force by firmly grasping the previously installed piles, and thus saves temporary works as shown in Figure 1. In addition, as it grasps piles not at the pile head but near the ground surface, it requires less headroom as shown in Figure 2.

There are four penetration techniques in the Press-in Method, as summarized in Figure 3. The basic one is "Standard Press-in" where no installation assistance such as water jets or augers are used. It is noted that "surging", a repeated penetration and extraction during installation, is not taken as an installation assistance and can be adopted in any of the four penetration techniques. The other three are "Press-in with Water Jetting" where water jets are used in the pile base, "Press-in with Augering" where soils near the pile base are excavated and temporarily lifted up, and "Rotary Cutting

Press-in" where piles with base cutting teeth are pushed and rotated at the same time. With the development of the latter two techniques, the applicability of the Press-in Method has been expanded to hard grounds including rocks and concretes.

More detailed information on the Press-in Method and its applications can be found in IPA (2016), IPA (2020) and Bolton *et al.* (2020).

Figure 1. Press-in piling system saving temporary works (JPA, 2017).

DOI: 10.1201/9781003215226-65

Figure 2. Press-in piling under a restriction of headroom (IPA, 2015).

1.2   Use of piling data in the Press-in Method

Generally, the construction and design of structures with piles are based on the subsurface information obtained by interpolating the limited number of subsurface investigation results. The spatial intervals between two adjacent points of subsurface investigations are typically several tens or one hundred meters. On the other hand, it is often the case that local variations can be seen in the actual ground. For examples, the geological structure may not be homogeneous horizontally, and weak soils or hard cobbles may exist locally. If the areas of such local variations are smaller than the spatial intervals of the subsurface investigation points, the prior information and the actual condition of the ground becomes different. This difference deteriorates the rationality of the construction and design.

In the Press-in Method, piling data can be obtained for every single pile. This feature is expected to be utilized as ICT (Information and Communication Technology) to provide a solution for the above-mentioned issue.

The press-in piling data includes the jacking force, torque, penetration depth, time, rotational number and so on. As shown in Figure 4, applications of the data are exemplified by the selection of press-in conditions, the estimation of subsurface information and the estimation of the pile performance.

In the selection of press-in conditions, the values for the press-in parameters such as the downward velocity, upward velocity, downward displacement, upward displacement and so on, are selected based on a judgement using the piling data. A good example is the automatic operation system where the press-in conditions selected based on the piling data

Standard Press-in	Press-in with Water Jetting	Press-in with Augering	Rotary Cutting Press-in
Press-in a pile without installation assistance	Press-in a pile wile applying water-jetting in the pile base	Press-in a pile while excavating the soil around the pile base	Rotate and press-in a pile equipped with base cutting teeth

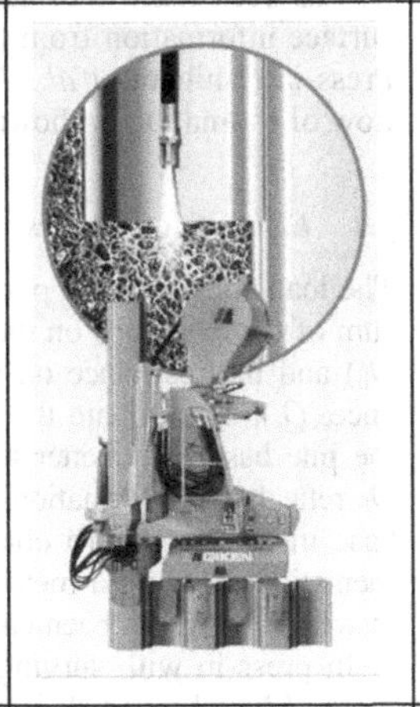

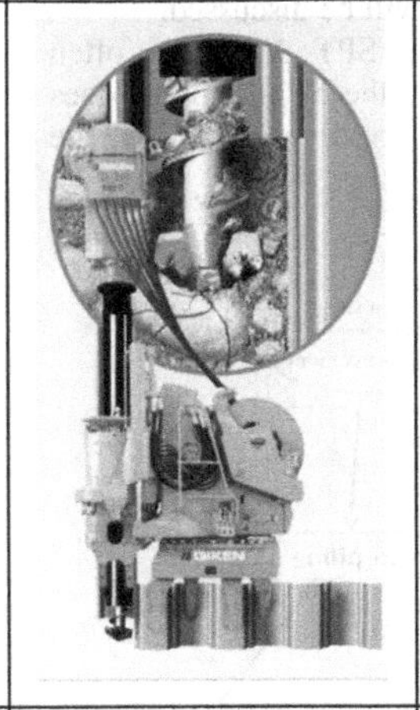

Figure 3. Four penetration techniques in the Press-in Method.

are feed-backed to the press-in machine continuously during the piling work (Ishihara, 2018).

The estimation of the subsurface information is to estimate the information (type and state) of the soil around the pile base, by interpreting the piling data. The estimated information would realize a more reliable termination control management (construction management at the end of installation of each pile), or provide contractors with objective materials for judging the necessity of changing the penetration techniques or the pile embedment depth.

The estimation of the pile performance means estimating the performance (such as the vertical capacity and the horizontal resistance) of the installed piles from the piling data. Although the confirmation of the pile performance is usually done by the load tests, they cannot be conducted for all the piles because of the additional time and cost required for them. If the pile performance is estimated from the piling data, it would become possible to assure a certain level of quality for all the piles without causing the issue of additional time and cost.

1.3 Objectives of this paper

Among the four applications of the piling data explained in the previous section, this paper introduces the methods of estimating the subsurface information, by re-structuring and adding recent findings to the contents in the IPA-TC2 technical material written in Japanese (IPA, 2017).

The penetration techniques dealt with in this paper will be Standard Press-in, Press-in with Augering and Rotary Cutting Press-in. For Standard Press-in, methods of estimating the cone resistance (q_c) of CPT (Cone Penetration Test), soil type and N value of SPT (Standard Penetration Test) will be discussed. For the other two penetration techniques, methods of estimating the SPT N value will be discussed.

In the hard ground, the SPT N values often exceeds 50 (more precisely, the SPT sampler does not penetrate into the virgin ground by the

designated value of 0.3 meters even when the blow count reaches 50). In such cases, this paper will adopt the converted N value as expressed by:

$$\text{Converted } N = 50 \times \frac{0.3 \text{ [m]}}{\delta z_{\text{SPT}(50)} \text{ [m]}} \quad (1)$$

where $\delta z_{\text{SPT}(50)}$ is the incremental penetration depth of the SPT sampler for the blow count of 50 (JGS, 2015).

There are several examples of the use of piling data in other piling methods. For bored piles or driven piles (using vibratory hammers), methods of confirming the bearing stratum based on the electric current values required for operating augers or vibratory hammers (Hashizume et al., 2002; JRA, 2015; JFCC & COPITA, 2017). For driven piles, methods of estimating a static vertical capacity of a pile based on the piling data obtained by using an instrumented pile (Likins, 1984; Rausche et al., 1985) are widely known. On the other hand, the methods in this paper are different from the above-mentioned existing methods in that they are based on the static loads during the piling work, which would allow simpler interpretation of the data, and that they do not require complicated instrumentation with piles (other than a device for measuring the length of the soil inside the pile, as explained in Section 4).

2 ESTIMATION FROM PILING DATA OBTAINED IN STANDARD PRESS-IN

In Standard Press-in, a pile is installed by a static jacking force without the use of any installation assistance, and the process of the penetration of the pile is similar to that of a cone in CPT. This similarity has been taken into account to estimate the subsurface information from the piling data in Standard Press-in (Ishihara et al., 2015a). The outline of the flow of estimation is shown in Figure 5.

2.1 Estimating base resistance and shaft resistance

The load applied to the pile head (head load, Q) is the sum of the resistance on the pile base (base resistance, Q_b) and the resistance on the pile shaft (shaft resistance, Q_s). To estimate the subsurface in formation at the pile base, it is better to use Q_b rather than Q, as Q_b reflects the information of the soil beneath the pile base more directly. To obtain Q_b without any instrumentation on piles, a method using the data of press-in with surging has been confirmed to be effective.

In press-in with surging, the downward displacement (l_d) and upward displacement (l_u) are applied to the pile alternately ($l_d > l_u$), as shown in Figure 6. Ogawa et al. (2012) conjectured that, as shown in Figure 7, the head load recorded when the pile base passes a certain depth for the first time (Q_1) is the sum of Q_b and Q_s, while the head load recorded

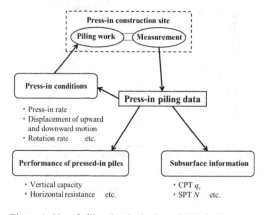

Figure 4. Use of piling data in the Press-in Method.

1) Load applied to a pile
⇒ Base resistance / Shaft resistance

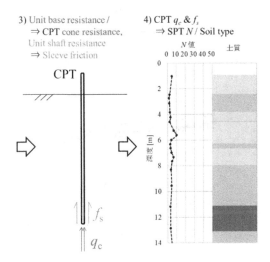

2) Base resistance /
⇒ Unit base resistance,
Shaft resistance
⇒ Unit shaft resistance

3) Unit base resistance /
⇒ CPT cone resistance,
Unit shaft resistance
⇒ Sleeve friction

4) CPT q_c & f_s
⇒ SPT N / Soil type

Figure 5. Flow of estimation in Standard Press-in.

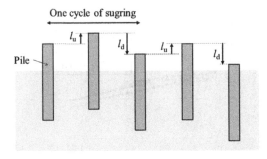

Figure 6. Process of surging.

when the pile base passes that depth for the second time (Q_2) consists only of Q_s, as expressed by the following two equations.

$$Q_1 = Q_b + Q_s \qquad (2)$$

$$Q_2 = Q_s \qquad (3)$$

From these equations, Q_b can be expressed as:

$$Q_b = Q_1 - Q_2 \qquad (4)$$

They argued that the values of Q_1 and Q_2 should be obtained shortly after the beginning of the downward motion in each cycle of surging, in order to avoid the influence of soils that may have collapsed into the void created beneath the pile base during the previous upward motion. Figure 8 shows the variation of Q and Q_s in one cycle of surging, which was obtained in C11 test series in a soft alluvial ground shown in Figure 9 by using a closed-ended pile with the outer diameter of 318.5 mm. The pile was equipped with a load cell in its base to measure Q_b, and Q_s was obtained by subtracting Q_b from Q. As can be confirmed in the figure, values of Q_s were similar in the first and the second penetrations. On the other hand, values of Q in the second penetration were identical with Q_s values when the downward displacement (increment in the penetration depth) in the second penetration was smaller than 0.1 m but gradually increased afterwards. Ishihara et al. (2015a) further analyzed the data obtained in C11 test series (C11-05) and confirmed that the difference between the estimated and measured Q_b values became larger with the increase in the second downward displacement (l_{d2}), regardless of the soil type or the penetration depth, as shown in Figure 10. Based on these, it is recommended to define the values of Q_1 and Q_2 as the arithmetic average of the Q values recorded in $0.1\,D_o < l_{d2} < 0.2\,D_o$, where D_o is the outer diameter of the pile.

Figure 11 Shows the comparison of Q_b measured by the load cell and Q_b estimated by Equation (4) in two tests in the C11 test series (C11-05 and C11-06). The two tests were conducted with the same condition ($l_d = 800$mm, $l_u = 400$mm) at different positions which were distant from each other by about 4 meters. Good agreement can be confirmed between the estimated and measured values in the depth

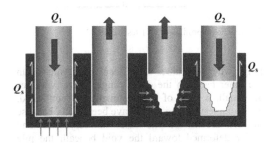

Figure 7. Forces acting on a pile during surging.

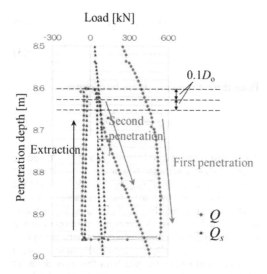

Figure 8. Variation of Q and Q_s in one cycle of surging.

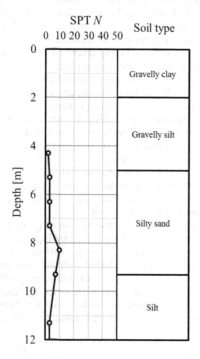

Figure 9. Site profiles of C11 test series.

range where N values show local peak values. On the other hand, in the depth range where N values are small, a trend of underestimation is confirmed. This underestimation might have been partly because the soil around the pile base was very weak and easily deformed toward the void beneath the pile base during the upward motion of the pile, leading to the increase in the Q_2 values.

By the way, according to Ogawa et al. (2011), it is not a better option to assume the extraction force in each cycle of surging as being identical with Q_s. This seems to be partly due to the difference in Q_b - Q_s interaction during the penetration and the extraction.

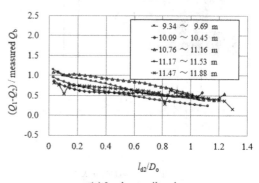

(a) In clay to silty clay

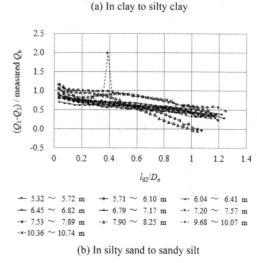

(b) In silty sand to sandy silt

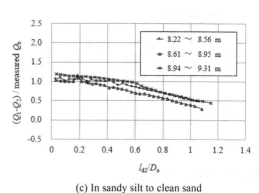

(c) In sandy silt to clean sand

Figure 10. The ratio of estimated Q_b (= Q_1 − Q_2) to measured Q_b plotted against l_{d2}/D_o in each cycle of surging.

2.2 Estimating unit base resistance and unit shaft resistance

This sub-section discusses the method to estimate the unit base resistance (q_b) and the unit shaft resistance (q_s) from Q_b and Q_s.

2.2.1 Estimating unit base resistance of open-ended tubular piles

Regarding the pile base, it is necessary to consider the plugging condition. For open-ended tubular piles, it is possible to assess the plugging condition based on the index called IFR (Incremental Filling Ratio) as expressed by the following equation (White & Deeks, 2007; Lehane et al., 2007; White et al., 2010).

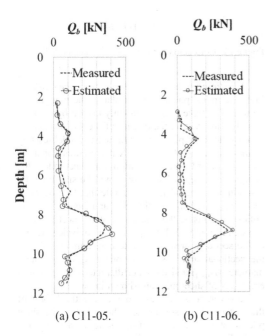

Q_b [kN] Q_b [kN]

(a) C11-05.   (b) C11-06.

Figure 11. Comparison of estimated and measured Q_b (Closed-ended, D_o = 318.5mm).

$$IFR = \frac{\delta h}{\delta z} \qquad (5)$$

Here, δz is the increment of the penetration depth, and δh is the increment of the length of the soil column inside the pile that is observed while the penetration depth increases by δz. The fully plugged condition, the fully unplugged condition and the partially plugged condition are represented by IFR = 0, IFR = 1 and 0 < IFR < 1 respectively.

For open-ended tubular piles, Q_b is expressed as the sum of the resistance on the bottom of the inner soil column (Q_{bi}) and the resistance on the pile base annulus (Q_{bp}):

$$Q_b = Q_{bi} + Q_{bp} \qquad (6)$$

Q_{bp} could be assumed as:

$$Q_{bp} = \left(\frac{\pi D_o^2}{4} - \frac{\pi D_i^2}{4}\right) \times q_{b,closed} \qquad (7)$$

where $q_{b,closed}$ is the unit base resistance of a closed-ended pile, and D_o and D_i are the outer and inner diameter of the pile respectively. On the other hand, the inner soil column receives not only Q_{bi} but also its self-weight (W_s) and the inner shaft resistance (Q_{si}) as shown in Figure 12, and these forces will satisfy the following equilibrium condition:

$$Q_{bi} = W_s + Q_{si} \qquad (8)$$

Kurashina (2016) conducted model tests to press-in a closed-ended pile (D_o = 101.6mm) or an open-ended pile (D_o = 101.6mm, D_i = 83.5mm) in a dry sand with the relative density being around 60%. The closed-ended pile was equipped with strain gauges in its base to measure Q_b, while the open-ended pile was equipped with a load cell in its head to measure Q_{bi} and a stroke sensor to obtain the length of the inner soil column, as shown in Figure 13. Figure 14 shows the correlation of (1 − IFR) and the right side of Equation (8) normalized by the potential push-up stress at the bottom of the inner soil column (= ($\pi D_i^2/4$) $q_{b,closed}$), which were obtained by analyzing the data of Kurashina (2016). Based on this figure, the following correlation is found:

$$\frac{W_s + Q_{si}}{\frac{\pi D_i^2}{4} \times q_{b,closed}} = \lambda \times (1 - IFR) \qquad (9)$$

which is basically in line with the findings of Lehane & Gavin (2001). This equation is converted into:

$$W_s + Q_{si} = \lambda \times (1 - IFR) \times \frac{\pi D_i^2}{4} \times q_{b,closed} \qquad (10)$$

Combining Equations (8) and (10),

$$Q_{bi} = \frac{\pi D_i^2}{4} \times \lambda \times (1 - IFR) \times q_{b,closed} \qquad (11)$$

47

The value of λ is difficult to be reliably determined, because the number of the database is limited. Tentatively, it will be assumed as unity (smaller than the experimental results), which will give a conservative estimation results (smaller SPT N values and soil types with smaller grain sizes) based on the methods explained later. Combining Equations (6), (7) and (11),

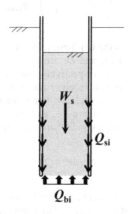

Figure 12. Forces acting on the soil column inside the pile.

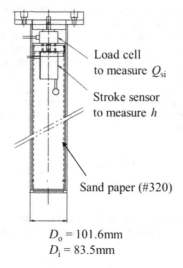

Load cell
to measure Q_{si}

Stroke sensor
to measure h

Sand paper (#320)

$D_o = 101.6\text{mm}$
$D_i = 83.5\text{mm}$

Figure 13. Open-ended model pile to measure Q_{si}.

$$q_{b,closed} = \frac{Q_b}{\frac{\pi D_o^2}{4} - IFR \times \frac{\pi D_i^2}{4}} \equiv \frac{Q_b}{A_{b,eff}} \quad (12)$$

where $A_{b,eff}$ is the effective base area reflecting the plugging condition.

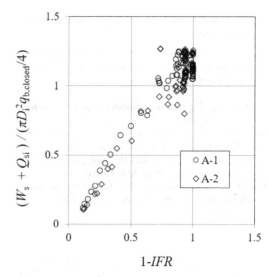

Figure 14. Correlation between $\frac{W_s + Q_{si}}{\frac{\pi D_i^2}{4} \times q_{b,closed}}$ and $(1\text{-}IFR)$.

2.2.2 Estimating unit base resistance of sheet piles

For sheet piles, phenomena similar to the plugging of tubular piles have been empirically known by contractors. The increase in the density of the soil around the base of the sheet pile was confirmed in model tests by Taenaka *et al.* (2006) using X-ray. In this paper, to consider the plugging condition for sheet piles, the complicated shape of sheet piles will be simplified by assuming an equivalent tubular pile that has the same sectional area both on the pile annulus and in the hollow part inside the pile as those of sheet piles, as shown in Figure 15. Example of the equivalent values of the sectional area of the pile base annulus ($A_{bp,eq}$), the sectional area inside the pile ($A_{bi,eq}$), the outer diameter ($D_{o,eq}$) and the inner diameter ($D_{i,eq}$) of several types of sheet piles are shown in Table 1.

However, it is not easy to measure the height of the surface of the soil in the hollow part of the sheet pile (to obtain the values of h) during the actual piling work. To cope with this difficulty, a constant plugging condition (i.e. the value of IFR) will be assumed for each type of sheet pile. Considering that the subsurface information estimated from the piling data will be more likely to be utilized for grasping the relatively hard layers, it will provide a practically reasonable IFR values if they are back-analyzed so that the local peak values of the N values estimated from the piling data match with the local peak values of the N values obtained by SPT. Table 2 shows the back-analyzed IFR values for three types of sheet piles.

Taenaka (2013) conducted a centrifuge model tests with the centrifugal acceleration being 20g, in which sheet piles were jacked into a dry sand. The height of the surface of the soil in the hollow part of the sheet pile was measuring by a steel bar, and the component of the surrounding soil collapsing into the hollow part was compensated by assuming an active failure. As a result, IFR was expressed as:

$$IFR = \min\left\{1.0, 1.12 \times \left(\eta_{sh} \times \frac{L}{W_b}\right)^{-0.45}\right\} \quad (13)$$

$$\eta_{sh} = \frac{\frac{1-\sin\theta}{\cos\theta} - \frac{W_b}{2 \times D_s} \times \left(\frac{\tau_b}{\tau_{si}} - 1\right)}{1 - \frac{D_s}{W_b} \times \tan\theta} \quad (14)$$

where η_{sh} is the Arching Strength Parameter (a parameter depending on the shape of the sheet pile), L is the penetration depth, W_b, D_s and θ are the parameters related to the shape (length or angle) of the sheet pile as shown in Figure 16, and τ_b and τ_{si} are the frictional stress at the soil-soil or soil-pile interfaces.

The frictional force at the soil-soil interface (acting reversely to the direction of penetration) is generated in reaction to the frictional force at the soil-pile interface. It will follow that the former does not exceed the latter, leading to:

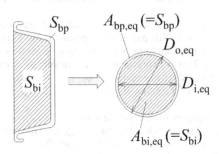

Figure 15. Conversion of a sheet pile into a tubular pile.

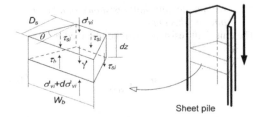

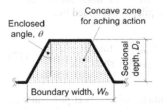

Figure 16. Explanation of symbols in Eqs. (13) and (14) (Taenaka, 2013).

$$\frac{\tau_b}{\tau_{si}} = \min\left\{\frac{\tan\phi}{\tan\delta}, 1 - 2 \times \frac{D_s}{W_b} \times \left(\frac{1}{\cos\theta} - \tan\theta\right)\right\} \quad (15)$$

Figure 17 shows the variation of IFR with depth for the three types of sheet piles, which were calculated by Equations (13), (14) and (15). Looking at the depth of 0.8 m, the IFR values for SP-10H, SP-IIIw and SP-III were 0.95, 0.81 and 0.60 respectively, which roughly corresponds to the values shown in Table 2 and suggests the validity of the back-analysis. However, considering that the IFR values sharply decreases with depth as can be seen in Figure 17, the IFR values in Table 2 might only be valid at a certain small penetration depth into a new layer.

Table 1. Example of $A_{bp,eq}$, $A_{bi,eq}$, $D_{o,eq}$ and $D_{i,eq}$ of sheet piles.

Type of sheet piles	$A_{bp,eq}$ [m²]	$A_{bi,eq}$ [m²]	$D_{o,eq}$ [m]	$D_{i,eq}$ [m]
SP-III	0.0076	0.0435	0.2552	0.2353
SP-IIIw	0.0104	0.0910	0.3593	0.3403
SP-10H	0.0110	0.1068	0.3873	0.3688

Table 2. Back-analyzed IFR values for sheet piles.

Type of sheet piles	IFR
SP-III	0.50
SP-IIIw	0.90
SP-10H	0.95

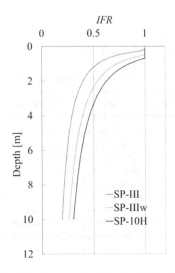

Figure 17. IFR values calculated by Eqs. (13) – (15).

49

2.2.3 Estimating unit shaft resistance of tubular piles or sheet piles

The unit shaft resistance will be obtained by:

$$q_s = \frac{Q_s}{A_{s,emb}} \qquad (16)$$

where $A_{s,emb}$ is the area of the outer surface of the embedded part of the pile, as expressed by:

$$A_{s,emb} = \begin{cases} \pi \times D_o \times z_{emb} \text{ (for tubular piles)} \\ L_p \times z_{emb} \text{ (for sheet piles)} \end{cases} \qquad (17)$$

where L_p is the perimeter of the sheet pile and z_{emb} is the embedment depth of the tubular pile or sheet pile.

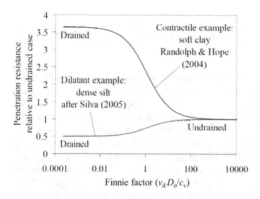

Figure 18. Expression of the rate effect by Finnie factor (after White et al., 2010).

The unit shaft resistance q_s obtained by these equations is the average over the whole embedment depth. On the other hand, the sleeve friction (f_s) obtained in CPT, into which q_s is to be converted based on the method explained in the next sub-section, is the local frictional stress near the cone. Considering that the local frictional stress on the pile shaft will be higher near the pile base than far above the pile base, the average unit shaft resistance (q_s) will be smaller than the local unit shaft resistance near the pile base. As a result, the subsurface information estimated by using q_s (instead of the local unit shaft resistance near the pile base) becomes conservative (the estimated SPT N values will be smaller and the particle size of the estimated soil type will be smaller).

2.3 Estimating CPT cone resistance and sleeve friction

The next step is to convert $q_{b,closed}$ and q_s into CPT q_c and f_s. The CPT cone has much smaller diameter and is installed at a lower penetration rate than the pile. These differences will lead to differences in the resistance on the base and the shaft of the penetrating materials (i.e. the CPT cone or the pile). Therefore, the conversion of $q_{b,closed}$ and q_s into CPT q_c and f_s requires the consideration of the effects of the penetration rate and the scale (diameter of the penetrating material).

2.3.1 Effect of penetration rate

As an index to capture the rate effect on $q_{b,closed}$, a dimensionless quantity "Finnie factor" $v_d D_o/c_v$, where v_d is the downward velocity and c_v is the coefficient of consolidation, is widely known (Finnie & Randolph, 1994; White et al., 2010). Figure 18 shows the example of explaining the rate effect by Finnie factor (White et al., 2010). Based on this framework, the rate effect can be ignored if v_d is controlled to give the same value of Finnie factor during installation as that of CPT.

The rate effect on q_s seems to have not been well understood. In this paper, the approach using Finnie factor is assumed to be applicable to q_s as well as to $q_{b,closed}$. This is expected to be acceptable, as q_s can be expressed as a function of $q_{b,closed}$ according to Jackson et al. (2008) and Ishihara et al. (2011).

2.3.2 Effect of scale (pile diameter)

To consider the scale effect on $q_{b,closed}$, the knowledge on the scale effect on the unit base capacity ($q_{bf,closed}$) will be referred to, by assuming $q_{b,closed} = q_{bf,closed}$. Generally, the set-up phenomenon (the increase in the resistance with time after the end of installation) is mainly seen in the shaft resistance (Komurka et al., 2003). It will follow that the effect of time after the end of installation on $q_{b,closed}$ could be ignored. Therefore, if the rate effect on $q_{b,closed}$ is ignored and the displacement to define $q_{bf,closed}$ is adequately chosen, the assumption of $q_{b,closed} = q_{bf,closed}$ would be acceptable.

It is known that $q_{bf,closed}$ becomes smaller as the outer diameter of the pile increases (Meyerhof, 1983; Jardine & Chow, 1996; Chow, 1997). In these researches $q_{bf,closed}$ is defined at a smaller displacement than what would give the "plunging load" (the load at a sufficiently large displacement, which reflects the fully mobilized strength of soils around the pile base). White & Bolton (2005) analyzed the database of Chow (1997) by defining $q_{bf,closed}$ by the plunging load, and confirmed that $q_{bf,closed}$ is not influenced by the pile diameter and can be linked with the averaged CPT cone resistance (q_{ca}) as:

$$q_{bf,closed} = \alpha \times q_{ca} \qquad (18)$$

$$\alpha = 0.9 \qquad (19)$$

As the base resistance during installation is mobilized in the process of the continuous penetration with a large displacement, it will be more appropriate to assume that $q_{b,closed}$ is comparable with the base capacity defined by the plunging load rather than the base capacity defined at a certain pile displacement. Then Equation (18) is re-written as:

$$q_{b,closed} = \alpha \times q_{ca} \qquad (20)$$

On the other hand, the scale effect on q_s seems to have not been well understood. Based on the linear correlation between $q_{b,closed}$ and $q_{s.}$, the scale effect on $q_{b,closed}$ (Eq. 20) will be assumed to be applicable to q_s as well, as expressed by:

$$q_s = \beta \times f_s \qquad (21)$$

$$\beta = 0.9 \qquad (22)$$

Ishihara et al. (2015) analyzed the piling data ($q_{b,closed}$ and q_s) obtained in a soft alluvial soil by using a closed-ended pile with $D_o = 318.5$ mm and the CPT results obtained in the same site. The back-analyzed values of α and β were 0.8 and 0.5 respectively, which are lower than the values shown in Equations (19) and (22). These lower values were possibly caused by the highly multilayered soil strata. Adopting Equations (19) and (22) will make the estimation results more conservative (the estimated SPT N values will be smaller and the particle size of the estimated soil type will be smaller).

2.4 Estimating SPT N and soil type

According to Robertson (1990), a value of an index I_c (Soil Behavior Type Index) is obtained from CPT q_c and f_s by:

$$I_c = \sqrt{(3.47 - \log Q_t)^2 + (1.22 + \log F_r)^2} \qquad (23)$$

$$Q_t = \frac{q_c - \sigma_{v0}}{\sigma_{v0}'} \qquad (24)$$

$$F_r = 100 \times \frac{f_s}{q_c - \sigma_{v0}} \qquad (25)$$

where σ_{v0} and σ_{v0}' are the overburden pressure and the effective overburden pressure respectively. The soil type can then be obtained based on the I_c value and the chart shown in Table 3. On the other hand,

according to Jefferies & Davies (1993), SPT N can be estimated from CPT q_c and f_s by:

$$N = \frac{\frac{q_c}{p_a}}{8.5 \times \left(1 - \frac{I_c}{4.6}\right)} \qquad (26)$$

where p_a is the atmospheric pressure ($= 100$ kPa).

Table 3. I_c values and soil types (Lunne et al. (1997)).

I_c	Soil type
$I_c < 1.31$	Gravelly sand
$1.31 < I_c < 2.05$	Sands - clean sand to silty sand
$2.05 < I_c < 2.60$	Sand mixtures - silty sand to sandy silt
$2.60 < I_c < 2.95$	Silt mixtures - clayey silt to silty clay
$2.95 < I_c < 3.60$	Clays
$3.60 < I_c$	Organic soils - peats

It is noted that it would be better to use the corrected cone resistance q_t, which compensates for the influence of pore water pressure on q_c, in the above equations. However, for simplicity, q_c will be consistently used in this paper. For soils other than soft fine soils, q_c and q_t can be taken as being comparable, according to Lunne et al. (1997). For soft fine soils, the excess pore water pressure would be positive and thus $q_c < q_t$. The consistent use of q_c (instead of q_t) will therefore make the estimation results more conservative (the estimated SPT N values will be smaller and the particle size of the estimated soil type will be smaller).

In CPT, the pore pressure (u) is usually measured just above the cone (Lunne et al., 1997) as shown in Figure 19. In this case, q_t is expressed as:

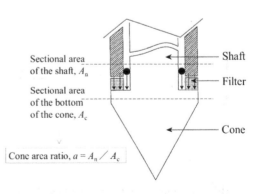

Cone area ratio, $a = A_n / A_c$

Figure 19. Typical shape of CPT cone (after JGS, 2013).

$$q_t = q_c + (1 - a) \times u = q_c + \left(1 - \frac{A_n}{A_c}\right) \times u \quad (27)$$

where a is the cone area ratio, which is the ratio of the sectional area of the shaft (A_n) to that of the widest part of the cone (A_c), as shown in Figure 19. However, JGS (2013a) recommends to obtain the value of a not based on the geometric information but by conducting an experiment in a pressure chamber.

Taking one example where the generated pore pressure (u) is comparable with q_c and the "a" value of the cone is 0.8 (VERTEK, 2017), Equation (27) is reduced to $q_t = 1.2 \, q_c$.

2.5 Verification through field tests

Figure 20 shows the comparison of the SPT results with the N values and the soil types estimated from the piling data obtained during the installation of

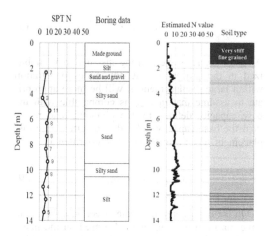

Figure 21. Comparison of N values and soil types obtained by SPT and estimated from CPT results.

a closed-ended tubular pile equipped with a base load cell (D_o = 318.5 mm, without surging, v_d = 2 mm/s). The value of Finnie factor was comparable to that of CPT so that the rate effect could be ignored. In the estimation, measured Q_b values were used instead of Equation (4). It can be confirmed that the trend of variation of the estimated N with depth agreed with the SPT results, while some of the local peak values were slightly overestimated. The differences in the depths where the local peak values were obtained would partly be because the positions of the SPT and the pile installation were different.

Figure 21 shows the comparison of the SPT results with the N values and the soil types estimated from CPT results. The N values were overestimated, as were in the previous case. It is suggested that this overestimating trend is the nature of the method of Equations (23) – (26). Miyasaka et al. (2009) conducted surveys at 36 sites in Japan and showed that the N values estimated from CPT results were greater than the N values obtained by SPT by 10% on average. One reason for this would be the influence of the energy efficiency in SPT. Equation (26) provides the N values with the energy efficiency of 60%, whereas the energy efficiency in SPT in Japan will usually be greater than 60% (JGS, 2013b), which would lead to smaller SPT N values in Japan.

Regarding the soil type, although the classification system in SPT is different from that in CPT, the soil types estimated from the piling data (Figure 20) or CPT data (Figure 21) based on the classification system in CPT roughly agreed with the SPT results.

Figure 22 shows the comparison of the SPT results with the N values and the soil types estimated from the piling data obtained during the installation of a closed-ended tubular pile equipped with a base load cell (D_o = 318.5 mm, without surging) at a more practical penetration rate (v_d) of 20 mm/s or 30 mm/s. In the estimation, measured Q_b values were used instead of Equation (4). As a result, the estimation results

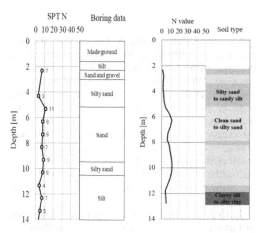

(a) Estimation using piling data in C07-03

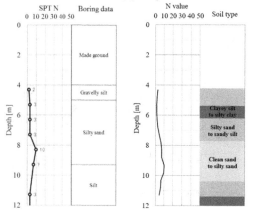

(b) Estimation using piling data in C09-14

Figure 20. Comparison of SPT results (left) and estimation results (right) (Closed-ended, D_o = 318.5 mm, v_d = 2 mm/s).

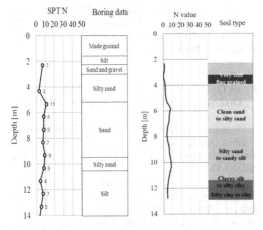

(a) Estimation using piling data in C07-17

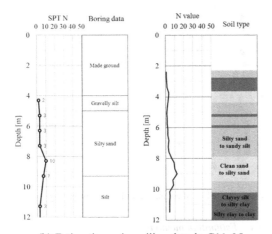

(b) Estimation using piling data in C11-05

Figure 22. Comparison of SPT results (left) and estimation results (right) (Closed-ended, D_o = 318.5 mm, v_d = 20-30 mm/s).

agreed well with the SPT results. It is suggested that this would be because the overestimating trends (Figures 20 and 21) were cancelled by the rate effect.

Figure 23 shows the comparison of the SPT results with the N values and the soil types estimated from the piling data obtained during the installation of an open-ended tubular pile equipped with base earth pressure transducers (D_o = 318.5 mm, D_i = 199.9 mm, without surging, v_d = 10 mm/s). The length of the inner soil column (h) was obtained by a stroke sensor attached inside the pile. In the estimation, the measured base earth pressure was taken as $q_{b,closed}$, and the shaft resistance was obtained by subtracting Q_b from the head load, where Q_b was obtained by Equation (12) from q_b and IFR. As with the previous case (Figure 21), the estimation results agreed well with the SPT results.

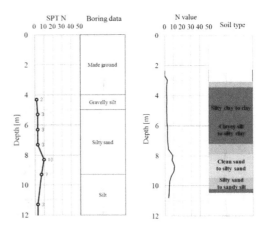

Figure 23. Comparison of SPT results (left) and estimation results (right) (C08-02: Open-ended, D_o = 318.5 mm, D_i = 199.9 mm, v_d = 10 mm/s).

Figure 24 shows the comparison of the SPT results with the N values and the soil types estimated from the piling data obtained during the installation of a closed-ended tubular pile (D_o = 318.5 mm, with surging, v_d = 20 mm/s). In the estimation, Q_b values were obtained by Equation (4). As a result, the estimation results agreed well with the SPT results. Looking at the figure more closely, the N values were underestimated at the depth where SPT N values were small. This will be because of the underestimating tendency of Q_b in the depth range where SPT N values are small, as discussed in Section 2.1.

Figure 25 shows the comparison of the SPT results with the N values and the soil types estimated from the piling data obtained during the installation of a sheet pile with the width of 600 mm (SP-IIIw,

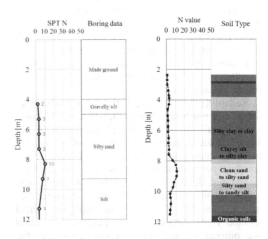

Figure 24. Comparison of SPT results (left) and estimation results (right) (C11-05: closed-ended, D_o = 318.5 mm, v_d = 20 mm/s).

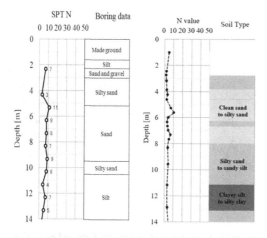

Figure 25. Comparison of SPT results (left) and estimation results (right) (J0617-05: sheet pile with the width of 600mm, v_d = 30 mm/s).

with surging, v_d = 30 mm/s). In the estimation, Q_b values were obtained by Equation (4). At the depth where local peak values of SPT N were recorded, the estimation results agreed well with the SPT results. At other depths (especially deeper than 6m), the estimation results were conservative. This will partly be because of the underestimating tendency of Q_b in the depth range where SPT N values are small, as discussed in Section 2.1. Another cause might be the assumption of *IFR* values being constant, although they should vary with depth in reality. If the penetration process in reality was more unplugged (having larger *IFR* values) than what is assumed in the estimation, the estimation results will become conservative.

Based on the above case studies, it can be said that the penetration rate and the surging stroke (downward and upward displacement in surging) do not influence the estimation results very much while the plugging condition has a significant influence on them. Especially for sheet piles, the plugging condition is assumed as being constant with depth, which can make the estimated trend of variation of N values with depth different from the actual SPT results in some cases.

Regarding the rate effect, the outer diameter of the pile is much larger than that of the CPT cone, and it could be the case that the value of the Finnie factor in press-in piling cannot be reduced to being comparable to that of CPT even if the smallest penetration rate of the press-in machine is adopted. Even so, if the values of the Finnie factor in piling and in CPT fall in between 0.1 and 30, the rate effect will be negligibly small (Bolton *et al.*, 2013). If the Finnie factor values are outside of this range, considering that the Standard Press-in is usually adopted in soft soils that might be expected to be contractile (rather than dilatant), and that the value of Finnie factor

in piling is greater than that in CPT, the unit base resistance in piling will become smaller than the tip stress in CPT, and consequently the estimation results will become conservative.

3 ESTIMATION FROM PILING DATA OBTAINED IN PRESS-IN WITH AUGERING

In Press-in with Augering, a sheet pile is installed by a static jacking force with the aid of an augering device consisting of an auger head, auger screw and casing. The similarity in the augering process and the drilling of rocks has been taken into account to estimate the subsurface information from the piling data in Press-in with Augering (Ishihara *et al.*, 2015a). The outline of the flow of estimation is shown in Figure 26.

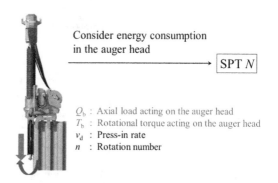

Figure 26. Flow of estimation in Press-in with Augering.

3.1 *Estimation methods*

In the field of rock drilling, it is known that the parameter T/d_c, where T is the effective rotational torque and d_c is the depth of cut, will be proportional to the unconfined compressive strength of a rock (Nishimatsu, 1972; Fujimoto *et al.*, 2005). Fukui *et al.* (1996) pointed out that the parameter $T/(d_c)^\gamma$ has a better correlation with the strength of the rock than T/d_c, where γ is a constant. Assuming that the similar relationship exists for the ground other than rocks, and that the unconfined compressive strength of the ground is linearly correlated with its SPT N value as was found for clays with SPT N being in between 10 and 100 (JGS, 2013c), the parameter $T/(d_c)^\gamma$ in Press-in with Augering is expected to linearly correlates with SPT N. It follows that the SPT N is estimated by:

$$N = A \times \frac{T}{\left(\frac{v_d}{n}\right)^\gamma} \tag{28}$$

where A is a constant, v_d is the downward velocity of the pile and n is the rotational revolution of the auger.

54

On the other hand, there is a technique called MWD (Measurement While Drilling) in the field of drilling, which estimate the hardness of the ground from the energy required for drilling it. In this technique, the SPT N is obtained by the following equation (JGS, 2004):

$$N = C_n \times (Q_b + \frac{2 \times \pi \times n \times T}{v_d}) \quad (29)$$

where C_n is a constant. Considering the similarity of the process of drilling to which MWD is applied and that of augering in Press-in with Augering, it is expected that this equation can be applied to Press-in with Augering by adjusting the value of C_n.

For a reliable estimation, the constants in Equations (28) and (29) should be obtained by the back-analysis based on the sets of SPT results and the piling data, so that the estimated N values agree with the SPT results.

As explained in Section 1.3, the converted N value will be adopted for hard grounds with SPT N exceeding 50 in this paper. It has to be noted that the reliability of the converted N value is limited and several methods of correction have been proposed (SN-EC, 2004). The higher the values of the converted N, the lower the reliability of the converted N would be. Considering this, it would be desirable to conduct the back-analysis of the parameters in Equations (28) and (29) based on the database of the piling data and the SPT results that are obtained in the ground where the converted N values are smaller than a certain value.

3.2 Verification through case studies

In this section, the estimated N values will be compared with the SPT results, based on the data obtained in piling construction sites. In the estimation, the values of the constants (A, γ and C_n) were obtained by the back-analysis based on the database with the converted N values smaller than 100. The values of the constants were common in all the cases. The press-in piling data were obtained during the pre-augering process using a standard type of the auger head, which has 450 mm outer diameter and 3 wings as shown in Figure 27.

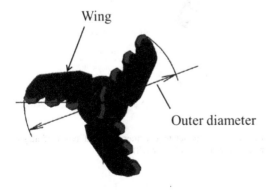

Figure 27. Standard type of auger head.

Figures 28 – 31 shows the comparison of the N values obtained by SPT and those estimated from the piling data based on the two methods explained in the previous section. In the both estimation methods, the trends of variation of N with depth were well estimated in general. However, significant differences in the values of N were found at 2 m in P0902-02 and at 7 – 9 m and 9 – 11 m in P0916-01. Reading carefully what is written in the "commentary" column in the boring log in the SPT results, the existence of gravels or cobble stones with its diameter being larger than 80 mm was commonly confirmed at these depths, which is believed to be the cause of the difference between the estimated and investigated N values at these depths. On the other hand, the differences were also found at 6 – 7 m in P0902-02 and at 12 – 13 m in P0908-02. There is a possibility that these differences

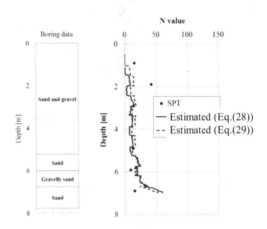

Figure 28. Comparison of SPT results and estimated N values (J0902-02).

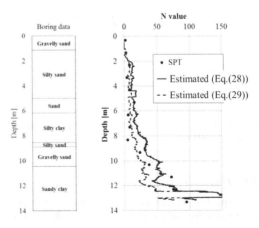

Figure 29. Comparison of SPT results and estimated N values (J0908-02).

are suggesting the existence of thin hard layers or cobbles that were not depicted by SPT which is conducted intermittently with depth (usually at every 1 m).

Comparing the estimation results obtained by the two methods, there seems to be no trend that the results obtained by one method is consistently larger or smaller than those obtained by the other. In practice, it is recommended to conduct the estimation both by the two methods and compare the results.

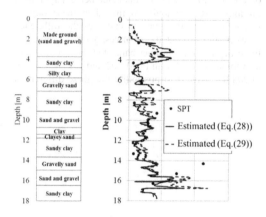

Figure 30. Comparison of SPT results and estimated N values (J0912-01).

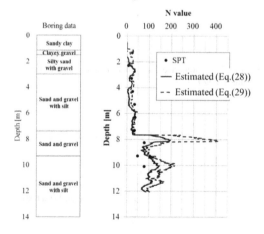

Figure 31. Comparison of SPT results and estimated N values (J0916-01).

3.3   Points to be noted

As explained in the previous section, the back-analyzed values of the constants A, γ and C_n were

obtained based on a specific type of the auger head (with 450 mm outer diameter and 3 wings), and the values cannot be directly adopted for the piling data obtained by using other types of auger heads. In the IPA-TC2 technical material, this point was reflected by restricting the use of different types of auger heads. More recently, Okada *et al.* (2018) proposed a method to allow the use of the auger heads with different diameters in the two estimation methods, by modifying the torque on the varieties of auger heads into what would be experienced on the specific type of the auger head (with 450 mm outer diameter and 3 wings) based on the model shown in Figure 32. They confirmed that their proposed method was effective in mitigating the dispersion of the estimation results that seemed to have been caused by the difference in the diameter of the auger head, as shown in Figure 33. The investigation on the applicability of this modification method has been continued using a variety of auger heads in several piling sites.

On the other hand, it should be kept in mind that the reliability of the SPT results in the hard ground would be limited. The diameter of the SPT sampler is around 50 mm, and it seems that the sampler could sometimes penetrate into the ground without hitting the gravels or cobbles that are actually contained in the ground, as reported by Ogawa *et al.* (2013). In addition, according to Mitsuhashi (1995), the cobbles usually "lie" in the ground, with their longer axis being in the horizontal plane as shown in Figure 34, and as a result the size of the cobbles written in the "commentary" column in the boring log (representing the shorter axis of the cobbles) can be smaller than one-third of the actual size of the cobbles.

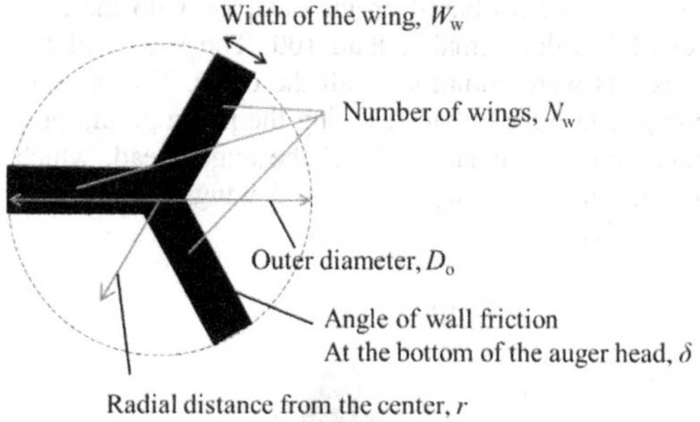

$$\text{Torque} = \int_0^{\frac{D_o}{2}} \left[\left\{ q_{b.closed} \times \tan\delta \times (W_w \times dr \times N_w) \times r \right\} \right]$$
$$= \frac{1}{8} \times D_o^2 \times W_w \times N_w \times \tan\delta \times q_{b.closed}$$

Figure 32. A model to consider different types of auger heads (after Okada *et al.*, 2018).

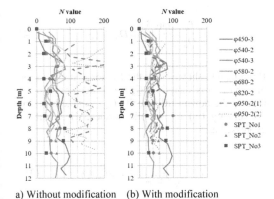

a) Without modification   (b) With modification

Figure 33. Comparison of N values estimated with or without the modification of torque.

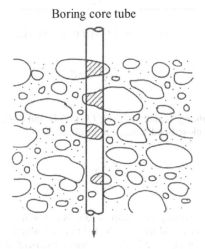

Figure 34. Boring core and cobble stones (after Mitsuhashi, 1995).

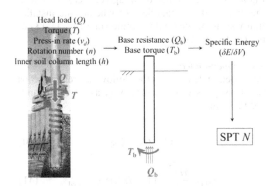

Figure 35. Flow of estimation in Rotary Cutting Press-in.

4 ESTIMATION FROM PILING DATA OBTAINED IN ROTARY CUTTING PRESS-IN

In Rotary Cutting Press-in, a tubular pile equipped with base cutting teeth is installed by the combination of vertical and rotational static jacking forces. The energy consumed for installing the pile has been considered for estimating SPT N from the piling data in Rotary Cutting Press-in, by referring to the knowledge in the field of rock drilling (Ishihara et al., 2015b). The flow of estimation is outlined in Figure 35.

4.1 Estimating base resistance and base torque

4.1.1 Closed-ended piles
As discussed in Section 2.1, when estimating the subsurface in formation at the pile base, it is better to use the resistance on the pile base rather than the load applied to the pile, as the former reflects the information of the soil beneath the pile base more directly. In Rotary Cutting Press-in, the pile receives the vertical load (Q) and the torque (T) from the press-in machine, and in reaction to these it receives the base resistance (Q_b), shaft resistance (Q_s), base torque (T_b) and the shaft torque (T_s) from the ground, as shown in Figure 36 and expressed by Equations (30) and (31).

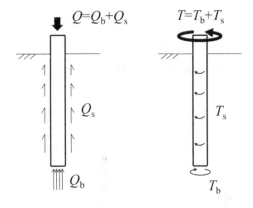

Figure 36. Forces acting on a pile during Rotary Cutting.

$$Q = Q_b + Q_s \tag{30}$$

$$T = T_b + T_s \tag{31}$$

For closed-ended piles, Q_b and T_b will be expressed as a function of the unit base resistance on the closed-ended pile ($q_{b,closed}$):

$$Q_b = \frac{\pi D_o{}^2}{4} \times q_{b,closed} \tag{32}$$

$$\begin{aligned} T_b &= \int_0^{\frac{D_o}{2}} \{(q_{b,closed} \times \tan \delta_{sp} \times 2\pi r \times dr) \times r\} \\ &= \frac{\tan \delta_{sp} \times \pi \times D_o{}^3}{12} \, q_{b,closed} \end{aligned} \tag{33}$$

where r is the distance from the center of the pile in the radial direction and δ_{sp} is the frictional angle at the soil-pile interface. From these equations, $q_{b,closed}$ can be deleted to give:

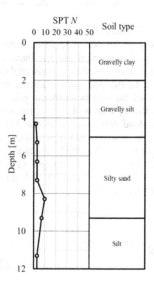

Figure 37. Site profile in C11 test series.

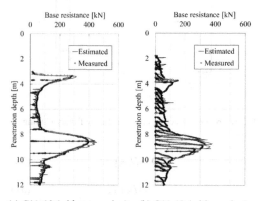

(a) C11-10 (without surging)   (b) C11-13 (with surging)

Figure 38. Comparison of estimated and measured Q_b (Closed-ended, D_o = 318.5mm).

$$\frac{T_b}{Q_b} = \frac{\tan \delta_{sp}}{3} \times D_o \equiv \xi_C \tag{34}$$

Bond (2011) showed that the frictional stress on the pile shaft (f) is shared by Q_s and T_s according to the ratio of the downward velocity (v_d) and the horizontal velocity (v_r) of the pile surface. This will be expressed as:

$$Q_s = \frac{1}{\sqrt{1+\mu^2}} \times f \times \pi D_o \times z \tag{35}$$

$$T_s = \frac{\mu}{\sqrt{1+\mu^2}} \times f \times \pi D_o \times z \times \frac{D_o}{2} \tag{36}$$

$$\mu = \frac{v_r}{v_d} \tag{37}$$

From Equations (35) and (36), f can be deleted as:

$$\frac{T_s}{Q_s} = \frac{\mu \times D_o}{2} \equiv \zeta \tag{38}$$

Combining Equations (30), (31), (34) and (38), Q_b of the closed-ended pile can be obtained by:

$$Q_b = \frac{T - \zeta \times Q}{\xi_C - \zeta} \tag{39}$$

The validity of Equation (39) was confirmed by a series of field tests conducted in a soft alluvial soil shown in Figure 37. The pile was closed-ended and was equipped with a base load cell to measure Q_b. From the piling data (Q and T), Q_b was estimated by Equation (39), and was compared with the values measured by the load cell. As can be confirmed in Figure 38, the estimated Q_b agreed very well with the measured Q_b regardless of whether the installation was associated with surging (in Test C11-13) or not (in Test C11-10).

4.1.2 Open-ended piles
For open-ended piles, the following relationships will be supposed in the same way as were in Standard Press-in:

$$Q_b = Q_{bp} + Q_{bi} \tag{40}$$

$$Q_{bi} = \frac{\pi D_i{}^2}{4} \times (1 - IFR) \times q_{b,closed} \tag{41}$$

Considering that there are cutting teeth on the pile base, Q_{bp} would be expressed as:

$$Q_{bp}=t_T \times w_T \times n_T \times q_{b,closed} \qquad (42)$$

where t_T and w_T are the thickness and the width of each cutting tooth and n_T is the number of the cutting teeth. In the same way as Q_b, the base torque (T_b) will be expressed as:

$$T = T_{bp} + T_{bi} \qquad (43)$$

where T_{bp} and T_{bi} are respectively the torque on the pile base annulus and at the bottom of the soil column inside the pile, that arise as a reaction from the soil beneath the pile base. Assuming that a slip plane is created at the pile base, T_{bi} could be written as:

$$T_{bi}= \int_0^{\frac{D_i}{2}} \left\{ (1 - IFR) \times q_{b,closed} \times \tan\phi \times 2\pi r \times dr \times r \right\} \qquad (44)$$

where ϕ is the internal friction angle of the soil around the pile base. On the other hand, assuming that a uniform stress of $q_{b,closed}$ acts both on the vertical and the horizontal planes in each cutting tooth as shown in Figure 39, T_{bp} will be expressed as:

$$T_{bp}=t_T \times d_c \times n_T \times q_{b,closed} \times \frac{D_o+D_i}{4} \qquad (45)$$

$$d_c= \min(h_T, \frac{\pi \times (D_o+D_i)}{2 \times n_T \times v}) \qquad (46)$$

where d_c is the depth of cut and h_T is the height of the cutting tooth. If IFR and $q_{b,closed}$ are independent of r, Equations (40) – (46) can be combined to give:

$$\frac{T_b}{Q_b}=\frac{3\pi t_T(D_o+D_i)^2+(1 - IFR) \times \tan\phi \times 2\pi v D_{in}^3}{24 v n_T t_T w_T+(1 - IFR) \times 6\pi v D_{in}^2} \equiv \zeta_{0,T} \qquad (47)$$

On the other hand, the relationship between Q_s and T_s would be comparable with that for closed-ended piles, as expressed by Equation (38). Combining Equations (30), (31), (38) and (47), Q_b of the open-ended pile can be obtained by:

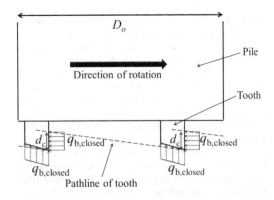

Figure 39. Assumption of the resistance on the cutting tooth.

$$Q_b=\frac{T - \zeta \times Q}{\zeta_{0,T} - \zeta} \qquad (48)$$

4.2 Estimating SPT N

When a penetrating material penetrates into the ground by an incremental depth of δz, the soil beneath the base of the penetrating material will receive an incremental energy of δE and will deform by a volume of δV. In the field of rock drilling, the index $\delta E/\delta V$ is called as the Specific Energy (SE), and has been widely used to represent the performance of the drilling machines (Teale, 1965; Hughes, 1972).

According to Hughes (1972) and Li & Itakura (2012), SE in rock drilling and the unconfined compressive strength of the rock are linearly correlated. Expanding this knowledge to assume that SE required for a penetrating material to penetrate into the ground has a linear relationship with the strength of the ground, it would follow that the SE required in Rotary Cutting Press-in and the SE required for an SPT sampler to penetrate into the ground are linearly correlated, which would be written as:

$$\frac{Q_b \times \delta z + 2 \times \pi \times n \times \delta t \times T_b}{A_{b,eff} \times \delta z} = \chi \times \frac{m_w \times g \times h_w \times e \times N}{a_{b,eff} \times \delta z^*}_{SPT} \qquad (49)$$

where the left side is the SE in Rotary Cutting Press-in and the fractional part in the right side is the SE in SPT. n is the revolution number, t is the time, m_w and h_w are the mass and the drop height of the weight, g is

59

the gravitational acceleration, e is the energy efficiency in SPT, $a_{b,eff}$ is the effective base area of the SPT sampler and δz^*_{SPT} is the reference incremental penetration depth of the sampler ($= 0.3$ m). χ is the parameter to represent the piling efficiency in terms of the energy consumption. The value of χ will become greater than unity when unnecessary energy is consumed in the piling process. This could happen when the pile is excessively extracted, inducing a drop of soils around the pile base into the cavity created by the extraction and the subsequent increase in Q_b. Another case where the energy is unnecessarily consumed could be encountered in a multilayered ground. If a high revolution number required for a pile to penetrate through hard layers is maintained in soft layers where the pile can penetrate without rotation, the energy associated with the rotation will be excessive in the soft layers.

From Equation (49), the estimated SPT N can be obtained by:

$$N = \frac{a_{b,eff} \times \delta z^*_{SPT} \times (Q_b \times \delta z + 2\pi n \times \delta t \times T_b)}{\chi \times e \times m_w g h_w \times A_{b,eff} \times \delta z} \quad (50)$$

In Rotary Cutting Press-in, the installation is often associated with surging (repeated penetration and extraction) in order to reduce the resistance and smoothen the piling process. As a machine setting, the motion of surging is displacement-controlled. If the values of Q or T exceed certain values, it will be conducted in a load-controlled manner. This is because it is necessary to maintain the values of Q and T sufficiently smaller than the reaction obtained from the pull-out resistance of the previously installed piles, the weight of the press-in machine and so on, for a smooth piling process.

Table 4. Specification of piles.

	D_o [mm]	D_i [mm]	n_T	t_T [mm]	w_T [mm]	h_T [mm]
J1001	800	776	6	40	65	80
C12	800	776	4	40	65	200
J1404	1000	976	6	40	65	200
J1501	1000	976	6	40	65	200

If the installation is associated with surging, the pile base passes a certain depth several times. The energy consumption in the surging process would be considered by integrating Equation (50) as:

$$N = \frac{a_{b,eff} \delta z^*_{SPT}}{\chi e m_w g h_w} \int \left(\frac{Q_b \delta z + 2\pi n \delta t T_b}{A_{b,eff} \delta z} dz \right) \quad (51)$$

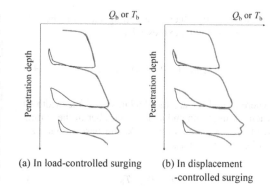

(a) In load-controlled surging   (b) In displacement-controlled surging

Figure 40. Data used for the integration in Equation (51).

Through the application of the piling data obtained in Site N1 (mentioned later), it was confirmed that a better agreement in the N values estimated by Equation (51) and those obtained by SPT was found if the integration in Equation (51) was conducted using the data recorded in the penetration into the fresh ground (Figure 40a) when the surging motion was controlled by displacement, while it was better to use the data recorded when the pile was moving downward (Figure 40b) if the surging was conducted in a load-controlled manner.

4.3 Verification through field tests

Field tests were conducted in two different sites to confirm the validity of the estimation method for open-ended piles with base teeth. The specification of the piles and the press-in conditions are summarized in Tables 4 and 5, where v_u is the upward velocity of the pile, f_w is the flowrate of water injected in the pile base, l_u is the upward displacement in each cycle of surging, and Q_{UL} and T_{UL} are the manually-set upper-limit values of Q and T respectively. The symbol "-" means that no specific values were set (i.e. arbitrary values were adopted).

The site profile in the first site (Site A) is shown in Figure 41. It is multilayered and inhomogeneous in plan in the depth range from 3 m to 8 m below the ground surface. The positional relationship of the SPT and the pile installation is shown in Figure 42.

In the estimation, χ was assumed as unity, and PLR (Plug Length Ratio, defined by Equation (52)) (Xu et al., 2005) was adopted in place of IFR.

$$PLR = \frac{h_{EOI}}{z_{EOI}} \quad (52)$$

Here, h_{EOI} and z_{EOI} are the length of the soil column inside the pile and the penetration depth of the pile at the end of installation. It is believed that the use of

Table 5. Press-in conditions.

	v_d	v_u	v_r	Q_{UL}	T_{UL}	l_u	f_w	
	[mm/s]	[mm/s]	[mm/s]	[kN]	[kNm]	[mm]	[l/min.]	Site
J1001-1	12-16	22	240	400	-	60	30	A
J1001-4	12	22	240	500	-	40	30	A
C12-21	8	6	150	600	-	40	90	N1
C12-22	8	18	110	600	-	40	90	N1
J1404-5	10	30	340	600	500	40	60	N1
J1501-3	-	-	-	600	-	20-80	60	N1

PLR instead of *IFR* did not deteriorate the validity of the estimation in Site A, as the pile installation proceeded in a fully unplugged manner (*IFR* = *PLR* = 1).

Figure 43 shows the *N* values estimated from two sets of piling data (jacking force *Q*, torque *T*, time *t*, penetration depth *z* and the length of the inner soil column *h*) obtained in Site A. The estimated *N* values agreed well with the SPT results, showing relatively small values (between 10 and 15) in 0 m < *z* < 10 m and a steep increase to more than 30 in 10 m < *z* < 12 m. On the other hand, some differences were found in the two estimated values in 3 m < *z* < 4 m. These differences would be due to the inhomogeneity of the ground, judging from the differences in the four SPT results in Figure 41.

The site profile in the second site (Site N1) is shown in Figure 44. It consists of a sand layer and a sand and gravel layer. The sand and gravel layer seems dense and have large *N* values (exceeding 50 at several depths). The positional relationship of the SPT and the pile installation is shown in Figure 45.

Figure 46 shows the *N* values estimated from three sets of piling data obtained in Site N1. In the estimation, χ was taken as unity and *PLR* was used instead of *IFR*. The estimated *N* values gradually increased with depth in *z* < 8 m, and exceeded 50 at several depths in 10 m < *z* < 12 m. These trends agree with the SPT results as a whole. Significant overestimations were found at 8.5 m in C12-22 and at 7 m in J1404-5. These would be because large l_u values (around 500mm in both cases) were adopted around these depths to cope with the deterioration of the piling efficiency caused by the phenomenon of plugging or the sudden increase in the frictional stress called "water binding" (Stevens, 2015). On the other hand, some overestimations were confirmed in 8 m < *z* < 12 m. This will mainly be because *PLR* was adopted instead of *IFR* in the estimation. This point will be further discussed in the next paragraph.

It has been understood that the transition of the plugging condition during the pile penetration can be explained by the equilibrium of forces that act on the

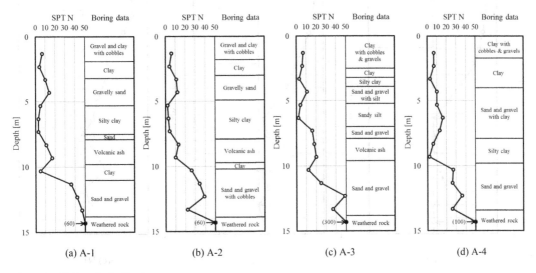

(a) A-1 (b) A-2 (c) A-3 (d) A-4

Figure 41. SPT results in Site A.

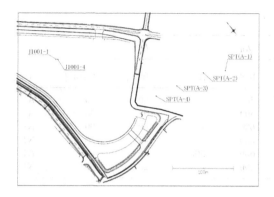

Figure 42. Positional relationship of SPT and pile installation in Site A.

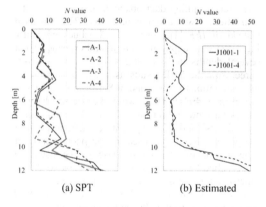

(a) SPT   (b) Estimated

Figure 43. Comparison of N values obtained by SPT and estimated from piling data in Site A.

soil column inside the pile (White *et al.*, 2000; Okada & Ishihara, 2012). The pile will become plugged where N values decrease with depth, because the push-up stress on the bottom of the soil column will decrease with depth. At such depths, *PLR* becomes larger (closer to 1) than *IFR*, leading to smaller $A_{b,eff}$ based on Equation (12). As a result, N values estimated by Equation (51) becomes larger. Figure 47 shows the comparison of the N values estimated by using *PLR* or *IFR*, and the N values obtained by SPT. Note that *IFR* values were obtained as the average over the depth range of 0.5 m, to cope with the insufficient frequency of measuring h relative to the frequency of surging, and the N values were estimated by the following equation:

$$N = \frac{a_{b,\,\mathrm{eff}}\delta z^* {}_{\mathrm{SPT}\,\delta z^*}(Q_b\delta z + 2\pi n\delta t T_b)dz}{\chi em_w g h_w\,A_{b,\mathrm{eff}}\delta z^*} \quad (53)$$

where δz^* is the reference value of the incremental penetration depth (= 0.5 m). In 10 m $< z <$ 12 m, N values were significantly overestimated if *PLR* was used, but the overestimating trend was mitigated if *IFR* was used. As can be confirmed in Figure 48, the penetration became almost fully plugged (i.e. *IFR* was nearly zero) in deeper than 10 m. However, although the significant overestimating trend in $z >$ 10 m was mitigated if *IFR* was used, the variation of N with depth became too eminent. This might have been caused by the use of *IFR* values averaged over 0.5 m.

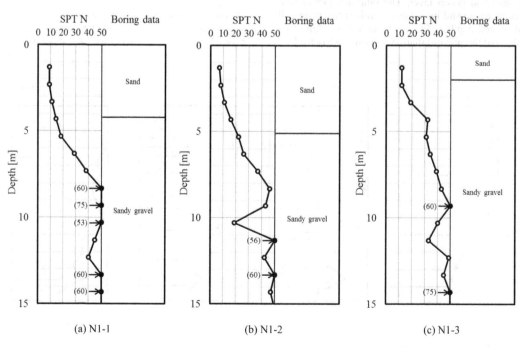

(a) N1-1   (b) N1-2   (c) N1-3

Figure 44. SPT results in Site N1.

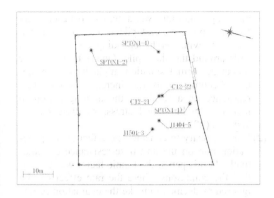

Figure 45. Positional relationship of SPT and pile installation in Site N1.

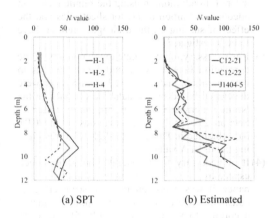

(a) SPT   (b) Estimated

Figure 46. Comparison of N values obtained by SPT and estimated from piling data in Site N1.

From these case studies, it can be said that the penetration and the downward and upward displacement in surging have small influence on the estimation results, while the plugging condition (values of IFR or PLR) influences significantly. If the plugging condition varies with depth to some extent, the estimation results based on PLR will become less reliable, and it is recommended to use IFR obtained by a continuous measurement of the length of the inner soil column (h). In addition, the sampling rate (frequency of the measurement) of h should be sufficiently high, so that the IFR values in each cycle of surging can be used to conduct the integration in Equation (51).

5 POSSIBLE UTILIZATION OF THE ESTIMATION METHODS AND REMAINING ISSUES

5.1 *Possible utilization of the estimation methods*

As discussed in the previous sections, the subsurface information estimated by the methods in this paper

agreed as a whole with the information obtained by the subsurface investigation techniques, but showed some disagreement under certain conditions. Such disagreements would sometimes be caused by the difference in the mechanical conditions such as the size and the shape of the pile, the CPT cone or the SPT sampler, and sometimes by the difference in the geological conditions due to the inhomogeneity of the ground. On the other hand, the subsurface investigations conducted several times in one site can yield results that are not consistent with each other. Considering these points, it would be recommended at the present stage that the subsurface information estimated by the methods in this paper be limitedly used as a reference when managing the termination of piling or judging the alteration of the penetration techniques, as exemplified below:

(1) Contractors confirm the bearing stratum by grasping the variation of the estimated N with depth, and assure an adequate embedment depth of the pile.
(2) Contractors adequately alter the press-in conditions required for the operation of the press-in machine and attempt to improve the piling efficiency, based on the estimated subsurface information.
(3) When encountering the unexpected ground conditions, contractors put together the estimated subsurface information and present them to the owners, and propose the alteration of the construction plan such as the re-selection of the installation assistance.

If the estimated information significantly disagrees with the subsurface investigation results and its validity needs to be carefully investigated, it is recommended to conduct additional subsurface investigations by the other methods. If the

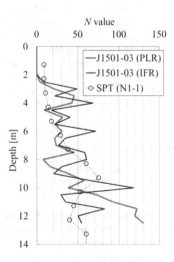

Figure 47. Differences in N values estimated by using PLR or IFR.

63

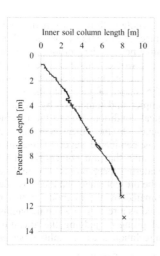

Figure 48. Variation of the inner soil column length with depth in J1501-03.

cause for the disagreement is expected to be the existence of gravels, cobbles or underground obstacles, it is desirable to avoid the use of SPT as an additional investigation method, considering the limitation of its applicability to such conditions as discussed in Section 3.3. The most reliable method would be the trial digging, where a backhoe could be used for a shallow excavation and a set of a casing tube and a hammer grab could be effective if the target depth is deep. The large-diameter core tube sampling (Watanabe et al., 2006) might be another option. On the other hand, the ground-penetrating radar could be a choice if its difficulty in distinguishing cobbles from cavities (Kimura et al., 2000) are overcome. The surface-wave method could also be adopted if its resolution is improved.

In Rotary Cutting Press-in, the reliability of the estimation results will be deteriorated if the plugging condition significantly varies with depth and the measurement of the inner soil column length is omitted or conducted with an insufficient sampling rate. It is recommended to measure the inner soil length with a sufficient sampling rate to grasp its variation during surging and conduct the integration in Equation (51).

5.2 Remaining issues

For Standard Press-in, the following points would be effective in improving the validity of the estimation methods:

(1) It is necessary to refine the method of considering the effective base area or the plugging condition of the sheet piles. Although the equivalent tubular pile was introduced in this paper, the actual direction of the shear stress on the sheet pile is different from that on the tubular pile. In addition, the

plugging condition was assumed to be constant for each type of sheet pile in this paper, but it will vary with depth depending on the state of soils around the sheet pile. One option to better reflect the actual situation might be to consider the phenomenon as the increase in the shaft resistance as a result of the arching action to increase the horizontal stress, as discussed by Taenaka (2013).

(2) It is necessary to consider the effect of the penetration rate on the unit base resistance quantitatively. In this paper, the Finnie factor was used to find the conditions where the rate effect can be ignored or deemed to make the estimation results conservative. If the variation of the unit base resistance is expressed as a function of the Finnie factor, the rate effect can be more directly considered. In addition, in using the Finnie factor, an adequate definition of D_o for sheet piles and the effect of surging on the drainage condition have to be investigated.

(3) It is necessary to investigate the regionality of the correlation between CPT and SPT. Methods to correlate them that were used in this paper were developed based on the database obtained in the North American Continent, and as discussed in Section 2.5, applying the methods to the CPT data obtained in Japan led to slightly overestimating trends of N values.

(4) It is necessary to investigate the scale effect, especially for large-diameter piles. In this paper, it was supposed that the scale effect is constant regardless of the pile diameter, according to White & Bolton (2005). However, the database was limited to piles having the diameter of about 1000 mm or smaller, and for larger piles it still remains possible that the unit base resistance becomes smaller as the pile diameter becomes larger, as have been widely understood. If so, applying the estimation methods in this paper to large-diameter piles will make the estimation results conservative.

For Press-in with Water Jetting, no methods have been developed. It is necessary to collect the piling data before establishing an estimation method.

In Press-in with Augering, the following points would be effective in improving the validity or expand the applicability of the estimation methods:

(1) It is necessary to continue collecting the data sets of SPT results and the piling data and conduct the back-analysis of the parameters to increase the versatility of the parameters. It is desirable that the positions of the SPT and the press-in piling are as close as possible.

(2) It is necessary to investigate the values of the parameters for hard grounds including rocks, to expand the applicability of the estimation methods.

In Rotary Cutting Press-in, the following points would be effective in improving the validity of the estimation methods:

(1) It is necessary to further investigate the relationship between the plugging condition and the inner shaft resistance, particularly the validity of Equation (44).
(2) It is necessary to confirm the validity of the assumption of the force acting on the cutting teeth (Equation (45)).
(3) It is necessary to develop a device to measure the length of the inner soil column at a sufficient sampling rate without difficulty.

6 CONCLUSIONS

This paper introduced the methods of estimating the subsurface information from the data obtained in Standard Press-in, Press-in with Augering and Rotary Cutting Press-in, most of which were organized into a technical material in Japanese under the activity of IPA-TC2 in 2017. The validity of these estimation methods was assessed by conducting the field tests or collecting the data in the piling construction sites. As a result, it was confirmed that the estimation results agreed to a certain extent with the SPT results. It was also confirmed that the plugging condition influenced the estimation results significantly.

Based on the above assessment, the subsurface information estimated by the methods in this paper is recommended to be limitedly used as a reference when managing the termination of piling or judging the alteration of the penetration techniques, at this present stage. Further researches are expected to improve the validity or expand the applicability of the estimation methods.

ACKNOWLEDGEMENTS

The authors appreciate the advices and discussions on the contents in this paper, which were made by the following members of IPA-TC2 (Technical Committee on Estimation of Subsurface Information Using Press-in Data): Dr. Yoshiaki Kikuchi, Dr. Junichi Koseki, Dr. Yoshito Maeda, Dr. Tatsunori Matsumoto, Dr. Masaaki Terashi, Ms. Nanase Ogawa and Mr. Koichi Okada.

REFERENCES

Bolton, M. D., Haigh, S. K., Shepley, P. and Burali d'Arezzo, F. 2013. Identifying ground interaction mechanisms for press-in piles. *Press-in Engineering 2013: Proceedings of 4th IPA International Workshop in Singapore*: 84–95.

Bolton, M. D., Kitamura, A., Kusakabe, O. and Terashi, M. 2020. *New Horizons in Piling*, CRC Press: 170p.

Bond, T. 2011. Rotary jacking of tubular piles. *M.Eng. Project Report, Cambridge University Department of Engineering*: 50p.

Chow, F. C. 1997. Investigations into the behavior of displacement piles for offshore foundations. *Ph.D. thesis, University of London (Imperial College)*.

Finnie, I. M. S. and Randolph, M. F. 1994. Punch-through and liquefaction-induced failure of shallow foundations on calcareous sediments. *Proceedings of International Conference on Behaviour of Offshore Structures, BOSS'94*: 217–230.

Fujimoto, A., Takenouchi, Y., Ogura, Y., Kobayashi, S. and Haino, H. 2005. Development of rational construction method for road tunnel junctions. *JSCE Proceedings of Tunnel Engineering*, Vol. 15: 323–330. (in Japanese)

Fukui, K., Okubo, S. and Homma, N. 1996. Estimation of rock strength with TBM cutting force and site investigation at Niken-goya tunnel. *Journal of MMIJ*, Vol. 112: 303–308. (in Japanese)

Hashizume, Y., Uchida, K. and Kiya, Y. 2002. Construction management of bored concrete pile. *Proceedings of the 37th Japan National Conference on Geotechnical Engineering*: 1391–1392. (in Japanese)

Hughes, H. M. 1972. Some aspects of rock machining. *International Journal of Rock Mechanics & Mining Sciences*, Vol. 9: 205–211.

International Press-in Association (IPA). 2016. *Press-in retaining structures: a handbook*, First edition 2016: 520p.

International Press-in Association (IPA). 2017. *Technical Material on the Use of Piling Data in the Press-in Method, I. Estimation of Subsurface Information*: 63p. (in Japanese)

International Press-in Association (IPA). 2020. *Press-in Piling Case History Volume 1*: 198p.

Ishihara, Y., Okada, K., Nishigawa, M., Ogawa, N., Horikawa, Y. and Kitamura, A. 2011. Estimating PPT data via CPT-based design method. *Proceedings of the 3rd IPA International Workshop in Shanghai, Press-in Engineering 2011*: 84–94.

Ishihara, Y., Ogawa, N., Okada, K. and Kitamura, A. 2015a. Estimating subsurface information from data in press-in piling. *Proceedings of the 5th IPA International Workshop in Ho Chi Minh, Press-in Engineering 2015*: 53–67.

Ishihara, Y., Haigh, S. and Bolton, M. D. 2015b. Estimating base resistance and N value in rotary press-in. *Soils and Foundations*, Vol. 55, No. 4: 788–797.

Ishihara, Y. 2018. Use of press-in piling data for automatic operation of press-in machines and estimation of subsurface information. *Proceedings of the First International Conference on Press-in Engineering 2018, Kochi*: pp. 651–660.

Jackson, A. M., White, D. J., Bolton, M. D. and Nagayama, T. 2008. Pore pressure effects in sand and silt during pile jacking. *Proceedings of the 2nd BGA International Conference on Foundations*, CD: 575–586.

Japan Federation of Construction Contractors (JFCC) and Concrete Pile Installation Technology Association (COPITA). 2017. Methods to confirm the bearing stratum in pile construction management: 21p. (in Japanese)

The Japanese Geotechnical Society (JGS). 2004. Other soundings. *Japanese Standards for Geotechnical and Geoenvironmental Investigation Methods*: 329-337. (in Japanese)

The Japanese Geotechnical Society (JGS). 2013a. Electric cone penetration test. *Japanese Standards and Explanations of Geotechnical and Geoenvironmental Investigation Methods*, Vol. 1: 392. (in Japanese)

The Japanese Geotechnical Society (JGS). 2013b. Standard penetration test. *Japanese Standards and Explanations of Geotechnical and Geoenvironmental Investigation Methods*, Vol. 1: 279–316. (in Japanese)

The Japanese Geotechnical Society (JGS). 2013c. Standard penetration test. *Japanese Standards and Explanations of Geotechnical and Geoenvironmental Investigation Methods*, Vol. 1, p. 311. (in Japanese)

The Japanese Geotechnical Society (JGS). 2015. Method for standard penetration test. *Geotechnical and Geoenvironmental Investigation Methods, Japanese Geotechnical Society Standards*, Vol. 1: 20p.

Japan Press-in Association (JPA). 2017. Non-staging construction method: 9p. (in Japanese)

Japan Road Association (JRA). 2015. *Handbook on Pile Construction*: 371p. (in Japanese)

Jardine, R. J. and Chow, F. C. 1996. *New Design Methods for Offshore Piles, Marine Technology Directorate*: 48p.

Jefferies, M. G. and Davies, M. P. 1993. Use of CPTu to estimate equivalent SPT N_{60}. *Geotechnical Testing Journal*, GTJODJ, Vol. 16, No. 4: 458–468.

Kimura, M., Nakamura, H. and Abe, H. 2000. The case to have confirmed the cavity which is in the cobblestone mixture gravel bed using the underground radar. *Proceedings of the 35th Japan National Conference on Geotechnical Engineering*, Vol. 35, No. 1: 427–428. (in Japanese)

Komurka, V. E., Wagner, A. B. and Edil, T. B. 2003. Estimating soil/pile set-up. *Wisconsin Highway Research Program 0092-00-14*, Final Report.

Kurashina, T. 2016. Model pile penetration test on the mechanism of mobilization of base capacity of an open-ended pile. *Master's thesis, Tokyo University of Science*: 106p. (in Japanese)

Lehane, B. M. and Gavin, K. G. 2001. Base resistance of jacked pipe piles in sand. *ASCE Journal of Geotechnical and Geoenvironmental Engineering*, Vol. 127, No. 6: 473–480.

Lehane, B. M., Schneider, J. A. and Xu, X. 2007. CPT-based design of displacement piles in siliceous sands. *Advances in Deep Foundations*: 69–86.

Likins, G. E. 1984. Field measurements and the pile driving analyzer. *Second International Conference on the Application of Stress Wave Theory on Piles*: 296–305.

Li, Z. and Itakura, K. 2012. An analytical drilling model of drag bits for evaluation of rock strength. *Soils and Foundations*, Vol. 52(2): 206–227.

Lunne, T., Robertson, P. K. and Powell, J. J. M. 1997. *Cone Penetration Testing in Geotechnical Practice*. Spon Press: 312p.

Meyerhof, G. G. 1983. Scale effects of ultimate capacity. *ASCE Journal of Geotechnical Engineering*, Vol. 109, Issue 6: 797–806.

Mitsuhashi, K. 1995. Know-how of reading and utilizing boring data to prevent troubles: 208p., Kindaitosho. (in Japanese)

Nishimatsu, Y. 1972. The mechanics of rock cutting. *International Journal of Rock Mechanics and Mining Sciences & Geomechanics*, Vol. 9, Issue 2: 261–270.

Ogawa, N., Okada, K. and Ishihara, Y. 2011. Fundamental model tests on shaft resistance during press-in and extraction of closed-ended piles. *Proceedings of the Japanese Geotechnical Society Shikoku Branch Annual Meeting*: 57–58. (in Japanese)

Ogawa, N., Nishigawa, M. and Ishihara, Y. 2012. Estimation of soil type and N-value from data in press-in piling construction. *Testing and Design Methods for Deep Foundations, IS-Kanazawa 2012*: 597–604.

Ogawa, N., Okada, K. and Ishihara, Y. 2013. Estimation of N value using PPT data in press-in with augering. *Proceedings of the Japanese Geotechnical Society Shikoku Branch Annual Meeting*: 97–98. (in Japanese)

Okada, K. and Ishihara, Y. 2012. Estimating bearing capacity and jacking force for rotary jacking. *Testing and Design Methods for Deep Foundations, IS-Kanazawa 2012*: 605–614.

Okada, K., Ogawa, N. and Ishihara, Y. 2018. Case study on estimation of ground information with the use of construction data in press-in method. *Proceedings of the First International Conference on Press-in Engineering 2018, Kochi*: 371–378.

Rauche, F., Goble, G. G. and Likins, G. E. 1985. Dynamic determination of pile capacity. *Journal of Geotechnical Engineering, Vol. 111, Issue 3*: 367–383.

Robertson, P. K. 1990. Soil classification using the cone penetration test. *Canadian Geotechnical Journal*, 27: 151–158.

Stevens, G. 2015. Mechanism of water binding during press-in in sand. *M.Eng. Project Report, Cambridge University Department of Engineering*: 50p.

Story of N value Editorial Committee (SN-EC). 2004. *Story of N Value*: 29-37, Rikoh Tosho, ISBN 978-4-5446-0697. (in Japanese)

Taenaka, S., Otani, J., Tatsuta, M. and Nishiumi, K. 2006. Vertical bearing capacity of steel sheet piles. *Proceedings of the Sixth International Conference on Physical Modelling in Geotechnics*, Vol. 1: 881–888.

Taenaka, S. 2013. Development and optimisation of steel piled foundations. *Ph. D. thesis, The University of Western Australia*: 289p.

Teale, R. 1965. The concept of specific energy in rock drilling. *International Journal of Rock Mechanics & Mining Sciences*, Vol. 2: 57–73.

VERTEK. 2017. *ConePlot for CPTSND - processing and graphing software for CPT data acquired using CPTSND or CPTDAS*: 4–5.

Watanabe, K., Araki, K. and Endo, M. 2006. Cobble or pebble survey method. *Micro Sampling Method Research Organization*: 10p. (in Japanese) http://www.microsampling.org/download/report_2006_JSTT.pdf

White, D. J., Sidhu, H. K., Finlay, T. C. R., Bolton, M. D. and Nagayama, T. 2000. Press-in piling: the influence of plugging on driveability. *Proceedings of the 8th International Conference of the Deep Foundations Institute*: 299–310.

White, D. J. and Bolton, M. D. 2005. Comparing CPT and pile base resistance in sand. Proceedings of the Institution of Civil Engineers, *Geotechnical Engineering* 158: 3–14.

White, D. J. and Deeks, A. D. 2007. Recent research into the behaviour of jacked foundation piles. *Advances in Deep Foundations*: pp. 3–26.

White, D. J., Deeks, A. D. and Ishihara, Y. 2010. Novel piling: axial and rotary jacking. *Proceedings of the 11th International Conference on Geotechnical Challenges in Urban Regeneration, London, UK, CD*: 24p.

Xu, X., Lehane, B. M. and Schneider, J. A. 2005. Evaluation of end-bearing capacity of open-ended piles driven in sand from CPT data. *Proceedings of the International Symposium on Frontiers in Offshore Geotechnics*: 725–731.

Proceedings of the Second International Conference on
Press-in Engineering 2021, Kochi, Japan – Matsumoto et al (eds)
© *2021 Taylor & Francis Group, London, ISBN 978-1-032-10414-0*

State of the art report on steel sheet pile method in geotechnical engineering -development of PFS method

J. Otani
Kumamoto University, Kumamoto, Japan

ABSTRACT: Steel sheet piles have been used mainly for the purpose of temporary works such as excavation and pre-construction structures. However, recently those application fields have been expanded to the permanent structures such as foundation structures. This paper is the state of the art report on steel sheet pile method as earth works. Recently, so called PFS method has been developed and the application fields, such as the countermeasure methods for any kinds of natural disasters including heavy rain and earthquake, are expanding. Based on these current situations, the recent development of the sheet pile method is summarized including the activities of Technical Committee No.3 under International Press-in Association. Finally, one of the recent developments on the construction technique, which is press-in method, is briefly summarized.

1 INTRODUCTION

When excavations or retaining works are done safely, some countermeasure techniques against ground failures are needed and sheet piles have been used not only in Japan but also around the world. Most of the sheet piles are made of steel and its shape is not plain plate but has some special shape. Figure 1 (Steel Sheet pile manual, 2014) shows some of the sheet piles in Japan and it has been standardized depending on the regions and countries. Sheet pile is one of the construction methods and had been often used as a temporary work such as for the case of excavation to stabilize cutting slope and often used as a wall with flexible stiffness in the soils to resist earth or water pressures. The applications in coastal engineering as a quay walls at the port and harbor are also quite familiar for the sheet piles. The functions of the sheet piles are categorized by following three topics (Steel sheet pile manual, 2014)

1) retaining structures including free-standing method, retractable one, and cut and saw,
2) impermeable structures, and
3) stress shut down.

The expected effectiveness can also be categorized as any countermeasures such as,

1) subsidence of soft ground,
2) stability with seismic actions,
3) reinforcement of main structures, and
4) reduction of environmental impacts.

Recently, the applications as permanent structures have been increased and those are as foundations or countermeasures for any natural disasters. In fact, the author has done a series of studies on the development of so called PFS (Partially Floating Sheet pile) method for the countermeasure of ground subsidence due to river embankment construction and based on those studies, the PFS method has been implemented with its design concept (PFS method, Technical Manual, 2005)

In this paper, firstly, the steel sheet pile method in geotechnical engineering is summarized and details on the development of PFS method is introduced. Then, activities of the technical committee (TC3) on the topic of the steel sheet pile method, which was established under International Press-in Association (IPA) from 2017 to 2019, are briefly presented with the studies made on effectiveness of the method as countermeasures against natural disasters such as earthquakes. Summary of this TC activities presented in this paper includes not only the effectiveness of the method itself but also some practical construction techniques and design methodologies well associated with the press-in technique.

2 DEVELOPMENT OF THE PFS METHOD

2.1 Overview

The sheet pile method has been used as a permanent structure for the prevention technique on slope failures. For the case of embankments constructed along rivers, sheet piles have been sometimes employed at the toe of the embankment for the purpose of stress cut off from the surrounding ground to reduce subsidence of the area where the private houses are closely

DOI: 10.1201/9781003215226-2

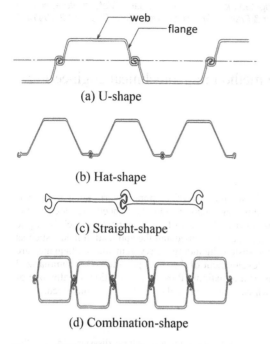

(a) U-shape

(b) Hat-shape

(c) Straight-shape

(d) Combination-shape

Figure 1. Different types of sheet piles (Steel sheet pile manual, JASPP, 2014).

located along rivers. Such a situation is frequently encountered when the embankment is constructed on soft ground. In such a case, the ground not only under the embankment but also adjacent ground may suffer a serious subsidence problem. Then, some counter-measures have to be considered. A steel sheet pile method is one of the candidates for this type of problem as shown in Figure 2. However, it has a cost problem especially when the area and depth of soft ground are wider and deeper. Now, a development of new sheet pile method is strongly awaited. In 1975, a collaborative study was started between Kyushu University and the Ministry of Construction (Ministry of Land, Infrastructure, Transportation and Tourism at present) in Japan. Under this collaboration, a series of in-situ full scale tests were conducted in Kyushu area. Based on those activities, a technical committee for developing a new sheet pile method was established in 2003 in which the chair was Prof. Hidetoshi Ochiai, Professor Emeritus of Kyushu University, Japan. In 2005, a new sheet pile method called PFS method was proposed under the activities of this com-mittee. In this method, the end bearing sheet pile and that of floating type were combined to deal with its effectiveness and cost as shown in Figure 3. Figure 4 shows the details of this structure (PFS Method, Tech-nical Manual, 2005).

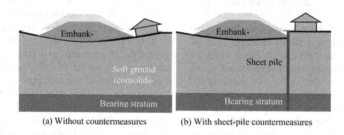

(a) Without countermeasures   (b) With sheet-pile countermeasures

Figure 2. Sheet-pile countermeasures.

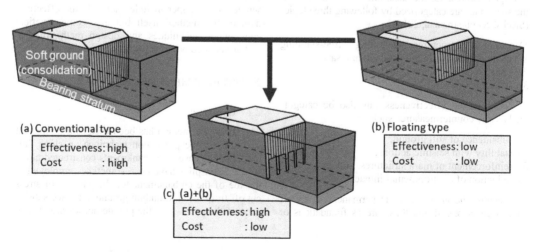

(a) Conventional type

Effectiveness: high
Cost : high

(b) Floating type

Effectiveness: low
Cost : low

(c) (a)+(b)

Effectiveness: high
Cost : low

Figure 3. Idea of PFS method.

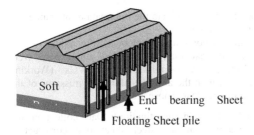

Figure 4. PFS method.

Photo 1. Photo. after the PFS construction.

2.2 In situ experiments

A large number of in-situ full scale tests for PFS method were conducted in Kumamoto City, Japan. This area is well known as a region of Ariake clay which is highly sensitive clay and its depth is up to 40m. Figure 5 shows the soil profile at the site of in-situ test for PFS method. In this case, one end bearing sheet pile with five floating sheet piles were constructed as one unit of the sheet piles. Figure 6 shows the results of measurement for the subsidence at the site and Photo 1 shows the view of the site after PFS construction. Since a large volume of sheet pile materials were reduced, the cost of the PFS method is obvious and the construction time was also highly reduced because of the less volume of the sheet piles.

2.3 Concept of design

Under the activities of research committee of the PFS method, a simple design method was proposed in 2003 and this idea is shown in Figure 7 which is the combination of spring with beam elements. It is noted that the basic idea, in the beginning, is the consideration of only vertical displacement.

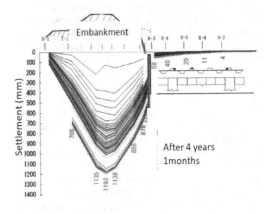

Figure 6. Results of measurements.

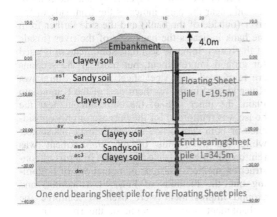

Figure 5. Ground condition at the site.

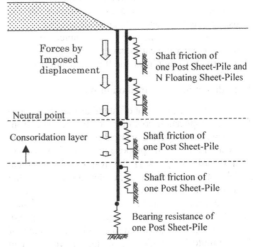

Figure 7. Simplified design model.

69

3 CURRENT ACTIVITIES ON THE DEVELOPMENT OF THE STEEL SHEET PILE

3.1 Introduction of technical committee No.3 under IPA

Steel sheet pile method has long been used as a temporary construction work but in recent years, it has been used as a permanent structure including the applications on the port and harbor structure. The PFS method is one of the methods to construct near the toe of the embankment as a measure to settle the embankment construction on soft ground, and this method is a partially floating sheet piles with the combination of the end bearing sheet piles and its cost effectiveness and construction feasibility can be easily recognized. Regarding the steel sheet pile method, the effectiveness under the earthquake motion has been also reported in the recent 2011 off the Pacific Coast of Tohoku Earthquake and the 2016 Kumamoto Earthquake in Japan but there are still the needs for the quantitative discussion to clarify the effectiveness of the method.

The objective of this committee is to propose the quantitative scope of application such as the quantitative discussion on the lateral displacement of the ground due to embankment construction and also to discuss its effectiveness and performance under the earthquake. In fact, the idea of starting this technical committee was based on the "PFS method technical manual" published in March, 2005 by the PFS research committee. Although it summarizes the policy on design and construction, those were based on the limited field data, so that it needs to include the general condition such as earthquake performance. Specifically, as an activity for three years from 2017 to 2019, the following terms of reference were shown:

1) To reconsider the quantification of the lateral displacement of the ground as a condition for the use of countermeasures against subsidence;
2) To discuss the precise performance of PFS method under the earthquake;
3) To propose guidelines for the application of the PFS method, and
4) To disseminate this method in Asia.

First, for 1), the concept of lateral displacement conditions as application conditions of this method from both experimental and numerical analysis is summarized with aiming at the propose of the design manual. With regard to 2), using the results of centrifugal model tests and dynamic finite element analyses, the behavior under dynamic loading of this method is confirmed and the earthquake resistance should be validated. For 3), the design manual draft is prepared. Lastly, for 4), a series of seminar with the contents of this committee's activities are organized every year in Asia. In addition, securing young talent in the construction industry today is widely concerned. Therefore, the committee will promote younger engineers and researchers to support their activities. The other point is to disseminate the information internationally on our technology.

In this committee, members from industry, government, academia, stakeholders and experts were selected. We established five WGs (Working Groups) with the aim of making the missions of the committee clear. Those are

1) WG1 (Field investigation): Data collection and analysis of the steel sheet pile construction method in Kumamoto after the 2016 Kumamoto earthquake;
2) WG2 (Laboratory experiments): Preliminary experiments and basic behaviors by centrifugal model tests;
3) WG3 (Numerical analysis): Determination of analytical condition (analysis case) and numerical modeling;
4) WG4 (Design): Confirmation of the basic matters concerning the design of the PFS method; and
5) WG5 (Overseas): Seminar on steel sheet piling method in Asia

Those WG activities are summarized in this paper.

3.2 WG1: Field investigation

In this section, research outcomes of WG1 (field investigation) will be outlined. Details of some of these researches can be found in Yamamoto et al. (2018a), Yamamoto et al. (2018b), Yamamoto (2019), Kasama et al. (2019) and Kasama et al. (2020).

3.2.1 Application of the steel sheet pile method in Kumamoto Prefecture

Figure 8 shows the location and purpose of the steel sheet pile construction method in the Shirakawa, Midorikawa, Hamato and Kase Rivers in the Kumamoto Plain. The left and right sides of the river are indicated in two colors, respectively, and the color of the side near the bank indicates the front side of the river (outside of the bank) and the side farther from the bank indicates the back side of the river (inside of the bank). In some sections, solidification is used for ground improvement, but most of the sections are reinforced with steel sheet piles. In Shirakawa River located in the northern part of the Kumamoto Plain, the inner side of the embankment near the house was reinforced for the purpose of only subsidence or both subsidence and earthquake resistance, while the outer side of the embankment was reinforced for earthquake resistance. In terms of construction methods, various combinations of bottoming out, the methods of PFS, FL (Floating) and ground improvement were used to reinforce the embankment near the mouth of the river. This is because field tests were conducted between 1995 and 2000 to investigate the effects of combinations of these methods. The FL method was also used at a point 9 km from the mouth of the Shirakawa River.

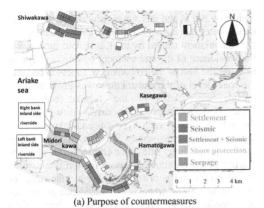

(a) Purpose of countermeasures

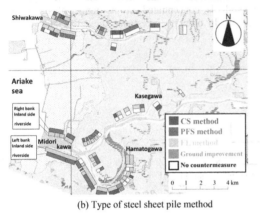

(b) Type of steel sheet pile method

Figure 8. Location of steel sheet pile method.

Table 1. The statistics of sheet pile length.

	No.	Mean (m)	Mode (m)	COV^*	Min (m)	Max (m)
CS method	35	34.2	37	0.24	14	42
FL method	121	14.6	15	0.41	8	30
PFS method						
End supporting	99	38.7	40.5	0.13	28	53
Floating	99	25.5	25.5	0.22	11.5	36.5
Ratio[**]	99	0.66	0.86	0.20	0.27	0.90

* The coefficient of variation
** The ratio of floating sheet pile and end supporting sheet pile

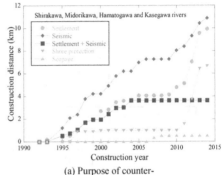

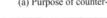

(a) Purpose of counter-

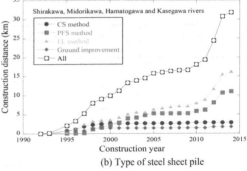

(b) Type of steel sheet pile

Figure 9. Construction history.

A reinforcement of the embankment was carried out on the inner side of the Midorikawa and Hamato Rivers to prevent subsidence, and on the outer side of the embankment, reinforcement of the embankment was carried out for the purpose of constructing a revetment using earthquake resistance measures or sheet pile method. Most of the river embankments were reinforced by the PFS and FL methods on the inner and outer sides of the embankment, respectively. In the case of Kase River, which is in the middle of Midori River, no countermeasures were taken on the outer side of the bank, and the inner side of the bank was partially reinforced by the embankment by PFS and FL methods. Table 1 shows the statistics of the sheet pile lengths used for each method in the rivers. The column of PFS method shows the sheet pile lengths of both end bearing and floating sheet piles used and their sheet pile length ratios. The average sheet pile lengths for end bearing (CS method) and floating method (FL method) are 34.2 m and 14.6 m, respectively, indicating that the end bearing sheet pile is about twice as long as the FL method. In the meantime, the average end bearing and floating sheet pile lengths of the PFS method are 38.7 m and 25.5 m, respectively. Figure 9 (a) shows the trend of construction period

of steel sheet pile method and ground improvement classified by purpose and construction method. The construction period of the sheet pile method and sheet pile revetment has been increasing at a constant rate and the construction period of the sheet pile revetment has been increasing rapidly since 2010. On the other hand, the volume of steel sheet pile construction method, which is the objective of both subsidence control and earthquake resistance, has remained unchanged since 2004. Figure 9 (b) shows the relationship between the year of construction and the length of construction of each sheet pile method in the four rivers of the Kumamoto Plain. Here, it is easily realized that the number of construction of the PFS and FL methods has increased rapidly after 2010. It is noted that Figure 10 shows the flow chart of the design method for the steel sheet pile due to earthquake motion. This idea is based on the liquefaction problem on river embankment construction.

3.2.2 *Adaptation of the steel sheet pile method in Kumamoto Prefecture*

The subsidence of the river embankment after the Kumamoto earthquake was calculated based on the cross-sectional survey at the top of the embankment conducted by the Kyushu Regional Bureau of the Ministry of Land, Infrastructure, Transport and Tourism at a pitch of 200 m and this survey was conducted immediately after the earthquake. In the case of damage to the embankment caused by the Kumamoto earthquake, the subsidence of the embankment was not so large as to cause the loss of the embankment function, because the water level of the embankment was lower than the outside water level of the Tohoku earthquake. Figure 11 shows the probability density distributions of the subsidence of each type of embankment. The legend in the figure. shows the type of sheet piling method for the inner side of the embankment. The subsidence of the no countermeasure section is measured at 551 points and is distributed over a wide range of -1.28 m to 1.56 m, whereas the subsidence of the section reinforced by the various steel sheet piling methods is concentrated in the range of -0.08 to 0.39 m. In order to investigate the effect of various

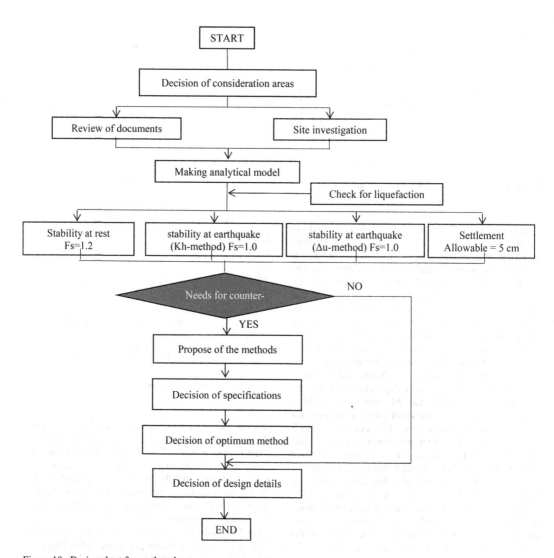

Figure 10. Design chart for earthquake.

72

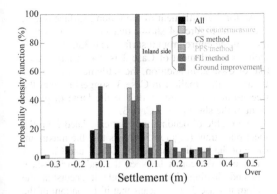

Figure 11. Probability density function of Settlements.

methods and combinations of methods on subsidence control, Table 2 summarizes the statistics of subsidence caused by earthquakes. There are 14, 49 and 15 points in the embankment where the CS, PFS and FL method were applied, respectively, and combination of FL method, ground improvement and no countermeasure was applied on the outer side of the embankment. In addition, there were 15 points where FL method was applied on the outer side of the embankment and no countermeasure was applied on the inner side of the embankment, and there was one point where both the inner and outer sides of the embankment were improved. When the inner side of the embankment was reinforced by the embankment, the average subsidence of the embankment was the largest and the range between the maximum and the minimum values varied widely from -0.08 m to 0.39 m for the embankment method and the FL method. The mean subsidence of

the embankment with the CS method (inside the embankment) and the ground improvement method (outside the embankment), and the no countermeasure (outside the embankment) was 0.03 m and 0.01 m, respectively, which is almost zero. When the inner side of the embankment was reinforced by the PFS method, the average subsidence of the embankment reinforced by the FL method (outside of the embankment), ground improvement (outside of the embankment), and no countermeasure (outside of the embankment) was 0.11 m, 0.08 m, and 0.04 m, respectively, which is a small value. When the embankment was reinforced by the FL method, the mean subsidence was slightly higher than 0.16 m and 0.13 m for either the inner or outer side of the embankment, respectively. In addition, the subsidence of the embankment reinforced on the outside of the embankment tended to be larger than that of the other methods, for example, 0.39 m was observed at the point where the embankment was reinforced by the FL method. The average subsidence of the combination of FL reinforcement and ground improvement was 0.08 m and 0.09 m, respectively.

3.3 WG2: Laboratory experiments

In this section, research outcomes of WG2 (laboratory experiments) will be outlined. Details of some of these researches can be found in Hizen et al. (2018), Kijima (2018), Inoue (2019), Iwasaki (2019), Kashiwagi (2019), Akimoto (2020), Kijima (2020) and Oka (2020).

3.3.1 Earthquake resistance of PFS in liquefied ground: Centrifuge Model Test

The behavior of the PFS in liquefied ground was investigated by centrifuge model tests. The equipment used was a beam-type centrifuge with an effective radius of 1.5 m owned by Kansai University. The scale of the model was set to 1/50 (centrifugal acceleration 50g, g is the acc. due to gravity), and the ground was a two-layered ground (10.0 m thick, 22.5 m wide, and 10.0 m deep in prototype scale) consisting of loosely packed saturated sand (Toyoura standard sand) and clay (kaolin + gypsum) (Figure 12). The thickness of the PFS model (model scale: 1.2 mm) was determined so that the bending stiffness was the equivalent to that

Table 2. Statistics of seismic settlement.

Island side	River side	No.	Mean	COV*	Max (m)	Min (m)
All		645	0.10	2.24	1.56	-1.28
No Countermeasure		551	0.10	2.36	1.56	-1.28
Countermeasure		64	0.09	1.06	0.39	-0.08
CS method	FL method	4	0.15	1.44	0.39	-0.08
	GI**	2	0.03	1.41	0.07	0.00
	No	8	0.01	5.06	0.13	-0.07
PFS method	FL method	29	0.11	0.84	0.38	0.00
	GI**	3	0.08	1.09	0.15	-0.02
	No	17	0.04	0.90	0.14	-0.01
FL method	FL method	4	0.08	1.11	0.15	-0.04
	GI**	3	0.09	0.73	0.16	0.05
	No	8	0.16	0.39	0.26	0.08
No	FL method	15	0.13	1.00	0.39	-0.06
GI**	GI**	1	0.05	-	0.05	0.05

* : The coefficient of variation
** : Ground improvement

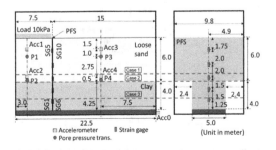

Figure 12. Cross sectional view of model test (for Case 2).

of the U-shaped sheet pile type 3. The PFS model consisted of two parts; one is the end-bearing sheet pile with 10 m in length and 9.8 m depth, and the other is floating sheet piles with 6 m in length and 5 m depth. The bottom edge of the end bearing sheet pile model was fixed for both rotation and displacement. A stainless steel plate as the embankment load with a distributed load of 10 kPa was applied to the ground surface. A half-section model was used for all experiments, with the symmetry axis at the center of the embankment cross-section. The test was conducted in three cases as shown in Table 3, and the embedment depth of the floating sheet pile was varied by changing the thickness of the upper sand layer. As shown in Figures 13(a)-(c), the input acceleration is a tapered sinusoidal wave lasting for 20 seconds. 5 accelerometers (Acc 0~Acc 4), 4 pore water pressure transducers (P1~P4), 10 strain gages on the sheet pile (SG1~SG10) were installed in the model. Also settlements of ground surface, and lateral displacements of the steel sheet pile were measured by hand.

3.3.2 Results of centrifuge tests
The acceleration responses under the embankment (Figures 13(d)-(f)) show a large decay in amplitude in all cases, reflecting the liquefaction of the ground. In Case 2, there is a tendency of one-sided oscillation, which may be due to the dilatancy spike caused by the displacement of the sheet pile.

As for the settlement of the ground surface due to excitation, Figure 14 shows that the average settlement of the entire ground surface is 0.33m to 0.35m, while the settlement of Case 3 is 0.49m larger than that of Case 2. In addition, the settlement of the free ground is the smallest in Case 3, suggesting that the free ground was uplifted due to the large displacement of the sheet pile in Case 3.

The profile of bending moment calculated from the sheet pile strain (Figure 15) shows that the maximum value of the bending moment in Case 1 and Case 2 is found near the corner between the end-bearing sheet pile and floating sheet pile, while it occurs at the bottom in Case 3. This means that if the bottom of the floating sheet pile is not rooted in the non-liquefied layer (Case 3), the sheet pile may be deformed significantly by the lateral flow of sand due to liquefaction. The amount of deformation depends on the bending stiffness of the bottom of the sheet pile. However, it is necessary to root the bottom of the floating sheet pile into the non-liquefied layer when liquefaction is expected.

3.4 WG3: Numerical analysis

In this section, research outcomes of WG3 (numerical analysis) will be outlined. Details of some of these researches can be found in Nakai et al. (2018), Fujiwara et al. (2019) and Fujiyama (2020).

Table 3. Test cases.

	Layer thickness (m)		Dr (%) of sand	Input PGA (m/s²)
	Sand	Clay		
Case 1	5.0	5.0	60.3	1.7
Case 2	6.0	4.0	51.9	1.0
Case 3	8.0	2.0	45.4	1.0

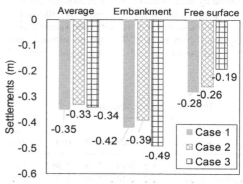

Figure 14. Ground subsidence due to earthquake.

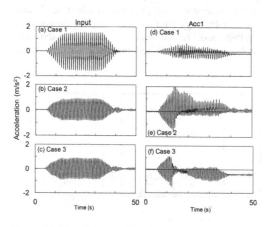

Figure 13. Time history of earthquake record (input and accl).

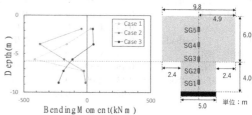

Figure 15. Distribution of bending moment at the maximum stage.

3.4.1 Verification of lateral displacement suppression effect during embankment construction

When constructing embankments on soft ground, the PFS method is effective in suppressing subsidence of the embankment due to the edge cutting effect, without the need for continuous placement of steel sheet piles up to the support layer as in the past. On the other hand, the effect of suppressing lateral displacement to the embankment land has not been fully verified. Therefore, in this section, we verified the lateral displacement suppression effect of the PFS method during embankment construction, paying particular attention to the stratum composition. Figure 16 shows the finite element mesh used in the analysis. The analysis code is the 3D finite element method program FEMtij-3D, and a linear elastic body is used for the sand layer and an elasto-plastic body (Sekiguchi-Ohta model) is used for the clay layer as the constitutive model of the ground material. The embankment height was 3.4 m and the crown width was 6.0 m, referring to the actual river embankment in the Kikuchi River, Kumamoto Prefecture. The material parameters were determined based on the results of the ground survey conducted at the same location. Figure 17 shows a model diagram of the PFS sheet pile. In this study, the spacing between the end bearing sheet piles was fixed at 5, and the embedment length of the PFS sheet piles was changed in 3 ways. Table 4 shows the stratum composition. The total stratum thickness was fixed at 30 m, and the clay layer thickness and sand layer thickness were systematically changed.

(1) Lateral displacement in unmeasured ground

Figure 18 shows the distribution of lateral displacement during embankment construction on unmeasured ground, and Figure 19 shows the relationship between the maximum lateral displacement and the sand thickness ratio. Looking at Figure 18, it can be seen that the thinner the sand layer on the surface is the larger the lateral displacement becomes. When there is no sand layer on the surface and only the clay layer, the lateral displacement is particularly large and reaches up to 80 cm. It was also found that the maximum value of lateral displacement occurs near the boundary between the sand layer and the clay layer.

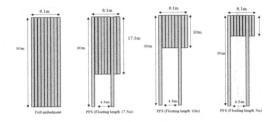

Figure 17. Examination model of PFS method.

Table 4. Stratum composition used for the study.

Clay	30m	29.5m	27.5m	25m	20m	15m
Sand	0m	0.5m	2.5m	5m	10m	15m
Total	30m					

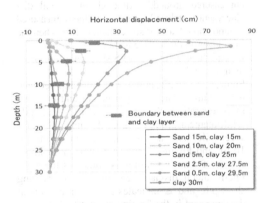

Figure 18. Lateral displacement distribution with respect to the depth direction in case of non-countermeasure.

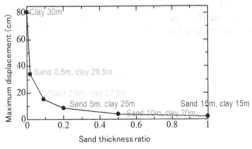

Figure 19. Relationship between maximum lateral displacement and stratum composition in case of non-countermeasure.

(2) Lateral displacement in sheet pile reinforcement ground

Figure 20 shows the relationship between the maximum lateral displacement and the sand thickness ratio with respect to the all-landing sheet pile

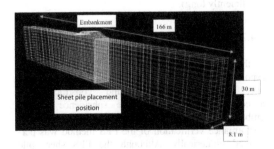

Figure 16. Finite element mesh.

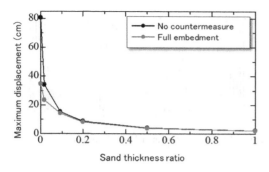

Figure 20. Relationship between maximum lateral displacement and stratum composition in case of full-embedment countermeasure.

reinforcement ground (the figure is omitted, but the maximum value of the lateral displacement occurs near the boundary between the sand layer and the clay layer, which is common to the unmeasured ground.). Although the lateral displacement is reduced compared to the unmeasured ground, large lateral displacement occurs when the sand layer on the surface is thin.

(3) Lateral displacement in PFS sheet pile reinforcement ground

Figures 21 and 22 show the relationship between the maximum lateral displacement and the sand thickness ratio with respect to the PFS sheet pile reinforcement ground. Figure 21 shows the PFS sheet pile with a long embedment length of 10 m, and Figure 22 shows a short embedment length of 5 m. In addition, ● indicates the maximum lateral displacement in the end bearing sheet pile, and ▲ indicates the maximum lateral displacement in the floating sheet pile. When the floating sheet pile length is long (Figure 21), the countermeasure effect is almost the same as that of all-landing sheet pile reinforcement ground, and the lateral displacement suppression effect of the PFS method when there is a sand layer on the surface layer can be confirmed. On the other

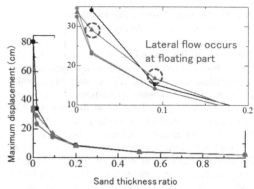

Figure 22. Relationship between maximum lateral displacement and stratum composition in case of PFS method (embedment depth of floating piles are as shallow as 5m).

hand, when the floating sheet pile length is short (Figure 22), the lateral displacement is large in the region where the sand thickness ratio is small. This is due to the soil slipping through and squeezing out from the floating sheet pile, suggesting that the PFS sheet pile reinforced ground does not have a sufficient lateral displacement suppression effect when the floating sheet pile length is short. In this section, using 3D FEM analysis, the lateral displacement suppression effect of the PFS method in sand-clay alternating layer ground was verified by focusing on the sand thickness ratio. As a result, the following conclusions were obtained. If there is not enough sand layer on the surface (sand thickness ratio is 0.2 or less), even if the steel sheet pile method (total landing/PFS method) is used, the amount of lateral displacement during embankment construction cannot be sufficiently suppressed. However, if there is a sand layer on the surface and the sand thickness ratio is 0.2 or more, the lateral displacement can be suppressed by the steel sheet pile method. When using the PFS method, if the floating sheet pile length is short, the ground will slip through and squeeze out in the floating part, and a sufficient lateral displacement suppression effect cannot be obtained. Therefore, the floating sheet pile length should be set appropriately (sufficiently long).

3.4.2 Seismic performance verification in the event of an earthquake

In the Kumamoto earthquake (2016), there are many reports that the seismic effect was exhibited in the PFS sheet pile reinforced ground constructed as a countermeasure against the subsidence of the embankment. Therefore, in this section, the seismic performance verification of the PFS method was performed numerically. Although the PFS sheet pile reinforcement ground is a three-dimensional problem,

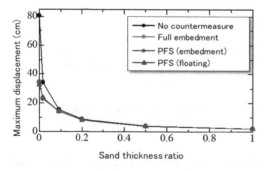

Figure 21. Relationship between maximum lateral displacement and stratum composition in case of PFS method (embedment depth of floating piles are as deep as 10m).

it is difficult to carry out a three-dimensional analysis of the actual ground on a regular basis. Therefore, first, we propose a simple two-dimensional modeling method for the PFS method. After that, seismic performance verification of PFS sheet pile reinforced ground in sand-clay alternating layer ground was carried out.

3.4.3 Numerical analysis of ground slip-through behavior between end bearing sheet piles and floating sheet piles

A seismic response analysis using a simple model was performed to understand the ground slip-through behavior between the end bearing sheet pile and the floating sheet pile. The analysis code used is the liquefaction analysis program LIQCA (Oka et al., 1994; Oka et al.,1999). Figure 23 shows the finite element mesh used in the analysis. A 1/4 cross-section model assuming symmetry was used. 8 m from the surface layer is assumed as a liquefied layer with a relative density of about 40% and 4 m below is assumed as a non-liquefied layer with a relative density of 90%. A sine wave with an amplitude of 9.0 m/s² and an input frequency of 3.0 Hz was applied to all nodes at the lower end for 10 seconds. Figures 24 and 25 show the distribution of lateral flow (horizontal displacement) due to liquefaction at the piles and floating sheet piles when the floating sheet pile length is 7m and 1m, respectively (the pile spacing is 1.0, 2.5, 5.0 and 10.0). The amount of lateral flow of the PFS sheet pile method is less than that of the unmeasured ground. On the other hand, in the floating sheet pile section, lateral flow at the lower part of the floating sheet pile is larger when the distance between the piles is wider or the embedment length of the floating sheet pile is shorter. This trend is similar to that of the lateral displacement during embankment construction shown in Section 1.1. Figure 26 is a cross-sectional view of the horizontal displacement at a depth of 7 m when the floating sheet pile length is 1 m. An arcuate lateralflow is generated starting from the end

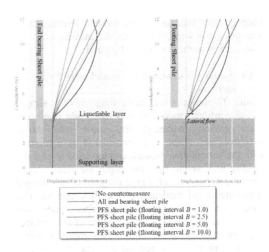

Figure 24. Distribution of lateral flow due to liquefaction in PFS method with its floating pile embedment 7m.

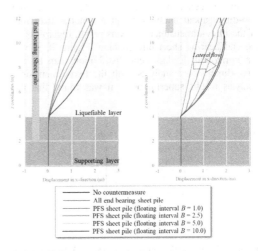

Figure 25. Distribution of lateral flow due to liquefaction in PFS method with its floating pile embedment 1m.

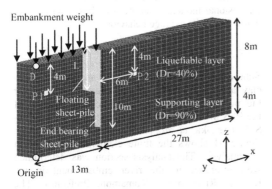

Figure 23. Finite element mesh.

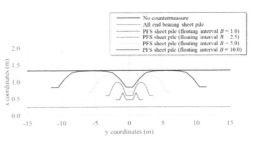

Figure 26. Cross section of horizontal Displacement.

bearing sheet pile, and when the end bearing sheet pile spacing is wide, it matches the maximum lateral flow amount in the unmeasured ground.

3.4.4 Simple two-dimensional modeling of PFS method

Figure 27 shows a schematic diagram of a simple two-dimensional model of the sheet pile reinforcement ground. Although the sheet pile shows a complicated shape, it is considered as a mixture with the surrounding ground, and modeled so that the rigidity and weight per unit depth are equal. Next, Figure 28 shows a schematic diagram of the simple two-dimensional modeling of the PFS method. The space between the end bearing sheet piles was regarded as one unit, and the mixture of the replacement sheet pile and the surrounding ground existing in one unit was modeled so that the rigidity and weight per the same depth would be equal. Figure 29 shows a comparison between the 3D analysis when the floating sheet pile length is 4 m and the 2D plane strain analysis results using the simple 2D modeling method of the PFS method proposed in this section. It can be seen that the lateral displacement of the two-dimensional analysis result is smaller than that of the three-dimensional analysis result, regardless of the end bearing sheet pile spacing. Figure 30 shows the horizontal displacement distribution map by the three-dimensional analysis with the two-dimensional analysis results superimposed. It was found that the

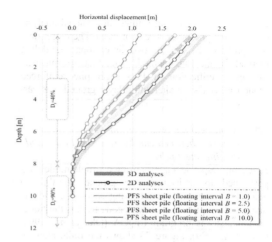

Figure 29. Comparison of lateral flow between 2D and 3D analysis.

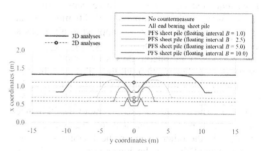

Figure 30. Comparison of lateral flow between 2D and 3D analysis (cross section).

simple two-dimensional model of the PFS method proposed in this section is just the average value of the lateral flow amount generated in an arc shape between the end bearing sheet piles. In reality, it shows a three-dimensional expansion of the lateral flow amount, but it was confirmed that a simple two-dimensional modeling can be used based on the fact that it shows its average behavior.

3.4.5 Seismic performance verification of PFS method in sand-clay alternating layer ground

Subsequently, a seismic response analysis of sand-clay alternating layers ground was carried out for the purpose of verifying seismic performance of the PFS method. The analysis code used is GEOA-SIA (Asaoka et al., 2002; Noda et al. 2008). Figure 31 shows the finite element mesh used in the analysis. The analysis section was determined with reference to the river embankment in the Kikuchi River basin, Kumamoto Prefecture. The elasto-plastic parameters used in the analysis are determined based on the results of various mechanical tests of undisturbed samples collected in the

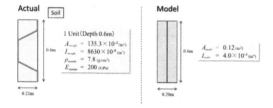

Figure 27. Schematic figure of simple 2D Modelling of sheet pile.

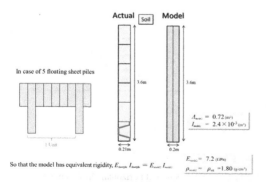

Figure 28. Schematic figure of simple 2D modelling of PFS method.

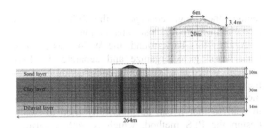

Figure 31. Finite element mesh.

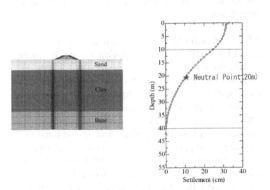

Figure 32. Determination of neutral point (floating pile length).

field. Figure 32 shows the distribution of horizontal displacement at the embankment toe at the time of embankment. Similar to the actual observational data, it shows an inverted S-shaped curve in the depth direction. According to the PFS method design and construction manual, the floating sheet pile length was set as its neutral point of 20 m. Figure 33 shows the ground conditions examined in this section. On the left side of the drawing, it was assumed that there were all landing sheet piles up to the base, and the right side was carried out in four ways: (a) no measures, (b) all landing, (c) floating, and (d) PFS. The method of 3.4.4 was used to model the sheet pile. Figure 34 shows the input seismic motion used in the

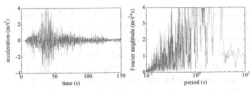

Figure 34. Input seismic motion.

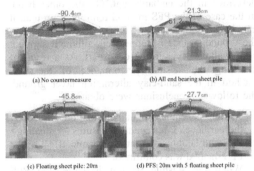

Figure 35. Comparison of embankment Deformation.

analysis. This is an assumed Nankai Trough earthquake with a long duration and long-period components. Figure 35 shows the shear strain distribution 10 years after the occurrence of the earthquake (enlarged display around the embankment) along with the amount of subsidence and horizontal displacement at the top of the embankment. In the case of unmeasured ground, the deformation of the embankment is very large, but in the case of all landing reinforcement ground, the effect of suppressing deformation during an earthquake can be confirmed, especially in horizontal displacement. Although not as much as the all landing, floating sheet pile and PFS method can also confirm the suppression effect. In addition to the smaller amount of subsidence and horizontal displacement compared to the floating method, the PFS method has a deformation suppression effect close to that of the end bearing method, confirming the seismic performance of the PFS method. Figure 36 shows the depth distribution of the lateral displacement in

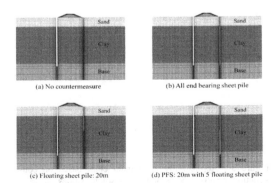

Figure 33. Analysis condition.

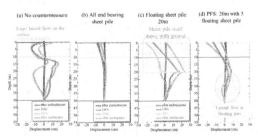

Figure 36. Lateral flow distributions of each Countermeasure.

79

the sheet pile. In the case of unmeasured ground, it can be seen that it sways greatly from side to side during the earthquake from shallow to deep, but in the case of all landing reinforcement ground, there is almost no lateral flow in the deep part, and the ground surface. In the case of floating sheet pile reinforced ground, since the floating sheet pile flows together with the ground, the amount of lateral flow near the ground surface becomes large, and there is a concern that the ground will be deformed in the embankment. On the other hand, in the case of the PFS method reinforced ground, it can be seen that the amount of lateral flow near the ground surface is suppressed, although slipping and swelling occur in the deep part. As a result of checking the seismic performance of the PFS method in the sand-clay alternating layer ground, the following conclusions were obtained.

- As for the embankment deformation caused by the earthquake, no countermeasures ≫ floating> PFS> all landing, so that the seismic effect of the PFS method was confirmed.
- When the seismic motion is large, the PFS method is considered to be more effective in suppressing lateral flow, especially near the ground surface, than the floating sheet pile (the floating sheet pile may flow along with the ground).

3.5 WG4: Design

In this section, research outcomes of WG4 (design) will be outlined.

3.5.1 Introduction
The PFS method itself was developed as "a construction method to prevent the settlement of the surrounding ground due to new embankment construction or bulky embankment construction" in "the ground which has a sand layer at the surface and where consolidation settlement is more dominant than horizontal displacement in response to load increase". However, there is a concern that the sand layer at the surface may liquefy during an earthquake, so liquefaction countermeasures are inevitably required in the ground conditions where the PFS method may be applied. As for the seismic reinforcement of embankment structures on liquefied ground, the technical standards are being developed by the management departments of each structure. For example, in the case of river embankments in Japan, the Public Works Research Institute (PWRI) has published a manual on design methods (PWRI, 2016). For example, for a river embankment, a manual from the PWRI describes the design method of the embankment, in which steel sheet piles are rooted into the non-liquefiable layer, and the horizontal resistance of the sheet piles determines the lateral flow of the ground under the embankment. On the other hand, these technical standards for seismic reinforcement do not

include the design concept of the PFS method as a countermeasure for close construction.

Against this background, the WG4 (design) has been working to organize and separate the PFS method, which is a "proximity construction countermeasure," from the "seismic reinforcement countermeasure" in each field, and to revise the design manual so that it can be practically used to easily design the PFS method, which is both a seismic reinforcement countermeasure and a proximity construction countermeasure. As a summary of the results, the main revisions of the design manual are shown below.

3.5.2 Determination of applicable conditions
The PFS method can be applied to the following ground conditions:

When the load increase caused by the embankment makes it necessary to take countermeasures against ground subsidence due to consolidation of the clay layer, but when direct countermeasures are not necessary because the displacement in the horizontal direction is suppressed by the surface sand layer. It was necessary to have a simple method to determine this from limited practical information. In this manual, the following three methods are presented.

1) Empirical method
   In this method, the applicability is judged based on the regional ground characteristics and construction results. For example, the Kumamoto area falls under this category.
2) Statistical method
   In this method, the required thickness of sand layer is determined based on the statistical analysis of past construction results, as shown in Figure 37 and Figure 38. In this method, the required thickness of the sand layer is determined based on the statistical analysis of past construction results, as

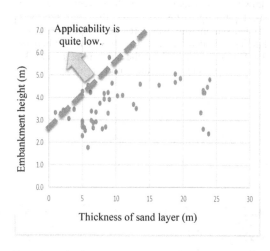

Figure 37. Applicability assessment chart based on construction results (Evaluation by absolute values of sand layer thickness and embankment height).

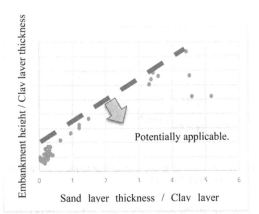

Figure 38. Applicability assessment chart based on construction results (Evaluation by ratio of sand layer thickness and embankment height to clay layer thickness).

shown in those figures, reflecting the results of WG1 (Field investigation).
3) Analytical methodIn this method, the nonlinear finite element method studied by WG (Numerical analysis) is used for evaluation.

3.5.3 *Design concept of seismic reinforcement measures*

The design concept of the seismic reinforcement of the PFS method is usually the same as that of the steel sheet pile method (design concept of the seismic reinforcement), which ignores the end bearing sheet pile in various fields. It is clear from the results of WG2 (Laboratory experiments) and WG3 (Numerical analysis) that this approach is at least on the safe side. It is also mentioned in the manual that it is possible to design with the resistance of end bearing sheet piles by using detailed analysis such as the 3D nonlinear finite element method, which can take into account the effect of ground slip between the piles.

3.5.4 *Determinants of the main design parameters*

Based on the analysis of the design concept of each field and the results of trial design, the determinants of the main design specifications of the PFS method are analyzed as follows.

1) Floating sheet pile lengthDetermined by the neutral axis depth determined from the allowable displacement.
2) Type of steel sheet pile (thickness and sectional stiffness)Regardless of the purpose of seismic countermeasure, it is necessary to have a type of sheet pile that can be driven to the bearing stratum (the deeper the bearing stratum and the stronger the stratum above the bearing stratum, the larger the type of steel sheet pile (i.e., the larger the thickness and the sectional rigidity)).
3) Sheet pile length and space (pitch) of end bearing sheet pileThe length of the sheet pile should be at

Table 5. Table of contents for design manual.

1. General remarks
   1.1 Objectives
   1.2 Positioning of the construction method
   1.3 Scope of application
   1.4 About this manual
   1.5 Related standards
   1.6 Basic concepts of PFS method
   1.7 Investigation, design, and construction procedures
2. Site investigation
   2.1 Objectives
   2.2 Contents and items
   2.3 Planning
3. Design
   3.1 Basic concept
   3.2 Determination of applicability
   3.3 Setting basic specifications
   3.4 Consideration of earthquake
   3.5 Consideration of countermeasures for surrounding ground
   3.6 sheet pile head treatment
4. Construction
   4.1 Installation of sheet pile
   4.2 Construction procedure
   4.3 Points to note for the use of water jet

least longer than the depth of the bearing layer determined by the soil conditions. The length of the sheet pile should be at least longer than the depth of the bearing layer determined by the ground conditions. The space (pitch) between the end bearing sheet piles should be determined so that the vertical bearing capacity is sufficient to withstand the negative friction.

3.5.5 *Table of contents of the design manual*

The table of contents of the revised design manual is shown in Table 5. The main feature of the manual is that the flow is designed to minimize the number of design trials.

In the design of civil engineering structures in Japan, seismic design is very important and has a great influence on the specifications of the structures, but the design methods are becoming more sophisticated. The design of civil engineering structures in Japan is very important and has a great impact on the specification of structures. However, the design methods are becoming more and more sophisticated. Therefore, the seismic design work in practice requires a large amount of human resources, and the development of a simple seismic design method is very important for the diffusion of the technology and the efficient development of social infrastructure. The design manual will contribute to improving the efficiency of the practical work when the PFS method is used together with the seismic reinforcement method.

4 PRESS-IN TECHNIQUE

The Press-in Method, originated in Kochi, is a piling method that installs piles with a static jacking force while gaining a reaction force from previously installed piles. Since it is a static method, it generates less noise and vibration (White *et al.*, 2002; White & Deeks, 2007). In addition to the function of a press-in machine to gain a reaction force from previously installed piles, the press-in machine and its related devices (power unit, crane and so on) can walk on top of the previously installed piles. Owing to these functions, this piling method can save temporary works or large construction areas for reaction weights or conventional types of cranes by the spatially-efficient piling system (called "GRB System") as shown in Figure 39.

The resistance applied to the pile from the ground during installation has to be controlled smaller than the sum of the reaction forces obtained mainly from the previously installed piles. In Standard Press-in (press-in without installation assistance), the resistance will be controlled by a repetition of penetration and extraction (called "surging") or by the penetration rate. There are several researches that confirmed the effect of surging to reduce the shaft resistance (Burali d'Arrezzo *et al.*, 2013). Increasing the penetration rate is effective in reducing the penetration resistance if the soil is contractile (White *et al.*, 2010). If the ground condition is too hard for Standard Press-in, the use of installation assistance is

Figure 39. Press-in piling system that requires no temporary works (GRB System).

effective. In Press-in with Water Jetting, the penetration resistance is reduced by the erosion of the soil and the build-up of the pore water pressure around the pile base (Gillow *et al.*, 2018). In Press-in with Augering, the reduction of the penetration resistance is attained by excavation and temporary lift-up of the soil beneath the pile base. In Rotary Cutting Press-in where a tubular pile with cutting teeth on its base is pushed and rotated at the same time to be installed into the ground, the vertical penetration resistance is reduced by the cut of the soil (or a rock or a concrete) beneath the pile base as well as by the decomposition of the frictional stress acting on the pile surface into vertical and horizontal directions. The above-mentioned penetration techniques in the Press-in Method are summarized in Figure 40.

Standard Press-in	Press-in with Water Jetting	Press-in with Augering	Rotary Cutting Press-in
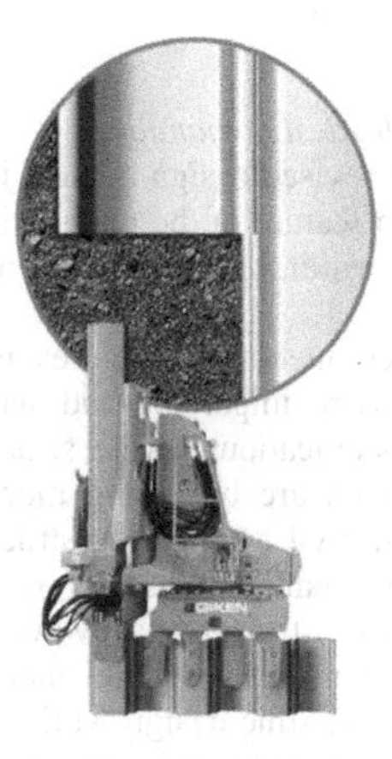	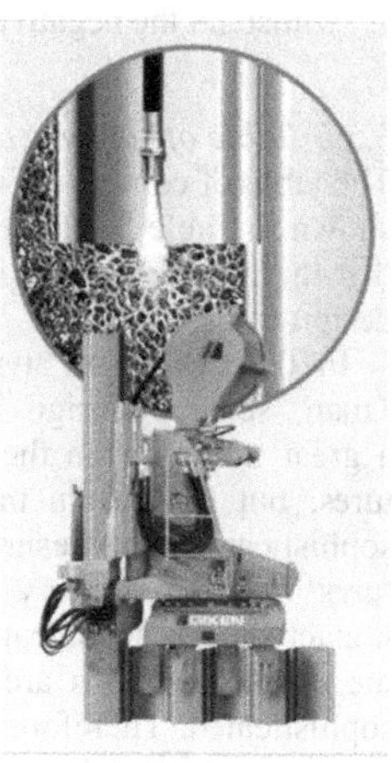		
Press-in a pile without installation assistance	Press-in a pile wile applying water-jetting in the pile base	Press-in a pile while excavating the soil around the pile base	Rotate and press-in a pile equipped with base cutting teeth

Figure 40. Penetration techniques in the Press-in Method.

One of the case histories that typically summarize the features of the Press-in Method is shown in Figure 41 (White *et al.*, 2010). This was the project of renovating the river levee in Tokyo. Under a strict spatial restriction due to the highway bridge above the levee and the buildings along the river, steel tubular piles were installed directly into the ground by penetrating though the concrete structures of the old levee, while doing away with the necessity of the removal work of the old levees and the temporary works for a crane that affects the river flow.

Another feature of the Press-in Method is the acquisition of the piling data by the automatic system equipped in the press-in machine. The data includes the time, penetration depth, jacking force, rotational torque and so on. Recent researches have developed methods of utilizing the press-in piling data for estimating subsurface information, which are summarized in IPA (2017) that provides recommendations on their applicability to real projects based on the validity of the estimation results at the moment.

On the other hand, the pile installed by the Press-in Method has a feature to effectively utilize the resistance of the ground. White & Deeks (2007) showed that the capacity of the pressed-in pile is higher than that of driven or bored piles, mainly due to the higher extent of soil plugging which leads to greater volume of soil displacement and to the smaller extent of friction reduction due to the cyclic motion of the pile (friction fatigue). Deeks & White (2007) demonstrated that the stiffness of the base response of the pressed-in pile is also higher than

(a) Single pile wall   (b) Double sheet pile wall

Figure 42. Application of pile walls to coastal levees in Kochi Coast as liquefaction countermeasure.

that of driven or bored piles, mainly due to the loading history of the pile (static loading and unloading at the end of installation of the pressed-in pile). Li (2010) confirmed in his centrifuge model tests that the pressed-in pile exhibited higher horizontal stiffness under horizontal loads than the bored pile which were embedded in dense sands.

It has been expected that the above-mentioned advantages in the performance of the pressed-in piles be reflected in the performance of structures with piles. Such structures are exemplified by retaining walls, coastal levees, circular or rectangular walls and so on. Figure 42 shows a project in Kochi Coast, which was conducted in direct control of Ministry of Land, Infrastructure, Transport and Tourism (MLIT), to improve the seismic performance of the coastal levees (Ishihara *et al.*, 2020). Pile walls were installed by the Press-in Method in the levee body, so that the height of the levee is maintained even when the ground is liquefied by the significant seismic motion expected in the coming Nankai Trough earthquake. In addition to the performance for liquefaction, the structure is expected to show tenacity for tsunami loads (Furuichi *et al.*, 2015). Varieties of other case histories related with the Press-in Method can be found in IPA (2016) and IPA (2019).

5 CONCLUSIONS

Steel sheet pile method was discussed as a state of the arts report. This method had long been used as temporary works in geotechnical engineering but recently more applications to a permanent structure have been extended. In this paper, the development of PFS method which was originally proposed as a countermeasure for soft soil subsidence due to river embankment constructions. And recently, this method has been tried to discuss on the applications to earthquake. This was done by Technical Committee under IPA. There are a large number of natural disasters such as heavy rains and earthquakes and

Figure 41. Renovation of river levee in Tokyo by Rotary Cutting Press-in Method.

the sheet piles can be an effective countermeasure such as the stability of the river embankments. A construction technique called press-in method has accelerated the application of the sheet pile method. Finally, more wide varieties of the usage of the sheet pile can be expected not only as the construction technique but also as the countermeasure method for any kinds of natural disasters related to geotechnical engineering problems.

ACKNOWLEDGEMENT

The contents of this paper are based on the activities by two committees which are the original Technical Committee on PFS method chaired by Emeritus Prof. Hidetoshi Ochiai of Kyushu University in 2003 and the TC3 of IPA. The author was deeply involved for both two committees. The author would like to express his deep gratitude to those technical members including Emeritus Prof. Hidetoshi Ochiai of Kyushu University, Japan and Prof. Osamu Kusakabe who is the immediate past President of IPA for their kind supports to the development of steel sheet pile method. In addition, I would like to thank the Kyushu Chapter of Ministry of Land, Infrastructures, Transportation and Tourism for providing us with a lot of valuable in-situ data. We would not have been able to carry out our activities without this data. Finally, the leading members of TC3, Dr. Shinj Taenaka, Prof. Kiyonobu Kasama, Prof. Tetsuo Tobita, Prof. Kentaro Nakai, Prof. Hidetoshi Nishioka and Dr. Yukihiro Ishihara, who have done a great support preparing the contents of this paper, are highly appreciated.

REFERENCES

Akimoto, T. 2020. Dynamic centrifuge model tests on the deformation of sheet piles installed at the foot of the embankment on a soft clay. *Bachelor's Thesis*, Tokushima University: 97pp. (in Japanese).

Asaoka, A., Noda, T., Yamada, E., Kaneda, K. & Nakano, M. 2002. An elasto-plastic description of two distinct volume change mechanisms of soils, *Soils and Foundations*, 42(5): 47–57.

Burali d'Arezzo, F., Haigh, S. K. and Ishihara, Y. 2013. Cyclic jacking of piles in silt and sand. Installation Effects in Geotechnical Engineering – *Proceedings of the International Conference on Installation Effects in Geotechnical Engineering*: 86–91.

Deeks, A. D. & White, D. J. 2007. Centrifuge modelling of the base response of closed-ended jacked piles. *Advances in Deep Foundations*: 241–251.

Fujiwara, K., Nakai, K. & Ogawa, N. 2019. Quantitative evaluation of PFS (Partial Floating Sheet-pile) Method under liquefaction. *International Conference on Geotechnics for Sustainable Infrastructure Development*: 467–472.

Fujiyama, H. 2020. A study on the evaluation of soil deformation suppression effect of PFS Method under different ground conditions using 3D FEM analysis. *Master's Thesis*, Kumamoto University: 20pp. (in Japanese).

Furuichi, H., Hara, T., Tani, M., Nishi, T., Otsushi, K. & Toda, K. 2015. Study on reinforcement method of dykes by steel sheet-pile against earthquake and tsunami disasters. *Japanese Geotechnical Journal*, Vol. 10, Issue 4: 583–594.

Gillow, M., Haigh, S., Bolton, M., Ishihara, Y., Ogawa, N. & Okada, K. 2018. Water Jetting for Sheet Piling. *Proceedings of the First International Conference on Press-in Engineering 2018*, Kochi: 335-342.

Hizen, D., Kijima, N. & Ueno, K. 2018. Centrifuge model tests and image analysis of a levee with partial floating sheet-pile method. *Proceedings of the First International Conference on Press-in Engineering 2018*, Kochi: 215-220.

Inoue, N. 2019. Research on the effectiveness of PFS Method in a ground consisting of clay and sand during earthquakes. *Bachelor's Thesis*, Kansai University: 10pp. (in Japanese).

IPA (International Press-in Association). 2016. *Press-in retaining structures: a handbook*, First edition 2016: 520pp.

IPA (International Press-in Association). 2017. *Technical Material on the Use of Piling Data in the Press-in Method, I.* Estimation of Subsurface Information: 63pp. (in Japanese)

IPA (International Press-in Association). 2019. *Press-in Piling Case History Volume 1*, 2019: 198pp.

Ishihara, Y., Yasuoka, H. & Shintaku, S. 2020. Application of Press-in Method to coastal levees in Kochi Coast as countermeasures against liquefaction. Geotechnical Engineering *Journal of the SEAGS & AGSSEA*: 10pp.

Iwasaki, T. 2019. Investigations into the effect of reducing the settlement of embankment by dynamic centrifuge model tests. *Bachelor's Thesis*, Tokushima University: 56pp. (in Japanese).

Kasama, K., Ohno, M., Tsukamoto, S. & Tanaka, J. 2019. Seismic damage investigation for river levees reinforced by steel sheet piling method due to the 2016 Kumamoto earthquake, *International Conference on Geotechnics for Sustainable Infrastructure Development*.

Kasama, K., Yamamoto, S., Ohno, M., Mori H. & Tsukamoto S., Tanaka J. 2020. Seismic damage analysis on the river levees reinforced with steel sheet pile by the 2016 Kumamoto earthquake. *Japanese Geotechnical Journal*, Vol. 15, No. 2: 395–404. (in Japanese).

Kashiwagi, K. 2019. Research on the dynamic behavior of PFS Method in clay. *Master's Thesis*, Kansai University: 9pp. (in Japanese).

Kijima, N. 2018. Centrifuge model tests and image analyses on the embankment with partial floating sheet-piles (PFS). *Bachelor's Thesis*, Tokushima University: 61pp. (in Japanese).

Kijima, N. 2020. Effect of partial floating sheet-pile (PFS) on dynamic deformation behavior of embankment. *Master's Thesis*, Tokushima University: 70pp. (in Japanese).

Li, Z. 2010. Piled foundations subjected to cyclic loads or earthquakes. *Ph.D. Thesis*, University of Cambridge: 290pp.

Nakai, K., Noda, T., Taenaka, S., Ishihara, Y. & Ogawa, N. 2018. Seismic assessment of steel sheet pile reinforcement effect on river embankment constructed on a soft clay ground, *Proceedings of the First International Conference on Press-in Engineering 2018*, Kochi: 221–226.

Noda, T., Asaoka, A. and Nakano, M. 2008. Soil-water coupled finite deformation analysis based on a rate-type equation of motion incorporating the SYS Cam-clay model, *Soils and Foundations*, 48(6): 771–790.

Oka, F., Yashima, A., Shibata, T., Kato., M. & Uzuoka, R. 1994. FEM-FDM coupled liquefaction analysis of a porous soil using an elasto-plastic model. *Applied Scientific Re-search*, 52: 209–245.

Oka, F., Yashima, A., Tateishi, A., Taguchi, Y. & Yama-shita, A. 1999. A cyclic elasto-plastic constitutive model for sand considering a plastic-strain dependence of the shear modulus. *Geotechnique*, 49 (5): 661–680.

Oka, R. 2020. Dynamic centrifuge model tests on the deformation behavior of a soft clay ground reinforced by sheet piles. *Bachelor's Thesis*, Tokushima University: 76pp. (in Japanese).

PFS Method, Technical Manual, PFS Technical Committee, 2005 (in Japanese).

Public Works Research Institute (PWRI). 2016. Guidelines for Liquefaction Countermeasures for River Embankments, *TECHNICAL NOTE of PWRI*, No.4332: 62–82. ((in Japanese).

Steel Sheet Pile Manual, JASSP, Japan 2014 (in Japanese).

White, D. J., Finlay, T., Bolton, M. & Bearss, G. 2002. Press-in piling: ground vibration and noise during pile installation. International Deep Foundations Congress, *ASCE, Special Publication 116*: 363–371.

White, D. J. and Deeks, A. D. 2007. Recent research into the behavior of jacked foundation piles. *Advances in Deep Foundations*: 3–26.

White, D. J., Deeks, A. D. & Ishihara, Y. 2010. Novel piling: axial and rotary jacking. *Proceedings of the 11th International Conference on Geotechnical Challenges in Urban Regeneration*, London, UK, CD: 24pp.

Yamamoto, S., Kasama, K., Ohno, M. & Tanabe, Y. 2018a. Seismic behavior of the river embankment improved with the steel sheet piling method. *Proceedings of the First International Conference on Press-in Engineering 2018*, Kochi: 227–232.

Yamamoto, S., Kasama, K., Ohno, M., Tsukamoto, S. & Tanaka J. 2018b. Seismic behavior of the river embankment improved with various steel sheet piling methods by the 2016 Kumamoto Earthquake instruction. *The 15th Japan Earthquake Engineering Symposium*: 10pp. (in Japanese).

Yamamoto, S., 2019. Seismic behavior of the river embankment improved with steel sheet piling methods by the 2016 Kumamoto Earthquake, *Master's Thesis*, Kyushu University: 124pp. (in Japanese).

Session A: Pile performance / Piling process

Proceedings of the Second International Conference on
Press-in Engineering 2021, Kochi, Japan – Matsumoto et al (eds)
© 2021 Taylor & Francis Group, London, ISBN 978-1-032-10414-0

Size effect of footing in ultimate bearing capacity of intermediate soil

T. Iqbal, S. Ohtsuka & Y. Fukumoto
Nagaoka University of Technology, Nagaoka, Japan

K. Isobe
Hokkaido University, Hokkaido, Japan

ABSTRACT: In geotechnical engineering, bearing capacity of rigid footings is perhaps common and complex problem in numerous facets. Although, a lot of researches have already been conducted pertaining to the bearing capacity but hardly a few researches present the size effect of foundation on ultimate bearing capacity (UBC) of intermediate soil using non-linear shear strength characteristics. The prime objective of this study is to analyze the effect of footing sizes on UBC by using the finite element analysis and assess the effectiveness of Architectural Institute of Japan's (AIJ) semi-empirical bearing capacity equation. Rigid plastic finite element method (RPFEM) using nonlinear shear strength characteristics of the soil is employed to evaluate the UBC of footing against the centric vertical load. The analysis results are compared with that of conventional bearing capacity formulae and a new bearing capacity equation is proposed with dimensional correction factor in cohesive part of AIJ's bearing capacity equation.

1 BACKGROUND

The precise estimation of the ultimate bearing capacity of foundation is the first and foremost step in order to ensure the stable footing-soil system. Its importance increases significantly especially in designing buildings or structures. A German engineer Ludwig Prandtl (1921) is globally admitted to be the torchbearer in development of primitive bearing capacity theory, where he utilized the theory of plasticity to understand the punching failure pattern of thick metals. The aforementioned researcher considered a weightless and infinite half-space just below the footing, to have strength characteristics "cohesion *c*" and "angle of internal friction ϕ" to illustrate the kinematic failure mode. The theory was further extended by Reissner (1924) who considered the perfectly frictional soil (*c*=0) loaded by adjacent uniform surcharge load. The closed form solution using the hyperbolic functions, resulted in the bearing capacity factor N_q. The ultimate bearing capacity equation developed by Terzaghi (1943) considering the effects of cohesion, material weight and surcharge load was a remarkable accomplishment in the concepts for laying the foundation of modern bearing capacity theories. The ultimate bearing capacity equation proposed by Terzaghi (1943) is as follows:

$$q = cN_c + 0.5\gamma B N_\gamma + \gamma D_f N_q \tag{1}$$

In the above equation; N_c, N_γ and N_q denote the soil bearing capacity factors. These factors depend upon the angle of internal friction of the material, ϕ. Other parameters given in the above equation are given below:

γ : Unit weight of soil (KN/m^3),
c : Soil cohesion (KN/m^2),
B : Width of foundation (m) and
D_f : Foundation depth (m)

A lot of researches have been conducted over the period of time for estimation of bearing capacity factors. By using the concept of moment equilibrium the boundary value solutions for N_q and N_c can be obtained, as proposed by Prandtl (1921) and Reissner (1924):

$$N_q = e^{\pi \tan \phi} \tan^2 \left(45 + \frac{\phi}{2}\right) \tag{2}$$

$$N_c = (N_q - 1) \cot \phi \tag{3}$$

Similarly, several researches have been conducted pertaining to the bearing capacity factor N_γ and relations are proposed accordingly. The research conducted

DOI: 10.1201/9781003215226-3

by Meyerhof (1963) resulted in the following mathematical equation:

$$N_\gamma = (N_q - 1)\tan(1.4\phi) \qquad (4)$$

The bearing capacity theory was further extended by Meyerhof (1963) by introducing depth and inclination factors for the situations where a line of action of applied load is inclined to the vertical plane. The following equation was proposed by the said researcher:

$$q = cN_c d_c i_c + 0.5\gamma BN_\gamma d_\gamma i_\gamma + \gamma D_f N_q d_q i_q \qquad (5)$$

$$i_c = i_q = \left(1 - \frac{\theta°}{90°}\right)^2 \qquad (6)$$

$$i_\gamma = \left(1 - \frac{\theta°}{\phi°}\right)^2 \qquad (7)$$

$$d_c = 1 + 0.2 \cdot \sqrt{k_p} \cdot \frac{D_f}{B} \qquad (8)$$

$$d_q = d_\gamma = 1 + 0.1 \cdot \sqrt{k_p} \cdot \frac{D_f}{B} \qquad (9)$$

Here in the above equations, θ represents the degree of inclined load with reference to the vertical axis. Coefficient of passive earth pressure is given as follows:

$$k_p = \tan^2\left(45 + \frac{\phi}{2}\right) \qquad (10)$$

The Architectural Institute of Japan (AIJ, 1988, 2001) proposed the semi-experimentally devised ultimate bearing capacity formula. This equation is being used all across Japan for the ultimate bearing capacity estimation. The ultimate bearing capacity formula of AIJ in terms of bearing capacity factors N_c, N_γ and N_q can be written as follows:

$$q = i_c \alpha c N_c + i_\gamma \gamma \beta B \eta N_\gamma + i_q \gamma D_f N_q \qquad (11)$$

Here, α and β denote the shape coefficients while η is the foundation size effect factor. De Beer (1970) used empirical or semi empirical techniques to propose shape modifiers i.e. $\alpha=1$ and $\beta=0.5$. Relationship for the foundation size effect factor is expressed as follows:

$$\eta = \left(\frac{B}{B_o}\right)^m \qquad (12)$$

$B_0 = $ 1m(Reference value in the footing width)

Based on the experimental considerations, $m= -1/3$ is recommended in engineering practices.

In contrast with the conventional bearing capacity formulations, AIJ equation considers the size effect of foundation on the ultimate bearing capacity. Therefore, the traditional approach results in the overvaluation of the calculated results, as the foundation width increased. The extent up to which foundation size influences the bearing capacity, needs to be carefully evaluated. The non-linear finite element method was also used by Ueno et al. (1998) to predict the confining stress dependence of material strength parameters i.e. "cohesion c" and "angle of internal friction ϕ" and consequently the shear failure criteria. Their research results showed that mean stress beneath the foundation varied from $2\gamma B$ to $10\gamma B$ in the case of strip footing and has considerable effect on material strength characteristics and hence the bearing capacity while taking into account the size effect. In this study primary focus is laid on evaluation of ultimate bearing capacity of foundation placed on intermediate soil subjected to centric vertical load. Moreover, this research also analyzes the size effect of footing in the bearing capacity for proposing the size effect factor in cohesive part of the AIJ bearing capacity formula. Rigid plastic finite element method (RPFEM) using the non-linear shear strength envelope against the confining stress is employed for the finite element analysis (FEA). Recently, use of finite element analysis is becoming increasingly common in almost all fields of engineering because of accuracy of obtained results and saving in terms of time and cost. The applicability of FEA in geotechnical engineering can be witnessed from the prominent bearing capacity studies conducted by various researchers namely Griffiths (1982), Sloan and Randolph (1982), Frydman and Burd (1997), Hoshina et al. (2012), Nguyen et al. (2016) and Pham et al. (2019). In light of analysis abnormality and resulting variability in the material stress-strain relationships very close to the shear failure state (De Borst and Vermeer, 1984), the rigid plastic finite element method was developed by Tamura et al. (1984) to analyze the response of soil structure in the limit state. Previous researches, namely to illustrate Mehdi et al. (2014) have made it clear that the flow rule in considerable effect in the results obtained from bearing capacity analysis. Similarly, some experimental studies conducted by Tatsuoka et al. (1986)

have well depicted the influence of confining stress on material "friction angle ϕ" in the case of frictional materials. The rigid plastic finite element method is eminent in terms of ease of introducing the non-associated flow rule to the material properties for diminishing the effect of dilatancy. Therefore, in order to present the actual failure pattern of soil underneath the foundation upon application of load, the non-linear shear strength model is used in this research in contrast with the conventional Mohr-Coulomb or Drucker-Prager criteria to ascertain the size effect of foundation on the ultimate bearing capacity of intermediate soil. The resemblance in the results obtained from the non-linear RPFEM and the AIJ method indicates that RPFEM is not only suitable for analyzing the soil response in the limit state rather and it also well accounts for the size effect of foundation in bearing capacity. Furthermore, the failure modes obtained for the soil mass portray the unerring contoured distribution of equivalent plastic strain rate and velocity vectors. Therefore, obtained results illustrate that RPFEM using non-linear shear strength parameters can better envisage the ultimate bearing capacity as compared to the ordinary bearing capacity formulas currently in practice.

2 CONSTITUTIVE EQUATION FOR RIGID PLASTIC FINITE ELEMENT ANALYSIS

The rigid plastic finite element method (RPFEM) was initially derived using the concept of upper bound theorem of plasticity theory, which in fact applies the upper bound on the actual limit load to be worked out. Reissner (1924) proposed the rigid plastic constitutive equation and authenticated that the results match well with those by upper bound limit analysis. The rigid plastic finite element analysis technique works well with both the linear and non-linear analysis of soil against the confining stresses. The rigid plastic constitutive equation in respect of Drucker-Prager yielding criteria is presented in the section below. Hoshina et al. (2011) introduced the rigid plastic constitutive equation by applying the constraint on dilatancy condition using the penalty method.

2.1 Drucker-Prager yield criteria and rigid plastic constitutive equation

Tamura et al. (1987) proposed the stress-strain rate relationship of Drucker-Prager type frictional materials by assuming that the associated flow rule holds. Drucker-Prager criteria describes the linear relationship between shear stress and normal stress in the limit state through material constants. It can be said that Drucker-Prager criteria is generalization of Mohr-Coulomb failure theory. The yield surface of Drucker-Prager criteria can be written as follows:

$$f(\sigma) = aI_1 + \sqrt{J_2} = b \qquad (13)$$

Here in the above equation I_1 denotes the first invariant of stress tensor σ_{ij}, J_2 is the second invariant of deviatoric stress tensor s_{ij}. Moreover, a and b represent the material properties i.e. internal friction and cohesion respectively under plane strain.

$$I_1 = tr(\sigma_{ij}) \qquad (14)$$

$$J_2 = \frac{1}{2} s_{ij} s_{ij} \qquad (15)$$

$$a = \frac{\tan\phi}{\sqrt{9 + 12\tan^2\phi}} \qquad (16)$$

$$b = \frac{3c}{\sqrt{9 + 12\tan^2\phi}} \qquad (17)$$

The expression for volumetric strain rate is given below:

$$\dot{\varepsilon}_v = tr(\dot{\varepsilon}) = tr\left(\lambda\left(aI + \frac{s}{2\sqrt{J_2}}\right)\right) = \frac{3a}{\sqrt{3a^2 + 1/2}}\dot{e} \qquad (18)$$

Here λ is the intermediate plastic multiplier and $\dot{e}$ represents the equivalent strain rate. The unit and deviatoric stress tensors are shown by I and s respectively. The strain rate $\dot{\varepsilon}$ is a perfectly plastic component, which should satisfy the following volumetric constraint condition against the dilation property of soil to be compatible with the Drucker-Prager failure surface:

$$h(\dot{\varepsilon}) = \dot{\varepsilon}_v - \frac{3a}{\sqrt{3a^2 + 1/2}}\dot{e} = \dot{\varepsilon}_v - \eta\dot{e} = 0 \qquad (19)$$

The stress vector can be resolved in two component vectors as given below. The first term accounts for the stress vector which is determined for the yielding function while the second component determines the indeterminate stress having direction parallel to the one of the side of conical Drucker-Prager yield surface.

$$\sigma = \sigma^1 + \sigma^2 = \frac{b}{\sqrt{3a^2 + 1/2}}\frac{\dot{\varepsilon}}{\dot{e}} + \beta\left(I - \frac{3a}{\sqrt{3a^2 + 1/2}}\frac{\dot{\varepsilon}}{\dot{e}}\right) \qquad (20)$$

Here β is the undetermined stress characteristic which remains unknown or undetermined until the boundary value problem satisfying the volumetric constraint condition is solved.

The particular analysis methodology adopted in this study involves the incorporation of constraint condition on the equivalent strain rate through penalty constant in the constitutive equation. The penalty method was introduced by the Hoshina et al. (2011).

$$\sigma = \frac{b}{\sqrt{3a^2 + 1/2}\,\dot{e}}\dot{\bar{\varepsilon}} + k(\dot{\varepsilon}_v - \eta\dot{e})\left(I - \frac{3a}{\sqrt{3a^2 + 1/2}\,\dot{e}}\dot{\bar{\varepsilon}}\right)$$

(21)

The above equation and the finite element method using the concept of upper bound theorem in plasticity as formulated by Tamura et al. (1987) is also given. This methodology is termed as RPFEM in the current research. In the rigid plastic finite element method spurious deformation of finite elements as a result of zero energy modes have been witnessed during the analysis. While, rigid plastic constitutive equation using the penalty constant stabilizes the analysis and hence avoids zero energy modes.

2.2 Ultimate bearing capacity analysis of footing

In this research study, the finite element analysis is performed for the strip foundation subjected to centric vertical loading and placed on the uniform soil mass. A set of input shear strength parameters have been used for carrying out the bearing capacity analysis and results have been compared with those obtained by conventional bearing capacity formulas being practiced by the engineering community. The comparison of obtained results with the existing formulations is used to assess the efficacy of the method employed in this study. The strength parameters of foundation are set sufficient enough to be rigid. The boundary conditions are set large enough to simulate an infinite soil mass. The typical finite element mesh, boundary conditions and loading arrangements are shown in the Figure 1.

The ultimate bearing capacity analysis is performed for varied foundation widths i.e. 1, 5, 10, 30 and 50m using a set of shear strength parameters i.e. angle of internal friction $\phi=30°$, $40°$ and cohesive shear strength $c=0$, 10 and 50 kPa. The obtained results showing the velocity field and equivalent strain rate distribution in case of B=10m at $\phi=30°$ and $c=10$ kPa is shown in the Figure 2. The strain rate distribution is shown by the colored contours for values ranging from $\dot{e}_{max}$ to $\dot{e}_{min}$ (0). The relative distribution and magnitude of $\dot{e}$ determines the magnitude of ultimate bearing capacity.

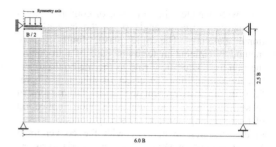

Figure 1. Finite element mesh and boundary condition for foundation width (B=10m).

Figure 2. Strain rate distribution for foundation width (B=10m) in case of $\phi=30°$ and $c=10$ kPa using Drucker-Prager criteria.

The failure pattern of soil underneath the foundation is similar to that of general failure theories. The maximum horizontal extent of failure mode from the edge of footing is 2.55B and depth is 1.05B with an ultimate bearing capacity of 1955.9 kPa. The failure mode also makes it clear that stress is concentrated on the edge of rigid foundation. This stress concentration seems to depict the problem of singularity in stress distribution, which is addressed by suitable meshing of elements. The efficacy of RPFEM for ultimate bearing capacity analysis is judged by comparing the results with conventional bearing capacity theories. The obtained results have made it clear that in spite of slight singularity in stress the results obtained are well matched with those of the past theories. The bearing capacity results of intermediate soil using Drucker-Prager yield criteria have been obtained for all the cases.

The above graph in Figure 3. shows that ultimate bearing capacity results in case Terzaghi and rigid plastic finite element method with Drucker Prager formulation are close to each other for all foundation widths. But, the results by using AIJ formula are less than others specially in case of larger foundations. As the AIJ formula is based on semi-experimental technique therefore it takes into account the size effect of footing. It infers that RPFEM should be devised in such a way that it can better depict the size effect of foundation in ultimate bearing capacity assessment.

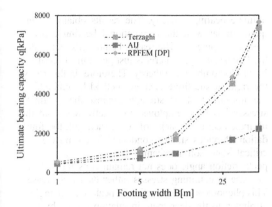

Figure 3. Size effect of foundation on ultimate bearing capacity in case of $\phi=30°$ and $c=10$ kPa.

2.3 Rigid plastic constitutive equation by using non-linear shear strength against confining pressure

The analysis in this research is based on the following higher order yield function by taking into account the non-linear shear strength of soil.

$$f(\sigma) = aI_1 + (J_2)^n = b \qquad (22)$$

Here a and b are the material constants representing the angle of internal friction and cohesion respectively while n depicts the extent of non-linearity in the shear strength of soil against the first stress invariant i.e. I_1. The above equation takes the form of Drucker-Prager yield function for $n=1/2$. The non-linear parameters have been identified for a series of analyses and comparing the results with AIJ formula which envisage the size effect of foundation. By assuming that the associated flow rule holds, the relationship for the strain rate for non-linear yield function can be expressed as follows:

$$\dot{\varepsilon} = \lambda \frac{\partial f(\sigma)}{\partial(\sigma)} = \lambda \frac{\partial}{\partial(\sigma)}(aI_1 + (J_2)^n - b) = \lambda(aI + nJ_2^{n-1}s) \qquad (23)$$

Here λ denotes the plastic multiplier. The strain rate being perfectly plastic component should satisfy the following volumetric constraint condition against the dilation property to figure out the non-linear behavior of soil mass:

$$\dot{\varepsilon}_v = tr\left(\lambda(aI + nJ_2^{n-1}s)\right) = \frac{3a}{\sqrt{3a^2 + 2n^2(b - aI_1)^{2-n^{-1}}}} \dot{e} \qquad (24)$$

From Equation (22) and Equation (24) the relationship for the first stress invariant can be easily obtained. The rigid plastic constitutive equation using the non-linear shear strength characteristics against the confining pressure is proposed by Nguyen et al. (2016) as follows:

$$\sigma = \frac{3a}{n}\left[\frac{1}{2n^2}\left\{\left(3a\frac{\dot{e}}{\dot{\varepsilon}_v}\right)^2 - 3a^2\right\}\right]^{\frac{1-n}{2n-1}}\frac{\dot{\varepsilon}}{\dot{\varepsilon}_v} + \left[\frac{b}{3a} - \frac{1}{3a}\left\{\frac{1}{2n^2}\left(3a\frac{\dot{e}}{\dot{\varepsilon}_v}\right)^2 - 3a^2\right\}^{\frac{n}{2n-1}} - \frac{a}{n}\left\{\frac{1}{2n^2}\left(3a\frac{\dot{e}}{\dot{\varepsilon}_v}\right)^2 - 3a^2\right\}^{\frac{1-n}{2n-1}}\right]I \qquad (25)$$

The value of stress obtained by using the above equation (25) is different from that obtained by using the Drucker-Prager yield function. From Figure 4 it can be seen that non-linear parameter n substantially affects that non-linearity in shear strength against the confining pressure. The ultimate bearing capacity results obtained in Figure 5 by using the multiple non-linear shear strength parameter n with rigid plastic constitutive equation indicates that the results obtained with $n=0.54$ are well matched with the semi-empirical formula in practice. Nguyen et al. (2016) indicated the

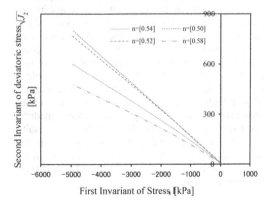

Figure 4. Effect of non-linear parameter n on non-linear shear strength property of soil in case of $\phi=30°$ and $c=10$ kPa.

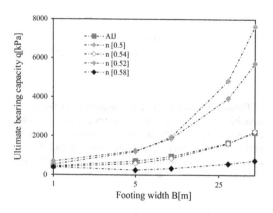

Figure 5. Effect of non-linear parameter n on ultimate bearing capacity of soil in case of $\phi=30°$ and $c=10$ kPa.

Table 1. Material characteristics data for analyses.

ϕ	c	a	b	n
	kPa		kPa	
30°	0	0.20	0	0.54
	10	0.20	9.9	0.54
	50	0.21	56	0.54
40°	0	0.25	0	0.525
	10	0.25	7.6	0.525
	50	0.25	43	0.525

coincidence in ultimate bearing capacity between AIJ formula and the computed results by RPFEM employing the non-linear shear strength of sandy soil by using the shear strength property of Toyoura sand. Based on the previous study, this manuscript is attempted to investigate UBC of the intermediate soil within the framework of AIJ formula and the non-linear shear strength of soil is set to fit the AIJ formula. The strength parameters thus obtained presented a good agreement of bearing capacity results with those of AIJ bearing capacity equation in the case of cohesionless soil. Keeping restraint on non-linearity and internal friction angle, the effect of variance in cohesion was then analyzed on the bearing capacity of intermediate soil having given strength characteristics. The results of ultimate bearing capacity obtained by reviewing the non-linearity have been obtained in this way and non-linear parameters a, b and n given in Table 1 have been set for intermediate soil.

3 SIZE EFFECT OF FOUNDATION ON THE ULTIMATE BEARING CAPACITY OF INTERMEDIATE SOIL USING NON-LINEAR SHEAR STRENGTH MODEL

RPFEM by using the Drucker-Prager yield criteria does not estimate the size effect of foundation on the

ultimate bearing capacity. The results obtained were quite similar with the conventional bearing capacity theories in practice. The reason behind this fact is that the Drucker-Prager criteria is just generalization of the Mohr-Coulomb failure theory. Therefore, in this study, the rigid plastic finite element method by embedding the non-linear shear strength against the confining stresses has been employed to exactly work out the ultimate bearing capacity of soil underneath the foundation. Previous studies conducted in the purview of critical state soil mechanics have revealed that the peak friction angle does not remain constant with the increase in confining stresses rather than it decreases. This phenomenon of reduction in peak friction angle is attributed to the decrement in dilation caused by high confining stresses.

In the case of ultimate bearing capacity of foundation resting on an infinite soil mass, size of foundation is a factor directly affecting the confining stresses. Therefore, as the footing size increases confining stresses also increase resulting in reduction of peak friction angle. The rigid plastic finite element method with non-linear failure envelope considers the variability in the internal friction angle.

The effect of change in cohesion at the same internal friction angle and non-linearity in shear strength was also analyzed for varied foundation sizes; the typical results of which are shown in Figures 6-8. The failure modes in the case of RPFEM (Higher order) are quite similar to those obtained by RPFEM (Drucker-Prager) but in the case of higher order analysis area deformed under applied load is smaller than that of linear case. For instance, in the case of Figure 7, B=10m at $\phi=30°$ and $c=10$ kPa the size of failure mode from the edge of footing is 1.82B having a depth of 0.88B and the bearing capacity of 852.5 kPa, which is in fact smaller than as computed in Figure 2 using the Drucker-Prager failure theory. Moreover, the obtained results are broadly in well accordance with the AIJ bearing capacity equation implying the efficacy of technique employed. In cohesionless case a small value of cohesion i.e. ($c=0.5$ kPa) is imparted in analysis to avoid the instability in computation process but the overall effect on results is found to be negligible.

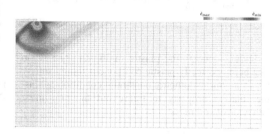

Figure 6. Strain rate distribution for foundation width (B=10m) in case of $\phi=30°$ and $c=0$ kPa using RPFEM (Higher order).

Figure 7. Strain rate distribution for foundation width (B=10m) in case of ϕ=30° and c=10 kPa using RPFEM (Higher order).

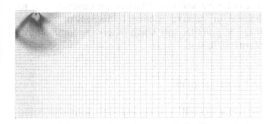

Figure 8. Strain rate distribution for foundation width (B=10m) in case of ϕ=30° and c=50 kPa using RPFEM (Higher order).

This study also indicated that upon intrusion of cohesion, the reduction in cohesive shear strength due to confining pressure was also witnessed. It's compound effect on bearing capacity is found to be very small. The intricate non-linear frictional response of soil just below the footing surface can be considered as a logical reason for this decrease in cohesive shear strength. Moreover, this reduction in cohesive shear strength is also a result of particle crushing, negative dilatancy and modified grain size distribution at high confining pressure resulting from large foundation sizes. The mechanical response of soil being investigated depends upon several index physical properties namely to illustrate particle size, shape and hardness. The phenomenon can be better studied by discussing the particle size, shape, crushing index and void ratio before as well as after the tri-axial tests on soil specimens, which is beyond the scope of this study. Although the effect of confining pressure resulting from large foundation sizes on cohesive shear strength of intermediate soils is found to be very small but still considering that effect a modified size effect factor is proposed in the AIJ bearing capacity equation. The size effect factor is calculated from the dispersion of finite element analysis results from the AIJ bearing capacity equation in relation to cohesive shear strength.

The representative bearing capacity results in the Figures 9-10 obtained by non-linear RPFEM are well matched with the AIJ with small discrepancies which in this study is attributed to the size effect of

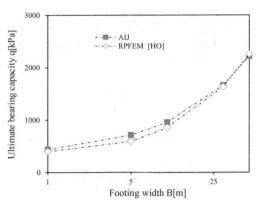

Figure 9. Estimation of UBC in case of ϕ=30° and c=10 kPa.

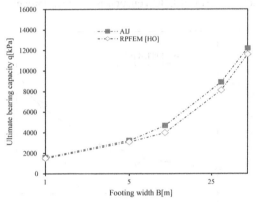

Figure 10. Estimation of UBC in case of ϕ=40° and c=10 kPa.

foundation on cohesive shear strength of soil. This effect on cohesive shear strength becomes conspicuous in the case of intermediate soils possessing higher internal friction angle. Therefore, a size effect factor is proposed in the cohesive part of the AIJ ultimate bearing capacity equation. The proposed equation better represents the behavior of intermediate soils which engineers usually come across in practical circumstances. Moreover, the reduction in ultimate bearing capacity due to the effect of confining stresses on cohesive shear strength produces conservative results.

In case of centric vertical load and absence of surcharge the AIJ ultimate bearing capacity equation given in Equation (11) can be rewritten as follows:

$$q = \alpha\eta_c c N_c + \gamma\beta B \left(\frac{B}{B_o}\right)^{\frac{-1}{3}} N_\gamma \qquad (26)$$

From the analysis results, the size effect factor η_c is computed as given below. The value of $m=-1/14$ is calculated from the obtained results. Hence, the final equation takes the following form:

$$\eta_c = \left(\frac{B}{B_o}\right)^m$$

$B_0 = 1\text{m}$(Reference value in the footing width)

$$q = \left(\frac{B}{B_o}\right)^{\frac{-1}{14}} cN_c + 0.5\gamma B\left(\frac{B}{B_o}\right)^{\frac{-1}{3}} N_\gamma \qquad (27)$$

The typical results reaped from proposed equation are plotted in comparison with the RPFEM (HO) analysis and AIJ results in Figure 11-12:

The exponential factor for the size effect term in adhesion part of UBC equation is obtained through rigorous trial and error mathematical computations based on complete set of analysis results. The obtained correction factor shows that there is a marginal effect of footing size on the cohesive shear strength of intermediate soil. The errors in obtained results remain within 3% on conservative side. The results fetched by using Equation (27) express that the proposed equation better represents the RPFEM analysis results using non-linear shear strength parameters against the confining stresses. It implies that the size effect of foundation also slightly governs the cohesive shear strength in case of intermediate soil. The applicability of Equation (27) is limited to intermediate soil only, as in the case of pure cohesive soil size effect is not observed due to absence of relative frictional mechanism between fines and granular soil which otherwise dominates in intermediate soil.

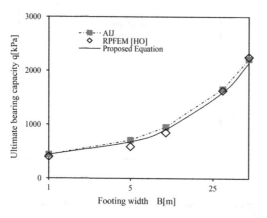

Figure 11. Comparison of results obtained from RPFEM (HO) and proposed equation (27) in case of $\phi=30°$ and $c=10$ kPa.

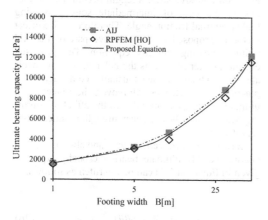

Figure 12. Comparison of results obtained from RPFEM (HO) and proposed equation (27) in case of $\phi=40°$ and $c=10$ kPa.

4 CONCLUSION

The conventional bearing capacity formulas being used by engineering community have very limited applicability due to a couple of disadvantages integral to the theories on the basis of which formulation is done. RPFEM is convenient in its use for analyzing the footing soil system because of its flexibility to employ in multiplex situations in terms of soil strata and footing shapes. The conclusions of this study are recapitulated as follows:

(1) RPFEM has well analyzed the footing soil system in the case of intermediate soils by using non-linear shear strength against the confining stresses.

(2) The results obtained from non-linear RPFEM better estimate the size effect of footing on the UBC.

(3) This study clarified that intermediate soils also have nearly the similar effect of footing size on bearing capacity as that of cohesionless soils.

(4) RPFEM using non-linear shear strength clarified the effect of footing size on cohesive shear strength of intermediate soils.

(5) The effect of non-linearity in shear strength of intermediate soils is envisaged by working out the relationship between the first stress invariant and the second invariant of deviatoric stress.

(6) The effectiveness of RPFEM for UBC was assessed through a set of footing widths, internal friction angle and cohesive shear strength of soil.

(7) The depth of failure modes is interestingly found to be nearly proportionate to footing size.

(8) The broad effectiveness of the AIJ UBC formula is confirmed as discrepancies are minimal.

(9) A thorough investigation of material physical properties before and after the strength tests is necessary to better predict the mechanical properties and their effect on the shear strength.

REFERENCES

AIJ, 2001. *Recommendations for design of building foundations*. Tokyo: Architectural Institute of Japan.

De Beer, 1970. Experimental determination of the shape factors and the ultimate bearing capacity factors of sand. *Geotechnique* 20 (4): 387–411.

De Borst, R., Vermeer, P.A., 1984. Possibilities and limitations of finite elements for limit analysis. *Geotechnique* 34 (2): 199–210.

Du, N.L., Ohtsuka, S., Hoshina, T., Isobe, K., 2016. Discussion on size effect of footing in ultimate bearing capacity of sandy soil using rigid plastic finite element method. *Soils and Foundations* 56 (1): 93–103.

Frydman, S., Burd, H.J., 1997. Numerical studies of ultimate bearing capacity factor N_γ. *Journal of Geotechnical and Geoenvironmental Engineering* 123 (1): 20–29.

Griffiths, D.V., 1982. Computation of bearing capacity on layered soil. *Proceedings of the 4th International Conference on Numerical Methods in Geomechanics* 1: 163–170.

Hoshina, T., Ohtsuka, S., Isobe, K., 2011. Ultimate bearing capacity of ground by rigid plastic finite element method taking account stress dependent non-linear strength property. *Journal of Applied Mechanics* 68 (2): I_327-I_336.

Meyerhof, G.G., 1963. Some recent research on the ultimate bearing capacity of foundations. *Canadian Geotechnical Journal* 1 (1): 243–256.

Maeda, K., Miura, K., 1999. Confining stress dependency of mechanical properties of sands. *Soils and Foundations* 39 (1): 53–67.

Maeda, K., Miura, K., 1999. Relative density dependency of mechanical properties of sands. *Soils and Foundations* 39 (1): 69–79.

Mehdi, V., Jyant, K., Fatemeh, V., 2014. Effect of the flow rule on the bearing capacity of strip foundations on sand by the upper-bound limit analysis and slip lines. *International Journal of Geomechanics* 14 (3): 04014008.

Prandtl, L., 1921. Über die Eindringungsfestigkeit (Härte) plastischer Baustoffe und die Festigkeit von Schneiden. *Journal of Applied Mathematics and Mechanics* 1: 15–20.

Pham, Q.N., Ohtsuka, S., Isobe, K., Fukumoto, Y., 2019. Ultimate bearing capacity of rigid footing under eccentric vertical load. *Soils and Foundations* 59 (6): 1980–1991.

Reissner, H., 1924. Zum erddruck problem. *1st International Conference of Applied Mechanics*: 295–311.

Sloan, S.W., Randolph, M.F., 1982. Numerical prediction of collapse loads using finite element methods. *International Journal of Numerical and Analytical Methods in Geomechanics* 6: 47–76.

Tamura, T., Kobayashi, S., Sumi, T., 1987. Rigid plastic finite element method for frictional materials. *Soils and Foundations* 27 (3): 1–12.

Tamura, T., Kobayashi, S., Sumi, T., 1987. Limit analysis of soil structure by rigid plastic finite element method. *Soils and Foundations* 24 (1): 34–42.

Tatsuoka, F., Goto, S., Sakamoto, M., 1986c. Effects of some factors on strength and deformation characteristics of sand at low pressures. *Soils and Foundations* 26 (4): 79–97.

Tatsuoka, F., Sakamoto, M., Kawamura, T., Fukushima, S., 1986. Strength and deformation characteristics of sand in plane strain compression at extremely low pressures. *Soils and Foundations* 26 (1): 65–84.

Terzaghi, K., 1943. *Theoretical Soil Mechanics*. New Jersey: John Wiley and Sons Ltd.

Ueno, K., Miura, K., Maeda, Y., 1998. Prediction of ultimate bearing capacity of surface footings with regard to size effects. *Soils and Foundations* 38 (3): 165–178.

Proceedings of the Second International Conference on
Press-in Engineering 2021, Kochi, Japan – Matsumoto et al (eds)
© 2021 Taylor & Francis Group, London, ISBN 978-1-032-10414-0

Influence of end geometry on aged behavior of segmental jacked pipe piles in clay

A.J. Lutenegger
University of Massachusetts, Amherst, USA

ABSTRACT: Small diameter segmental jacked pipe piles are one option for underpinning lightly structures to reduce settlement and restore the structure. Short sections of steel pipe are jacked into the ground through poor soils to achieve the desired load capacity from end bearing on a firm layer using the mass of the structure as the reaction. At the present there are few design guidelines available for jacked piles. A study was conducted to evaluate the influence of end configuration on the installation jacking force and load behavior on jacked segmental pipe piles in clay. Piles with diameters of 73 mm and 89 mm were used. During installation the jacking force was measured to relate jacking force to load capacity. Compression tests were conducted one day after installation to determine short-term axial capacity and then repeated on over a period of 10 days to 750 days to evaluate aging behavior.

1 INTRODUCTION

Underpinning of distressed structures often requires an innovative approach so that the structure can be economically restored to a serviceable state. In the 1890's, Breuchaud developed an underpinning system using sections of steel pipe jacked into the ground using a hydraulic jack and the existing structure as reaction. The system was used extensively in New York to underpin large buildings during construction of the subway system. This system is still used today to underpin light and medium loaded structures, such as residential and light commercial buildings. In many cases there is little to no design performed prior to installation. The selection of pile size, length and number is usually left to the installation contractor who essentially performs installation until the structure begins to lift. The load is then locked off.

Jacked piles have a number of advantages over other available underpinning systems. They are minimally intrusive, they are easy to handle, they can be installed in low headroom and directly adjacent to a structure, the required installation equipment is light-weight, a continuous measurement of jacking force can be obtained. In fact, a jacked pile is the only foundation element that can provide a direct measurement of the axial compression capacity, although the installation force is an instantaneous measurement and may not be equal to the long-term capacity. Generally the installation force is lower than the aged ultimate capacity and therefore is a conservative value.

Steel pipe pile sections are typically installed open ended and internal sleeves are used between sections to give a flush external connection, although in some cases an external sleeve can also be used. Often contactors attach an oversized donut ring to the tip of the lead pile section to reduce the required installation force. The piles can be installed as end bearing elements if oversized couplings or an oversized end is used and the pile is installed until a firm bearing strata is reached to support the required load. If an external flush connection is used and no oversized end ring is used, the pile acts as a side resistance element, with only a nominal end bearing component.

In the current work described in this paper a set of segmental jacked pipe piles were installed and load tested in axial compression to evaluate: 1) any relationship between installation jacking force and load capacity; 2) aging behavior; and 3) influence of pipe tip geometry on both load capacity and aging. The piles described in this paper are a subset of a larger number of jacked piles installed at the test site used.

2 TEST PROGRAM

Tests were performed at a well characterized research site adjacent to the University of Massachusetts, Amherst, Ma. Concrete bridge piers from a previous research project were used as reaction to advance the piles and during load testing. Two different diameter steel pipe piles were installed and tested having

DOI: 10.1201/9781003215226-4

Table 1. Summary of Piles Tested.

Set	Pile No.	O.D. (mm)	L (m)	End
1	P-1	73	7.3	Ring
1	P-2	73	7.3	Flat
1	P-3	73	7.3	Cone
1	P-6	89	7.3	Ring
1	P-7	89	7.3	Flat
2	P-10	73	3	Ring
2	P-13	73	3	Flat
2	P-12	73	3	Cone

O.D. of 73 mm and 89 mm and wall thickness of 6.3 mm. These sizes represent common commercial size piles used for underpinning. Pile sections were equipped with internal sleeves between sections so that the external surface was flush. The initial set of piles had an embedded length of 7.3 m. A second set of piles was installed and had a length of 3 m to be embedded only in the upper overconsolidated fill and crust. Lengths were chosen to give a range of load capacity to compare with installation forces.

End configurations consisted of: 1) an open end fitted with an oversized steel donut ring "friction reducer"; 2) a closed end using either a flush welded flat steel plate; and 3) a closed end with a flush welded steel cone tip with a 60° apex (similar to a CPT). The oversized donut end ring used on the 73 mm piles had an O.D. of 89 mm and an I.D. of 64 mm giving a slight relief on the inside. The donut end ring used on the 89 mm pile had an O.D. of 105 mm and an I. D. of 80 mm also giving a slight relief on the inside. Flat ends and cone ends were used to investigate whether the end configuration influence the capacity since the piles are dominated by side resistance because of the large length/diameter ratios. Table 1 gives a summary of the piles described in this paper.

3 SITE GEOTECHNICS

Tests were performed at the Geotechnical Experimentation Site located at the University of Massachusetts – Amherst. The soils consist of approximately 1.5 m of stiff silty-clay overconsolidated fill overlying a thick deposit of lacustrine varved clay which extends to a depth of about 25 m. This deposit is locally known as Connecticut Valley Varved Clay (CVVC) and is a lacustrine deposit composed of alternating layers of silt and clay as a result of deposition into glacial Lake Hitchcock. At the location of the test piles, the fill consists of CVVC placed about 40 years ago after excavations at the Town of Amherst Wastewater Treatment Plant, adjacent to the site. Below the clay fill, the CVVC has a well-developed stiff overconsolidated crust that grades into near normally consolidated clay at about 5 m. The site has been previously used for other field

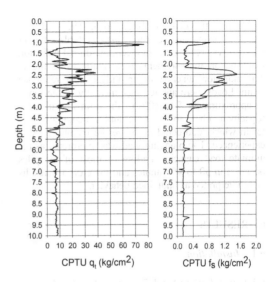

Figure 1. Results from CPTU profile at the test site.

studies (Lutenegger 2008, 2009). Figure 1 shows typical piezocone (CPTU-Type 2) results obtained in the upper 10 m. Tests were stated at a depth of 1 m in a small prebored hole filled with water to insure saturation of the porous element. These results show the upper 1.5 m of clay fill, the original surface of the overconsolidated CVVC crust and the transition to softer, near normally consolidated CVVC at around 4.8 m. Below 4.8 m the clay is very soft. Results of field vane tests performed at the site show that the sensitivity of the clay is on the order of 4 to 6.

4 PILE INSTALLATION

Piles were installed in 0.9 m sections using a specially fabricated jacking frame attached to the foundation of the piers using anchor bolts. The dead load of the pier and concrete footing provided the reaction for pile installation. The jacking force was measured using an in-line calibrated electronic load cell placed between the top of the pile and the hydraulic drive cylinder. An average jacking rate of approximately 0.15 m/min. was used. During installation, in addition to measuring the jacking force for each 0.15 m. of pile advance, measurements of the length of soil plug inside the open end piles were recorded after each pile extension section was advanced.

5 LOAD TESTS

Load tests were performed using the incremental maintained load method using the general procedures described in ASTM D1143-14 *Standard Test Method for Deep Foundations Under Static Axial Compressive Load*. Load was applied by a single acting hollow ram 250kN hydraulic jack placed on a small load frame attached to the concrete bridge pier. The load was measured using an electronic load cell placed over pile and was read using an electronic digital indicator. Deformation measurements were made using digital dial indicator attached to an independent reference beam. Loads were applied incrementally in the range of approximately 5 to 10% of the estimated ultimate capacity and held for 2.5 min. until a plunging behavior in which load could no longer be maintained was experienced. The plunging load was interpreted as the ultimate load capacity. Failure of most of the piles occurred in 30 to 45 minutes and therefore the results were interpreted as undrained behavior.

6 RESULTS

6.1 Installation

6.1.1 Installation: Set-1 7.3 m flat closed end vs. open ring end

Figure 2 shows results of the installation force profile for the two different pile sizes for the open pile with oversized end ring and flat closed end; Piles P-1, P-2, P-6 and P-7. As expected, the piles with the oversized ring showed lower installation force as compared to a flush closed flat end pile of the same diameter. The flush end pile accumulates side resistance along the outside of the pile as penetration proceeded. From measurements of the plug length taken inside the open piles during installation, both open end piles became fully plugged after about 6 diameters, despite the fact that the inside diameter of the end ring was slightly smaller than the inside diameter of the pile.

The installation force for the closed end piles consists of both side resistance and end bearing, while the installation force for the piles with oversized end rings largely consists of end bearing, especially after becoming plugged. However, even with an oversized end ring initially producing a small gap between the outside of the pile shaft and soil it is likely that some side resistance would develop along the pile shaft during installation as the soil "rolls" back onto the shaft. The soil along the sides of the pile would be remolded and not necessarily in full contact with the pile along the length. In the upper 2 m the penetration forces are similar.

The profiles of jacking force are similar is shape (but of course different magnitude) to both the CPTU cone tip resistance and sleeve resistance profiles shown in Figure 1 and show the buildup of jacking force through the upper overconsolidated zone and then the reduction as the pile tip passes into the lower softer clay. The piles with an oversized ring show a more pronounced difference in jacking force in the softer clay but show nearly the same behavior as the closed end piles in the upper stiff clay. The jacking force provides an indication of changes in soil layers

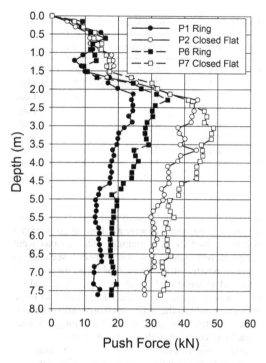

Figure 2. Measured jacking force for oversized ring and closed flat end piles.

just like the CPTU, even though the CPTU only reflects end bearing.

6.1.2 Installation: Set-1 7.3 m flat closed end vs. cone closed end

Figure 3 shows a comparison of the jacking force measured during installation of the 73 mm piles with both a flush flat end and a flush 60° cone end; Piles P-2 and P-3. From these two tests it appears that there is a difference in the two piles. Most literature would generally suggest that the end bearing would be similar for both piles, which means that the difference in installation force might be attributed largely to a difference in side resistance created by the different geometry of the pile tips. There is actually very little information available on how the geometry of the pile tip might influence the side resistance in clays.

6.1.3 Installation: Set-2 3 m flat closed end vs. open ring end

Based on the observations of the jacking force obtained for the 7.3 m long piles with closed flat and cone ends, another set of 73 mm piles having closed flat and cone ends along with a pile with an oversized end ring was installed to a depth of just 3 m, fully embedded in the stiff overconsolidated zone, Piles P-10 and P-13. The jacking force profile is shown in Figure 4 which shows the substantial difference between the two piles.

100

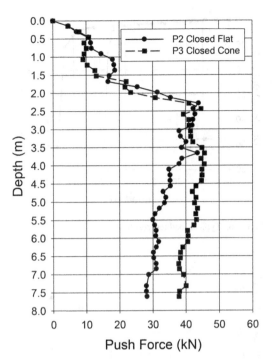

Figure 3. Measured jacking force for closed flat end and closed cone end piles.

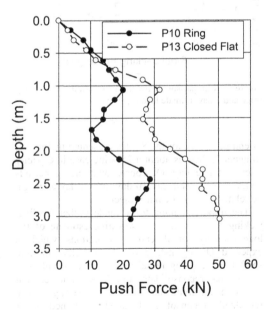

Figure 4. Measured jacking force for shallow oversized ring and closed flat end piles.

6.1.4 *Installation: Set-2 3 m flat closed end vs. cone closed end*

Results of the measured jacking force for the flat closed end and cone closed end for piles embedded to

3 m, Piles P-12 and P-13, are shown in Figure 5. These results show similar results to the longer piles, with the largest jacking force measured on the pile fitted with a cone.

A comparison of Figures 2 and 3 shows that as the piles pass through the heavily overconsolidated zone the jacking force on comparable size and end configuration piles decreases. This suggests that even though the pile is accumulating side resistance with increasing embedment length, the end bearing is decreasing rapidly so that the total jacking force is lower for piles embedded deeper in the softer clay than the shorter piles embedded fully within the overconsolidated soil. Table 2 gives a summary of the measured final jacking force for all piles.

6.2 *Load tests*

A summary of load tests for each of the piles at different ages after installation is given in Table 3. Ages of tests were not exactly the same for all piles

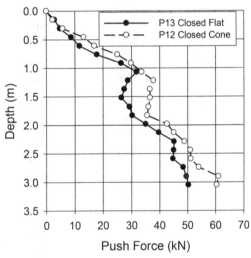

Figure 5. Measured jacking force for shallow closed flat end pile and closed cone end pile.

Table 2. Summary of final jacking force.

Set	Pile No.	Final Jacking Force (kN)
1	P-1	14.0
1	P-2	28.0
1	P-3	36.9
1	P-6	18.1
1	P-7	34.9
2	P-10	22.5
2	P-13	50.3
2	P-12	60.4

because of scheduling and weather, especially after about 100 days. In most cases, the piles exhibited plunging failure at very low displacements, typically on the order of 5 to 10 mm. Figure 6 shows typical results.

6.2.1 Installation vs. load capacity

Figure 7 shows a comparison between the measured final jacking force and the measured ultimate load capacity for the first load test conducted after 1 day. The 1 day capacity was selected simply for consistency and to illustrate the capacity that might be gained after just one day of rest. The results show a more-or-less consistent relationship, independent

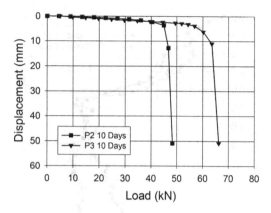

Figure 6. Typical load test results.

Table 3. Summary of load tests.

Set	Pile No.	Age (Days)	Q^{ult} (kN)
1	P-1	1	19.1
		10	23.3
		100	42.7
		300	42.9
		676	41.7
	P-2	1	44.3
		10	45.8
		100	72.7
		300	58.9
		689	71.8
	P-3	2	45.4
		10	63.5
		157	106.5
		360	95.0
		745	68.5
	P-6	1	27.4
		10	32.6
		100	35.8
		300	36.4
		674	39.4
	P-7	1	46.0
		10	51.7
		100	76.3
		300	84.4
		689	92.5
2	P-10	1	25.8
		10	25.8
		100	29.7
		361	38.3
		768	30.7
	P-13	1	61.6
		10	60.6
		100	59.5
		334	53.3
		741	53.9
	P-12	1	66.3
		10	81.3
		100	78.1
		346	70.7
		755	58.9

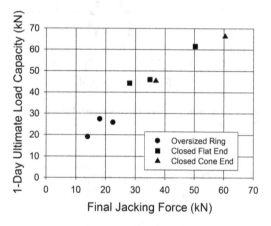

Figure 7. Comparison between final installation jacking force and 1 day ultimate load capacity.

of end geometry for the 8 piles presented in Table 1. In general, there is about a 25% increase in capacity after just one day. Of course other aging periods could also be used, say 10 day or 100 day as might be relative to a particular project.

The important point to be made is that the final jacking force gives a conservative estimate of the load capacity for all end configurations, which is expected. During installation, the soil along the pile shaft is undergoing initial remolding and with age, the soil has reconsolidated to a lower water content giving a different undrained shear strength. The remolded, reconsolidated, aged undrained shear strength may be either higher or lower than the initial undisturbed undrained shear strength depending on the sensitivity ratio of the clay and thixotropic behavior.

6.2.2 Aging

Figure 8 shows the aging behavior of the 73 mm piles with a length of 7.3 m (P-1, P-2, P-3). During

102

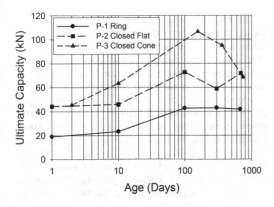

Figure 8. Aging behavior of 73 mm 7.3 m long piles.

the first 150 days, the piles all showed an increase in capacity with repeat loading. Even the pile with the oversized end ring showed an increase in capacity, likely some increase in side resistance as the soil along the shaft reconnected to the pile. It is doubtful that the change in capacity of any of the piles was affected by and substantial increase in end bearing since the pile tips were all located in the lower soft clay. After 150 days all of the piles showed either a decrease in load capacity or stabilized behavior from 150 days to about 750 days. As expected, the pipe with the oversized end ring showed the lowest capacity which is consistent with the lower installation force.

Figure 9 shows the results of the two 89 mm piles with a length of 7.3 m (P-6, P-7). In both cases, the piles showed a consistent continual increase in capacity through 750 days, with the closed flat end pile showing a much larger initial capacity and a large increase in capacity with time. However, even in this case, the pile with the oversized ring also showed an increase in capacity, similar to the 73 mm pile (P-1).

The tests reported herein were performed over a much longer period of time than many previously reported pile tests in stiff clay. For example the results presented by Karlsrud & Haugen (1985) were only performed over 40 days. Fresh tests on different piles at the same site sometimes show relaxation and then an increase in capacity over longer periods of time (e.g. Karlsrud et al. 2014).

Figure 10 shows the aging behavior of the shorter piles, embedded in just the upper stiff clay. The pile with the oversized end ring showed an increase in capacity up to 350 days but then showed a decrease at 750 days. Initially, the piles with the oversized ring showed very little increase in capacity within the first 10 days after installation. With longer aging periods the increase appeared to be more significant.

In this case the closed cone end pile (P-12) in the upper stiff clay showed only a short term increase in capacity but after 10 days showed a steady decrease in capacity. The closed flat end pile (P-13) showed no increase in capacity and actually showed a gradual decrease in capacity after the initial test. This suggests a steady degradation of the side resistance with time.

6.2.3 *Closed cone end vs. closed flat end*

The results shown in Figure 10 also show that while the flat end and cone end closed piles initially showed similar capacity, the pile with the cone end showed a much larger increase in capacity up to about 150 days. This is consistent with the longer piles, shown in Figure 8, at least up to about 150 days. Both sets of tests suggest that the end geometry of a pipe pile in clay may have some influence on the development of side resistance. The author could locate no published data comparing pile capacity of the same diameter and length pipe piles with both a flat tip and a cone tip.

For the two sets of Flat end and cone end piles presented here, there appears to be more than just a random difference in behavior. This difference may help explain, at least in part, some of the observed scatter in empirical correlations between undrained shear strength and undrained side

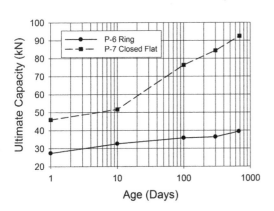

Figure 9. Aging behavior of 89 mm 7.3 m long piles.

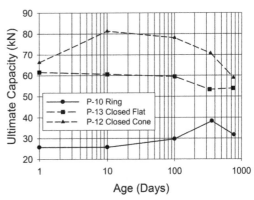

Figure 10. Aging behavior of 73 mm 3 m long piles.

103

resistance often seen in various design charts. That is, the Alpha value ($\alpha = f_s/s_u$) may be influenced by pile end geometry. This needs additional verification. Pile load tests at other sites using piles with both a flat tip and cone tip are needed.

7 SUMMARY

The results of axial compression tests on small diameter segmental jacked piles in clay showed:

1) Using an oversized end ring on an open end pile requires less installation force than a closed end pile but also provides less load capacity for the same amount of steel;
2) The increase in load capacity of flush closed end pipe piles appears to be primarily the result of additional side resistance developed in the softer clay over time, given the length/diameter ratios of the piles tested;
3) Piles with a 60° apex cone tip gave higher aged capacity as compared to piles with a flush closed flat end, at least up to 150 days; this may be related to difference in the movement of soil away from the pile tip during pile penetration and the degree of remolding experienced by the soil along the pile shaft;
4) No aging occurred with a flat end closed pile in the upper stiff clay; this behavior is contrary to previously reported results and at the present is unexplained;
5) Piles with an oversized end ring showed a small increase in capacity with aging which may be the result of remolded clay reconsolidating against the pile shaft;
6) Several of the piles showed relaxation behavior after repeat loading tests after 150 days. This is likely related to the degradation of shear strength with repeat loading;
7) The results showed a clear relationship between installation force and 1 day load capacity with the load being approximately 20% higher than the installation force after 1 day.

The results presented in this paper suggest that for jacked piles in clay a design based on final jacking force would leave unused load capacity, even for short term (1 day) aging. The use of a "friction ring" may appear to make it easier for the contractor to install piles but these piles developed about 50% of the load capacity of a comparable size and length pile with a closed end, at least for the clays investigated. Results may be different for sands. Closed end piles are clearly more economical in terms of the load capacity developed with the same quantity of steel, but have the added expense of having a tip welded onto the pile. For underpinning applications, the piles are require the same amount of time to install and unless there is a strong layer at relatively shallow depth so that the pile can act an as end bearing element, there is little need to use an oversized "friction ring" on the pile.

The results presented herein are based on repeat tests on the same pile. Aging may be influenced by the "preshearing" of the previous load test since all tests were taken to failure. There is some suggestion in the literature that aging behavior of one time "fresh" load tests on identical piles aged for different periods of time may show greater aging capacity.

REFERENCES

Karlsrud, K. and Haugen, 1985. Axial static capacity of steel model piles in overconsolidated clay. Proceedings of the 11th International Conference on Soil Mechanics and Foundation Engineering, 3, 1401–1406.

Karlsrud, K., Jensen, T., Lied, E., Nowaki, F. and Simonsen, A., 2014. Significant aging effects for axially loaded piles in sand and clay verified by new field tests. *Proceedings of the Offshore Technology Conference*, Paper OTC 25197-MS, 17 pp.

Lutenegger, A.J., 2008. Tension tests on single-helix screw-piles in clay. *Proceedings of the 2nd British Geotechnical Association International Conference on Foundations*, 201–212.

Lutenegger, A.J., 2009. Cylindrical shear or plate bearing? – uplift behavior of multi-helix screw anchors in clay. *Contemporary Issues in Deep Foundations*, ASCE, GSP 185, 456–463.

*Proceedings of the Second International Conference on
Press-in Engineering 2021, Kochi, Japan – Matsumoto et al (eds)*
© *2021 Taylor & Francis Group, London, ISBN 978-1-032-10414-0*

Feedback on static axial pile load tests for better planning and analysis

F. Szymkiewicz & A. Le Kouby
Université Gustave Eiffel, Champs/Marne, France

T. Sanagawa
Railway Technical Research Institute, Kunitachi, Japan

H. Nishioka
Chuo University, Tokyo, Japan

ABSTRACT: Static load tests are usually carried out to either aliment extended databases from which design bearing capacity of piles are derived, control the design, or develop new piling methods. However, testing practices have evolved with time, usually with the expansion of the scope of the standards regulating these tests, but also because of the cost involved with such tests, making it difficult to achieve representative results and maintain a certain continuity of the results over the years. Therefore, it is most important to carry out and analyse such tests in ways that are adapted to their purposes, and reproducible. This publication aims to provide some feedback about the way to plan for and to conduct static axial pile load tests, having in mind that when planning and then performing a static pile load test, many factors can and will impact its preparation, the protocol followed to carry it out, and the results and their subsequent analysis.

1 INTRODUCTION

Static pile load tests are the most reliable way to assess correctly the ultimate bearing capacity of piles and their behavior.

However, nowadays, as they are usually time consuming, difficult, and expensive to carry out, the design bearing capacity of piles is often derived from the analysis of extended databases, (which need to be constantly alimented with new results) of piles that were statically loaded to the failure.

Yet, these tests are still conducted routinely, for design and control purposes alike. Furthermore, for the development and validation of new tools or new piling process such as pressed-in piles/sheet piles or Gyropress Method (GIKEN LTD, 2018), static pile load tests are mandatory.

At the same time, testing practice have evolved over the years, leading to testing standards being more inclusive and therefore in a certain way more permissive.

Hence, the precise purpose of the test should be always defined in advance and known from every actor, as they will define the preparation of the test as well as the testing method and load steps sequence and duration. Also, thorough soil investigations and detailed planning, preparation and execution are necessary, to ensure that the tests provide results that can be exploited, given their actual purposes.

Furthermore, analysis should be done by experienced engineers, following a method that will ensure the proper interpretation of the results.

However, there are few technical papers that systematically summarize this information, as until now it has been treated as the know-how of the engineers conducting such tests. Therefore, throughout the whole paper, a number of static axial pile load tests carried out or analysed by Université Gustave Eiffel and Railway Technical Research Institute are analysed through a new light, in order to illustrate in details the most important points to focus on for a static axial pile load test in terms of organisation and execution, and to serve as feedback for future pile load tests.

First, the different purposes that may be the reason for carrying out a static axial pile load test are described in details, as they will most probably impact the setup and execution method of the test pile.

Depending on the predefined objective of the test, the load steps sequence and duration are defined, and the pile may be instrumented or not. Thus, the consequences of the choices made for the loading sequence and step duration as well as for the nature and position of this instrumentation are studied.

DOI: 10.1201/9781003215226-5

Then, the possible impacts of the chosen time-frame (or planning) are assessed, taking into account not only the nature and state of the soil, but also the nature of the material constitutive of the pile.

A discussion is also made about the way to interpret measures.

Finally, conclusions are drawn from this detailed study for the planning of a static axial pile load test, its setup and protocol as well as for the analysis of the data achieved during this test, taking into account all these background parameters.

2 REASONS FOR CARRYING OUT A STATIC PILE LOAD TEST

2.1 Control tests: Verifying the overall behavior at Serviceability Limit States (SLS)

The kind of test carried out to control the behavior of a pile under a certain load (usually under SLS load, or slightly higher) is often called a control test: its purpose is therefore only to observe the displacement of the head of the pile under this predefined load, and to compare this measured displacement to the calculated one on one hand and to the acceptable displacement for this given project on the other hand.

Therefore, this kind of test provides the first part of the load-settlement curve.

The only other information given by this kind of test is the time-displacement curve and the load-creep rate curve up to the maximum load applied.

Sometimes, the step under the SLS load is maintained for a longer time, to expressly observe the evolution of the creep rate of the pile under this load.

2.2 Conformity tests: Validating a design value or bearing capacity

The conformity tests are usually carried out to validate a design value, by carrying out a static load test on a pile by loading said pile up to its geotechnical resistance (or at least up to its theoretical failure).

These tests are usually carried out on instrumented piles, as the level of load applied is sufficient to determine the unit shaft friction mobilization for each level, and even the mobilization of the base resistance.

In some countries, conformity tests can be used as control tests (AFNOR, 2012).

2.3 Tests carried out for the development of new methods, and for the creation or alimentation of a database

These tests usually take place outside the scope of an actual project. Their main purposes are to determine the specific base resistance and unit shaft friction that can be mobilized in a given soil, whose state and nature are well documented. The piles are therefore instrumented.

In order to build a database on which reliable and sound design rules will be based, it is absolutely necessary to perform all the tests the same way, as well as to analyze them with the same procedure (Baguelin et al., 2012 and Burlon et al., 2014).

3 PREPARATION AND PLANNING OF A TEST PILE

3.1 Definition of the purpose of the test

Defining the clear purpose of the test will bring the engineers to choose whether the pile shall be instrumented or not, to define loading sequence and the steps durations, as well as to choose to build the test pile in a certain way so as to ensure that the goals of the test are performed.

The first three topics will be covered in the following paragraphs.

The last is also very important: planning for a test pile whose purpose is to determine the geotechnical resistance of the pile or the pile base resistance, especially on large diameter piles in strong soils, it may be necessary to apply great efforts (higher than 15-20 MN) to mobilize its overall resistance.

In this case, it may be possible to perform a test on a pile of slightly smaller dimensions (AFNOR, 2005b). On the other hand, if the aim of the test is to validate a base resistance, the pile may be over-drilled or tubed over a certain length, so as to decrease the shaft friction on the upper part of the pile, reducing at the same time the overall effort to apply on the test pile. If the last solution is retained, it is important to note that it will have a direct impact on the measured strains and on the interpretation of the results, as can be seen on Figure 1 (see paragraph 3.3.2).

Furthermore, when planning for a load test, if there are multiple purposes for it, it shall be studied if these purposes are compatible with each other. Indeed, given the cost of such a test, engineers could be tempted to optimize this cost by trying to use it for multiple purposes such as the determination of the resistance through a dynamic load test and a static load test (Figures 2 and 3) or such as the study of the creep rate under an extra long step as well as the determination of the creep load of the pile (see paragraph below).

It can be seen on Figure 2 that the repeated impacts induced deformations greater than 3500 µdef each time. Eurocode 2 (AFNOR, 2005a) fixes the ultimate relative deformation of concrete in compression ε_{cu} at 3.5/1000. Here, this deformation is significantly exceeded during each impact. These levels of deformation therefore very probably damaged the pile head (which was confined by a ferrule), and even more the top of the shaft of the pile (under the head ferrule) at the place where the maximum deformations were felt, causing a cracking and irreversible damage to the upper part of the pile, followed by deeper plastic deformations.

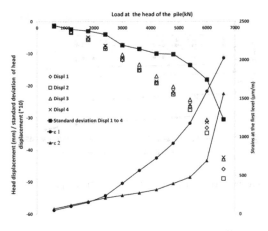

Figure 3. Head displacement and strain measures at the top level during a static load test carried out after a dynamic load test.

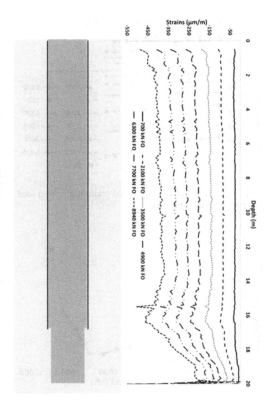

Figure 1. Longitudinal section of a pile with variable geometry and strain measures along the shaft, for different load steps.

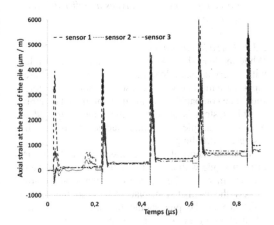

Figure 2. Strains-time relationship during a dynamic load test.

In addition, it can be seen in Figure 3 that the measures of strains (at the first level at 0.3 meters deep) and displacements at the head show a notable and increasing dispersion from the level at 3000 kN: this reflects an inclination of the head increasing hand in hand with the increase of the applied load,

and therefore a displacement of the head of the pile which is not representative of the shaft.

3.2 Determination of the loading sequence and load steps durations

3.2.1 Loading sequence

The loading sequence has a direct impact on the results of the test, and more precisely on the precision of the derived values such as the overall resistance of the pile or its creep load. Many procedures exist that specify different loading sequences (Szymkiewicz et al, 2020).

Usually, when performing a static pile load test, each load increment is of equal magnitude, and this load increment is chosen so as to reach the calculated resistance in eight or 10 steps. This is usually enough to determine the resistance or creep load cited above.

However, if for any reason the equal magnitude between steps is not respected, it can have a detrimental impact on the precision of these values, as seen on Figures 4 and 5.

On these Figures 4 and 5, it can be seen that the change in step magnitude between 6000 kN and 7500 kN occured at a crucial time, and that the creep rate – axial load curve (drawn following the French practice) presents a gap not allowing to assess with precision the creep load.

The same could be applied to the determination of the overall resistance, which may be determined on the load-settlement curve following a criteria different from country to country.

Some testing procedures allows for the decreasing of the magnitude of the steps when approaching the failure load (AFNOR, 2018), with the express purpose to determine more accurately the behavior of the pile to refine the determination of the pile resistance.

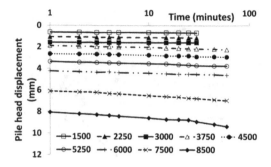

Figure 4. Head displacement – time relationship.

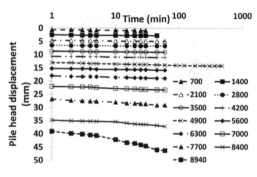

Figure 6. Head displacement – time relationship for a load test with an extra long step.

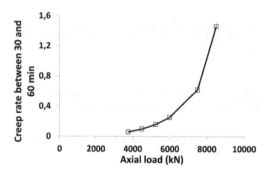

Figure 5. Evolution of the creep rate with the axial load.

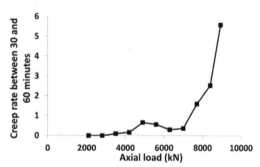

Figure 7. Evolution of the creep rate with the axial load for a load test with an extra long step.

Under these high loads (compared to the resistance of the pile), and particularly in some soils allowing for a very important creep, this particular adaptation may on the contrary decrease the resistance of the pile.

3.2.2 Steps duration

Step duration also has an impact on the results of the test.

While the measure of strains along the shaft does not vary so much with time, meaning that the mobilized shaft friction and base resistance do not evolve with time, pile displacement evolve with time: creep rate and displacement are indeed criteria which are scrutinized when deciding if the load step can be shortened or must be lengthened, depending on the local practice.

Obviously, lengthening a step under loads higher than the creep load may have an impact on the overall resistance, as seen in the previous paragraph.

Furthermore, it can also have an impact on the displacement behavior of the pile for the few next steps. Figures 6 and 7 show the results of a test pile for which an extra long step under the estimated service load was performed. As a result, it caused strain hardening of the soil, and it can be seen that the creep coefficient decreased for the next few steps.

3.3 Force input system and types and distribution of sensors

3.3.1 Force input system

While the load shall, whenever possible, be controlled with a very accurate load cell, the choice of the force input system that will be used to apply the force should be compatible with the maximum load to apply.

This system is almost all the time composed of one or multiple hydraulic jacks controlled by a single hydraulic pump.

While in theory it may be possible to test a 1 MN resistance pile with a 10 MN system, this would be to the detriment of the accuracy of the load control. Hence, the estimated displacements and the overall resistance of the pile should be taken into account when choosing the jack and the pump. The jack section and the estimated displacements for the first few load steps should be correlated to the debit of the pump, so as to ensure to maintain a load closest to the target value.

Figure 8 shows the load-time curve of a test performed on a very small pile (R_c = 230 kN) performed with a system design for the test up to 5000 kN: the applied load is clearly not constant and often higher than the target load.

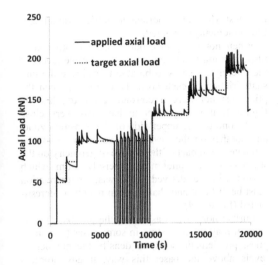

Figure 8. Comparison of the target load and the applied load during a load test using a manual pump.

Furthermore, whenever possible, the pump shall be an automatic one, allowing to maintain a constant load without the intervention of the operator, who may not be always close to the pump for the whole duration of the test for different reasons.

Results showed in Figure 8 were achieved with a manual pump, without any automatic regulation.

Figure 9 shows parts of the load-time curve of a test where transversal and axial loads were applied at the same time. The axial load was applied with a manual 2500 kN system, while the transversal load was applied with an automatic 1000 kN system. It can be seen that the transversal load is clearly more constant than the vertical load.

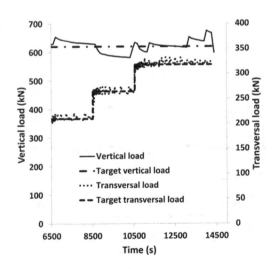

Figure 9. Comparison of the target loads and the applied loads during a combined load test.

3.3.2 Types and distribution of sensors

Regarding the displacement sensors localized at the head of the pile, their number should never be less than three, and preferably four.

Indeed, these numbers ensure that if a moment occurs, because the load is applied with any eccentricity or for any other reason, it could be seen instantly during the test, and precautions could then be taken (Figures 3 and 10).

Concerning the instrumentation embedded or inserted in the pile, while its choice should in theory be transparent in regards to the analysis method and therefore to the results, it should also be linked to the purpose of the test and to the geotechnical context, as it will have an impact on the planning (see paragraph above) and on the precision of the measures and could very well have an impact on the overall results of the pile (Szymkiewicz et al., 2021).

While embedded sensors of different technologies (mostly vibrating wire strain gauges, resistance strain gauges and optical fiber) are providing almost identical measures (Figure 11), it is not the case for retrievable extensometers (Figure 12).

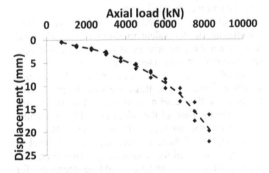

Figure 10. Evolution of the dispersion of measured pile displacements during a static load test performed on a pile with an eccentricity.

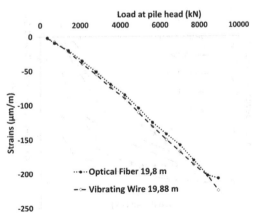

Figure 11. Comparison of measured strains in a bored pile with a optical fiber and vibrating wire strain gauges.

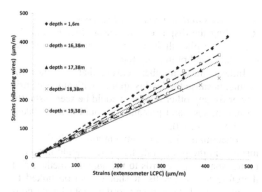

Figure 12. Comparison of measured strains at different depths in a bored pile with a retrievable extensometer and vibrating wire strain gauges.

The embedded sensors are measuring the strains over a length equal to their own length, while the retrievable extensometers measures strains between two anchors assumed to be fixed, and moving with the pile itself, making the strain measurement 'global' compared to the 'local' measurements achieved with embedded sensors.

However, the achieved results will be very comparable, as the equivalent modulus of the pile used to interpret the data will be adapted to the different levels of deformations (see paragraph 4).

Distribution of the sensors along the shaft is also a very important topic. Regardless of the purpose of the test, the first level of sensors shall be positioned just under the head of the pile, but under the ferrule, so as to estimate the real stiffness of the shaft.

A number of three to four sensors per level is necessary to achieve redundancy. However, the number of sensors per level could be decreased for deeper levels, as moments should not occur at these depths, as can be seen in Figure 13, presenting the standard deviations of deformations at each level,

for a test where an important eccentricity (and therefore a moment) was noted.

When not using distributed sensors like optical fiber, it is important to place sensors at the levels of the interlayers, so as to be able to estimate the unit shaft friction of each layer. Furthermore, when the pile does not have a constant geometry, like on Figure 1, it is important to have two very close levels, one on the upper side of the transition zone, and another on the lower side, so as to be able to estimate the impact of the change of geometry on the stiffness of the pile more precisely, from which efforts will be derived. Otherwise, interpretation must be difficult and shaft friction may be underestimated (Figure 14).

Furthermore, when assessing the pile base resistance, it may be interesting in some cases (very long piles, problematic soils) to densify the number of levels above the base: this way, if any problem occurred during the concreting phase (collapsed walls for example), the sensors will provide useful information (Figure 15) and allow the engineer to understand what happened and not to interpret blindly the data.

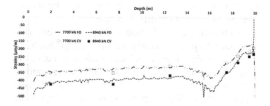

Figure 14. Comparison of strains measured along the shaft with optical fiber (FO) and vibrating wires strain gauges (CV) on a pile with a variable geometry.

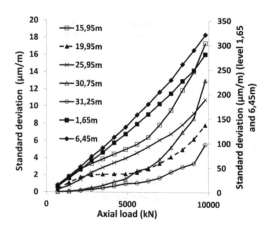

Figure 13. Evolution of the standard deviation of the strains at different depths.

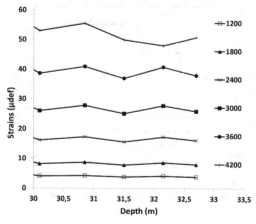

Figure 15. Strain measures at the base of a pile with a concreting default.

110

3.4 Preparation

Preparation of the pile is necessarily linked to the purpose(s) of the test, as well as to the choices of sensors and their distribution.

For example, a retrievable extensometer is particularly well adapted for the instrumentation of driven piles (but also for CFA piles), as the sensors are not present during the critical phases and therefore cannot suffer from it. However a reservation needs to be placed beforehand on the pile.

On the contrary, the use of embedded sensors implies the intervention of a team of external workers during the realization phase, as well as extra care during the manipulation phases of the cages.

Concrete formulation shall also be chosen with precaution, to ensure that the test can be carried out up to the maximum target load, but also, in the case of CFA piles notably, that the reinforced cage (playing the role of the instrumentation support) can be inserted in the fresh concrete without too much vibrations, which could damage the sensors.

Furthermore, the pile head shall be, whenever it is possible, encased in a ferrule, so as to increase the strength of the concrete, especially when high stresses (20 MPa or more) should be reached during the last load steps.

3.5 Planning

Depending on the type of pile (displacement pile of bored pile), the nature of the soil and the material constitutive of the pile, the waiting time between the realization and the test of the pile shall differ.

For steel piles and concrete driven piles, the main factor influencing the length of the waiting period is the soil set up (Augustesen et al., 2006, and Jardine et al., 2006): during this phase, excess pore water pressures induced during the setup of the pile will dissipate. This duration is therefore strongly dependent on the permeability of soils.

Concerning cast-in place piles, concrete strength on the day of the test should be enough to accept the maximum target load.

Furthermore, the modulus of the concrete shall also be taken into account, especially when working with special formulations and long piles. Performing a test on such a pile at a young age (or with a formulation different from the formulation which will be used during the actual project) could lead to the overestimation of the pile head displacement, and could therefore lead to overdesign of the piles for a project.

The example presented in Figure 16 shows that a difference in modulus from 20 to 50 GPa induce

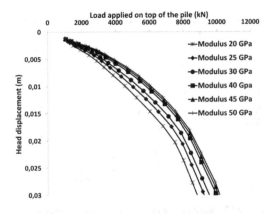

Figure 16. Load-settlement curves for a same pile with different modulus.

a difference in displacement of about five millimeters under a load between 4000 and 6000 kN, which for this particular project was the estimated SLS load.

This is particularly important to take this aspect into account when planning to perform a test whose objectives are to determine the settlement under SLS load and to determine the resistance of the pile at the same time, as the second objectives may necessitate the use of a special formulation in prevision of the very high loads applied at the head of the pile.

4 INTERPRETATIONS OF THE MEASURES

As said in paragraph 3.3.2, the choice of the sensor impacts (among other things) the level of deformations that will be reached during the test.

Nevertheless this just implies that the modulus used to interpret the data will differ from one type of instrumentation to another.

4.1 Modulus determination and load distribution

Except when embedding a load cell at the base of the pile, the usual information given by the instrumentation distributed along the pile is measures of deformations.

These deformations are then just multiplied by the cross sectional area of the pile and the modulus of the pile to obtain efforts (or loads), from which unit shaft friction are calculated (by subtraction between two levels), and base resistance as well.

In a first approach, some codes such as the French National application standard for the implementation of Eurocode 7 (AFNOR, 2012) relative to deep

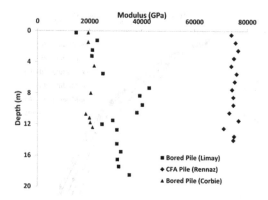

Figure 17. Variability of the modulus in cast-in place piles.

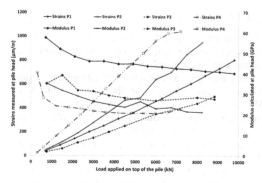

Figure 18. Evolution of strains and modulus during load test for four different piles.

foundations give values for the concrete modulus: in this case, 20 GPa.

However, for concrete piles, the modulus values can be very variable, depending on the age of the concrete (as discussed in paragraph 3.5) and its formulation.

Therefore, it is necessary to determine this modulus for each project. Figure 17 (partly issued from Bustamante and Doix, 1980) shows the great variability of the modulus of pile concrete, through the results of modulus determination tests carried out on specimens cored from piles.

It is also necessary to take into account the reinforcing steel bars of the cage inside the pile. The French standard NF P 94-150-1 (AFNOR, 1999) proposes the equivalent modulus method, taking into account the modulus of each material in the pile as well as their respective cross sectional area.

This method is often non-representative of the stiffness of the pile. Indeed, the values of the concrete modulus are given for a unique strain level, generally very small, leading to modulus values higher than in reality (see for example the value for the CFA Pile, on Figure 17) and, if used as it is, to incorrect interpretations, as the base resistance would be greatly overestimated in this case.

Therefore, it is mandatory to take into account the strain-dependency of the modulus.

To do so, multiple authors proposed different methods, summarized by Lam and Jefferis (2011).

Of these methods, the one proposed by Fellenius (2001) seems one of the most commonly used.

However, a simpler method would be to use, as stated in paragraph 3.3.2, the first level of strain gauges to assess the stiffness of the pile, in function of the applied load. Then, knowing the cross sectional area of the pile, the modulus can be easily estimated (Figure 18).

From there, it is possible by approximation by segments or polynomial approximation, to determine the modulus for each strain level, and then the load corresponding. This is the approximation by segment of the P4 case from Figure 18 that is presented in Table 1: the two boxes (Load 1800 kN at 1 m depth and Load 7200 kN at 17.65 m depth) showing the almost same exact measured strains, albeit for different load steps and at different depth, result in the same load.

If the pile body is a composite of two materials, it is better to consider the effect of breaking the adhesion between the two materials as necessary. One example is a steel pipe soil cement pile made by inserting a ribbed steel pipe into a mixed soil cement as shown in Figure 19 (JASPP, 2017). The ribs of this pile are attached to the entire outside and the inside near the tip, where ground resistance is required. On the other hand, most of the inner surface that does not contribute to skin friction has no ribs attached. Therefore, as the vertical load increases, the adhesion between the soil cement and the inner surface of the steel pipe will break off on the inner surface, and the axial rigidity of the entire pile will decrease.

Figure 20 shows the relationship between the output of the strain gauge of the steel pipe at the pile head and the axial rigidity calculated back from the applied load in the loading test with a steel pipe diameter of 800 mm and a soil cement diameter of 1200 mm (Nihei et al., 2011). At the initial stage of loading, it has the rigidity calculated under the condition that the adhesion between the soil-cement and the inside of the steel pipe is not broken. However, as the strain increases, the back-calculated axial stiffness decreases to the rigidity of the steel pipe alone. By considering the strain level dependency of such a composite material, the axial force distribution can be evaluated properly.

Table 1. Example of the analysis of the data, linking the measured strains and the stress level to the loads.

		load applied on top (kN)												
	Depth (m)	300	600	1200	1800	2400	3000	3600	4200	4800	5400	6000	6600	7200
strains (µdef)	1	25	70	164	252	346	454	560	666	764	858	960	1024	1039
	2	23	62	147	227	317	417	515	618	717	816	932	1022	1173
	3.75	23	51	116	183	253	333	411	488	563	637	722	804	881
	6.5	16	37	81	125	174	228	281	335	390	452	518	583	640
	9.5	9	20	47	77	110	143	177	215	254	294	342	389	436
	12.5	7	16	35	59	88	121	159	193	231	266	307	347	388
	15.08	5	10	18	31	48	71	97	126	155	186	221	257	299
	16.65	5	9	17	27	42	63	87	111	138	164	194	228	266
	17.65	5	9	16	26	40	60	83	105	130	152	183	214	251
	18.65	5	8	16	25	38	57	79	99	122	140	163	187	223

		load applied on top (kN)												
	Depth (m)	300	600	1200	1800	2400	3000	3600	4200	4800	5400	6000	6600	7200
modulus (Gpa)	1	40	28	24	24	23	22	21	21	21	21	21	21	21
	2	40	30	25	24	23	22	22	21	21	21	21	21	21
	3.75	40	33	26	24	24	23	22	22	21	21	21	21	21
	6.5	40	37	28	26	24	24	23	23	23	22	22	21	21
	9.5	40	40	34	28	27	25	24	24	24	23	23	23	22
	12.5	40	40	38	29	28	26	24	24	24	23	23	23	23
	15.08	40	40	40	39	34	28	27	26	25	24	24	24	23
	16.65	40	40	40	40	36	30	28	27	25	24	24	24	23
	17.65	40	40	40	40	36	31	28	27	26	25	24	24	24
	18.65	40	40	40	40	37	32	28	27	26	25	24	24	24

		load applied on top (kN)												
	Depth (m)	300	600	1200	1800	2400	3000	3600	4200	4800	5400	6000	6600	7200
load (kN)	1	297	600	1197	1795	2392	3000	3600	4200	4800	5400	6000	6600	7200
	2	282	571	1107	1632	2215	2823	3377	3914	4543	5171	5912	6490	7200
	3.75	277	517	921	1328	1798	2315	2783	3226	3633	4011	4576	5101	5584
	6.5	199	412	679	983	1268	1635	1984	2327	2662	3022	3395	3735	4059
	9.5	111	245	490	656	882	1086	1289	1547	1804	2067	2369	2656	2933
	12.5	90	192	402	513	731	958	1174	1401	1653	1887	2151	2397	2647
	15.08	66	121	220	362	498	608	793	984	1153	1350	1591	1827	2098
	16.65	56	111	200	321	457	574	729	893	1059	1201	1407	1637	1886
	17.65	55	105	194	310	442	561	697	851	1012	1136	1327	1539	1790
	18.65	55	99	188	298	426	546	664	808	964	1068	1194	1359	1601

5 CONCLUSIONS

Static pile load tests are expensive, long tests, the results of which can considerably impact the design of the foundations of a structure or building.

Because of these high stakes, it is a priority to plan thoroughly these tests, and to carry them out in the conditions.

However, the actual testing standards and recommendations are now tending to be more and more inclusive in terms of loading procedures, and do not describe in details the planning and the different steps of the analyses one has to follow when performing such a test. This can have a great impact on the overall results of the test on one hand, and on the other hand on the consistency and reliability of the database they are part of.

Hence, throughout this communication, all the most important topics relative to the preparation and the performing of such tests have been discussed, in order to help plan and carry out future tests:

- the main reasons for carrying out these tests,
- the need for a clear definition of the objectives of the tests,

113

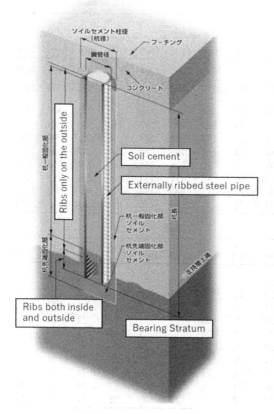

Figure 19. Outline of steel pipe soil cement piles (JASPP, 2017).

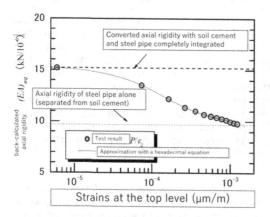

Figure 20. Relationship between strain of steel pipe at pile head and axial rigidity by composite of steel pipe and soil-cement.

– the importance of the choice of sensors, if the pile is to be instrumented, as well as their number and distribution along the shaft, as this choice will impact the planning of the test as well as the realization of the said pile,

– the realization of the pile (ranging from the pertinent choice of its constitutive materials to its geometry and reinforcement), which may in certain ways dictates that the test will run smoothly, or not,

– the waiting time between the realization of the pile and the day of the test,

– the importance of the loading sequence, load step magnitude and durations, as they will have direct consequence on the achieved results,

– the choice of the force input system.

Furthermore, some information was given to ensure that the interpretation of the strains measures is done correctly, with a focus on the importance of the modulus in the analysis of these tests, its variability, and a simple solution to assess it.

For more complex or longer load tests, carried on geothermal piles for examples, or on piles from an actual building, it would also be interesting and certainly necessary to study the impact of weather (temperature, sun, wind) and creep of the concrete on the behavior of the pile.

REFERENCES

AFNOR. 1999. NF P94-150-1: Essai statique de pieu isolé sous effort axial en compression, *test standard*.

AFNOR. 2005a. NF EN 1992-1-1: Eurocode 2 – Design of concrete structures – General rules and rules for buildings, *design standard*.

AFNOR. 2005b. NF EN 1997: Eurocode 2 – Geotechnical design, *design standard*.

AFNOR. 2012. NF P94-262: Justification of geotechnical work - National application standards for the implementation of Eurocode 7 - Deep foundations, *design standard*.

AFNOR. 2018. *NF EN ISO 22477–1*: Geotechnical investigation and testing - Testing of geotechnical structures – Part 1: Testing of piles: static compression load testing, *testing standard*.

Augustesen A.H., Andersen L. and Sørensen C.S. 2006. Assessment of Time Functions for Piles Driven in Clay, *DCE Technical Memorandum No. 1. Aalborg University*, 22p.

Baguelin F., Burlon S., Bustamante M., Frank R., Gianeselli L., Habert J. and Legrand S. 2012. Justification de la portance des pieux avec la norme "Fondations profondes" NF P 94–262 et le pressiomètre. *JNGG2012, Bordeaux, 4–6 juillet*, pp 577–584.

Burlon S., Frank R., Baguelin F., Habert J., Legrand S. 2014. Model factor for the bearing capacity of piles from pressuremeter test results – Eurocode 7 approach. *Géotechnique* 64(7):513–525.

Bustamante, M., Combarieu, O. and Gianeselli, L. 1980. Portance des pieux dans la craie altérée, *Annales de l'ITBTP* n°388.

Fellenius, B.H. 2001. From strain measurements to load in an instrumented pile, *Geotechnical News Magazine*, 19(1): 35–38.

GIKEN LTD. 2018. *Gyropress Method*, Kochi: https://www.giken.com/en/wp-content/uploads/press-in_gyro press.pdf

Japanese Technical Association for Steel Pipe Piles and Sheet Piles (JASPP). 2017. *Construction management procedure of steel pipe soil cement pile method*, Edition 1, p.6.

Jardine, R. J., Standing J. R. and Chow F. C. 2006. *Some observations of the effects of time on the capacity of piles driven in sand.* Géotechnique 56, No. 4, pp 227–244.

Laboratoire Central des Ponts et Chaussées. 2001. *La mesure des déformations à l'aide des extensomètres amovibles LPC: Méthodes d'essai LPC n°34,* 17 (in French)

Lam, C. and Jefferis, S.A. 2011. Critical assessment of pile modulus determination methods, *Canadian Geotechnical Journal,* 48, 1433–1448.

Nihei T., Nishioka H., Kawamura C., Nishimura M., Edamatsu M. and Koda M. 2011. A study of displacement-level dependency of vertical stiffness of pile - comparisons between static loading test and measurements during train passing, *Journal of Japanese society of civil engineers, ser. C (geosphere engineering),* 67(1), pp. 78–97.

Szymkiewicz F, Sanagawa T and Nishioka H. 2020. Static Pile Load Test: International Practice Review And Discussion About The European And Japanese Standards, *International Journal of GEOMATE,* 18(66), pp. 76–83.

Szymkiewicz F, Minatchy C. and Reiffsteck R. 2021. Static pile load tests: contribution of the measurement of strains by optical fiber, *International Journal of GEOMATE* (accepted)

Proceedings of the Second International Conference on
Press-in Engineering 2021, Kochi, Japan – Matsumoto et al (eds)
© *2021 Taylor & Francis Group, London, ISBN 978-1-032-10414-0*

Study of bearing capacity of tubular piles with diaphragm under pressing loads

M.P. Doubrovsky & V.O. Dubravina
Odessa National Maritime University, Odessa, Ukraine

ABSTRACT: Marine structures often include steel tubular piles of essential length (80-100 m and more) that should provide high bearing capacity in case of external axial loads application. One of the interesting peculiarities of long tubular piles' behavior is the formation of soil plug at the piles' tip. To increase piles bearing capacity under static pressing load, such an additional element as the internal diaphragm has been applied in some practical cases. Presented research aimed to study two connected processes during steel tubular pile driving: soil plug formation at the tip of the open-end pile and soil behavior under the internal diaphragm fixed inside the tubular pile's shaft. Obtained results of internal diaphragm application may be useful to provide an increase of pile's bearing capacity (in case of bearing capacity deficit) or to justify pile length reduction.

1   INTRODUCTION

Modern marine transportation and offshore structures such as deep-water port's berths, oil and gas platforms, raid and offshore fixed single point moorings, submerged stores and others often include steel tubular piles of essential length (80-100 m and more) as main bearing elements. Some examples of these structures are presented on Figures 1-3 (Doubrovsky & Pereiras 2014).

Such tubular piles should provide high bearing capacity in case of external axial loads application (Doubrovsky & Pereiras 2014, Doubrovsky et al. 2017, Doubrovsky et al. 2018, Tomlinson & Woodward 2008).

One of the interesting peculiarities of long tubular open-end piles behavior is the formation of soil plug at the piles' tip (Randolph et al. 1991, White et al. 2010).

From this point of view, we support known opinion that it is important to study the influence of the soil plug not only on the pile's tip bearing capacity but also on soil behavior inside the tubular shaft (Tomlinson & Woodward 2008).

In case of necessity the bearing capacity of long tubular piles may be increased by different methods:

- by driving a pile with a closed end to develop increased end-bearing resistance (but it requires application of too powerful hammers)
- by installing the pile at larger depth in order to reach the bearing soil strata (but such penetrations are often much greater than those required for fixity against lateral loading)
- by grouting beneath the pile toe (but the operations of cleaning-out the pile and grouting are slow and relatively costly)
- by welding a steel plate diaphragm across the interior of the pile in order to increase bearing capacity by use of soil reaction under the diaphragm (such method demonstrated good results on some marine projects and got positive references (Tomlinson & Woodward 2008).

Regarding that in many cases large diameter tubular piles of shelf structures are installed without plugging effect (so called "fully coring mode" (Gudavalli et al. 2013) or with partial plugging, the last approach (closure of the pile's shaft) looks rather attractive for deep water port, marine and offshore engineering but needs detailed consideration and study aiming to determine method's peculiarity, appropriate sphere of application, details of diaphragm construction and proper location along the pile's shaft.

2   APPLICATION OF INTERNAL DIAPHRAGM IN STEEL TUBULAR PILES

Recommended technology to install the internal diaphragm was described by Tomlinson & Woodward 2008 (Figure 4).

A hole is necessary in the diaphragm for release of water pressure in the soil plug and to allow expulsion of silt. Stresses on the underside of the diaphragm are high during driving and radial stiffeners are needed.

DOI: 10.1201/9781003215226-6

Figure 1. Single point mooring fixed to the sea bottom by long steel tubular piles (LNG carrier service).

Figure 2. Piled cluster applied to fix mooring device of the tanker.

Figure 3. Use of large-diameter bearing monopile as fixed single point mooring for offshore industry.

According to Tomlinson & Woodward 2008, the minimum depth above the pile toe for locating the diaphragm is the penetration below sea bed required for fixity against lateral loading. There

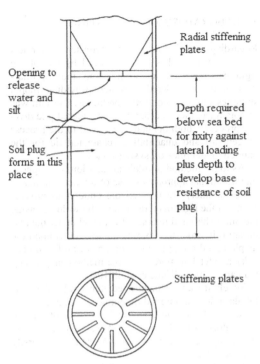

Figure 4. Steel tubular pile with diaphragm.

are formulas in some norms allowing determination of the fixity's depth depending on soil properties and pile's bending ridigity; roughly this depth may be determined in the interval of (5-7)d. However a further penetration is necessary to form the soil plug under the diaphragm by compacting the soil within the plug and to develop the necessary base resistance. Thus mentioned authors considered two locations for two soil plugs formation during the tubular pile driving: at the open end of the pile and under the internal diaphragm.

As an example of diaphragm's practical application, we may refer to the piling works at the Hadera coal unloading terminal near Haifa (Tomlinson & Woodward 2008). Open-end piles 1424- and 1524-mm OD were proposed but initial trial driving showed that very deep penetrations, as much as 70m below sea bed in calcareous sands, would be needed to develop the required axial resistance. The blow count diagram showed quite low resistance at 36m below sea bed. Trials were then made of the diaphragm method. A diaphragm with a 600 mm hole giving 83% closure of the cross-section was inserted 20m above the toe. This increased the driving resistance at 39m below sea bed and another trial with a 300 mm hole (95% closure) gave a higher resistance at 37m. It was supposed that such improvement of piles bearing capacity was stipulated by soil plug formation below the mentioned diaphragms.

3 LABORATORY MODEL TESTING

Regarding that obvious effect (increase of the pile's bearing capacity) has been achieved by use of the rigid diaphragm, our intention was to study peculiarities of the considered approach providing model static tests in the laboratory conditions. Our aim was to obtain parameters describing considered pile driving process – both qualitative (related to the process in general) and quantitative (characteristic for the applied model pile-soil system) ones.

As to the method of pile's installation, we suppose that traditional approaches (use of impact hammer or vibro hammer) are not reliable enough to provide safety of the rigid diaphragm fixed by welding inside the pile's shaft and interacted with soil under the diaphragm. In order to avoid dynamic actions upon the diaphragm during pile penetration we prefer to consider a safer but more effective method of pressing load application (White et al. 2010).

To clarify above mentioned items related to the tubular pile with internal diaphragm, we have started a series of experimental studies in Geotechnical Laboratory of the Department "Sea, River Ports and Waterways" at Odessa National Maritime University (Odessa, Ukraine).

For pile testing we used soil box of dimensions: width 600 mm, length 750 mm, depth 1100 mm (Figures 5-8). For the model of tubular open-end pile, we apply steel pipe d=50 mm external diameter, wall pipe thickness 1 mm, *l*=800 mm length. To drive the pipe into fine sand mechanical jack has been applied.

For experimental studies we used fine sand with the following characteristics: internal friction angle 33°; density 14.5 kN/m³; void ration 0.71; moisture 0.07%; Young modulus 16 MPa, Poisson's ratio 0.3.

The first series of the experiment was aimed to determine conditions of the soil plug formation at the tip of open-end pile model.

Measured parameters (at each stage of load application) were:

- applied vertical axial pressing force (named *Load, F* on the diagrams)
- pile's penetration depth (named *Displacement, t* on the diagrams)
- soil level inside the pile's model (the initial position was fixed when soil levels inside and outside the pile were equal).

In order to describe the process of the model tubular pile plugging we applied:

- earlier proposed IFR (Incremental Filling Ratio) and PLR (Plug Length Ratio) characteristics (for example, recommendations Gudavalli et al. 2013, Brucy et al. 1991, Paik & Salgado 2003, Lehane & Gavin 2004 and others) presented in Figures 9, 10
- assumption that as an evidence of completing of the process of soil plug formation we may

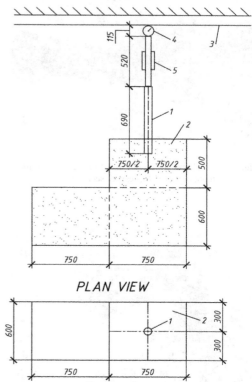

Figure 5. Scheme of the experiment.

1 – pile model; 2 – sand box; 3 – bearing beam; 4 – force gauge (dynamometer); 5 – jack loading system (telescopic) (all sizes in millimeters)

consider equal pile's vertical displacements and related settlements of the sand surface inside the shaft at each stage of further axial load application (related diagrams are presented in Figures 11, 12).

According to the PLR and IFR diagrams (Figures 9, 10) at the initial stage of pile installation [till t=(1.5-2)d] PLR=IFR=100%, i.e. pile is driven according to the "fully coring mode". From the penetration depth of some (4-5)d soil plug length is almost unchanged; average IFR value is stable too. Some IFR fluctuations below the penetration depth=20 cm may be explained by technical reasons: resetting of the jack each 20 cм of the penetration process and corresponding installation of extension tubes.

According to the PLR and IFR diagrams (Figures 9, 10) at the initial stage of pile installation [till t=(1.5-2)d] PLR=IFR=100%, i.e. pile is driven according to the "fully coring mode". From the penetration depth of some (4-5)d soil plug length is almost unchanged; average IFR value is stable too. Some IFR fluctuations below the

Figure 6. Experimental system: side view.
1 – soil box; 2 – model pile; 3 – mechanical jack;
4 – dynamometer; 5 – displacement gauge.

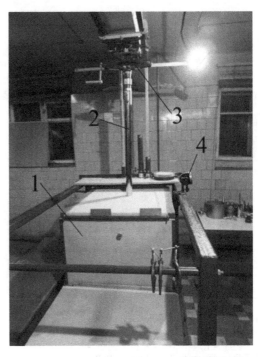

Figure 7. Experimental system: front view.
1 – soil box; 2 – model pile; 3 – mechanical jack;
4 – displacement gauge.

penetration depth=20 cm may be explained by technical reasons: resetting of the jack each 20 см of the penetration process and corresponding installation of extension tubes.

Similar conclusion may be done on the base of diagrams in Figures 11, 12. Soil level inside the pile's shaft become stable at the relative penetration depth (normalized pile displacement) approx. t/d=5-6 Figure 11). Dependencies for t/d and S/d describing pile and soil displacements also become parallel starting from the t/d=5-6 and till reaching the pile bearing capacity =5-6 Figure 12).

Some important results of the first series of the laboratory experiments may be formulated in the following way:

- in fine sandy soil the plug is formed at the comparatively early stage of pile's driving (in the considered case – at the penetration depth of around 4-5 pile's diameters)
- if to locate the internal diaphragm at the recommended depth required for fixity against lateral loading as described above (approx. at the penetration depth of around 5-7 pile's diameters), we may meet the situation of no contact between the diaphragm and the soil inside the shaft (clearance space); i.e., the diaphragm does not catch up with soil.

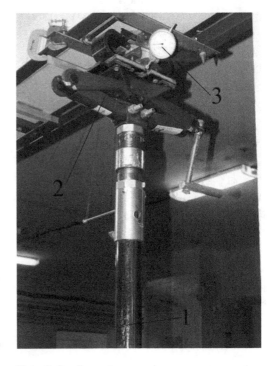

Figure 8. Loading system.
1 – model pile; 2 – mechanical jack; 3 – dynamometer.

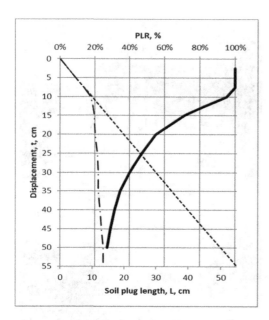

Figure 9. Typical dependencies for the tests of the first series.

------- - 1;   —·— - 2;   ——— - 3
1- fully coring mode; 2 – L=L(d); 3 - PLR

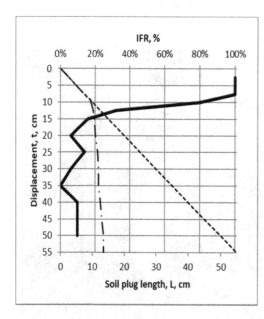

Figure 10. Typical dependencies for the tests of the first series.

------- - 1;   —·— - 2;   ——— - 3
1- fully coring mode; 2 – L=L(d); 3 - IFR

The second series of the experiment was devoted to clarification of the role and contribution of the internal

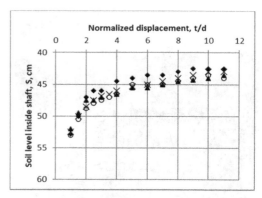

Figure 11. Results of open-ended model pile jacking (4 similar tests): S – distance between pile's top and soil in the shaft.

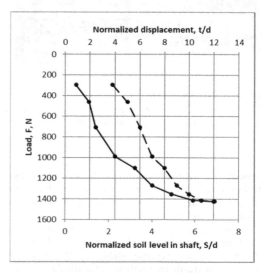

Figure 12. Results of open-ended model pile jacking (average points of 4 similar tests).

—•—·· - t/d=t/d(F);   ——•— - S/d=S/d(F)

diaphragm. For model pile the diaphragm was produced as a circular steel plate (4 mm thickness), with its diameter corresponded to the inner diameter of the pile.

By use of the rigid steel bar (located in the pile's shaft) the diaphragm was connected with the pile head. Varying the length of the mentioned rigid bar we had the possibility to locate the internal diaphragm at different places along the model pile's shaft.

At the second series of the experiment internal diaphragm was fixed at several positions by changing the distance from the tip of the model pile: 0 (closed end); 3d; 6d; 9d (total length of the pile was equal to 16d).

It has been discovered that due to sand settlements inside the pile shaft during pile installation, there is an empty space under the diaphragm and, correspondingly, no contact between soil and diaphragm.

In order to avoid clearance space under the diaphragm and to provide constant contact of the sand with underside of the diaphragm, we applied the diaphragm with several small holes allowing sand filling into the space under the diaphragm (Figure 13). For above mentioned options of the diaphragm location the sand was filled after the pile driving on the depth 3d, 6d and 9d correspondingly (i.e., at the moment of the first potential contact of the soil surface with underside of the diaphragm).

The volume of the filled sand was calculated to provide required diaphragm-sand contact during the whole process of model pile jacking.

As it is demonstrated by diagrams presented on Figures 14-17, application of the internal diaphragm provides increasing open-end pile bearing capacity. The degree of such increase depends on the diaphragm location. For the considered options of the diaphragm fixing point, the minimal increment of the open-end pile bearing capacity relates to the 3d distance between diaphragm and the pile tip (Figures 14, 17) and the maximal increment is measured at the 9d distance (Figures 16, 17).

Perhaps mentioned circumstances may be commented by the following way. The upper plug under the diaphragm may be formed if there is the proper base reaction developed inside the shaft. Such situation may occur if the upper plug (being in the process of formation) meets already formed lower plug. The last transfers additional pressure to the soil under the toe and provokes additional base reaction. Thus, additional external force acts on the plug and increases soil density in it. In fact, after that stage two plugs are combined and work as one large plug between the diaphragm and pile's toe. Obviously that creation of the mentioned large plug and its effective contribution to

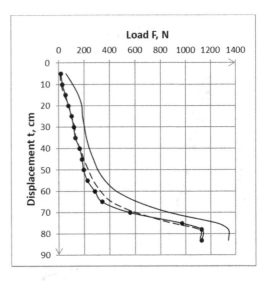

Figure 14. Dependencies between vertical axial load upon the model pile and its displacements.

 – – – · - open-end pile; ———— – closed end;
 —•— - diaphragm at 3d distance from the pile tip.

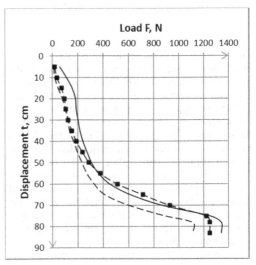

Figure 15. Dependencies between vertical axial load upon the model pile and its displacements.

 – – – · - open-end pile; ———— – closed end;
 – – – · - diaphragm at 6d distance from the pile tip.

Figure 13. Underside of the internal diaphragm with peripheral holes for sand filling.

the pile bearing capacity may be provided only in case of "right" location of the diaphragm (not too low and not too high). For our model tests the maximal bearing capacity of the open-ended pile was measured in case of 9d distance of the diaphragm from the pile tip.

It may be explained, particularly, by the fact that for the considered test conditions, approximate

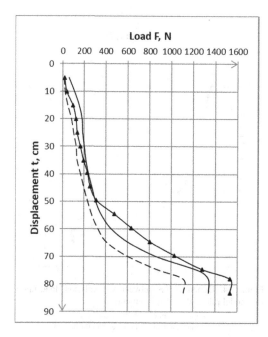

Figure 16. Dependencies between vertical axial load upon the model pile and its displacements.

$----$ - open-end pile; $———$ – closed end;
$———$ - diaphragm at 9d distance from the pile tip.

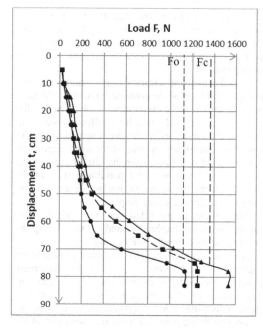

Figure 17. Dependencies between vertical axial load upon the model open-end pile and its displacements.

$—•—$ - diaphragm at 3d distance from the pile tip;
$—■—·$ - diaphragm at 6d distance from the pile tip;
$—▲—$- diaphragm at 9d distance from the pile tip;
F_o – open-end pile bearing capacity without diaphragm;
F_c – closed-end pile bearing capacity.

driving depth t=(4-5)d at the initial stage of pile installation is needed to dense a soil due to the development of the friction forces inside the pile's shaft and to form a lower soil plug at the pile tip. If then to apply similar consideration for the follow-on stage of driving process – compaction of the soil under the diaphragm due to the similar friction forces, required penetration depth for this stage to form the upper plug may be of similar value (4-5)d. So total distance between pile toe and the diaphragm may be considered as sum of these two parts of the penetration depth, i.e. approx. (8-10)d. Such location of the diaphragm may be optimal to form two plugs consecutively and to combine them in one large plug.

Regarding quantitate parameters of open-end pile bearing capacity (Figure 17), we would like to note that due to the diaphragm's contribution, pile bearing capacity may be increased (in our tests up to 15-20%). Another effect consists in possibility to decrease pile driving depth (10-15%). Obviously mentioned figures should be considered with regard to possible experimental errors stipulated by differences in the reproducibility of the model ground preparation as well as to measurement inaccuracy (perhaps up to 10% in total).

Regarding scale-effects for the considered problem, it should be noted the following. From the point of view of so called "direct modelling" (Florin 1961, Ivanov 1985), dependencies between limit axial

force in the pile $N_{\lim,p}$ (as well as related displacements $U_{\lim,p}$) and similar parameters of the model may be presented as

$$\frac{N_{\lim,p}}{N_{\lim,m}} = C_L^n,$$

$$\frac{U_{\lim,p}}{U_{\lim,m}} = C_L^m,$$

where C_L – sizes (scales) correlation between prototype and the model;

n, m – parameters depending on soil properties and pile dimensions.

For the conditions of our laboratory model testing (skipping the details of intermediate conversions and calculations) it was determined that related prototype is a tubular pile of diameter 1.0 m driven up to 10 m into similar sandy soil. Its bearing capacity (sum of the toe and shaft bearing capacities) is 1723 kN. For comparison: calculated value of the prototype bearing capacity according to the recommendation of the related Ukrainian code occurred to be 2020 kN (some 15% difference).

Also, for plugging effect assessment we have to consider scale-effects stipulated by influence of internal pile diameter. This aspect is subject to a study for the further investigations.

4 CONCLUSION

As obtained from the presented initial series of our experimental studies, rigid diaphragm inside the tubular open-end pile may be a useful element for increasing of the pile's bearing capacity.

Consecutive formation of two soil plugs (lower one formed just at the pile tip and then upper one formed under the diaphragm) leading to their partial or full integration is most effective when the optimal location of the diaphragm inside the pile shaft is provided. From the point of view of pile bearing capacity under axial compressive load and for the considered experimental conditions, such proper distance between the pile tip and internal diaphragm occurred to be around 9d (d – pile diameter).

As demonstrated by our tests, during pile jacking process there is a possibility of clearance space between the diaphragm and soil inside the pile's shaft (no contact situation). For real construction site conditions and inhomogeneous soil base it is complex task to check proper diaphragm-soil contact and their interaction or to determine clearance space formation under the diaphragm. That's why it is proposed (and checked by our tests) the technological improvement based on sand filling into space under the internal diaphragm to provide constant diaphragm-soil contact and related soil resistance. Such approach guarantees force interaction between the internal diaphragm and the soil inside the shaft via filled sand and may simplify calculation scheme of such interaction (the last is a task of future development of the considered problem).

Improvement of the pile effectiveness determined by experimental modeling for the above-mentioned pile-soil conditions provides 15-20% increasing of the bearing capacity or 10-15% reduction of the driving depth.

Presented experimental studies should be continued and developed in order to study installation peculiarities of piles with diaphragm by pressing technologies. Also, it looks prospective to investigate influence of internal diaphragm's design (different from the considered flat plate option).

ACKNOWLEDGEMENT

Authors would like to express gratitude to colleagues from the Geotechnical Laboratory of the Department "Sea, River Ports and Waterways" at Odessa National Maritime University for their valuable support and assistance in model tests arrangement and fulfillment.

REFERENCES

Brucy, F., Meunier, J., Nauroy, J.F. (1991). Behavior of pile plug in sandy soils during and after driving. In: Proceedings of 23rd Annual Offshore Technology Conference, Houston, 1, pp. 145–154.

Doubrovsky M., Pereiras R. (2014). Single point fixed moorings for Ukrainian shelf and seaports: problems and prospects of development. *Herald of the Odessa National Maritime University.* No. 41, pp. 123–133.

Doubrovsky M., Gerashchenko A., Dobrov I., Dubrovska O. (2017). Piled structures for marine transportation facilities: innovative structures and technologies. *Proceedings of the Second International Conference «Challenges in Geotechnical Engineering 2017»,* Kiyv, Ukraine, pp. 104–105.

Doubrovsky M., Gerashchenko A., Dobrov I., Dubrovska O. (2018). Innovative Design and Technology Solutions for Development of Port and Offshore Pressed-in Piled Structures, *Proc. of the First International Conference on Press-in Engineering,* Kochi, Japan, pp. 91–99.

Florin V.A. Fundamentals of Soil Mechanics: Volume 1 General Relationships and State of Stress Caused By Foundation Loads. Moscow, National Technical Information Service, 1961. 357 p.

Gudavalli S.R., Safaqah O., Seo H. (2013). Effect of soil plugging on Axial Capacity of Open-Ended Pipe Piles in Sands. *Proceedings of the 18-th International Conference on Soil Mechanics and Geotechnical Engineering,* Paris, France, pp. 1487–1490.

Ivanov P.L. Soils and bases of the hydraulic structures. Textbook. Moscow, High School. 1985. 352 p.

Lehane, B. M. and Gavin, K. G. (2004). Discussion of "Determination of bearing capacity of open-ended piles in sand". Journal of Geotechnical and Geoenvironmental Engineering, Vol. 130, No. 6, pp. 656–658.

Paik, K. and Salgado, R. (2003). Determination of bearing capacity of open-ended piles in sand. Journal of Geotechnical and Geoenvironmental Engineering, 129(1), pp. 46–57.

Randolph, M.F., Leong, E.C. and Houlsby, G.T. (1991). One-dimensional analysis of soil plugs in pipe piles, *Geotechnique,* Vol.41, No.4, pp. 587–598.

Tomlinson M. and Woodward J. (2008). *Pile design and Construction Practice,* Fifth edition, Taylor & Francis, N.Y.

White, D.J., Deeks, A.D. and Ishihara, Y. (2010). Novel piling: axial and rotary jacking, *Proc. of the 11th International Conference on Geotechnical Challenges in Urban Regeneration,* London, UK, CD, 24 p.

Proceedings of the Second International Conference on
Press-in Engineering 2021, Kochi, Japan – Matsumoto et al (eds)
© 2021 Taylor & Francis Group, London, ISBN 978-1-032-10414-0

Results of static vertical load tests on tubular piles installed by Standard Press-in and Rotary Cutting Press-in

K. Okada, K. Inomata & Y. Ishihara
GIKEN LTD., Kochi, Japan

ABSTRACT: The Press-in Method, being started with Standard Press-in (SP, press-in without installation assistance), has expanded its applicability to hard ground conditions with the development of Press-in with Augering and Rotary Cutting Press-in (RCP, press-in with the use of rotational forces onto a pile with cutting teeth on its base). Although the piles installed by these methods are usually used for retaining walls, there have recently been an increasing number of cases where SP piles and RCP piles are used for foundations. To understand the performance of these piles and establish rational design methods, it is necessary to accumulate load test results with different pile specifications and in different ground conditions. This paper reports the records of installation and the results of static vertical load tests on two piles with the outer diameter of 1000mm and the thickness of 12mm, which were installed by SP or RCP in the same site.

1 INTRODUCTION

Among several installation techniques in the Press-in Method (IPA, 2016), Standard Press-in (SP) installs a pile without any installation assistance such as water jetting and augering, and it is suitable for soft ground conditions. The applicability of the Press-in Method to hard grounds has been significantly improved by the development of penetration techniques such as Press-in with Augering and Rotary Cutting Press-in Method (RCP, with the use of vertical and rotational forces onto a pile with cutting teeth on its base). Although piles installed by these installation techniques are usually used for retaining walls, there have recently been some cases where RCP piles are used for foundations. On the other hand, SP piles are beginning to be used as non-temporary structures (e.g. Tanaka et al., 2018), in which they are supposed to resist to vertical loads. To understand the performance of these piles and establish a rational design method, it is necessary to accumulate load test results with different pile specifications and in different ground conditions.

This paper reports the records of installation and the results of static vertical load tests on two piles with the outer diameter of 1000mm and the thickness of 12mm, which were installed by Standard Press-in or Rotary Cutting Press-in.

2 METHOD OF FIELD TESTS

2.1 Site profile and test piles

Figure 1 shows the results of SPT in the test site. The ground consists of soft alluvial soils. For the depth (z) from 0m to 4m, the ground contains gravels, which is reflected in relatively high SPT N values. In $4m < z < 23m$, it consists of soft sand and silt with SPT N being smaller than 10, except for a relatively large value of 19 at around $z = 8m$. Below 23m is a hard sand and gravel layer with SPT N exceeding 50.

Two piles (A and B) with the outer diameter of 1000mm and the thickness of 12mm were used as test piles in the field tests. The specification of the test piles is shown in Table 1. Pile A had its embedment depth of 24m and was installed by Rotary Cutting Press-in into the sand and gravel layer with SPT N value exceeding 50. It was instrumented with strain gauges at several sections for measuring axial strains, as shown in Figure 1. Pile B had its embedment depth of 8m and was installed by Standard Press-in into a silty sand layer with SPT N being around 19.

2.2 Test layout and procedures

The test layout is shown in Figure 2. Pile A was positioned in between the two sheet pile walls with the embedment depth of 14.4m, which were used as reaction piles during static vertical load test. Pile B was

DOI: 10.1201/9781003215226-7

Table 1. Specification of test piles.

		Pile A	Pile B
Installation method		RCP	SP
Outer diameter (D_o)	mm	1000	1000
Pile thickness	mm	12	12
Pile length	m	25	9
Final installation depth	m	24	8
Number of teeth		6	6

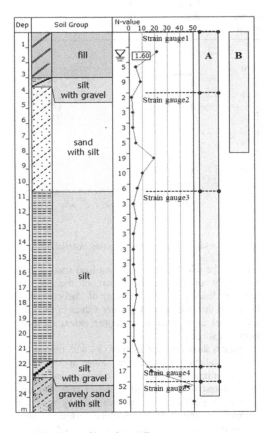

Figure 1. Site profile and test piles.

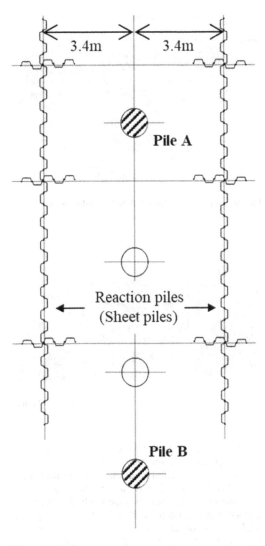

Figure 2. Test layout.

positioned slightly away from the sheet pile walls compared with pile A. The distance between the test piles and the sheet pile walls were greater than $3D_o$ (where D_o is the pile outer diameter) to avoid the influence of the walls on the pile capacity (JGS, 2002).

The field test was conducted with the following procedures.

Firstly, Pile A was installed by Rotary Cutting Press-in using a press-in machine (F401) while gaining a reaction force from a self-weight system as shown in Figure 3. The installation was associated with load-controlled surging, in which the pile was extracted when the jacking force applied by the press-in machine (Q') or the torque applied by the press-in machine (T') reached their upper-bound values (Q'_{UL} or T'_{UL}). Water was injected in the pile base, with the flowrate (f_w) being maintained at 15 liters per minute in $0m < z < 23m$ and being minimized in deeper than 23m. At the end of installation, termination load (Q_T) was applied for a certain period of time, without rotation and water injection.

Secondly, Pile B was installed by Standard Press-in using a press-in machine (F401) while gaining a reaction force from the same self-weight system as was used for Pile A. The installation was associated with load-controlled surging except for the last 1m, where it was conducted monotonically (without extraction). Water injection was not used. At the end of installation, termination load (Q'_T) was applied for a certain period of time without water injection. The jacking force and the penetration depth were measured by an automatic measurement system equipped in the press-in machine.

125

Figure 3. Press-in machine with Self-weight system.

Figure 4. Displacement measurement for inner soil column.

Table 2. Test conditions.

		Pile A	Pile B
Installation method		RCP	SP
l_d	mm	Arbitrary (manual)	
l_u	mm	Arbitrary (manual)	
v_d	m/min	0.5-0.7	0.7-2.1
v_u	m/min	1.8	1.8-2.6
v_r	rpm	7-11	-
f_w	L/min	15	-
Curing period	day	57	25

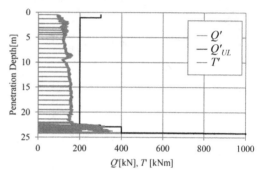

Figure 5. Jacking force and torque during installation.

Thirdly, the sheet piles were installed by Standard Press-in using a press-in machine (F111).

Finally, static vertical load tests were conducted on Piles B and A when 25 days and 57 days had passed since the end of installation respectively, based on the method recommended by JGS (2002). Strain gauge readings were offset to be zero just before the start of the load test. During the load test, the load and displacement at the pile head were measured by a load cell and displacement transducers. The depth of the surface of the soil inside the pile (z_i) was measured at the center of the pile, using a steel bar and displacement transducers as shown in Figure 4.

The press-in conditions and load test conditions for Piles A and B are summarized in Tables 2, where l_d and l_u is the downward and upward displacement in one cycle of surging, v_d and v_u are the downward and upward velocity, and v_r is the peripheral velocity of the pile.

3 RESULTS OF FIELD TESTS

3.1 RCP pile (pile A)

Figure 5 shows Q' and T' during installation of Pile A. The manually-set limitation of the jacking force (Q'_{UL}) was maintained very low when the pile was penetrating through a soft layer, with Q' being mostly

zero (i.e. the pile was installed by a load smaller than the weight of the chucking part of the press-in machine), leading to little necessity of load-controlled surging. It then increased to greater values in the hard sand-gravel layer, leading to higher extent of load-controlled surging.

Figure 6 shows the variation of Q'_T with time at the end of installation. Q'_T was maintained at 950kN for about 15 minutes. The pile head displacement during this load test was measured to be 112mm, but this value is unreliable as it was measured by a sensor equipped in the press-in machine, which cannot remove the additional displacement associated with the

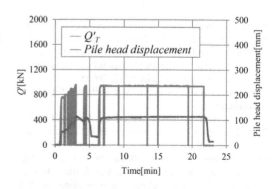

Figure 6. Axial force at the end of installation.

inclination of the press-in machine under a large jacking force.

Figure 7 shows the variation of the head load (Q) with time during the vertical static load test. As can be seen from this figure, the loading was conducted by multi-cycles and step sequence (with 5 cycles and 11 steps).

Figure 8 shows the variations of Q, base resistance (Q_b) and shaft resistance (Q_s) with the pile base displacement (δ_b) during the load test. Q_s showed its peak when δ_b reached around 10mm (= 1/100 D_o), and slightly decreased to its residual value until δ_b reached 80mm. Q_b continuously increased with the increasing δ_b. This continuous increase may partly be due to the small embedment length into the hard layer (bearing stratum). Defining the pile capacities as the values recorded when δ_b became 1/10 of D_o (JGS, 2002), the total capacity (Q_f), base capacity (Q_{bf}) and shaft capacity (Q_{sf}) were obtained as 5000kN, 3102kN and 1898kN, respectively. On the other hand, according to JGS (2002), the first-limit- resistance (Q_y, yielding load) can be obtained by plotting the relationship between head load and pile head displacement as shown in Figure 9 and by judging as the load giving the maximum curvature. As a result, Q_y of Pile A can be judged as 3000kN. The obtained values of Q_f, Q_{bf}, Q_{sf} and Q_y are summarized in Table 3.

Figure 10 shows the distribution of the axial load and the averaged unit shaft resistance (fs) with

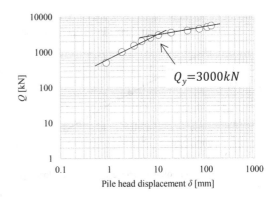

Figure 9. Determination of first-limit resistance.

Table 3. Result of load test on Pile A.

	kN
Q_f	5000
Q_{bf}	3102
Q_{sf}	1898
Q_y	3000

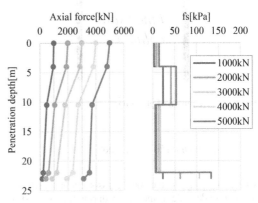

Figure 10. Distribution of axial force and fs.

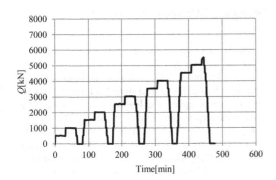

Figure 7. Loading record during load test.

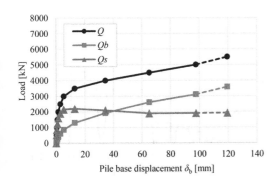

Figure 8. Load-Displacement relationship during load test.

depth, recorded at different load steps. It is clearly seen that the major part of the shaft resistance was generated in soil layers which contains sands or sand and gravels.

Figure 11 shows the increment of the inner soil column surface depth (z_i, as illustrated in Figure 12) plotted against the pile head displacement (δ) during the load test. It can be confirmed that the pile was fully plugged during the load test, if the plugging condition is defined by Incremental Filling Ratio (the ratio of the increment in z_i to the increment in δ). By the way, according to AIJ (2019), Q_{bf} of a closed-ended pile in the smallest case (for bored piles) is obtained by:

127

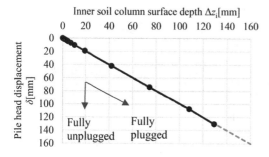

Figure 11. Inner soil column surface depth during loading test.

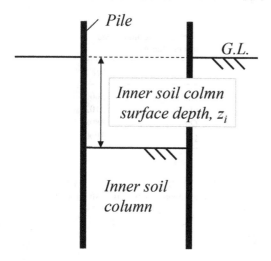

Figure 12. Definition of plugging.

$$Q_{bf} = min(150 \times N_D, 9000) \times A_{b,closed} \quad (1)$$

where $A_{b,closed}$ is the net cross-sectional area of a closed-ended pile. If the Pile A during the load test is assumed to be closed-ended based on the observation in Figure 11, Eq. (1) provides a significant overestimation (Q_{bf} = 7065kN). It is suggested that a compression of the inner soil is occurring during the load test.

3.2 SP pile (Pile B)

Figure 13 shows Q' during installation of Pile B. The pile was installed with surging manually. The manually-set limitation of the jacking force (Q'_{UL}) was maintained very low. It then increased to greater values in the sand layer, but a higher extent of load-controlled surging was exerted.

Figure 14 shows the variation of Q'_T with time at the end of installation. Q'_T was maintained at 650kN for about 8 minutes, and the pile head displacement during this maintained load was about 275mm,

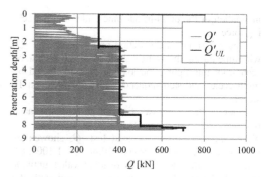

Figure 13. Jacking force during installation.

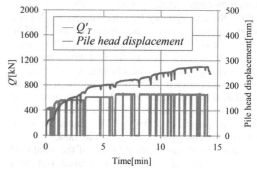

Figure 14. Axial loading at the end of installation.

which is unreliable due to the same reason as explained for Pile A.

Figure 15 shows the variation of the head load (Q) with time during the vertical static load test. As can be seen from this figure, the loading was conducted by multi-cycle step loading (with 6 cycles and 20 steps).

Figure 16 shows the variation of Q with the pile head displacement (δ) during the load test. Defining the total capacity (Q_f) as the load recorded when the pile head (not base, due to the lack of base displacement measurement) displacement reached 1-/10 of D_o, Q can be obtained as 1300kN.

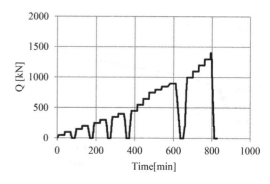

Figure 15. Loading record during load test.

128

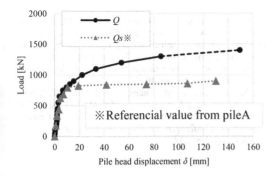

Figure 16. Load-Displacement relationship during load test.

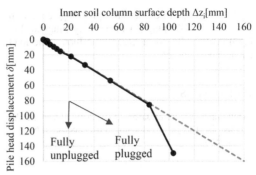

Figure 18. Inner soil column surface depth during loading test.

The base and shaft resistances during the load test on Pile B are unknown, as this pile was not instrumented with axial strain gauges. Here, it is attempted that the shaft resistance (Q_s) on Pile B is roughly estimated based on the results on Pile A in Figure 10, ignoring the effect of differences in the piling method and the pile length. Figure 16 shows the variation of Q_s on Pile A in the depth range of 0m < z < 8m (corresponding to the embedment depth of Pile B), plotted against δ (not δ_b). It is seen that Q_s became constant at 860kN when δ is greater than 20mm. Adopting this Q_s value as the shaft capacity of Pile B, Q_b of Pile B can be estimated as 440kN. On the other hand, Figure 17 shows the relationship between the head load and the rate of pile head displacement recorded at different loading steps during the load test, by which Q_y can be judged as 1000kN (as the load giving the maximum curvature). The obtained values of Q_f, Q_{bf}, Q_{sf} and Q_y are summarized in Table 4.

Figure 18 shows the increment of the inner soil column surface depth (z_i, as illustrated in Figure 12) plotted against the pile head displacement (δ) during the load test. It can be confirmed that the plugging condition of Pile B shifted from a fully-plugged condition (for δ < 80mm) to a partially plugged condition (for δ > 80mm). This value of δ ($\fallingdotseq$ 80mm) corresponds to the point of time when the head load

Table 4. Result of load test on Pile B.

	kN
Q_f	1300
Q_{bf}	440
Q_{sf}	860
Q_y	1000

became unmaintainable as expressed by a dotted line in Figure 16, but the shift of the plugging condition does not appear to be reflected in this load-displacement curve (i.e. the load appears to be increasing continuously).

4 CONCLUSIONS

The records of installation and the results of static vertical load tests on two piles with the outer diameter of 1000mm and the thickness of 12mm were reported.

Pile A had its embedment depth of 24m and was installed by Rotary Cutting Press-in into a bearing stratum consisting of a sand and gravel layer with SPT N value exceeding 50. It was instrumented with strain gauges at several sections for measuring axial strains. Axial jacking force of 950kN was applied at the end of installation without rotation. The load test was conducted 57 days after the end of installation. The obtained base and shaft capacities were approximately 3000kN and 2000kN respectively. The plugging condition, judged by IFR (Incremental Filling Ratio), was fully plugged during the load test.

Pile B had its embedment depth of 8m and was installed by Standard Press-in into a loose silty sand layer with SPT N value around 19. Axial jacking force of 600kN was applied at the end of installation. The load test was conducted 25 days after the end of installation. The obtained total capacity was around 1300kN. Shaft capacity was estimated to be 860kN, from the axial loads recorded in the corresponding

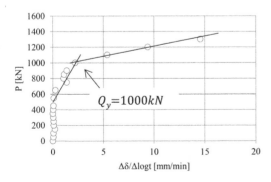

Figure 17. Determination of First-limit resistance

depth range in Pile A. The plugging condition shifted from fully plugged to partially plugged, but this transition did not apparently influence the load displacement curve.

REFERENCES

Architectural Institute of Japan (AIJ). 2019. Recommendations for Design of Building Foundations, 196p. (in Japanese)

International Press-in Association (IPA). 2016. *Press-in retaining structures: a handbook, First edition 2016*, 520p.

The Japanese Geotechnical Society (JGS). 2002. Method for static axial compressive load test of single piles. *Standards of Japanese Geotechnical Society for Vertical Load Tests of Piles*, pp. 49–53.

Tanaka, K., Kimizu, M., Otani, J. and Nakai, T. 2018. Evaluation of effectiveness of PFS Method using 3D finite element method. *Proceedings of the First International Conference on Press-in Engineering 2018, Kochi*, pp. 209–214.

*Proceedings of the Second International Conference on
Press-in Engineering 2021, Kochi, Japan – Matsumoto et al (eds)*
© 2021 Taylor & Francis Group, London, ISBN 978-1-032-10414-0

The inner friction resistance and the resistance of an actual part of open-ended piles by the double-pipe model pile experiment

H. Yamazaki
Nippon steel corporation, Chiba, Japan

Y. Kikuchi, S. Noda & M. Saotome
Tokyo University of Science, Chiba, Japan

M. Nonaka
Tokyo Institute University, Tokyo, Japan

ABSTRACT: A penetration experiment was conducted using a double-pipe model pile capable of measuring the inner friction resistance in isolation. The objective was accurately measuring the bearing capacity of open-ended piles and determining the range of machining that needs to be performed on the toe of the piles in order to plugging. The friction coefficient of the model pile was changed to verify the relationship between the inner friction resistance and the *IFR**, and how plugging at the toe is affected by the range and position of the where machining area. From the results, it can be concluded that a superior outcome cannot be achieved solely by increasing the friction coefficient around $0.5D$ from the end of the pile. To achieve effective plugging, the friction coefficient should be increased not only at the toe of the pile but across an area of at least $1.0D$ from the toe of the pile.

1 INTRODUCTION

Pile foundations have been used for a long time in the port facilities in Japan. In recent years, as port facilities increase in size and their structural types change, large-diameter steel pipe piles are being widely used to support large vertical loads. Therefore, steel pipe piles with a large diameter, including up to $\phi 2{,}000$ mm, are also used. For open-ended piles, such as steel pipe piles, it is the problem that the plugging effect decreases as the pile diameter increases.

According to Kikuchi (2011), there is a significant difference in the estimated bearing capacity depending on the difference in the estimation equations from areas where the pile penetration depth exceeds 60 m when the bearing capacity of the pile is estimated using internationally applied estimation equations.

According to the construction standards of Japan (Architectural Institute of Japan, 2001), as shown in Figure 1, the toe bearing resistance of an open-ended pile, which can be separated into the resistance acting on the annular part, R_{out}, and the inner friction between the pile wall and soil, R_{in}, can be theoretically represented as $R_{open} = R_{out} + R_{in}$ (Figure 1).

Considering this, the plugging effect η which is the ratio of the resistance of an open-ended pile and a closed-ended pile, is written as equation (1). However, it is difficult to measure R_{in} and R_{out} separately

in actual piles, and it has not been possible to discuss the plugging effect quantitatively.

$$\eta = \frac{R_{open}}{R_{close}} = \frac{R_{out} + R_{in}}{R_{close}} \qquad (1)$$

The authors considered an estimation of the internal friction resistance R_{in} of open-ended piles to be a serious problem in the estimation of R_{open}. Therefore, we previously devised a method to estimate the inner friction resistance R_{in} of open-ended piles through a model pile penetration experiment on sandy ground.

If the inner friction resistance is not exerted in comparison with R_{close} with unplugging at the pile toe, the plugging effect may be insufficient. In such cases, a special pile is considered to promote plugging by machining the pile toe with cross-ribs. However, neither has a method to promote plugging by machining around the pile toe been systematized, nor has a rational method been developed.

Therefore, it is important to accurately estimate the bearing capacity of a pile and to clarify the method of machining the toe of a pile tip to promote plugging. In this study, penetration experiments were conducted using a double-pipe model pile, which can measure the inner friction resistance and the

DOI: 10.1201/9781003215226-8

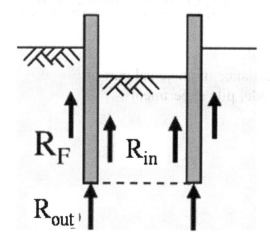

Figure 1. Schematic of the bearing capacity.

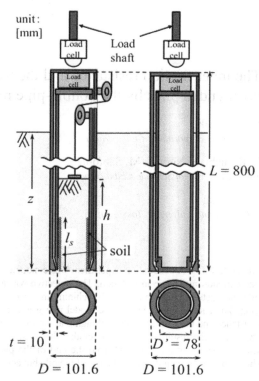

Figure 2. Diagram of the model piles.

resistance acting on the ring of the pile toe. Furthermore, the relationship between the inner friction resistance and IFR*, which was defined as $1 - IFR$, was investigated. Definition of IFR is shown in equation (2). IFR* represents the state of plugging by the inner friction coefficients of the inner surface of a double-pipe model pile that are varied by attaching sand to the inner surface of the pile. The effects of the machining area and position of the sand on the acceleration of plugging were also studied.

2   MATERIAL AND TEST METHODS

The model piles used in this series of experiments are shown in Figure 2. Each model pile is comprised of both outer and inner pipes. The dimensions of the outer pipe of the model pile, shown in red, are as follows: outer diameter $D = 101.6$ mm, total length $L = 800$ mm, and pipe thickness $t = 10$ mm. The toes of the piles were formed as shown in Figure 2. The ratio of the outer diameter to the pipe thickness of the outer pile (D/t) was approximately 10.2. There were two types of inner pipes, open-ended and closed-ended. They are shown in blue in Figure 2. The outer and inner diameters of the open-ended pile were 88 mm and 80 mm, respectively, whereas the tip of the closed-ended pile had a diameter of $D' = 78$ mm. Both pipes were 4 mm thick and made of stainless steel. In the closed-ended pile, the attachment was screwed to the pile toe. When it was combined with an outer pipe, the positions of the pile toes were aligned. The inside of the outer piles was tapered at the pile shaft to prevent the inner and outer pipes from protruding from each other. Adhesive tape was applied to the pile tip clearances to prevent sand from entering between the piles.

Two load cells were used to measure the resistance of each pile individually. A load cell with

a capacity of 30 kN was inserted between the upper part of the inner pipe and the outer pipe, through which the outer and inner pipes were rigidly connected. A load cell with a capacity of 49 kN was set between the loading shaft and top of the outer pile. The penetration resistance R_{in} acting on the inner pipe was measured directly by the load cell set between the piles. The penetration resistance acting on the outer pipe was calculated from the total resistance measured by the load cell set between the loading shaft and the top of the outer pipe and subtracting the load measured by the load cell set between the pipes. The penetration resistance acting on the outer pipe was the sum of the tip resistance of the outer pipe R_{out} and outer frictional resistance R_f. The outer frictional resistance R_f was estimated based on a previous study[2].

In this study, sand particles of Tohoku silica sand #5 (Table 1), which was used for the model ground, adhered to the inner surface of the inner pipe with epoxy adhesive, as shown in Figure 2. The friction coefficient of the inner surface was partially changed by this method in this series of experiments. The lengths of the attached sand particles, l_s, were 0, 20, 50, and 100 mm. The distance from the pile toe to the lower end of the sand-adhered part was either 0 or 50 mm. Table 2 shows the relationship between the experimental case names and the positions of the sand-adhered part.

Table 1. Soil properties of test sand.

Soil particle name	Tohoku silica sand #5
ρ_s	2.658 g/cm^3
ρ_{dmax}	1.718 g/cm^3
ρ_{dmin}	1.479 g/cm^3
D_{50}	0.6 mm
U_c	1.9
D_r	65 %

Table 2. Summary of sand-adhered part.

		Sand-adhered part		
Type of pile	Case name	Lower end from toe (mm)	Upper end from toe (mm)	Length l_s (mm)
Open-ended	S0-0	0	0	0
	S0-20	0	20	20
	S0-50	0	50	50
	S0-100	0	100	100
	S50-100	50	100	50
Closed-ended	Closed			

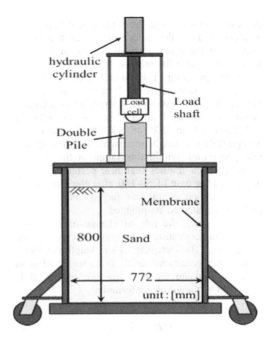

Figure 3. Diagram of the soil tank.

The model ground was prepared in a soil tank with an inner diameter of 772 mm, as depicted in Figure 3, and was made of dry Tohoku silica sand #5 (Table 2). The model ground was prepared with a relative density of 65 % and a ground height of 800 mm using an air pluviation method. The internal friction angle of the sand was $\phi_d = 35°$ when the relative density was 65 %.

The model pile penetrated the model ground statically at a speed of 30 mm/min, and the data measured included the total penetration resistance R, inner friction resistance R_{in}, and penetration depth z. The penetration depth was measured continuously using a displacement transducer, the capacity of which was 1000 mm. In the case of an open-ended pile, the displacement of the inner soil surface, h', was measured continuously using a displacement transducer, the capacity of which was 2000 mm. The inner soil height, h ($h = z - h'$), is the length from the pile tip to the soil surface inside the inner pile.

3 RESULTS AND DISCUSSION

3.1 Inner soil height

Figure 4 depicts the relationship between the inner soil height h and the penetration depth z in the open-ended pile penetration experiments. The values on both the X and Y axes are normalised by the outer diameter of the outer pile. When the increase in inner soil height h was almost equal to the increase in penetration depth, the soil could be easily intruded into the pile. Thus, it can be considered that the pile was not plugged. When the increment rate of the inner soil height was smaller than the pile penetration, the soil was less likely to enter the pile. In this case, it can be considered that the pile was partially plugged. When the inner soil height did not change during the penetration of the pile, the soil was never intruded into the pile. The pile was then perfectly plugged.

From Figure 4, the results suggest that no plugging occurs throughout the pile penetration

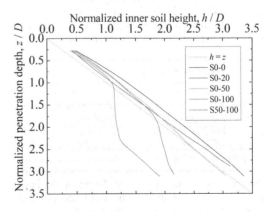

Figure 4. Relationship between h/D - z/D.

process in cases where the sand-adhered area was a small area of the inner surface of the pipes, such as in the cases of S0-0, S0-20, and S0-50. In the cases of S50-100 and S0-100, partially or perfectly plugging phenomena were observed during the pile penetration. Plugging can be considered to occur when the normalized penetration depth z/D ranged from 1.9 to 2.8 for S50-100 and from 1.1 to 2.3 for S0-100. In particular, a negligible soil height increment occurred during pile penetration within this range in the case of S0-100, which implies that near-perfect plugging occurs. In the case of S0-100 after the normalized pile penetration depth was further than 2.8, the pile was considered un-plugged.

In this study, the rate of change of inner soil height with penetration was compared with the state of plugging. The increment of soil length per increment of pile penetration depth, IFR^*, was used as an index to evaluate the plugging phenomena. It is expressed by the following formula:

$$IFR^* = 1 - \frac{\Delta h}{\Delta z} \qquad (2)$$

The parameters Δh and Δz indicate the increment in inner soil height and pile penetration depth, respectively. $IFR^* = 1$ indicates that no soil enters the pile during penetration and plugging occurs. $IFR^* = 0$ indicates that the amount of soil equivalent to the penetration depth penetrates the inner pipe, and no plugging occurs. Therefore, a small IFR^* value indicates less plugging, and a high value indicates more plugging.

There are other indices for examining pile plugging, such as PLR = h/z. Usually, the pile is not plugged; then, $PLR = 1$. When this index was applied to the case S0-100, the PLR was less than one at the pile penetration depth deeper than $z/D > 1.1$. As seen in Figure 5, the plug could be destroyed after the penetration depth of $z/D > 2.3$. At that condition, PLR was still smaller than one. Thus, PLR is a less important parameter for the plugging index than IFR or IFR^*.

In this report, the relationship between $q_{\text{in-open}}$/ $q_{\text{in-close}}$ and IFR^* is considered. The effects of the formation situation of plugging on the inner friction resistance are also discussed in this report.

3.2 Penetration resistance

$R_{\text{in-open}}$ is the resistance of the inner friction of the open-ended pile, and $R_{\text{in-close}}$ is the resistance acting on the inner pipe of the closed-ended pile. In the case of the closed-ended pile, $R_{\text{in-close}}$ is the resistance at a radius of 39 mm from the centre of the pile tip. It is not possible to compare the resistance measured on the inner pipe between $R_{\text{in-open}}$ and $R_{\text{in-close}}$ because of the different cross-sectional areas at the tip of the inner pipe. Therefore, to discuss the resistance per unit cross-sectional area, $q_{\text{in-open}}$ is defined as $R_{\text{in-open}}$

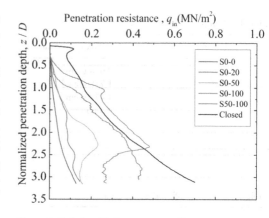

Figure 5. Relationship between q_{in} - z/D.

divided by the internal area of the inner pipe (i.e., $(D-2t)^2/4$), and $q_{\text{in-close}}$ is defined as $R_{\text{in-close}}$ divided by the cross-sectional area at the end of the inner pipe (i.e., $\pi(D')^2/4$). By converting the resistance per unit cross-sectional area using this approach, it is possible to compare and discuss them. The relationship between the unnormalized penetration resistance and depth ($R_{\text{in-open}} - z$, $R_{\text{close}} - z$) has also been described in a previous paper[3].

Figure 5 depicts the relationship between the inner friction resistance per unit area, $q_{\text{in-open}}$ or $q_{\text{in-close}}$, and the normalized penetration depth, z/D. In Figure 5, although the value of $q_{\text{in-close}}$ was relatively larger than that of $q_{\text{in-open}}$ at the beginning of the pile penetration. When the soil adhered area was large, $q_{\text{in-open}}$ tended to increase rapidly after the penetration depth was greater than $z/D = 0.5$. However, a significant difference in $q_{\text{in-open}}$ was observed between S0-50 and S50-100, where the soil adhered area was the same size. In S0-100, $q_{\text{in-open}}$ sharply increased up to the pile penetration depth of $z/D = 1.1$. From Figure 4, it was evident that the pile was plugged at this point. In the case of S0-100 $q_{\text{in-open}}$ was larger when the pile penetration depth ranged from 1.1 to 2.3. The $q_{\text{in-open}}$ of this case sharply decreased at the pile penetration depth of $z/D = 2.3$. At this point, the pile plug was destroyed and IFR^* was almost zero. In S50-100, the plugging occurs in a similar manner, $q_{\text{in-open}}$ sharply increases from $z/D = 0.8$ until $z/D = 1.2$, and once $q_{\text{in-open}}$ matches $q_{\text{in-close}}$, it remains approximately the same as $q_{\text{in-close}}$. In this pile penetration range, $z/D = 0.8 - 1.2$, plugging did not occur, as depicted in Figure 4. In addition, after z/D exceeds 1.7, plugging is formed and $q_{\text{in-open}}$ is relativity smaller to $q_{\text{in-close}}$. The difference in the relationship between the formation situation of plugging viewed from the behaviour of inner soil height and $q_{\text{in-open}}$ and $q_{\text{in-close}}$ is complex.

Comparing S0-20 and S0-50, the $q_{\text{in-open}}$ were almost the same until the pile penetration depth was $z/D = 1.5$. After $z/D = 1.5$, the $q_{\text{in-open}}$ of S0-50 was larger than that of S0-20, and both values were almost the same when the pile penetration depth z/D was close to 3.0. In these cases, the inner soil heights were almost the same for the full pile penetration depth. When the plugging effect was small, IFR^* was not an effective parameter for presenting the plugging situation. The authors considered that the effect of the change in friction angle in the inner pile along the pile axis was not even along the pile axis and was more effective in increasing the inner friction if the friction angle of the area near the pile toe increased. The value of $q_{in\text{-}open}/q_{in\text{-}close}$ also increases when h/D was nearly 1.8. Figure 5 does not show a sharp decrease in $q_{in\text{-}open}$ after the disintegration of the plugin S0-100.

Figure 6 shows the relationship between the ratio of inner soil resistance ($q_{in\text{-}open}/q_{in\text{-}close}$) and the normalized inner soil height (h/D), which is normalized by the outer diameter of the outer pile. For S0-100, where complete plugging occurred, and the $q_{in\text{-}open}/q_{in\text{-}close}$ value was the highest shortly after the point where the inner soil height (h/D) reached 1.0, that was, the point at which complete plugging occurred. Then, the value of $q_{in\text{-}open}/q_{in\text{-}close}$ sharply decreased as the inner soil height (h/D) increased when the plug was destroyed. A similar phenomenon occurred in the case of S50-100, which was the other case where plugging occurred. In this case, $q_{in\text{-}open}/q_{in\text{-}close}$ reached almost 1.0 when h/D was almost 1.3. At that h/D, plugging occurred. Thereafter, as h/D increased, a smaller change in the ratio occurred up to 1.8. h/D.

Figure 6 shows a sharp decrease relative to the increase in h/D immediately after the highest value is reached in the case of S0-100, but the resistance ratio was almost constant from h/D of 1.3 to 1.8 in the case of S50-100. The reason for this difference in the behaviour of $q_{in\text{-}open}/q_{in\text{-}close}$ between the cases that were both plugged in this manner was that S0-100 plugged perfectly, but S0-50 was incompletely plugged.

Figure 7 depicts the relationship between $q_{in\text{-}open}/q_{in\text{-}close}$ and IFR^*. Although there was variance in the relationship in each case, $q_{in\text{-}open}/q_{in\text{-}close}$ began to

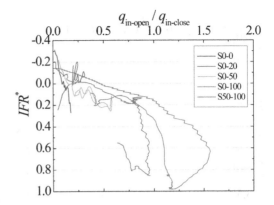

Figure 7. Relationship between $q_{in\text{-}open}/q_{in\text{-}close}$ and IFR^*.

increase after IFR^* reached approximately –0.1. The figure shows that until IFR^* reached 0.2, $q_{in\text{-}open}/q_{in\text{-}close}$ increased relative to IFR^* increase. IFR^* reached more than 0.2 in only two of the cases, S0-100 and S50-100. In these cases, $q_{in\text{-}open}/q_{in\text{-}close} > 0.9$ was maintained, although the values of $q_{in\text{-}open}/q_{in\text{-}close}$ varied. In the cases of S0-100 and S50-100, IFR^* decreased to some extent in the pile penetration depth. From these experimental results, the authors found that the values of $q_{in\text{-}open}/q_{in\text{-}close}$ were different whether IFR^* increased or decreased.

Even if the values of IFR^* were similar, the ratio of inner friction resistance was different; thus, further studies are required to estimate the ratio of the inner friction resistance circumference when IFR^* is above a certain value.

The relationship between the ratio of inner friction resistance and plugging at the pile toe was discussed. Therefore, the effect of the change in the range or position wherein the friction coefficient of the inner surface of the pile was changed on the IFR^* and the inner friction resistance must be considered.

Figure 8 depicts the relationship between IFR^* and the penetration depth when normalized by the outer diameter of the pile (z/D). In S0-0, S0-20, and S0-50, the results show that processing the end of the pile had most no effect on IFR^*. For example, S0-0, wherein no processing was performed, exhibited a higher value below $z/D = 1.0$ than that of S0-20, wherein processing was performed across an area extending 20 mm from the end. In S0-100, the case with the largest processed area, the IFR^* value increased sharply immediately after penetration, demonstrating the effect of processing. In S50-100, IFR^* began to increase sharply only after z/D exceeds 1.5, despite having a processed area of the same size as that of S0-50. From Figure 4, it can be observed that the inner soil height, h/D, was the same at $z/D = 1.5$, as it was previously, from which it can be presumed that the outcome was affected by processing an area that is at a distance from the pile end. Past research has indicated that the inner friction resistance of an

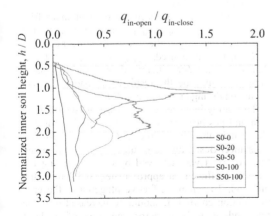

Figure 6. Relationship between $q_{in\text{-}open}/q_{in\text{-}close} - h/D$.

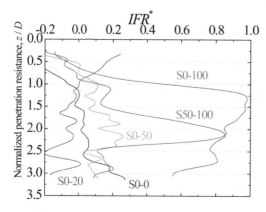

Figure 8. Relationship between IFR^* and z/D.

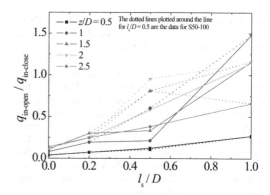

Figure 10. Relationship between l_s^*/D and $q_{\text{in-open}}/q_{\text{in-close}}$.

open-ended pile is noticeable near the pile end; thus, it can be presumed that it is important to change the friction coefficient at a distance from the pile end in some cases and the end in other cases for easy plugging.

Figure 9 depicts the relationship between l_s/D (the area of the inner surface of the pile that is processed, l_s, divided by the outer diameter of the pile, D) and h (the inner soil height when a specific penetration depth is reached). The dotted lines plotted around the line for $l_s/D = 0.5$ are the data for S50-100. From Figure 9, it can be observed that plugging was not facilitated until z/D reached $0.5 - 1.0$, regardless of the processed area. There is no difference in the state between $l_s/D = 0.0$ and $l_s/D = 0.2$ after $z/D = 1.5$ but plugging occurs at the end. Note that when $l_s/D = 0.5$ for S0-50, the state remains the same as that at $l_s/D = 0.2$ until $z/D = 2.0$; however, a notable degree of plugging occurs at the same l_s/D value (i.e., $l_s/D = 0.5$) for S50-100.

Figure 10 depicts the relationship between l_s/D and the ratio of inner friction resistance, $q_{\text{in-open}}/q_{\text{in-close}}$. From Figure 10, the difference between $q_{\text{in-open}}/$

$q_{\text{in-close}}$ was small when the relative penetration depth was less than $z/D = 0.5$. However, as the relative penetration depth increased, the $q_{\text{in-open}}/q_{\text{in-close}}$ increased only in the case of $l_s/D = 1$. It can be observed that the $q_{\text{in-open}}/q_{\text{in-close}}$ values are larger for $l_s/D = 0.5$ in the case of S50-100, as depicted by the dotted line, in comparison with that in the case of S0-50 with the same $l_s/D = 0.5$. Thus, it can be presumed that increasing the coefficient of inner friction in the section of approximately $1D$ from the pile toe or $0.5D$ from the pile toe to $1D$ may affect the acceleration of plugging.

4 CONCLUSIONS

In this study, a penetration experiment was conducted using a double-pipe model pile to measure the inner friction resistance, $q_{\text{in-open}}$, and the resistance acting on the ring of the pile toe. The sand was applied to the inner surface of the double-pipe model pile to change the friction coefficient, and the relationship between the inner soil height and IFR^*, which represents the plugging effect, was verified. Furthermore, the effect of the area and position where the sand was applied on the plugging at the pile end was verified.

The relationship between the ratio of inner friction resistance, $q_{\text{in-open}}/q_{\text{in-close}}$, and IFR^* indicated that when the plug at the pile end disintegrated and IFR^* sharply decreased, $q_{\text{in-open}}/q_{\text{in-close}}$ was lower than that when IFR^* increased as a result of plugging. This may be the result of a historical effect occurring during an increase in IFR^*.

This study investigated the effect of changing the area and position where the friction coefficient was changed in the pile on the IFR^* and inner friction resistance. Based on the results, it can be presumed that no effect was achieved by only increasing the friction coefficient at approximately $0.5D$ from the pile end. To achieve effective plugging at the end, the friction coefficient should be increased near the pile end as well as across an area of at least

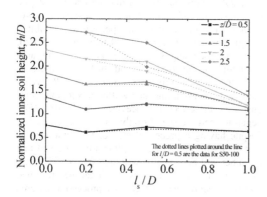

Figure 9. Relationship between l_s^*/D and h/D.

1.0D from the end. It was also found that processing the area from 0.5 to 1.0D from the pile end facilitates plugging more effectively than processing the area from 0 to 0.5D.

Future research is necessary to determine whether the same phenomenon occurs during plug reformation. Furthermore, it would be useful to quantify indices, such as the friction coefficient in the pile, inner friction resistance, and IFR^*.

There were many differences in the actual field and this kind of small-scale experiment. For example, the pile diameter to pile wall thickness ratio D/t used in the experiment was quite large compared to the actual field. The ratio of pile diameter to soil particle must be an important parameter, but the authors did not discuss this point. The stress level was significantly different between the field and this experiment. Furthermore, the maximum penetration depth was only approximately 3D; thus, it is not established whether a plug forms again after the initial plug disintegrates.

According to these limitations in laboratory experiments, the conclusions extracted from this research cannot be applied directly to the field. However, the conclusions might provide useful information for planning larger-scale model experiments or field experiments. The authors believe that this research provides useful information to readers preparing further research.

ACKNOWLEDGEMENTS

The financial supports for the research works provided by the Japan Iron and Steel Federation (JISF) is gratefully acknowledged.

REFERENCES

Architectural Institute of Japan, 2001. *Recommendations for design of building Foundations*, p. 207 (in Japanese).

Kikuchi, Y. 2011. Mechanism of inner friction of an open-ended pile. *Proceedings of 3rd IPA International Workshop, Press-in Engineering 2011*: 65–83.

Yamazaki, H., Kikuchi, Y., Noda, S., Sakimoto, K. Matsuoka, H. 20193 Mechanism of the bearing capacity of the open-ended piles using double-pipe model pile, *Journal of Japan Society of Civil Engineers, Ser. B3 (Ocean Engineering), Vol. 75*, 2: I_462-I_467.

Yamazaki, H., Kikuchi, Y., Noda, Saotome, M., Nonaka, M. 2020. Unit inner friction resistance and unit resistance of actual part of open-ended piles based on the double-pipe model pile experiment, *Journal of Japan Society of Civil Engineers, Ser. B3(Ocean Engineering), Vol. 76*, 2: I_450-I_455.

Proceedings of the Second International Conference on
Press-in Engineering 2021, Kochi, Japan – Matsumoto et al (eds)
© 2021 Taylor & Francis Group, London, ISBN 978-1-032-10414-0

The vertical and horizontal performance of pressed-in sheet piles

T. Zheng, S.K. Haigh, A. Dobrisan & F. Willcocks
University of Cambridge, Cambridge, UK

Y. Ishihara, K. Okada & M. Eguchi
GIKEN Ltd., Kochi, Japan

ABSTRACT: This paper aims to investigate the vertical and horizontal performance of pressed-in sheet piles, focusing on the effect of the installation process. An instrumented pile was installed four times (once as part of a three-wide wall) in sand under varying installation conditions and subjected to vertical and horizontal load tests. A novel liquefiable sand tank was used in which 'boiling' of the sand between each pile installation allowed the sand to return to a loose, undisturbed state. In the vertical direction, the performance was found to vary significantly with the installation conditions, and an accurate, straightforward method for assessing the pile capacity retroactively from the installation loads is presented. The horizontal performance was found to be relatively independent of the installation conditions, and a simple modification to the *P-y* method is proposed to model the behaviour, to account for existing methods being suited to circular rather than flat cross sections.

1 INTRODUCTION

Sheet piles are a type of deep foundation which are reusable, lightweight and easy to install. Additionally, interlocking geometries at their section ends al- low for the construction of continuous walls. These key advantages have led to their widespread adoption in certain applications, such as retaining walls and temporary excavation support. The press-in method is an installation procedure which uses a static jacking force to push a pile into the ground (International Press-in Association 2016). The reaction force required to keep the piling equipment grounded can be provided by a ballast, or conveniently by the rig anchoring itself to previously in- stalled piles. The latter provides an incredibly efficient way to install lines of piles. While the maximum achievable pile depths are generally more limited than for traditional methods such as driving or vibration, the press-in method causes no disturbance (noise and vibration) to the surrounding environment – another significant advantage.

Both the behaviour of sheet piles and the effect of the press-in method on pile perform- ance are sparsely researched. This paper investi- gates the response of sheet piles subjected to vertical and horizontal loads, with particular focus on how the behaviour is affected by the press-in process. The findings are then applied to the prospect of how a potential design method could be developed.

2 METHODOLOGY

Multiple full-scale load tests were carried out on an instrumented 8.15 m long SP-25H steel sheet pile, installed in a 7 × 7 × 10 m tank of silica sand. The geotechnical properties of the sand are summarised in Table 1. G_s, e_{min}, e_{max} and φ'_{cs}, which represent the specific gravity, minimum and maximum voids ratios and critical state friction angle respectively, have been derived from density and triaxial laboratory testing. D_r, the relative density, has been estimated from in-situ CPT data using the empirical correlation pro- posed by Jamiolkowski et al. (2001). γ_{dry} and γ'_{sub}, the dry and submerged unit weights, have been calcu- lated using the aforementioned parameters.

A key feature of the tank is that it can be liquified. Water is pumped and circulated through the tank at a high rate, causing a build-up of pore pressure, reducing the effective stress of the soil everywhere within the tank to zero (Ogawa et al. 2018). As a result, the sand returns to a loose, undisturbed state. By liquefying the tank between each test, the ground conditions for all tests can be considered equivalent, as the liquefaction process removes the influence of the previous test's stress history.

The structural properties of the pile are summar- ised in Table 2, and a cross-section is shown in Figure 1. D_{eq} represents the outer diameter of an equivalent tubular pile, with equal cross-sectional area and equal enclosed soil area (based on the cen- tral trapezium). I, A_s, E and σ represent the second

DOI: 10.1201/9781003215226-9

Table 1. Sand properties.

G_s	e_{min}	e_{max}	φ'_{cs}	D_r	γ_{dry}	γ'_{sub}
-	-	-	°	-	kN/m³	kN/m³
2.634	0.531	0.891	30.7	0.4	14.8	9.2

Table 2. Pile properties.

H	W	t	D_{eq}	I	A_s	E	σ
mm	mm	mm	mm	cm⁴	cm²	GPa	MPa
300	900	13.2	438	22000	144.4	210	295

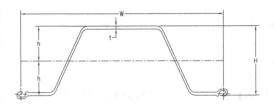

Figure 1. Pile cross-section.

moment of area, cross-sectional area, Young's modulus and yield strength of the pile respectively.

The test suite comprised 4 cases. For each case the pile was installed using the press-in process. Following this, a vertical load test was carried out using a hydraulic jack to apply the load. A horizontal load test was then carried out with the jack attached approximately 1.3 m above the soil surface. In both directions, the load level was increased linearly, at rates of approximately 400 and 200 kN/hr for the vertical and horizontal tests respectively. Once the peak loads were reached, unloading commenced at twice the loading rate. The rates of displacement at the pile head varied between 5 and 70 mm/hr, slow enough for pseudo-static, undrained behaviour to be assumed.

The maximum peak loads were determined by either of two criteria, whichever was met first.

A strength criteria limited the load level to below the ultimate capacity of the pile (i.e. when load no longer increases with displacement), while a stiffness criteria limited the deflection to $D_{eq}/10$ in the vertical case, as specified in the Japanese standard for axial pile testing (JGS 2002), and to $D_{eq}/30$ at the ground surface in the horizontal case.

Load cells, displacement sensors, inclinometers, earth pressure sensors, and the piling machine's internal data logger were used to record data during testing.

The test set-up is illustrated in Figure 2, and the differences between each test are presented in Table 3, where L_{down} and L_{up} are the surge stroke lengths – two parameters related to the installation condition. During the press-in process the pile is cyclically moved upwards and downwards as it progresses through the soil (known as surging), and the surge stroke lengths determine the size of these cyclic movements.

3 VERTICAL PERFORMANCE

3.1 Results

Force-displacement curves of the vertical load tests for cases 1-3 are presented in Figure 3. In case 1, the test was terminated prematurely due to equipment failure, hence the results are incomplete.

Force-penetration depth curves of the installation processes have been plotted for cases 1-3 in Figure 4. These give an indication of the evolution of vertical capacity with pile length, as the penetration strokes are themselves are vertical load applications. The negative spikes seen are anomalous readings due to the piling machine behaviour only and are not reflective of the soil resistance experienced by the pile – they occur when the piler releases the pile to progress the clamp position. The final penetration depths indicated by the installation data are slightly smaller than the actual achieved depths. This inaccuracy again arises from result of clamping mechanism – when unclamped, at shallow depths the pile occasionally 'slipped' downwards due to the low soil resistance, resulting in unrecorded penetration.

Table 3. Test cases.

Case	Number of Piles	z_w m	L_{down} / L_{up} mm	F_{max} kN	L m	e_H m
1	1	7	400 / 200	800	6.16	1.10
2	1	1	400 / 200	800	6.04	1.30
3	1	1	100 / 50	800	6.19	1.15
4	3*	1	100 / 50	300 to 350**	5.84	1.50

* Two neighbouring piles were installed either side of the instrumented pile in a wall formation
** Maximum install force was increased during installation after pile refusal occurred

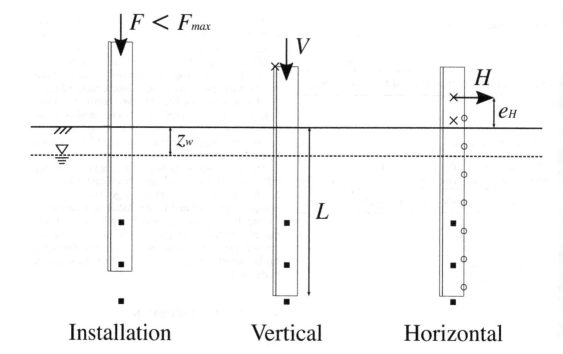

Figure 2. Test set-up.

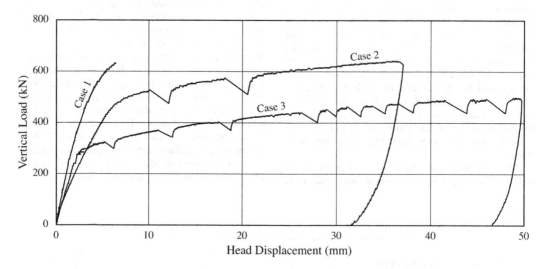

Figure 3. Vertical load test results.

3.2 Interpretation of vertical load tests

The three force-displacement curves in Figure 3 demonstrate significant variability. However, they appear to all follow a similar form, consisting of two distinct regions: an initial response characterised by high pile-head stiffness; and a subsequent 'yielding' characterised by low pile-head stiffness and

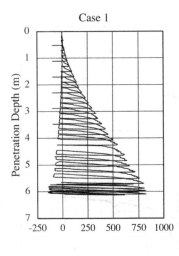
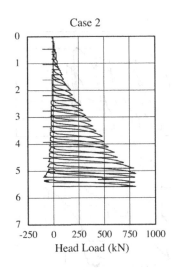
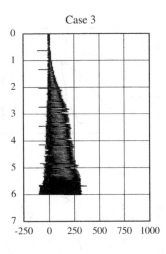

Figure 4. Installation data.

instability (the piles underwent multiple sudden, large deformations). These regions are clearly observed in cases 2 and 3, and the available data from case 1 seems to follow this characterisation reasonably well. It is therefore assumed that this shape is typical of a pressed-in sheet pile installed in loose sand.

The initial region can be considered to represent elastic deformation, and the 'yielded' region to represent plastic deformation. Not only is this intuitive, it is supported by the displacements regained during unloading (5.5 and 3.2 mm for cases 2 and 3 respectively) being approximately equal to the displacements at the end of the initial region.

It can be inferred from Figure 3 that the installation process is highly influential towards the scaling of this overall form. The piles of cases 2 and 3 only differ significantly in terms of their installation characteristics, yet their force-displacement curves are notably different. In order to characterise a given installation regime, a rate of surging (in spatial terms) can be defined as:

$$R_{surge} = 1 / (L_{down} - L_{up}) \qquad (1)$$

where R_{surge} is equal to the number of cycles per unit length of net pile progress, and hence represents the intensity of the surging regime. Case 3 has a much higher R_{surge} than case 2 (20 cyc/mm compared to 5 cyc/mm respectively), and correspondingly a much lower yield load and ultimate capacity.

This apparent detrimental effect of heavier surging on pile capacity is not unexpected in the context of existing literature. Friction fatigue is a known phenomenon whereby cyclic loading of sand at a pile-soil interface leads to a reduction in skin friction. Repeated shearing densifies the near-field soil and leads to a relaxing of horizontal pressure, and hence a reduction of the available skin friction (White & Bolton 2002, Burali d'Arezzo et al. 2015). It is therefore reasonable to assume that higher levels of surging (greater values of R_{surge}) induce more friction fatigue, causing greater reductions in skin friction and hence vertical capacity. Quantifying this effect, however, is more challenging.

3.3 Back analysis of installation data

Given that the vertical load tests were performed after unloading from the final penetration stroke of the installation, they can be considered the next load cycle. It is therefore useful to plot the force-penetration data from the final installation cycles against the vertical load test data for each case. This comparison is shown in Figure 5.

In all three cases, there is a strong correlation between the yield load in the vertical load test data and the loads at which consecutive cycles cross in the installation data (hereafter referred to as a cycle intersection). While there is some scatter, the yield loads generally align closely with the last few cycle intersections, as indicated by the dashed lines. This is a key result, which can be explained by considering the physical significance of the cycle intersection:

At the end of any surge cycle, a pile has reached its maximum depth d. Crucially, the pile must be in the plastic region; otherwise, no net penetration would have occurred during the cycle. It is then unloaded during the up-stroke. Now, for the pile to progress further than d it must re-enter the plastic region i.e. yield. From Figure 5, the cycle intersection corresponds to the point at which the pile returns to d after a surge cycle (it is slightly less than

141

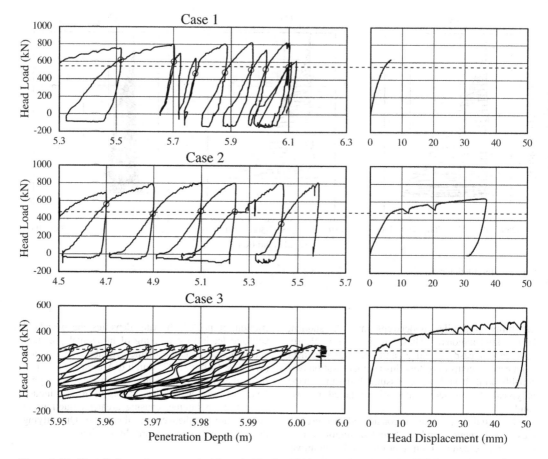

Figure 5. Final installation cycles compared with vertical load tests data (cycle intersections circled).

d in reality but the difference is small since the unloading curve is steep), and as such, represents the yield point. The force at d is now lower than the previous cycle due to the effect of the surging, but regardless, the soil must yield for further plastic deformation (i.e. penetration) to occur.

The ultimate capacity of the pile, however, is not well predicted by the installation data. In case 3, the ultimate capacity is never reached during installation, as the yield point/cycle intersection is only just exceeded. In case 2, the ultimate capacity achieved during installation exceeds that of the vertical load test. The difference is due to the high penetration rates during installation resulting in undrained behaviour. In this case, negative excess pore pressures were generated, leading to higher effective stresses and increased resistance.

It is worth noting that generally during elastic deformation, very little shearing and volume change occurs, meaning little excess pore pressure is generated regardless of penetration rate. As a result, the cycle intersections, which occur at the limit of elastic deformation, are unaffected by the penetration rate,

and hence remain valid as an indicator of the yield load in any case.

3.4 Consequences for design

It is proposed that for design purposes, the transition between the elastic and plastic regions - the yield point - is an appropriate point for design. This would satisfy strength (capacity) requirements, as the load would never exceed the yield load and hence the ultimate capacity. It would satisfy stiffness (deflection) requirements – in all cases the maximum displacements within the elastic region (approximately 1.6% and 0.7% of D_{eq} for cases 2 and 3) were much lower than the deflection limit of $D_{eq}/10$. Lastly, it would ensure robustness, by preventing cumulative plastic deformation from repeated loading, and eliminating the possibility of potentially damaging sudden, large settlements, which were observed within the plastic region. The key design problem therefore lies in predicting the load at which yielding occurs.

In light of this, the observations made in Section 3.3 lead to an incredibly significant result. Once installed, the yield (i.e. design) load of a pressed-in sheet pile can be readily assessed from the installation data by observing the final cycle intersections. Since the required data is recorded automatically by the piler, 'zero-effort' experimental verification of the design resistance is carried out by default.

The observations also show that there is a clear trade-off between ease of installation and high axial capacity, as the two are inherently coupled. While this can be seen as a downside (both are desired), it does give the designer a level of control over the vertical capacity via the choice surging regime. Of course, these findings only relate to post-installation verification. A method of design in the first instance would also be necessary. Then, a designer could choose the pile size, length and installation regime with much flexibility to provide the required capacity.

No such design process has been presented in this paper; there is too small a data set to assert such a method with confidence. However, the data does present insight into the behaviour of a sheet pile un- der vertical loading, which is a subject rarely studied in detail.

It can be seen from Figure 4 that all cases demonstrate a much larger penetration force (envelope of positive peaks) than extraction force (envelope of negative peaks). Typically, the former represents the sum of the base resistance and shaft friction, and the latter solely the shaft friction (which should be equal in both directions). For a thin-walled sheet pile, the much higher penetration force is surprising as the base area is negligible compared to the shaft area. The pile must therefore be able to utilise some of its gross base area on downward penetration.

Plugging of up to 2 m was observed during the later stages of installation for all piles (Figure 6), which supports this hypothesis. Further evidence is provided by the lateral earth pressure measurements (example in Figure 7), which show large spikes during installation. Note that there are slight discrepancies between the peaks and sensor depths due to the inaccuracies in penetration depth as explained section 3.1. In standard bearing/skin friction failure, the lateral soil pressures generated would be minimal with such a small displaced soil volume. It is therefore likely that an effect similar to the exponential stress increase seen in tubular piles - known as vertical arching (Randolph et al. 1991) - is taking place, but with open-sided soil columns, meaning the horizontal stress is transferred to the external soil.

The complexity of the stress fields due to such a mechanism, combined with the highly complicated installation induced stress changes, suggest that an empirical approach may be best suited to

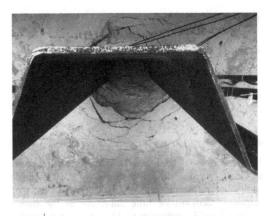

Figure 6. Plugging with open-sided soil plug.

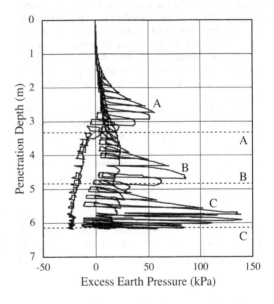

Figure 7. Recorded horizontal earth pressures from 3 sensors (A, B and C) during Case 1 installation (sensor depths shown by dashed lines).

a design process. Only a rough estimate of capacity would be required, provided the installation regime can be monitored and adjusted live (using the cycle intersections) in order to end with the right capacity.

Again, too little data was collected in this study to reliably inform any empirical process, though from the study it is likely that the key factors in determining vertical performance are R_{surge}, L, the pile surface perimeter and the gross cross-sectional area, with the latter two influencing the vertical arching mechanism. These soil properties (e.g. peak friction angle and unit weight) will also likely be influential.

4 HORIZONTAL PERFORMANCE

4.1 *Results*

Data from the horizontal load tests for all cases is shown in Figure 7. The ground surface displacement has been extrapolated linearly from data recorded at two positions above ground.

4.2 *Interpretation of horizontal load tests*

Unlike during the axial tests, here the piles were seen to reach specified deflection limits ($D_{eq}/30$) significantly before the ultimate capacity. In all cases, the responses demonstrate little curvature as the maximum allowable deflection of 14.6 mm is reached or approached, suggesting that the ultimate capacities are much higher than the maximum loads applied during the tests. As such, the horizontal performance of sheet piles can be considered as predominantly governed by stiffness (deflection) criteria and not by strength (capacity), other than in situations where excessive deflection is tolerable. As sheet piles tend to be much more flexible than tubular piles, it is not surprising that deflection criteria dominates over strength criteria.

Case 4 represents the group case, and hence the response was significantly stiffer (higher applied load for the same displacement) than the other three cases. In case 1, the water table was below the pile toe, meaning the effective stresses within the soil were higher than in cases 2 and 3 (due to the absence of hydrostatic pore pressure), and hence the response was also stiffer.

The piles in cases 2 and 3 are useful to compare as they were approximately equivalent in all respects other than the rate of surging applied during installation. It can be seen from Figure 8

that both behave very similarly, despite large differences in their installation characteristics. There are slight differences in load eccentricity and pile length, but these should be of minor significance, particularly as the displacement has been interpreted at the ground surface and not the point of load application. This suggests that performance under lateral loading is not significantly influenced by the installation process, unlike for the axial case.

Interestingly, most of the discrepancy between the case 2 and 3 load-displacement curves arises at small displacements. At a surface displacement of 4 mm, the applied load is 3.5 kN higher for case 3 than 2, yet at a displacement of 16 mm, it is still only 3.5 kN higher. The curves run approximately parallel after the initial region.

4.3 *P-y analysis*

Inclinometer and displacement data was collected at various points along the pile length, as illustrated in Figure 2. Using this data, the pile deflected shapes can be inferred via numerical integration (see Figure 9). Additionally, using beam bending relationships, successive differentiation can be used to relate this data to the internal bending moments, shear forces and external loads along the pile. However, differentiation typically amplifies noise, meaning simple numerical differentiation did not yield reliable results. Instead, a program called 'Probfit' (Dobrisan et al., in prep.) was used to process the data. The software uses recorded deflections and rotations along with boundary conditions to create a compatible family of curves pertaining to the deflection, rotation, bending moment, shear and soil pressure along the pile length which have the best fit to the data, allowing for measurement error.

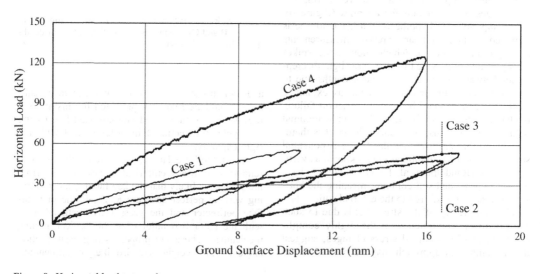

Figure 8. Horizontal load test results.

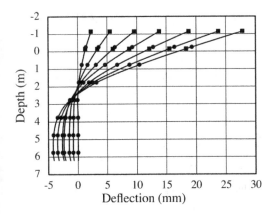

Figure 9. Integrated deflected shapes during horizontal loading for case 3. Displacement sensor and inclinometer positions are shown with square and circular markers respectively.

'Probfit' analysis was carried out at seven different loads for cases 2-4, and P-y curves were subsequently derived from the deflection and soil pressure results. For comparison, theoretical P-y curves have been produced based on established correlations within the literature. In this study the hyperbolic form proposed by O'Neill & Murchison (1983) is used, in which the curves are characterised by two parameters, k_{py} and p_u. This formulation is currently adopted in the API (2000) and DNV (2014) design guidelines. k_{py} and p_u are estimated via correlations derived from large amounts of empirical data. However, all of this data was obtained using tubular piles, as they are primarily what P-y analysis is used for. The sheet pile deformation mechanism for lateral movement through sand is likely to be different to that of a circular section. As such, there is no reason to assume that these empirical correlations for k_{py} and p_u should be applicable to sheet pile behaviour. For this reason, modified P-y curves have also been produced, where modification factors F_k and F_p have been applied to k_{py} and p_u (as demonstrated in Figure 10), in order to provide a better fit to the experimental data.

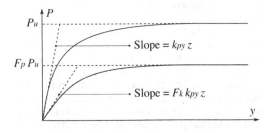

Figure 10. O'Neill & Murchison (1983) and modified P-y curves, where z represents the depth below the surface.

The experimental, O'Neill & Murchison, and modified P-y curves are all presented in Figure 11 for cases 2-4 at three different depths. The depths extend to 2 m only as beyond this the deflections were found to be small.

Considering the individual pile cases, from Figure 11 it can be deduced that the O'Neill & Murchison curves, which are widely adopted for analysis of tubular piles, do not fit the experimental results well. They tend to over-estimate the soil pressures experienced by the sheet pile, and would therefore be unconservative in design.

The modified curves have been developed as a best fit to the experimental data, with $F_k = 0.5$ and $F_p = 0.75$ for cases 2 and 3. It became evident, however, that the hyperbolic curves could not be perfectly fitted to the experimental data, as the fundamental shapes differed. Instead, they run parallel to and slightly below the experimental curves after an initial high stiffness region which could not be accurately modelled (this high stiffness region can be observed as a 'hump' in the data). This correlation can be clearly observed in all 6 plots for cases 2 and 3. In this way, the modified curves represent a conservative lower bound to the true P-y curves, excluding the effects of the initial 'hump'.

It is argued that the initial high stiffness is the result of densification of the near-field soil due to friction fatigue during installation. This densification process itself is a widely accepted phenomenon, and it is also well established that a denser sand provides a stiffer response (k_{py} is correlated with relative density). At small displacements, only the near-field, denser soil is mobilised, leading to a high initial stiffness. As the pile deforms further, the zone of influence grows beyond this region, and the increases in load are determined by the mobilization of the looser, undisturbed, further-field soil.

This hypothesis is further evidenced by the results of Figure 8. As mentioned previously, most of the discrepancy between cases 2 and 3 (which only differ in their installation characteristics), occurs at small displacements i.e when only the near-field soil is mobilised. In case 3, where heavier surging was used (leading to more friction fatigue/densification), the initial stiffness was higher. At larger displacements, the further-field undisturbed soil is mobilised, which is the same for cases 2 and 3, thus they undergo similar increases in load.

Since the modified P-y curves model the response excluding the initial 'hump', this hypothesis suggests that they are effectively modelling the response of the sheet pile excluding the effects of installation, which are beneficial, in contrast to the axial case.

4.4 Accounting for pile width

The proposed new modification must be able to account for varying pile/pile wall sizes to be of

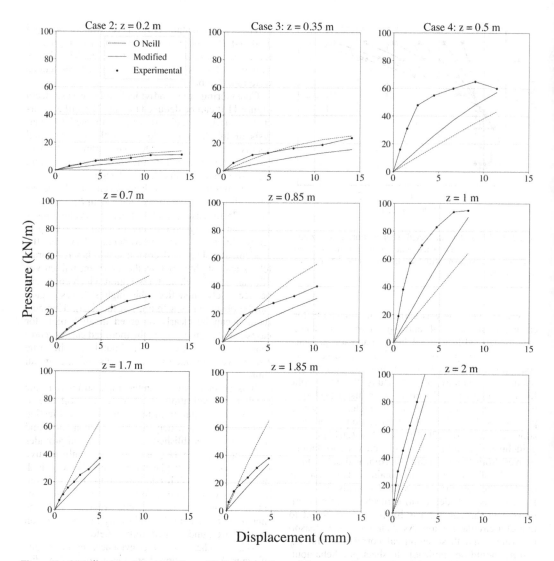

Figure 11. O'Neill & Murchison (1983), modified and experimental *P-y* curves for cases 2-4 (left to right).

any practical use. The variation of F_k and F_p with width W must therefore be determined.

In O'Neill and Murchison (1983), the function for determining P_u has an inherent linear dependency on D. This may be a reasonable assumption for a circular cross section, as the failure mechanism is likely to be the same. For a sheet pile or wall, however, the mechanism will be different for an individual pile and a long planar wall due to end effects, with P_u/W being on average lower for the longer case. How sensitive P_u/W is with respect to width is more difficult to know.

k_{py} on the other hand is usually defined as independent of the size. This has been disputed for tubular piles (Sorensen and Augustesen 2016), and Figure 8 shows that this is clearly unrealistic for sheet piles as the initial gradients of case 4 (group

of 3 piles) are far higher than those of cases 2 and 3 (individual piles). It would be expected that a sheet pile of double width would receive approximately double the soil resistance at small strains (or slightly less due to relatively reduced end effects).

Assuming both P_u and k_{py} are approximately linearly dependent on width, $F_p = 0.75$ and $F_k = 0.5 \ W/0.9 = W/1.8$ m. These expressions have been used to produce the modified curves for case 4 in Figure 11. The correlation between modified and experimental curves does not appear to be as strong as for the individual cases, indicating that further refinement of the modification factors is needed for larger widths. However, the initial 'hump' is still clearly seen (although is much larger here), and beyond this the correlation does improve as expected.

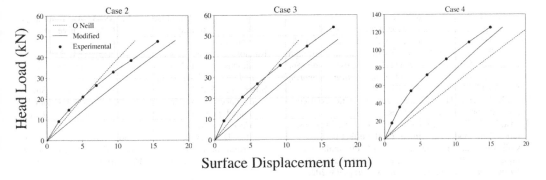

Figure 12. Numerical pile simulation results for cases 2-4.

4.5 Consequences for design

It is proposed that by applying F_k and F_p, the P-y method can be used for the design of sheet piles as well as tubular piles. Based on the limited data, factors of $F_p = 0.75$ and $F_k = W/1.8$ m are suggested. It is acknowledged however, that both are likely to decrease from the above as W increases, though not enough data was collected to assert a more accurate relationship. A numerical simulation of the installed pile can then be performed using the modified P-y curves.

This has been carried out via a finite difference method for cases 2-4, using both O'Neill & Murchison and modified formulations. The resulting global force-displacement curves are presented, along with the experimental data (via 'Probfit'), in Figure 12.

In cases 2 to 3, as for the P-y curves, the simulated force-displacement curve using the O'Neill & Murchison formulation poorly predicts the true behaviour, and is unconservative in its nature. The modified curves, again, run parallel to and below the true curve beyond an initial 'hump'. As mentioned previously, the hump is postulated to be the result of densification due to friction fatigue during installation – this is supported by the fact that it is larger for case 3 than case 2, which utilised significantly more surging. This modified curve therefore predicts the behaviour excluding the effect of installation. Since this effect is relatively small, and slightly beneficial, design using this method would fairly well predict the true pile behaviour, while always remaining conservative.

In the group case, the predicted gradient using the modified method is slightly higher than the true gradient beyond the initial 'hump'. This is expected, as the linear F_k and F_p relationships do not account for reducing end effects as width increases. However, the difference is modest and the result remains conservative within the tested range of displacements due to the initial installation-induced 'hump'.

5 CONCLUSIONS

Through full-scale pile testing, a number of key insights into the axial and lateral behaviour of sheet piles, particularly regarding the effects of the installation process, have been deduced.

The level of surging during installation is found to be heavily influential towards the vertical capacity, with heavier surging leading to greater reduction in capacity. The load observed during installation has been found to be inherently coupled to the vertical performance, to such an extent that the installation data can even serve as a pile test indicating the pile's performance (the cycle intersections toward the end of installation inform the yield point/design strength). Additionally, despite not having a closed section, gross area effects such as vertical arching and plugging are still very prevalent for sheet piles, and central to their behaviour.

The lateral response is only modestly affected by installation conditions, and this effect appears to be a slight beneficial increase in initial stiffness (due to densification of the soil close to the pile), so can conservatively be ignored. The P-y method can be applied, but with modified parameters to account for non-circular geometry. Multiplying k_{py} and P_u by correction factors $F_k = W/1.8$ m and $F_p = 0.75$ respectively is deemed a reasonable approach for modelling isolated piles/small pile groups, provided excessive deflections are not expected.

REFERENCES

API 2000. Recommended practice for planning, designing and constructing fixed offshore platforms – working stress design.

Burali d'Arezzo, F., Haigh, S.K., Talesnick, M. & Ishihara, Y. 2015. Measuring horizontal stresses during jacked pile installation. *ICE Proceedings: Geotechnical Engineering* vol. 168(4): 306–318.

DNV 2014. DNV-OS-J101 – Design of offshore wind turbine structures.

Dobrisan, A., Haigh, S.K. & Deng, C. 2020. In prep.

International Press-in Association (IPA) 2016. Press-in retaining structures: a handbook, First edition: 520.

Jamiolkowski, M., Lo Presti, D.C.F. & Manassero, M. 2001. Evaluation of Relative Density and Shear Strength of Sands from CPT and DMT. *Symposium on Soil Behavior and Soft Ground Construction Honoring Charles C. "Chuck" Ladd*: 201–238.

JGS 2002. JGS 1811-2002 – Method for static axial compressive load test of single piles.

Ogawa, N., Ishihara, Y., Ono, K. and Hamada, M. 2018. A large-scale model experiment on the effect of sheet pile wall on reducing the damage of oil tank due to liquefaction. *Proceedings of the First International Conference on Press-in Engineering, Kochi*: p193–202.

O'Neill, M.W., Murchison, J.M. 1983. An Evaluation of P-y Relationships in Sands. University of Houston.

Randolph, M. F., Leong, E.C., Houlsby, G.T. 1991. One dimensional analysis of soil plugs in pipe piles. *Géotechnique* vol. 41(4): p578–598.

Sørensen, S.P.H., Augustesen, A.H. 2016. Small-displacement soil-structure interaction for horizontally loaded piles in sand. *Proc. 17th Nordic Geotechnical Meeting, Reykjavic*: 775–786.

White, D.J., Bolton, M.D. 2002. Observing friction fatigue on a jacked pile. *Proc. Workshop on Constitutive and Centrifuge Modelling: Two Extremes*: 347–354.

Proceedings of the Second International Conference on
Press-in Engineering 2021, Kochi, Japan – Matsumoto et al (eds)
© 2021 Taylor & Francis Group, London, ISBN 978-1-032-10414-0

Performance of pressed-in piles in saturated clayey ground: Experimental and numerical investigations

L.T. Hoang
Thuyloi University, Hanoi, Vietnam

X. Xiong & T. Matsumoto
Kanazawa University, Ishikawa, Japan

ABSTRACT: This paper investigates performance of pressed-in piles in saturated clay ground under isolated (SP) condition and pile group (PG) condition via both experimental and numerical methods. In the experiments, static load test (SLT) was conducted on one SP immediately and on another SP and PGs 24 hrs after the installation to investigate the consolidation effect on the pile capacity. The numerical analyses were conducted following the procedure of the model tests using PLAXIS 3D. The constitutive soil model called "soft soil creep model" was employed to describe the soil behaviors with parameters mainly obtained from laboratory element tests. The effect of the pile installation process on the ground stresses is simulated by cylindrical expansion. Both the measured and calculated results indicate that the ground consolidation significantly increases pile capacity, and the simulations results show a good agreement with the experimental results in terms of initial stiffness and pile shaft resistance.

1 INTRODUCTION

In foundation engineering, piles are well utilized as a foundation solution to support heavy structures or structures located on soft grounds, of which the settlements of shallow foundations are excessive for the allowable values or the bearing capacities of shallow foundations do not meet the design load. According to the construction method, piles are classified into two types: non-displacement piles, such as bored piles and pre-augered piles, and displacement piles, such as jacked-in piles and driven piles. It is widely accepted that the installation process makes a great difference in bearing capacity between two pile types (Deeks et al., 2005; Dijkstra, 2009).

The effects of the installation process on pile behavior were observed in many previous studies. When a pile is jacked into the ground, the soil will be displaced outwards from the pile with a volume equal to the pile volume, and the soil near the pile shaft is completely remodeled. The behavior of soil surrounding the pile depends on many factors such as the initial density, degree of saturation of the soil, or grain sizes. If the soil is a saturated cohesive soil with low permeability, the total stresses generated by the installation process in the soil surrounding the pile basically transfer to the pore water pressure (PWP) first. PWP generated during the installation period may equal or exceed the total overburden

pressure and requires a duration after the installation process to dissipate and transfer into effective stress (Flaate, 1971; Bozozuk et al., 1978). A plastic zone is developed around the pile where the mobilized shear stress exceeds the original undrained shear strength of the soil. At the zone near the pile shaft, as the PWP dissipates with time during and after the installation process, the soil strength and stiffness increase, and the shear strength of the soil recover to an even higher value in magnitude than the initial value before piling (referred to as "side shear set-up" and "set-up effect"), which results in the change of pile performance with time (Cooke et al., 1979; Konrad and Roy, 1987; Whittle and Sutabutr, 1999; Svinkin and Skov, 2000; Bullock et al., 2005; Yan and Yuen, 2010; Basu et al., 2014). When such behavior of displacement piles is predicted more accurately, the design and construction of these piles will be more economically effective and safe.

The installation effect on the capacity of a pile in clay ground is currently analyzed by several methods. The total stress approach and the effective stress approach are the simplest methods, and they are commonly used in the current design. For these methods, pile resistance is calculated by an equation related to initial undrained shear strength s_u (Tomlinson, 1957), radial effective stress (Chow, 1997) or initial in-situ vertical effective stress σ'_{vo} (API-RP2A 1969) or a combination of both s_u and σ'_{vo}

DOI: 10.1201/9781003215226-10

(Ladd et al., 1977; Karlsrud et al., 2005). Since the fact that soil stiffness and soil strength change with time during and after the installation process, these calculated methods cannot take into account these changes. Another method for pile capacity prediction is based on cone penetration test (CPT)/piezocone penetration tests (CPTu). By this method, the pile shaft resistance is calculated by semi-empirical equations linked to the measurement results of cone tip resistance, PWP, and sleeve friction through a reduction factor (Nottingham, 1975; Almeida et al., 1996; Li et al., 2020). The reduction factor depends on pile shape, pile material, cone type, embedment ratio (Nottingham, 1975; Bustamante and Giaeselli, 1982), or measured cone shaft resistance (Clarke et al., 1993), or cone tip resistance and initial effective overburden stress (Almeida et al., 1996), or using an empirical value based on data from case histories (Eslami and Fellenius, 1997). Li et al. (2020) indicated that CPTu measurements could be a feasible approach to the estimation of the time-dependent bearing performance of jacked piles. In general, CPT/CPTu method has indicated as a reasonable approach to predict the bearing capacity of displacement piles. However, it should be noted that the empirical factors have been usually derived for particular regions (particular soil).

In recent times, a finite element analysis (FEA) is seen as an advanced and promising approach to model the pile installation effects. Some special code programs, such as Solid Nonlinear Analysis Code (SNAC) and Material Point Method (MPM), are developed to simulate the jacking and loading behavior of a pile; and the results showed reasonable agreements with measured results if appropriate soil parameters and constitutive model were selected (Basu et al., 2014; Lorenzo et al., 2018; Phuong, 2019). One disadvantage of these methods is that such programs are relatively complicated and uncommon. Therefore it is still difficult for designers to use these programs practically and straight-forwardly. Some commercially available programs, such as PLAXIS and ABAQUS, are employed in several studies to model the behavior of jacked-in piles in clay ground in recent years (Engin, 2013; Engin et al., 2015; Lim et al., 2018; Khanmoham-madi and Fakharian, 2019). In these studies, the pile installation process in clay ground was simulated under fully undrained conditions with a hypoplastic soil model (Engin, 2013; Engin et al., 2015) or Modified Cam-Clay model (Lim et al., 2018; Khan-mohammadi and Fakharian, 2019). The above-mentioned studies paid attention to simulation techniques, parametric studies, or comparison with other calculation methods, however, a very limited number of studies utilized experimental results or field measurement data to validate the modeling.

In this study, the behaviors of jacked-in piles in saturated clay ground were investigated through both physical modeling and numerical modeling. For the numerical modeling, a standard available commercial package PLAXIS 3D was employed. The main aim of this study is to simulate the behavior of jacked-in piles using available software with simple techniques, to achieve practical and simple designs. The piles were investigated not only in isolated conditions but also in group conditions, and the results were compared with the corresponding experimental results. The focuses are on: i) load-settlement behavior of a single pile with and without considering the ground consolidation caused by the installation process; ii) load-settlement behavior of jacked piles in group conditions; iii) axial forces distributing along the pile. Furthermore, one more attempt for the modeling approach in this study is that almost the input soil parameters for the soil constitutive model were obtained directly from laboratory soil tests.

2 EXPERIMENTAL DESCRIPTION

2.1 Model grounds

Clay ground was prepared in a cylindrical chamber with a height of 420 mm and a diameter of 420 mm. The soil used for the model ground was a mixture of Kasaoka clay and Silica sand #6.

The model ground was prepared as follows: Firstly, Silica sand #3 was saturated and compacted in the chamber for a bottom drainage layer with a height of 50 mm, this drainage layer could be regarded as a rigid layer. Secondly, dry Kasaoka clay powder and Silica sand #6 were mixed at a mass ratio of 1:1 (K50S50) in a rectangular basin. Water was then added to the mixed soil to obtain a soil slurry with a water content of 1.3 LL (LL: liquid limit). This soil slurry was poured into the soil chamber to an initial thickness of 370 mm. The soil was left to consolidate under its self-weight for two days. After that, another Silica sand layer was placed on the clay to provide the top drainage layer, as shown in Figure 1. Next, a rigid circular loading plate was placed on the top drainage layer, and the vertical load on the loading plate was increased to consolidate the soil one-dimensionally in several steps up to vertical stress of 100 kPa. Each load step was maintained until the degree of consolidation reached 90% following Terzaghi's one-dimensional consolidation theory. The final load step was kept for one more week to reach a higher degree of consolidation. Finally, the consolidation pressure was removed and the ground was allowed for the swelling process in 10 days. After the swelling process, the thickness of the clay layer was 297 mm.

T-bar tests, cone penetration tests (CPTs), and unconfined compression tests (UCTs) were carried out immediately after completion of the load test on the piles to obtain properties of the model ground and to confirm the consistency between model grounds. The more details of the ground preparation, T-bar tests, CPTs, and UCTs were described in Hoang and

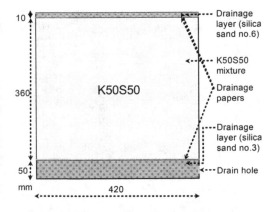

Figure 1. Preparation of model ground: longitudinal cross-sectional view before applying consolidation pressure.

Matsumoto (2020). A series of laboratory soil tests such as oedometer test, consolidated undrained (CU) triaxial compression test, Atterberg limits, density of soil particles were also carried out to obtain ground properties. Table 1 shows the properties of the model grounds. It is noted that effective cohesion c', tensile strength σ_t, and Poisson's ratio for unloading/reloading ν_{ur} are not obtained from the above-mentioned soil tests (they are estimated/assumed values). However, they are also shown in the table because they are later used for the numerical analyses, together with the parameters obtained from the laboratory soil tests.

2.2 Model foundations

Model piles used in the experiments were ABS (Acrylonitrile Butadiene Styrene) solid bars with a diameter D of 10 mm and a length L of 150 mm, as shown in Figure 2(a). Young's modulus E_p and Poisson's ratio v of the model piles are 2920 N/mm^2 and 0.406, respectively. To measure axial forces along each pile, strain gages were attached on the pile shaft at different levels as shown in Figure 2(b). Model raft was a square aluminum plate with a thickness of 12 mm and a width B of 125 mm. The raft was regarded as a rigid raft.

In the experiments of single piles (SPs), the piles with 4 levels of strain gages were used. In the experiments of three group piles (PGs) with 4, 9, or 16 piles, the piles were arranged as shown in Figure 2 (c)-(e).

2.3 Test procedure

For the load test of SPs: Two SPs were jacked into the ground one by one with a center-to-center pile spacing of $20D$ (D: pile diameter) until the pile tip reached 140 mm below the ground surface. Static load test (SLT) was then conducted on one SP immediately and on another SP 24 hrs after the installation process to investigate the consolidation effect on the pile capacity. The rest period of 24 hrs is the necessary duration for the PWP generated during the installation process to dissipate (according to Hoang and Matsumoto, 2020).

For the load test of PGs: The load tests on three PGs with 4, 9, and 16 piles respectively and the same center-to-center pile spacing of $3D$ were carried out. In each PG, the piles were jacked into the ground one by one, after the pile installations, the raft was placed on the pile heads with a gap between the raft base and the ground surface of around 5 mm. The SLTs of PGs were then conducted 24 hrs after the completion of pile installation.

Figure 3 shows the set-up of an experiment during the pile installation process and the SLT of a PG.

3 NUMERICAL SIMULATION

3.1 Soil constitutive model

Among available soil models in PLAXIS 3D (V20-CONNECTION), the Soft Soil Creep (SSC) model is the suitable model to describe the behavior of overconsolidated clayey soil considering the time effect. The detail of the SSC model is described in the material models manual of PLAXIS (PLAXIS 3D, 2018). To validate the soil model as well as to determine appropriate input soil parameters, simulation of CU triaxial test was conducted first. Almost

Table 1. Properties of model ground.

Soil parameter	Value
Density of soil particle ρ_s	2.653 Mg/m^3/g/cm^3
Plastic limit PL	13.6 %
Liquid limit LL	33.9 %
Plastic index PI	20.3 %
Compression index C_c	0.291
Swelling index C_s	0.055
Secondary compression index (creep index)C_α	0.00125
Effective cohesion c'	0.005 N/mm^2/5 kPa (estimated from simulations of CU test)
Friction angleϕ'	34.8 degrees
Tensile strength σ_t	0 (defaut)
Poisson's ratio for unloading/reloading ν_{ur}	0.17 (assumed)
Pre Overburden pressure POP	0.1 N/mm^2/100 kPa
Initial void ratio e_0	0.703
Permeability k	0.00038 mm/min
Change of permeability c_k	0.425
Unsaturated unit weight γ_{unsat}	0. 000019 N/mm^3
Saturated unit weight γ_{sat}	0. 000019 7 N/mm^3
Unit weight of water γ_{water}	0.00001 N/mm^3

* After completion of ground preparation.

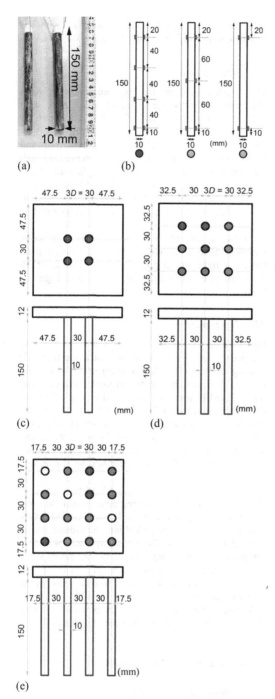

(a)

(b)

(c)

(d)

(e)

Figure 2. Model piles, raft, and pile foundation: (a) model piles; (b) locations of strain gages; (c) dimensions and arrangement of 4-pile pile group 4P-PG; (d) dimensions and arrangement of 9-pile pile group 9P-PG; (c) dimensions and arrangement o16-pile pile group 16P-PG.

all the input soil parameters were obtained from laboratory element tests, except for c', σ_t, and ν_{ur}

(a)

(b)

Figure 3. Set-up of an experiment during: (a) pile installation process; (b) SLT of a PG.

which were assumed/estimated values or set as default values. Table 1 shows the input soil parameters of the SSC model for the simulation of CU triaxial test $(C_c; C_s; C_\alpha; c'; \phi'; \sigma_t; \nu_{ur}; POP; e_0; k; c_k; \gamma_{unsat}; \gamma_{sat}; \gamma_{water})$. All geometry dimensions, drainage condition, consolidation time, shearing rate for the simulation followed the CU test in the laboratory. The analysis results of the CU triaxial test are shown and compared with measured results in Figures 4 and 5.

It is seen from Figure 4 that, in the consolidation stage, the trend of changes in volume strain was simulated quite well. Compared to the measured result, the changes in volume strain were a little higher at the initial consolidation process of each step. At the end of the consolidation stage, the

Triaxial test of K50S50_consolidation stage

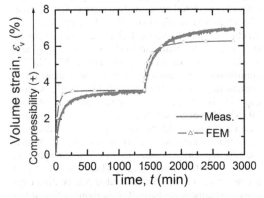

Figure 4. Change of volume strain during consolidation stage of CU triaxial test.

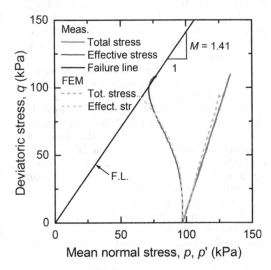

Figure 5. Normal stress versus deviatoric stress during axial compression stage of CU test.

calculated volume strain reached about 90% of the measured result.

Figure 5 shows the comparisons of total stress paths and effective stress paths between simulated and measured results in the undrained shearing stage. The relationship between effective mean normal stress p' and deviatoric stress q was simulated well from the start of the shearing stage until the failure state was nearly reached.

In general, the SSC model with the input parameters basically obtained from laboratory element tests can reasonably simulate the CU triaxial test. Therefore, in the next step, the SSC model and the same soil parameters are employed to simulate pile load tests.

3.2 Numerical setup

The FEM program PLAXIS 3D V20 CONNECTION implements a fully automatic generation of finite element meshes. The basic soil elements are the 10-node tetrahedral elements. For beam elements, 3-node line elements are used. Figure 6 shows the mesh for the calculations.

For the piles, the pile body is considered as linear elastic non-porous material (L.E.). A hybrid model of which beam elements surrounded by solid volume elements was employed to model the piles, according to Kimura and Zhang (2000). The main advantage of the hybrid pile modeling is that it is easier to obtain axial forces and bending moments along piles. In this research, the beam element of the hybrid pile model carried 90% of the bending stiffness EI and axial stiffness EA of the pile. The stiffness of the surrounding solid volume elements of the hybrid pile was reduced to 10% of the actual value, however, it was still much higher than the stiffness of the soil surrounding the pile. Table 2 shows the hybrid-pile model parameters.

Interface elements were set between the pile and the surrounding soil to model the pile-soil interaction. The properties of the interface elements are described by the strength properties of the surrounding soil with the application of an interface reduction factor R_{inter}. To determine an appropriate magnitude of R_{inter}, a series of simple experiments on piles slipping on the air-dry K50S50 ground was conducted; and the results show that the piles started to slip on the ground at an average incline angle of $31.2°(= 0.9$ times of $34.8°$ of internal effective friction angle of K50S50 ground). The value $R_{\text{inter}} = 0.9$, therefore, was used for the analyses. It is noted that in the model test, the ground soil is saturated. However, R_{inter} was determined from the tests on dry ground because R_{inter} is defined in terms of effective stress in PLAXIS. Hence, the slip test on the dry ground is reasonable.

The geometry followed experiments, however, in numerical modeling, only 1/4 of the physical modeling was simulated owing to symmetric conditions. A groundwater table was set on the ground surface which is the same as the experiments. Figure 6 shows the geometry of the analyses.

In this paper, the pile installation effect was modeled by the cavity expansion method. This

Table 2. Parameters of the elastic elements.

Description	Beam	Solid pile	Raft
Material model		L.E.	L.E.
Unit weight γ (N/mm^3)	10.52×10^{-6}	1.169×10^{-6}	78.8×10^{-6}
Young's modulus, E (N/mm^2)	2628	292	200×10^3
Poisson's ratio ν	-	0.406	0.3

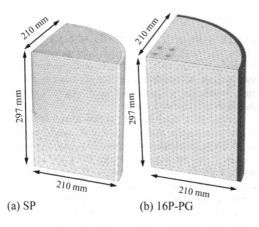

(a) SP   (b) 16P-PG

Figure 6. Mesh and geometry: (a) modeling of a single pile; (b) modeling of a pile group 16P-PG.

method has been previously used with PLAXIS by Broere and van Tol (2006) to model the bearing capacity of displacement piles in sand. The effect of the pile installation process on the ground stresses was simulated simply by prescribing volumetric strains on volume elements in the area representing the pile (see Broere and van Tol (2006) for more details of the simulation method). Several simulations with different values of volume strains were conducted first on SPs to select an appropriate strain value. This value is then used to simulate PGs. One another simulation scheme in this study is combining cylindrical cavity expansion with an amount of vertical displacement of $0.1D$ at the same time. The aim of using vertical displacement is to increase pile tip resistance due to pre-vertical pressure.

The analyses of pile load tests include: (i) SLT of SP was conducted immediately after installation, using volume expansion alone; (ii) SLT of SP was conducted immediately after installation, using volume expansion combine with vertical displacement; (iii) SLT of SP was conducted 24 hrs after installation, using volume expansion alone; (iv) SLT of SP was conducted 24 hrs after installation, using the combination of volume expansion with vertical displacement; (v) SLT of PGs was conducted 24 hrs after installation, using both volume expansion alone and combination of volume expansion with vertical displacement.

4 RESULTS AND DISCUSSIONS

4.1 Load-settlement behavior of single piles in SLT conducted immediately after installation

Figure 7 shows the load-settlement behavior during SLT of SP, of which SLT was conducted immediately after pile installation. Both measured and calculated results are shown in the figure.

Looking at Figure 7(a) first, the pile installation effect is simulated by the volume expansion scheme. The measured result shows that the pile head load reached a peak of about 125 N at a settlement of 0.75 mm. After that pile load reduced slightly and reached a residual capacity of around 121 N. The simulation result shows that when no expansion is applied ($\varepsilon_{xx} = \varepsilon_{yy} = 0\%$), the calculated pile bearing capacity is far below the measured result (about 70 N). However, when a small value of volume strain $\varepsilon_{xx} = \varepsilon_{yy} = 2.5\%$ was applied, the calculated pile capacity increases effectively to about 97 N, and the pile capacity increases to 104 N when applying $\varepsilon_{xx} = \varepsilon_{yy} = 5.0\%$. Interestingly, when strains (ε_{xx}, ε_{yy}) are larger than 5.0%, the calculated pile bearing capacity was almost unchanged. It is thought that at the initial state just after volume expansion (the consolidation has not taken place), a small amount of lateral strain (ε_{xx}, $\varepsilon_{yy} > 5.0\%$) is enough to fully mobilize the shear strength of soil at the zone surrounding the pile shaft. Therefore, the calculated pile bearing capacity (of which SLT is carried out immediately after expansion) does not increase when applying larger lateral strain.

Figure 7(b) shows the results for the case of which the combination between volume expansion and vertical displacement of $0.1D$ was used for simulating the pile installation effect. Compared to Figure 7(a), at the same lateral strain, the pile capacity was almost unchanged when the vertical displacement is added, however, the initial stiffness of the load-settlement curve increased slightly. In both schemes shown in Figure 7, the calculated results were still smaller than the measured one.

In general, when simulating the installation effect of piles in saturated clay ground without consolidation after installation, a small amount of volume strain could increase pile resistance effectively and pile resistance becomes much closer to the measured result, in comparison with no expansion case.

4.2 Load-settlement behavior of single piles in SLT conducted 24 hrs after installation

Figure 8 shows the load-settlement behavior during SLT of SP, of which SLT was conducted 24 hrs after pile installation. The period of 24 hrs is necessary for the PWP generated during the installation process to completely dissipate (according to Hoang and Masumoto, 2020). Both measured and calculated results are shown in the figure for comparison.

The measured result in Figure 8 shows that the pile head load reached a peak of about 170 N at a settlement of 0.75 mm. After that pile load reduced slightly and reached a residual capacity of 150 N. Compared with the pile resistance at the time immediately after installation (Figure 7), the pile resistance after consolidation significantly increases by 25% (from 121 N to 150 N). The pile resistance increased because the PWP dissipated with time, the strength and stiffness of soil increased, and the shear

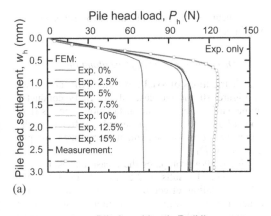

(a)

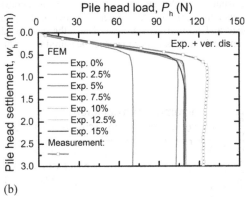

(b)

Figure 7. Load-settlement behavior of SP immediately after pile installation: (a) volume expansion alone; (b) combination of volume expansion with vertical displacement.

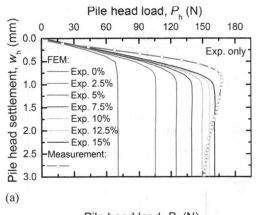

(a)

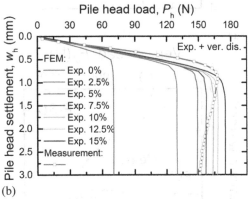

(b)

Figure 8. Load-settlement behavior of SP 24 hrs after pile installation: (a) volume expansion alone; (b) combination of volume expansion with vertical displacement.

strength of the soil recovered. This is commonly known as the set-up effect.

Figure 8(a) shows the results of SLTs of SP when the installation effect was simulated by volume expansion alone. Obviously, there is no change in pile resistance in case of no expansion. For the cases where the volume expansion was applied, the pile resistance increased significantly after the consolidation process (compared with Figure 7(a)), and the pile resistance increases as the magnitude of volume expansion increases. The value $\varepsilon_{xx} = \varepsilon_{yy} = 12.5\%$ gives a good agreement between measured and calculated results at residual state.

Figure 8(b) shows the results of SLTs of SP when the installation effect is simulated by combining volume expansion (ε_{xx}, ε_{yy}) with vertical displacement (u_z). When vertical displacement was added, the pile resistance increased significantly. The combination of $\varepsilon_{xx} = \varepsilon_{yy} = 7.5\%$ and $u_z = 0.1D$ gives a good agreement with measured result. The initial stiffness of load-settlement behavior also increases when u_z is added.

In general, both the experimental and numerical results indicated that the pile resistance increases significantly due to the soil consolidation caused by the pile installation. The addition of vertical displacement has a clear influence on the cases of simulation of SLTs after the consolidation, although it shows negligible influence on the cases of SLTs immediately after the pile installation.

4.3 Load-settlement behavior of pile groups

Figure 9 shows the load-settlement behavior during SLT of three PGs, of which SLTs were conducted 24 hrs after the completion of pile installation. In each figure, the measured result and the calculated results with both schemes are shown.

The results of all three PGs show that the initial stiffnesses are simulated quite well, both schemes give quite similar results. Regarding the magnitude of the group resistance, the resistance of 4P-PG is calculated reasonably. However, when the number of piles in the group increased, the calculated resistance was under the estimated one. This phenomenon may be explained by the differences in measured results and calculated results of load distributions between the pile tip and pile shaft. Further study is needed for this aspect. It is seen from Figure 9(a)-(c) that the differences between

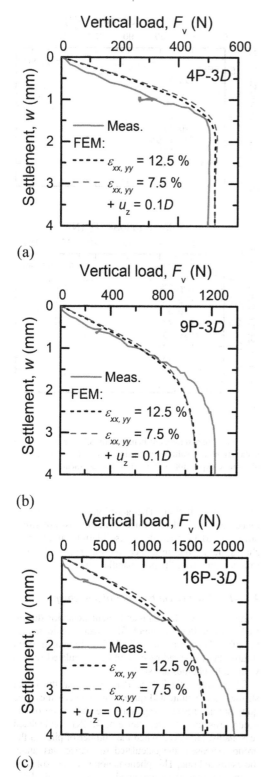

(a)

(b)

(c)

Figure 9. Load-settlement behavior of three PGs: (a) 4P-PG; (b) 9P-PG; (c) 16P-PG.

the calculated results of the two schemes are negligible.

4.4 Distributions of axial force along piles

The distributions of axial force along SPs are shown in Figure 10 for the case of the SLTs conducted immediately after the pile installation. The distributions of axial forces are presented at different settlements (in form of settlement normalized by pile diameter w/D).

Figure 10(a) compares the measured results with the calculated results in the case of simulating the pile installation effect by using only the volume expansion. It is seen from the figure that the shaft resistances of the SP are simulated well at depths deeper than 60 mm. At the depths between the pile head and 60 mm from the pile head, the calculated shaft resistance is larger than the measured one. The

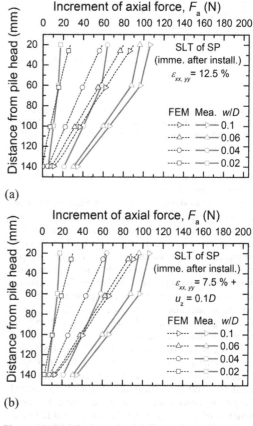

(a)

(b)

Figure 10. Distribution of axial force along piles when SLTs are conducted immediately after pile installation: (a) pile installation effect is simulated by volume expansion method; (b) pile installation effect is simulated by a combination of volume expansion method with vertical displacement.

calculated tip resistance is noticeably smaller than the measured values.

Figure 10(b) compares the measured results with the calculated results in the case of simulating the pile installation effect by using the combination of volume expansion with vertical displacement. It is found that the results in Figure 10(b) are very similar to the results in Figure 10(a). The addition of vertical displacement seems to have a slight effect on pile resistance as well as distributions of axial forces along the pile.

Figure 11 shows the distributions of axial force along SPs for the case of the SLTs conducted 24 hrs after pile installation in the cases: (a) pile installation effect is simulated by volume expansion method and (b) pile installation effect is simulated by a combination of volume expansion method with vertical displacement. It is found from the figure that at the small settlement ($w/D = 0.02$), the calculated axial force is larger than the measured one.

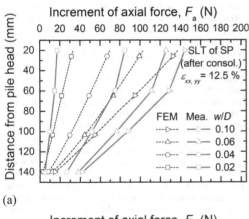

(a)

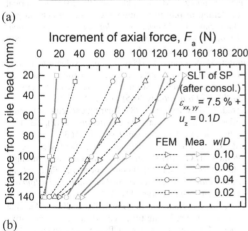

(b)

Figure 11. Distribution of axial force along piles when SLTs were conducted 24 hrs after pile installation: (a) pile installation effect is simulated by volume expansion method; (b) pile installation effect is simulated by a combination of volume expansion method with vertical displacement.

The results also show that shaft resistances are simulated well at depths deeper than 60 mm meanwhile the pile shaft resistances were overestimated at the top sections. The calculated tip resistance is noticeably smaller than the measured one. The tip resistance in the case of addition of vertical displacement (Figure 11(b)) is slightly higher, compared to the case of cylindrical cavity expansion alone (Figure 11(a)).

Comparing the measured axial forces along piles obtained from SLTs conducted before (Figure 10) and after (Figure 11) consolidation, the increment of pile resistance is mainly due to the increment of pile shaft resistance. The increment of tip resistance is minor.

The calculated results of two schemes show that the addition of vertical displacement has influences on pile shaft resistance rather than on pile tip resistance (as the amount of lateral strain reduces from 12.5 % to 7.5 %, however, the tip resistance does not increase much). Based on the measured results, several previous studies also pointed out that the installation process of a pile in saturated clay ground affects pile shaft resistance rather than pile tip resistance (Attwooll et al., 1999; Bullock et al., 2005). Therefore the reason for the underestimation of tip resistance and overestimation of shaft resistance near the pile top might be the use of a constant effective cohesion c' in analyses. If c' increases with depth, the more reasonable analysis results may be obtained.

5 CONCLUSION

This paper investigates behaviors of pressed-in piles in saturated clay ground under isolated and group conditions via both experimental and numerical methods. The following remarks are derived from the study:

The pile shaft resistance increased significantly after the soil consolidation, in comparison with the pile shaft resistance obtained from SLT conducted immediately after pile installation without consolidation.

Simulating the pile installation effect by the volume expansion method: When conducting SLT immediately after pile installation, a small value of lateral strain (reference value in this study: $\varepsilon_{xx} = \varepsilon_{yy} \geq 5.0$ %) is enough to fully mobilize the pile shaft resistance, and the pile resistance does not increase as the lateral strains increase. When conducting SLT after the consolidation process, the pile resistance increases as the lateral strains increase, and $\varepsilon_{xx} = \varepsilon_{yy} \approx 12.5$ % are appropriate strains for the case in this study.

Simulating pile installation effect by combination of volume expansion with vertical displacement: When conducting SLT immediately after pile installation, the addition of vertical displacement makes a slight increase in initial stiffness, however, the residual resistance of pile is almost unchanged, compared with the case of applying volume expansion alone. When conducting SLT after the consolidation process, the pile resistance increases as the lateral strains increase and $\varepsilon_{xx} = \varepsilon_{yy} \approx 7.5$ % are appropriate strains for the case in this study. The addition of vertical displacement in the simulation of the pile installation effect increases pile resistance noticeably, in comparison with applying volume expansion purely. Interestingly, a large amount of the increment of pile resistance caused by the addition of vertical displacement is the increment of pile shaft resistance, not pile tip resistance. Other schemes should be done to increase the resistance of the pile tip.

In general, the installation effect could be simply simulated by applying volume strains before SLTs, although there are some differences between calculated and measured results. The pile set-up effects are also simulated reasonably by this method.

REFERENCES

Almeida, M.S.S., Danziger, F.A.B. & Lunne, T. 1996. Use of the piezocone test to predict the axial capacity of driven and jacked piles in clay. *Canadian Geotechnical Journal* 33(1),23–41. https://doi.org/10.1139/t96-022.

API (1969) Recommended practice for planning, designing, and constructing fixed offshore platforms, *API RP2A*, 1st edn. American Petroleum Institute, Washington

Attwooll, W.J., Holloway, D.M., Rollins, K.M., Esrig, M.I., Sakhai, S. & Hemenway, D. 1999. Measured pile setup during load testing and production piling: I-15 corridor reconstruction project in Salt lake city, Utah. Transportation research record, *Journal of the Transportation Research* 1663(1). https://doi.org/10.3141/1663-01

Basu, P., Prezzi, M., Salgado, R. & Chakraborty, T. 2014. Shaft resistance and setup factors for piles jacked in clay. *Journal of Geotechnical and Geoenvironmental Engineering* 140(3), 04013026. https://doi.org/10.1061/(ASCE)GT.1943-5606.0001018

Bozozuk, M., Fellenius, B.H. & Samson, L. 1978. Soil disturbance from pile driving in sensitive clay. *Canadian geotechnical journal* 15(3),346–361. https://doi.org/10.1139/t78-032

Broere, W. & van Tol, A. F. 2006. Modelling the bearing capacity of displacement piles in sand. *In Proc. of the Institution of Civil Engineers, Geotechnical Engineering* 159 (ICE, GE3), 1–13.

Bullock, P.J., Schmertmann, J.H., McVay, M.C. & Townsend, F.C. 2005. Side shear setup. I: Test piles driven in Florida. *Journal of Geotechnical and Geoenvironmental Engineering* 131(3),292–300. https://doi.org/10.1061/(ASCE)1090-0241(2005)131: 3

Bustamante, M. & Gianeselli, L. 1982. Pile bearing capacity prediction by means of static penetrometer CPT. *In Proc. of the 2nd European symposium on penetration testing*, Amsterdam, 493–500

Chow, F.C. 1997. Investigations into the behavior of displacement piles for offshore foundations. *PhD Thesis , Imperial College London, United Kingdom.*

Cooke, R.W., Price, G. & Tarr, K. 1979. Jacked piles in London Clay: a study of load transfer and settlement under working conditions. *Geotechnique* 29 (2), 113–147. https://doi.org/10.1680/geot.1979.29.2.113

Clarke, J., Long, M.M. & Hamilton, J. 1993. The axial tension test of an instrumented pile in overconsolidated clay at Tilbrook Grange. *In book: Large-scale pile tests in clay. Thomas Telford*, London, 362–380

Deeks, A., White, D. & Bolton, M. 2005. A comparison of jacked, driven and bored piles in sand. *In Proc. of the 16th International Conference on Soil Mechanics and Geotechnical Engineering* 4, pp. 2103–2106.

Dijkstra, J. 2009. On the modeling of pile installation. *Phd thesis of Technical University Delft,Netherlands.*

Eslami, A. & Fellenius, B.H. 1997. Pile capacity by direct CPT and CPTu methods applied to 102 case histories. *Can Geotech J* 34(6),886–904. https://doi.org/10.1139/t97-056

Engin, H.K. 2013. Modelling Pile Installation Effects – A Numerical Approach. *Ph. D. thesis, Geo-Engineering Section Delft University of Technology, The Netherlands.* DOI: 10.4233/uuid:3e8cc9e2-b70c-403a-b800-f68d65e6ea85

Engin, H.K., Brinkgreve, R.B.J. & van Tol, A.F. 2015. Simplified numerical modelling of pile penetration – the Press-Replace technique. *International Journal for Numerical and Analytical Methods in Geomechanics* 39 (15),1713–1734. https://doi.org/10.1002/nag.2376

Flaate, K. 1971. Effects of Pile Driving in Clays. *Canadian Geotechnical Journal* 9(1), pp. 81–88. https://doi.org/10.1139/t72-006

Hoang, L.T. & Matsumoto, T. 2020. Long-term behavior of piled raft foundation models supported by jacked-in piles on saturated clay. *Soils and Foundations* 60(1),198–217. https://doi.org/10.1016/j.sandf.2020.02.005

Karlsrud, K., Clausen, C.J.F. & Aas, P.M. 2005. Bearing capacity of driven piles in clay, the NGI approach. *In: Proc. of frontiers in offshore geotechnics: ISFOG, Perth*, pp 775–782. https://doi.org/10.1201/NOE0415390637.ch88

Khanmohammadi, M. & Fakharian, K. 2019. Numerical modelling of pile installation and set-up effects on pile shaft capacity. *International Journal of Geotechnical Engineering* 13(5),484–498. https://doi.org/10.1080/19386362.2017.1368185

Kimura, M. & Zhang, F. 2000. Seismic evaluations of pile foundations with three different methods based on three-dimensional elasto-plastic finite element analysis. *Soils and Foundations* 40(5),113–132.

Konrad, J. M. & Roy, M. 1987. Bearing capacity of friction piles in marine clay. *Geotechnique* 37(2),163–175. https://doi.org/10.1680/geot.1987.37.2.163

Ladd, C.C., Foott, R., Ishihara, K., Schlosser, F. & Poulos, H.G. 1977. Stress-deformation and strength characteristics, state-of the-art report. *In: Proc. of the International conference on soil mechanics and foundation engineering*, Tokyo, 421–494.

Li, L., Li, J., Sun, D. & Gong, W. 2020. A feasible approach to predicting time-dependent bearing performance of jacked piles from CPTu measurements. *Acta Geotechnica* 15, 1935–1952. https://doi.org/10.1007/s11440-019-00875-x

Lim, Y.X., Tan, S.A. & Phoon, K.K. 2018. Numerical study of pile setup for displacement piles in cohesive soils. *In: Cardoso et al. (Eds) Numerical Methods in Geotechnical Engineering IX*, Volume 1, 451–456, Taylor & Francis

Group, London. https://doi.org/10.1201/9780429446931 (ISBN 978-1-138-33198-3, viewing online on google book, no hard coppy or pdf version)

Lorenzo, R., Cunha, R.P., Cordao-Neto, M.P. & Naim, J.A. 2018. Numerical simulation of installation of jacked piles in sand using material point method. *Canadian Geotechnical Journal* 55(1),131–146. https://doi.org/10.1139/cgj-2016-0455

Nottingham, L.C. 1975. Use of quasi-static friction cone penetrometer data to estimate capacity of displacement piles. *PhD thesis of Department of Civil Engineering, University of Florida*, Gainesville, America.

Phuong, N.T.V. 2019. Numerical Modelling of Pile Installation Using Material Point Method. *PhD thesis, Delft University of Technology*, https://doi.org/10.4233/uuid:5580c747-d8a1-464c-8db8-e16eb2a499f7

PLAXIS 3D 2018 - Material Models Manual, PLAXIS, https://www.plaxis.com/support/manuals/plaxis-3d-manuals/

Svinkin, M. R. & Skov, R. 2000. Set-up effect of cohesive soils in pile capacity. *Proc., 6th Int. Conf. Application of Stress Wave Theory to Piles, S. Niyama and J. Beim, eds.*, Balkema, Rotterdam, Netherlands, 107–111.

Tomlinson, M.J. 1957. The adhesion of piles driven in clay soils. *In: Proc. of the 4th international conference on soil mechanics and foundation engineering*, London.

Whittle, A.J. & Sutabutr, T. 1999. Prediction of pile setup in clay. *Journal of the Transportation Research Board* 1663(1),33–40. https://doi.org/10.3141/1663-05

Yan, W.M. & Yuen, K.V. 2010. Prediction of pile set-up in clays and sands. *IOP conf. series: Materials Science and Engineering* 10. https://doi.org/10.1088/1757-899X/10/1/012104

Proceedings of the Second International Conference on
Press-in Engineering 2021, Kochi, Japan – Matsumoto et al (eds)
© *2021 Taylor & Francis Group, London, ISBN 978-1-032-10414-0*

Proposal of vertical design bearing capacity estimation formula of Gyropress method based on Japanese railway standard

T. Ozaki & T. Sanagawa
Railway Technical Research Institute, Tokyo, Japan

Y. Kimura & N. Suzuki
Japan Press-in Association, Tokyo, Japan

ABSTRACT: In recent years, Gyropress method (Self-Walking Rotary Press-in Method for Tubular Piles with Tip Bit) has become popular as one of the methods for installation of piles. In the previous Japanese railway design standard, the limit state design method have been used. However, in 2012, it was revised to adopt the concept of the performance-based design method. In this paper, the authors propose a design vertical bearing capacity estimation formula and a partial resistance factor for Gyropress method based on the concept of the performance-based design method.

1 INTRODUCTION

1.1 *Gyropress method*

Damage to social infrastructure has occurred in various parts of Japan due to the torrential rains caused by the recent abnormal weather. In addition, an occurance of large-scale earthquake is also predicted in the near future. Therefore, it is also necessary to implement countermeasures against large-scale earthquakes and torrential rains.

In recent years, Gyropress method (Self-Walking Rotary Press-in Method for Tubular Piles with Tip Bit) has become widespread as one of the construction methods for steel pipe piles. Gyropress method is one of pile installation methods, a steel pipe pile with a ring bit consisting of a steel ring and a bit for cutting hard ground is pressed-in by rotary cutting at the tip (Figure 1). Gyropress method is expected to be applied, especially in places where it is difficult to construct structures, such as hard grounds and restrictions on the use of upper space. Moreover, it has been used for seismic reinforcement and scours measures for existing railway bridge piers (Figures 2-3).

1.2 *Current design vertical bearing capacity of usual construction method and Gyropress method*

In the Japanese design standard of railway foundation structures, the design bearing capacity of piles R_{vd} is expressed as follows Eqs.(1)-(3).

$$R_{vd} = f_{rf} R_f^k + f_{rt} R_t^k \qquad (1)$$

$$R_f^k = U \sum r_{fi}^k l_i \qquad (2)$$

$$R_t^k = q_t^k A_t \qquad (3)$$

where R_{vd} = total design vertical bearing capacity; R_f^k = characteristic value of the ultimate bearing capacity of shaft friction; R_t^k = characteristic value of the ultimate bearing capacity of the tip resistance; f_{rf} = partial resistance factor of the shaft friction; f_{rt} = partial resistance factor of the tip resistance; U = circumferential length of the pile; r_{fi}^k = characteristic value of the intensity of the ultimate shaft friction at i th layer; l_i = thickness of i th layer; q_t^k = characteristic value of the intensity of the ultimate bearing capacity of the tip resistance; and A_t = the area of the pile tip.

In this paper, the value with the superscript notation k denotes the characteristic values, the subscript f denotes shaft friction resistance, and the subscript t means tip resistance.

The partial resistance factor of the shaft friction f_{rf} and that of the tip resistance f_{rt} are distinct, as shown in Eq. (1). Hence the design vertical bearing capacity can be takes into consideration the effects of the ratio of the tip resistance to the shaft friction. The characteristic values r_{fi}^k and q_t^k shown in Eqs. (2)-(3) are defined as the unit ultimate resistance when the settlement of the focused part of the pile (pile at i th layer for r_f^k, pile tip for q_t^k) reaches 10% of the

DOI: 10.1201/9781003215226-11

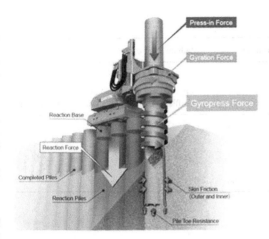

Figure 1. Mechanism of Gyropress method (GIKEN LTD. 2015).

Figure 2. Overview state of Gyropress method (Kimura et al. 2019).

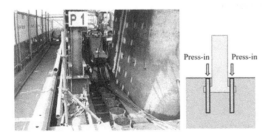

Figure 3. Detail state of Gyropress method (Kimura et al. 2019).

pile diameter. This definition follows the standard for the pile loading tests of Japanese Geotechnical Society (JGS 2007).

The characteristic values r_{fi}^k, q_t^k, and the partial resistance factors f_{rf}, f_{rt} of the bearing capacity of the pile should be determined based on the results of the loading tests with a full-scale pile at a specific construction site for each project. However, in terms of cost, it is not ractial to conduct the pile loading tests for design calculation at all construction sites. For this reason, the Japanese railway design standard presents the estimation formulas of the characteristic values r_f^k, q_t^k, and the partial resistance factors f_{rf}, f_{rt} derived on the basis of the soil investigation results. In fact, almost all the structures have been designed using only the soil investigation results (N-value measured by standard penetration test is most prevalent in Japan).

On the other hand, in Gyropress method, with the estimation formulas of the characteristic value r_f^k, q_t^k, and partial resistance factors f_{rf}, f_{rt} are not presented in the Japanese railway design standard. This is because the track record of vertical loading test results by Gyropress method is insufficient. Therefore, the design vertical bearing capacity has been currently designed with a safety margin that is more sufficient.

On the basis of the above, in this paper, we propose the estimation formulas of the characteristic values r_f^k, q_t^k of Gyropress method in Chapter 3, and the partial resistance factors of that f_{rf}, f_{rt} in Chapter 4 with the derivation method.

2 OUTLINE OF VERTICAL STATIC LOADING TEST OF PILE INSTALLED BY GYROPRESS METHOD

First, the authors collect the data of the vertical loading test by Gyropress method.

We evaluate 5 cases shown in Table 1, which are a multi-stage static loading test results following the standard of JGS. The converted rooting depth in the support layer is longer than the outer diameter of steel pipe D, which is 800 to 1000 mm.

Figure 4 shows the relationship between the pile tip displacement and the load. The y-axis load is the value when the pile tip displacement reaches 10% of the pile diameter, and the displacement in Figure 4 (b) is normalized by pile diameter. For reference, the reference displacement in the Japanese railway design standard is shown in Figure 4.

The resistance of the shaft friction reaches the limit with a displacement of about 5 to 20 mm. Since the resistance of the shaft friction in case of general method reaches the limit when the pile head displacement is about 10 to 20 mm, we find the tendency of Gyropress method is the same as genral construction methods. Besides, regarding the pile head bearing capacity, a route exerts a large bearing capacity from the initial stage of displacement is drawn. The variation of between in each case is small.

Table 1. Outline of loading test data of Gyropress method.

Test name	Diameter	Thickness of steel pipe pile	Length of pile	Normalized depth into support layer	Support layer Soil clasiffication	N value	Medium layer Soil clasiffication	N value	Soil clasiffication	N value
I1	800mm	16mm	19.65m	2.5m	G	86	G	43~66	-	-
T1	800mm	16mm	17.5m	2.0m	G	36	S	3	S	9
F1	1000mm	16mm	15.0m	3.1m	S	58	S	7	S	58
A1	800mm	12mm	4.7m	0.8m	S	15	C	3	-	-
N1	1000mm	12mm	25.0m	2.4m	G	60	S	6	C	4

Soil classification G:gravel S:sand C:clay (Suzuki et al.2019)

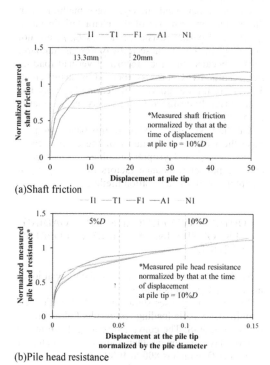

(a)Shaft friction

(b)Pile head resistance

Figure 4. Relationship between pile tip displacement and normarized bearing capacity (Suzuki et al. 2019).

3 ESTIMATION FORMULAS FOR VERTICAL SUBGRADE REACTION OF PILE INSTALLED BY GYROPRESS METHOD

3.1 Proposed formulas for coefficient of vertical subgrade reaction

3.1.1 Coefficient of vertical subgrade reaction at pile tip

The estimation formula of the coefficient of vertical subgrade reaction at the tip of the pile k_{tv} is established as follows. The closed area is used as the pile tip area when adpoting Eqs. (4)-(5) for calculating the vertical ground spring constant K_{tv} at the pile tip.

$$k_{tv} = 1.4 \rho_{gk} E_d D^{-3/4} \qquad (4)$$

$$K_{tv} = k_{tv} A_t \qquad (5)$$

where k_{tv} = coefficient of the vertical subgrade reaction at the pile tip, ρ_{gk} = geotechnical modification factor, E_d = deformation modulus of the ground, and K_{tv} = vertical ground spring constant.

Figure 5 shows the relationship between the coefficient of the vertical subgrade reaction force and the deformation modulus of the ground estimated from the N value at the pile tip in each loading test case. The measured coefficient of the vertical subgrade reaction is obtained by dividing the ground spring constant at the tip of the pile by the tip area A_t and the effect of the load duration time is corrected to $\rho_{gk} = 1.0$.

The dotted line in Figure5 shows the result by the proposed estimation formula. From Figure 5, it is seen that the proposed estimation formula can evaluate the loading test results roughly equivalent to the average value.

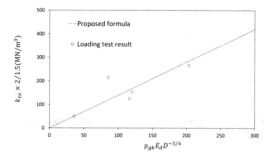

Figure 5. Relationship between coefficient of vertical subgrade reaction force at pile tip and deformation modulus of ground (Kimura et al. 2019).

3.1.2 Coefficient of vertical subgrade reaction on pile shaft

The estimation formula of the coefficient of vertical subgrade reaction of shaft friction k_{fv} is established as following Eq. (6).

$$k_{fv} = \min(0.3\rho_{gk}E_d, 6000) \qquad (6)$$

Although the upper limit of the coefficient of the vertical subgrade reaction force is not set, the upper limit is set in consideration of the loading test results in Gyropress method. It is presumed that this is because Gyropress method often targets the hard support ground and the dependence on the original strength of the ground becomes smaller due to the effect of rotary cutting during press-fitting into the hard support ground.

Figure 6 shows the relationship between the coefficient of the vertical subgrade reaction and the deformation modulus of the ground on the shaft of the pile. In Figure6, the measured coefficient of the vertical subgrade reaction force is calculated back from the loading test result, and the effect of the load duration time is increased by 1.33 times so that it is equivalent to $\rho_{gk} = 1.0$, the same as one of the tip resistance.

3.2 Proposed estimation formula for reference bearing capacity

3.2.1 Reference bearing capacity at pile tip

Figure 7 shows the relationship between the reference bearing capacity at pile tip and the N value. The reference bearing capacity at pile tip is obtained by dividing the measured value of the reference tip bearing capacity R_t of the loading test by the tip area of the closed steel pipe. The Japanese railway design standard indicates to use "the minimum N value in $1D$ above and $3D$ from the pile tip in the depth direction" as the tip N value. However, if the minimum N value is used in this study, the coefficient of

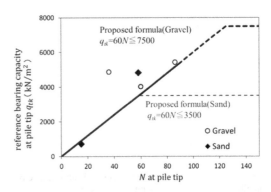

Figure 7. Relationship between reference bearing capacity at tip and N value (Kimura et al. 2019).

variation during statistical processing (described later) is large. Therefore, "average N value in $1D$ above and $3D$ from the pile tip in the depth direction "is used. However, it should be notedit is better to design on the safe side in actual operation by using the minimum N value as in other pile construction.

As for the estimation formula of the reference bearing capacity at pile tip of the pile installed by Gyropress method, we propose the following Eq. (7). The proposed formula, which is shown by a solid line, is derived from Figure 7 as a model equivalent to the lower limit value as in other pile construction methods. Since the loading test range is limited, the range confirmed by the loading test is shown by the solid line, and the dotted line shows the outside of the range.

$$q_{tk} = \begin{cases} \min(60N, 3500): \text{sand} \\ \min(60N, 7500): \text{gravel} \end{cases} \qquad (7)$$

The proposed estimation formula is equivalent to the cast-in-place pile formula presented in the Japanese railway design standard. On the other hand, in the driven pile method with a pile diameter of 800 mm or less, which is the same open-ended steel pipe pile method, $q_{tk} = 35\ (L/D)\ N$ is presented. In the case where the embedding ratio L/D in the support layer is 1 to 2, $q_{tk} = 35$ to $70\ N$, which is almost the same as the proposed estimation formula. The mechanism of the Gyropress method is considered to be closer to one of the driven pile rather than the cast-in-place pile in the process of constructing piles near the support layer.

However, for the upper limit, we apply mutatis mutandis the value of cast-in-place pile, which has the smallest estimated value of bearing capacity at the tip in the pile construction method. This value is likely to be changed by increasing the number of loading tests in the future.

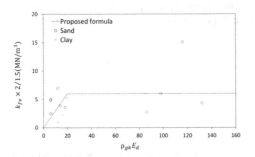

Figure 6. Relationship between coefficient of vertical subgrade reaction force on pile shaft and deformation modulus of ground (Kimura et al. 2019).

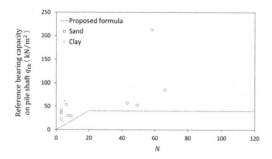

Figure 8. Relationship between reference bearing capacity on pile shaft and N value (Kimura et al. 2019).

3.2.2 Reference bearing capacity on pile shaft

Figure 8 shows the relationship between the reference bearing capacity on pile shaft obtained from the loading test and the N value. Here, since there are few data on clay, we propose the following Eq. (8) which is the same reference bearing capacity of pile shaft estimation formula for sand and clay. The proposed formula is shown by dotted lines in Figure 8.

$$r_{fk} = \min(2N, 40) \qquad (8)$$

From Figure 8, it is confirmed that the proposed estimation formula evaluates the loading test results at the lower limit, as with other pile construction methods.

Moreover, estimated reference bearing capacity including both tip resistance and shaft friction is calculated by Eqs. (1)-(3), (7)-(8) . The relationship between estimation value and measured value is shown in Figure 9. The characteristic values estimated from Figure 9 are generally smaller than the measured values, indicating that the estimation formula exists in the safe side.

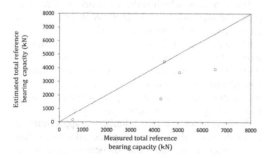

Figure 9. Relationship between measured value and estimated value of total reference bearing capacity (Kimura et al.2019).

4 CALCULATION OF PARTIAL RESISTANT FACTOR BY RELIABILITY ANALYSIS

4.1 Calculation method of partial resistance factors

4.1.1 Definition of partial resistance factor

In this section, we dfine the partial resistance factors f_{rf} and f_{rt} using the method in the Japanese railway design standard.

First, we calculate the normalized resistance force of pile (NRF), which is the measured shaft friction and the tip resistance of each limit state divided by the characteristic value of total resistance R_k ($=R_{fk}+R_{tk}$). Moreover, the database of NRF for Gyropress method for each limit state is created. In this procedure, singular values are excluded by Grubbs' test (Grubbs 1950).

The partial resistance factors f_{rt} and f_{rf} are calculated using FORM as follows on the assumption that the NRF database follows the normal distribution:

$$f_{rt} = \mu_{Xt} - \beta_a \alpha_t \sigma_{Xt} \qquad (9)$$

$$f_{rf} = \mu_{Xf} - \beta_a \alpha_f \sigma_{Xf} \qquad (10)$$

$$\alpha_t = R_t^k \, \sigma_{Xt} \, / \sqrt{\left(R_t^k \, \sigma_{Xt}\right)^2 + \left(R_f^k \, \sigma_{Xf}\right)^2} \qquad (11)$$

$$\alpha_f = R_f^k \, \sigma_{Xf} \, / \sqrt{\left(R_t^k \, \sigma_{Xt}\right)^2 + \left(R_f^k \, \sigma_{Xf}\right)^2} \qquad (12)$$

where β_a = target reliability index; α_t, α_f = sensitivity coefficients; μ_{Xt}, μ_{Xf} = mean values; σ_{Xt}, σ_{Xf} = standard deviations of NRF database.

However, the calculated mean values μ_{Xt} and μ_{Xf} are low with a confidence level of 0.75 because the number of data is small. The partial resistance factor of the shaft friction f_{rf} and that of the tip resistance f_{rt} can be calculated by Eqs. (9)-(12).

Moreover, the partial resistance factor of the total resistance f_r is defined as the function of the ratio of pile tip bearing capacity p_t ($=R_{tk}/(R_{tk}+R_{fk})$) as following Eqs. (13)-(14) from a practical perspective.

$$R_{vd} = f_r \left(R_t^k + R_f^k\right) \qquad (13)$$

$$f_r = \left(f_{rt}R_t^k + f_{rf}R_f^k\right) / \left(R_t^k + R_f^k\right) = p_t f_{rt} + (1 - p_t)f_{rf} \qquad (14)$$

However, in the case where μ_X is small and σ_X is large, some of the calculated results become very small (or sometimes become negative). In the case, the actual phenomena and the calculation result are incompatible.

Therefore, we adopt the calculation method by Sanagawa et al. in order to approximate NRF distribution more highly in the next section. In the method, NRF database follows the lognormal distribution.

4.1.2 Approximation by lognormal distribution
1) Evaluation by lognormal distribution

It is assumed that the NRF database follows the lognormal distribution as shown in Figure 10 (a). After the x-axis, which denotes the probability density function of NRF, is logarithmically transformed as shown in Figure 10 (b), the mean $\mu_{\ln X}$ and the standard deviation $\sigma_{\ln X}$ are calculated. The statistic parameters $\mu_{\ln X}$ and $\sigma_{\ln X}$ can also be calculated from the mean μ_X and the standard deviation σ_X of NRF as following Eqs. (15)-(16) (μ_X and σ_X are the statistic parameter, without logarithmic transformation).

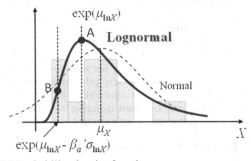

(a) Probability density function

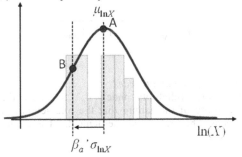

(b) Logarithmically transformed

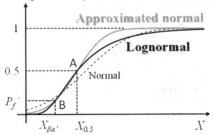

(c) Cumulative distribution function

Figure 10. Outline of approximation of distribution (Sanagawa et al. 2019).

$$\mu_{\ln X} = \ln(\mu_X) - 1/2\sigma_{\ln X}{}^2 \qquad (15)$$

$$\sigma_{\ln X} = \sqrt{\ln\left(1+(\sigma_X/\mu_X)^2\right)}$$
$$= \sqrt{\ln\left(1+V^2\right)} \qquad (16)$$

where V = design coefficient of variation.

2) Set the part for normal distribution approximation

The part of the lognormal distribution, which considerably influences on the partial resistance factors f_r is approximated by the normal distribution. The approximated part is set as the part whose cumulative probabilities is between P_f and 0.5 shown in Figure 10 (c). The random variable of NRF (X), whose cumulative probability are from 0.5 ($X_{0.5}$) to P_f' ($X_{\beta a}$') can be calculated by $\mu_{\ln X}$ and $\sigma_{\ln X}$ as following Eqs. (17)-(18).

$$X_{0.5} = \exp(\mu_{\ln X}) \qquad (17)$$

: Cumulative probability of failure is 0.5

$$X_{\beta_a'} = \exp(\mu_{\ln X} - \beta_a'\sigma_{\ln X}) \qquad (18)$$

: Cumulative probability of failure is Pf'

3) Approximation by normal distribution

The cumulative distribution function of the lognormal distribution is approximated by that of the normal distribution which passes through two points, A ($X_{0.5}$, 0.5) and B ($X_{\beta a}$', P_f') as shown in Figure 10 (c).

The mean value μ_X' and the standard deviation σ_X' of the approximated normal distribution are given as following Eqs. (19)-(20).

$$\mu_X = X_{0.5} = \exp(\mu_{\ln X}) \qquad (19)$$

$$\beta_a'\sigma_X' = X_{0.5} - X_{\beta a'}$$
$$= \exp(\mu_{\ln X}) - \exp(\mu_{\ln X} - \beta_a'\sigma_{\ln X}) \qquad (20)$$

4.1.3 Definition of reference settlement for each limit state

In order to calculate the statistic parameter of NRF for each limit state from the loading test database, it is necessary to define the reference settlement for each limit state. We determine the reference settlement so as to correspond to the limit state in the Japanese railway design standard.

Table 2 shows an example. These reference settlements are empirically determined considering the

Table 2. Example of limit state and reference settlement for each limit state for railway structures (Sanagawa et al. 2019).

Limit state	Reference settlement for each limit state s	Assumed damage condition	Action
Serviceability	20mm	Cracks on superstructure by differential settlement	Variable (Train load)
Restorability	Min. (50mm, 0.05D)	Need to restore track because of residual settlement	Accidental (Level 1 earthquake)
Ultimate	0.1D	Need to restore structure because of large settlement	Accidental (Level 2 earthquake)

limit states for maintenance (serviceability), restorability, and safety for the structures (restorability and iltimate).

4.1.4 *Subdividing of design coefficient of variation*
In in Eq. (18), the design coefficient of variation V ieffects of variations are considered as differences in the quality of the pile construction, variations when calculating the coefficients of the ground and variation due to diversion of soil investigation results. In this section, the design coefficient of variation V is divided into three factors as following Eq. (21).

$$V = \sqrt{V_1^2/n + V_2^2 + V_3^2} \qquad (21)$$

where V_1 = coefficient of variation depending on the pile construction method, V_2 = coefficient of

variation of estimation ground strength from N value, V_3 = coefficient of spatial variation of stiffness and strength of soil, and n = number of piles integrated as a group pile.

Although V_2 is originally considered to be variable, we assume V_2 as a constant value of $V_2 = 10\%$ (conversion error from N value to internal friction angle φ) with reference to (Nishioka et al. 2017) V_3 is estimated by linearly interpolating the minimum value $V_{3min} = 18\%$ to the maximum value $V_{3min} = 45\%$ according to the distance ΔL from the ground survey position with reference to the research by (Otake et al. 2014) Here, the minimum value V_{3min}, the value where $\Delta L \le 5$ m, corresponds to the design by the boring nearby the pile construction spot. On the other hand, the maximum value V_{3max}, the value where $\Delta L \ge 50$ m, corresponds to the design by boring diversion.

We assume that V when $n = 1$, $V_2 = 10\%$, and $V_3 = V_{3min}$ are assigned to Eq. (21) is the same as the coefficient of variation V_{test}. Then, V_{test} is obtained by convering into consideration of the lognormal distribution after statistically processing the loading test database. Here, $n = 1$ because the loading test is conducted with a single pile; $V_3 = V_{3min}$ because the loading test is conducted by the test pile. Moreover, V_1 is considered to be a unique value for each pile construction method corresponding to the construction quality and is calculated by the following Eq. (22).

The result of each design coefficient of variation is shown in Table 3.

$$V_1 = \sqrt{V_{test}^2/n - V_2^2 - V_{3min}^2} \qquad (22)$$

4.1.5 *Target reliability index*
The target reliability index β_a' indicates the probability of the failure that the pile settlement exceeds the reference settlement. There are

Table 3. Each coefficient of variation about Gyropress method loading test with normal distribution.

Limit state		Serviceability		Restorability		Ultimate	
Reference settlement for each limit state s		20mm		Min. (50mm, 0.05D)		10%D	
Resistant part		Tip resistance	Shaft friction	Tip resistance	Shaft friction	Tip resistance	Shaft friction
Subdivision of coefficient of variation	Actual coefficient of variation V_{test}	0.54	0.53	0.48	0.55	0.49	0.49
	variation depending on construction method $V_1 = \sqrt{(V_{test}^2 - V_2^2 - V_3^2)}$	0.50	0.49	0.43	0.51	0.44	0.44
	Coefficient of variation using in design: V	0.68	0.67	0.69	0.64	0.64	0.69

Table 4. Example of target reliability index β_a' and probability of failure P_f'.

Approximated normal distribution	Limit state		
	Serviceability	Restorability	Ultimate
Target reliabilit index β_a'	0.85	0.4	0.1
Probability of failure P_f'	20%	34%	46%

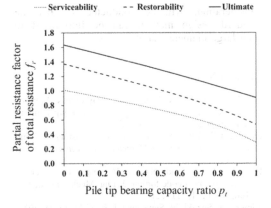

Figure 11. Partial resistance factor f_r for each limit state.

various methods to determine the values of β_a'. The Japanese railway design standard determine β_a' by code calibration. The code calibration is conducted based on the reliability of the design bearing capacity of driven pile in the current standard, so that it becomes equal to that of the previous one about driven pile (Railway Technical Research Institute 1997). And all the construction methods adopt the same target reliability indexes. Target reliability index β_a' and the corresponding probability of failure P_f' for each limit state are shown in Table 4 (Sanagawa et al. 2020).

4.2 Calculation of partial resistance factor of Gyropress method

Table 5 shows the statistic parameter of NRF for each limit state calculated from the NRF database for Gyropress method.

The partial resistance factor was calculated using the formulas in section 4.1 and values in Table 4 and Table 5. Figure 11 shows the relationship between the partial resistance factor of the total resistance f_r and the pile tip bearing capacity ratio p_t.

5 CONCLUSION

In this paper, on the basis of the collected loading test results of piles constructed by Gyropress method, the authors proposed the design method for Gyropress method on the concept of reliability design of the Japanese railway design standard. In addition, the authors proposed the partial resistance factor. As described above, the number of loading tests is not large enough. Therefore, the proposal is made with enough safety margin. In other words, it is highly possible that it is improved by increasing the number of loading test databases in the future.

The Gyropress method was initially used for the revetment repair of urban rivers. After that, it has been also used for installing steel pipe piles into hard rock and rubble mounds, and as of the end of August 2018, there are more than 400 construction records (of which 25 are expected to have bearing capacity).

Construction machines for Gyropress method applying for large diameter tubular piles up to $\varphi 2500$ mm have been used. Various

Table 5. Statical parameter of Gyropress method for each limit state.

Limit state		Serviceability		Restorability		Ultimate	
Reference settlement for each limit state s		20 mm		50 mm or 5%D		10%D	
Resistant part		Tip resistance	Shaft friction	Tip resistance	Shaft friction	Tip resistance	Shaft friction
Static parameter (Approximated normal distribution)	Valid data number n	5	5	5	5	5	5
	Mean value μ	0.49	1.70	0.68	1.76	0.96	1.73
	Coefficient of variation using in design: V	0.48	0.48	0.52	0.55	0.57	0.57

considerations will be conducted so that the standard design method can be introduced to such large diameter tubular piles.

REFERENCES

GIKEN LTD. Update/Brochures - Gyropress Method. retrieved from https://www.giken.com/en/wp-content/uploads/press-in_gyropress.pdf

Grubbs, F.E. 1950. Sample criteria for testing outlying observations. *Annals of Mathematical Statistics*. 21 (1). 27–58.

Hasfer, A.M. & Lind, N.C. 1974. Exact and invariant second-moment code format. J. Eng. Mech., *ASCE*, Vol. 100, No. EM1, 111–121.

Kimura, Y. Suzuki, N. Sanagawa T & Nishioka, H. 2019. Design vertical bearing capacity of Gyropress method used for railway structures. *Kisokou*. Vol. 47. No.8: 40–43 (in Japanese).

Nishioka, H. Koda, M. Shinoda. M & Tateyama. M. 2009. Method for calculating design bearing capacity of piles under various construction method by statistical analysis of loading test database. *Proc., Performance-Based Design in Earthquake Geotechnical Engineering*. Tokyo. 1667–1671.

Nishioka, H. Sanagawa, T. Koda, M. Suzuki, K. Takisawa S. & Fujiwara, T. 2017. Verification to the evaluation method of vertical bearing capacity of the pile for railway structure. *52nd Japan Natural Conference one Geotechnical Engineering*.

Otake, Y & Honjo, Y. 2014.Characterization of transformation error in geotechnical structural design. *Journal of JSCE C. vol,70.No.2*, 186–198 (in Japanese).

Railway Technical research Institute. 1997. Design Standard for Railway Structures and Commentary (Foundation Structures).

Railway Technical research Institute. 2012. Design Standard for Railway Structures and Commentary (Foundation Structures).

Sanagawa, T & Nishioka, H. 2020. Calculation results of partial factor of vertical resistance of pile based on reliability analysis. 62th Geotechnical Symposium (in Japanese).

Sanagawa, T. Nishioka, H & Kasahara, K. 2019. Proposal of partial factors of bearing capacity of piles for Japanese design standard of railway structures. *7th International Symposium on Geotechnical Safety and Risk (ISGSR 2019)*. Taipei. Taiwan. 11–13.

Proceedings of the Second International Conference on
Press-in Engineering 2021, Kochi, Japan – Matsumoto et al (eds)
© 2021 Taylor & Francis Group, London, ISBN 978-1-032-10414-0

Comparison of SPT-based design methods for vertical capacity of piles installed by Rotary Cutting Press-in

K. Toda & Y. Ishihara
Giken LTD, Kochi, Japan

N. Suzuki
Giken LTD, Tokyo, Japan

ABSTRACT: The applicability of the Press-in Method to hard grounds has been significantly improved by the development of Rotary Cutting Press-in (RCP), where vertical and rotational forces are simultaneously applied to a pile with base cutting teeth. Although RCP piles are usually used for retaining walls, there have been some cases where they are used for foundations. Several SPT-based design methods for RCP piles have been prepared so far. Some of these are based on existing design codes in Japan for piles installed by other piling methods, which are intended for specific construction fields (roads and railways). This paper overviews these existing SPT-based design methods for RCP piles, and proposes another method by making adjustments on the existing design method for the field of buildings in Japan, based on the results of five static load tests on RCP piles.

1 INTRODUCTION

The applicability of the Press-in Method to hard grounds has been significantly improved by the recent development of installation techniques such as Press-in with Augering and Rotary Cutting Press-in (RCP, which installs piles by applying vertical and rotational forces onto a pile with base cutting teeth). Although piles installed by these installation techniques are usually used for retaining walls, there have been some cases where RCP piles are used for foundations. Several design methods for RCP piles have been prepared so far (IPA, 2014, Suzuki et al., 2019; JSCE, 2020), based on some existing design codes in Japan for piles installed by other piling methods. These methods are intended for specific construction fields (roads and railways), to fit for the conventional style in Japan where different design codes based on different design concepts are adopted in different construction fields. There is another research (Ishihara et al., 2020) providing a method of estimating the base capacity of RCP piles either from CPT or SPT results, which was obtained by introducing a method to estimate the plugging condition (*IFR*, Incremental Filling Ratio) into the framework of CPT-based UWA-05 design method (Lehane et al., 2005).

This paper firstly overviews some of the above-mentioned SPT-based design methods for RCP piles. After that, another method will be proposed by making adjustments on the existing code for the field of buildings in Japan, based on the results of five static load tests on RCP piles.

2 EXISTING DESIGN METHODS FOR VERTICAL CAPACITY OF RCP PILES

2.1 IPA (2014)

There had been no design methods for the vertical capacity of RCP piles until the method of IPA (2014) was prepared, by adjusting the SPT-based method for driven piles (JRA, 2012) based on three load test results including that of Hirata et al. (2009).

According to JRA (2012), the base capacity (Q_{bf}) of a driven pile is expressed as:

$$Q_{bf} = q_{bf} \times A_{b,closed} \tag{1}$$

$$q_{bf} = 300 \times \min\left(\frac{z_{bs}}{5D_o}, 1\right) \times \min(N_{D0}, 40) \ [kPa] \tag{2}$$

where q_{bf} is the unit base capacity, $A_{b,closed}$ is the cross-sectional area of a fully plugged (closed-ended) pile, z_{bs} is the embedment length into a bearing stratum (defined as the layer with SPT N not smaller than 30 in JRA (2012)), D_o is the outer diameter of the pile and N_{D0} is the "bearing stratum N value" determined by

DOI: 10.1201/9781003215226-12

the SPT N values in the depth range from the pile base to 4 D_o above the pile base. IPA (2014) recommended the value of z_{bs} for RCP piles to be 1 D_o, aiming for assuring the efficiency of piling work by avoiding excessive time for installing a pile into a hard layer while securing a certain level of the vertical performance of the pile. As a result, in IPA (2014), the unit base capacity is obtained by:

$$q_{bf}= \min(60 \times N_{D1}, 2400) \ [kpa] \quad (3)$$

Note that the averaging method for N is different from that in JRA (2012), with N_{D1} being the N value averaged from the pile base to 1 D_o above the pile base. It is also noted that the bearing stratum is defined in IPA (2014) as the layer with SPT N not being smaller than 40.

On the other hand, in IPA (2014), the shaft capacity (Q_{sf}) is expressed as:

$$Q_{sf}= \int (q_{sf} \times \pi D_o) dz \quad (4)$$

$$q_{sf}= \begin{cases} \min(2 \times N, 100) & [kPa] \ \text{(for sand)} \\ \min(8 \times N, 100) & [kPa] \ \text{(for clay)} \end{cases} \quad (5)$$

where q_{sf} is the unit shaft capacity.

The ratio of design load to the ultimate capacity ($A_{D/f}$) is prescribed in IPA (2014) as:

$$\frac{Q_D}{Q_f} \equiv A_{D/f} = \frac{1}{3} \quad (6)$$

$$Q_f = Q_{bf} + Q_{sf} \quad (7)$$

where Q_D is the design load and Q_f is the ultimate capacity.

2.2 JSCE (2020)

The method of JSCE (2020) was prepared in response to the revision of JRA (2012) into JRA (2017). In JRA (2017), the concept of reducing q_{bf} for smaller z_{bs} (Eq. (2)) was removed, and as a result, the following simpler expression has newly been provided for piles having z_{bs} values equal to or greater than 2 D_o.

$$q_{bf}= \begin{cases} \min(130 \times N_{D2}, 6500) & [kPa] \ \text{(for sand)} \\ \min(90 \times N_{D2}, 4500) & [kPa] \ \text{(for clay)} \end{cases} \quad (8)$$

where N_{D2} is the arithmetic average of SPT N value from the pile base to 3 D_o below the pile base.

In JSCE (2020), JRA (2017) was adjusted based on five load test results introduced later in Section 3.2, and Q_{bf} and Q_{sf} are estimated by:

$$Q_{bf} = q_{bf} \times A_{b,closed} \quad (= Eq.(1)) \quad (9)$$

$$q_{bf}= 4500 \ [kPa] \ \text{(for gravelorsand, } N_{D2} \geq 40) \quad (10)$$

$$Q_{sf}= \int (q_{sf} \times \pi D_o) dz \quad (= Eq.(4)) \quad (11)$$

$$q_{sf}= \begin{cases} \min(5 \times N, 50) & [kPa] \ \text{(for sand)} \\ \min(6 \times N, 50) & [kPa] \ \text{(for clay)} \end{cases} \quad (12)$$

The main point in the revision of JRA (2012) into JRA (2017) was the introduction of the partial factor design method and the limit state design method, which has been reflected in a more sophisticated (subdivided) expression of the ratio of design load to the ultimate capacity ($A_{D/f}$) as follows:

$$A_{D/f} = A_{D/y} \times A_{y/f} \quad (13)$$

$$A_{D/y} = \xi_1 \phi_r \lambda_f \quad (14)$$

where $A_{D/y}$ is the ratio of design load (Q_D) to yield load (Q_y, the first-limit-resistance), $A_{y/f}$ is the ratio of yield load (Q_D) to ultimate capacity (Q_f), ξ is the parameter called "investigation and analysis factor", Φ_y is the parameter called "resistance factor" and λ_f is the parameter to consider the effect of the pile type (e.g. end-supported pile or friction pile). JSCE (2020) provides each value of these parameters, but this paper will only quote the values of $A_{D/y}$ and $A_{y/f}$ for simplicity, later in Table 3.

2.3 Suzuki et Al. (2019)

Suzuki et al. (2019) proposed a design method for RCP piles, by analyzing the five load test results introduced later in Section 3.2 based on statistical processing and reliability analysis, as instructed in RTRI (2012) and RTRI (2018). Their method is expressed as:

$$Q_{bf} = q_{bf} \times A_{b,closed} \quad (= Eq.(1)) \quad (15)$$

$$q_{bf}= \begin{cases} \min(60 \times N_{D3}, 3500) & [kPa] \ \text{(for sand)} \\ \min(60 \times N_{D3}, 7500) & [kPa] \ \text{(for gravel)} \end{cases} \quad (16)$$

$$Q_{sf}= \int (q_{sf} \times \pi D_o) dz \quad (= Eq.(4)) \quad (17)$$

$$q_{sf} = \min(2 \times N, 40) \ [kPa] \ \text{for sand, clay} \quad (18)$$

$$\frac{Q_D}{Q_f} \equiv A_{D/f} \qquad (19)$$

$$Q_f = Q_{bf} + Q_{sf} \quad (= \text{Eq.}(7)) \qquad (20)$$

where N_{D3} is the smallest N value in the depth range from 1 D_o above to 3 D_o below the pile base. $A_{D/f}$ can be obtained by a chart shown in Figure 1.

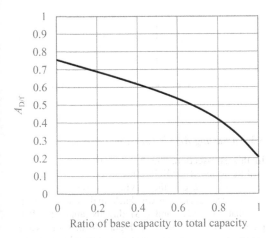

Figure 1. Ratio of design load to the ultimate capacity (after Suzuki et al., 2019).

3 PROPOSAL OF A DESIGN METHOD FOR RCP PILES IN THE FIELD OF BUISLINGS

3.1 Framework of AIJ (2019) design method

The latest design method for piles in the field of buildings in Japan is provided in AIJ (2019). Q_{bf} and Q_{sf} of driven piles are expressed as:

$$Q_{bf} = \eta \times q_{bf} \times A_{b,closed} \qquad (21)$$

$$\eta = \begin{cases} \min\left(0.16 \times \frac{z_{bs}}{d_i}, 0.8\right) & \text{(for open - ended, } \frac{z_{bs}}{d_i} \geq 2) \\ 1 & \text{(for closed - ended)} \end{cases} \qquad (22)$$

$$q_{bf} = \begin{cases} \min(300 \times N_{D4}, 18000) & \text{[kPa] (for sand)} \\ \min(6 \times c, 18000) & \text{[kPa] (for clay)} \end{cases} \qquad (23)$$

$$Q_{sf} = \int (q_{sf} \times \pi D_o) dz \quad (= \text{Eq.}(4)) \qquad (24)$$

$$q_{sf} = \begin{cases} \min(2 \times N, 100) & \text{[kPa] (for sand)} \\ \min(0.8 \times c, 100) & \text{[kPa] (for clay)} \end{cases} \qquad (25)$$

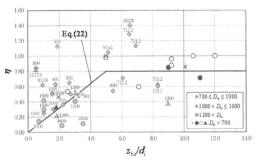

Figure 2. Plug efficiency (after AIJ (2019)).

Here, in general, the base capacity will increase as the open-ended pile becomes more plugged, and the pile will become more plugged as the embedment into the bearing stratum becomes longer. In AIJ (2019), these relationships are considered by introducing the plug efficiency η (Eq. (22), Figure 2), d_i is the inner diameter of the pile, N_{D4} is the SPT N value averaged from 4 D_o above to 1 D_o below the pile base, and c is the cohesion of the soil. The bearing stratum is defined as the layer having the SPT N values greater than 50. The way of considering the plugging condition is similar to that in JRA (2012), which can be confirmed by comparing q_{bf} obtained by Eq. (2) and $\eta \times q_{bf}$ obtained by Eqs. (22) and (23).

AIJ (2019) is a limit-state design method. Piles in their ultimate limit state, damage limit state and serviceability limit state are supposed to exhibit Q_f, two-thirds of Q_f and one-third of Q_f respectively, where Q_f is the ultimate capacity.

3.2 Load test database

In this paper, information of five static load tests on RCP piles were collected, as summarized in Table 1. Tests I2006, T2007 and F2008 were conducted by Hirata et al. (2009) and outlined in IPA (2014). Detailed information on A2016 and N2017 can be found in Ishihara et al. (2016) and Okada et al. (2021) respectively. Site profiles and the press-in conditions for these tests are shown in Figure 3 and Table 2 respectively. Load test conditions were basically in compliance with JGS standard (JGS, 2002), except for the shorter period from the end of installation to the start of load test in A2016 and the existence of adjacent piles within the horizontal distance of 3 D_o from the center of the test pile in I2006, T2007 and F2008.

I2006 was conducted at an initial stage of the development of this piling method. The pile accidentally experienced strong plugging during its installation, and it was extracted fully (with its inner soil column being stuck to the pile) and installed again (after removing the plugged soil). In A2016, the pile was installed by RCP down to 2m BGL and by Standard Press-in (press-in without any installation assistance) in deeper than 2m.

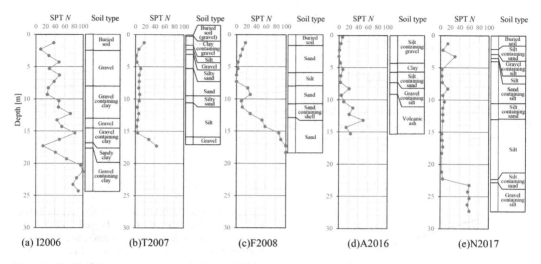

Figure 3. Site profiles.

Table 1. Load test databases.

(a) Test conditions

Test name	Pile Diameter D_o [mm]	Pile Thickness t [mm]	Embedment depth z [m]	Embedment length z_{bs} [m]	Curing period t_c [day]
I2006	800	16	19.65	0.9	18
T2007	800	16	17.5	0	16
F2008	1000	12	15	1.7	15
A2016	800	12	4.7	0	1
N2017	1000	12	24	0.8	57

(b) Test results

Test name	Total capacity Q_f [kN]	Base capacity Q_{bf} [kN]	Shaft capacity Q_{sf} [kN]
I2006	4168	2548	1620
T2007	4060	2368	1692
F2008	6363	3576	2787
A2016	578	363	215
N2017	5000	3102	1898

Table 2. Press-in conditions.

Test name	Jacking force [kN]	Flowrate of water injection [L/min]	Number of base teeth
I2006	500~900	30	4→8
T2007	400	10~30	4
F2008	500	20~30	5
A2016	300	0	4
N2017	300	15	6

3.3 Adjusting AIJ (2019) based on load test results

In this sub-section, the AIJ (2019) design method will be adjusted for RCP piles based on the load test results introduced in the previous sub-section.

For base capacity, η was back-analyzed by assuming that Eqs. (21) and (23) directly apply to RCP piles. It can be found in Figure 4 that the same expression as Eq. (22) provides the lower limit for the load test databases. However, the correlation between the plots of load test results and the estimation line given by Eq. (23) is very weak. One major reason was thought to be the effect of averaging of SPT

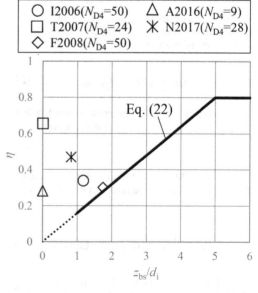

Figure 4. Comparison of η back-analyzed by load test data and estimated by Eq.(22).

172

N (averaged from 4 D_o above to 1 D_o below the pile base in Figure 4). Figure 5 shows the results of the same analyses with different averaging methods (averaged from 0 to 3 D_o below the pile base in Figure 5 (a) and averaged from 1 D_o above to 1 D_o below the pile base in Figure 5(b)). Better correlations between the load test results and the estimation lines can be found in Figure 5 compared with those in Figure 4. Figure 6 shows the comparison of measured and estimated base capacities. It can be confirmed that all the estimated values are conservative, and better matches can be obtained when the SPT N values are averaged either from 0 to 3 D_o below the pile base or from 1 D_o above to 1 D_o below the pile base.

For shaft capacity, values of constants in Eq. (25) were back-analyzed based on the q_{sf} values obtained by the strain gauge readings in the load tests and the SPT N values averaged over the corresponding depth ranges. For sands, as shown in Figure 7(a), q_{sf} = $3.5N$ with its upper limit (q_{sf}^{UL}) being 100kPa provides a reasonable threshold, with the excess ratio (ratio of load test data plots exceeding the estimation line) being greater than 75% as instructed in AIJ

(2019). For clays, as shown in Figure 7(b), the threshold of q_{sf} = $6N$ gives the excess ratio of 75%, while the q_{sf}^{UL} value is difficult to be judged. Figure 8 is the comparison of the measured and estimated Q_{sf}, with estimation provided based on different q_{sf}^{UL} values. It can be confirmed that q_{sf}^{UL} = 100kPa (as adopted in Eq. (25)) provides a slightly conservative results than the average trends in the load test results, with slight overestimation for two of the four load test results (excluding I2006 due to the irregularity experienced in its installation as previously explained).

3.4 Comparing the SPT-based design methods

The SPT-based design methods reviewed in Section 2 and proposed in Section 3.3 are summarized in Table 3. Here, the ultimate limit state, damage limit state and serviceability limit state in AIJ (2019) were interpreted by the authors as corresponding to the states where a pile shows its ultimate capacity, yield load and design load respectively. The subdivided expression of $A_{D/y}$ in JSCE (2020) was simplified in this table to enable an easier comparison. It should also be noted

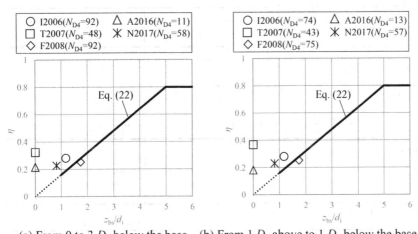

(a) From 0 to 3 D_o below the base   (b) From 1 D_o above to 1 D_o below the base

Figure 5. Effect of different averaging methods on back-analyzed η.

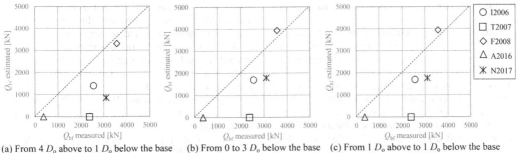

(a) From 4 D_o above to 1 D_o below the base   (b) From 0 to 3 D_o below the base   (c) From 1 D_o above to 1 D_o below the base

Figure 6. Comparison of Q_{bf} obtained in load test and estimated by the proposed method with different averaging methods.

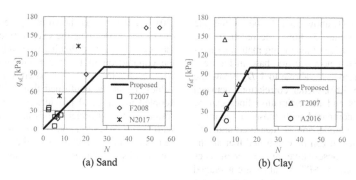

Figure 7. Correlation between unit shaft capacity and SPT N values.

Table 3. Comparison of SPT-based design methods for RCP piles in Japan.

	Code		IPA (2014)	JSCE (2020)	Suzuki et al. (2019)	Proposed
	Instalation method		RCP	RCP	RCP	RCP
(1)	Ratio of design load to ultimate capacity, $A_{D/f}(=A_{D/y} \times A_{y/f})$		1/3	0.32	Figure 1	-
(2)	Ratio of design load to yield load, $A_{D/y}$			0.72		-
(3)	Ratio of design load to ultimate capacity, $A_{y/f}$			0.45		-
(4)	q_{bf}	clay				
		sand	$60N_b \leqq 2400$	$4500\ (N_b \geqq 40)$	$60N_b \leqq 3500$	$300N_b \leqq 18000$
		gravel	$60N_b \leqq 2400$	$4500\ (N_b \geqq 40)$	$60N_b \leqq 7500$	
(5)	η		1	1	1	Eq.(22)
(6)	q_{sf}	clay	$8N \leqq 100$	$6N \leqq 50$	$2N \leqq 40$	$6N \leqq 100$
		sand	$2N \leqq 100$	$5N \leqq 50$	$2N \leqq 40$	$3.5N \leqq 100$
(7)	Definition of bearing stratum		$N \geqq 40$	$N \geqq 40$	$N \geqq 30$	$N \geqq 50$
	Reqired z_{bs} value		$1D_o$	$1D_o$	$1D_o$	$1D_o$
	Range of N considered for q_{bf}	Bearing stratum Embedded depth	above $1D_o$		above $1D_o$	above $1D_o$
		N_b	average			belw $1D_o$ average
				below $3D_o$ average	below $3D_o$ minimum	
(8)	Target performance of estimation		Lower limit of load test database	Average of load test database	Lower limit of load test database	base : Excess ratio 50% shaft : Excess ratio 75%

that N_{D4} in the proposed method was obtained by averaging SPT N values from 1 D_o above to 1 D_o below the pile base, to reflect the discussion in the previous section.

Figure 9 shows the comparison of Q_{bf}, Q_{sf} and Q_f obtained in the load tests and those estimated by different design methods. Q_{bf} was estimated reasonably, except for the significant underestimations by IPA (2014) (as intended in this design method as explained in Section 2.1). Q_{bf} remained almost constant with the value of z_{bs} in IPA (2014), Suzuki et al. (2019) and JSCE (2020), while it increased significantly with z_{bs} in the proposed method due to the effect of η. This can lead to a significant overestimation if the pile is embedded into the bearing stratum deeper than the z_{bs}

values of the load test databases in this paper (at most about 2 D_o). For Q_{sf}, the result in I2006 was overestimated by all the methods, due to the irregular extraction to remove the plugged soil. In the other tests, significant (about half) underestimations were found in Suzuki et al. (2019) (as intended in this design method as explained in Section 2.3), while a slight overestimating trend was found in IPA (2014). For Q_f, good agreement with the load test results were found in the proposed method and JSCE (2020), while conservative estimation results were yielded by the other two methods.

Figure 10 shows the comparison of design loads (Q_D) obtained by IPA (2014), Suzuki et al. (2019) and JSCE (2020). It can be seen that similar Q_D

174

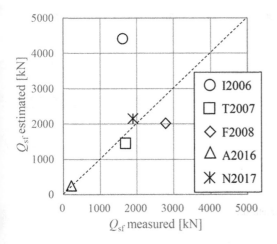

Figure 8. Comparison of measured and estimated Q_s.

values were obtained by Suzuki *et al.* (2019) and JSCE (2020) despite their difference in estimating Q_f values, and these values were greater than what was obtained by IPA (2014). It is necessary to obtain a value of $A_{D/f}$ for the proposed method in the future.

4 CONCLUSIONS

Existing SPT-based design methods for RCP piles were reviewed, and another method was proposed by making adjustments on the existing design method for the field of buildings in Japan, based on the results of four static load tests on RCP piles. Comparing the estimation results and the field test results, it was confirmed that the ultimate capacity obtained by each code was different from each other but the design load obtained by each code was more

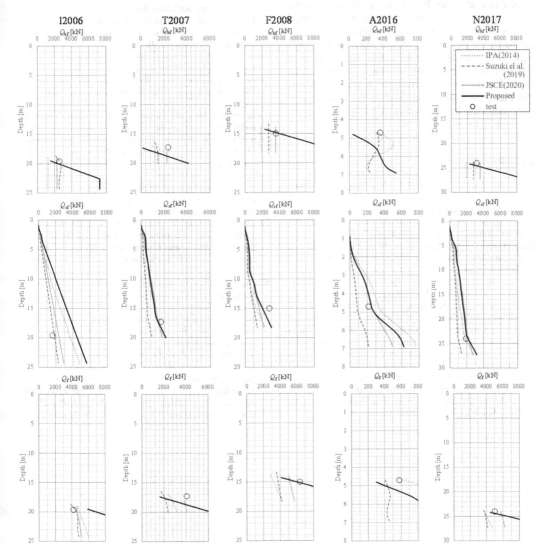

Figure 9. Comparison of Q_{bf}, Q_{sf} and Q_f obtained in load tests and estimated by SPT-based design methods.

175

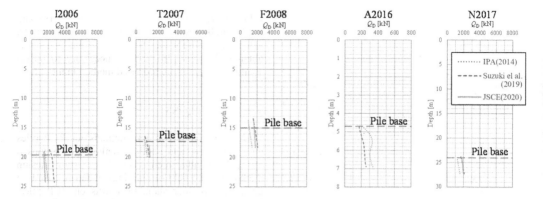

Figure 10. Comparison of Q_D estimated by SPT-based design methods.

similar to each other. It was also confirmed that the base capacity estimated by the proposed method sharply increased in the bearing stratum due to the consideration of the plugging condition, which could lead to significant overestimation for RCP piles installed into the bearing stratum by greater than about 2 times its outer diameter. It was recognized that the ratio of the design load to the ultimate capacity obtained by the proposed method has to be clarified in the future.

REFERENCES

Hirata, H., Suzuki, T., Matsui, N. and Yasuoka, H. 2009. Bearing capacity performance of rotary press-in method (Gyropress Method): part 1 (vertical capacity). *Proceedings of the 65th JSCE (Japanese Society of Civil Engineers) Annual Meeting*, pp. 41–42. (in Japanese).

International Press-in Association (IPA). 2014. *Design and Construction Manual for Earth Retaining Walls with Tubular Piles Installed by Gyropress (Rotary Cutting Press-in) Method*, 143p. (in Japanese).

Ishihara, Y., Ogawa, N., Okada, K., Inomata, K., Yamane, T. and Kitamura, A. 2016. Model test and full-scale field test on vertical and horizontal resistance of hatted tubular pile. *Proceedings of the Third International Conference Geotec Hanoi 2016 - Geotechnics for Sustainable Infrastructure development*, pp. 131–139.

Ishihara, Y., Haigh, S. and Koseki, J. 2020. Assessment of base capacity of open-ended tubular piles installed by the Rotary Cutting Press-in method. Soils and Foundations, Vol. 60, pp. 1189–1201.

The Japanese Geotechnical Society (JGS). 2002. Method for static axial compressive load test of single piles. *Standards of Japanese Geotechnical Society for Vertical Load Tests of Piles*, pp. 49–53.

Japan Road Association (JRA). 2012. *Specifications for Highway Bridges: Part 4 Substructures*, 586p.

Japan Road Association (JRA). 2017. *Specifications for Highway Bridges: Part 4 Substructures*, 569p. (in Japanese).

Japan Society for Civil Engineers (JSCE). 2020. Report of technology assessment of design and construction method for Rotary Cutting Press-in (Gyropress) Method. *Technology Promotion Library*, No. 25, 113p. (in Japanese).

Lehane, B.M., Schneider, J.A. and Xu, X. 2005. The UWA-05 method for prediction of axial capacity of driven piles in sand. *International Symposium on Frontiers in Offshore Geotechnics*, pp. 683–689.

Okada, K., Inomata, K. and Ishihara, Y. 2021. Results of static vertical load tests on tubular piles installed by Standard Press-in and Rotary Cutting Press-in. *Proceedings of the second International Conference on Press-in Engineering*. (submitted).

Railway Technical Research Institute (RTRI). 2012. *Design Standards for Railway Structures and Commentary (Foundation Structure), Maruzen*, 608p. (in Japanese)

Railway Technical Research Institute (RTRI). 2018. *Design Standards for Railway Structures and Commentary (Foundation Structure) 2012: Guidance of Performance Assessment of Foundation Structure, Kenyusya*. (in Japanese).

Suzuki, N., Kimura, Y., Sanagawa, T., Nishioka, H. 2019. Modeling of vertical bearing capacity of rotary Press-in pile used for a railway structure. *Journal of Railway Engineering Research 23*, pp.217–222. (in Japanese).

Proceedings of the Second International Conference on
Press-in Engineering 2021, Kochi, Japan – Matsumoto et al (eds)
© 2021 Taylor & Francis Group, London, ISBN 978-1-032-10414-0

An investigation into vertical capacity of steel sheet piles installed by the Standard Press-in method

K. Toda & Y. Ishihara

Giken LTD, Kochi, Japan

ABSTRACT: Steel sheet piles have long been used for temporary retaining structures. Nowadays, they are increasingly applied to non-temporary structures, but the number of load tests on sheet piles conducted so far is limited, and methods of estimating vertical and horizontal performance of sheet piles have not been well developed, possibly leading to some conservatism in design of sheet piles. Accumulation of load test results on sheet piles is essential for understanding the performance of sheet piles and rationalizing their design method. This paper focuses on sheet piles installed by Standard Press-in, and introduces three cases of static vertical load tests conducted in the field, which are collected from published sources. The results were used for back-analyzing the values of coefficients in an SPT-based design method. As a result, it was found out that the total capacity of the sheet pile can be safely estimated.

1 INTRODUCTION

Steel sheet piles have long been used for temporary retaining structures. Nowadays, they are increasingly applied to non-temporary structures such as coastal levees (Ishihara *et al.*, 2020a), railway bridge foundations (Kasahara *et al.*, 2018), stress cut-off walls (Tanaka *et al.*, 2018), liquefaction countermeasures for buildings (Kato *et al.*, 2014) and so on. However, the number of load tests on sheet piles conducted so far is limited, and methods of estimating vertical and horizontal performance of sheet piles have not been well developed, possibly leading to some conservatism in design of sheet piles. Accumulation of load test results on sheet piles is essential for understanding the performance of sheet piles and rationalizing their design method.

This paper introduces results of static vertical load tests on sheet piles installed by Standard Press-in (a press-in technique that does not use the installation assistance such as water jetting or augering but may use the technique of repeated penetration and extraction (surging)) from published sources as well as from our original experiments, and attempts to make modifications in an existing SPT-based design method based on the collected load test results.

2 COLLECTING INFORMATION ON STATIC LOAD TESTS ON SHEET PILES INSTALLED BY STANDARD PRESS-IN

2.1 *Is20 test series: U-shaped 400-millimeter-wide sheet pile (SP-III) in soft ground layered by clay and sand*

The first case was taken from a published source (Ishihara *et al.*, 2020b). As shown in Figure 1, the ground consists of soft alluvial soils. A relatively hard layer of fine sand, with SPT N exceeding 10 at the depth (z) from 5.5m to 9m, was sandwiched by soft layers of silty sand and silty clay. As shown in Figure 2, a total of 14 U-shaped sheet piles with the width of 400mm (SP-III) were installed by Standard Press-in, based on the press-in conditions summarized in Table 1. Piles No. 2, 5, 6, 9 and 10 were load tested by a simplified test method using the press-in machine, with different time after the end of installation (t_{LT}).

As shown in Figure 3, the set-up ratio increased linearly with t_{LT}. The greater values of No. 5, 9 compared than No. 6, 10 is thought to be due to the effect of the existence of piles in both sides of the test piles during the load test.

In this test series, the sheet piles were not instrumented with strain gauges. Instead, pull-out tests were conducted immediately after the compressive

DOI: 10.1201/9781003215226-13

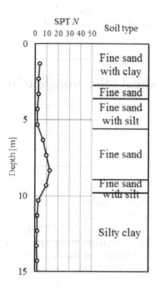

Figure 1. Site profile (Is20 test series).

the ground consists of soft sand layer and a soft sandy silt layer underlaying the sand layer. Hat-shaped sheet piles with the width of 900mm (SP-50H) were used as test piles. As shown in Figure 6, three test piles (A-1, A-2, A-3) were installed by Standard Press-in and then load tested, without their interlocks being connected to the adjacent pile. The other piles were installed by a vibro-hammer method and were used as reaction piles during press-in piling.

The test piles were instrumented with axial strain gauges and a measurement device for the base displacement, as shown in Figure 5. Q_{bf} and Q_{sf} were defined as the values recorded when the pile base displacement reached $0.1W_{sp}$, where W_{sp} is the width of the sheet pile. Q_{bf}, Q_{sf} and the unit shaft capacity are summarized in Table 3.

2.3 Eg21 test series: U-shaped 600-millimeter-wide sheet pile (SP-IIIw) in soft ground layered by sand and silt

The third case was our original experiment. As shown in Figure 7, the ground consists of soft alluvial silt with SPT N values being smaller than 5, underlain by a relatively hard (but still soft) silty sand and gravel layer with SPT N values exceeding 10. The test pile was a U-shaped sheet pile with the width of 600mm (SP-IIIw), and was instrumented with axial strain gauges as shown in Figure 8.

The test layout is shown in Figure 9. The test pile was installed by Standard Press-in based on the press-in conditions summarized in Table 4, using a press-in machine (F201) which was grasping the piles No. A, B, C and D. Pile A was very short and had been welded to the pile B, without being embedded into the ground. The other piles (B, C, D) were embedded by 6.4m.

Figure 10 shows the load displacement curves obtained in the load test. Q_{bf} and Q_{sf}, determined as the values at the pile head displacement of $0.1D_{o,eq}$, are summarized in Table 5.

load test, and the pull-out resistance during the pull-out test was assumed to be equal to the shaft resistance during the compressive load test. The base capacity (Q_{bf}) and the shaft capacity (Q_{sf}) were defined as the values recorded when the pile head displacement reached $0.1D_{o,eq}$, where $D_{o,eq}$ is the outer diameter of an imaginary tubular pile that has the same areas of the annulus and the hollow part of the pile as shown in Figure 4 (IPA, 2017). As a result, the base capacity (Q_{bf}) and the shaft capacity (Q_{sf}) of pile No. 10 was estimated to be 140kN and 140kN respectively, as summarized in Table 2.

2.2 Om19 test series: Hat-shaped 900-millimeter-wide sheet pile (SP-50H) in soft ground layered by sand and silt

The second case was taken from another published source (Omura et al., 2019). As shown in Figure 5,

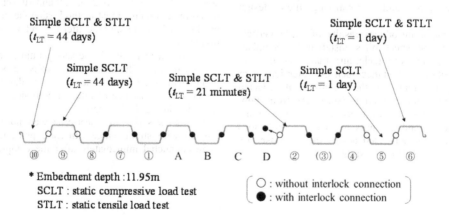

Figure 2. Layout of the field test (Is20 test series) (after Ishihara et al., 2020).

178

Table 1. Test conditions (Is20 test series).

(a) Press-in piling.

Pile No.	Displacement of penetration and extraction (l_d and l_u) [mm]	Rate of penetration and extraction [mm/s]	Interlock connection	Press-in machine (WC in kN)
2	l_d=400, l_u =200	30	None	F111 (19.7)
5	l_d=400, l_u =200	30	None	F111 (19.7)
6	l_d=400, l_u =200	30	None	F111 (19.7)
9	l_d=400, l_u =200	30	None	F111 (19.7)
10	l_d=400, l_u =200	30	None	F111 (19.7)

(b) Load test.

Pile No.	Tensile loading history	Curing period t_{LT}
2	None	21 mintues
5	Yes	1 day
6	None	1 day
9	Yes	44 days
10	None	44 days

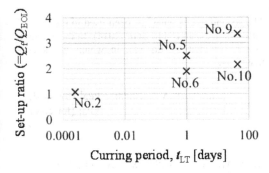

Figure 3. Effect of curing period on pile capacity (Ishihara et al., 2020).

3 SPT-BASED DESIGN METHODS FOR VERTICAL CAPACITY OF SHEET PILES INSTALLED BY STANDARD PRESS-IN

3.1 Existing SPT-based design methods in Japan

In this sub-section, two SPT-based methods to estimate Q_{bf} and Q_{sf} of piles installed by Standard Press-in will be introduced.

The first one is what has been used in the field of roads in Japan (JRA, 1999). In this method, Q_{bf} is estimated by:

$$Q_{bf} = q_{bf} \times A_{bp} \qquad (1)$$

Type of sheet pile	$A_{o.eq}$ [m^2]	$A_{i.eq}$ [m^2]	$D_{o.eq}$ [m]	$D_{i.eq}$ [m]
SP-III	0.007642	0.043496	0.2552	0.2353
SP-IIIw	0.010390	0.090971	0.3593	0.3403
SP-10H	0.011000	0.106836	0.3873	0.3688

Figure 4. The outer diameter of an imaginary tubular pile (IPA, 2017).

Table 2. Summary of the load test results (Is20 test series).

Pile No.	Curing period t_{LT}	End of installation capacity Q_{EOI} [kN]	Total capacity Q_f [kN]	Base capacity Q_{bf} [kN]	Shaft capacity Q_{sf} [kN]	$\frac{Q_f}{Q_{EOI}}$
2	21 minutes	150	160	70	90	1.07
5	1 day	100	250	-	-	2.50
6	1 day	120	25	100	125	1.88
9	44 days	110	370	-	-	3.37
10	44 days	130	280	140	140	2.15

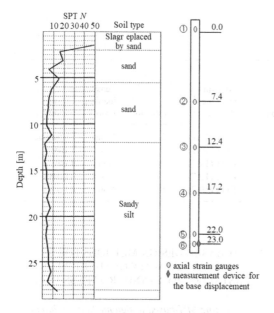

Figure 5. Site soil profile, and distribution of strain gauges and measurement device (Om19 test series) (after Omura et al., 2019).

where q_{bf} is the unit base resistance obtained by Eq. (2) and A_{bp} is the net cross-sectional area of the pile.

$$
q_{bf} = \begin{cases} \min(200 \times N_{D1}, 8000) & \text{[kPa] (for sand)} \\ - & \text{[kPa] (for clay)} \end{cases}
$$
(2)

Here, N_{D1} is the SPT N value averaged from 0 to 2 meters above the pile base. On the other hand, Q_{sf} can be estimated by:

$$
Q_{sf} = \int (q_{sf} \times W_{sp}) dz
$$
(3)

where q_{sf} is the unit shaft resistance obtained by Eq. (4) and z represents the depth.

$$
q_{sf} = \begin{cases} \min(2 \times N, 100) & \text{[kPa] (for sand, } N>2) \\ \min(10 \times N, 150) & \text{[kPa] (for clay, } N>2) \\ 0 & \text{[kPa] (for, } N \le 2) \end{cases}
$$
(4)

The second one is what has been used in the field of railways in Japan (RTRI et al. 2014). In this method, Q_{bf} is obtained by Eq. (1), in which q_{bf} is estimated by:

$$
q_{bf} = \begin{cases} \min(210 \times N_{D2}, 10000) & \text{[kPa] (for sand)} \\ - & \text{[kPa] (for clay)} \end{cases}
$$
(5)

where N_{D2} is the smallest SPT N value from 0 to 3 meters below the pile base. On the other hand, Q_{sf} is estimated by:

$$
Q_{sf} = \int (q_{sf} \times P_{sp}) dz
$$
(6)

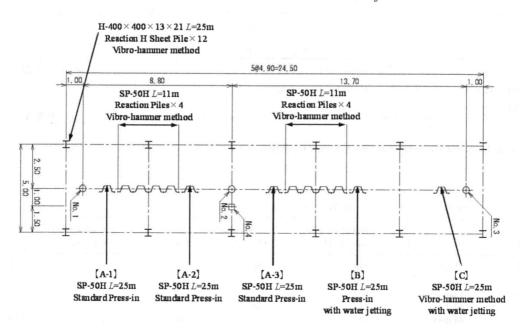

Figure 6. Layout of the field test (Om19 test series) (after Omura et al., 2019).

180

Table 3. Summary of the load test results (Om19 test series).

(a) Capacity

Pile No.	Curing period t_{LT}	Total capacity Q_f [kN]	Base capacity Q_{bf} [kN]	Shaft capasity Q_{st} [kN]
A-1	14 days	1500	78	1422
A-2	14 days	*	124	*
A-3	14 days	*	117	*

(b) Unit shaft capacity, q_{sf} [kPa]

Gauges No.	① - ②	② - ③	③ - ④	④ - ⑤	⑤ - ⑥
Soil type	sand	sand	silt	silt	silt
N value	10.5	4.8	3.6	4.8	3.3
A-1	21.8	7.6	22.0	30.6	25.5
A-2	*	*	14.5	24.4	27.0
A-3	*	*	40.4	37.5	15.4

* Data not indicated in Omura *et al.*, 2019

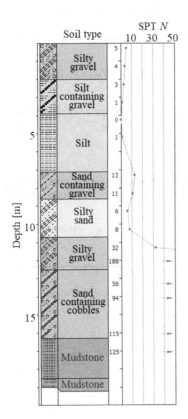

Figure 7. Site soil profile (Eg21 test series) (after Eguchi *et al.*, 2021).

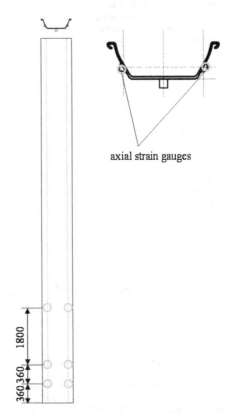

axial strain gauges

Figure 8. Distribution of strain gauges (Eg21 test series).

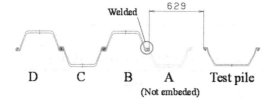

Welded

629

D C B A Test pile

(Not embeded)

Figure 9. Layout of the field test (Eg21 test series).

Table 4. Press-in conditions (Om19 test series).

z [m]	l_d [mm]	l_u [mm]	v_d [mm]	v_u [mm]	Q'_{UL} [KN]	Memo
0~5	400	200	160	122	200	
5~6.35	400	200	160	122	300	
6.35~6.4	400	200	67	-	300	No extraction

z : Depth
l_d : Displacement of penetration in each cycle of surging
l_u : Displacement of extraction in each cycle of surging
v_d : Penetration velocity in each cycle of surging
v_u : Extraction Velocity in each cycle of surging
Q'_{UL} : Upper – limit of jacking foce

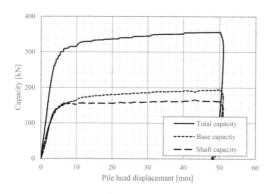

Figure 10. A result of the load displacement curves (Eg21 test series).

Table 5. Summary of the load test results (Eg21 test series).

Curing period t_{LT}	Total capacity Q_f [kN]	Base capasity Q_{bf} [kN]	Shaft capacity Q_{sf} [kN]
20 days	350	189	161

$$q_{sf} = \begin{cases} \min(3 \times N, 120) \ \text{[kPa] (for sand)} \\ \min(6 \times N, 120) \ \text{[kPa] (for clay)} \end{cases} \quad (7)$$

where P_{sp} is the perimeter of the sheet pile.

3.2 Adjusting the SPT-based design methods for Standard Press-in

The two methods introduced in the previous sub-section have similar structure in their formulae to estimate Q_{bf} and Q_{sf}. In this section, Eqs. (8) – (11) will be adopted based on these formulae, where N_D is the smallest SPT N value from 0 to 3 meters below the pile base (= N_{D2}), α_b, $\alpha_{s(s)}$ and $\alpha_{s(c)}$ are the coefficients and q_{bf}^{UL}, $q_{sf}^{UL(s)}$ and $q_{sf}^{UL(c)}$ are the upper limits for q_{bf} or q_{sf}. These parameters will be back-analyzed based on the load test results in Om19 test series, as the sheet piles in this test series were highly instrumented and thus the information of measured q_{sf} values at different soil layers are available.

$$Q_{bf} = q_{bf} \times A_{bp} \ (= \text{Eq.}(1)) \quad (8)$$

$$q_{bf} = \min(\alpha_b \times N_D, q_{bf}^{UL}) \ \text{[kPa]} \ (= \text{Eq.}(6)) \quad (9)$$

$$Q_{sf} = \int (q_{sf} \times P_{sp}) dz \quad (10)$$

$$q_{sf} = \begin{cases} \min(\alpha_{s(s)} \times N, q_{sf}^{UL(s)}) \ \text{[kPa] (for sand)} \\ \min(\alpha_{s(c)} \times N, q_{sf}^{UL(c)}) \ \text{[kPa] (for clay)} \end{cases} \quad (11)$$

Figure 11 shows the correlation between q_{bf} and N_{D2} in Om19 test series, together with the lines

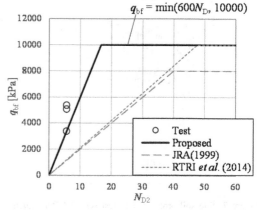

Figure 11. q_{bf} and N_{D2} in Om19 test series and design methods.

determined by each design method. Based on this figure, the α_b value can be determined to be 600 as a lower limit value. It can be confirmed that the existing methods (JRA (1999) and RTRI *et al.* (2014)) provide more than two-fold underestimation in this test series.

Figure 12 shows the correlation between q_{sf} and N in Om19 test series, together with the lines determined by each design method. N was averaged in each soil layer (i.e. for each data plots in this figure). Based on this figure, the $\alpha_{s(s)}$ and $\alpha_{s(c)}$ values can be determined to be 1.9 and 6.6 respectively as average values. It can be confirmed that JRA (1999) provides values comparable to the average of the database and RTRI *et al.* (2014) provides values larger than the average for sand, and vice versa for clay, in this test series.

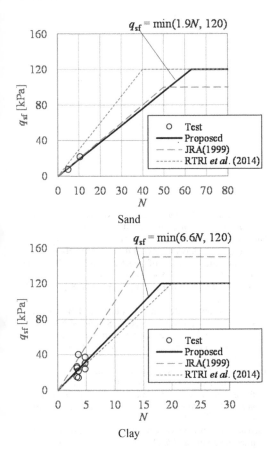

Figure 12. q_{sf} and N in Om19 test series.

Figure 13 shows the comparison of the measured and estimated total capacity (Q_f), Q_{bf} and Q_{sf} for three test series introduced in the previous section. In the estimation, q_{sf} below the depth of the lowest strain gauges (0m, 1m and 0.36m above the pile base in Is20, Om19 and Eg21 test series, respectively) were eliminated. It can be confirmed that the total capacities are safely estimated by all the methods, and that the proposed method provides better estimation than the other two methods for all the three test series. However, the base capacities are significantly underestimated in all the three methods, whereas shaft capacities are slightly underestimated in the proposed method. One reason for the significant underestimation of Q_{bf} would be the effect of plugging phenomenon in the narrow side of the sheet pile, as pointed out by Taenaka *et al.* (2013). It might be that the sheet piles in Om19 test series were less plugged than those in Is20 and Eg21 test series, leading to smaller back-analyzed α_b value which yields much smaller estimation for Q_{bf} in Is20 and Eg21 test series.

It should be noted that the base displacement to determine the capacity was not consistent in the three test series introduced in Section 2. The base displacement determined based on the definition in Om19 test series (= $0.1W_{sp}$) will be larger than the base displacement determined as one-tenth of the outer diameter of the equivalent tubular pile as adopted in the other two test series (Is20, Eg21), and consequently the capacity determined at $0.1W_{sp}$ will be larger. This paper ignored this effect when analyzing the data in this section. The error caused by this can be judged as being small, being less than 5% in Eg21 test series for example if judged from Figure 10. One reason for this small error might be that the sheet piles in these test series were friction piles, and the resistance (both on base and shaft) did not vary significantly with the base displacement after the yielding point.

4 CONCLUSIONS

Three cases of static vertical load tests on sheet piles installed by Standard Press-in were collected from published sources. An SPT-based design method for the vertical capacity of sheet piles was proposed, by adjusting the existing methods in Japan based on the collected load test results. As a result, the total capacity of the sheet pile was safely estimated if the unit base resistance on the net sectional area was taken as $600N$ in kPa and the unit shaft capacity on the net surface area was taken as $1.9N$ in sandy soil and $6.6N$ in clayey soil. However, the base capacity was significantly underestimated while the shaft capacity was slightly overestimated. The significant underestimation of the base capacity may be due to the effect of the plugging condition of the sheet piles in the field load tests.

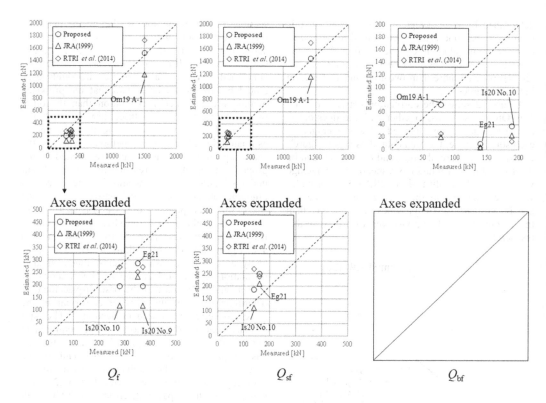

Figure 13. Comparison of measured and estimated capacity.

REFERENCES

International Press-in Association (IPA). 2017. Technical Material on the Use of Piling Data in the Press-in Method, I. Estimation of Subsurface Information, 63p. (in Japanese).

Ishihara, Y., Yasuoka, H. and Shintaku, S. 2020a. Application of Press-in Method to coastal levees in Kochi Coast as countermeasures against liquefaction. *Geotechnical Engineering Journal of the SEAGS & AGSSEA*, 10p.

Ishihara, Y., Ogawa, N., Mori, Y., Haigh, S. and Matsumoto, T. 2020. Simplified static vertical loading test on sheet piles using press-in piling machine. *Japanese Geotechnical Society Special Publication, 8th Japan-China Geotechnical Symposium*, pp. 245–250.

Japan Road Association (JRA). 1999. *Highway Earthwork Series: Manual for Retaining Walls*, 378p.

Kasahara, K., Sanagawa, T., Nishioka, H., Sasaoka, R. and Nakata, Y. 2018. Seismic reinforcement for foundation utilizing sheet piles and soil improvement. *Proceedings of the First International Conference on Press-in Engineering 2018, Kochi*, pp. 555–562.

Kato, I., Hamada, M., Higuchi, S., Kimura, H. and Kimura, Y. 2014. Effectiveness of the sheet pile wall against house subsidence and tilting induced by liquefaction. *Journal of Japan Association for Earthquake Engineering*, 14(4), pp. 4_35-4_49.

Omura, A., Hamaguchi, M., Uetani, O., Oghi, S., Ueda, T., Yokoyama, N. and Osaki, H. 2019. Vertical bearing capacity of sheet-pile by loading tests. *Journal of Japan Society of Civil Engineers, B3 (Ocean Engineering)*, 75(2), pp. I_438-I_443. (in Japanese).

Railway Technical Research Institute (RTRI), Obayashi Corporation, and Nippon Steel & Sumitomo Metal Corporation. 2014. *Design and Construction Manual for Sheet Pile Foundations Applied to Rrailway Structures (draft)*, Version 3, 253p.

Tanaka, K., Kimizu, M., Otani, J. and Nakai, T. 2018. Evaluation of effectiveness of PFS Method using 3D finite element method. *Proceedings of the First International Conference on Press-in Engineering 2018, Kochi*, pp. 209–214.

Taenaka, S. 2013. Development and optimisation of steel piled foundations. *Ph. D. thesis, The University of Western Australia*, 289p.

The Japanese Geotechnical Society (JGS). 2002. Method for static axial compressive load test of single piles. *Standards of Japanese Geotechnical Society for Vertical Load Tests of Piles*, pp. 49–53.

Proceedings of the Second International Conference on
Press-in Engineering 2021, Kochi, Japan – Matsumoto et al (eds)
© *2021 Taylor & Francis Group, London, ISBN 978-1-032-10414-0*

Predicting the capacity of push and rotate piles using offshore design techniques and CPT tests

M.J. Brown
University of Dundee, Dundee, Scotland, UK

Yukihiro Ishihara
Giken Limited, Kochi, Japan

ABSTRACT: Design methods for pile installation e.g. cast insitu or full displacement driven piles are often used as the basis for the design of more advanced installation approaches such as push and rotate piles with cutting shoes (RCP). These approaches, though, do not reflect the variation in installation and the effects it may have on capacity when applied to RCP piles. Although continuous development and improvement of RCP pile design is ongoing, it was seen as valuable to explore current insitu test based design of offshore piles. This paper compares offshore CPT based pile design techniques with field measurements of pile performance. Results suggest improved prediction of shaft capacity when compared to SPT based methods but with mixed performance for end bearing prediction. Attempts to predict installation torque using a method developed from DEM modelling showed some success irrespective of the complexity of installation using water injection and surging.

1 INTRODUCTION

1.1 *Background*

Design of the more advanced push and rotate piles is often undertaken using methodologies designed to predict the capacity of cast insitu piles or full displacement piles (i.e. driven piles). Such approaches though do not reflect the differences in installation approaches between the different pile installation techniques and the likely affect installation may have on in service capacity.

Refinement of a design and reduction in conservatism for push and rotate piles is an ongoing area of research. Although this is being looked into specifically, it was also decided to look to other pile design methodologies from the offshore sector which have much greater reliance on insitu test measurement (e.g. CPT rather than onshore SPT). It was decided to look to this sector not only for the use of insitu tests but also because this is an area that sees continuous refinement and development of approaches and there has been significant advances in capturing installation effects and how they affect in service capacity (e.g. Lehane et al. 2005, Jardine et al. 2005 and as outlined in API RP2 Geo, 2011).

1.2 *Approach*

The approach used here was to take the results of site investigation (CPT and SPT only), pile installation and subsequent load testing for three field test sites at different locations in Japan (Giken test sites Akaoka, Nunoshida, Takasu). These particular sites were chosen due to the presence of CPT test which may not be all that common in Japan with SPT normally being used in design. They were also chosen because they included instrumentation on the piles allowing separation of skin and tip resistance.

This information was then used to retrospectively calculate the pile capacity using offshore techniques which in the main use CPT data as their input and compare this with the empirically based classical API offshore design technique and previously developed SPT approaches (IPA, 2014) for RCP piles. The design methods used are referred to herein as Sand 05 (Kolk et al. 2005), Sand ICP (Jardine et al. 2005), Sand UWA (Lehane et al. 2005) and Sand API (API RP2 GEO, 2007). The approach to design was to follow what might be considered an industry based designed approach based purely on the codes and as outlined in the help manual of the OPile software developed by Cathie (Cathie, 2020). For example, the API code takes a simplified approach to design and assumes pure coring throughout installation such that the incremental filling ratio (IFR) equals 1. Which in

DOI: 10.1201/9781003215226-14

the case studies used here seems appropriate as the piles appeared to core throughout and have final filling ratios (FFR) close to one. Although the OPile software is referred to above, the actual calculations were undertaken in a spreadsheet form rather than using proprietary software.

2 CASE STUDY SITES AND INPUT DATA

2.1 Case study site A2016

A typical site profile for the site A2016 is shown in Figure 1. As well as a description of the encountered stratigraphy. Figure 1 also shows the result of SPT testing. Details of the pile installation and other key information is shown in Table 1.

It can be seen that this pile was only installed to a shallow depth of 4.8 m below ground level and has been included to contrast the results with those of the longer piles of the other sites. Also, the methods developed for offshore pile design considered are in the main intended for longer piles than those considered here although the origins of the API method appear to come from short onshore piles. The pile included shaft strain gauges that were installed at 0, 0.5, 1.0, 2.0, 3.0 and 4.3 mbgl (meters below ground level). The lower gauges for each pile were used to

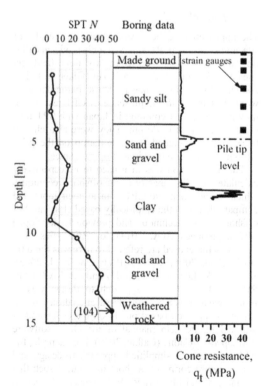

Figure 1. Soil stratigraphy and results of SPT and CPT testing for site A2016.

Table 1. Details of pile installation and key site information, A2016.

Parameter	Value
Pile outer diameter, D_o (m)	0.800
Pile outer diameter, D_i (m)	0.776
Installed length, L (mbgl)	4.8
Groundwater table (mbgl)	0.06
Assumed interface friction angle, δ (°)	24

infer the pile base resistance and it is acknowledged that there will be some influence of the short shaft zone on these values which may have resulted in slightly enhanced tip resistance values. Corrections were applied to the tip resistance values to remove the affects of the additional skin friction zone based upon the shaft resistances determined above. Installation was unusual in that the pile was pushed and rotated to 2 mbgl and then purely pushed or jacked to the final installation depth. The pile was also "surged" to reduce installation requirements (maintained below a pre decided vertical force and torque during installation). This involved effectively lifting the pile up under load control during installation. The pile incorporated 4 teeth at the base resulting in a cutting action which is obviously different to an offshore driven pile. Water injection was not used during installation.

A static axial maintained load test was conducted based on JGS (Japanese Geotechnical Society) standard (JGS, 2002), except for the condition of the wait period prior to testing. JGS (2002) requires t_{LT} to be greater than 7 days for sands and 14 days for clays. In this load test, t_{LT} was 26 hours, as one of the objectives of this load test was to confirm the short-term performance of the pile. The piles were observed to be fully coring throughout with internal soil material at the same level throughout or close to this.

2.2 Case study site N2017

Again, the site profiles for this case study site are shown in Figure 2 with key information in Table 2. In this case the pile was pushed and rotated throughout installation. Shaft strain gauges were installed at 0, 4.0, 10.5, 22.0 and 23.0 mbgl. The pile tip incorporated 6 cutting teeth and water injection of 15 litres per minute was used at the base although this was minimized on approaching the final pile installation depth ($0.9D_o$). Water injection was used rather than jetting where the injection rate was 1/5th of that associated with jetting (greater than 300 liters/minute) which may lead to soil transport.

The pile was also "surged" to reduce installation requirements. Pile testing was carried out in a similar manner to A2016 with a period of 57 days between installation and testing.

186

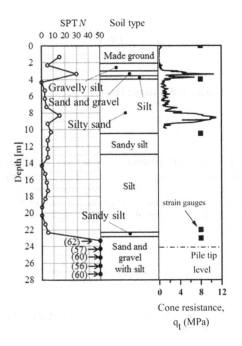

Figure 2. Soil stratigraphy and results of SPT and CPT testing for site N2017.

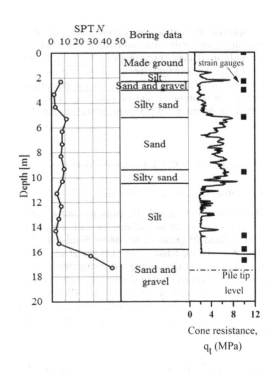

Figure 3. Soil stratigraphy and results of SPT and CPT testing for site T2007.

Table 2. Details of pile installation and key site information, N2017.

Parameter	Value
Pile outer diameter, D_o (m)	1.000
Pile outer diameter, D_i (m)	0.976
Installed length, L (mbgl)	24.1
Groundwater table (mbgl)	1.6
Assumed interface friction angle, δ (°)	24

Table 3. Details of pile installation and key site information, T2007.

Parameter	Value
Pile outer diameter, D_o (m)	0.8
Pile outer diameter, D_i (m)	0.768
Installed length, L (mbgl)	17.5
Groundwater table (mbgl)	1.1
Assumed interface friction angle, δ (°)	24

2.3 Case study site T2007

As per pile N2017, pile T2007 was installed to far greater depth than A2016 with these two deeper case studies more aligned with the depth that the offshore pile designed methods were developed for. In the case of both pile N2017 and T2007, the piles were designed to generate significant capacity through tips founded in competent sand and gravel layers (Fig. 2-3). Above this relatively little bearing capacity would be developed due the presence of low density silt and sand layers. Shaft strain gauges were again incorporated at 0, 2.3, 3.0, 5.2, 9.6, 14.7, 15.8 and 16.7 mbgl. Water injection of 18 to 24 litres per minute at the tip was used although this was minimized over the final $0.5D_o$. Rotation was also stopped about 4 mm above the final installation depth and the pile only pushed or jacked into final position. The pile tip incorporated 4 cutting teeth to aid installation.

2.4 CPT data

The CPT data used for the three case study sites is shown in Figures 1, 2 and 3. Unfortunately, the CPTs for N2017 and T2007 did not extend to sufficient depth to allow for continuous design based upon CPT. In this case the missing CPT results were inferred from the SPT results using the conversion from SPT to CPT proposed by Jefferies & Davies (1993). Soil unit weight was determined based upon the method proposed by Robertson & Cabal (2010). The relative density of the soil to allow parameter selection as part of the API methodology was determined based upon the approach set out by Jamiolkowski et al. (2001). All CPT results used to determine pile end bearing were average $1.5D_o$ above and below the tip position but

187

no specific averaging was applied for pile shaft resistance determination (although some of the methods used may propose this). It is noted that due to the low strength/density of the soil over the some of the pile lengths, CPT derived relative density was not appropriate and SPT readings were converted to determine input parameters for the API method.

3 CALCULATION OF PILE CAPACITY

3.1 Offshore CPT based methods

The offshore CPT based pile capacity design methodologies are those as outlined in API RP2 GEO (2007): Sand 05 (Kolk et al. 2005), Sand ICP (Jardine et al. 2005), Sand UWA (Lehane et al. 2005). To aid consistent design and aid comparison within spreadsheet, approaches to design the methods have been summarized by a single equation (Equation 1) for shaft resistance where parameters are varied depending on the specific methodology used (Table 4).

$$\tau_s = u.q_c \left(\frac{\sigma'_v}{p_a}\right)^a .A_r^b . \max\left(\frac{L-z}{D_o}, v\right)^{-c} .\tan(\delta_{cv})^d$$
$$\min\left(\frac{L-z}{D_o}.\frac{1}{v}, 1\right)^e$$
$$\tag{1}$$

Where q_c is the cone resistance, p_a is the atmospheric pressure (taken as 100 kPa), A_r is an area ratio $(1-D_i^2/D_o^2)$, L is the final embedded length of the pile, z is the depth of the pile during installation, δ_{cv} is the pile interface friction angle and all other symbols are defined in Table 4.

Although the shaft friction determination for the methods is based upon a semi-unified approach, the end bearing resistance varies from method to method. For example, the base resistance for the Sand 05 method is represented by Equation 2:

$$q_b = 8.5.\left[p_a.\left(\frac{q_c}{p_a}\right)^{0.5}\right].A_r^{0.25} \tag{2}$$

The Sand UWA method uses:

$$q_b = q_c.(0.15 + 0.45.A_r) \tag{3}$$

For the ICP method the pile was considered plugged or unplugged. Where the pile was considered to be coring the base resistance was calculated based upon the cone resistance multiplied by the annular base area. When plugged Equation 4 was used:

$$q_b = q_c.\left(0.5 - 0.25.\log\left(\frac{D_o}{D_{CPT}}, 10\right)\right) \tag{4}$$

Where D_{CPT} is the diameter of the CPT taken here as 36 mm.

3.2 Results from the CPT based methodologies

The results for the analysis applied to pile A2016 are shown in Figure 4-6 for total capacity, shaft resistance and end bearing capacity respectively. Where Figures 5, 8 & 11 refer to coring this means that both internal and external skin friction are considered and that the tip resistance was calculated using the annular area. When plugged, only the external skin friction was considered with a fully plugged base (total base area). The UWA method was assumed to be coring during installation and plugged during testing (external shaft friction only) with the full base area used in Equation 3 (as this is modified by A_r and IFR). Figures 5, 8 & 10, plugged refers to external skin friction only, x2 means this

Table 4. Parameter values for Equation 1.

Parameter	Method		
	Sand05	SandUWA	Sand ICP
a	0.05	0	0.1
b	0.45	0.3	0.2
c	0.90	0.5	0.4
d	0	1	1
e	1	0	0
u	0.043	0.030	0.023
v	$A_r^{0.5}$	2	$A_r^{0.25}$

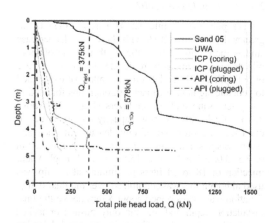

Figure 4. Total pile capacity calculated and compared to that measured after installation for Pile A2016.

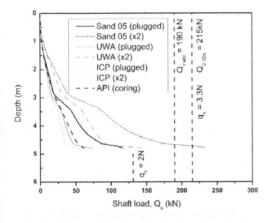

Figure 5. Shaft resistance calculated and compared to that measured after installation for Pile A2016.

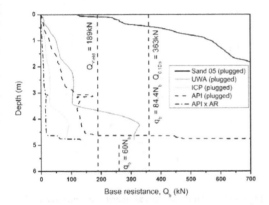

Figure 6. Base resistance calculated and compared to that measured after installation for Pile A2016.

has been doubled to show the effect of the coring state where in both cases A_r is unmodified by the IFR i.e. IFR is set at 1 throughout. In this case internal and external skin friction are calculated in the same manner.

Pile A2016 is obviously relatively short at 4.8m installed final length compared with the other two piles and the length of piles the offshore design methods were designed for. It would appear that the Sand 05 method significantly overpredicts the total pile capacity (Figure 5) with a similar result for pile N2017 (Figure 8) although the over prediction is less significant for pile T2007 (Figure 11). This appears to be attributed to overprediction of end bearing resistance. The Figures show capacity measured from the pile tests for the first yield, Q_{yield} (if present or easily identifiable) and at a displacement of 10% of diameter, $Q_{0.2Do}$ where piles are assumed to be rigid.

In terms of shaft resistance for A2016, all of the methods appear to underpredict capacity for this short

pile (Figure 5) even where allowance has been made for the piles to be coring (by doubling the calculated shaft resistance). Only the Sand 05 method appears close when doubled up but manages to overpredict slightly. It is noted though that as well as being a very short pile the pile was also tested relatively quickly after installation which may have had an influence on the effective stress state at the shaft. Again, this may reflect the origins of the offshore methods for longer piles. Comparison with the SPT based design method as outlined in IPA (2014) is shown on the shaft ($q_s = 2N$) and base resistance figures ($q_b = 60N_b$), where N is the SPT number and N_b the average one diameter above the base or tip. In both cases the SPT based approach underpredicts capacity (by 39% & 29% respectively) as it was originally designed to be conservative (to give results lower than the lower limit of the small testing database).

Due to the relatively low skin friction developed for the three piles here based upon the nature of the ground i.e. low density upper layers (Figures 1-3) and substantial layers (Figure 1 & 3) of silt, the end bearing resistance developed is probably of most interest here. Figure 6 shows best performance for the Sand UWA method with good prediction at large pile deformation whereas Sand 05 and Sand API significantly overpredict capacity.

Considering pile N2017, which is installed to 24.1 m, in terms of total capacity, the UWA (assumed plugged), ICP plugged and the API plugged methods do a relatively good job of predicting total capacity whereas the Sand05 method seems to significantly overpredict again (Figure 7). Better predictions of final shaft resistance are achieved if the methods are applied to the internal and external shaft (Figure 8) which is consistent with the full coring behavior noted in the field for all piles (FFR generally close to 1). Again, the UWA method performs well when doubled for coring and so does the ICP and API method which is surprising (not directly CPT derived). The SPT based method seems to again underpredict shaft resistance by 28%

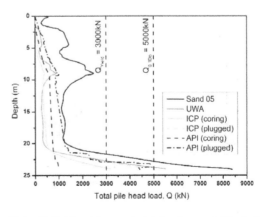

Figure 7. Total resistance calculated and compared to that measured after installation for Pile N2017.

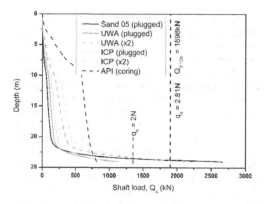

Figure 8. Shaft resistance calculated and compared to that measured after installation for Pile N2017.

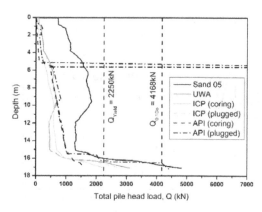

Figure 10. Total resistance calculated and compared to that measured after installation for Pile T2007.

suggesting there is potentially better performance for the offshore methods. Unfortunately, again, though end bearing predictions are not particularly satisfactory in that in general quite significant over prediction is encountered (Figure 9), whereas the SPT method again underpredicts by 24%. Based upon the shaft resistances predicted for piles N2017 and T2007 (Figure 11) it would appear that the ICP and UWA methods do a good job of predicting shaft resistance if they are allowed to core and the shaft resistance is doubled internally and externally. It is not clear if the shaft friction fatigue they were derived for was designed to be applied internally and externally but it appears to work well here. Maybe it is also surprising that the RCP installed piles with water injection appear to have such similar shaft performance to that predicted for long offshore driven piles. This may be particularly remarkable where the RCP piles effectively have a cutting shoe and water injection during installation is used. This may be as a result of the nature of the soil where significant depths of silt are encountered although it is unlikely that sand-based methods used here where

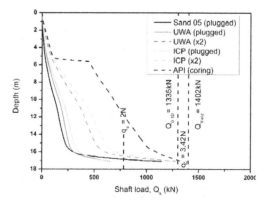

Figure 11. Shaft resistance calculated and compared to that measured after installation for Pile T2007.

exposed to significant data from silt sites during their development. The results, though, would suggest that there is potential for further development and investigation of the CPT based shaft resistance methods outlined where RCP piles have significant length and rely on this component of pile resistance for capacity as these appear to be an improvement over SPT based approaches.

For pile T2007 the results are more mixed in terms of both total capacity and end bearing resistance there was also a tendency for reducing shaft capacity with displacement (Figure 11) as the final resistance was lower than that at the yield point (not shown herein). The total capacity results tend to span the difference in capacity measured at yield and at a displacement equivalent to 10% of the pile diameter ($0.1D_o$) with the API and Sand05 methods closest to the ultimate resistance and the ICP and UWA method close to yield. For tip resistance the UWA method underpredicts whilst the Sand 05 and API approaches overpredict to a similar degree.

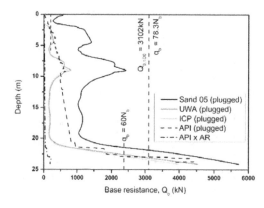

Figure 9. Base resistance calculated and compared to that measured after installation for Pile N2017.

190

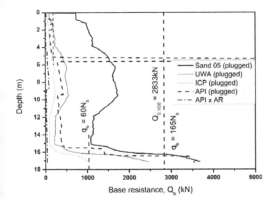

Figure 12. Base resistance calculated and compared to that measured after installation for Pile T2007.

Again, the SPT methods underpredict capacity in terms of shaft and tip capacity by 42% & 63% respectively. The SPT based performance of based resistance seems particular poor here due to averaging of the SPT values above the pile tip where it would seem more appropriate to consider the SPT resistance above and below the tip as adopted in the CPT methods.

3.3 Discussion of results

Based upon the results presented and discussed above, it would appear based upon the limited data set that there is some potential in using the UWA and ICP based CPT methods to predict pile shaft capacity for RCP piles in the sand/silt horizons encountered, although some apparent scatter is created with site A2016 (short pile). White & Deeks (2007) and Okada & Ishihara (2012) have suggested reducing the magnitude of term c in Table 4 which would reduce the effects of friction fatigue which has been observed for push and rotate piles. This would not seem necessary for sites considered here except maybe site A2016 but this is most likely due to the relatively short nature of this pile i.e. any cycles leading to friction fatigue are limited. Further reduction may also not be seen to be required as the piles here are not only push and rotate but also have cutting teeth and surged during installation which has led to considerable numbers of cycles of stress reversal on the shaft such that these piles may behave more like long driven piles (White & Lehane, 2004). In all cases interface friction angles were kept constant for all of the analysis and set at 24° based upon the author's experience of lab characterization of sand-steel interfaces. This assumption though is also in line with the guidance in API RP2GEO. It is noted that pile shaft resistance may be sensitive to this value but that as the paper focuses on relative performance, and the piles here are more reliant on end bearing resistance, this

would seem a lesser concern in this situation although accurate interface friction angles should be determined for the specific case under consideration where possible.

Final end bearing capacity prediction, which is potentially more important for the sites considered, is less satisfactory and further investigation of other methods (CPT based outside of those considered here) is suggested. It is noted, though, that White & Deeks (2007) and Okada & Ishihara (2012) have previously modified the stand UWA method (increasing the magnitude of the final term in Equation 3) in an attempt to improve the situation but that was not considered directly here. It would in this case improve the results from sites A2016 and T2007 but increase the over prediction in N2017. The overall results may not be unsurprising in that the methods investigated in general have their origins in long slender driven offshore piles where more recent development has gone into capturing the friction fatigue effects with respect to shaft resistance. Therefore, in an offshore environment, more reliance may be placed on developing shaft capacity as significant pile deflection may be required to mobilize large tip resistances which could lead to serviceability failure. For example, in Equation 3 q_b can only reach a maximum of $0.6q_c$ which may be included to reduce the reliance on end bearing at serviceability level deflections (in Okada & Ishihara, 2012 this is increased to $0.9q_c$) whereas when coring the ICP method allows the use of full q_c on the annulus only. On this basis it would seem appropriate to investigate other CPT based end bearing capacity methods and look at any previous variations proposed for RCP piles with a wider database of tests.

Performance of the existing SPT based methods consistently underpredict performance which ranges from 28-42% for shaft resistance and to a greater degree for tip resistance (39-63%) which in part seems attributable to how average N values are captured around the tip in the IPA (2014) method. This, though, has the potential to have a significant effect on efficiency and cost when design loads may be reduced by a further one third over ultimate capacity suggesting further work is required to refine the SPT based design approach for these pile types. A simple comparison of performance of the various methods is presented in Table 5 with Qs or Qb here referring to the ultimate capacity proven in static load testing.

It is noted that only a limited data set of RCP pile case studies have been compared with the offshore CPT based methods due to a lack of high-quality instrumented field studies with the inclusion of instrumentation and subsequent load testing. For more conclusive analysis a wider data set in a variety of soils would be required.

3.4 Installation torque prediction

As well as investigating capacity prediction it was also decided to see if a recently developed CPT

Table 5. Simple comparison of performance.

Method		Site		
		A2016	N2017	T2007
Sand 05	Base[1]	3.87Q_b	1.84Q_b	1.29Q_b
	Shaft[1]	1.07Q_s	1.39Q_s	0.88Q_s
UWA	Base[1]	0.85Q_b	1.45Q_b	0.82Q_b
	Shaft[2]	0.68Q_s	1.18Q_s	0.98Q_s
ICP	Base[1]	0.24Q_b	1.19Q_b	0.37Q_b
	Shaft[2]	0.46Q_s	1.04Q_s	0.93Q_s
API	Base[1]	2.67Q_b	1.26Q_b	1.25Q_b
	Shaft[2]	0.28Q_s	0.43Q_s	1.02Q_s
IPA	Base	0.71Q_b	0.76Q_b	0.37Q_b
	Shaft	0.61Q_s	0.72Q_s	0.58Q_s

[1]plugged,[2]coring,

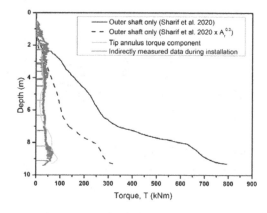

Figure 13. Comparison of calculated torque and that predicted based upon methodology proposed by Sharif et al. (2020) for N2017.

based methodology to predict installation torque based upon centrifuge modelling and DEM simulation (Sharif et al. 2020) was applicable for RCP piles. This methodology (Eq. 5 & 9) has not been testing outside the laboratory and was developed for solid or plugged piles with conical tips of different apex angles (Sharif et al. 2020).

$$T_s = \sum_{\Delta xi=1}^{\Delta xi=L} a\overline{q_c} \ tan\delta\pi\Delta D_o \frac{P_i}{\sqrt{1+P_i^2}} f \quad (5)$$

$$f = 0.63D_r + 0.52 \quad (6)$$

$$a = \frac{F_r}{tan\delta} \quad (7)$$

$$P_i = \frac{\dot{\theta}D_o}{2\dot{w}} \quad (8)$$

Where T_s is the torque developed on the shaft, f is a correction for the initial relative density (D_r), a is a stress drop index on the CPT, θ is the rate of angular rotation of the pile and w the rate of vertical displacement, P_i is the pitch of the rotating pile, δ is an interface friction angle.

The torque prediction compared with field measurements for pile N2007 is shown in Figure 13. The results of the prediction from Equation 5 are for the external shaft only (Ts) and do not consider soil inside the shaft. The results are also limited to 9.2 m in depth as this is the extent of CPT data available at this site (Figure 3). For the torque prediction is it not possible to use the equivalent CPT data derived from SPT as the torque prediction requires friction ratio as an input. It is clear that the method (Figure 13) overpredicts the torque seen in the field associated with the pile shaft and appears to increase with depth. The results are also modified by the area ratio to see the effects that coring may have on the shaft resistance using the modification

proposed as part of the UWA method but again there is still significant overprediction from the outer shaft component. When the torque is determined for the tip annulus alone (Eq. 9) the measured torque is relatively well predicted. This suggests that the majority of the torque is generated by the tip with little input from the shaft. It also highlights how effective the water injection and surging are at reducing the shaft resistance during installation. Without instrumentation to measure local torque on the pile this is difficult to decipher further, though. This may suggest that there is some merit in investigating and developing the tip torque method proposed by Sharif et al. (2020) further but there is still work to do with respect to shaft prediction especially where water injection and surging is deployed. Investigation of the coring mechanism for a pile using DEM is challenging and time consuming due to the size of the elements required to avoid scaling issues in the DEM which increases computational time and practicality of investigation. Further development would have to be undertaken to try and capture surging or water injection, but surging could be implemented in an equivalent dry soil currently without further development.

$$T_b = \overline{q_c}S\pi \frac{D_o^3}{12} \frac{tan\beta tan\delta}{sin\beta tan\beta + tan\beta}$$
$$\times \left(\left[1 - 2\left(\frac{cos\beta}{P_i}\right)^2\right]\sqrt{1 + \left(\frac{cos\beta}{P_i}\right)^2} + 2\left(\frac{cos\beta}{P_i}\right)^3 \right)$$
$$(9)$$

$$S = 0.013\beta + 0.52 \quad (10)$$

where T_b is the base torque, β is the apex angle of a solid base conical tip on the pile (here assumed to be 90 degrees as a flat tip), S is a base angle correction factor.

4 CONCLUSIONS

This paper makes a simplistic attempt to compare the results of RCP pile installation and testing case study site data from three sites in Japan to offshore CPT based pile design techniques to identify if similar techniques can be used and applied to RCP design. The results suggest that the methods referred to as Sand UWA and Sand ICP do a relatively good job of predicting pile shaft resistances but unfortunately pile tip resistance is generally overpredicted or returns mixed performance where for the end bearing piles considered here this is most important. It would seem that the end bearing resistance predicted based upon IPA (2014) is conservative to varying degrees which may be exacerbated by how the SPT values are averaged around the pile tip which may be an area for further improvement.

Attempts to predict torque installation requirements for one of the piles using a recently developed methodology from DEM simulation of solid piles with conical tips had some success when applied to the open coring piles considered here. Shaft torque was significantly over predicted suggesting that the approach to installation may have significantly reduced this component during installation and whereas the tip torque prediction apparently worked well. Interpretation of the results is complicated though by the differences in the situation modelled e.g. solid piles and the presence of cutting teeth, jetting and surging in the field. Further development with DEM could be made to consider coring piles, with surging and including cutting teeth.

It is noted that the case study data set used here is only very limited and that there is a continuing need for high quality instrumented RCP pile testing campaigns to inform improved and efficient design approaches where the results here suggest current approaches used may be overly conservative.

REFERENCES

API. 2007. *Planning, designing and constructing fixed off-shore plat-forms—working stress design* (API RP2). USA, American Petroleum Institute.

API. 2011. *Geotechnical and foundation design considerations*. ANSI/API recommended practice 2Geo (API RP2GEO). USA, American Petroleum Institute.

Cathie 2020. OPile user manual. Cathie, Belgium.

International Press-in Association (IPA), 2014. design and construction manual for earth retaining walls with tubular piles installed by Gyropress (rotary cutting press-in) method. 143p. (in Japanese)

Jamiolkowski, M. Lo Presti, D.C.F., & Manassero, M. 2001. Evaluation of relative density and shear strength of sands from CPT and DMT. In Germaine, J.T., Sheahan, T.C. & Whitman, R.V. (eds.), *Soil Behavior and Soft Ground Construction, ASCE Geotechnical Special Publication, 119, American Society of Civil Engineers (ASCE)*, Reston, Virginia: 201–238

Jardine, R., Chow, F., Overy, R. & Standing, J. 2005. *ICP design methods for driven piles in sands and clays*. Imperial College, London: Thomas Telford Publishing.

Jefferies, M.G., & Davies, M.P., 1993. Use of CPTU to estimate equivalent SPT N60. *Geotechnical Testing Journal*, ASTM, 16(4): 458–468.

Kolk, H. J., Baaijens, A. E. & Senders, M. 2005. Design criteria for pipe piles in silica sands. In Gourvenec, S. & Cassidy, M. (eds), *ISFOG 2005: Proc. 1st Int. Symp. On Frontiers in Offshore Geotechnics, University of Western Australia, Perth, 19–21 September 2005*. London: Taylor & Francis: 711–716.

Lehane, B.M., Schneider, J.A. & Xu, X. 2005. The UWA-05 method for prediction of axial capacity of driven piles in sand. In Gourvenec, S. & Cassidy, M. (eds), *ISFOG 2005: Proc. 1st Int. Symp. On Frontiers in Offshore Geotechnics, University of Western Australia, Perth, 19–21 September 2005*. London: Taylor & Francis: 683–689.

Okada, K. & Ishihara, Y. 2012. Estimating bearing capacity and jacking force for rotary jacking. *Testing and Design Methods for Deep Foundations. IS-Kanazawa, 2012*: 605–614.

Robertson, P.K. & Cabal, K.L. 2010. Estimating soil unit weight from CPT. *2nd International Symposium on Cone Penetration Testing, Huntington Beach, CA, USA, May 2010*.

Sharif, Y.U., Brown, M.J., Ciantia, M.O., Cerfontaine. B., Knappett, J.A., Davidson, C., Meijer, G.J. & Ball, J. 2020. Using DEM to create a CPT based method to estimate the installation requirements of rotary installed piles in sand. *Canadian Geotechnical Journal*. Published online 19/08/20. DOI: doi.org/10.1139/cgj-2020-0017.

The Japanese Geotechnical Society (JGS), 2002. Method for static axial compressive load test of single piles. *Standards of Japanese Geotechnical Society for Vertical Load Tests of Piles*: 49–53.

White, D.J. & Deeks A.D., 2007. Recent research into the behaviour of jacked foundation piles. In Kikuchi, Y., Otani, J. Kimura, M., Morikawa, Y. (eds.), *Advances in Deep Foundations: International Workshop on Recent Advances of Deep Foundations (IWDPF07)*, CRC Press, Japan: 3–26.

White, D.J. & Lehane, B.M. 2004. Friction fatigue on displacement piles in sand. *Geotechnique* 54, No. 10: 645–658.

Proceedings of the Second International Conference on Press-in Engineering 2021, Kochi, Japan – Matsumoto et al (eds)
© 2021 Taylor & Francis Group, London, ISBN 978-1-032-10414-0

Behaviour of three types of model pile foundation under vertical and horizontal loading

W.T. Guo
Graduate School of Natural Science and Technology, Kanazawa University, Kanazawa, Japan

Y. Honda
Graduate School of Hokkaido University, Sapporo, Japan

X. Xiong & T. Matsumoto
Graduate School of Natural Science and Technology, Kanazawa University, Kanazawa, Japan

Y. Ishihara
Scientific Research Section, GIKEN LTD., Kochi, Japan

ABSTRACT: In recent years, efficient installation methods of piles have been developed. A target in this research is to show a possibility to use steel sheet piles for permanent pile foundations, because time and cost of construction for sheet piles could be lower than those for pipe piles. In this study, a series of experiments were conducted to investigate the load transfer behaviours of model foundations supported by three different types of piles in dry sand ground subjected to vertical and horizontal loading. According to experiment results, PPF (Plate Pile Foundation) can carry almost the same loads with OPF (Open-ended pipe pile foundation) and larger loads than those of BPF (Box pile foundation) under both vertical and horizontal conditions. Hence, sheet pile foundation would be a promising alternative to conventional round pipe pile foundation, especially in high-seismic areas where foundations will experience both vertical and horizontal loading.

1 INTRODUCTION

In recent years, efficient installation methods of piles have been developed (e.g. IPA, 2016; Proc. of the 1st Int. Conf. on Press-in Eng., 2018). Nowadays, a challenge in piling engineering is to reduce costs including transportation and construction costs, and at the same time keep safety of foundation structures. Hence, a target in this research is to show a possibility to use steel sheet piles for pile foundations of permanent structures, because time and cost of construction of sheet piles could be lower than those of steel pipe piles.

In this study, a series of experiments were conducted to investigate the load transfer behaviours of model foundations supported by open-ended pipe piles, plate piles, or square box pile in dry sand ground subjected to vertical and horizontal loading. Load-displacement relationships and load sharing between the raft and the piles of three types of foundations during vertical loading and horizontal loading are presented and discussed.

2 EXPERIMENT DESCRIPTION

2.1 *Model ground*

The sand used as model ground was Silica sand #6. Table 1 shows the physical properties of the sand. The model ground was prepared in a rectangular box with dimensions of 500 mm in width × 800 mm in length × 530 mm in height. The model ground was prepared with 10 layers of 50 mm thick and one top layer of 30 mm thick. The sand of each layer was poured into the soil box and compacted by hand tamping to get a target relative density, $D_r = 82\%$ ($\rho_d = 1.533$ ton/m^3).

2.2 *Model foundations*

The model piles used in all experiments were made of aluminium round pipes, plates representing sheet piles or square box pipe. Three different types of model pile foundations were used in the experiments, as shown in Figure 1. The first one is

DOI: 10.1201/9781003215226-15

Table 1. Physical properties of Silica sand #6 (after Vu et al, 2018).

Property	Value
Soil particle density, ρ_s (ton/m³)	2.679
Minimum dry density, ρ_{dmin} (ton/m³)	1.268
Maximum dry density, ρ_{dmax} (ton/m³)	1.604
Maximum void ratio, e_{max}	1.089
Minimum void ratio, e_{min}	0.652
Model ground relative density, D_r (%)	82.5
Model ground density ρ_d (ton/m³)	1.533

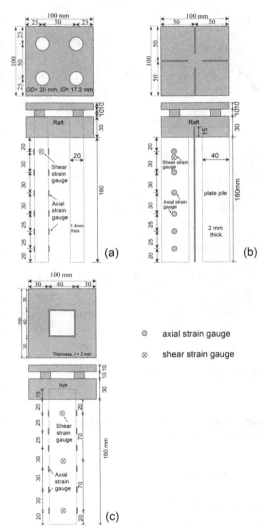

Figure 1. Model pile foundations: (a) Pipe pile foundation; (b) Plate pile foundation; (c) Box pile foundation.

a foundation supported by open-ended pipe piles with an outer diameter of 20 mm, an inner diameter of 17.2 mm, a wall thickness of 1.4 mm and a length of 210 mm (called OPF, Figure 1a). The 2nd is a foundation supported by plate piles with a width of 40 mm, a length of 195 mm and a thickness of 2 mm (called PPF, Figure 1b). The 3rd is the one supported by square box pile with a width of 40 mm, a length of 195 mm and a thickness of 2 mm (called BPF, Figure 1c). The geometrical and mechanical properties of the model piles are listed in Table 2.

It was intended to use the same volume of pile material for the three model foundations with the same length. Note that one BP was used, while 4 OPs or 4PPs were used. Hence, the total volume of piles of each foundation was almost same, as shown in Table 2.

Fourteen strain gauges were attached on opposite sides of each pile of PP and OP. And eighteen strain gauges were attached on sides of BP.

The model square raft had a side length of 100 mm and a thickness of 30 mm, as shown in Figure 1. Pile heads were rigidly connected to the raft in all the foundations.

2.3 Experimental procedure

For each model foundation, two cases were carried out. The first case was aimed to obtain the penetration resistance during PPT (Pile Penetration Test) and the bearing capacity of the foundation in VLT (Vertical Load Test). The second case was mainly aimed to obtain the performance of the foundation subjected to horizontal loading. In the 2nd case, HLT (Horizontal Load Test) was carried out after PPT.

In PPT, the foundation was penetrated in the model ground using a screw jack (Figure 4a) until the pile embedment length reached 170 mm. After that, in the 1st case, vertical load test (VLT) was conducted with the raft base being untouched to the ground surface (called PG stage or PG condition, PG: Pile group) as shown in Figure 2. After the raft base touched the ground surface (called PR stage or PR condition, PR: Pile raft), VLT was again conducted. In the 2nd case, after the raft base touched the ground surface, the vertical load by screw jack was unloaded. HLT was carried out immediately after PPT. A death

Table 2. Properties of model piles.

	OP	PP	BP
Length from raft base, L (mm)	180	180	180
Cross sectional area, A (mm)	81.8	80	304
Wall thickness, t (mm)	1.4	2	2
Young's modulus, E (GPa)	71.3	69.5	75.6
Poisson's ratio, v	0.343	0.297	0.356
Bending rigidity, EI (MNmm²)	253.7	1.85	5546.1
Bending rigidity, EI (MNmm²) (strong axis)	—	742	—

195

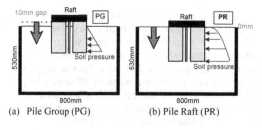

(a) Pile Group (PG)   (b) Pile Raft (PR)

Figure 2. Illustrations of load tests of PG and PR.

weight of 1000 N was placed on the foundation prior to the start of HLT, as shown in Figure 4 (b).

Locations of open-ended pipe piles and plate piles are shown in Figure 3. In OPF, OP2 and OP3 are "front pile" while OP1 and OP4 are "back pile". In PPF, PP2 is called "front pile", PP1 and PP3 are called "middle pile", and PP4 is called "back pile".

A total of 6 experiments were carried out. After the completion of each experiment, cone penetration tests (CPTs) were conducted in the model ground. The cone used in CPTs had a diameter of 20.5 mm and an apex angle of 60 degrees. The diameter of the cone was similar with that of the model pipe pile.

2.4 Measurement

During the load test, the horizontal load applied to the foundation was measured by means of a load cell attached between the raft and the winch. Horizontal displacement and vertical displacements of the raft were measured by means of dial gauges (Figure 4b). And the inclination of the raft was obtained from an inclinometer. Axial forces and bending moments of the model piles were estimated from the measured axial strains, and the horizontal load of each pile was estimated from the shear strain gauges (cross-gauges) near the pile top.

In HLT, distribution of horizontal displacements along the pile shaft was calculated from the measured distribution of bending moments, the horizontal displacement at the loading point and the raft inclination. It is noticed that the inclination of the pile top was equal to the raft inclination as the piles were rigidly connected to the rigid raft. Unfortunately,

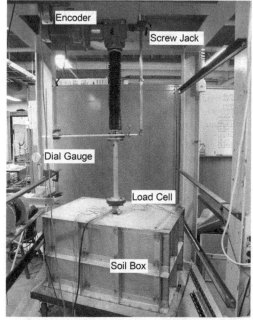

(a) VLT

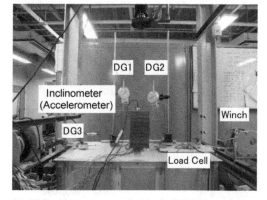

(b) HLT with a constant vertical load of 1000 N

Figure 4. Experimental set-up.

some data of BPF was not obtained due to a technical problem.

3 EXPERIMENT RESULTS

3.1 Results of CPTs

Figure 5 shows the locations of CPTs. As shown in Figure 6, CPT tip resistant q_c increased almost linearly with depth from the ground level. And, distributions of q_c with depth are almost uniform among the cases. That is, all the experiments were conducted under the same ground condition.

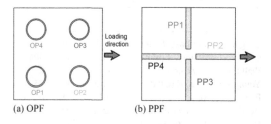

(a) OPF   (b) PPF

Figure 3. Locations of piles in two foundations (top view).

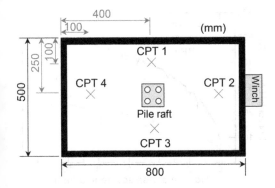

Figure 5. Location of CPTs.

3.2 Results of PPT and VLT

Figure 7 shows the relationships of the vertical load P and the settlement w of three model foundations during the PPT and VLT in the first cases. The results show that the load of open-ended pipe piles is around two times that of plate piles in the PPT and PG stages. However, in the PR stage, the loads of OP and PP are almost the same. This will be discussed in detail later.

3.3 Results of VLT on PG and PR

Figure 8 shows comparisons of vertical load-settlement relationships of the foundations in stages of PG and PR. The loads of piles were calculated from axial strain gauges. The loads carried by 4 OPs, 4 PPs and BP are also shown by the dashed lines. The difference between the total load and the pile load is the load carried by the raft. Note that touch down level of each foundation is different. It was intended to leave a gap about 10 mm between the raft and the soil surface after the end of the PPT. However, due to the limited precision of the instrument, there was a difference about 3 mm of the gap between three types of foundation prior to the start of VLT. The vertical displacement, w, was zeroed at the start of VLT in Figure 8.

In the PG condition, of course, the load carried by piles was almost equal to the total load. The load of OPs was around two times that of PPs and 1.5 times that of BP.

In the PR condition, the total loads of the foundations increased rapidly when the experimental stage turned to PR condition after the raft base touched the ground surface. It is interesting to notice that the loads of 4 OPs, 4 PPs and BP continued to increase in PR condition. However, the load of 4 PPs increased significantly faster than that of 4 OPs or BP. The load of 4 PPs is only around 1000 N in the PG condition, while in the PR condition, this value became about 2.5 times when the settlement of the foundation reached about 27 mm.

A possible reason for this result is as follows. In the PR condition, because a part of vertical load is

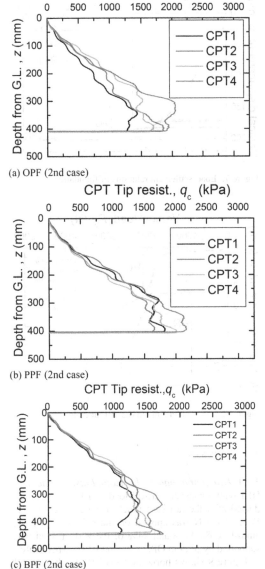

(a) OPF (2nd case)

(b) PPF (2nd case)

(c) BPF (2nd case)

Figure 6. Results of CPTs in model ground.

transferred to the ground through the raft base, stress levels in the soil surrounding the piles increase. Hence, the unit shaft resistance of the piles in PR condition increases compared to that in PG condition. Moreover, since the shaft area of PPs is the largest among the three model foundations, the increase in the shaft resistance (= the unit shaft resistance × the shaft area) of PPs is greater than that of OPs and BP in PR stage. As a result, the load carried by 4 PPs increased significantly faster than that carried by 4 OPs and BP under PR condition.

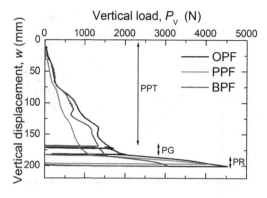

Figure 7. Load-settlement relations of foundations.

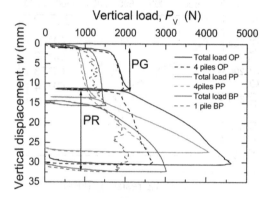

Figure 8. Load-settlement curves of foundations.

3.4 Results of HLT

3.4.1 Horizontal load vs. horizontal displacement

HLT with a constant vertical load of 1000 N was conducted after the raft base touched the ground surface. When 1000 N was applied on the raft prior to HLT, almost 100%, 93% and 94% of the vertical load was supported by piles in OPF, PPF and BPF, respectively.

Figure 9 shows horizontal load P_H vs. horizontal displacement u of OPF, PPF and BPF. It can be clearly seen from the figure that P_H of OPF is larger than that of PPF and BPF, and reached peak at $u = 12$ mm. On the other hand, PPF carried the almost same horizontal load with OPF after u reached 12 mm and kept increasing. Among three model foundations, the load of BPF was smallest.

3.4.2 Inclination of raft

Figure 10 shows horizontal displacement u vs. inclination θ of the raft in cases of OPF, PPF and BPF. It can be seen that the inclination of the raft in PPF is much smaller than that in OPF and BPF. It is thought that a high value of bending rigidity EI of PP2 and PP4 (strong axis, see Table 2 and Figure 3) contributes to supressing the inclination. Another reason is considered that even the bending rigidity EI of the

whole (four) PPs including two PPs in weak axis and another two PPs in strong axis is smaller than that of BP, the greater distance between the front and the back edges of PPs can also contribute to preventing the foundation from rotating, as shown in Figure 3.

3.4.3 Axial force on each pile

Figure 11 shows the changes of the axial force ΔF_a on each pile during HLT. In both cases of OPF and

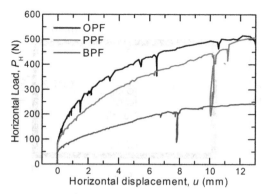

Figure 9. Horizontal displacement vs. displacement load.

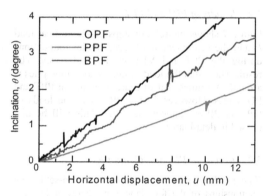

Figure 10. Horizontal displacement vs. inclination.

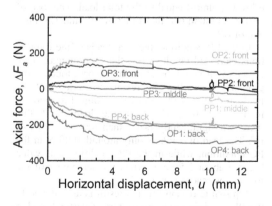

Figure 11. Axial force on each pile during HLT.

PPF, compression forces were generated on the front piles (OP2, OP3, PP2), while tension forces were generated on the back piles (OP1, OP4, PP4). The compression force of PP2 was smaller than that of OP2 and OP3, while the tension force of PP4 was almost equal to that of OP1. It is noticed that ΔF_a of the middle piles in PPF (PP1 and PP3) remained almost unchanged.

3.4.4 Horizontal load sharing

Figure 12 shows the relationship between horizontal displacement u and horizontal resistances between different piles in OPF, PPF and BPF. As for PP2 and

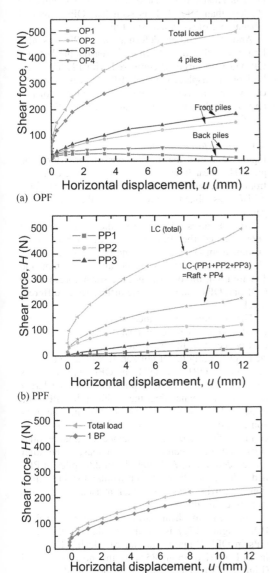

(a) OPF

(b) PPF

(c) BPF

Figure 12. Horizontal resistance vs. horizontal displacement.

PP4, the shear force near the pile head was estimated from the cross-gauges near the pile head (see Figure 1). However, since the cross-gauges of PP4 were damaged during HLT, the data of shear force in PP4 is not available.

In OPF (Figure 12a), four OPs carried around 80% of the total horizontal load. It is clearly seen that the front piles (OP2 and OP3) carried larger horizontal resistance than the back piles (OP1 and OP4). It is interesting that the horizontal resistance of the front piles (OP2 and OP3) continued to increase with increasing u, while the horizontal resistance of the back piles (OP1 and OP4) showed a softening behaviour.

One mechanism for the smaller horizontal resistance of the back piles would be the existence of the front piles. This may be related to larger inclination of OPF (see Figure 10). The soil in front of the foundation was compressed during the HLT, and the horizontal earth pressure on the front OPs increased. On the other hand, because of the inclination of OPF, the earth pressure on the back OPs decreased, and the shear forces of the back OPs increased slightly and then decreased gradually. As a result, there is a larger difference of the horizontal resistance between the front OPs and the back OPs, as pointed out by e.g. Horikoshi et al. (2003) and Vu et al (2018).

In addition, the bending moment distributions of different piles in the same pile group is also different (see Figure 13). Although the pile head displacement of front OPs is the same as that of the back OPs (see Figure 14), the change of bending moment of the front OPs is more obious and the peak bending moment is higher. As a result, they are more likely to suffer damage.

In PPF (Figure 12b), the front pile (PP2) carried larger horizontal resistance than the middle piles (PP1 and PP3). Since the width (thickness) of PP2 in the loading direction is only 2 mm, the horizontal resistance of PP2 is mainly the friction resistance acting on two side walls. The behavior of the PP4 might be similar to the PP2, although the horizontal resistance was not obtained due to the damage of the cross-strain gauges.

In BPF (Figure 12c), BP carried around 85% of the total horizontal load.

3.4.5 Bending moments in piles

Figure 13 shows the distributions of bending moment in piles for three types of foundations. It was difficult to obtain bending moments in PP2 and PP4 in PPF, because of very large bending stiffness of these piles in strong axis. Unfortunately, strain data of BPF deeper than 80 mm was not obtained due to a technical problem.

When the horizontal load reached 500 N, the maximum bending moment of PP3 is about 1/3 of that of OP3, as shown in Figure 13(b) and (f). However, the bending stress in PP3 is much larger than that of OP2, because EI of PP (weak axis) is only 1/135 of that of OP (Table 2).

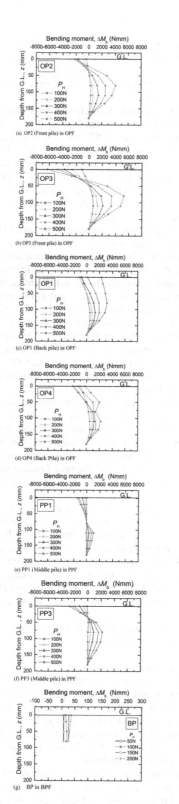

(a) OP2 (Front pile) in OPF

(b) OP3 (Front pile) in OPF

(c) OP1 (Back pile) in OPF

(d) OP4 (Back Pile) in OPF

(e) PP1 (Middle pile) in PPF

(f) PP3 (Middle pile) in PPF

(g) BP in BPF

Figure 13. Distribution of bending moment of piles.

Under the horizontal loading condition, the performance of PP is better than that of OP.

As shown in Figure 13(g), the bending moment of BP is much smaller than that of OPs and PPs at the same horizontal load level. To explain this phenomena, further study is necessary.

3.4.6 *Horizontal displacements of piles*

Figure 14 shows the distributions of horizontal displacements of the piles with depth for different P_H. Local horizontal displacements were estimated from the measured bending moments, pile head displacement, and inclination of the pile head.

In general, the horizontal displacement of the PP head is greater than that of OP head especially for P_H less than 500 N. It is seen that OPs exhibited a behaviour of so-called "short pile".

And, the bending deformation of PP was also larger than that of OP, which is reasonable because OP had very large EI compared with PP. The local horizontal displacement of BP decreases significantly with increase in z, compared with that of PPs and OPs at the same horizontal load level ($P_H = 150$ N, 200 N).

3.4.7 *Shear forces in piles*

Figure 15 shows the distributions of shear forces in piles. The shear forces of OPs and PPs were estimated from the measured bending moments. The shear force of BP was obtained directly from the shear strain gauges.

From Figures 15(a) to (d), the shear forces in OPs changed the direction at a depth z of about 90 mm from ground level. The largest shear forces along OP2 and OP3 occurred at $z = 35$ mm and 145 mm. On the other hand, the shear forces along OP1 and OP4 showed a little difference at different levels.

The horizontal resistance of each pile is almost equal to the shear force near the pile head. It is clearly seen that the front piles (OP2 and OP3) carried significantly larger horizontal resistance than the back piles (OP1 and OP4) in OPF. Moreover, as shown in Figure 12, front pile (PP2) carried larger horizontal resistance than the middle piles (PP1 and PP3) in PPF, but when u exceeded 6 mm, the horizontal resistance of PP2 levelled off at 100 N. In contrast, the horizontal resistance of middle piles (PP1 and PP3) continued to increase even after u exceeded 6 mm. Although the middle piles in PPF had a larger width in the horizontal loading direction, it carried less horizontal resistance than the front piles in OPF.

The trend of shear force distribution of BP in BPF (Figure 15g) is similar to that of the front OPs in OPF (Figure 15a and b). During HLT, BP in BPF carried about 85% of the horizontal load (Figure 15g). This result can be found also in Figure 12.

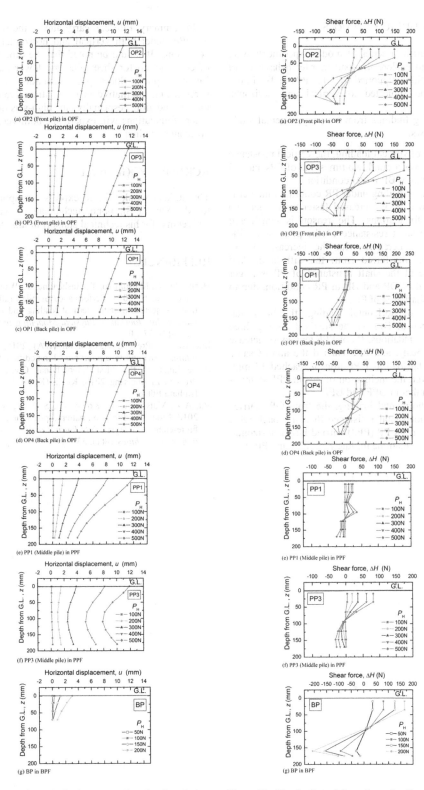

Figure 14. Horizontal displacement of each piles during HLT.

Figure 15. Distribution of shear forces in piles.

4 CONCLUDING REMARKS

In this research, a series of model experiments were conducted to explore the load transfer behaviours of pile foundations supported by OPs, PPs or BP in dry sand ground subjected to vertical and horizontal loading.

Interesting findings from this experimental study are as follows:

(1) In case of vertical loading, the vertical resistance of OPs was around two times that of PPs and 1.5 times that of BP in PG condition.

(2) In vertical loading under PR condition, the vertical load of PPF increased faster than other two model foundations. The reason is considered that the shaft resistance of the piles under PR condition increases with the increase in soil pressure transferred from the raft base. And increase of the shaft resistance of PP is larger than that of OP and BP in PR condition, due to larger shaft area of PP.

(3) In case of horizontal loading, the horizontal resistance of BPF is the smallest among three model foundations. P_H of PPF tended to increase continuously, and PPF carried the almost same horizontal load as OPF after u reached 12 mm. Moreover, the inclination of raft in PPF is smaller than that in OPF and BPF.

(4) Under horizontal loading, the front OPs carries larger horizontal resistance than the back OPs. And the front PP (PP2) carried larger horizontal resistance than the middle PPs (PP1 and PP3).

In summary, PPF can carry almost the same load as OPF under both vertical and horizontal loading conditions. It should be noted that the time and cost of construction for sheet piles are lower than those for pipe piles. Sheet pile foundation would be a promising alternative to conventional round pipe pile foundation, especially in high-seismic areas where foundations will experience both vertical and horizontal loading.

ACKNOWLEDGEMENT

The authors would like to express our appreciation to Mr. Shinya Shimono, technician of Kanazawa University, for his technical support in this study.

REFERENCES

Horikoshi, K., Matsumoto, T., Hashizume, Y., Watanabe, T., Fukuyama, H. 2003. Performance of piled raft foundations subjected to static vertical loading and horizontal loading, *Int. Journal of Physical Modelling in Geotechnics*, 3(2): 37–50.

International Press-in Association (IPA) 2016. *Press-in retaining structures: a hand book* First edition 2016.

Proceedings of the First International Conference on Press-in Engineering 2018, Kochi (Kusakabe, Ueno & Ishihara Edt.),

Vu, A.T., Matsumoto T., Kobayashi, S. & Nguyen, T. L. 2018. Model load tests on battered pile foundations and finite-element analysis. *Int. Journal of Physical Modelling in Geotechnics* 18(1): 33–54.

Proceedings of the Second International Conference on
Press-in Engineering 2021, Kochi, Japan – Matsumoto et al (eds)
© 2021 Taylor & Francis Group, London, ISBN 978-1-032-10414-0

Experimental study on the pile group effect in the horizontal resistance of spiral piles

N. Ohnishi
Graduate School of Science and Engineering, Chuo University, Tokyo, Japan

H. Nishioka
Department of Civil and Environmental Engineering, Chuo University, Tokyo, Japan

ABSTRACT: Here, horizontal loading tests on spiral piles and cylindrical piles were conducted. The model ground was made using dry sand. Model piles were arranged in a row on the model ground; a wire was attached to the pile head and loaded monotonically in one direction. The experiment was conducted for five cases, with the center-to-center spacing of the piles changing from 1.5 times to 8 times the pile diameter. The results revealed that the tendency of the pile group effect of the spiral pile varied vastly depending on the evaluated load level. At the initial loading stage, spiral piles were more affected by the pile group effect than ordinary cylindrical piles; the relatively large initial displacement should be considered while designing spiral piles. However, as the loading progressed, spiral piles tended to have a smaller reduction in bearing capacity than the cylindrical piles owing to the pile group effect.

1 INTRODUCTION

1.1 Background and objectives of the project

Spiral piles are small-scale piles constructed by twisting flat steel plates that are used in the construction of pile foundations by manually drilling them into the earth's surface (Figure 1). It is very convenient in a site where the use of heavy construction machinery is arduous owing to a lack of space. An example of this could be a platform door foundation developed for railways using spiral piles (Nonaka et al. 2019), as shown in Figure 2.

When a pile group foundation is designed, it is necessary to consider the pile group effect fully. In practice, it is common to set the center-to-center interval of piles at a pile diameter (D) of approximately 2.5–3.0. However, most of the sites where spiral piles are used are narrow, and the area in which the foundation can be installed is limited. In the case of a pile group structure that uses spiral piles, a rational and economical design will be possible even if the pile spacing is smaller than the conventional pile spacing. However, only a few examples of horizontal loading tests on pile groups comprising spiral piles are available, and there are no studies in which the pile group effect was quantitatively investigated.

In the present study, horizontal loading tests on spiral and cylindrical piles were conducted, and the results were compared to examine the pile group effect of spiral piles.

2 EXPERIMENT OUTLINE

2.1 Model pile

A spiral pile made of phosphor bronze with a width of 20 mm, plate thickness of 2 mm, and length of 400 mm was utilized. The plate was twisted four times to make the pile (Figure 3). That is, the pitch is 50 mm. A separate (i.e., without sand) bending test demonstrated that the bending stiffness of this pile was $EI = 2613.3$ kN–mm^2. Figure 4 shows the bending test of a spiral pile. It can be confirmed that the localization of deformation has not occurred. Therefore, it assumes that the entire pile has uniform bending stiffness. In this experiment, all the spiral piles were grabbed by the fixing jig at pile head and fixed in the weak axis direction at the loading point, so the bending test was also fixed in the weak axis direction at the fixed part of the cantilever. A similar experiment was performed using an acrylic pile, which followed a conventional cylindrical pile model, with an

DOI: 10.1201/9781003215226-16

Figure 1. Construction status of spiral piles (Nonaka et al. 2019).

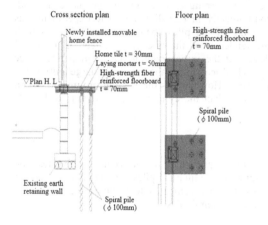

Figure 2. Example of foundation structure using spiral piles and high-strength fiber reinforced floorboards (Platform door foundation on embankment type platform) (Nonaka et al. 2019).

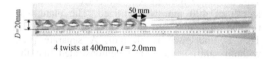

Figure 3. Model spiral pile.

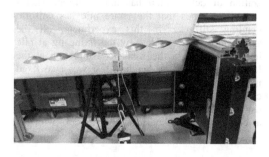

Figure 4. Bending test for spiral pile.

outer diameter of 20 mm and a plate thickness of 2 mm. The bending stiffness was found to be EI = 14838 kN–mm^2. The specifications of these models were set based on the following concepts. First, for the model acrylic pile, the pile diameter and plate thickness were set to satisfy $\beta l > 3$, a condition that can be generally treated as a semi-infinite length pile. β is the characteristic value of the pile represented by Equation (1). (Japanese Geotechnical Society. 2010)

$$\beta = [k_h \cdot D/(4EI)]^{1/4} \qquad (1)$$

where k_h is the coefficient of the horizontal subgrade reaction, E is Young's modulus, and I is the moment of inertia of the area.

For k_h, the value obtained from a horizontal loading test result of a single pile conducted in advance was used. For the model spiral pile, the plate width was set to be the same as the outer diameter of the model acrylic pile. The plate thickness was set in proportion to the ratio of the plate width to the wall thickness of the actual spiral pile.

2.2 Experimental device

The model ground was made using dry sand. The sand was dropped into a soil tank with a width of 1000 mm and a depth of 400 mm using the multiple sieving method. The relative density of the sand was approximately 80%.

In the model pile installation method, the spiral piles were individually drilled into the model ground to a total penetration depth of $l = 390$ mm. The acrylic pile penetrated 150 mm when 200 mm of sand was poured; then, the remaining sand was poured. (Therefore, the penetration depth, l, was 350 mm).

The spiral pile was constructed using the jig shown in Figure 5, but due to the influence of the problem in the design, blurring and over-rotation inevitably occurred during construction.

Model piles were arranged in a row on the model ground, and a wire was attached to the pile head and loaded monotonically in one direction (Figure 6,7). The experimental condition is that the pile head and tip are free.

2.3 Experimental pattern

The experiment was conducted for five cases, with the center-to-center spacing of the piles changing from 1.5 times to 8 times the pile diameter D (Figure 8). For spiral piles, D was set to the width of the plate.

As shown in the figure, all spiral piles were constructed to be in the weak axis direction at the loading point.

To improve the reliability of the results, acrylic piles and spiral piles were tested three times in all cases.

Figure 5. Construction jig for spiral pile.

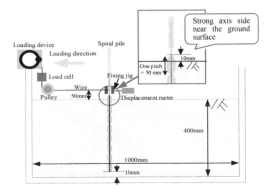

Figure 6. Overview of the test soil tank and loading device.

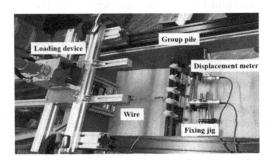

Figure 7. Installation status of loading equipment and measuring equipment.

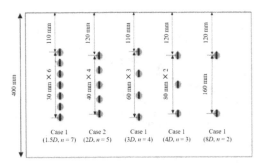

Figure 8. Experimental case pile placement.

3 EXPERIMENTAL RESULT

3.1 Acrylic pile

Figure 9 shows the load–displacement relationship of the acrylic piles. Although it differs slightly from test to test, the resistance decreases as the pile spacing reduces, owing to the pile group effect.

3.2 Spiral pile

Conversely, from the load–displacement relationship curves of the spiral piles shown in Figure 10, in some cases, the resistance per pile obtained from the experiment conducted at $L = 1.5D$ exceeded the resistance obtained at $L = 8D$; the resistance did not decrease even if the pile spacing was reduced. We will discuss these differences in chapter 4.

3.3 Pile characteristic value βl

For the test results of $L = 8D$ of the acrylic pile, the value of βl was calculated from the average value of the load when the horizontal displacement was 0.2 mm. The Chang formula (Japanese Geotechnical Society. 2010) used is shown in Equation (2).

$$y_t = \frac{(1 + \beta h)^3 + 1/2}{3EI\beta^3} P \qquad (2)$$

where y_t is the horizontal displacement, β is the characteristic value of the pile, h is the height of the loading point, E is Young's modulus, I is the moment of inertia of the area, and P is the load.

From the equation, $\beta l = 3.89$ ($l = 350$ mm) was obtained. It was found that the setting concept of the model specification ($\beta l > 3$) was satisfied when the coefficient of the horizontal subgrade reaction at this time was calculated for $k_h = 55609$ kN/m³.

Similarly, βl and k_h for the spiral pile were obtained to be $\beta l = 7.72$ ($l = 400$ mm) and $k_h = 72510$ kN/m³. It was observed that the flexural rigidity was lower than that of acrylic piles and βl was larger owing to the longer l, but k_h itself was higher

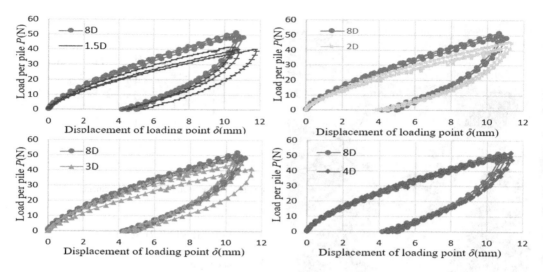

Figure 9. Acrylic pile load–displacement relationship.

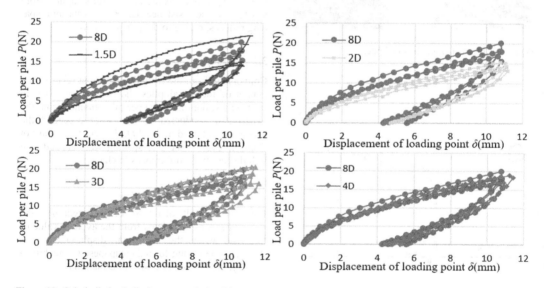

Figure 10. Spiral pile load–displacement relationship.

than that of acrylic piles owing to the influence of load-width dependence. (Japanese Geotechnical. 2017)

4 DISCUSSION OF EXPERIMENTAL RESULTS

4.1 Large load

The behavior when the ground resistance was sufficiently plasticized was compared (hereinafter referred to as "large load"). Table 1 shows the experimental results and coefficient of variation of the acrylic pile under a 40 N load. Table 2 shows the

experimental results and coefficient of variation of the spiral pile when 14 N is loaded.

Table 1 shows that the experimental results of the acrylic piles are relatively stable, although there is some variation at the pile center spacing $L = 3D$.

Conversely, Table 2 shows that the results of spiral piles are more inconsistent than those of acrylic piles. The possible causes could be the construction failure of the pile and disturbance of the ground around the pile during rotational penetration. Notably, for spiral piles, slight differences in construction conditions, such as the amount of force applied during rotational penetration and the slight deviation of the pile center, may significantly affect the

Table 1. Displacement of the acrylic pile under 40 N load.

Pile center spacing	P = Displacement when 40 N is loaded (mm)			Average	Coefficient of variation
$L = 8D$	7.79	8.17	7.36	7.77	4%
$L = 4D$	7.96	7.69	7.42	7.69	3%
$L = 3D$	9.00	8.01	10.96	9.32	13%
$L = 2D$	10.42	10.18	9.27	9.96	5%
$L = 1.5D$	9.66	10.70	11.49	10.61	7%

Table 2. Displacement of the spiral pile under 14 N load.

Pile center spacing	P = Displacement when 14 N is loaded (mm)			Average	Coefficient of variation
$L = 8D$	8.37	5.96	7.39	7.24	14%
$L = 4D$	6.47	7.72	7.00	7.06	7%
$L = 3D$	5.77	5.69	8.95	6.80	19%
$L = 2D$	9.45	10.74	11.42	10.54	8%
$L = 1.5D$	4.45	9.62	10.10	8.06	32%

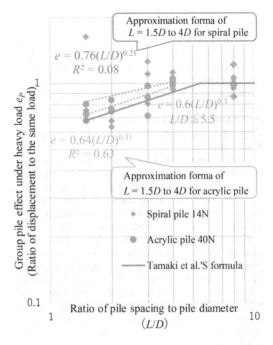

Figure 11. Relationship between pile spacing and group pile effect e_p under large load.

development of resistance. Therefore, there is a possibility that the disorder of construction when using the construction jig described in Chapter 2 had a significant influence. Remarkably, at $L = 1.5D$, which has the closest pile spacing, the results varied widely; the peripheral ground of the pile was strongly compacted during the rotational penetration, and the resistance increased.

However, it is difficult to eliminate this blurring and over-rotation even in the actual field. Therefore, although it is necessary to consider the uncertainty associated with construction in the design, it is important for research to separate construction variation from other factors.

Figure 11 shows the relationship between pile spacing and pile diameter, the L/D ratio, obtained in this experiment, and the pile group effect, e_P, under a large load. The pile group effect e_P on the vertical axis in Figure 11 is the value obtained by dividing the displacement of a single pile under the same load by the displacement of the pile group, referring to the definition by Tamaki et al (1971). However, in this experiment, the average of the cases with the pile center-to-center spacing $L = 8D$ was set to be the same as that of a single pile without the pile group effect. Figure 9 shows the pile group effect when 14 N is loaded on the spiral pile and 40 N is loaded on the acrylic pile. Further, Figure 11 also illustrates the relational expression between the pile spacing and the pile group effect in the case of the free pile head given by Tamaki et al (1971). shown in Equation (3).

$$e = 0.6(L/D)^{0.3}$$
$$5.5 > (L/D) > 1.5 \qquad (3)$$

where e is the pile group effect, L is the pile center spacing, and D is the pile diameter. In addition, in Figure 9, the results of approximating the $L = 1.5D$ to $4D$ data in both cases with a power function are shown according to Equation. (3).

From Figure 11, we understand that the result of the acrylic pile is relatively close to the relational expression of Tamaki et al (1971). Conversely, the spiral pile has a larger average e_P value of the pile group effect than the acrylic pile. Furthermore, because the spiral pile has a smaller index value in the approximation formula than the acrylic pile, it was found that the effect of the pile interval on the pile group effect was relatively small when evaluated under a large load. However, the value of the coefficient of determination, R^2, is very low owing to the large variation in the results.

The reason for this difference could be that the stress distribution range in the ground near the spiral pile is smaller than that of the conventional cylindrical pile. It is predicted that this is because there are points where the effective width of the spiral pile when pushed into the ground is reduced to the plate thickness instead of the full width. Moreover, since the projected area of the loading direction of the $1/\beta$ section of the spiral pile is smaller than the cylindrical pile, D can be evaluated to be smaller than the plate width, and as a result, there is a possibility that a difference in the group pile effect appears.

4.2 Initial stage of loading

Next, to consider the difference in tendency when the load level to be evaluated is reduced, the pile group effect at a displacement of 1% (0.2 mm) of the pile diameter is evaluated (Figure 12,13). This displacement level is generally treated as an equivalent linear range in design practice. Therefore, we focus on the horizontal spring constant K obtained by dividing the horizontal load by the horizontal displacement.

Table 3 shows the horizontal spring constant of the acrylic pile, and Table 4 shows the horizontal spring constant of the spiral pile when displaced by 0.2 mm. There is no significant difference in the variation in the results of the acrylic pile and the spiral pile, unlike in the case of a large load.

Figure 14 shows the relationship between pile spacing L/D and horizontal spring constant pile group effect e_K when the pile diameter is displaced by 1% (0.2 mm). Pile group effect e_K on the vertical axis is the ratio of the horizontal spring constant at 0.2 mm displacement of $L = 8D$ to the average of all three cases with the other horizontal spring constants, where the L values are $4D$, $3D$, $2D$, and $1.5D$.

Figure 14 shows that the acrylic pile has a greater pile group effect than the spiral pile. In particular, for acrylic piles, the effect was very large at the initial stage of loading.

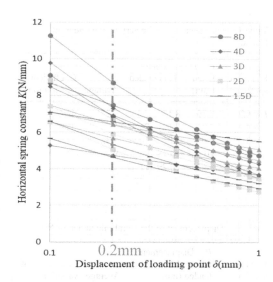

Figure 13. Relationship between displacement and horizontal spring constant of spiral pile.

Table 3. Horizontal spring constant at 0.2 mm displacement of acrylic pile.

Pile center spacing	Horizontal spring constant at 0.2 mm displacement K (N/mm)			Average	Coefficient of variation
$L = 8D$	15.56	14.11	18.58	16.08	12%
$L = 4D$	17.30	16.70	16.63	16.88	2%
$L = 3D$	15.88	17.21	11.09	14.73	18%
$L = 2D$	15.15	14.47	16.37	15.33	5%
$L = 1.5D$	15.57	13.60	13.26	14.14	7%

Table 4. Horizontal spring constant at 0.2 mm displacement of spiral pile.

Pile center spacing	Horizontal spring constant at 0.2 mm displacement K (N/mm)			Average	Coefficient of variation
$L = 8D$	8.68	6.58	6.83	7.36	13%
$L = 4D$	4.72	7.29	6.90	6.30	18%
$L = 3D$	6.39	6.39	5.67	6.15	6%
$L = 2D$	5.88	6.64	5.04	5.85	11%
$L = 1.5D$	6.55	4.63	5.30	5.49	14%

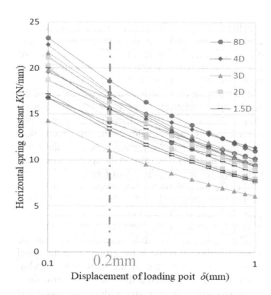

Figure 12. Relationship between displacement and horizontal spring constant of acrylic pile.

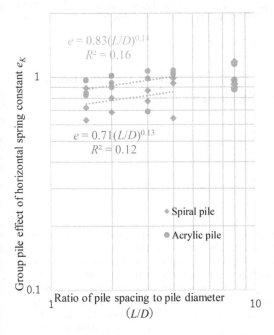

Figure 14. Relationship between pile spacing and horizontal spring constant group pile effect e_K.

Figure 15. New construction jig for spiral pile.

5 CONCLUSIONS

Examination of the pile group effect of spiral piles revealed that the effect varied significantly depending on the evaluated load level. Designs must consider that at the initial stage of loading, spiral piles are more affected by the pile group effect than ordinary cylindrical piles, and the initial displacement is relatively large. However, as the loading progressed, it was found that spiral piles tended to have a larger pile group effect than the cylindrical acrylic piles.

The test results for spiral piles vary widely, and slight differences in conditions at the time of penetration may significantly affect the resistance. Therefore, great care should be taken when constructing spiral piles.

In our future studies, we will use the new construction jig shown in Figure 15 and conduct experiments to reduce construction disturbances. In addition, we will conduct experiments that simulate actual construction conditions, such as horizontal load tests with multiple rows of pile groups that

match the bending rigidity of spiral piles and acrylic piles, to explain the pile group effect of spiral piles further.

REFERENCES

Japanese Geotechnical Society: *Method for Lateral Load Test of Piles*, pp32 Maruzen Publishing, 2010. (in Japanese)

Japanese Geotechnical Society: *Soil and Foundation Design Calculation Exercises Corresponding to New Design Methods (2017 edition)*, pp75 Maruzen Publishing, 2017. (in Japanese)

Nonaka, T., Tsushima, F., Maehara, S., Nishioka, H., Otsuka, K., Inoyae, Y., Sugawara, T., Fujita, M. 2019. Proposal of home door foundation structure using spiral pile and high strength fiber reinforced slab. *JSCE the 74th Annual Lecture, VI-851*. (in Japanese)

Tamaki, S., Mitsuhashi, K., Imai, T. 1971. Horizontal resistance of a pile group subjected to lateral load. *Proceedings of JSCE 192: 79–89.* (in Japanese)

Proceedings of the Second International Conference on
Press-in Engineering 2021, Kochi, Japan – Matsumoto et al (eds)
© 2021 Taylor & Francis Group, London, ISBN 978-1-032-10414-0

Experimental observation on the ultimate lateral capacity of vertical-batter screw pile under monotonic loading in cohesionless soil

A. Jugdernamjil & N. Yasufuku
Kyushu University, Fukuoka-shi, Japan

Y. Tani, T. Kurokawa & M. Nagata
HINODE Co., Ltd, Fukuoka-shi, Japan

ABSTRACT: 1g model experimental studies were carried out in the laboratory to evaluate vertical perform-ance with varying angles battered pile in dense sand. Relatively rigid piles with slenderness ratios 9 and 13.5 were undertaken in a sand-filled tank under strain-controlled lateral loading. The tank dimensions were designed so that boundary effects could be minimized and earth pressure measured using transducers at the front and bottom, and rear in each case. The relative density of 90% was achieved to simulate a more precise field condition in all test cases using the tamper compaction method. In order to determine the efficiency of the ultimate bearing capacity of screw pile configurations in lateral loading, the plate and pipe pile configur-ations were chosen for comparison. Totally 26 experimental cases were performed, which included vertical-batter combinations. The results indicated that the load-displacement characteristic was nonlinear under lateral loading. The case of 45 degree demonstrated higher ultimate lateral resistance.

1 INTRODUCTION

In recent years, in agriculture and civil engineering, the screw pile is becoming popular in supporting foundations designated in soft ground. A screw pile is made by twisting a flat steel bar. It has a lower bending stiffness than other conventional piles because of its lower cross-section area (Figure 1). Preceding research on the bearing and pull-out cap-acity of the screw pile was studied by Sato et al. 2014, and Wang et al. 2018. Wang et al. 2018 found out that the optimal pitch width ratio (p/w) under a pull and pushing test is 4.5, selected in this experimental study. In this experimental test, the pitch and width of the screw model pile were taken as 72 mm and 16 mm, respectively, as shown in Figure 1. The pitch width ratio is equal with divid-ing pitch length by width length. As mentioned above, the screw piles and anchors are implemented in the foundation of greenhouses, solar panel farms, wind turbines, road signs, and guide rails because of their applicable axial loading and simple installation.

The present study focuses on the foundation of guide rails for protection from a vehicle accident that gives impact loading to them. The resultant lat-eral impact is made over lateral displacement in a single screw foundation due to less frontal and side soil resistance than a conventional pipe pile.

Therefore, additional reinforcement is required for improving lateral performance. The screw shape, which has a limitation that the lateral resistance is less than a steel pipe pile even having the same diameter, caused less bending rigidity concerning the lateral direction loading.

In the literature, as a method of improving the horizontal resistance, increasing pile dimensions such as length and diameter can be mentioned. How-ever, about a screw shape, there is a limitation in the manufacturing and installation environment. Thus, the coupled piles may be one of the improvement techniques in the lateral capacity, consisting of verti-cal and batter piles. In the decades, quantitative empirical and theoretical studies have been done on the lateral capacity of the conventional rigid pile such as tapered pile, belled-type pile, and steel H-pile in cohesionless soil using approaches such as analysis based on limiting equilibrium or plastic theory, analysis based on elastic theory, and non-linear analysis. Furthermore, one of the representa-tives of screw type of pile is the helical pile. The theoretical and empirical studies on the helical pile can be found in the literature. However, the study on the pile made by a flat bar twisting into a screw is scarce.

The aim of this research is to observe the lateral capacity of a single screw pile by a 1-g model test and comparing it with conventional model pile types

DOI: 10.1201/9781003215226-17

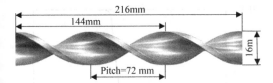

216mm
144mm
16m
Pitch=72 mm

Figure 1. View of screw pile.

such as a flat bar and a pipe. Additionally, in order to find out an optimal and rational angle between the vertical and batter piles for reinforcing a single screw pile, the cases of 30, 45, and 60 degrees are performed and evaluated.

2 METHODOLOGY

2.1 Experimental test apparatus

A macro-scale 1g model test was performed to determine the behavior of the lateral capacity of the screw pile and its combination. Figure 2 shows the experimental test setup conducted in this study. Lateral monotonic displacement was applied to the top of the pile. Pile head displacement of 20mm was applied, and the allowable pile capacity was taken as the load corresponding to 0.2D (3.2mm) of pile head deflection (Narasimha Rao et al. 1998, Chandrasekaran et al. 2010). The lateral load was applied at the pile head as free with an eccentricity of 40mm.

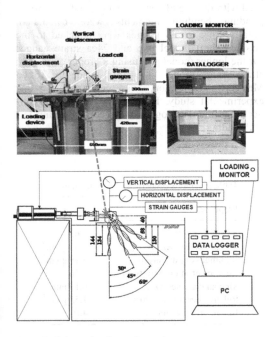

Figure 2. Schematic of experimental apparatus.

2.2 Model pile

Three types of model piles were tested, such as a steel screw, a flat bar, a pipe in this study for comparing lateral capacities. The screw pile was made by twisting a flat bar. The dimensions of the flat bar and a screw were the same: the thickness was 3mm, and the width was 16mm, respectively. The outer diameter of the pipe was equal to 16mm and 3mm thickness. The Young's modulus of the used steel model piles was $2 \times 10^{11} \text{N m}^{-2}$. The scaling of the model pile was adjusted for the chamber dimension that minimized its boundary effect. Considering the influence range as five times of pile diameter (5D), the lengths of the pile were set to 144mm and 216mm. Slenderness ratios (L/D) of the model pile were 9 and 13.5, respectively.

The scaling of each dimension is shown in Table 1. Dimensions of the model pile of the pipe and the flat bar were settled the same as the model of the screw pile. The scaling of the prototype represented by the model pile was calculated using the following formula (Wood et al. 2002):

$$\frac{E_m I_m}{E_p I_p} = \frac{1}{F^{4.5}} \qquad (1)$$

where F = scale factor: $E_m I_m$=flexural rigidity of model pile; and $E_p I_p$ = flexural rigidity of prototype. For an assumed cast iron prototype screw pile of diameter 220mm, the scaling factor is estimated at 10.

A pile in cohesionless soil can be considered to be rigid for practical purposes if the following condition is satisfied:

$$\frac{L}{T} \leq 2 \qquad (2)$$

In which $T = \sqrt[5]{\frac{EI}{n}}$

Table 1. Properties of model piles.

	Model		Prototype	
Material	Steel		Cast Iron	
E [GPa]	200		170	
I_x [m^4]	3.60E-11		1.34E-06	
I_y [m^4]	1.02E-09		3.81E-05	
EI_x [N·m2]	0.01		227.68	
EI_y [N·m2]	0.20		6476.34	
Length [m]	0.14	0.216	2	3
Diameter [m]	0.016		0.22	
Thickness [m]	0.003		0.04	
Scale factor [x axis]	1		10	
Scale factor [y axis]	1		10	

211

L = length of the pile; EI = flexural rigidity; and n = constant of horizontal subgrade reaction.

Figure 3 illustrates a comparison of the pile flexural rigidity with a depth between a screw, a flat bar, and a pipe. The moment of inertia of the screw varied along with the actual depth. For simplification, a four-point bending test was performed in screw piles for obtaining moment of inertia, which was close to 15 degrees tilted flat bar (I=20.4 Pa•m^4) using the following equation (2):

$$I = \frac{bh \cdot \left(h^2 \cdot cos^2\theta + b^2 \cdot sin^2\theta\right)}{12} \quad (2)$$

Where θ is the angle of a tilted flat bar with the axis.

2.3 Soil setup

Tests were conducted in a rectangular chamber; width 30cm, length 60cm, height 42cm, respectively. The sand used in the experiment was dried at the room temperature with a uniformity coefficient of 1.76 and a specific gravity of 2.63. Figure 4 shows the grain size distribution of soil samples used in this study. The height of the chamber was divided into 12 layers of thickness. For the required density of soil corresponding to 90% relative density, the weight of soil to be filled in each layer was calculated and poured into the chamber. The wooden stick was used to compact for satisfying uniformity all over the soil medium. This preparation method was approximate; however, it was conducted with as much care as possible. Relative density may be varied by 3% (Dr = 87-90%). Triaxial tests gave a friction angle of 42° at a dense density, γ = 15.2kN/m^3.

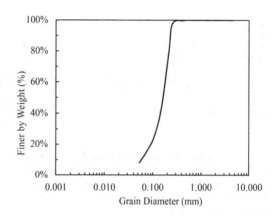

Figure 4. Grain size distribution curve of K7.

Initially, the sand was filled up to the tip of the model pile, and after that model pile was kept in its position, the sand was filled again. After running the test, the sand medium was removed, and model piles were detached. Lateral load on the pile was applied with servo cylinder and was measured with load cell attached to the cylinder head with loading rate 0.01mm/sec. A data

Overall, 26 cases were conducted with dense sand (Table 3).

In order to obtain bending moment through pile length, the strain gauges were attached to both front and rear sides, as shown in Figure 5. Cables of strain gauges were managed by bonding on the pile surface. The strain along the pile was measured with the use of a strain gauge, and the corresponding bending moments were calculated. Totally, 16ea strain gauges were attached to the coupled short model pile and 24ea strain gauges for the longer model pile using a proper bond. The strain gauge model FLA-200-3 made by Tokyo Measuring Laboratory was conducted in this experimental study. Each strain gauge was connected to the data logger and recorded during lateral loading.

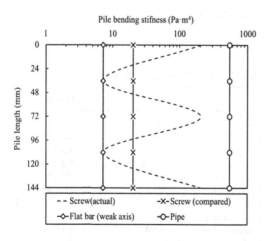

Figure 3. Variation of pile bending stiffness with depth.

Table 2. Index properties of soil medium.

Properties	Value
Specific gravity, Gs	2.63
Maximum dry density, ρ_{max}	1.56 g/cm^3
Minimum dry density, ρ_{min}	1.19 g/cm^3
Coefficient of uniformity, U_c	1.76
Median diameter, D_{50}	0.17 mm
Relative density, Dr=90%	1.53 g/cm^3
Internal friction angle, °	42°

Table 3. Experimental test conditions.

Test ID	Batter angle (°)	No. Pitch	L/D	y [mm]
S-Screw144	0	2	9	3.2
30-Screw144	30	2	9	3.2
45-Screw144	45	2	9	3.2
60-Screw144	60	2	9	3.2
S-Pipe144	0	-	9	3.2
30-Pipe 144	30	-	9	3.2
45-Pipe 144	45	-	9	3.2
60-Pipe 144	60	-	9	3.2
S-Flatbar144	0	-	9	3.2
30-Flatbar144	30	-	9	3.2
45-Flatbar 144	45	-	9	3.2
60-Flatbar 144	60	-	9	3.2
S-Screw216	0	3	13.5	3.2
30-Screw216	30	3	13.5	3.2
45-Screw216	45	3	13.5	3.2
60-Screw216	60	3	13.5	3.2
S- Pipe216	0	-	13.5	3.2
30-Pipe216	30	-	13.5	3.2
45- Pipe216	45	-	13.5	3.2
60- Pipe216	60	-	13.5	3.2
S-Flatbar216	0	-	13.5	3.2
30-Flatbar216	30	-	13.5	3.2
45-Flatbar216	45	-	13.5	3.2
60-Flatbar216	60	-	13.5	3.2
45-Screw144-216	45	2;3	-	3.2
45-Screw216-144	45	3;2	-	3.2

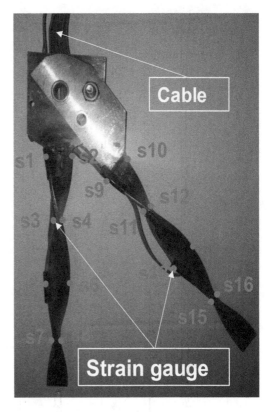

Figure 5. Strain gauge attachment.

3 RESULTS AND DISCUSSIONS

3.1 Bending moment of the model pile

The yield stress at 10N and 30N was selected in respective shorter and longer model piles for collecting strain values. In order to measure precise bending moment along with the pile, the pure bending strain value was calculated in respective model piles considering front and rear strain data. The bending moment along the length of the pile was calculated using the following equation:

$$M_y = S \times \sigma_y \qquad (3)$$

Where S = section modulus of the model pile, σ_y = yield stress.

Figure 6-7 illustrates the variation of bending moment with the depth of the screw, the flat bar, and the pipe on single and coupled piles. The left-hand figures describe the vertical pile bending moment. Meanwhile, right-hand ones describe the bending moment of the batter pile (or reinforcement pile).

Bending moment in single piles is shown relatively higher than vertical pile in coupled piles. The trend of the curves showed a similar pattern in the screw, flat bar, and pipe. Results indicated a rather small strain along with the pile, such as less than 1kNmm. Bending moments in a single pile of 216 mm cases were reduced much comparing with the 144 mm after reinforcing by batter pile. Bending moments in 216 mm shows the amount that less than 4kNmm in vertical piles. The maximum bending moments in the configurations were shown in the same depth (d = 62 mm) aside from the case of pipe pile in 216 mm (d = 98 mm).

Based on the bending moment results, after reinforcement, the bending moments were transferred from the vertical piles to batter piles effectively.

3.2 The ultimate lateral capacity of piles

Nonlinear load-deflection curves of the model pile were measured from the monotonic lateral load test shown in Figure 8. In the literature, various criteria have been proposed to consider the ultimate lateral capacity of model piles from the lateral load-deflection curve. Meyerhof et al. (1981), Rao & Prasad (1993), Georgiadis & Georgiadis (2010), Lee et al. (2010) suggested that the ultimate pile capacity

213

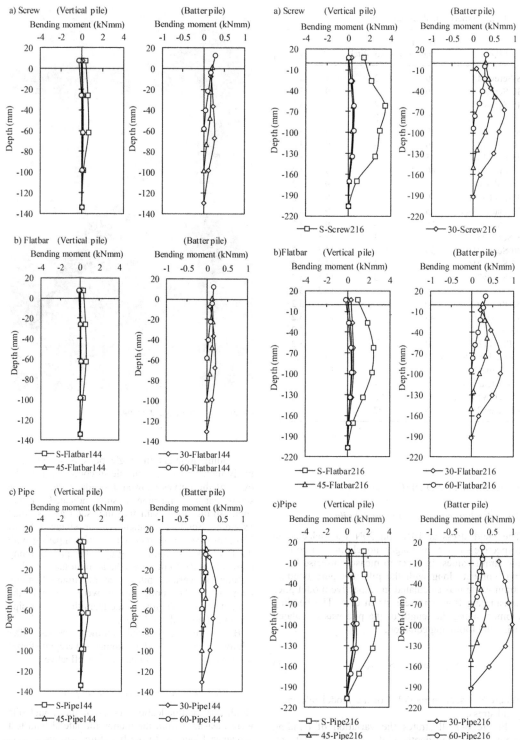

Figure 6. Variation of bending moment with a depth of 144mm model pile a) Screw b) Flat bar c) Pipe.

Figure 7. Variation of bending moment with a depth of 216mm model pile a) Screw b) Flat bar c) Pipe.

214

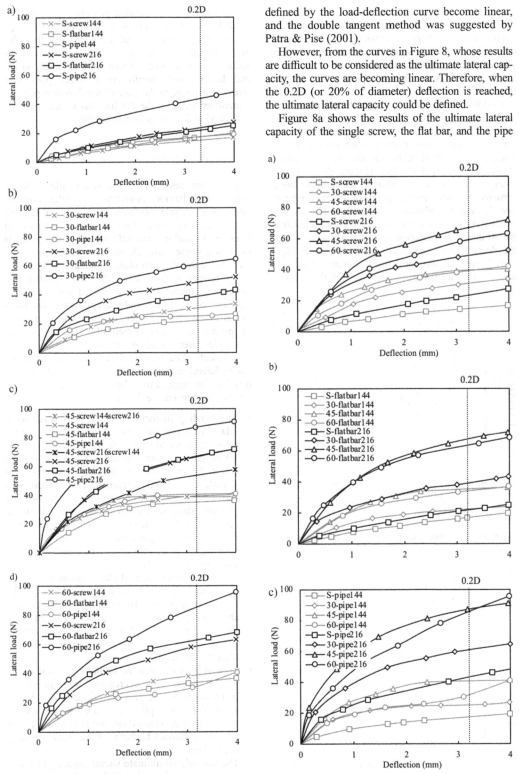

defined by the load-deflection curve become linear, and the double tangent method was suggested by Patra & Pise (2001).

However, from the curves in Figure 8, whose results are difficult to be considered as the ultimate lateral capacity, the curves are becoming linear. Therefore, when the 0.2D (or 20% of diameter) deflection is reached, the ultimate lateral capacity could be defined.

Figure 8a shows the results of the ultimate lateral capacity of the single screw, the flat bar, and the pipe

Figure 8. Comparison of ultimate capacity of piles with deflection a) single pile b) 30 degree combination c) 45 degree combination d) 60 degree combination.

Figure 9. Variation of ultimate lateral capacity with batter angles in case of a) screw b) flat bar c) pipe.

215

model piles, respectively. The result of a case of 144 mm shows not much difference between screw, flat bar, and pipe. However, in the case of 216 mm, the flat bar and screw pile were shown a similar amount of ultimate lateral capacity aside pipe and can be found an obvious difference between pipe and screw pile (Figure 8).

Results indicated that the ultimate lateral capacity of the screw was decreasing with increasing batter angle. In the case of a 45 degree combination, the screw pile was shown close amount with a flat bar. The pipe pile combination was shown higher ultimate lateral capacity than others (Figure 8b, c, d).

Figure 9 shows the variation of the ultimate lateral capacity of batter angle increment in each type of pile. 45 degree combination indicated the higher value in ultimate lateral capacity in every pile. Meanwhile, the 30 degree combination indicated a lower value in the ultimate lateral capacity, as shown in Figure 10.

In Figure 11, the normalized ultimate lateral capacity was evaluated using the ratio of the ultimate

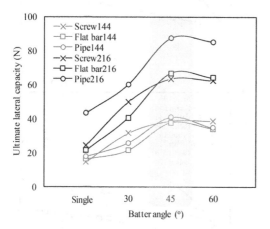

Figure 10. The ultimate lateral capacity with batter angle.

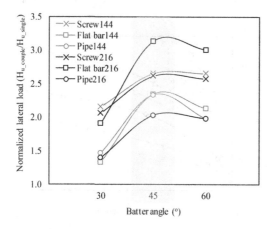

Figure 11. Variation of the normalized ultimate lateral capacity and the ultimate lateral capacity in different angles.

lateral capacity of a coupled pile (H_{u_degree}) and the ultimate lateral capacity of a single pile (H_{u_single}). Screw configuration showed a higher value than that of the flat bar and the pipe, which indicates that the screw pile has a more efficient working ability in lateral loading with reinforcement batter piles than others. However, in the case of the 216 mm pile, the normalized value of the flat bar was increased in the case of 45 and 60 degrees dramatically.

4 CONCLUSIONS

The aim for research on the screw pile with the batter pile behavior comparing with other conventional types of pile foundation under lateral loading is addressed by conducting experiments on model piles. Soil medium was selected as sand, which is used as a dense state. Dimensions of the pile were modeled carefully, avoiding boundary effects and using proper scaling down for the experimental chamber. Experiments were carried out for single piles and coupled piles under lateral monotonic loading to investigate the effect of the screw to find out the proper angle in combined piles. The pile capacities under lateral loading were interpreted from the experiments. The pile head displacement/deflection during lateral monotonic was also measured and analyzed in this study. From the test results, the following conclusions can be drawn:

- The bending moment in single piles was shown relatively higher than the vertical pile in coupled piles. After reinforcing with batter piles as 30, 45, 60 degrees, the bending moments of the vertical pile in the screw, flat bar, and pipe were transferred to batter piles. Bending moments in the model piles showed the nearly same pattern in the bending moment along with the pile depth aside from the case of the 216 mm pile. The maximum bending moments in configurations were measured at the same depth as 62 mm in 144 mm and 92 mm in 216 mm, respectively.
- The model piles showed close values comparing with each other in a single case aside from the case of pipe in the 216 mm pile. The screw pile showed slightly less ultimate lateral capacity than the other two. The ultimate lateral capacity of 216 mm of single screw pile was shown 1.5 times higher value than 144mm pile.
- Results indicated that the ultimate lateral capacity of the screw was decreasing with increasing batter angle. In the case of a 45 degree combination, the screw pile was shown close amount with a flat bar. Meanwhile, the pipe pile combination was shown a higher ultimate lateral capacity than others.
- The normalized ultimate lateral capacity of screw configuration showed a higher value than a flat bar and the pipe, which indicates that the screw pile has the more efficient working ability in

lateral loading with reinforcement batter piles than others. However, in the case of 216 mm of the pile, the normalized value of the flat bar was increased in the case of 45degrees degree dramatically.

ACKNOWLEDGMENTS

The authors would like to thank Mr. Michio Nakajima, Technician of Geotechnical Engineering Laboratory, Kyushu University, for providing technical support in this research.

REFERENCES

Amarbayar, J., Yasufuku N., Ishikura, R., Tani, Y., Nagata, M. & Kurokawa, T. 2020. Experimental studies on the behavior of screw vertical-batter pile under lateral loading in the sand. *The 55th Geotechnical Research Presentation*.

Chandrasekaran, S. S., A. Boominathan, & G. R. Dodagoudar. 2010. Group interaction effects on laterally loaded piles in clay. *J. Geotech. Geoenviron. Eng*. 136 (4): 573–582.

Georgiadis, K & Georgiadis, M. 2010. Undrained lateral pile response in sloping ground. *J. Geotech. Geoenviron. Eng*: 1489–1500

Hirata, A., Kokaji, S., Kang, S, S. & Goto, T. 2005. Study on the Estimation of Axial Resistance of Spiral Bar Based on Interaction with Ground. *Shigen-to-Sozai*, 121. 8. (in Japanese) (submitted).

Kurokawa, T., Tani, Y., Nagata, M. & Nagasaki, R. 2020. Tension and Four-Point Bending Test of Spiral Piles (Twisting a Strip Flat Steel). *The 55th Geotechnical Research Presentation, online* (in Japanese).

Lee, J., Kim, M & Kyung, D. 2010. Estimation of lateral load capacity of rigid short piles in sands using CPT result. *J. Geotech. Geoenviron. Eng*: 48–56.

Meyerhof, G. G. 1981. The bearing capacity of rigid piles and pile groups under inclined load in clay. *Can. Geotech. J*: 18(2). 159–170.

Narasimha Rao, S., V. G. S. T. Ramakrishna, & M. Babu Rao. 1998. Influence of rigidity on laterally loaded pile groups in marine clay. *J. Geotech. Geo-environ. Eng*. 124 (6): 542–549.

Patra, N. R & Pise, P. J. 2001. Ultimate lateral resistance of pile groups in the sand. *J. Geotech. Geoenviron. Eng*: 481–487.

Rao, S. N & Prasad, Y. V. S. N. 1993. Uplift behavior of pile anchors subjected to lateral cyclic loading. *J. Geotech. Eng*: 786–790.

Sato, T., Harada, T., Iwasa, N., Hayashi, S. & Otani, J. 2010. Effect of shaft rotation of spiral piles under its installation on vertical bearing capacity. *Japanese Geotechnical Journal*, 10. 2. 253–265.

Tani, Y., Wang, K., Yasufuku N., Ishikura, R., Fujimoto, H. & Nagata, M. 2019. Model test for bearing capacity of the spiral pile in sandy ground focused on pitch-width ratio. *The 54th Geotechnical Research Presentation, Saitama (in Japanese) (submitted)*.

Tani, Y., Amarbayar, J., Yasufuku N., Ishikura, R., Kurokawa, T. & Nagata, M. 2020. Horizontal Loading Test of Spiral Piles in Sand. *The 55th Geotechnical Research Presentation, online (in Japanese)*.

Wang, K., Tani, Y., Yasufuku N., Ishikura, R., Fujimoto, H. & Nagata, M. 2019. Bearing capacity characteristics of the spiral pile in sandy ground focused on pitch-width ratio. *The 54th Geotechnical Research Presentation, Saitama*.

Wood, D. M. 2004. *Geotechnical modeling*. New-York: Spon Press, Taylor & Francis Group.

Wood. D. M., Crewe, A., & Taylor, C. 2002. Shaking table testing of geotechnical models. *Int. J. Phys. Model. Geotech*: 2(1). 1–13.

Proceedings of the Second International Conference on
Press-in Engineering 2021, Kochi, Japan – Matsumoto et al (eds)
© 2021 Taylor & Francis Group, London, ISBN 978-1-032-10414-0

Experimental evaluation of the lateral capacity of large jacked-in piles and comparison to existing design standards

A. Dobrisan & S.K. Haigh
Department of Engineering, University of Cambridge, Cambridge, UK

Y. Ishihara
GIKEN LTD., Kochi, Japan

ABSTRACT: The devastation caused by the 2011 Tōhoku earthquake and subsequent tsunami revealed the need for a rethink of seawall design. Along the Kochi coast a new generation of tsunami defences has been installed, consisting of large diameter adjoining steel piles with deep embedment. This paper presents the results of a full-scale lateral test on a pile identical to those used in the new seawalls. The lateral test is an optimal opportunity to check how well design codes, originally intended for smaller, flexible piles, scale-up to larger pile classes. A novel data analysis method is used to retrieve accurate p- y curves from experimental data. Results show good agreement with design p- y relationships at shallow depths, while below 3 depth the design curves significantly overpredict soil stiffness. The paper highlights the need for new, appropriate design specifications to account for large stiff piles and obtain better assessment of their lateral capacity.

1 INTRODUCTION

The devastating Tōhoku seismic event and tsunami brought about great loss of life and extensive socio-economic damage. For geotechnical engineers it highlighted the immediate need to reconsider design not just for a 100 or 200-year return period event, but for much larger, potentially never before seen, magnitudes of natural hazards.

In this light the traditional concrete caisson design of seawalls is ill suited for larger-than-planned-for events as the structure can overturn and provide no further protection to coastal areas, as happened in 2011, illustrated in Figure 1 from Kato et al. (2012).

A modern seawall design currently implemented along the Kochi coast consists of large diameter, jacked-in steel tubular piles embedded up to 15 into the ground. The large embedment, coupled with steel's capacity to dissipate energy when yielding, makes these piles suitable designs even against extremely large waves as they are likely to stay in place and reduce the tsunami's energy through yielding for overtopping, larger-than-designed-for waves.

Since the design of these seawall structures is mostly concerned with their capacity to withstand very powerful waves it can be considered a true Ultimate Limit State (ULS) design scenario. Currently evaluating the capacity of these seawalls employs design codes not specifically meant for this application.

By analysing the experimental data from a lateral loading test on a pile of the type used in the Kochi coast defences insight can be gained regarding how well the current design guides scale to large diameter stiff tubular structures.

2 CURRENT CODES OF PRACTICE

Modern design practice for predicting the lateral capacity of piles in sand is based on data from tests on flexible, 0:4m to 0:6m diameter piles published in the 1960's and 1970's in well known papers such as Broms (1964) and Reese et al. (1974).

Since soil displaces around a laterally loaded pile in a complex manner, lacking a known analytical expression, the general way to model soil-structure interaction has been to employ non-linear Winkler springs (Winkler 1867). These link the net soil pressure exerted on the pile at each depth $p(z)$ to the pile displacement $y(z)$ through a spring constant which varies with y, depth z, ground conditions and number of cycles. Knowing the $p - y$ relation is sufficient to derive the pile lateral response, including ultimate load and displacement.

An important requirement for the $p - y$ method is a good prediction of the ultimate soil pressure p_u. Most empirical p_u relations are of the form shown in eq.1.

DOI: 10.1201/9781003215226-18

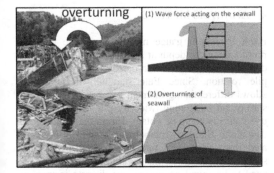

Figure 1. Seawall failure (Kato et al. 2012).

$$p_u = K_q \gamma' z D, \qquad (1)$$

where K_q is an earth pressure coefficient depending on Φ (friction angle), γ' is the effective unit weight of soil and D is pile width or diameter. Broms (1964) proposed the following expression for K_q:

$$K_q = 3 \times K_p, \qquad (2)$$

in which K_p is the passive earth pressure coefficient. Reese et al. (1974) advanced a different semi-empirical definition for K_q by assuming a wedge mechanism forming close to the ground surface and radial, 2D deformation dominating at significant depths:

$$K_q = K_p - K_a + \frac{z}{D}(K_p - K_0)\sqrt{K_p} \tan \alpha + \frac{z}{D} K_0 \sqrt{K_p} \left(\frac{1}{\cos \alpha} + 1\right) \tan \Phi \sin \beta \qquad (3)$$

near the surface and

$$K_q = K_p^3 + K_0 K_p^2 \tan \Phi - K_a \qquad (4)$$

well below ground surface. K_a and K_0 are the active and in-situ earth pressure coefficients, $\beta = 45° + \Phi/2$, α is wedge angle. The resulting K_q is then scaled by pure empirical parameters depending on z and D representing the deviation between eq.3-4 and the measured K_q in the tests conducted by Reese et al. (1974).

Two widely used design standards by Det Norske Veritas (DNV 1992) and the American Petroleum Institute(API 2000) feature a simplified version of Reese et al. (1974) based on a report by O'Neill and Murchinson (1983). The empirical factor A, defined by eq.5, is significantly simplified.

$$A = \left(3 - 0.8\frac{z}{D}\right)0.9 \qquad (5)$$

Finally, the $p - y$ relation as given in design codes (eq.(6)) is a hyperbolic function as opposed to the parabolic expression in Reese et al. (1974).

$$p = A p_u \tanh\left(\frac{kz}{A p_u} y\right) \qquad (6)$$

k is the initial modulus of subgrade reaction and it is derived in DNV (1992) and API (2000) by identical methods based purely on Φ.

Hence by evaluating the in-situ conditions of our lateral loading test and knowing pile diameter, length and stiffness the predicted $p - y$ can be calculated from the above equations. These curves can be integrated to give predictions of pile deflection, bending moment and net soil pressure distribution which may be compared with experimental findings.

3 EXPERIMENTAL SETUP

The tests were carried out at Giken Ltd's Nidahama trial site in Kochi, Japan. Next to the Pacific, and within a mile from where the new piled seawalls were being actively deployed in Kochi, Nidahama provided ground conditions similar to the ones the production piles would be installed in. Sieve analysis was used to grade the Nidahama soil according to ASTM (2017). The result was a silty sand with 20% fines by mass bearing the SM designation. Table 1 summarises the ground properties at the test site, with further data from Nidahama found in Gillow et al. (2018).

Table 2 provides a summary of key pile properties as specified by the manufacturer.

Table 1. Derived soil properties at Nidahama site (Dobrisan et al. 2018).

Water table depth	γ_t	γ'	R_d	Φ_{crit}	Φ_{peak}
m	kN/m³	kN/m³	%	Deg	Deg
7.45	20	12.4	70	32	40

Table 2. Key data for the tested pile of type SKK490 (Dobrisan et al. 2018).

Diameter	Section Thickness	Pile Length	Embedment	Yield Stress	Tensile Stress
m	mm	m	m	MPa	MPa
1.0	22	15	10	315	490

A hydraulic jack was used to load the pile up to forces equal in magnitude to those a tsunami would apply on a seawall. A reaction system made up of four steel piles similar to the one being tested was used (Figure 2). The deflection of the reaction system was monitored during the test to ensure it was stiff enough not to affect the experiment.

A schematic of the experiment setup and of the instruments used is shown in Figure 3. The hydraulic jack position of 2.4m above the ground level was chosen to be similar to the point of application of the equivalent static wave force as dictated by Japanese codes (Okada et al. 2006).

The pile was instrumented with ten strain gauge pairs, each coupled with an inclinometer. These were placed at 1 intervals below ground. LVDTs and wire displacement sensors were used to infer pile movement at the ground level.

The pile was subjected to two load cycles, the maximum applied force being 1836, which coincided with the observed onset of yielding in the strain gauges.

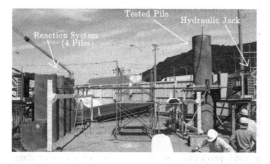

Figure 2. Lateral pile loading system (Dobrisan et al. 2018).

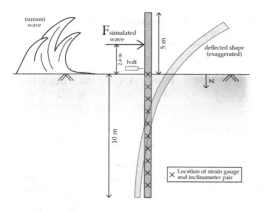

Figure 3. Schematic of pile loading and instrumentation. The pile head is free to move and rotate.

4 ANALYSIS METHOD

From the strain gauge measurements the moment profile along the depth at a given lateral load can be derived. The inclinometer data gives a measure of pile rotation. Since Euler-Bernoulli beam theory allows differentiating moment twice to get soil pressure p and integrating twice for pile displacement y, $p - y$ curves could be obtained from the strain gauge data alone. Similarly rotation data could be differentiated three times and integrated once to yield the same result. To differentiate and integrate the data fits need to be found connecting the discrete instrument measurements and giving continuous plots of M and θ (rotation) respectively. The errors in interpolating using standard techniques like polynomial or spline fitting compounded by measurement errors means obtaining $p - y$ results consistent between the strain gauge and inclinometer measurements could not be achieved. However, a different approach, named *multifit* (Dobrisan et al. 2020), enables obtaining reliable $p - y$ results from the noisy data set. In short, by specifying the errors for each instrument type, the method searches for the family of polynomial fits that satisfies all experimental measurements (strain gauges, inclinometers etc.) simultaneously within the given error bounds. Afterwards, the range of probable values for M, P, y etc. is taken as the interval between the 20th and 80th percentiles (Figure 4).

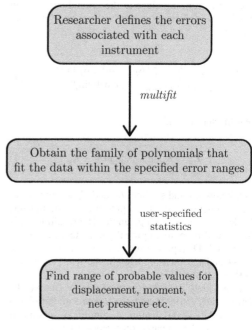

Figure 4. Diagram of *multifit* procedure (Dobrisan et al. 2020).

5 RESULTS. DISCUSSION

Figure 5 shows the measured load-displacement curve at the hydraulic jack location.

It can be seen that Reese et al. (1974) predicts the overall pile stiffnes to a high degree of accuracy with the API (2000) result predicting a somewhat stiffer response. Both numeric solution underpredict the ultimate load capacity of the pile by a considerable factor. To further compare the experimental results to the design code, *multifit* was used to derive the pile properties at each loading stage.

A conservative overall relative error of 10% was considered for both strain gauges and inclinometers alike. The results are shown in Figure 6 for a representative hydraulic jack loading of 984kN, with the other loading stages generating comparable plots.

A first observation is that, even though the Reese et al. (1974) deflection curve is almost identical to the *multifit* result for the top 5m in Figure 6a, the predicted behaviour is very different to the *multifit* one at larger depths. Both the design codes and Reese analyses predict virtually no deflection at the base of the pile, whereas the fitted data suggests there is mobilisation at this level, with a rotation point present in the pile just above 6m depth. Due to the high confining stress around the base of the pile this toe movement generates significant soil pressure as shown in Figure 6e. The numeric predictions show negligible p at the pile base ($z = 10$) consistent with the prediction of zero deflection at this point. However, the significant soil pressure at depth predicted by multifit aligns well with the measured moment data as the increased base resistance brings down the peak bending

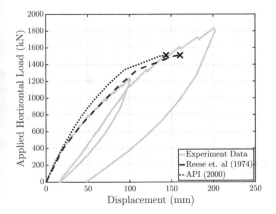

Figure 5. Load-displacement curve, experimental vs predicted.

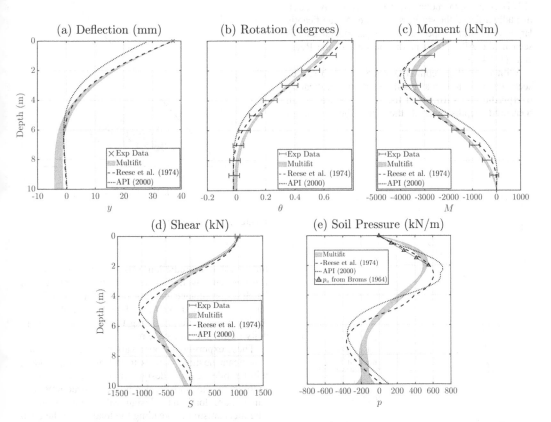

Figure 6. Comparison between multifit results and current codes at 984kN loading.

moment compared to the numeric results, with both overshooting the measured value. The shear plot in Figure 6d consolidates the observation that mobilising the deeper soil strata leads to a reduction of the bending stress on the pile, which may potentially explain the larger than predicted ultimate load measured. This mobilisation of deeper layers does not seem to be captured by current codes, potentially due to the fact that the flexible piles on which they were based lacked sufficient rigidity to move the toe of the pile.

Since most of the soil was not mobilised enough to reach p_u, comparing the K_q predictions is restricted to the very shallow top 1.5m of soil where failure occurred. It can be seen that both Broms (1964) and Reese et al. (1974) underpredict the ultimate capacity of the soil. However, the correction codes like DNV (1992) and API (2000) make on K_q through factor A (eq.5) seems to move the prediction significantly closer to the experimentally deduced gradient at shallow depths.

By using the p and y data from *multifit* across each loading stage, the experimental $p - y$ curves can be predicted. Figure 7 plots $p - y$ at 2m depth. p_u is not reached at this location and the experimental results show a slightly less stiff response than Reese et al. (1974) predicts. Even though both numeric results have the same initial stiffness and similar p_u the fact that API (2000) replaces the parabola in Reese et al. (1974) with a hyperbolic (eq. 6) seems to have detrimentally affected the stiffness estimate. A significantly larger error with respect to the experimental data is seen in the API (2000) prediction than in the Reese et al. (1974) one.

With increasing depth (Figures 8 and 9) it can be seen that the numerical results start to significantly overpredict the stiffness of the soil response. At all measured depths past 3 the same behaviour of

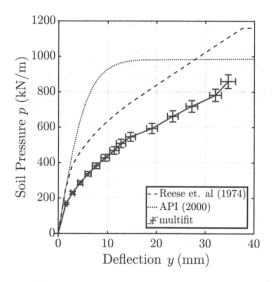

Figure 8. $p - y$ result at 3m depth.

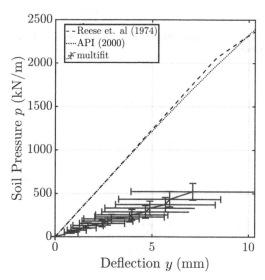

Figure 9. $p - y$ result at 8m depth.

numerical overprediction was found. Conceptually this correlates well with earlier remarks: a stiffer soil has less compliance and thus more of the lateral load has to be carried by the pile which leads to the design codes underpredicting the load carrying capacity of the pile-soil system.

These experimental observations suggest there are significant positive aspects to the highly embedded, tubular pile seawall design. It was shown that by mobilising the soil at depth these structures can carry more load than design calculations suggest. The mechanism of spreading the load out to the deep soil strata is especially encouraging since the

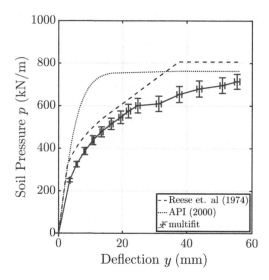

Figure 7. $p - y$ result at 2m depth.

incoming wave might erode some of the top soil layers and significantly decrease the load carrying capacity of the shallower layers.

6 CONCLUSIONS

The usual engineering practice of designing a structure for a given return period of a natural hazard might have to be rethought considering the extremely large, unpredictable hazards in recent memory such as the Tōhoku earthquake and tsunami. Further global climate change might make these super-events even more likely and a new mindset is required to prevent tragic losses. Engineering structures ought to behave well not only for designed scenarios, but should have a beneficial role even for loads much larger than those that could be reasonably expected.

A seawall design of stiff, large diameter, adjoined steel tubular piles was discussed as a possible replacement for traditional concrete caisson structures. Benefits include a higher embedment, piles being unlikely to be 'washed-off' by the wave, and capacity to dissipate wave energy through yielding.

Designing such a pile wall requires knowledge of the soil-structure interaction and design codes are shown to include empirical relations based on experiments on smaller and more flexible piles. A lateral loading test was discussed and the experimental results were compared with numerical predictions using design code equations. A novel fitting method, *multifit*, was used to get improved insight from the recorded data and allow for measurement errors.

It was found that design codes tend to predict large soil stiffness increase with depth and thus imply very little pile toe movement during deformation. This means the bulk of the load has to be carried in the shallower depths which translates to decreased load carrying capacity and potential issues during a tsunami event, as the wave may erode the top soil and weaken the shallow layers. However, the data analysis shows the soil stiffness is significantly less than the code predictions. This implies the deeper soil layers are mobilised and the pile toe can move. The result is significantly less loading on the pile and a higher measured ultimate capacity than prediction. The fact that the deeper soil is mobilised is encouraging since these layers are less likely to significantly weaken during a wave collision.

Overall it was found that the pile wall design has improved geotechnical properties for withstanding wave loading when compared to concrete caissons and that design codes might have to be adjusted to account for the larger diameter piles being used presently in seawalls, wind turbines etc.

7 MULTIFIT ROUTINE

The *multifit* procedure of analysing experimental data from retaining wall and lateral pile tests has been implemented in MATLAB and is freely accessible at gitlab.developers.cam.ac.uk/ad622/multifit. Documentation and worked examples are made available at the link above to facilitate the use of *multifit* in geotechnical research.

ACKNOWLEDGEMENTS

The authors are grateful to EPSRC for support in the form of a Doctoral Training Award to A.D, grant number EP/M508007/1. The support of Giken Ltd. in providing the data from the pile lateral loading experiment is gratefully acknowledged.

REFERENCES

API (2000). Recommended Practice for Planning, Designing and Constructing Fixed Offshore Platforms-Working Stress Design. Technical Report 2A-WSD.

ASTM (2017). Standard Practice for Classification of Soils for Engineering Purposes (Unified Soil Classification System). Technical Report D2487, ASTM International.

Broms, B. B. (1964). Lateral Resistance of Piles in Cohesionless Soils. *Journal of the Soil Mechanics and Foundations Division* 90(3), 123–158.

DNV (1992). Foundations. Technical Report 30.4, Det Norske Veritas.

Dobrisan, A., S. K. Haigh, C. Deng, & Y. Ishihara (2020). Analysis of the behaviour of retaining structures through a novel data interpretation approach. *Unpublished*.

Dobrisan, A., S. K. Haigh, & Y. Ishihara (2018). Evaluating the Efficiency of Jacked-in Piles as Tsunami Defences. In *The First International Conference on Press-in Engineering*, Volume 1, Kochi, Japan, pp. 289–296. International Press-In Association.

Gillow, M., S. K. Haigh, & Y. Ishihara (2018). Water Jetting for Sheet Piling. In *The First International Conference on Press-in Engineering*, Volume 1, Kochi, Japan, pp. 335–342. International Press-In Association.

Kato, F., Y. Suwa, K. Watanabe, & S. Hatogai (2012). Mechanisms of coastal dike failure induced by the Great East Japan Earthquake Tsunami. *Coastal Engineering Proceedings* 1(33).

Okada, T., T. Sugano, T. Ishikawa, S. Takai, & T. Tateno (2006). Tsunami loads and structural design of tsunami refuge buildings. *The Building Centre of Japan*.

O'Neill, M. & J. Murchinson (1983). Fan evaluation of p-y relationships in sands. Technical report, American Petroleum Institute.

Reese, L. C., W. R. Cox, & F. D. Koop (1974). Analysis of laterally loaded piles in sand. In *Sixth Annual Offshore Technological Conference*, Houston, Texas, pp. 473–483.

Winkler, E. (1867). *Theory of elasticity and strength*. Prague: Dominicus.

Proceedings of the Second International Conference on
Press-in Engineering 2021, Kochi, Japan – Matsumoto et al (eds)
© 2021 Taylor & Francis Group, London, ISBN 978-1-032-10414-0

A study on analysis of horizontal resistance of screw coupled foundation with vertical and battered piles in cohesionless soil

T. Kurokawa
Research and Development Center, HINODE, Ltd., Fukuoka, Japan
Department of Civil Engineering, Kyusyu University, Fukuoka, Japan

Y. Tani & M. Nagata
Research and Development Center, HINODE, Ltd., Fukuoka, Japan

A. Jugdernamjil & N. Yasufuku
Department of Civil Engineering, Kyusyu University, Fukuoka, Japan

ABSTRACT: The screw piles, made by twisting a strip of flat steel, are excellent for ease of driving and demonstrate high resistance to vertical loads. However, screw piles have a horizontal resistance is lower than that of steel pipe piles of the same diameter. In this study, the horizontal resistance of screw coupled piles, which are expected to improve, was tested by model tests in the sand tank. Based on the model test results, a static nonlinear analysis is carried out to rationalize the structural design. The analytical model was that the ground is a bilinear soil reaction spring and the pile is a beam model on the soil reaction spring. It was found out that this analysis allowed to reproduce the initial rigidity and the maximum load of the screw coupled piles structure, and an example of the optimization of the structure of vertical and battered piles was identified.

Keywords: screw pile, coupled pile, horizontal resistance, nonlinear analysis, model test

1 INTRODUCTION

The screw piles are made of flat steel twisted into a spiral shape, and are used as foundations for solar panels, sign poles and lighting columns. The spiral piles can penetrate into the ground without disturbing the surrounding ground and have high vertical resistance to vertical load (push-in and pull-out), so that small-diameter and short steel pipe piles and screw piles are being developed as a useful foundation method for soft ground. Especially in the foundation method using screw piles (Figure 1), Sato *et al.* (2014) studied the use of sand tank and ground for a more rational design. The bearing capacity characteristics were clarified by Hirata *et al.* (2005) and a bearing capacity formula was proposed and validated by elastoplastic analysis using the FEM method.

Steel screw piles are made of flat steel and have a uniform cross section. In order to optimize the shape of these piles for the external forces (moments) acting on the piles in the ground, machining is required, which is unrealistic in terms of cost. (Figure 2) is a screw pile made of ductile cast iron, allowing for shape optimization and connection to the superstructure. Kanno *et al.* (2017, 2019) have conducted test to confirm the performance of the screw piles by the driving method in a narrow area where heavy machinery is difficult to be carried in.

Ductile cast iron screw pile has already been applied as a foundation for pedestrian guard-pipes. And, it is also expected to be used as a foundation for vehicle guard-pipes, which require greater strength. Currently, design rationalization studies have also been conducted by (Wang *et al.* 2019; Tani *et al.* 2019) to determine the optimal pitch-width ratio (*pw*) for vertical bearing capacity and penetration of screw piles.

2 BACKGROUND

The horizontal resistance of screw piles is lower than that of steel pipe piles of the same diameter due to their low bending rigidity under horizontal loads. The methods to increase the horizontal resistance are to increase the pile diameter and pile length; however, there are some cases where it is difficult to do so due to manufacturing constraints and the installation environment (e.g. interference with underground

DOI: 10.1201/9781003215226-19

Figure 1. Steel screw pile ($\phi50 \times t12 \times L4200$mm with a flange).

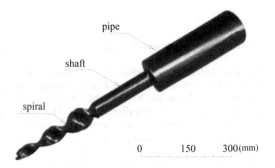

Figure 2. Ductile cast iron screw pile.

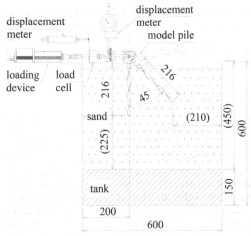

Figure 3. Experimental device.

Table 1. Sand tank.

Date	Detail	Note
Tank Size	300*600*600 (mm)	Width*Depth*Height
Sand Type	Silica Sand (K7)	Dry Sand

Table 2. Index properties of soil medium.

Properties	Value
Specific gravity, Gs	2.63
Maximum dry density, ρ_{max}	1.56 g/cm³
Minimum dry density, ρ_{min}	1.19 g/cm³
Coefficient of uniformity, U_c	1.76
Median diameter, D_{50}	0.17 mm
Relative density, Dr=90%	1.53 g/cm³
Internal friction angle, °	41.9°

Table 3. Loading device.

Date	Detail
Maximum Load	800 (N)
Loading Speed	0.01 (mm/s)
Maximum Stroke	50 (mm)

Table 4. Instrumentation.

Date	Detail	Note
Load Cell	1000 (N)	
Displacement Meter	25 (mm)	Contact Type

structures). The other method is to use two screw piles. The screw coupled piles are composed of vertical and battered pile, and the vertical resistance of the screw pile is utilized to improve the horizontal resistance of the structure as compared to the single pile. Tani, Amar *et al.* (2020, 2021) reported on the horizontal resistance characteristics of screw, plate and pipe single piles and screw coupled piles in an experimental sand tank and the effect of the angle of installation of battered piles. However, the rationalization of coupled pile structure design is still unclear.

In this study, the effects of the change in pile length on the horizontal resistance of vertical and battered piles were investigated by the model horizontal loading tests for rationalization of the screw coupled pile structure design. Based on the results, a static nonlinear analysis by FEM was performed to evaluate the horizontal resistance of the screw coupled piles structure by reproducing the model test and simulating full-scale.

3 HORIZONTAL LOAD MODEL PILE TESTS

3.1 Plan of experimental

The model test was conducted to confirm the effect of pile length change on the horizontal resistance of coupled piles. The experimental devices were the same as those used by Tani, Amra *et al.* (2020, 2021) as shown in Figure 3. The sand tanks, properties of soil medium, loading devices and instrumentation are outlined in Tables 1 to 4, and the sand tank conditions, loading conditions, and materials, dimensions, configuration and method of installation of the model piles are the same to compare with the existing results. The pw (pitch width ratio) of the screw piles, which affects bearing capacity and bending rigidity, was unified to 4.5.

3.2 Model piles

Table 5 shows the model test cases. No. 1 to 4 are for comparison with the conditions tested by Tani, Amra *et al.* (2020, 2021). Firstly, No.4 to 6 were prepared to confirm the change with the increase of the pile length in each case. In addition, the effect of the angle of the battered piles was also confirmed. Then, the effect of the structure of vertical piles and battered piles on the horizontal resistance was compared in the case of long battered piles (No.5) and long vertical piles (No.6) at an angle of 45°. The shapes of the model piles are shown in Figure 4.

3.3 Results

Figure 5 shows the load-horizontal displacement curves. Except for No. 6, the resistance gradually decreased with increasing displacement. No. 3 and 5 dropped to about the same level as No. 1 (144 mm single pile) and No. 4 to about the same level as No. 2 (216 mm single pile). When the load exceeded the maximum pull-out resistance of the vertical pile, the

Table 5. Model Test Case (Pile Length, Batter Pile Angle).

No	Type	L1(mm)	L2(mm)	θ(°)
1	Single*	144	-	-
2	Single*	216	-	-
3	Coupled*	144	144	45
4	Coupled*	216	216	45
5	Coupled	144	216	45
6	Coupled	216	144	45

* Tani, Amra *et al.* (2020, 2021).

a) Coupled piles No.5.   b) Coupled piles No.6.

Figure 4. Model piles.

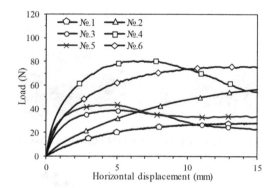

Figure 5. Load-horizontal displacement.

pile was pulled out. Therefore, it was assumed that the resistance was lost. The initial resistance and rigidity of No.4 was the largest, while No.6 had the largest resistance at 15mm displacement. The rigidity of No. 6 was small by the length of the battered pile. Therefore, a large amount of displacement was required before it was pulled out. The final result was estimated to be reduced to the same level. In the case of No.5, the initial rigidity of the No.5 pile was similar to that of No.6, but the resistance was lower. The results indicated that the horizontal resistance of the piles had a great effect on the pull-out resistance of vertical piles.

The ultimate bearing capacity (R_u) of a pile was the load at a displacement of 20% of the pile diameter. Figure 6 shows the results of changing the display range of the displacement amount, and Table 6 indicates the values. Also, a comparison with No. 1 and No. 3 is shown. The horizontal resistance was increased more than twice by changing from single piles to coupled piles (No.1 to No.3: 2.26, No.2 to No.4: 2.86). There was a difference of 22% in the rate of increase in resistance due to the extension of vertical and battered piles versus coupled piles. In addition, the resistance of No. 6 was still on an upward trend, while that of No. 5 had started to decrease from about 4 mm. The performance of No. 6 seems to be better.

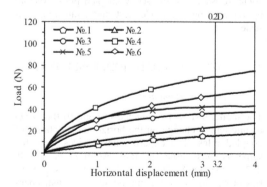

Figure 6. Load-horizontal displacement (20 % of Pile diameter).

Table 6. Ultimate horizontal bearing capacity.

No	Type	R_u (N)	Ratio No.1	Ratio No.3
1	S144	15.8	1.00	0.44
2	S216	23.6	1.49	0.66
3	V144-B144	35.8	2.26	1.00
4	V216-B216	67.4	4.25	1.88
5	V144-B216	41.9	2.65	1.17
6	V216-B144	51.2	3.23	1.43

4 STATIC NONLINEAR ANALYSIS OF HORIZONTAL LOAD TESTS OF MODEL PILES

4.1 Analysis methods

The results of the model test were reproduced by analysis, and the horizontal resistance was predicted under conditions that had not been tested. The same method of analysis was used as in Tani et al. (2020). In the analytical model, the ground was used as a non-linear soil reaction spring model with coefficients of horizontal and vertical subgrade reaction, and the pile was a beam model on the soil reaction spring, which was adjusted to the model test results. Figure 7 shows the analytical model image and Table 7 shows the boundary conditions. The pile diameter was reduced by 15% to 12 mm to match the projected area of a flat plate of the same

a) Single pile.   b) Coupled piles.

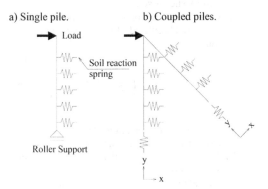

Figure 7. Soil reaction spring model.

Table 7. Fixed constraint.

Single			Coupled			
Displacement		Rotation	Displacement		Rotation	
X	free	X	fix	X	free	X	fix
Y	fix	Y	fix	Y	free	Y	fix
Z	fix	Z	free	Z	fix	Z	free

diameter. The results of the four-point bending test of flat plates and spirals in air conducted by the authors (2020) revealed the relationship between the bending rigidity EI (E: Young's modulus, I: Moment of inertia of area) of a spiral, and Amar et al. (2021) proposed a formula for calculating it. The I of the spiral was found to be $I=102(\text{mm}^4)$. From these results, the dimensions of the analytical model were determined to be $w=12$mm and $t=4.68$mm. Table 8 shows the material properties of the pile body and Table 9 shows the dimension properties of the pile body. The pile body was treated as a linear material since it was rigid enough against sand.

4.2 Estimation of horizontal subgrade reaction coefficient and deformation modulus for sand tanks

The horizontal subgrade reaction coefficient kh_0, which was the initial rigidity of the soil reaction spring, was calculated from the formula 1 according to the "Specifications for Highway Bridges, 2017 (Japan Road Association)".

$$kh_0 = \alpha \cdot \frac{E_0}{0.3} \tag{1}$$

E_0 : deformation modulus
α : conversion factor

Otake et al. (2017) reported a method for estimating soil deformation modulus for design. E_1 (deformation modulus at axial strain $\varepsilon_a=0.01(1\%)$ is called the base deformation modulus. This corresponds to the average value of E_{50} (deformation modulus calculated from 1/2 of the maximum principal stress difference in the triaxial compression test and the slope

Table 8. Material properties.

Date	Value	Note
E (GPa)	200	Yong's modulus
ν	0.3	Poisson's ratio

Table 9. Dimension properties.

Date	Actual size	Analysis size	Ratio
w (mm)	16	12	0.75
t (mm)	3	4.68	1.56
I (mm^4)	36	102	2.83

w : Width
t : Thickness
I : Moment of inertia of area ($I = w \cdot t^3/12$)

from the zero point). From this, the deformation modulus can be estimated for an arbitrary axial strain.

The results of the triaxial compression test of the sand used in this study are indicated in Figure 8. E_{50} and E_1 were plotted for each of the axial stresses in Figure 9, and the respective approximate equations were obtained. The soil pressure applied to the pile was calculated from

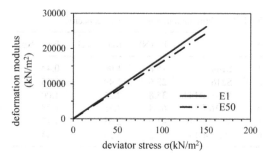

Figure 9. Deformation modulus.

a) 50kPa

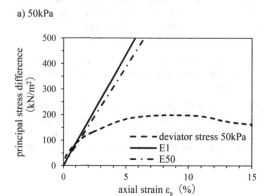

b) 100kPa

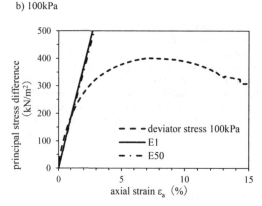

c) 150kPa

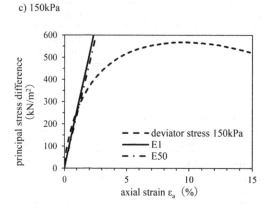

Figure 8. Triaxial compression test.

the installation depth of the pile and the unit volume weight of sand in the model test, and the deformation modulus was obtained from the approximate equation of E_{50}. In this study, the approximate equation of E_{50} was adopted because the design was made on the safety side. The value of α was 4 when the deformation modulus was obtained from the triaxial compression test. The horizontal subgrade reaction coefficient was estimated from these. The specific procedure for setting the soil reaction springs is described below.

• Setting procedure for bilinear soil reaction springs.
  I. Setting the number of springs.
    The mesh size was divided into 20 segments for a pile length of 144 mm and 30 segments for a pile length of 216 mm. One spring per two meshes is set.
  II. Calculation of coefficient of passive earth pressure. Calculated from Coulomb's formula 2.

$$K_p = tan^2\left(45° + \frac{\phi}{2}\right) \qquad (2)$$

ϕ : Internal friction angle, 41.9° (from triaxial compression test)
  III. Setting of passive soil pressure (maximum spring force).
    Calculated from Coulomb's formula 3. The passive soil pressure (kN/m²) is calculated by multiplying the passive earth pressure (kN/m²) by the mesh area (m², pile diameter x spring spacing) and this is the maximum spring force. According to Blombs (1964), there is a difference between the calculated and measured passive soil pressure, and correction is required. The correction factor (CF) varies depending on the ground conditions. In this analysis, corrections were made to the calculated passive soil pressure values and adjusted with the model tests.

228

$$P_p = K_p \cdot \gamma \cdot z \qquad (3)$$

γ : unit weight, 15.3(kN/m^3)
z : Installation depth of the spring (m) (optional)

IV. Setting of spring rigidity.
The coefficient of horizontal subgrade reaction (kN/m^3) is multiplied by the mesh area (m^2) to calculate the spring rigidity (kN/m).

V. Setting of displacement at the maximum spring force.
The displacement is calculated by dividing the maximum spring force (kN) by the spring stiffness (kN/m).
This is the displacement to reach the maximum spring force.

VI. Setting of bilinear soil reaction springs.
The bilinear soil reaction spring is obtained from the spring rigidity, the maximum spring force and the displacement. Since the maximum spring force depends on the depth of the spring, set the spring for each position. Figure 10.

4.3 Estimation of axial soil reaction springs of sand tank

Next, the axial spring was set up to reproduce the bearing capacity of the pile to compress and pull-out the pile. The vertical bearing capacity of a screw pile consists of the frictional force acting on the circular area of the pile body, which is determined from the shear force of the sand. However, in practice, it is assumed that soil getting into the twisting part of the screw pile would also add physical resistance, which would result in greater bearing capacity than this, but the detailed mechanism is not clear.

In this study, model tests were carried out to estimate the vertical bearing capacity and pull-out resistance of the screw pile. The test conditions and other details were identical to those reported by Tani, Wang et al. (2019). Screw piles of 25 and 44 widths with different pile diameters from the present tests were used for the model piles. Later, an empirical equation was obtained to allow estimation of the pile diameter and sand conditions as they changed (pw=4.5 is fixed). The results are shown in Figures 11 and 12. As the pile diameter increased, both resistances increased. In contrast to the horizontal resistance, the displacement did not progress almost at the beginning of loading, but gradually increased from a certain point to the maximum level. In the pull-out test, the resistance force decreased after reaching the maximum value. From the results, the axial soil reaction springs of this test is estimated. Based on the behavior of the test results and the analysis time, the trilinear type was set up in this study. Setting up a multilinear type seems to improve the accuracy more. The specific procedure for setting the springs is described below.

• Setting procedure for trilinear soil reaction springs.
I. Non-dimensionality of model test results (w25 and w44).
Figures 11 and 12 shows the result of bearing capacity divided by circumference area and displacement divided by pile diameter. In both cases, the tendency was almost the same at w25 and w44, and it was confirmed

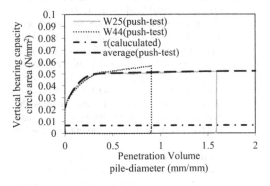

Figure 11. Static axial compressive load model test.

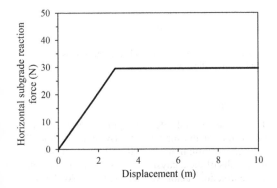

Figure 10. Bilinear soil reaction springs. (Example of a 216mm coupled piles 45deg).

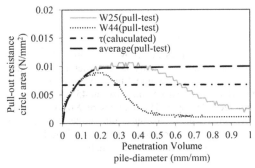

Figure 12. Axial tensile load model test.

that the bearing capacity depended on the pile size.

II. Calculation of the shear strength of sand (compressive, pull-out model test) and estimation of correction factors.

The shear strength of the sand in the tank was calculated from the following formula. The difference between the calculated value and the test results (Figures 11 and 12) was confirmed, and the correction factor was determined by the shape of the screw pile.

$$\tau = c + \sigma \tan \phi \qquad (4)$$

c : cohesion, 0
σ : earth pressure at rest (normal stress)
ϕ : 41.1° (From Kang et al. 2019 test results, sand type: K7, relative density: 75%)

$$\sigma = \sigma_c + K_0 \cdot \gamma \cdot z \qquad (5)$$

σ_c : overburden pressure, 12.6(kN/m^2)
K_0 : coefficient earth pressure at rest, 0.5
γ : 14.3(kN/m^3)
z : 0.7(m)

III. Calculation of the shear strength of sand (This horizontal load model test).

The bearing capacity of the horizontal load model test was calculated using the above equations and correction factors. The sand tank conditions and model pile conditions were changed (ϕ =41.9°, σ_c=0, D_r=90%, γ=15.3(kN/m^3), z : optional) and reflected in the calculations. Figures 13 and 14 show the average values of the model tests and the calculated values (with and without overburden pressure).

IV. Setting of trilinear soil reaction springs.

From these, the trilinear soil reaction spring was set up that connecting the zero-point, linear elastic range, and maximum value. The vertical and battered piles were set axially at their respective ends. An example is shown in Figure 15.

4.4 Analysis cases

The analysis cases are shown in Table 10. Firstly, No.1 to No.5 were analyzed in order to check the consistency with the model tests. Amar et al. (2021) reported that resistance increases at 45-60 degrees of battered pile angle. The analysis of No. 6 to 9 was conducted with the battered pile angle as a change parameter. The horizontal resistance of the battered pile angle is determined by the analysis.

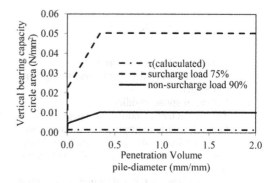

Figure 13. Static axial compressive load - non-dimensionless.

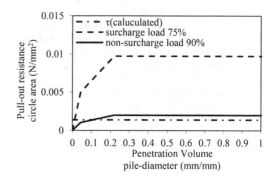

Figure 14. Axial tensile load - non-dimensionless.

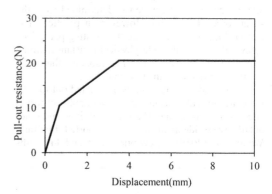

Figure 15. Trilinear pull-out soil reaction springs. (Example of a 216mm vertical pile).

4.5 Results

The results are indicated in Figure 16 together with the results of the model tests. By adjusting the CF for the maximum horizontal soil reaction springs, the results of the analysis are close to the test results. The correction factor was set to 30 for No. 1 and 3,

230

Table 10. Model test case (Pile Length, Batter Pile Angle).

No	Type	L1(mm)	L2(mm)	$\theta(°)$
1	Single	144	-	-
2	Single	216	-	-
3	Coupled	144	144	45
4	Coupled	216	216	45
5	Coupled	216	144	45
6	Coupled	216	216	50
7	Coupled	216	216	55
8	Coupled	216	216	60
9	Coupled	216	216	70

and 12 for all other cases, depending on the length of the vertical pile. Table 11 shows the difference between the ultimate bearing capacity and the model test, as well as the difference in the battered pile angle.

The error of the ultimate bearing capacity was less than 8% for No. 1 to No. 4 and 14% for No. 5. It is assumed that the correction factors will vary with the pile structure and ground conditions. The accuracy of the analysis would be improved if it is possible to set the correction factors for each case. However, more elemental experiments are needed to organize the correction factors logically and systematically. The load-displacement curves generally reproduced the trends of the model tests. This was made possible by setting up two types of nonlinear soil reaction springs.

Next, it was found that the horizontal resistance decreased gradually as the angle of the battered piles became larger than 45°. It is assumed that this is due to the shallow installation depth of the battered piles. However, the initial rigidity is higher at larger angles (Figure 16-g). It is assumed that this is because the axial spring of the battered pile is more likely to contribute to the horizontal resistance. It is necessary to investigate the optimum angle of the battered piles by unifying the installation depth of the battered piles and varying the angle of the piles.

In the case of coupled piles, some deviation was observed in the initial stage. It is assumed that this is due to the non-dimensionality of the estimation by setting the springs and so on. In this study, only the range of small displacement (about 25% of the pile diameter) was focused on, but it seems that the setting of the spring in the axial direction will make it possible to predict the deformation in a larger range.

5 STATIC NONLINEAR ANALYSIS OF PRODUCTS IN COHESIONLESS SOIL

5.1 Outline of analysis

Based on the previous model tests and analysis results, a study on the full-scale screw coupled pile foundations

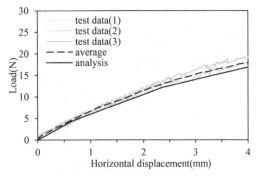

a) No.1 Single pile 144. *CF*=30.

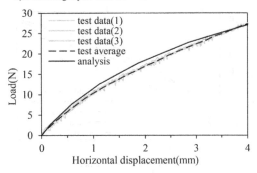

b) No.2 Single pile 216. *CF*=12.

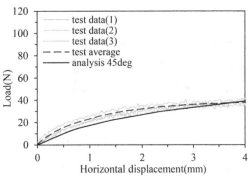

c) No.3 V144-B144. *CF*=30.

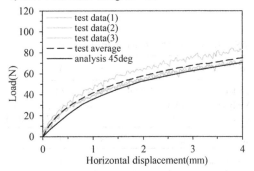

d) No.4 V216-B216-45deg. *CF*=12.

e) No.5 V216-B144. *CF*=12.

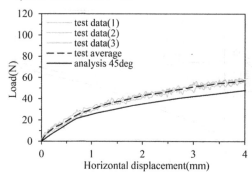

f) No.4 and No.6, 7, 8, 9. *CF*=12.

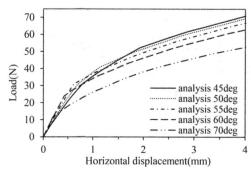

g) No.4 and No.6, 7, 8, 9. Less than 1.5mm range.

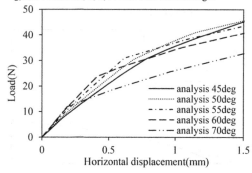

Figure 16. Load-horizontal displacement analysis.

was conducted. From "Standard Specifications of Vehicle Guard-pipe, 2014 (Japan Road Association)", a steel pipe pile $\phi 114.3 \times t4.5mm \times L1500mm$ was used as a comparison target for a foundation for a vehicle guard-pipe. In contrast, a foundation with a screw coupled pile angle of 45° was devised and analyzed. The screw piles were constructed with a tubular part and a spiral part is shown in Figure 17 with referring to the ductile cast iron pile structure shown in Figure 2. The vertical piles, the tubular part was $\phi 114.3 \times t4.5 \times L_p$ =400mm, the spiral part was $w150 \times t28 \times L_s$=350, 600, 850, 1100mm, and the total length was L=750, 1000, 1250, 1500mm. The battered piles, tubular part was $\phi 114.3 \times t4.5 \times L_p$=150mm, the spiral part was $w150 \times t28 \times L_s$=350mm, and the total length was L=500mm. The analysis of vertical pile only (L=1500mm) was also performed for comparison. The analysis cases were shown in Table 12 and the analysis model was shown in Figure 18. The tube is provided to ensure strength and to connect to the superstructure. Vertical piles and battered piles were connected with rigid beam elements at the pile head to be integrated. The piles and the guard-pipe poles were connected at the pile head and horizontal loads were applied to the poles.

The coefficients of horizontal subgrade reaction force and the deformation modulus were based on the values obtained for the sand tank condition in the

a) Steel pipe pile.   b) Screw coupled piles.

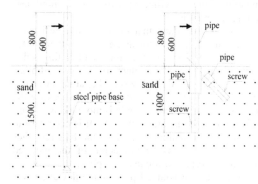

Figure 17. Piles foundation.

Table 11. Ultimate horizontal bearing capacity.

No	1	2	3	4	5
Model test	15.8	23.6	35.8	67.4	51.2
Analysis	14.6	24.1	35.0	65.2	44.2
Ratio test	0.92	1.02	0.98	0.97	0.86
No	6	7	8	9	
Model test	64.0	61.6	58.2	48.4	
Ratio No.4	0.98	0.95	0.89	0.74	

Table 12. Analysis Cases (Pile Length, Spiral part Length).

No	Type		L(mm)	L_p(mm)	L_s(mm)
1	Single	Pipe	1500	1500	-
2	Single	Screw	1500	400	1100
3	Coupled	Screw	1500	400	1100
4	Coupled	Screw	1250	400	850
5	Coupled	Screw	1000	400	600
6	Coupled	Screw	750	400	350

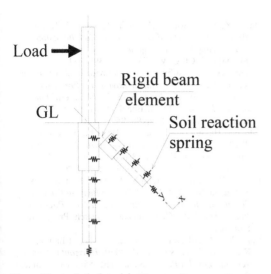

Load →

Rigid beam element

GL

Soil reaction spring

Figure 18. Analytical model.

previous chapter. The axial soil reaction springs was estimated from the results of the pull-out test conducted in the field. Otake *et al.* 2017 also discussed the case of estimating the design deformation modulus from the *N* value. In practice, it was also important to estimate from the *N* value. However, they reported that there is no data for the surface layer less than 2.5m, and setting the deformation modulus for the surface layer, such as the assumption in this case, is an important issue, which was not clear.

5.2 Results

The load-horizontal displacement curves were shown in Figure 19. *CF* for passive earth pressure were assumed to be 12. The rigidity of the vertical pile alone was inferior to that of the steel pipe pile due to the presence of the spiral parts. The rigidity and horizontal resistance of the coupled piles were higher than those of the steel pipe piles in No. 3, 4 and 5. The range of displacements less than 30 mm is shown in Figure 20. The ultimate horizontal bearing capacity was compared. The initial rigidity of

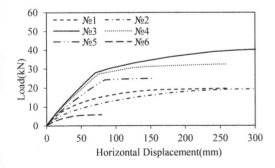

Figure 19. Load-horizontal displacement analysis.

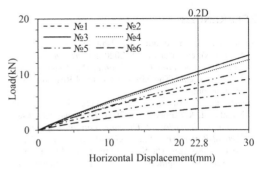

Figure 20. Load-horizontal displacement analysis (20% of Pile diameter).

No. 5 was almost the same as that of the steel pipe pile, and the ultimate horizontal bearing capacity was a little higher, about 1.1 times. In the future, it is necessary to investigate the stresses in the pile body, the manufacturability, the connection between vertical and diagonal piles, and the constructability of the pile. In addition, we would like to reproduce the behavior of the loss of pull-out resistance of vertical piles, rationally set the correction factor for the passive earth pressure (upper limit of spring), the influence of pile length in setting the axial soil reaction springs of screw piles, and confirm the validity by experiments on field. This will improve the accuracy of analysis and enable more rational design.

6 CONCLUSION

- A study was carried out to clarify the relationship between the screw coupled pile structure and the horizontal resistance force in cohesionless soil through model tests and FEM analysis. Next, FEM analysis was carried out on a full-scale in order to propose a practical coupled pile foundation.
- Based on the properties of soil medium and the results of previous model tests, horizontal subgrade reaction coefficients, deformation modulus, and axial soil reaction springs in the sand tank were determined to provide a sufficiently accurate FEM analysis.
- From this, we obtained a proposal for the structure of coupled piles with the same performance as steel pipe piles.

ACKNOWLEDGEMENTS

In this study, we would like to acknowledge Mr. M, Nakajima and Mr. Wang, K. technical staff, for their technical support and cooperation.

REFERENCES

Amarbayar, J., Yasufuku, N., Ishikura, R., Tani, Y., Nagata, M. & Kurokawa, T. 2020. *Experimental Studies on Behavior of Screw Vertical-Batter Pile Under Lateral Loading in Sand*. The 55th Geotechnical Research Presentation, online.

Amarbayar, J., Yasufuku, N., Ishikura, R., Tani, Y., Nagata, M. & Kurokawa, T. 2021. *Experimental Observation on Behavior of Ultimate Lateral Capacity of Vertical-Batter Screw Pile Under Monotonic Loading in Cohesionless Soil*. The Second International Conference on Press-in Engineering 2021, Kochi (submitting a paper).

Broms, B, B., 1964. *Lateral Resistance of Piles in Cohesionless Soils*. Journal of the soil mechanics and foundations division. Proceedings of the American Society of Civil Engineers. May, 1964.

Hirata, A., Kokaji, S., Kang, S, S. & Goto, T. 2005. Study on the *Estimation of Axial Resistance of Spiral Bar Based on Interaction with Ground*. Shigen-to-Sozai, Vol.121, No.8.

Kang, J. G., Yasufuku, N., Ishikura, R., & Purama, A. Y. 2019. *Prediction of Uplift Capacity of Belled-type Pile with Shallow Foundation in Sandy Ground*. Lowland Technology International 2019; 21 (2): 71–79.

Kanno, T. & Muraoka, H. 2017. *Traffic Safety Facility Foundations for Community Roads*. The 32th Japan Road Conference, Tokyo.

Kanno, T. & Kurokawa, T. 2019. *Study of Foundation Screw Piles for Guard-Pipes, Vehicle Stops, etc. for Community Roads*. The 33th Japan Road Conference, Tokyo.

Kurokawa, T., Tani, Y., Nagata, M. & Nagasaki, R. 2020. *Tension and Four-Point Bending Test of Spiral Piles (Twisting a Strip Flat Steel)*. The 55th Geotechnical Research Presentation, online.

Otake, Y., Nanazawa, Y., Honjo, Y., Kono, T. & Tanabe, A. 2017. *Improvement of Deformation Modulus Estimation Considering Soil Investigation Types and Strain Level*. Journal of JSCE C, Vol. 73, No. 4, 396–411.

Purama, A, Y., Yasufuku, N. & Ahmad, R. 2019. *Evaluation of Filler Material Behavior in Pre-Bored Pile Foundation System due to Slow Cyclic Lateral Loading in Sandy Soil*. International Journal of Geomate, June 2019, Vol.16, 58, pp.90–96.

Sato, T., Harada, T., Iwasa, N., Hayashi, S. & Otani, J. 2010. *Effect of Shaft Rotation of Spiral Piles Under its Installation on Vertical Bearing Capacity*. Japanese Geotechnical Journal, Vol. 10, No. 2, 253–265.

Tani, Y., Wang, K., Yasufuku N., Ishikura, R., Fujimoto, H. & Nagata, M. 2019. *Model Test for Bearing Capacity of Spiral Pile in Sandy Ground Focused on Pitch-Width Ratio*. The 54th Geotechnical Research Presentation, Saitama.

Tani, Y., Amarbayar, J., Yasufuku N., Ishikura, R., Kurokawa, T. & Nagata, M. 2020. *Horizontal Loading Test of Spiral Piles in Sand*. The 55th Geotechnical Research Presentation, online.

Wang, K., Tani, Y., Yasufuku N., Ishikura, R., Fujimoto, H. & Nagata, M. 2019. *Bearing Capacity Characteristics of Spiral Pile in Sandy Ground Focused on Pitch-Width Ratio*. The 54th Geotechnical Research Presentation, Saitama.

Yamagata, K., Ito, A., Yamada, T. & Tanaka, T. 1991. *Statistical Study on Ultimate Point Lord and Point Load Settlement Characteristics of Cast-in-Place Concrete Piles*. Journal of Struct. Constr. Engng, AIJ, No 423, May.

Yamagata, K., Ito, A., Tanaka, T. & Kuramoto, Y. 1992. Statistical Study on Ultimate Point Lord and Point Load Settlement Characteristics of Bored Precast Piles. Journal of Struct. Constr. Engng, AIJ, No 436, June.

Proceedings of the Second International Conference on
Press-in Engineering 2021, Kochi, Japan – Matsumoto et al (eds)
© 2021 Taylor & Francis Group, London, ISBN 978-1-032-10414-0

Influence of horizontal loading height on subgrade reaction behavior acting on a pile

A. Mohri
Graduate School of Tokyo University of Science, Chiba, Japan

Y. Kikuchi & S. Noda
Tokyo University of Science, Chiba, Japan

K. Sakimoto, Y. Sakoda & M. Okada
Graduate School of Tokyo University of Science, Chiba, Japan

S. Moriyasu
Nippon Steel Corporation, Chiba, Japan

S. Oikawa
Nippon Steel Corporation, Tokyo, Japan

ABSTRACT: Coastal areas along the Pacific Ocean suffered extensive damage due to the 2011 tsunami caused by the Tohoku earthquake. Installing a row of piles behind the caisson and filling the space in between with rubble are proposed as a method of reinforcement against tsunamis. Understanding the load conditions acting on a pile is necessary to determine the cross-section and embedment length of the pile used in the structure. Model experiments were conducted based on the bending moment distribution to estimate the external force acting as a distributed load on the offshore side of the pile. The behavior of the subgrade reaction on the horizontal resistance characteristics of a pile depends on its deformation mode. In this study, horizontal loading experiments were conducted on piles with different loading heights. The deformation mode of the pile changed the value of the subgrade reaction and its behavior with depth.

1 INTRODUCTION

The 2011 tsunami due to the Tohoku earthquake off the Pacific coast of Japan caused extensive damage to coastal areas. Several breakwaters that were structurally classified as gravity-type composite breakwaters were destroyed. The primary reasons for the damage to the gravity-type composite breakwaters were the presence of extensive horizontal forces caused by water level differences between the two sides of the caisson and erosion of the foundation at the rear side of the caisson caused by overflowing water transported by the tsunami (Arikawa et al. 2012). Therefore, tsunami-resistant breakwaters are necessary to reduce damage.

Moriyasu et al. (2016) proposed a method of reinforcing breakwaters by installing steel pipe piles behind a breakwater and filling the space in between with aggregates (Figure 1). The passive resistance exerted by the ground can be used to support the steel piles, because the horizontal forces acting on the caisson are effectively transferred to the steel piles through the rubble that fills the space between the piles and the caisson. The fundamental effectiveness of this reinforcement method was examined by performing static loading (Kikuchi et al. 2015, Suguro et al. 2015) and hydraulic flume (Arikawa et al. 2015) experiments.

To design the piles, the cross-section and embedment length of the steel pipe pile must be set in such a way that the stress level of the piles generated by the tsunami force acting on the caisson does not exceed the yield stress level. It is necessary to comprehensively understand the load conditions acting on the piles to determine the cross-section and embedment length.

Here, the loads acting on the piles used in the structure are classified into three types (Figure 2): (1) the pressure of the ground acting on the part of the piles protruding from the filling with the displacement of the caisson, (2) the pressure of the ground acting on the rooted part of the piles from the foundation due to the slope of the caisson, and (3) the subgrade reaction acting on the piles from the onshore side.

DOI: 10.1201/9781003215226-20

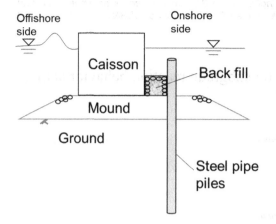

Figure 1. Schematic diagram of a reinforced gravity type breakwater (Moriyasu et al., 2016).

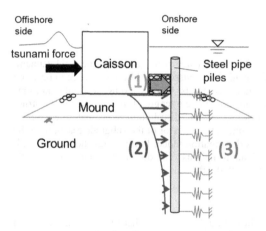

Figure 2. Classification of loads acting on the pile.

Loads (1) and (2) are external forces acting on the piles as distributed loads. Understanding the subgrade reaction received by the piles from the onshore side against the external forces is important for predicting the maximum bending moment and its depth of occurrence.

In this study, static model experiments were conducted to generalize the external force distribution acting on the piles from the offshore side of the structure (Hikichi et al. 2018, Mohri et al. 2018). In the experiment, the bending strain and bending moment distribution of the pile were calculated.

The deflection distribution was obtained by second-order integration of the bending moment distribution, and the load distribution acting on the pile was obtained by the second-order difference of the bending moment distribution. Here, the estimation of the external force distribution was studied by subtracting the assumed subgrade reaction

distribution based on the deflection distribution from the load distribution. The insufficient points regarding the external force distribution estimated using this method are due to a problem with the assumption of the coefficients of the subgrade reaction with depth, which was used to calculate the subgrade reaction distribution.

The behavior of the subgrade reaction on the horizontal resistance characteristics of a pile depends on the deformation mode of the pile (Iwata et al. 2007, Nakai 1985). In this study, horizontal loading experiments were conducted on piles with different loading heights, and the difference in the horizontal resistance of the piles was investigated.

2 MODEL LOADING TEST OF REINFORCED GRAVITY TYPE BREAKWATER

2.1 Experimental outline

A schematic diagram of the model setup is shown in Figure 3. A rigid container 1600 mm long, 800 mm high, and 400 mm wide was used to prepare the model ground (Figure 3). Tohoku silica sand #5 (ρ_s = 2.647 g/cm^3, D_{50} = 0.591 mm, e_{max} = 1.072, and e_{min} = 0.689) was used to prepare the model ground using a relative density of 90% and an air pluviation method.

The model caisson was 380 mm long, 300 mm high, and 300 mm wide and had a controlled weight of 752 N. The average contact pressure acting on the ground was 8.4 kN/m^2. The caisson model was designed at a 1/60 scale with a width of 300 mm.

Twelve steel piles (800 mm long, 30 mm wide, 2.3 mm thick, and E = 2.05 × 10^5 N/mm^2) were set 50 mm behind the caisson. The center-to-center distance of the piles was 33 mm, and the embedment length was 500 mm. A backfill was maintained at a height of 50 mm between the caisson and the steel piles (D_{50} ≈ 10 mm, G_s =2.65, γ_d = 12.1 kN/m^3).

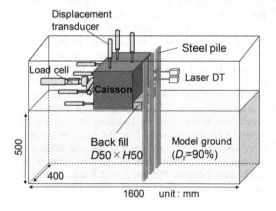

Figure 3. Schematic diagram of the experimental model set up.

The friction on the contact surface between the caisson and the loading blade was reduced using a Teflon sheet and silicone grease. Sandpaper #150 was attached to the bottom of the caisson. In the loading experiment, a tsunami force horizontal load was applied under displacement control at a height of 150 mm; the monotonous loading speed was 1 mm/min.

Twenty strain gauges were attached to both sides of the central pile and the pile at the end to measure the bending strains (Figure 4).

2.2 Experimental results

Figure 5 shows (a) bending moment distribution, (b) deflection distribution, and (c) load distribution in the central pile. The bending moment distribution was obtained by approximating the bending moment obtained from the strain gauges using the smoothing spline function. The deflection and load distributions were calculated as follows: the deflection distribution was obtained by the second-order integration of the bending moment distribution by the bending stiffness EI of the pile; the load distribution was obtained by the second-order difference of the bending moment distribution.

Figure 5 shows the results of applied loads of 670 and 800 N on the caisson. The maximum bending moment tends to increase as the applied load on the caisson increases. However, the generation depth $l_{m,max}$ and first zero-point depth of the bending moment l_{m1}—an important index for determining the embedment length of the pile—do not change significantly. Piles receiving horizontal force at the pile head generally experience increased $l_{m,max}$ and l_{m1} characteristics as the

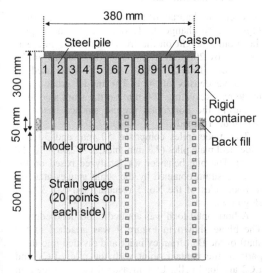

Figure 4. Schematic diagram of the pile condition (the view from the rear of the caisson).

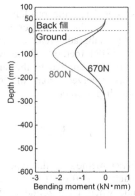
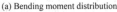

(a) Bending moment distribution

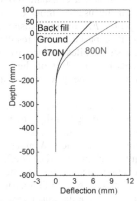

(b) Deflection distribution

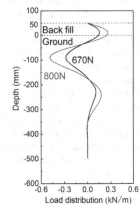

(c) Load distribution

Figure 5. Bending moment distribution and deflection distribution of the central pile.

load level increases. In this experiment, the horizontal force acting on the piles from the offshore side is considered a distributed load, which is different from the general tendency. The load distribution is obtained by second-order differentiation of the bending moment distribution of the central pile and is considered the resultant force, which is classified into three categories (Figure 2). Therefore, the subgrade reaction model (Kubo 1964) of the PHRI method of the type S model (Equation (1)), was applied to the subgrade reaction that was received from the onshore side owing to pile displacement.

$$p = k_s x y^{0.5} \qquad (1)$$

In the PHRI method of the type S model, the subgrade reaction coefficient is considered unique to the ground when a horizontal load acts on the pile head. Therefore, a constant subgrade reaction coefficient can be used without depending on the flexural rigidity, EI, or pile displacement. This model is derived based on the results of horizontal loading tests of piles under various conditions and represents the behavior of the pile by assuming nonlinearity in the p–y relationship (Kikuchi 2009).

The subgrade reaction behavior of a pile under a horizontal load generally depends on the deformation mode of the pile. Considering this phenomenon using the Winkler–Spring type model requires changing the coefficient of the subgrade reaction depending on the deformation mode of the pile. The depth distribution of the subgrade reaction coefficient used in the PHRI method of the type S model was modified based on the results obtained from the horizontal loading test of the pile. The subgrade reaction distribution was calculated using the deflection distribution shown in Figure 5 (b) and the corrected subgrade reaction coefficient.

The external force distribution acting on the pile from the offshore side was estimated by subtracting the subgrade reaction distribution based on the assumed load distribution. The external force distribution estimated in this experiment (Figure 6) presents discontinuous behavioral problems with depth and imbalance between the resultant force of the external force and the force of the load acting on the caisson.

The accuracy of the coefficient distribution of the subgrade reaction used in the assumption of the subgrade reaction distribution is assumed to be responsible for the insufficient points in the estimated external force distribution. As the behavior of the subgrade reaction varies depending on the deformation mode of the pile, the coefficient distribution of the subgrade reaction must be assumed to correspond to the deformation mode of the pile. Therefore, to change the deformation mode of the pile, loading experiments with different loading heights were conducted. The behavior of the subgrade reaction acting on the pile was examined based on the behavior of the pile.

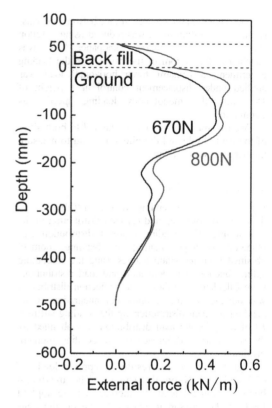

Figure 6. Estimated external force distribution.

3 HORIZONTAL LOADING EXPERIMENTS ON THE PILES

3.1 Experimental outline

A schematic diagram of the model setup is shown in Figure 7. To change the loading height on the pile, the ground height on the front side of the pile was lowered by 200 mm from the ground surface on the rear side. Tohoku silica sand #5 was used at a relative density of 90%. The model ground was made by excavating 200 mm on the front side after preparing the horizontal ground surface based on that of the rear side.

The model piles comprised 50 mm and 30 mm wide steel plate piles (2.3 mm thick, $E = 2.05 \times 10^5$ N/mm²). The pile behavior was analyzed based on the bending strain obtained by the strain gauges attached to both sides of the No. 5 (central) and No. 3 piles (Figure 8).

A horizontal load was applied under load control. The blade shown in Figure 9 was attached to the shaft of the Berofram cylinder and divided into three parts at the contact point with the piles. LC1 and LC2 are load cells. LC1 measured the total load on the piles arranged in a row, and LC2 measured the

238

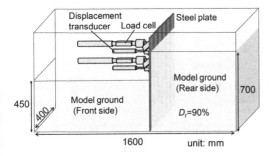

Figure 7. Schematic diagram of the experimental model set up.

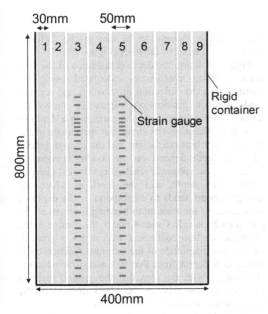

Figure 8. Schematic diagram of the pile condition.

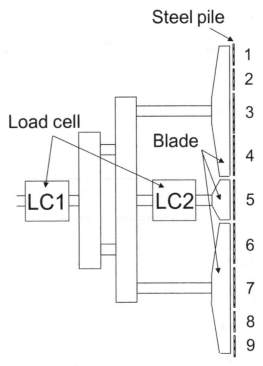

Figure 9. Structure of loading blade.

3.2 Experimental results

Figure 11 Shows the relationship between the horizontal load on the central pile and the displacement of the ground surface for each case. The horizontal load on the central pile is directly measured by LC2 (Figure 10). The height of the ground surface is that on the rear side of the piles.

The results show that the displacement of the ground surface of the pile with the same horizontal load differs greatly depending on the loading height. The displacement of the pile decreases with the same load for deeper external forces. However, Case-1 and Case-2, which have the same loading height and different ground conditions, do not differ significantly. Therefore, the effect of excavating 200 mm on the front side is small.

Figure 12 shows (a) bending moment distribution, (b) deflection distribution, and (c) subgrade reaction distribution for each case with a ground surface displacement of 4 mm. This result shows that the bending moment distribution varies greatly and that the deformation mode of the pile varies as the loading height changes.

When determining the subgrade reaction distribution, the bending moment distribution is approximated using a smoothing spline function (Kikuchi 2003). For Case-4 and Case-5, the bending moment distribution is not smoothed at the loading height. The smoothing spline function is applied separately above and below the loading

load acting on the central pile. Additionally, the horizontal displacement of the loading point was measured using a displacement meter fixed to the rigid container frame.

Figure 10 shows the experimental conditions. Case-1 is a horizontal ground condition where the ground height is the same on the front and rear sides of the pile. Cases-2–5 are ground conditions, in which the ground surface on the front side is lowered by 200 mm (Figure 7). The loading heights of each case differ by 0, −100, or −150 mm based on the ground surface on the rear side.

In this loading test, as the penetration length of the pile with respect to the horizontal force becomes sufficient, the difference in the ground height for each test case has a negligible effect on the pile behavior.

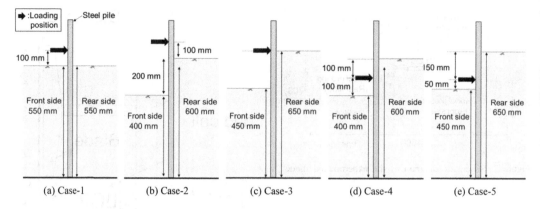

Figure 10. Details of the experimental cases.

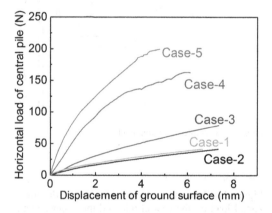

Figure 11. Relationship between horizontal load on central pile and displacement of ground surface.

height of the bending moment distribution (Morikawa et al. 2011). At this time, the discontinuity of the shear force at the loading height obtained by the differentiation of the bending moment distribution should be equal to the horizontal load. Additionally, the subgrade reaction obtained by the differentiation of the shear force distribution should be continuous at the loading height.

Figure 13 compares the p-y relationship at a depth of −60 mm and shows the relational lines based on the PHRI method of the type S model using several subgrade reaction coefficients. This result shows the differences in the p-y relationship depend on the loading height. The subgrade reaction at the same displacement is generally smaller for deeper loading heights. The subgrade coefficient in Case-5 is approximately 30% that of Cases 1–3. The difference (even when the ground conditions remain

unchanged) is thought to be due to the different deformation modes of the piles (Figure 13).

Figure 14 shows the depth distribution of the subgrade reaction coefficients calculated using the results of the (b) deflection distribution and (c) subgrade reaction distribution. Figure 12 shows the results of the section from the surface to a depth of −250 mm. These results show that the values of the subgrade reaction coefficient and the change behavior with depth depend on the loading height. In Cases-1–3, the subgrade reaction coefficient changes rapidly with depth, and the depth at which the change appears tends to correspond to the depth of occurrence of the maximum bending moment of the pile in each case. In Cases-4 and 5, the change in the subgrade reaction coefficient with depth is gradual. Further, the magnitude of the subgrade reaction coefficient is different from that in Cases-1–3, which is inconsistent with the idea that the ground reaction force coefficient used in the PHRI method of the type S model is unique to the ground.

Figure 15 shows the result of superimposing the deflection distribution at 670 N in Figure 5 (b) on the deflection distribution shown in Figure 12 (b). These results show that the deformation mode of the pile in the model experiment of the breakwater reinforcement structure is most similar to the pile deformation mode in Case-3. Multi-stage loading must be performed with an increased number of loading points to match the pile deformation mode in the model experiment. Assuming the distribution of the subgrade reaction coefficients in this experiment to be the same as those in Case-3 is necessary to obtain a subgrade reaction corresponding to the deformation mode of the pile in the reinforcement structure of the breakwater. Moreover, the external force acting on the pile may change depending on the difference in the

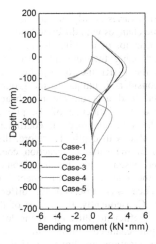

(a) Bending moment distribution

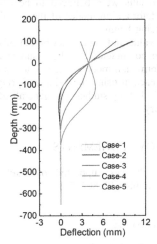

(b) Deflection distribution

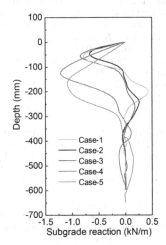

(c) Subgrade reaction

Figure 12. Pile behavior in each case at y_0=4mm.

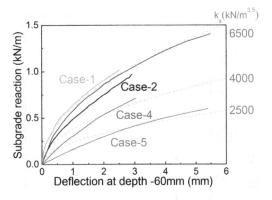

Figure 13. *P-y* relationship at a depth of -60 mm.

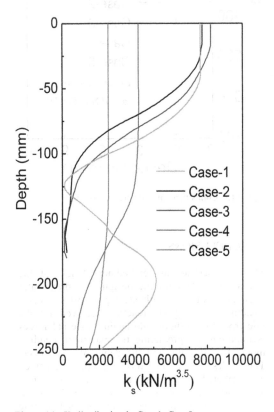

Figure 14. K_s distribution in Case1~Case5.

foundation ground and the rigidity of the pile. Figure 14 shows that the values and depth distributions of the subgrade reaction coefficients differ according to the deformation mode of the pile. Therefore, the subgrade reaction coefficient distribution corresponding to the deformation mode of the pile must be used.

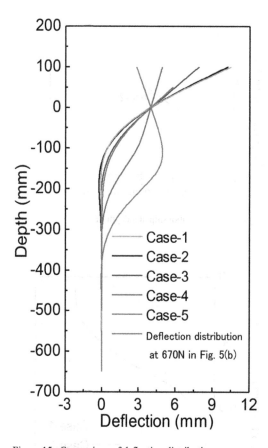

Figure 15. Comparison of deflection distribution.

4 CONCLUSIONS

The cross-section and embedment length of the pile are important for the design of a breakwater structure reinforced with steel pipe piles. Understanding the subgrade reaction received by the pile from the onshore side against external forces is important for predicting the maximum bending moment of the pile and the depth of occurrence. The piles used in this structure received a distributed load along the perpendicular pile axis direction from the offshore side. However, insufficient knowledge has been obtained on pile resistance for a load distributed in this manner. The change in the reaction force behavior cannot be considered. Therefore, the change in the subgrade reaction due to the difference in the deformation mode of the pile cannot be considered. In this study, horizontal loading experiments were conducted with different horizontal loading heights to examine the differences in the horizontal resistance characteristics of the piles.

The displacement of the pile decreases with the same load as the height of the external force moves deeper in the ground. The bending moment

distribution varies greatly, and the deformation mode of the pile changes with different loading heights.

The behavior of the piles in the experiment was compared using the subgrade reaction coefficient used in the PHRI method of the type S model, which considers it unique to the ground when the horizontal load acts on the pile head. However, because of the influence of the pile deformation mode, the subgrade reaction coefficient does not acquire value with depth. Therefore, the depth distribution of the subgrade reaction coefficient was calculated based on the bending moment distribution of the pile in the experimental results, and a difference was observed for different loading methods. Consequently, the values of the subgrade reaction coefficients differed as the loading height on the pile changed. Further, the behavior of the subgrade reaction coefficient varied with depth. These results were believed to be caused by different deformation modes of the pile depending on the loading height.

The value and depth distribution of the subgrade reaction coefficient must be assumed according to the deformation mode of the pile to express the subgrade reaction using the Winkler-Spring type PHRI method. Future studies on generalization will be necessary.

ACKNOWLEDGEMENTS

This work was supported by JSPS KAKENHI (Grant Number JP1804352).

REFERENCES

Arikawa, T., Sato, M., Shimosako, K., Tomita, T., Tatsumi, D., Yeom, G., Takahashi, T. 2012. Investigation of the Failure Mechanism of Kamaishi Breakwaters due to Tsunami - Initial Report Focusing on Hydraulic Characteristics-, *Technical note of Port and Airport Research Institute* 1251(52). (In Japanese)

Arikawa, T., Oikawa S., Moriyasu S., Okada K., Tanaka R., Mizutani T., Kikuchi Y., Yahiro A., and Shimosako K. 2015. Stability of the breakwater with steel pipe piles under tsunami overflow. *Technical note of Port and Airport Research Institute* 1298(44). (In Japanese)

Hikichi, K., Kikuchi, Y., Hyodo, T., Mohri, A., Akita, K., Shoji, N., Taenaka, S., Moriyasu, S., Oikawa, S. 2018. Estimation of external force acting on steel pile reinforced breakwater. *Proceedings of the 1st International Conference on Press-in Engineering 2018*: 281–288.

Iwata, N., Nakai, T., Zhang F, Inoue, T, Takei, H. 2007. Influence of 3D effects, Wall Deflection Process and Wall Deflection Mode in Retaining Wall Problems. *Soils and Foundations* 47(4): 685–699.

Kikuchi, Y. 2003. Lateral Resistance of soft landing moundless structure with piles. *Report of the Port and Airport Research Institute* 1039: 5–191. (In Japanese)

Kikuchi, Y. 2009. Horizontal Subgrade Reaction Model for Estimation of Lateral Resistance of Pile, *Report of the*

Port and Airport Research Institute 48(4): 3–22. (In Japanese)

Kikuchi, Y., Kawabe, S., Taenaka, S., and Moriyasu, S. 2015. Horizontal loading experiments on reinforced gravity type breakwater with steel walls. *Japanese Geotechnical Society Special Publication* 2: 1267–1272.

Kubo, K. 1964. A New Method for the Estimation of Lateral Resistance of Piles, *Report of the port and harbour research institute* 2(3): 13–15. (In Japanese)

Nakai, T. 1985. Finite Element Computations for Active and Passive Earth Pressure Problems of Retaining wall. *Soils and Foundations* 25(3): 98–112.

Mohri, A., Hikichi, K., Kikuchi, Y., Hyodo, T., Akita, K., Taenaka, S., Moriyasu, S., Oikawa, S. 2018. Estimation of external force acting on steel pile of reinforced breakwater. *Annual Journal of Civil Engineering in the Ocean* 74(2): 1_420-1_425. (In Japanese)

Moriyasu, S., Oikawa, S., Taenaka, S., Harata, N., Tanaka, R., Tsujii, M., Kubota, K. 2016. Development of New Type of Breakwater Rein-forced with Steel Piles against a Huge Tsunami, *Nippon Steel & Sumitomo Metal Technical Report* 113: 64–70. (In Japanese)

Morikawa, Y., Kikuchi, Y., Mizutani, T. 2011. Development of Design Method for Anchored Sheet Pile Wall Reinforced by Additional Anchorage Work, *Report of the Port and Airport Research Institute* 50(4): 107–131. (In Japanese)

Suguro, M., Kikuchi, Y., Kiko, M., Nagasawa, S., Moriyasu, S. 2015. Load-displacement properties of gravity type break-water reinforced by steel pipe piles. *Japanese Geotechnical Society Special Publication* 1(6): 1–5.

Proceedings of the Second International Conference on
Press-in Engineering 2021, Kochi, Japan – Matsumoto et al (eds)
© 2021 Taylor & Francis Group, London, ISBN 978-1-032-10414-0

Influence of different pile installation methods on vertical and horizontal resistances

S. Moriyasu
Nippon Steel Corporation, Tokyo, Japan

M. Ikeda
Kanazawa City Government, Kanazawa, Japan (former student of Kanazawa University)

T. Matsumoto, S. Kobayashi & S. Shimono
Kanazawa University, Kanazawa, Japan

ABSTRACT: This study focuses on the influence of the pile installation method on vertical and horizontal pile resistance. In a series of laboratory experiments, a model pile was installed using four types of pile installation methods: monotonic push-in, surging (repetitive push-in and pull-out), vibratory pile driving, and bored pile installation in dense dry sand ground. It was found that the cyclic shearing of surging or vibratory pile driving prevented soil dilation and decreased pile penetration resistance. During a static load test in the vertical direction, the pile installed using push-in, surging, or vibratory pile driving exhibited a higher vertical resistance in comparison with the bored pile installed in a similar manner. In the horizontal load tests, relatively high horizontal resistances were obtained in the surging and push-in cases in comparison with the bored pile, indicating that the effect of the displacement pile increases the horizontal soil resistance.

1 INTRODUCTION

In Japan, owing to the occurrence of many earthquakes, the assessment of the horizontal and vertical resistances (i.e., bearing capacity) of a pile is important. Although many types of pile installation methods exist, the difference in the horizontal resistance of a pile installed by each piling method has not been fully clarified yet. Therefore, this study investigates the influence of different piling methods on the horizontal pile resistance, and correlation between the horizontal resistance and the vertical resistance.

It is well known that the vertical resistance depends on the pile installation method. Some studies (e.g., Jack-in installation method: White & Deeks 2007, Ogawa et al. 2011; Vibratory pile driving: Holeyman et al. 1996, Vanden Berghe 2001) have clarified that pile movement during pile penetration causes soil dilation or contraction. Moriyasu et al. (2020) investigated the influence of the pile installation method on the vertical resistance of a model pile in sand ground through a series of laboratory experiments. It was discovered that while the pile installation force in surging or vibratory pile driving was lower than that in monotonic push-in, the bearing capacity of the pile installed in these

methods was the same or higher than that in the monotonic push-in. According to Moriyasu et al. (2020), soil contraction caused by cyclic pile movement during pile installation transforms into soil dilation under monotonic shearing during the static load test. However, Moriyasu et al. (2020) treated only the displacement pile installation methods. Therefore, this study adopted the bored pile method as a representative non-displacement pile installation method and compared the vertical resistance with those of the displacement pile installation methods. In a series of laboratory experiments, a model pile was installed using four different piling methods, i.e., push-in, surging, vibratory pile driving and boring (buried). After the pile installation process, vertical load test (VLT) and horizontal load test (HLT) were conducted.

2 EXPERIMENTAL DESCRIPTION

2.1 *Experimental cases*

Table 1 presents the experimental cases and conditions. Each experiment comprised a pile penetration test (PPT), VLT, and HLT. In Case 1, a model pile was installed using a monotonic push-in at a penetration

DOI: 10.1201/9781003215226-21

Table 1. Experimental cases and conditions.

Case No.	Case 1	Case 2	Case 3	Case 4
Relative density, D_r (%)	80	80	80	80
Penetration method	push-in	surging	vibration	bored pile
Penetration rate (mm/s)	0.15	0.15	-	-
Vibration frequency (Hz)	-	-	25 ~ 55	-
Test sequence	PPT[*1] ↓ VLT[*2] ↓ HLT[*3] ↓ CPT[*4]	PPT ↓ VLT ↓ HLT ↓ CPT	Static loading by V.H.[*5] ↓ Vibration ↓ VLT ↓ HLT ↓ CPT	Embedded with ground preparation ↓ VLT ↓ HLT ↓ CPT

* 1 PPT: pile penetration test, *2 VLT: vertical load test, *3 HLT: horizontal load test, *4 CPT: cone penetration test, *5 V.H.: vibratory hammer.

rate of 0.15 mm/s during a PPT. In Case 2, surging implies the repetition of 4 mm push-in (downward) and 2 mm pull-out (upward) strokes. In Case 3, a vibratory hammer model was employed to install a model pile. In Case 4, a model pile was embedded in the ground during model ground preparation. After the PPT, both the VLT and HLT were performed continuously.

2.2 Model pile

A closed-ended aluminium pipe pile with a diameter of 32 mm, wall thickness of 1.3 mm, and length of 600 mm was used for the model pile. As shown in Figure 1, two strain gauges were attached on opposite

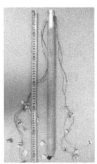

(a) Photograph of pile

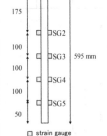

(b) Locations of strain gauges

Figure 1. Model pile.

faces at each level to calculate the axial strains and bending strains. The pile surface with the strain gauges was coated with an acrylic adhesive and glued with silica sand, i.e., the same material used for the model ground.

2.3 Model ground

Silica sand was used as the model ground material in all cases. Table 2 lists the physical properties of the sand. The model ground was prepared in a cylindrical soil container of diameter 566 mm and height 580 mm. The silica sand was poured into a container divided into 12 model ground layers (11 layers of height 50 mm and one layer of height 30 mm). The sand was compacted using a small tamper to adjust the relative density D_r to 80 %. The reason for choosing such a high relative density is to investigate the pile behaviour in a dense sand condition and obtain the soil resistance against the pile clearly within a short penetration depth.

2.4 Experimental apparatus and procedure

During the PPT, the model pile was installed into the model ground until the pile head displacement w_h reached approximately 420 mm by push-in, surging, and vibration. In the cases of push-in and surging, an electrical jack, shown in Figure 2(a), was employed. While the pile was installed monotonically at a pile penetration rate of 0.15 mm/s in the push-in case, it was installed with the repetitions of 4 mm push-in and 2 mm pull-out strokes at a pile movement speed of 0.15 mm/s in the surging case. The pile head load, P_h, was measured using a load cell between the pile head and jack. Figure 2(b) shows the vibratory hammer model employed in the vibration case. The vibratory hammer model weighed 275 N and had a maximum frequency of 60 Hz.

Vibration was generated by two electric motors with an eccentric mass, as shown in Figure 2(b). When the motors were rotated in opposite directions, the horizontal vibrations were negated, and the vertical vibrations of the two motors were harmonised. At the beginning of the PPT in the vibration case, the pile was installed by the weight of the vibratory hammer alone. When it became difficult to install the pile by the self-weight of the hammer, the vibration began. Until w_h reached approximately 420 mm, if the pile penetration stopped, the vibration frequency was

Table 2. Physical properties of silica sand.

Minimum dry density, ρ_{dmin}	(t/m^3)	1.37
Maximum dry density, ρ_{dmax}	(t/m^3)	1.63
Maximum void ratio, e_{max}		0.96
Minimum void ratio, e_{min}		0.65
Mean particle size, D_{50}	(mm)	0.51

(a) Motor jack   (b) Vibratory hammer model

Figure 2. Loading apparatus for pile penetration test and vertical load test.

increased to enhance the pile installation. The P_h was estimated from the strains near the pile top (i.e., SG1).

Subsequently, a VLT was performed. In all cases, the pile head was pushed by the electrical jack at a rate of 0.1 mm/s until w_h reached 440 mm. In the case of bored (buried) piles, the pile was embedded at a depth of 420 mm during the preparation stage of the model ground. During the VLT, the bored pile was pushed from a depth of 420 to 440 mm. In all cases, P_h was measured using the load cell between the pile head and jack.

After the VLT, an HLT was conducted. As shown in Figures 3 and 4, the pile head was pulled using a wire and a hand winch. An accelerometer was employed to obtain the slope-deflection at the pile top. The relationship between the inclined angle and acceleration was calibrated in advance.

Finally, cone penetration tests (CPTs) were performed to investigate the ground condition after the piling tests. A rod with a circular cone with a diameter of 20.05 mm at the tip was jacked at four

locations in the model ground using an electrical jack (see Figure 4). After the CPT, the pile was removed.

2.5 Monotonic and cyclic triaxial shear tests of sand

Consolidated monotonic and cyclic shear tests were performed to investigate the mechanical properties of the sand. The confining pressure of the tests was approximately 100 kPa. Figure 6 shows the results of the triaxial consolidated drained shear tests (CD tests) of dense sand ($D_r = 82\%$). For the result of the monotonic loading case, Figure 6(a) shows the softening behaviour after the axial strain, ε_a, exceeds 5%. The internal friction angle at the peak strength,

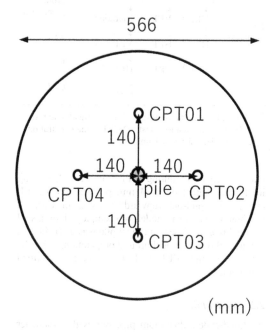

Figure 4. Locations of cone penetration tests.

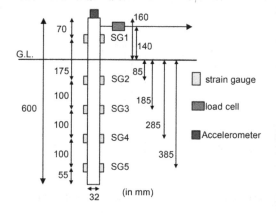

Figure 3. Loading apparatus for horizontal load test.

Figure 5. Schematic diagram of horizontal load test.

246

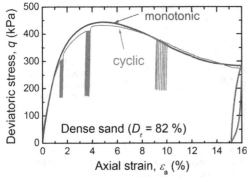

(a) Axial strain ε_a vs. deviatoric stress q

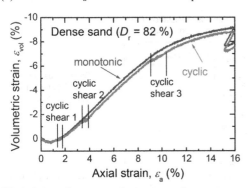

(b) Axial strain ε_a vs. volumetric strain ε_{vol}

Figure 6. Results of triaxial CD tests in dense ground.

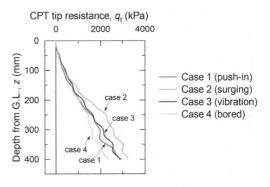

Figure 7. Results of cone penetration tests.

$\phi_p{}'$, was 42.8° and that at the residual state, $\phi_r{}'$, was approximately 35.0°. In the case of cyclic loading, cyclic shearing was applied when ε_a reached 1.3%, 3.4%, and 8.9%, separately, by changing the axial stress, σ_a, with a constant radial stress, σ_r, of 100 kPa. According to the volumetric strain in Figure 6 (b), the volume increase during the cyclic shearing stages was smaller than that of the monotonic shearing case. This indicates that cyclic shearing prevented soil dilation. When the cyclic shearing transformed to monotonic shearing, the q–ε_a and ε_a –ε_{vol} relationships were similar to the corresponding curves of the monotonic shearing case.

3 RESULTS

3.1 Results of cone penetration tests

Figure 7 shows the distributions of the CPT tip resistance, q_t, with depth, z, for all cases. Each line represents the average of q_t at four CPT locations for each case. As shown in Figure 7, the q_t of Case 4 (bored pile) was smaller than those of other cases. Because the q_t in Cases 1, 2, and 3 (push-in, surging, and vibration, respectively) was higher than that in

Case 4 (bored pile), the pile installation in Cases 1, 2, and 3 (displacement pile cases) would have increased the strength of soil around the pile.

3.2 Results of pile penetration tests and vertical load tests

Figure 8 shows the relationship between P_h and w_h for Cases 1, 2, and 3. At the end of the PPT ($w_h = 420$ mm), the P_h in Case 2 (surging) was smaller than that in Case 1 (push-in). Furthermore, the P_h in Case 3 (vibration) was much smaller than those in Cases 1 and 2. As mentioned above, cyclic CD tests show that cyclic shearing prevented soil dilation. If a similar behaviour of the soil occurred in Cases 2 and 3, the cyclic pile movement by surging or vibration would generate soil contraction around the pile. Because the cycle number, n, of the pile vibration in Case 3 was much higher than that in Case 2 (Case 2: $n = 210$, Case 3: $n \approx 21,600$), the effect of cyclic shearing may be significant in Case 3.

For the result of the VLT, Figure 9 shows the relationship among the w_h, P_h, pile base resistance, P_b, and pile shaft resistance, P_s. P_b is the axial force obtained from strain gauge SG5, whereas P_s is the difference between P_h and P_b. As shown in Figure 9(a), the P_h, in the cases of displacement pile (Cases 1, 2, and 3) was higher than that in Case 4 (bored pile). Figure 9(b) shows that P_b and its stiffness in Cases 1, 2, and 3 were much higher than those in Case 4. Meanwhile, as shown in Figure 9(c), P_s and its stiffness in Cases 1 and 2 were similar to those in Case 4. Regarding the vibration (Case 3), after P_s reached a peak value, it decreased significantly. Figure 10 shows the distribution of increment of unit shaft resistance, $\Delta\tau_s$, during VLT. $\Delta\tau_s$ equals zero at the beginning of VLT, and each line shows the increment of unit shaft resistance when P_s reaches each number illustrated in Figure 9(c). When P_s decreased significantly (#1 to #2), Figure 10 shows that $\Delta\tau_s$ from ground surface to $z = 78$ mm and $\Delta\tau_s$ from $z = 278$ mm to $z = 378$ mm decreased largely.

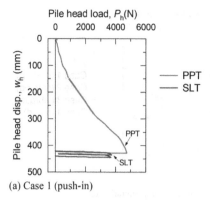

(a) Case 1 (push-in)

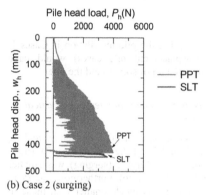

(b) Case 2 (surging)

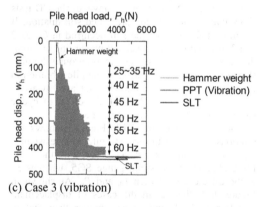

(c) Case 3 (vibration)

Figure 8. Relationship between pile head load, P_h, and pile head displacement, w_h, during PPT and VLT.

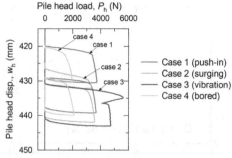

(a) Relationship between pile head load, P_h, and pile head displacement, w_h, during VLT.

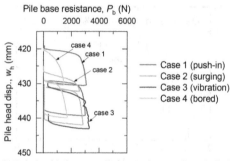

(b) Relationship between pile base resistance, P_b, and pile head displacement, w_h, during VLT.

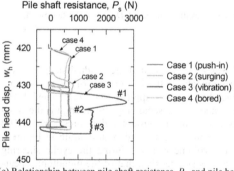

(c) Relationship between pile shaft resistance, P_s, and pile head

Figure 9. Results of vertical load test.

A possible reason why P_s decreased suddenly is that a hardening zone around the pile shaft formed during the PPT is sheared within a certain pile displacement during SLT. This hypothesis is based on a previous study which investigated a pile shaft resistance under a large number of quasi-static cyclic pile movements in a sand model ground (Bekki et al. 2013). As a result of the study, after the pile shaft friction was degraded by 300 cycles of pile movement, the friction kept increasing until the end of test (10^5 cycles). Bekki et al. (2013) pointed out that a large number of cycles progresses a densification of the sand within the interface zone around the pile. If a similar phenomenon occurred in the vibration case (Case 3), a large number of cyclic pile movement turns the aforementioned soil contraction to forming a hardening zone around the pile shaft. As a result, the P_h during PPT may increase gradually (see Figure 8 (c)). At the beginning of SLT, the hardening zone could increase P_s and its stiffness. When a certain monotonic pile displacement during SLT sheared the hardening zone, P_s decreased suddenly. Although such a softening occurred, as shown in

248

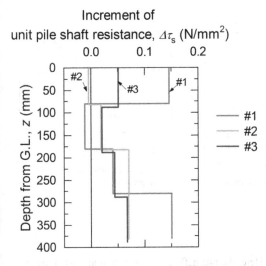

Figure 10. Increment of unit pile shaft resistance, $\Delta \tau_s$, during VLT in Case 3 (vibration).

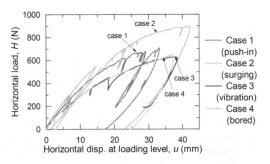

Figure 11. Relationship between pile head load, H, and horizontal displacement at loading level, u, during HLT.

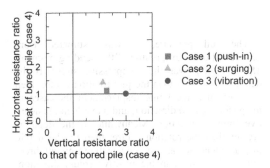

Figure 12. Pile resistance ratio of displacement pile cases to bored pile case in both VLT and HLT.

Figure 9, the P_s and P_h in the vibration remained higher than those in other cases.

While the P_h in Cases 2 and 3 (surging and vibration) during the PPT was smaller than that in Case 1 (push-in), the VLT indicated that the P_h in Cases 2 and 3 was the same or higher than that in Case 1. From the results of the cyclic triaxial CD test, it was presumed that the monotonic shearing phase after cyclic shearing enhanced soil dilation. When the cyclic pile movement during the PPT transformed into monotonic movement during the VLT, soil dilation occurred.

Returning to Figure 8, P_h during the VLT is lower than those during the PPT in Cases 1 and 2. A possible reason is the difference of the pile penetration rate between PPT and VLT. Watanabe and Kusakabe (2013) found that the failure strength of sand increases with increasing the loading rate. Because the pile penetration rate during VLT (0.1 mm/s) was slower than that during PPT (0.15 mm/s), P_h during the VLT was lower than those during the PPT.

3.3 Results of horizontal load tests

Figure 11 shows the relationship between the horizontal load, H, and horizontal displacement at the loading level, u. H was measured using a load cell connected between the pile head and the winch. As shown in Figure 11, whereas the H in Cases 1 and 2 (push-in and surging) was higher than that in Case 4 (bored pile), the H in Case 3 (vibration) was similar to that in Case 4.

Figure 12 presents the influence of the pile installation method on the pile head load during the VLT or HLT. In Figure 12, the horizontal axis represents the ratio of the pile head load in displacement pile cases (Cases 1, 2, and 3) to Case 4 (bored pile) during the VLT, and the vertical axis indicates the same ratio during the HLT. To estimate the ratio of the VLT, the yield load in each case was employed. Because the softening of the pile head load was observed in Case 3 (see Figure 9(a)), the converged value after the softening (4,200 N) was selected. Regarding the ratio in the HLT, when the horizontal displacement at the loading level, u, was 30 mm, the horizontal load was employed in each case. While the ratio in the VLT ranged from 2.1 to 3.0, that in the HLT ranged from 1.0 to 1.4. Therefore, as shown in Figure 12, the influence of the pile installation method on H was less prominent in comparison with that on the vertical load, P_h, in the VLT.

Figures 13(a)-(c) show the horizontal pile behaviours at representative horizontal displacements, u, in Case 1. Figure 13(a) shows the bending moment distribution of the pile, which was obtained from the pile's flexural stiffness and measured bending strains. Figure 13(b) shows the distributions of lateral displacements of the pile, which were calculated from the second-order integration of the bending moment distribution, pile head horizontal displacement, and slope-deflection at the pile head. Figure 13(c) shows the horizontal soil pressure, p, acting on the pile shaft at representative depths of $z = 85$ and 385 mm.

249

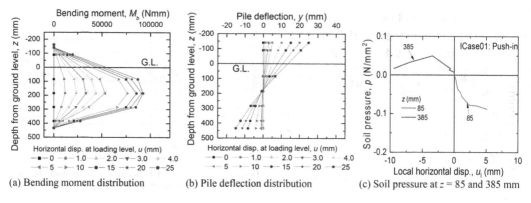

(a) Bending moment distribution   (b) Pile deflection distribution   (c) Soil pressure at $z = 85$ and 385 mm

Figure 13. Results of horizontal load test: Case 1 (push-in).

The soil pressure, p, was estimated from the second-order differential of the bending moment, M_b, and all p was the compressive stress. If the depth is shallower than the pile rotation centre, then the horizontal local pile displacement increases in the positive (plus) direction, and p exerts in the negative (minus) direction. Meanwhile, if the depth is greater than the rotation centre, then the local pile displacement increases in the negative (minus) direction, and p exerts in the positive (plus) direction. Figures 14 to 16 show the same pile behaviours for Cases 2 to 4.

The bending moment distributions ((a) in Figures 13 to 16) show that the M_b in Cases 1 and 2 was larger than that in Cases 3 and 4 at the same u. This order of M_b corresponds to the order of H, as shown in Figure 11.

As shown in (b) of Figures 13 to 16, the pile rotated with increasing u. While the centre of the pile rotation was at approximately $z = 200$ mm ($z =$ depth from the ground level) in Cases 1, 2, and 4, that in Case 3 was at $z = 300$ mm. The deeper centre of the pile rotation in Case 3 indicates that the horizontal earth pressure on the pile was lower than those in the other cases.

Here, the pile deflection distribution in Case 3 was not obtained after $u = 10$ mm because the accelerometer for measuring the pile head slope-deflection was not available owing to technical reasons.

While the pile shaft resistance in Case 3 (vibration) during VLT was higher than those in other cases, the horizontal soil pressure, p, during HLT was not higher than other cases. From this result, the soil condition in the vibration case is presumed as shown in Figure 17. As mentioned above, a large number of vibration during PPT may form a hardening zone surrounding the pile and increase the pile shaft resistance in the vertical direction (i.e., in VLT). Because the hardening associates with densification, the soil in the outer zone is loosened when the pile is vertically loaded statically after the installation. Therefore, when the pile is loaded in the horizontal direction (i.e., HLT), the outer loosened zone could have small horizontal pile resistance. More investigations are required to clarify this hypothesis.

In addition, in Case 2 (surging), whereas the pile penetration resistance during the PPT was smaller than that in Case 1 (push-in), the VLT and HLT demonstrated the same or higher resistances than that in

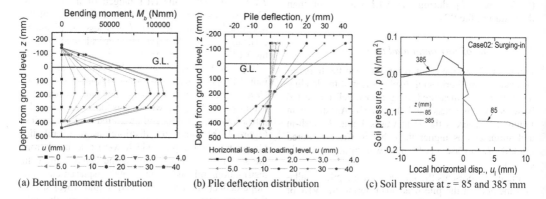

(a) Bending moment distribution   (b) Pile deflection distribution   (c) Soil pressure at $z = 85$ and 385 mm

Figure 14. Results of horizontal load test: Case 2 (surging).

250

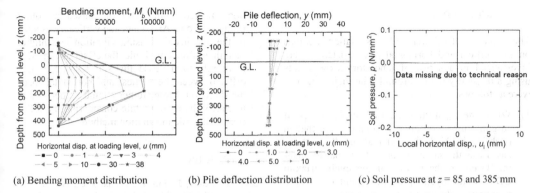

(a) Bending moment distribution

(b) Pile deflection distribution

(c) Soil pressure at $z = 85$ and 385 mm

Figure 15. Results of horizontal load test: Case 3 (vibration).

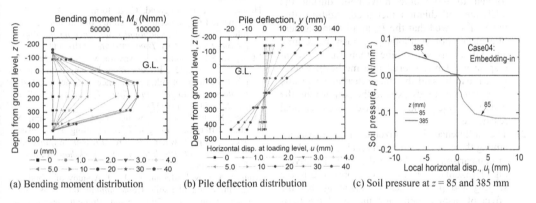

(a) Bending moment distribution

(b) Pile deflection distribution

(c) Soil pressure at $z = 85$ and 385 mm

Figure 16. Results of horizontal load test: Case 4 (bored pile).

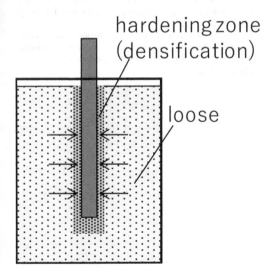

Figure 17. A concept of the soil condition in Case 3 (vibration).

Case 1. Settlements of the ground surface just after the end of installation in each test were not measured in the experiments. However, relative large settlement (2 to 4 mm by visual inspection) occurred at the ground surface within 2 or 3 mm from the pile shaft. Currently, the mechanism that yielded these results has not been clarified. Further studies are required to understand the relationship between the pile installation methods and the vertical or horizontal resistance.

Among (c) of Figures 13 to 16, p at the location of SG2 ($z = 85$ mm) in Case 2 was the highest at a specified local pile displacement, u_l, (i.e., the local pile displacement attached strain gauge SG2 in this case). This corresponded to the highest horizontal resistance shown in Figure 11. Meanwhile, the p at SG5 ($z = 385$ mm) was similar in all cases. Furthermore, as shown in (c) of Figures 13 to 16 that almost all of p–u_l relations showed a yield stress at a local pile displacement of approximately 2.0 mm, corresponding to $u = 10$ mm. It was presumed that the soil stress reached the yield stress, and that the residual stress state occurred by a local pile displacement of 2.0 mm.

4 CONCLUSIONS

The influence of different pile installation methods on the vertical and horizontal pile resistance were investigated in a series of laboratory experiments using a dense dry model ground. The major findings were as follows:

1) While the P_h during the PPT in the cases of surging and vibration was smaller than that in push-in, the VLT showed that the P_h in those cases was the same or higher than that in push-in. Based on the results of the cyclic triaxial CD test, the cyclic pile movement of surging or vibration during the PPT prevented soil dilation around the pile and increased P_h. When the cyclic pile movement during the PPT transformed to monotonic movement during the VLT, the soil dilation increased. Additionally, the VLT showed that the P_h in all displacement pile installation methods (i.e, push-in, surging and vibration) was higher than that of non-displacement pile (bored pile method). Therefore, it can be said that these displacement pile installation methods expand the soil around the installed pile and increase the vertical resistance.

2) The difference in horizontal pile resistance among the cases was not remarkable in the HLT in comparison with that in the VLT. Relatively high horizontal resistances were observed in the surging and push-in cases, indicating that the effect of displacement pile increased the horizontal soil resistance. Although the horizontal resistance of the pile caused by vibration was smaller than those of other cases, it was similar to that of the bored pile. The rotation centre of the pile caused by vibration was deeper than that caused by other pile installation methods. This implied that the distribution of the horizontal earth pressure on the pile shaft due to the vibration differed from those of caused by other piling methods. And, since the vertical resistance of the pile installed using vibration was higher than that of the bored pile, the effect of displacement pile to increase the soil resistance may differ between the vertical and horizontal loading direction. More studies should be performed to clarify the reason while considering the influence of the pile installation method.

REFERENCES

Bekki, H., Canou, J., Tali, B., Dupla, J. C., Bouafia, A., 2013. Evolution of local friction along a model pile shaft in a calibration chamber for a large number of loading cycles. *Comptes Rendus Mécanique* 341 (6): 499–507.

Holeyman, A. E., Legrand, C. & Rompaey, D. V. 1996. A method to predict the drivability of vibratory driven piles. *Proc. 5th International Conference on the Application of Stress-Wave Theory to Piles*: 1101–1112.

Moriyasu, S., Matsumoto, T., Aizawa, M., Kobayashi, S. & Shimono, S. 2020. Effects of cyclic behaviour during pile penetration on pile performance in model load tests. *Geotechnical Engineering Journal of the SEAGS & AGSSEA* 51(2): 150–158.

Ogawa, N., Ishihara, Y., Nishigawa, M. & Kitamura, A. 2011. Effect of surging in Press-in piling: shaft resistance and pore water pressure. *Press-in Engineering 2011, Proc. 3th IPA International Workshop in Shanghai*: 101–106.

Rodger, A.A. & Littlejohn, G. 1980. A study of vibratory driving in granular soils. *Geotechnique* 30 (3): 269–293.

Vanden Berghe, J.F. 2001. Sand strength degradation within the framework of vibratory pile driving. *Ph. D. thesis*, Universite catholique de Louvain.

White, D.J. & Deeks, A.D. 2007. Recent research into the behaviour of jacked foundation piles. *Advanced in Deep Foundations*: 3–26.

Watanabe, K. & Kusakabe, O., 2013. Reappraisal of loading rate effects on sand behaviour in view of seismic design for pile foundations. *Soils and Foundations*: 53 (2), 215–231.

Proceedings of the Second International Conference on
Press-in Engineering 2021, Kochi, Japan – Matsumoto et al (eds)
© 2021 Taylor & Francis Group, London, ISBN 978-1-032-10414-0

Stress changes in adjacent soils of tapered piles during installation into sand

Y. Ishihama, S. Taenaka & Y. Sugimura
Nippon Steel Corporation, Chiba, Japan

N. Ise
Nippon Steel Corporation, Hokkaido, Japan

ABSTRACT: Tapered piles with increasing pile diameter from the pile tip to the top can develop high vertical bearing capacity, especially a high shaft friction owing to tapering and wedging effects. Many researchers have reported these merits over conventional straight piles based on small model tests and furthermore the mechanism has been explained using a cylindrical cavity expansion theory. However, few researchers have examined the installation effect of tapered piles. There must be a certain installation effect even in a tapered pile, and its effect could be affected by its pile geometry. Therefore, this study investigates the radial and vertical pressure in adjacent soils between a straight pile and a tapered pile during installation into sand at 1g. The test results suggest that the profiles of pressure changes around the piles are quite different, leading to differences in the radial stress distribution between tapered and straight piles.

1   INTRODUCTION

Tapered piles with increasing pile diameter from the pile tip to the top can develop highly vertical bearing capacity, especially in high shaft friction owing to tapering wedge effect. Many studies have reported these merits over conventional straight piles. Experimental studies have demonstrated that the shaft friction under compression loading is higher than that of the straight piles through a comparison between different tapered angles, both 1g chamber tests and centrifuge tests (e.g. Wei & El Naggar, 1998; El Naggar & Sakr, 2000). Theoretical solutions under compression loads have also been provided based mainly on the cavity expansion theory by several researchers (e.g. Tominaga & Chen, 2006 and Manandhara et al., 2013). Other useful studies have already been carried out under different loading conditions for lateral loading (e.g. Sakr et al., 2005) and for uplift loading (e.g. El Naggar & Wei, 2000). In situ field testing was introduced for compression loading (e.g. Tominaga et. al., 2007a; Sato, et al., 2010) and lateral loading (Tominaga, et al., 2007b). These studies focused mainly on the behavior under the load-carrying stage. The mechanism behind the high performance of the tapered piles can be explained in a manner similar to that of non-displacement piles.

However, there are few studies discussing the installation effect for tapered piles. A few recent investigations were performed to model the pile installation process. Manandhar & Yasufuku (2013) measured the lateral soil pressure changes during pile installation at different distances from the pile center. They also have discussed the stress distribution where the stress was higher near the pile and reduced significantly from the center of the pile. Their results demonstrated that, when a tapered pile is installed into sand, radial displacement occurs in the adjacent soil owing to the increasing pile diameter along the pile shaft, leading to radial stress changes around the piles. These radial stress changes cannot be neglected when evaluating the performance of the tapered piles. There must be a certain installation effect even in a tapered pile, and its effect could be affected by the pile geometry of the tapered piles.

This paper discusses the installation effect of the tapered pile through the investigation of radial and vertical pressure changes in adjacent soils between a straight pile and a tapered pile during installation into sand. For this purposes, small model pile tests on medium dense dry sand ground at 1g were conducted for comparison between tapered and straight piles. Furthermore, the radial and vertical pressures were measured around the piles at different depth.

2   TEST METHODOLOGY

2.1   Model piles

Two types of piles were prepared for this study. One pile was a straight pile (STR) with a diameter

DOI: 10.1201/9781003215226-22

Table 1. Model piles of straight (STR) and taper (TPR).

	STR	TPR
Dia. at top D_1 (mm)	34.0	49.5
Dia. at tip D_2 (mm)	34.0	18.0
End bearing area (cm²)	9.1	2.5
ΔD (mm) (= $D_1 - D_2$)	0	31.5
Taper length L_{TPR} (mm)	620	620
Taper ratio $\Delta D/L_{TPR}$ (%)	0.0	5.08
Pile Volume (cm³)	2252	2380

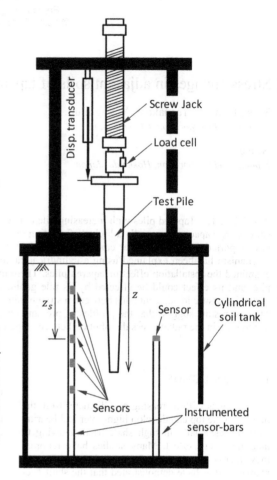

Figure 1. Experimental apparatus (soil tank, jack system and pressure sensors on instrumented bars).

of 34 mm. The other pile was a tapered pile (TPR) with an increasing pile diameter from the pile tip to the top. The maximum and minimum diameters of the tapered pile were determined to be similar volume to those of the straight pile at the final installation depth. This means that the soil volume moved by the pile installation was also similar between both piles, although their pile shapes were different. The detailed dimensions of both piles are summarized in Table 1.

2.2 Test conditions

The pile installation tests were carried out in a cylindrical steel tank of 700 mm in height and 520 mm in inner diameter at 1g with no surcharge pressure, as shown in Figure 1. Figure 2 shows the experimental apparatus and the tapered model pile. The ratio of the pile diameter to the diameter of the soil tank has been studies, and then Bolton et al. (1999) and Lee & Salgado (2000) has suggested less than 1/10 and 1/12 respectively. The ratio of the average diameter in the model piles (= 34 mm) to the soil tank (=520) was approximately 1/15 in this study, which satisfied their suggested range.

The soil, dry silica sand (D_{50}0.20 mm), was prepared by the free-fall method with vibration to satisfy the pre-defined relative density condition of D_r50 %. The soil strength at each depth was not clear in this study where the investigation hereby focused mainly on the comparison between two piles.

Then, such a test pile was installed monotonically from the soil surface at 10 mm/min until the pile tip reached around 620 mm in depth. The load and displacement at the pile head were measured during pile installation.

2.3 Data gathering

As introduced in the beginning, a particular emphasis is given to the stress changes in the tapered pile during installation compared with the straight pile. Experimental studies utilizing several types of pressure sensors have been reported in many researches (e.g., Klotz & Coop, 2001; White & Lehane, 2004; Lehane & White, 2005; Jardine et al., 2009; Taenaka et al., 2010), where pressure sensors were mounted on the model pile itself to measure the radial pressure directly acting on the pile shaft. Another method was conducted using pressure sensors placed in the soil. Using this method several studies measured pressure changes or stress distributions in the soil around model piles (e.g., Leung et al., 1996; Gavin & Lehane, 2003; Manandhar & Yasufuku, 2013).

In order to investigate pressure changes, several miniature pressure sensors (Tokyo Sokki; PDA-200KPA; 7.6mm in diameter and 2.0 mm in thickness) were used in this study. Five pressure sensors were placed on the shaft of the instrumented sensor

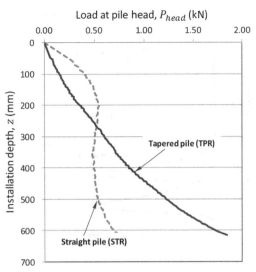

Figure 2. Photographs of the experimental apparatus (left) and the tapered model pile under testing (right).

Figure 3. Load displacement curves during pile installation for straight pile (STR) and tapered pile (TPR).

bar (square stainless bar with 10 mm in width) fixed at the bottom of the steel soil tank (the left bar in Figure 1), which could influence on the resistance of pile installation. Therefore, while the installation resistance was reference data in this study, main emphasis was given to changes in soil pressure during pile installation. The sensors were located at 130, 230, 330, 430 and 530 mm from the soil surface to measure lateral pressure changes in the radius direction 48 mm from the pile center (i.e., pressure sensors in soils) during pile installation. Unfortunately, the sensors at 230 mm did not work well, and these data were removed from this study. Another pressure sensor for vertical pressure changes was embedded on the top of the instrumented sensor bar (the right bar in Figure 1) at 330 mm from the soil surface. Utilizing each sensor, the pressure changes were measured during the installation of straight and tapered piles in medium-dense dry sand.

3 TEST RESULTS

3.1 Load displacement curves

Different load-displacement curves were obtained between two kinds of piles, straight pile (STR) and tapered pile (TPR) during installation, as shown in Figure 3. The load on the straight pile increased first and then remained steady with installation depth.

During installation at shallow depth, the load on the tapered pile was lower than that on the straight pile owing to its small pile tip area. On the other hand, the tapered pile developed at larger load at the pile as the installation depth increased. It is easy to guess that the tapered shaft increased the shaft resistance according to the tapered angle, even when the end bearing resistance was still lower. Many researchers have already discussed this higher shaft resistance of tapered piles and also explained the mechanism based mainly on the cavity expansion theory.

3.2 Radial pressure changes

This study is focusing on the changes measured on the pressure sensor in the soils. These pressure sensors were set to zero before pile installation, which means that the initial soil pressure was not included.

Figure 4 shows the radial pressure changes with pile installation depth as the straight pile tip was approached. It is clear that each peak pressure was recorded slightly before the pile tip reached each pressure sensor level. Then the pressure level rapidly dropped as the pile tip passed. Similar observations were reported in the 1g chamber test, centrifuge test, and in-situ field test by Gavin & Lehane (2003), Leung et al. (1996) and Chow (1996) respectively.

On the other hand, Figure 5 presents the radial pressure changes in the tapered pile installed into sand. The peak pressure was recorded at a level very close to the sensor level (i.e. almost at the same time), and the magnitudes of the pressure level were quite lower than those in the straight piles. This lower pressure level could be caused by the small end-bearing area.

3.3 Comparison of pressure changes

The radial pressures during installation were compared between the straight pile and the tapered pile on the sensor at 330 mm in depth in Figure 6. The radial pressure in the straight pile increased as the

255

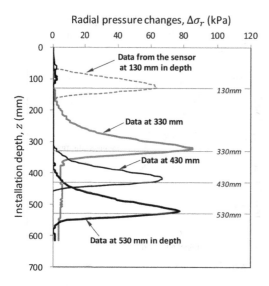

Figure 4. Radial pressure changes during pile installation of straight pile (STR) on four sensors at 48 mm from pile center.

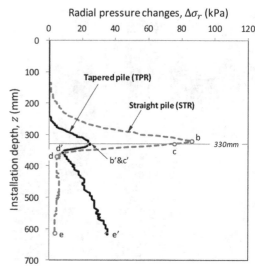

Figure 6. Comparison of radial pressure changes between straight pile (STR) and tapered pile (TPR) during installation.

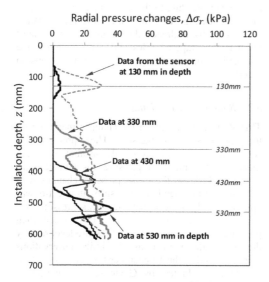

Figure 5. Radial pressure changes during pile installation of tapered pile (TPR) on four sensors at 48 mm from pile center.

pile was installed, and then increased rapidly to the peak at point [b] slightly before the pile tip reached the sensor level. Afterward, the pile tip passed through point [c] as the pressure dropped sharply. The radial pressure was almost bottomed out at point [d] a little bit above the pile tip. Finally, the soil pressure tended to decrease slightly as the pile was installed deeper. The pile installation test was completed at point [e].

The profile of the radial pressure changes in the tapered pile was different from that of the straight pile. The peak pressure of the tapered pile recorded

almost at the same time when the pile tip reached the sensor level (points [b'] & [c']). The magnitude of the peak pressure was much smaller than that of the straight pile, which is approximately 30% of the straight pile. Another feature of this profile was that the radial pressure tended to increase after the pressure drop passed through the sensor level (i.e., from point [d'] to [e']). These increasing radial pressures could be developed by the tapered shaft angle of the tapered piles.

The vertical pressure changes were recorded at the same sensor level (i.e., 330mm in depth). These measurements were plotted for the straight and tapered piles in Figure 7. The vertical pressure in the straight pile increased earlier than the radial pressure (compared to Figure 6). The peak of the vertical pressure at point [a] was also earlier before the pile tip approached. Then, the vertical pressure dropped rapidly to point [d], similar to the radial pressure, but the vertical pressure increased again as the pile was installed at point [e]. This increase could be caused by the drag force owing to the load transfer from the interface of the pile to sand during pile installation. The load transfer and drag force were pointed out by De Nicola (1996) and O'Neill (2001). On the other hand, the vertical pressure changes in the tapered pile were quite different from those in the straight pile. The peak pressure was not very high at point [a']. The profile of the pressure changes meandered, but increased approximately at a constant rate throughout the entire process.

3.4 Stress path in pile installation

Figure 8 shows the stress path from the vertical and radial pressure changes in the vertical and horizontal

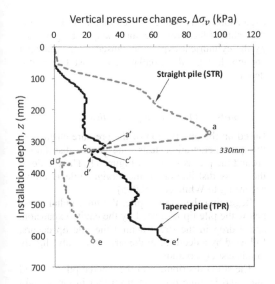

Figure 7. Comparison of vertical pressure changes between straight pile (STR) and tapered pile (TPR) during installation.

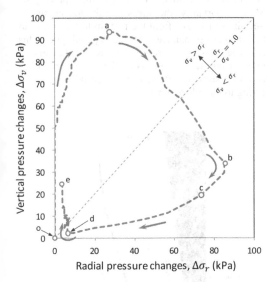

Figure 8. Stress path on pressure sensors at 330 mm in depth during installation of straight pile (STR).

directions, respectively, measured in the sensor at 330 mm during the straight pile installation for the entire process. The subscripts from [a] to [e] in the data plots are the same as in Figures 6 and 7.

In the first stage, the vertical and radial pressures increased to point [a], but more rapidly in the vertical pressure than in the radial one. Then, the radial pressure kept increasing with a decrease in the vertical pressure, reaching point [b] at which the radial pressure was at the peak. Point [c] indicates that the pile

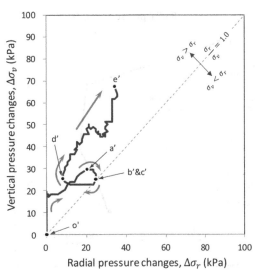

Figure 9. Stress path on pressure sensors at 330 mm in depth during installation of tapered pile (TPR).

tip was at the same level as the sensor in both radial and vertical pressures decreasing between points [b] to [d]. After passing through point [d], the vertical pressure increased again from point [d] to point [e], while the radial pressure still slightly decreased. The profile of this stress pass demonstrated dramatic changes in the pile installation, but the final position [e] was not so far from the initial point [o] despite of such a large loop of the stress path.

Similarly, the stress paths for the vertical and radial pressure changes in the tapered pile are shown in Figure 9. This profile of the stress path was clearly different from that in the straight pile. In the tapered pile, the vertical pressure reached point [a'] and then dropped to the point [b'] and [c'] while the radial pressure increased. Subsequently, a decrease was observed in the radial pressure from point [c'] to [d']. This loop of the stress path was quite small compared to the loop in the straight pile installation. Unlike the straight pile, both radial and vertical pressures increased in the tapered pile from point [d']. As a result, both pressures at the final point [e'] were pretty higher than the initial point [o']. In particular, the pressure level in the tapered pile at the final installation [f'] ($\Delta\sigma_v$ = 67.3 kPa, σ_r = 34.3 kPa) was much higher than the one in the straight pile ($\Delta\sigma_v$ = 24.0 kPa, $\Delta\sigma_r$ = 3.5 kPa).

4 STRESS FIELD AROUND PILES

4.1 Radial pressure distribution

The radial pressure distribution with sensors depth are shown in Figure 10, focusing on the final installation stage when the pile tip was penetrating the

257

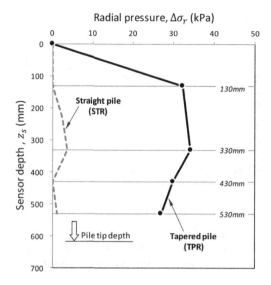

Figure 10. Comparison of radial pressure distribution with sensor depth at final installation stage between straight and tapered piles.

depth of 615mm. The pressure level was relatively small, as confirmed in Figure 8 in the straight pile.

On the other hand, the radial pressure level was highly developed along the pile shaft in the tapered pile. There was a tendency of reduction in the radial pressure under 330 mm in the sensor depth. This tendency likely depends on the ratio of the radial

expansion owing to the tapered diameter. Therefore, the value of the radial pressure likely to be governed by the cylindrical expansion caused by the diameter change during pile installation at least below this depth.

4.2 Stress field around tapered piles

Based on the observation of the pressure changes in the pile penetration tests, the stress distribution around the pile is shown in Figure 11. The profile of the stress distribution in the straight pile is drawn according to White et al. (2005).

Regarding the straight pile, the high radial stress below the pile tip is created by the cavity expansion, and a drop in the stress behind the pile tip occurs, followed by a decrease of the stress, but only slightly in this test observation.

The stress distribution in the tapered pile installed into sand is quite different from that in the straight pile. The radial stress around the pile tip level is smaller in the tapered pile than in the straight pile, because the expanded cavity below the pile tip is governed by the small pile area in the tapered pile. More importantly, the stress distribution along the pile shaft increases from the pile tip to top owing to the increasing diameter of the tapered pile. This stress level is much greater than that of the straight pile, leading to the conclusion that the tapered piles could be strongly affected by the installation effect owing to the pile geometry.

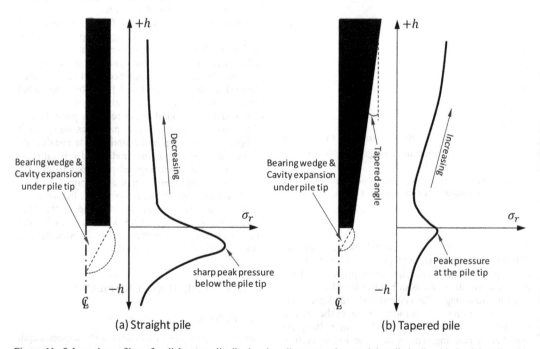

Figure 11. Schematic profiles of radial stress distribution in adjacent sand around installed pile; (a) Straight pile, (b) Tapered pile.

5 CONCLUSION

This study investigates the installation effect of tapered piles in sand in comparison to straight piles. The experimental results of this study indicate:

1. The peak pressure developed in the tapered pile around the similar level of the pile tip, although this peak pressure was not as great owing to cavity expansion below the pile tip as that of the straight pile.
2. The radial pressure increased behind the pile tip of the tapered pile as the pile was installed deeper owing to pile diameter expansion, while a slight decrease was observed in the straight pile.
3. The stress paths of the radial and vertical pressures were completely different between the tapered pile and the straight pile. The stress path in the straight pile drew a large loop, while the loop was small and the stress level tended to increase through the pile installation process in the tapered pile.
4. The stress field around the tapered pile was proposed conceptually based on the interpretation of the experimental results, suggesting that the installation effect could be much greater in a tapered pile than in a conventional straight pile.

Due to the page limitation, the data were reported for one geometry of the tapered pile with about five in the taper ratio in this study. Therefore, there is not enough data to discuss the phenomenon in detail, and it remains questionable whether this trend depends on the taper-ratio. Further discussion should be conducted to better understand the installation effect on tapered piles.

ACKNOWLEDGMENTS

This work was supported by Hironobu Matsumiya, formerly a researcher at Nippon Steel Corp. Without his contribution, the findings would not have been achieved in this paper. We would like to express our gratitude to him.

REFERENCES

Bolton, M.D., Gui, M.W., Garnier, J., Corte, J.F., Bagge, G., Laue, J. & Renzi, R. 1999. Centrifuge cone penetration tests in sand, *Géotehnique* 9, No.4, 43–552.

Chow, F. 1996. Investigations into the behaviour of displacement piles for offshore structures. *Ph.D. thesis , University of London (Imperial College)*, London, U.K.

De Nicola, A. 1996. The performance of pipe piles in sand. *PhD dissertation, the University of Western Australia*, Australia.

El Naggar, M.H. & Sakr, M. 2000. Evaluation of axial performance of tapered piles from centrifuge tests, *Canadian Geotechnical Journal* 37: 1295–1308.

El Naggar, M.H. & Wei, J.Q. 2000. Uplift behaviour of tapered piles established from model tests, *Canadian Geotechnical Journal* 37: 56–74.

Gavin, K.G. & Lehane, B.M. 2003. The shaft capacity of pipe piles in sand, *Canadian Geotechnical Journal*. 40: 36–45.

Jardine, R.J., Zhu, B., Foray, P. & Dalton, C.P. 2009. Experimental arrangements for investigation of soil stresses developed around a displacement pile. *Soils and Foundations* Vol. 49, No.5, 661–673.

Klotz, E.U. & Coop, M.R. 2001. An investigation of the effect of soil state on the capacity of driven piles in sands, *Géotehnique* 51, No.9, 733–751.

Lee, J.H. & Salgado, R. 2000. Analysis of calibration chamber plate load tests. *Canadian Geotechnical Journal* 37:14–25.

Lehane, B.M., Jardine, R.J., Bond, A.J. & Frank, R. 1993. Mechanisms of shaft friction in sand from instrumented pile tests, ASCE, *Journal of Geotechnical Engineering*, Vol. 119, No. 1.

Lehane, B.M. & White, D.J. 2005. Lateral stress changes and shaft riction for model displacement piles in sand. *Canadian Geotechnical Journal* 42: 1039–1052.

Leung, C.F. Lee, F.H. & Yet, N.S. 1996. The role of particle breakage in pile creep in sand, *Canadian Geotechnical Journal*. 33: 888–898.

Manandhar, S. & Yasufuku, N. 2013. Vertical bearing capacity of tapered piles in sands using cavity expansion theory, *Soils and Foundations*; 53(6):853–867

O'Neil, M.W. 2001. Side resistance in piles in sand and drilled shafts. *ASCCE Journal of Geotechnical and Geoenvironmental Engineering*, Vol. 127, No.1, 3–16.

Sakr, M. El Nagger, M.H. & Nehdi, M. 2005. Lateral behavior of composite tapered piles in dense sand. *Proc. Of the Institution of Civil Engineers, Geotechnical Engineering* 158: Issue GE3, 145–157.

Sato, T., Shinozawa, N. & Adachi, T. 2010. In-Situ test of the tapered steel pile used for small buildings, *AIJ J. Technol. Des*. Vol. 16, No. 32, 113–117.

Taenaka, S., White, D.J. & Randolph, M.F. 2010. The effect of pile shape on the horizontal shaft stress during installation in sand. Springman, Laue & Seward (eds), *Physical Modelling Geotechnics*, 835–840. Taylor & Francis Group, London.

Tominaga, K. & Chen, Q. 2006. An analysis for relationships between vertical load and settlement on a tapered-pile, *Journal Struct. Constr. Eng*., AIJ, No.603, 77–83.

Tominaga, K., Chen, Q., Tamura, M. & Wakai, A. 2007a. Some sorts of in-situ tests on steel pipe tapered piles installed by press-in method, *AIJ J. Technol. Des*. Vol. 13, No. 26, 487–490.

Tominaga, K., Chen, Q., Tamura, M. & Wakai, A. 2007b. Analytical method for tapered pile under lateral load – Compressions between predicted and experimental results- , *Journal Struct. Constr. Eng*., AIJ, No.622, 115–120.

Wei, J.Q. & El Naggar, M.H. 1998. Experimental study of axial behaviour of tapered piles, *Canadian Geotechnical Journal* 35: 641–654.

White, D.J. & Lehane, B.M. 2004. Friction fatigue on displacement piles in sand. *Géotehnique* 54, No.10, 645–658.

White D.J., Schneider, J.A. & Lehane, B.M. 2005. The influence of effective area ratio on shaft friction of displacement piles in sand. Gourvenec & Cassidy (eds), *Frontiers in Offshore Geotechnics: ISFOG 2005*, 741–747. Taylor & Francis Group, London.

Proceedings of the Second International Conference on
Press-in Engineering 2021, Kochi, Japan – Matsumoto et al (eds)
© 2021 Taylor & Francis Group, London, ISBN 978-1-032-10414-0

Comparison of penetration resistance and vertical capacity of short piles installed by Standard Press-in in loose sand

Y. Ishihara & M. Eguchi
GIKEN LTD., Kochi, Japan

M.J. Brown
University of Dundee, Dundee, UK

J. Koseki
University of Tokyo, Tokyo, Japan

ABSTRACT: In the Press-in Method, a vertical jacking force required for installing a pile is automatically measured for every single pile. This information is expected to be utilized to confirm the vertical performance of the pile. In this paper, results of a series of large-scale model tests are reported to compare the resistance during installation and during static load test. The test pile was a closed or open-ended tubular pile with an outer diameter of 318.5 mm, embedded in a loose sand to around 6 m. Phenomena of pile set-up or set-down were observed in some cases. Through literature review and detailed analyses of the experimental data, the cause of the set-down was concluded to be mainly due to the pile installation being terminated at a depth where the soil strength decreased with depth and to a lesser extent because negative pore water pressure was generated during installation.

1   INTRODUCTION

In the Press-in Method without any installation assistance such as water jetting or augering (Standard Press-in), a pile is installed by a static jacking force, and the piling data including the information on the jacking force can be monitored and recorded for every single pile. It has been suggested that the jacking force during installation (especially at the end of installation (EOI)) can be linked to the capacity of the pile. It is generally understood that the penetration resistance in a load test is usually larger than that during installation (Komurka et al., 2003; Gavin et al., 2015), which is known as "pile set-up". For pressed-in piles in cohesive soils, a two-fold or three-fold increase in total resistance with time after EOI has been confirmed in the field tests (Ishihara & Haigh, 2018; Ishihara et al., 2020). In such cases, the penetration resistance at EOI of a pile can be taken as the lower-bound for the capacity of that pile. To put this idea into practice, it is important to know under what circumstances the opposite trend (pile set-down) is encountered.

This paper reports the results of a series of large-scale model tests using a closed or open-ended tubular pile with the outer diameter of 318.5 mm, which were pressed-in with or without surging (applying downward and upward displacement l_d and l_u repeatedly as illustrated in Figure 1) in a loose sand to around 6 m. Phenomena of pile set-up or set-down were observed in some cases, and the causes for the set-down are discussed through literature review and detailed analyses on the experimental data.

2   POSSIBLE MECHANISMS OF PILE SET-DOWN

The difference in the penetration resistance at EOI and in a load test (i.e. either pile set-up or pile set-down) is partly caused by the differences in loading conditions during installation and in the load test. For example, the loading rate (or penetration rate) during installation (typically 20 mm/s or greater) is much higher than that in the load test (0.06 mm/s or lower for piles with the outer diameter of 1000 mm). The loading direction is often two-way during installation (i.e. installation is associated with surging) while it is consistently one-way in a load test. The condition

DOI: 10.1201/9781003215226-23

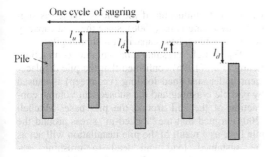

Figure 1. Process of surging.

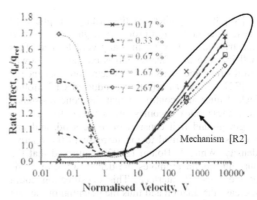

Figure 3. Effect of strain level on the rate effect (after Robinson & Brown, 2013).

of the soil around the pile is relatively "unstable" (highly variable with time) during installation, with excess pore water pressure being generated by the pile installation, while it is relatively "stable" at the start of the load test because there is usually a 5-days (for sand) or 14-days (for clay) interval after EOI to assure a complete dissipation of excess pore water pressure (JGS, 2002) prior to load testing. In this paper, possible mechanisms for pile set-down will be summarized in terms of (1) penetration rate, (2) surging and (3) others.

2.1 Effect of penetration rate

The penetration rate could be the cause of pile set-down in the following ways.

[R1] If the soil is dilatant, negative or reduced excess pore water pressure will be generated during installation, which will increase the penetration resistance (Silva & Bolton, 2005; Lauder et al. 2012), as illustrated in Figure 2 (after White et al., 2010) where v is the penetration depth, D is the pile diameter and c_v is the coefficient of consolidation. The higher penetration rate during installation than in the load tests will lead to greater negative excess pore water

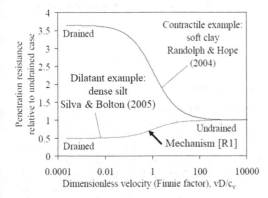

Figure 2. Rate effect on penetration resistance (after White et al., 2010).

pressure during installation, causing the pile set-down after a period of pore pressure equalization.

[R2] The higher penetration rate might increase the penetration resistance due to the effect of viscosity. Considering the findings of Robinson & Brown (2013) in clays as shown in Figure 3 where the definition of the horizontal axis is identical with that in Figure 2, with the higher penetration rate and the greater strain level during installation than in the load test, the resistance during installation might become larger than in the load test. On the other hand, this trend will be absent in loose contractile sands (Chow et al., 2020).

[R3] According to Tatsuoka et al. (2008) and Enomoto et al. (2009), unbound granular materials show four different responses due to the viscosity effect when sheared, as shown in Figure 4, depending on their particle shape, particle grading, particle size, particle crushability and so on. Until the local peak stress is reached, the effective stress increases when the strain rate is increased via four typical responses. After the local peak stress is experienced, soils behave differently. It might follow that, for the base resistance which may be associated with peak stresses, larger resistance would always be experienced at higher installation rates, while the shaft resistance may be increased or decreased according to the types of responses of the soil as it is associated with residual stresses.

[R4] Watanabe & Kusakabe (2013) conducted triaxial compression tests on dense Toyoura sand and confirmed that the internal friction angle and deformation modulus become greater when the strain rate is increased. This will lead to a larger resistance (both for base and shaft) during installation.

2.2 Effect of surging

The motion of surging (repeated penetration and extraction during installation) could be the cause of pile set-down in the following ways.

[S1] It is hypothesized that lifting the pile may cause a vacuum to form (or very negative pressures) that tries to draw water into any void formed. This will increase the effective stress on the next downward stroke and increase the penetration resistance.

[S2] Jeffery et al. (2016) conducted CPT around jacked piles and confirmed that the cone resistance near these piles becomes greater than that of the virgin ground if it consists of loose sand, and this increase is attributed to the increase in the relative density. White & Bolton (2002) confirmed that the surging motion promotes the densification of sands at the pile-soil interface. This densification could increase the dilation and thus the potential of the rate effect associated with the pore pressure effect (i.e. [R1]) during installation, while reducing the confinement stress from the far field (and the horizontal stress on the pile shaft) when loaded slowly, which is known as "friction fatigue".

2.3 Other effects

Other than the penetration rate and surging, the following ([O1], [O2], [O3]) would be possible influences on the mechanisms of pile set-down. In addition, mechanisms [O4], [O5] and [O6], which lead to the absence of set-up of shaft resistance, can be the indirect causes of set-down in total resistance when the set-down trend in base resistance due to other mechanisms are significant.

[O1] The variation of the soil strength (cone resistance or SPT) with depth. If the installation is stopped where the soil strength decreases with depth, the resistance in the load test will be lower than the resistance at EOI, as the depth of pile base is greater during and after load testing.

[O2] Lower displacements during the load test (to define the pile capacity) than during installation. This factor could be more influential if the stiffness of the soil around the pile is lower. Also, considering that a greater displacement is required to mobilize full resistance on pile base than on pile shaft, this factor

will be more influential if the pile is shorter as suggested by Zhang et al. (2006), since the total capacity depends more on base capacity for shorter piles.

[O3] Particle crushing and volume contraction. Leung et al. (1996) confirmed that the pile displacement under sustained loading (i.e. creep) is caused by particle crushing and the subsequent volume contraction of the soil around the pile base. Mitchell (2004) argued that the "locked-in" stress around the pile base as a result of the pile installation will act as the sustained load and lead to crushing, re-orientation and slip of particles, which causes a reduction in confining stress and base capacity.

[O4] Lack of increase in dilatancy with time after EOI. Based on the results of the measurement of horizontal stresses on the pile shaft after EOI and during load test, Axelsson (2000) proposed that the set-up in shaft resistance is explained by an increase in dilatancy of soils on the pile shaft with time after EOI, which would be caused by the rearrangement and the subsequent intrusion of soil particles into the rough surface of the pile ("constrained dilatancy"). Bowman & Soga (2005) conducted creep tests and confirmed that the increase in the volumetric strain (i.e. dilation) was apparent in dense sand but was absent in loose sand.

[O5] Lack of increase in horizontal stresses on the pile shaft. It is believed that a creation of arching (increased hoop stress) in the soil around the pile shaft during installation and the dispersion of the arching with time leading to the increase in horizontal stresses on the pile shaft after EOI (Chow et al., 1998). However, this mechanism was not clearly observed in the experiments of Axelsson (2000).

[O6] Lack of increase in plug strength (or inner shaft capacity). Randolph et al. (1991) showed that the plug strength of an open-ended pile in drained conditions is much higher than that in undrained conditions. This will lead to pile set-up if the pore water pressure inside the pile is sufficiently increased during installation. Alternatively, if the pore pressure increase is small, the set-up trend may be absent, and set-down trend could be observed especially when base or shaft resistance shows set-down behavior due to other mechanisms.

3 EXPERIMENTAL METHODS

3.1 Apparatus

Experiments were conducted in a large soil tank 7 m square and 9 m deep as shown in Figure 5. The soil tank is equipped with a water supply system at its base, to make the whole model ground liquefiable (boiling state) to allow preparation of a consistent soil bed. Silica sand NSK-40 (with its effective grain size D_{50} being 0.23 mm) was used as the material for the model ground (soil test bed). The relative density (D_r) of the model ground was roughly 50% on average (estimated from the weight of the sand put into the soil tank and the depth of the completed model ground),

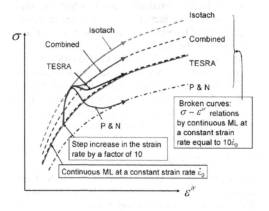

Figure 4. Four types of responses in terms of viscosity effect (Tatsuoka et al., 2008).

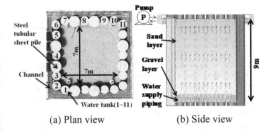

(a) Plan view	(b) Side view

Figure 5. Soil tank (Ogawa *et al.*, 2018).

although the site density tests conducted at shallow depths (surface and 1 m below the surface) showed that D_r was as small as 4 %. Details of this apparatus and the model ground can be seen in more detail in Ogawa *et al.* (2018).

A closed-ended pile or an open-ended pile with an outside diameter (D_o) of 318.5 mm was used as a test pile. The closed-ended pile was equipped with a load cell and strain gauges at its base to measure base resistance and base torque separately, and with earth pressure transducers and pore water pressure transducers on its surface to measure the total horizontal stresses and pore water pressures at 0.25 m, 1.5 m and 3.0 m above the pile base (σ_{hp-1}, σ_{hp-2}, σ_{hp-3}, u_{p-1}, u_{p-2} and u_{p-3}). The wall thickness of the open-ended pile was 10.3 mm.

Horizontal stresses in the model ground were measured by earth pressure transducers at the positions indicated in Figure 6. Horizontal stresses in radial direction (σ_{hr}) were measured at three levels at two positions, whereas those in circumferential direction ($\sigma_{h\theta}$) were measured at one position.

3.2 Procedures

The tests were conducted based on the following procedures. (1) The model ground was prepared by injecting water from the bottom of the soil tank inducing a "boiling" state throughout the

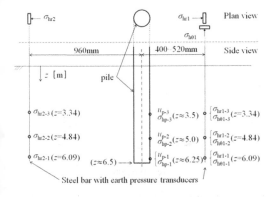

Figure 6. Positions of earth pressure measurement.

test bed, stopping the water injection and vacuum pumping the water from the bottom of the stand pipes (installed at four corners in the tank) to create the water level required, and waiting for several hours to attain static pore water pressure distribution in the model ground. (2) The test pile was installed at the center of the model ground by Standard Press-in by a press-in machine with or without surging as described in Table 1. In press-in without surging, the pile was unloaded at every 850mm penetration depth as the length of the cylinder of the piling machine was 850mm. To secure the reaction force, the press-in machine grasped the reaction beam which was fixed to the soil tank. (3) The test pile was left for 1 or 7 days. (4) The test pile was vertically load tested, based on JGS standard (JGS, 2002). (5) The test pile was extracted monotonically using the press-in machine.

During installation, head load (Q) was obtained by removing the weight of the chuck from the load applied by the press-in machine measured by pressure sensors (Q'). The penetration depth (z) was measured using a wire-type stroke sensor, except for J1902-04 where a stroke sensor in the press-in machine was used. During static load tests, head load was measured by a load cell placed on the pile head. The pile displacement was measured by two displacement transducers placed on the pile head.

3.3 Test cases

Test cases are summarized in Table 1, where v_d and v_u are the downward and upward velocity of the pile and t_{LT} is the time after the end of installation to the start of load test. The ground surface level and the water level were measured from the reference level in the soil tank (near the ground surface). Initially, three tests (J1902 test series) were conducted using the closed-ended pile. As discussed later in Section 4, pile set-down was observed in these tests, which was thought to be mainly due to the decreasing trend of the soil strength around the depth of the pile base as shown in Figure 7a. To cope with this issue, the model ground was carefully mixed by installing an H-shaped pile with water jetting hose and nozzle with the flowrate of around 300 liters per minute at nine different positions, in addition to injecting water from the bottom of the soil tank. Then additional tests (J2002 test series) were conducted in an improved soil condition with a shorter embedment depth as shown in Figure 7b, using an open-ended pile. In J2002-03, the installation was conducted with the upper limit of jacking force (Q'_{UL}), in which the pile was extracted by a certain amount (l_u) every time Q' reached Q'_{UL}.

It should be noted that CPT (Cone Penetration Test) in Figure 7 was not conducted in each test

Table 1.   Test cases.

Test No.	End condition	Ground Surface level [m]	Water level [m]	l_d [mm]	l_u [mm]	v_d [mm/s]	v_u [mm/s]	t_{LT}
J1092-4	Closed	-0.01	-7.5	400	200	45	48	7 days
J1092-5	Closed	-0.10	-1.0	400	200	45	48	7 days
J1092-6	Closed	-0.11	-1.1	850	0	45	-	7 days
J2002-01	Open	-0.20	-0.5	850	0	51	46	24 hrs
J2002-02-2	Open	-0.21	-7.9	850	0	51	46	24 hrs
J2002-02	Open	-0.25	-8.4	100	50	51	46	23.5 hrs

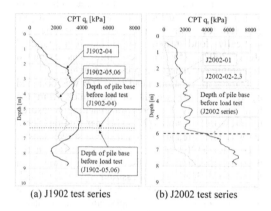

(a) J1902 test series   (b) J2002 test series

Figure 7. Ground condition and depth of pile base.

shown in Table 1, but conducted in a fresh ground with high water level once and in a fresh ground with low water level once, in each test series, at the center of the model ground.

4   EXPERIMENTAL RESULTS AND DISCUSSIONS

4.1   Resistances recorded in J1902 test series

Figure 8 shows the measured head load (Q), base resistance (Q_b) and shaft resistance (Q_s) obtained by $Q - Q_b$ in each test, where bold lines r epresent the values recorded in load tests. In J1902-05 and J1902-06 (with higher water levels), values of Q, Q_b and Q_s were lower than those in J1902-04 (with low water level). Comparing J1902-05 and J1902-06, both of Q_b and Q_s were reduced by surging during installation, and this reduction was more prominent in Q_s than in Q_b. Similar trends of reduction in Q_b and Q_s were found in the subsequent load test in these two tests.

In all the three tests, Q_b during load test was lower than that at EOI. On the other hand, such a reduction in Q_s was only found in J1902-04 (with low water level), and in the other two tests there was little reduction (and little increase) in Q_s values. As a result of these trends in Q_b and Q_s, Q was lower in the load test than at EOI in all of the three tests.

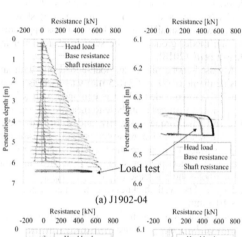

(a) J1902-04

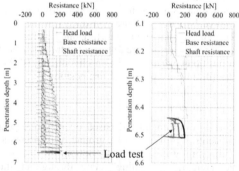

(b) J1902-05

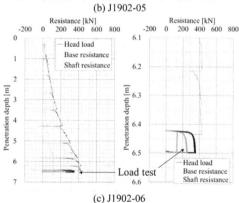

(c) J1902-06

Figure 8. Resistances during installation and load test (bold lines) in J1902 test series.

264

4.2 Narrowing down the mechanisms for the pile set-down observed in J1902 test series

4.2.1 Penetration rate: mechanism [R1]

Figure 9 shows the pore water pressure measured on the pile shaft (u_p) during installation in J1902-06 (closed-ended, without surging), together with the theoretical static pore water pressure. Reduced pressures were recorded in u_{p-2} and u_{p-3} by up to 5 kPa, which will lead to an increase in Q_s at EOI by up to around 10 kN. After EOI, values of u_{p-2} and u_{p-3} recovered to their static values immediately (in less than 10 seconds) as shown in Figure 10, where zero in the horizontal axis was taken as the point of time when Q became zero and z became constant (i.e. when the rebound finished).

Figure 11 shows the axial strain – volumetric strain relationships obtained in triaxial tests (CD) on saturated NSK-40 sand, with the relative density (D_r) at 4% which was confirmed as similar to 1 m BGL in the model ground. The cylindrical sample had a diameter and height of 50 mm and 10 mm respectively. The strain rate was 0.5 %/min. There was no tendency of positive dilatancy under the confinement stresses of 50, 100 and 200 kPa. These stresses are lower than the base resistance recorded during press-in, and it is not clear whether

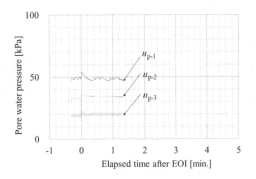

Figure 10. Variation of pore water pressure with time at EOI in J1902-06.

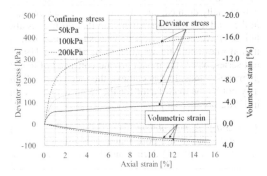

Figure 11. Results of triaxial test on NSK-40 sand.

this soil shows positive dilatancy under the higher confining stresses during installation. However, judging from the experimental data, the reduction in u_p values (which might have been caused by dilation) was recorded at levels away from the pile base while the set-down was observed not in Q_s but only in Q_b. This suggests that mechanism [R1] is not the reason for the set-down observed in J1902-06, although it has to be noted that this discussion is limited to pore pressure measurement on the pile shaft as there was no pore pressure information below the pile base.

4.2.2 Penetration rate: mechanism [R2]

It was confirmed by a consolidation test on NSK-40 that the coefficient of consolidation (c_v) at the average consolidation pressure of around 3.5 MPa, which roughly corresponds to the unit base resistance during installation, is around 3.8×10^3 cm^2/day. On the other hand, the penetration rate (v_d) in J1902-06 varied from around 40 to 50 mm/s as a whole. It follows that the normalized velocity V ($= v_d D_o/c_v$) in this experiment was approximately 3300 in installation. Considering the loading record in the load test in Figure 12, V was about 0.3 in the load test. According to Figure 3 which applies to clays, these V values mean that the

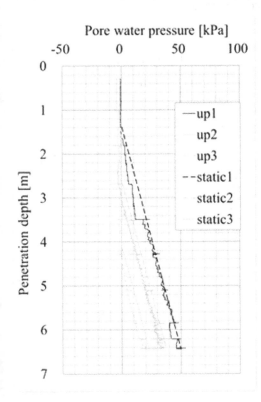

Figure 9. Pore water pressure on pile shaft during installation in J1902-06.

265

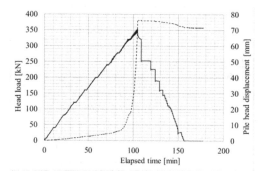

Figure 12. Loading record in J1902-06 load test.

installation was fully undrained while the load test was partially drained, and the unit base resistance (q_b) in the load test will be smaller than that during installation. However, this effect of viscosity at high penetration rates could be assumed to be absent in J1902 test series, where the model ground consisted of loose permeable sands. In this context, the mechanism [R2] may not be the main factor for the set-down observed in J1902-06.

4.2.3 Surging: mechanism [S1]

Figure 13 shows the pore water pressure measured on the pile shaft (u_p) during installation in J1902-05

(with high water level). The excess pore water pressure was negative during the upward motion, while it was positive during the downward motion. However, these values were small regardless of their sign conventions, resulting in little variation after EOI as shown in Figure 14. Therefore, the mechanism [S1] is not thought to be the main factor for the set-down observed in J1902-05.

On the other hand, in J1902-04 (closed-ended, with low water level), a greater extent of negative excess pore pressure was recorded near the pile base (u_{p-1} and u_{p-2}) than in J1902-05 (closed-ended, with high water level), as can be confirmed in Figure 15.

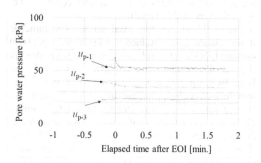

Figure 14. Variation of pore water pressure with time at EOI in J1902-05.

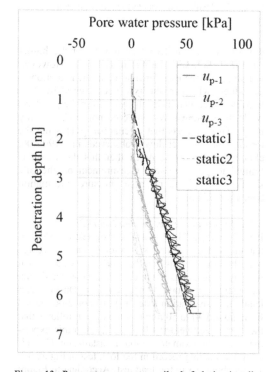

Figure 13. Pore water pressure on pile shaft during installation in J1902-05.

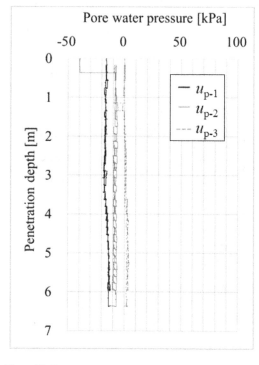

Figure 15. Pore water pressure on pile shaft during installation in J1902-04.

266

This suggests that the suction generated by the creation of a void beneath the pile base when the pile was moved upwards continued to exist during installation because of the lack of u_p building up in the subsequent downward motion of the pile. As well as this, the greater suction (because of the lower water level) may have led to a reduced volume of soil collapsing into the void when the pile was moved upwards and thus increased the extent of the additional suction induced by the pile's upward motion. As shown in Figure 16, values of u_{p-1} and u_{p-2} increased during the period from EOI to the start of load test by $5 \sim 15$ kPa. This will cause $10 \sim 30$ kN decrease in Q_s, which partly explains the reduction in Q_s in Figure 8. However, u_{p-3} decreased from a positive value by around 5kPa after EOI. It would be concluded that the mechanism [S1] could be a minor cause for the observed set-down in Q_s in J1902-04.

4.2.4 Surging: mechanism [S2]

Looking at Figure 8, Q_s during installation were lower in J1902-05 (closed-ended, with surging) than that in J1902-06 (closed-ended, without surging). This suggests either that there was little promotion of densification (and dilatancy) of soils around the pile shaft due to surging, or that the densification was promoted but the reduction in the confining stress (and resultant friction fatigue) was more influential than the increase in dilation even during installation. Therefore, [S2] is not thought to be the main mechanism for the set-down observed in J1902-05.

4.2.5 Others: mechanism [O1]

Figure 17 is the comparison of q_b and CPT corrected cone resistance (q_t). In J1902-04 (closed-ended, with low water level), the depth of EOI was where q_t tends to decrease with depth. This suggests that the mechanism [O1] is the cause of the set-down observed in J1902-04. On the other hand, in J1902-05 and J1902-06 (closed-ended, with high water level), at the depth of EOI, q_t tends to increase with depth. However, it is generally understood that Q_b reflects the strength of the soils not only at the depth

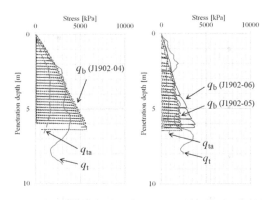

Figure 17. Comparison of q_b and CPT q_t in J1902 test series.

of the pile base but also those around the pile base. If q_t is averaged by the Dutch method (Lehane et al., 2005) to consider the strength of the soils around the pile base, the averaged q_t (denoted by q_{ta}) tends to decrease slightly with depth, reflecting the decrease in q_t values at greater depth. In light of this, [O1] could be the mechanism for the set-down in J1902-05 and J1902-06 as well.

4.2.6 Others: mechanism [O2]

Looking at Figure 18, Q_b almost reached its residual values during load test, at the pile head displacement of 30 mm which is lower than the displacement defining the capacity. This suggests that the mechanism [O2] would not be the reason for the observed set-down in the three tests.

4.2.7 Others: mechanism [O3]

Figure 19 shows the base stress just after EOI (i.e. after rebound) and just before the start of load test in each test. Stresses were almost constant (with a slight increase) after EOI except for J1902-05, with the maximum stress level being around 700 kPa. This does not seem to be high enough to induce particle breakage, which was confirmed by taking images of NSK-40 sand particles (Figure 20) before and after applying a constant stress of 735 kPa for 24 hours in a simple

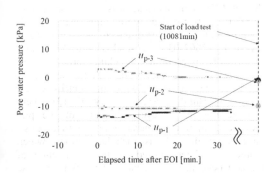

Figure 16. Variation of pore water pressure with time after EOI in J1902-04.

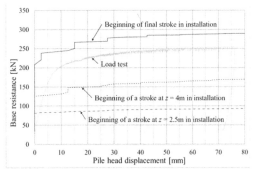

Figure 18. Comparison of base resistance at the beginning of each stroke in installation and in load test in J1902-06.

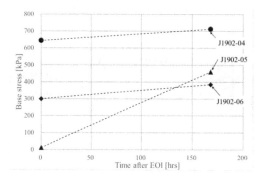

Figure 19. Base stress after EOI and before load test.

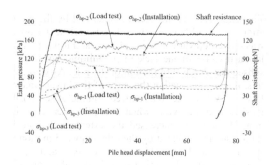

Figure 22. Earth pressure on pile shaft in the final stroke in installation and in load test (J1902-06).

(a) Before test   (b) After test

Figure 20. Sand particles before and after being subjected to a constant stress.

testing apparatus shown in Figure 21. Mechanism [O3] does not appear to be the reason for the observed set-down in the three tests.

4.2.8 Others: mechanisms [O4] and [O5]

Axelsson (2000) argued that the increase in horizontal stresses measured on the pile shaft (σ_{hp}) in load tests becomes definitive due to the "confined dilatancy". Figure 22 shows the comparison of σ_{hp} during the initial loading in the final stroke in installation and in the load test in J1902-06. In

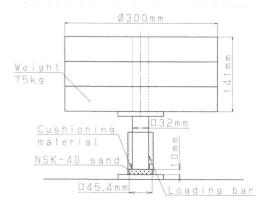

Figure 21. Apparatus to apply a constant stress to sand particles.

the beginning of the load test, σ_{hp} values decreased with displacement, and were lower than those at the beginning of the final stroke of installation at the displacement when the peak shaft resistance was recorded. In other words, the trend of confined dilatancy was not seen at the displacements lower than those at the peak shaft resistance. This suggests the possibility of the mechanism [O4] in J1902-06.

Figure 23 shows the variation of σ_{hp}, σ_{hr} and $\sigma_{h\theta}$ with time after EOI in J1902-06. It was found that $\sigma_{h\theta}$ slightly increased shortly (in less than 30 seconds) after EOI and remained constant until the start of the load test, and the opposite trend was found for σ_{hr}. σ_{hp} showed a similar trend as σ_{hr}, decreasing shortly after EOI and remain almost constant until the start of the load test. These trends suggest that mechanism [O5] could contribute to the observed set-down in J1902-06.

4.3 Additional experiments and discussions

As explained in Section 3, additional tests (J2002 test series) were conducted with an improved soil condition using an open-ended pile. Figure 24 shows the variation of resistances in these tests. Values of Q at larger displacements in the load test were comparable to or slightly greater than that at EOI. This increase was more apparent in J2002-01 (open-ended, with high water level) than in J2002-02-2 (open-ended, with low water level). In J2002-03 (open-ended, with low water level, with upper limit of jacking force (Q'_{UL})), this increase was more significant. This result seems to be consistent with the centrifuge test results by Burali d'Arezzo et al. (2013) where a large number of cyclic motions of the pile induced a significant increase in the base resistance when the pile is installed further, which would have been caused by the compaction of the soil beneath the pile base. It is often the case that the press-in piling is conducted with Q'_{UL}, and this seems to lead to greater pile set-up than

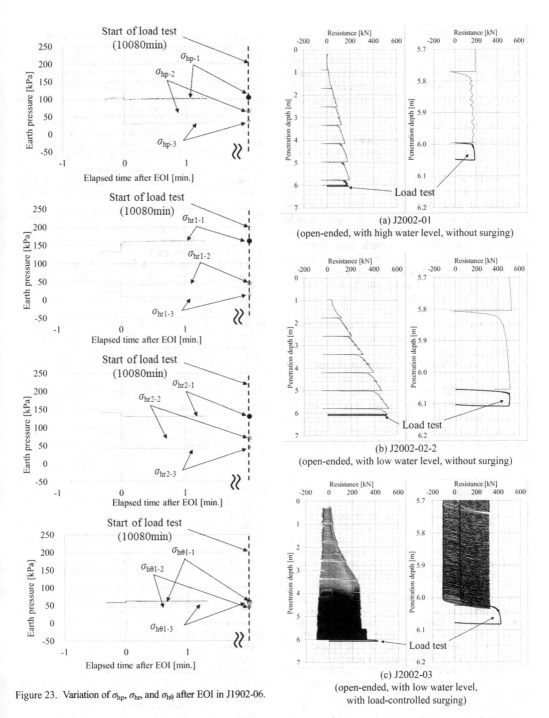

Figure 23. Variation of σ_{hp}, σ_{hr}, and $\sigma_{h\theta}$ after EOI in J1902-06.

(a) J2002-01
(open-ended, with high water level, without surging)

(b) J2002-02-2
(open-ended, with low water level, without surging)

(c) J2002-03
(open-ended, with low water level,
with load-controlled surging)

Figure 24. Resistances in installation and in load test (bold lines) in J2002 test series.

what would be experienced in a pile installed without Q'_{UL}.

From the experimental evidence that the set-up trend was found consistently in three tests in J2020 test series, and that the differences of test conditions in J2020 and J1902 test series were the ground condition and the pile base condition (open or closed), it is suggested that the observed set-down in J1902 test series

was caused mainly by mechanism [O1], and to a lesser extent by [O4], [O5] and [S1] as discussed in Section 4.2, and additionally by [O6].

269

5 CONCLUSIONS

Large scale model tests were conducted to compare the penetration resistance between installation and load tests in loose permeable sands. A trend of pile set-down was observed in some tests using closed-ended piles. Conversely, in additional model tests conducted in an improved model ground using an open-ended pile, a trend of pile set-up (or no trend of pile set-down) was observed.

Through literature review of the mechanisms of pile set-down and detailed analyses on the experimental data, the set-down trends observed in tests using a closed-ended pile were thought to be caused by (1) mechanism [O1], the decreasing trend of soil strength with depth at around the pile base, leading to the set-down in base resistance (Q_b); (2) to a lesser extent mechanisms [O4], [O5] and [O6], the lack of factors that lead to set-up in shaft resistance (Q_s) or in plug strength; and (3) similarly mechanism [S1], negative pore water pressure during installation, induced by surging in unsaturated soils, leading to minor set-down in Q_s.

Further model tests in an improved ground using a closed-ended pile, as well as triaxial tests with higher strain rates, will be effective in assessing these observations and suggested mechanisms. In addition, investigating the pore water pressure behavior beneath the pile base will allow a more reliable discussion on the causes of set-up or set-down of base resistance.

REFERENCES

Axelsson, G. 2000. Long-term set-up of driven piles in sand. *Doctoral Thesis, Department of Civil and Environmental Engineering, Royal Institute of Technology*, 194p.

Bowman, E. T. & Soga, K. 2005. Mechanisms of setup of displacement piles in sand: laboratory creep tests. *Canadian Geotechnical Journal*, Vol. 42, No. 5, pp. 1391–1407.

Burali d'Arezzo, F., Haigh, S. K. and Ishihara, Y. 2013. Cyclic jacking of piles in silt and sand. *Proceedings of the International Conference on Installation Effects in Geotechnical Engineering, ICIEGE 2013*, pp. 86–91.

Chow, F. C., Jardine, R. J., Brucy, F. and Nauroy, J. F. 1998. Effects of time on capacity of pipe piles in dense marine sand. *Journal of Geotechnical and Geoenvironmental Engineering*, Vol. 124, No. 3, pp. 254–264.

Chow, S. H., Diambra, A., O'Loughlin, C. D., Gaudin, C. and Randolph, M. F. 2020. Consolidation effects on monotonic and cyclic capacity of plate anchors in sand. *Geotechnique*, Vol. 70, Issue 8, pp. 720–731.

Enomoto, T., Kawabe, S., Tatsuoka, F., Benedetto, H. D., Hayashi, T. and Duttine, A. 2009. Effects of particle characteristics on the viscous properties of granular materials in shear. *Soils and Foundations*, Vol. 49, No. 1, pp. 25–49.

Gavin, K., Jardine, R., Karlsrud, K. and Lehane, B. M. 2015. The effects of pile ageing on the shaft capacity of offshore piles in sand. *Frontiers in Offshore Geotechnics III – Proceedings of the 3rd International Symposium on Frontiers in Offshore Geotechnics, ISFOG 2015*, pp. 129–151.

Ishihara, Y. and Haigh, S. 2018. Cambridge-Giken collaborative working on pile-soil interaction mechanisms. *Proceedings of the First International Conference on Press-in Engineering 2018, Kochi*, pp. 23–45.

Ishihara, Y., Ogawa, N., Mori, Y., Haigh, S. and Matsumoto, T. 2020. Simplified static vertical loading test on sheet piles using press-in piling machine. *Japanese Geotechnical Society Special Publication, 8th Japan-China Geotechnical Symposium*, pp. 245–250.

The Japanese -Geotechnical Society (JGS). 2002. Method for static axial compressive load test of single piles. *Standards of Japanese Geotechnical Society for Vertical Load Tests of Piles*, pp. 49–53.

Jeffrey, J. R., Brown, M. J., Knappett, J. A., Ball, J. D. and Causis, K. 2016. CHD pile performance: physical modelling. *Proceedings of the Institution of Civil Engineers, Geotechnical Engineering*, Volume 169, Issue GE5, pp. 421–435.

Komurka, V. E., Wagner, A. B. and Edil, T. 2003. Estimating soil/pile set-up. *Final Report, Wisconsin Highway Research Program* #0092-00-14, 43p.

Lauder, K, Brown, M.J., Bransby, M.F. & Gooding, S. 2012. The variation of tow force with velocity during offshore ploughing in granular materials. *Canadian Geotechnical Journal*. Vol. 49, No. 11: 1244–1255.

Lehane, B. M., Schneider, J. A. and Xu, X. 2005. CPT based design of driven piles in sand for offshore structures. *Report, the University of Western Australia*, GEO: 05345, 46p.

Leung, C. F., Lee, F. H. and Yet, N. S. 1996. The role of particle breakage in pile creep in sand. *Canadian Geotechnical Journal*, Vol. 33, pp. 888–898.

Mitchell, P. W. 2004. Jacked piling in a soil subject to relaxation. *Australian Geomechanics*, Vol. 39, No. 4, pp. 25–31.

Ogawa, N., Ishihara, Y., Ono, K. and Hamada, M. 2018. A large-scale model experiment on the effect of sheet pile wall on reducing the damage of oil tank due to liquefaction. *Proceedings of the First International Conference on Press-in Engineering 2018, Kochi*, pp. 193–202.

Randolph, M. F., Leong, E. C. and Houlsby, G. T. 1991. One-dimensional analysis of soil plugs in pipe piles. *Geotechnique*, Vol. 41, No. 4, pp. 587–598.

Randolph, M. F. and Hope, S. 2004. Effect of cone velocity on cone resistance and excess pore pressures. *Proceedings of IS Osaka – Engineering Practice and Performance of Soft Deposits*, pp. 147–152.

Robinson, S. and Brown, M. J. 2013. Towards a framework for the prediction of installation rate effects. *Installation Effects in Geotechnical Engineering* – Hicks et al. (eds), Taylor & Francis Group, London, pp. 128–134.

Silva, M. F. and Bolton, M. D. 2005. Interpretation of centrifuge piezocone tests in dilatant, low plasticity silts. *Proceedings of International Conference on Problematic Soils*, 8p.

Tatsuoka, F., Di Benedetto, H., Enomoto, T., Kawabe, S. and Kongkitkul, W. 2008. Various viscosity types of geomaterials in shear and their mathematical expression. *Soils and Foundations*, Vol. 48, Issue 1, pp. 41–60.

Watanabe, K. and Kusakabe, O. 2013. Reappraisal of loading rate effects on sand behavior in view of seismic design for pile foundation. *Soils and Foundations*, Vol. 53, Issue 2, pp. 215–231.

White, D. J. and Bolton, M. D. 2002. Observing friction fatigue on a jacked pile. *Centrifuge and Constitutive Modelling: Two Extremes*, pp. 347–354.

White, D. J., Deeks, A. D. and Ishihara, Y. 2010. Novel piling: axial and rotary jacking. *Proceedings of the 11th International Conference on Geotechnical Challenges in Urban Regeneration*, London, UK, CD, 24p.

Zhang, L. M., Ng, C. W. W., Chan, F. and Pang, H. W. 2006. Termination criteria for jacked pile construction and load transfer in weathered soils. *Journal of Geotechnical and Geoenvironmental Engineering*, Vol. 132, No. 7, pp. 819–829.

Proceedings of the Second International Conference on
Press-in Engineering 2021, Kochi, Japan – Matsumoto et al (eds)
© 2021 Taylor & Francis Group, London, ISBN 978-1-032-10414-0

Performance comparison of close-ended pressed-in steel pipe piles with helical pile in dense sand: An experimental study

M.A. Saleem, K. Takeuchi, N. Hirayama, A.A. Malik & J. Kuwano

Graduate School of Civil and Environmental Engineering, Saitama University, Saitama, Japan

ABSTRACT: Bearing behaviour of steel pipe piles and helical piles has been investigated in the past. However, their performance, considering similar pile tip diameter in dense ground condition is not well understood. Based on the above, the present study was focused on the model study of closed-ended steel pipe piles and single helix piles having similar tip diameters under dense ground conditions. Test results showed that steel pipe piles require 332% and 417% higher installation force than helical piles having equivalent tip diameters $D_{PS} = D_H = 43mm$ and 60mm respectively. However, the installation effort of screw pile in terms of power consumption was on average 29% higher than steel pipe pile. Also, helical piles exhibit 27% less ultimate bearing capacity than steel pipe piles having equivalent tip diameters. It was also observed that the installation force required to install the steel pipe pile is quite close to the ultimate bearing capacity of pile.

1 INTRODUCTION

Deep foundations are preferred worldwide when a shallow stratum does not offer sufficient resistance required to carry the superstructure load. In the evolution of pilling industry, various types of piles and construction methods have been developed and implemented at construction sites. This advancement includes driven/displacement piles, and are preferred both for onshore and offshore structures. The reason behind the preference of driven piles (steel pipe piles, helical piles, etc.) over classic non-displacement piles is attributed to legislation and restrictions on the allowable noise, generated during the installation of deep foundations, especially in urban environments. Araki (2013), Sato et al. (2015), Hirata et al. (2005) acknowledged the use of small-diameter steel pipe piles and spiral piles in solar power generations projects. The press-in piling technique is used for the installation of steel pipe piles (Deeks & White 2007). Whereas, the rotatory press-in method is used for helical pile installation (Lutenegger 2009). These installation methods are often labelled as "the silent method" owing to the reduced noise level and minimized vibration to abutting structure. Close-ended steel pipe piles cause minimized displacement compared to driven concrete piles (Leppanen 2000). Installation methods have different effects on the surrounding ground and the ultimate bearing capacity of the pile. According to Phuong et al. (2016), installation of jacked piles densify the surrounding ground and thus result in

increased static bearing capacity. It was investigated through acoustic emission that breakage of sand particles occurred in the shear zone area when the close-ended pile is driven. Moreover, the sand below the pile tip, i.e. within the compression zone showed insignificant breakage (Mao et al. 2020). According to Perko (2009), crowd (axial) force should be applied for ensuring the advancement of the helical pile in the ground and should be equal to at least 80 percent of the blade pitch during each revolution. The geometry of helical elements, soil properties and depth of installation effect the installation torque (Ghaly 1991). Malik (2019) investigated that the thickness of helix also affects helical pile performance under dense ground conditions if it is deflected or deformed.

A lot of researches have been conducted on the axial capacity of driven pile in dense sand but still, it is the most arguable area with high uncertainty in foundation design (Randolph et al. 1994). Also, previous research studies were mainly focused on understanding individual behaviour of steel pipe piles and helical piles. Therefore, the current study was focused on the performance of close-ended steel pipe piles and single helix helical piles under similar pile tip diameter and ground conditions. In the current study, steel pipe piles were installed in dense ground using the pressing method of installation. Whereas, helical piles were installed in dense ground with a combination of pressing and rotation mechanism. In the case of helical piles, the pressing rate and rotation were adjusted in such a way that 1 pitch penetration was achieved in 1

DOI: 10.1201/9781003215226-24

rotation of helix. Steel pipe piles having different shaft diameters and helical piles having different helix diameters were considered in this study. Installation force was measured during the installation of steel pipe piles and helical piles. Whereas, the installation torque was also measured for helical piles. Subsequently, the bearing capacities, including shaft resistances as well as base capacities were observed. Installation effort (installation force, installation torque) required for installing the piles and bearing capacities of closed-ended piles (steel pipe pile and helical pile) having similar tip diameters were compared.

2 TESTING EQUIPMENT

In this study, model close-ended steel pipe piles and close-ended single helix helical piles were used. Length of the pile (600mm) was identical in both types of piles. Steel pipe piles having different shaft diameters (D_{PS} = 21.7mm, 43mm and 60mm) were used in this study. Dimension details of steel pipe piles are shown in Figure 1. On the other hand, helical piles having an equivalent shaft diameter (D_{HS} = 21.7mm) but different helix diameters (D_H = 43mm and 60mm) were used (see Figure 1). To achieve the objectives of this study, equivalent tip diameters were used for both steel pipe piles and helical piles ($D_{PS} = D_H$ = 43 and 60mm). Consistent helix to pitch ratio (D_H/P = 3.6) were considered in this study. The pitch of helix was measured from the inner edge of the upper and lower helix blades. Both steel pipe piles and helical piles comprised of a hollow central shaft having a flat end attached at the bottom.

Strain gauges were also fixed on the bottom of the inner side of the pile wall to measure the shaft resistance during the pile load test. Preparation of model ground was accomplished in a steel container using Toyoura sand in dry condition. Toyoura sand was compacted to a relative density of 70% to assure homogeneity in all tests. The properties of Toyoura sand used

are: specific gravity = 2.65, D_{50} (50% pass particle size) = 0.20mm, maximum void ratio (e_{max}) = 0.98 and minimum void ratio (e_{min}) = 0.60. Both pile types were designed to nullify the impact of pile size and soil on measured data.

According to Dickin (1983) and Abdoun et al. (2008), in the model-scale test, the size of the buried structure should be 48 times greater than the D_{50} value to eliminate the size effect. In this study, the shaft diameter of the steel pipe pile (D_{PS}) was ranging from 21.7mm ~ 60mm and the D_{50} value of the soil was 0.2mm. Thus, ratio D_S/D_{50} was ranging between 109 ~ 300. Helical piles used in this study also share the similar range because the steel pipe pile shaft diameter (D_{PS}) and the helix diameter (D_H) are consistent in this study. Hence, it is believed that the results were not affected by the size of the pile and the soil particles.

Also, the size of the model container is crucial owing to its influence on the pile capacity measurement. Previous research shows that the loading influence zone ranges from 3 to 8 times the pile diameter (Kishida 1963, Robinsky 1964). Yang (2006) proposed an influence zone in clean sand above the pile tip to be 1.5 to 2.5 times the shaft diameter and 3.5 to 5.5 times the shaft diameter below the pile tip. The dimensions of steel container were carefully selected in an attempt to avoid the influence of boundary conditions on the data measurement during installation and pile load testing. In this study, the cylindrical steel container having a diameter of 1000mm and a height of 1100mm was used. The container diameter was $15D_{PS}$ and the vertical clearance beneath the pile was $10D_{PS}$ for the pile having a maximum diameter (D_{PS} = 60mm). The displacement control loading system was used to install the pile. Whereas, the rotation system was used to rotate the pile during the installation. Installation force and torque were monitored with the help of load and torque transducers. The pile penetration during the installation and the pile load test were measured with the help of displacement transducer. The data logger was used to record all the data (Figure 2).

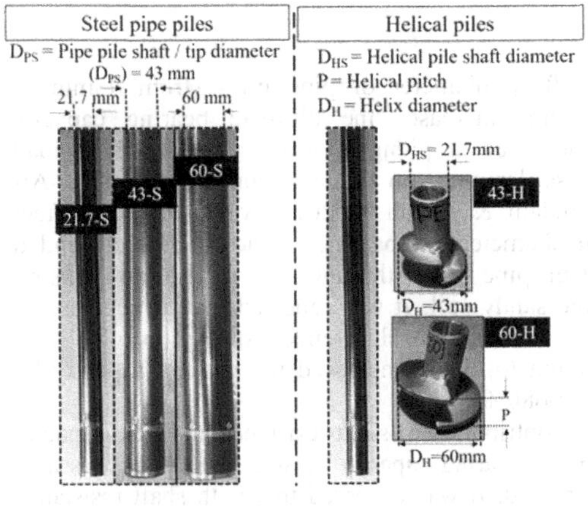

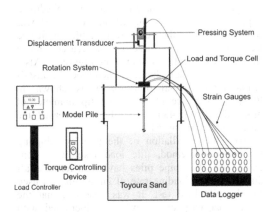

Figure 1. Model steel pipe piles and helical piles.

Figure 2. Schematic illustration of testing equipment.

273

3 TESTING PROCEDURE

The model ground was prepared using Toyoura sand, compacted using rammer in such a way that each compacted layer had a thickness of 100mm. The total depth of the compacted model ground was 1000mm. The relative density of each compacted layer was 70%. Uniformity of the model ground was maintained in all tests. Subsequently, model steel pipe piles and helical piles were installed in the model ground. Steel pipe piles were installed using the pressing method of installation. Whereas, helical piles were installed with a combination of pressing and rotation mechanism. The penetration rate for installing both types of piles was 15mm/min. In the case of helical piles, the rotation rate was adjusted in such a way that 1 pitch penetration was achieved in 1 rotation of helix. The embedment depth (E_d) of the piles during installation was measured with the help of displacement transducer. Whereas, the installation force (F) and the installation torque (T) were measured with the help of load and torque cell. Installation of piles was followed by pile load tests. During pile load testing, the penetration rate of the pile was reduced to 2mm/min. The settlement (S) of the pile during pile load test was measured with the help of displacement transducer. Whereas, the compressive force (P) was measured using the load cell.

4 RESULTS AND DISCUSSION

4.1 *Steel pipe piles*

In this test series, steel pipe piles having an equivalent pile length but different shaft diameters were installed by the pressing method. The pressing rate of 15mm/min was considered during the installation of piles. Steel pipe piles having three different shaft diameters (D_{PS} = 21.7mm, 43mm and 60mm) were installed in the dense ground.

Pipe piles were installed to an embedment depth of 400mm. The installation force (F) was recorded during the test. Test results indicated that the installation force was increased almost linearly with depth as shown in Figure 3. Also, an increase in shaft diameter increased the installation effort (installation force) as shown in Figure 4. Increase of 40% shaft/tip diameter of pipe piles (from 43mm to 60mm) increased the installation force by 61%.

After the installation of the pile by adopting the pressing method, pile load tests were conducted for steel pipe piles having different shaft diameters. The loading rate of 2mm/min was used for the pile load test. It was observed that the bearing capacity of the pile was increased with increase in pile shaft diameter. Increase of 40%

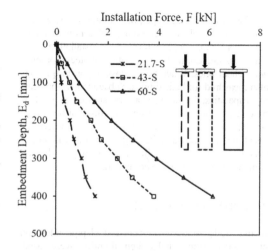

Figure 3. Effect of steel pipe pile shaft diameter on the installation force in dense sand.

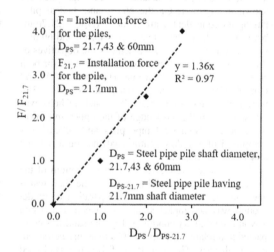

Figure 4. Relation of normalized pile shaft diameter with normalized installation force.

shaft/tip diameter of pipe piles (from 43mm to 60mm) increases the ultimate bearing capacity (measured at plunging state; state at which load to settlement ratio becomes constant) by 64%. Al-Soudani & Fattah (2020), investigated the effect of diameter on bearing capacity of close-ended steel pipe piles through the model study using fine sandy soil. It was reported that an increase in pile diameter of close-ended steel pipe piles from 20mm to 40mm increased the bearing capacity by 320-680%.

Contribution of shaft resistance and base capacity in the bearing capacity of steel pipe piles was also explored. It was observed that both shaft resistance and base capacities were increased with an increase

274

in shaft diameter (Figure 5). Chow (1995) and Randolph et al. (1994) investigated that the unit skin friction capacity (shaft resistance) of a pile may be influenced by lateral effective stress at the pile-soil interface. Increase in shaft resistance can be attributed to increased surface area (increases with the diameter of the pile). Similarly, an increase in base capacity can be attributed to the increased bearing area for large tip diameter piles compared to piles having small tip diameters. It was also observed from test results that there was a nominal difference between the maximum installation force and the ultimate bearing capacity (considered at settlement equal to 15% of pipe pile shaft diameter) of steel pipe piles (Figure 6).

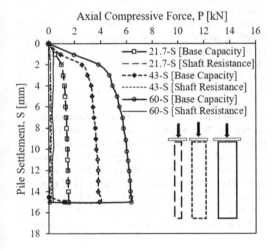

Figure 5. Load-settlement curves of steel pipe piles having different shaft diameter in dense sand.

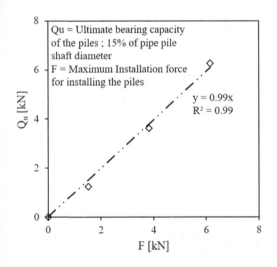

Figure 6. Relation of maximum installation force and ultimate bearing capacity of steel pipe piles in dense sand.

4.2 Helical piles

In this test series, helical piles having an equivalent pile length were installed by the pressing and rotation method. The pressing rate of 15mm/min was considered during the installation of piles. The pressing rate and rotation were adjusted in such a way that 1 pitch penetration was achieved in 1 rotation of helix (recommended by Perko (2009)). Helical piles having a similar shaft diameter (D_{HS} = 21.7mm) but with different helix diameters (D_H = 43mm and 60mm) were installed in the dense ground. Helical piles were installed to an embedment depth of 400mm.

Installation force (F) and installation torque (T) were recorded during the tests. It was observed that the installation force was increased with depth in all tests. It was also revealed by the test results that increase in helix diameter increased the installation force (Figure 7).

Installation torque (T) also experienced an increase with depth (Figure 8). This increase in installation torque with depth is in line with the study conducted by Ghaly et al. (1991). Ghaly et al. (1991) investigated that installation torque increases with an increase in soil strength parameters and/or installation depth. It was observed that the pile having a greater helix diameter (60mm) required greater torque for installation compared to that having a small helix, 43mm (Figure 8). The discrepancy in torque requirements, by helical piles having different helix diameters, for achieving the final installation depth is in line with findings of Ghaly et al. (1991). Ghaly et al. (1991) support the increment of installation torque requirements with the increase in shaft to helix diameter.

Pile load tests were also conducted for helical piles using 2mm/min loading rate (also adopted by

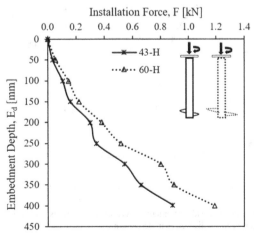

Figure 7. Effect of helix diameter of the helical pile on installation force in dense sand.

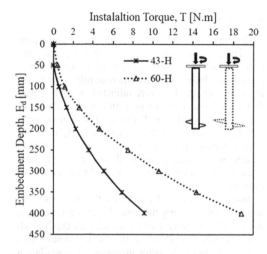

Figure 8. Effect of helix diameter of the helical pile on installation torque.

Matsumiya et al. 2015). The test results showed that a pile having a large helix exhibit enhanced bearing capacity compared to a pile having a small helix (Figure 9).

It was observed that the ultimate bearing capacity (considered at settlement equal to 15% of helical pile shaft diameter) of helical piles increased by 60% by increasing the helix to shaft diameters by 40% (D_H/D_{HS} = 1.98 to 2.76) as shown in Figure 9. This increase in the ultimate bearing capacity owing to increased helix diameters is in line with the study by Sakr (2011). Sakr (2011) identified that trimming

the pile helices results in reduction of axial capacity of piles because of reduced bearing area.

4.3 Comparison of steel pipe piles and helical piles having equivalent tip diameter

Steel pipe piles and helical piles having equivalent tip diameters were also compared for their installation effort requirement and bearing behaviour. The installation force of steel pipe piles having tip diameters (D_{PS}) of 43mm and 60mm were compared with helical piles having similar helix diameters (D_H). It was observed that, in general, the axial force requirement for installing steel pipe piles is quite high compared to helical piles having equivalent helix diameters. This discrepancy in axial force required for installation is owing to different installation mechanism; steel pipe piles are installed using the pressing method. Whereas, the rotation also accompanied pressing for installing helical piles. The comparison of steel pipe piles and helical piles having the equivalent tip diameter shows that the installation force of steel pipe pile was increased by 332% for D_{PS} = D_H = 43mm and 417% for D_{PS} = D_H = 60mm (see Figure 10). However, the installation effort of screw pile (installation force and torque) in terms of power consumption was on average 29% higher than steel pipe pile. The comparison of load-settlement curves of steel pipe piles and helical piles having the equivalent tip diameter shows that steel pipe pile bearing capacities are more than helical piles as shown in Figure 11. Figure 12 showed that the ultimate bearing capacity of the helical pile is 27% less than a steel pipe pile having a similar pile tip area under dense ground conditions.

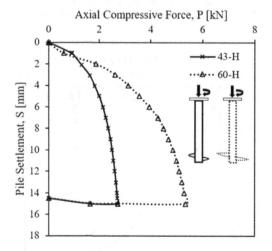

Figure 9. Load-settlement curves of helical piles with different helix diameters in dense sand.

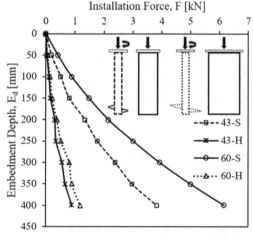

Figure 10. Comparison of installation force of steel pipe piles and helical piles having equivalent tip diameter.

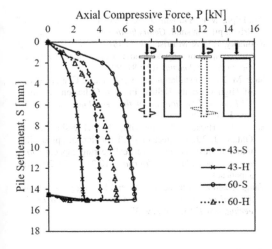

Figure 11. Comparison of load-settlement curves of steel pipe piles and helical piles having equivalent tip diameter.

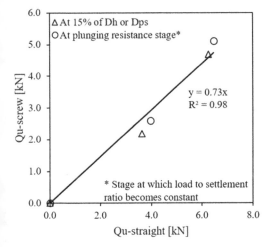

Figure 12. Relationship of ultimate bearing capacity between steel pipe pile and helical piles in dense sand.

5 CONCLUSIONS

This study was based on the comparison of steel pipe piles and helical piles having equivalent tip diameters. Following conclusions are drawn from the test results:

1. Steel pipe piles require 332% and 458% higher installation force than helical piles having equivalent tip diameters $D_{PS} = D_H = 43$mm and 60mm respectively. However, the installation effort of screw pile (installation force and torque) in terms of power consumption was on average 29% higher than steel pipe pile.

2. Helical piles exhibit 27% less ultimate bearing capacity than steel pipe piles having equivalent tip diameters.

3. In the case of steel pipe piles, the installation force required to install the pile is quite close to the ultimate bearing capacity of the pile (considered at settlement equal to 15% of pipe pile shaft diameter). The ultimate bearing capacity of steel pipe piles is 0.99 times the maximum force required for their installation.

This study is particularly useful for design engineers involved in decision making regarding the type of driven deep foundations to be used for a construction project. To decide which pile type is more efficient in dense ground conditions, it is recommended that the effect of stress level and lateral ground disturbance due to pile installation should be studied in future.

REFERENCES

Al-Soudani, W. & Fattah M.Y. 2020. Effect of diameter on the load carrying capacity of closed-open ended pipe piles. *IOP Conference Series: Materials Science and Engineering* 737(1)

Araki, K. 2013. Simplified foundation method subject to small-scale structure "PILE FOUNDATION METHOD". *The Japanese Geotechnical Society* 61(8): 32–33. (in Japanese)

Chow, F. 1995. Field measurements of stress interactions between displacement piles in sand. Ground Engineering. July/August: 36–40.

Deeks, A.D., & White, D.J. 2007. Centrifuge modelling of the base response of closed-ended jacked piles. *Advances in deep foundations-proceeding of the international workshop on recent advances of deep foundations, IWDPF07*: 241–251.

Dickin, E.A., & Leung, C.F. 1983. Centrifugal model tests on vertical anchor plates. *Journal of Geotechnical Engineering* 109(12): 1503–25.

Foray, P., Balachowski, L., & Colliat, J.L. 1998. Bearing capacity of model piles driven into dense overconsolidated sands. *Canadian Geotechnical Journal* 35: 374–85.

Ghaly, A., Hanna, A., & Hanna, M. 1991. Installation torque of screw anchors in dry sand. *Soils and Foundations* 31(2): 77–92.

Ha, D., Abdoun, T. H., O'Rourke, M. J., Symans, M. D., O'Rourke, T. D., Palmer, M. C., & Stewart, H. E. 2008. Buried high-density polyethylene pipelines subjected to normal and strike-slip faulting—a centrifuge investigation. *Canadian Geotechnical Journal* 45(12): 1733–1742.

Hirata, A., Kokaji, S., & Goto, T. 2005. Study on the estimation of the axial resistance of spiral bar based on interaction with ground. *Shigen-to-Sozaiitle* 121: 370–377. (in Japanese)

Kishida, H. 1963. Stress distribution of model piles in sand. *Soils and foundations* 4(1): 1–23.

Leppanen, M. (ed.) 2000. *Steel pipe piles*. Helsinki: Finnish national road administration.

Lutenegger, A.J. 2009. Cylindrical shear or plate bearing-uplift behaviour of multi-helix screw anchors in clay. *Proceedings of international foundation congress and equipment expo (Orlando, Florida: United States/ASCE)* (185): 456–463.

Malik, A.A., Kuwano, J., Tachibana, S., & Maejima, T. 2019. Effect of helix bending deflection on load settlement behaviour of screw pile. *Acta Geotechnica* 14(5): 1527–43.

Mao, W., Aoyama, S., & Towhata, I. 2020. A study on particle breakage during pile penetration process using acoustic emission source location. *Geoscience Frontiers* 11(2): 413–427.

Matsumiya, H., Ishihama, Y., & Taenaka, S. 2015. A study on the bearing capacity of steel pipe piles with tapered tips. 6th Japan-China Geotechnical Symposium, SJGS 2015: 47–52.

Perko, H.A. 2009. *Bhelical piles: A practical guide to design and installation.* New Jersey: John Wiley & Sons, Inc.

Phuong, N.T.V., Van Tol, A.F., Elkadi, A.S.K., & Rohe, A. 2016. Numerical investigation of pile installation effects in and using material point method. *Computers and Geotechnics* 73: 58–71.

Randolph, M.F., Dolwin, J., & Beck, R. 1994. Design of driven piles in sand. *Geotechnique* 44 (3): 427–448.

Robinsky, E.I., & Morrison, C.F. 1964. Sand displacement and compaction around model friction piles. *Canadian Geotechnical Journal* 1(2): 81–93.

Sakr, M. 2011. Installation and Performance Characteristics of High Capacity Helical Piles in Cohesionless Soils. *DFI Journal-the journal of the deep foundations institute* 5(1): 39–57.

Sato, T., Harada, T., Iwasaki, N., Hayashi, S., & Ohtani, J. 2015. Effect of shaft rotation of spiral piles under its installation on vertical bearing capacity. *Japanese Geotechnical Journal* 10(2): 253–26. (in Japanese)

Yang, J. 2006. Influence zone of end bearing piles in sand. *Journal of Geotechnical and Geoenvironmental Engineering, ASCE* 132(9): 229–37.

Proceedings of the Second International Conference on
Press-in Engineering 2021, Kochi, Japan – Matsumoto et al (eds)
© 2021 Taylor & Francis Group, London, ISBN 978-1-032-10414-0

Discrete element modelling of silent piling group installation for offshore wind turbine foundations

B. Cerfontaine
University of Southampton, Southampton, UK

M.J. Brown & M. Ciantia
University of Dundee, Dundee, UK

M. Huisman & M. Ottolini
Heerema Marine Contractors, Leiden, The Netherlands

ABSTRACT: Offshore wind farms are now built in deeper water and bigger foundations are required to stabilise wind turbines of increasing sizes. Pile driving is the most widespread foundation installation method, but more stringent environmental regulations necessitate costly mitigation methods to reduce underwater noise emissions. The silent piling (push-in) concept presented in this work is composed of a cluster of four piles, progressively installed by successive jacking sequences. During one sequence, each pile is moved downward by 0.5m stroke, while the other piles are used as reaction. This paper presents the results of Discrete Element Method (DEM) of the installation process. This work identifies the main features of the push-in installation method, such as pile interaction, progressive plugging and loss of efficiency as a function of depth. It is shown that the cluster capacity can reach six times the weight of the tool necessary to silently install the piles.

1 INTRODUCTION

Pile driving is one of the main offshore installation methods for large monopiles or smaller piles supporting the corner of jacket structures. One of the disadvantages of this method is the large amount of underwater noise generated by the repeated impact of the hammer on the pile, which can be harmful for marine inhabitants (Bailey *et al.*, 2010). Mitigation methods such as bubble curtains (Koschinski & Lüdemann, 2013), can be very expensive with an importantcarbon footprint. Subsequently, there is a need for innovative and silent piling installation methods.

Pile jacking generates very low noise during installation, as it does not require impact. However, large reaction force is necessary to install a pile to a target depth. The press-in piling method overcomes this hurdle by using previously installed piles as reaction piles (White *et al.*, 2002) to create a retaining wall made of piles. The push-in concept presented here follows the same rationale. This concept replaces a traditional single open tubular pile with a cluster of four smaller diameter open tubular piles (Huisman et al., 2020), see Figure 1. In a number of strokes, each of the piles in this cluster is statically pushed into the soil, with two or three piles of the cluster providing the uplift resistance

required to push in a third pile, with a tool gripping on the "uplift" piles and pushing down onto the pile that is penetrating. By sequentially pushing in each of the piles while holding on to two or three others, the cluster as a whole is penetrated into the seabed. A novelty in the push-in pile concept is that the installation method makes use of force equilibrium (uplift loads equal compression loads) without moment equilibrium (one side of the cluster is pushed down with more force than the opposite side). The moment equilibrium is therefore reached by introducing bending into the piles, something that the installation tool is specifically designed for.

The Discrete Element Method (DEM) represents the soil as an assembly of rigid particles that obey Newton's laws of motion and interact between each other by means of contact laws (O'Sullivan, 2011). The DEM is often employed to investigate soil behaviour at the element scale as it offers readily accessible information at the micro-scale, which may be used to uncover relevant micromechanics.

One of the main advantages of the DEM is to easily handle large displacement and large deformation problems. For this reason Arroyo *et al.*, (2011) started to use DEM to simulate CPT in calibration chambers. Ciantia *et al.* (2016) extended this approach to investigate crushing effects on CPT and pile jacking

DOI: 10.1201/9781003215226-25

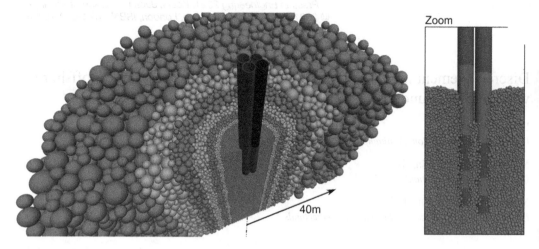

Figure 1. Half of the DEM sample, together with the four piles during installation (looking down from above).

respectively. Sharif *et al.* (2020) demonstrated the efficiency or rotary installation of piles to reduce the necessary crowd force by using DEM as a virtual centrifuge. Liu *et al.* (2019) simulated jacked open-ended piles in 2D. They showed that the plugging mechanism was related to some particle arching inside the pile. The closed-ended pile penetration mechanism was also investigated by Zhang and Wang (2015) in 3D.

This paper uses the DEM to simulate the push-in installation of a four pile cluster in a sandy material. The main objective is to demonstrate the feasibility of the technique and to investigate the main physical mechanisms during installation. The problem was first simplified as a displacement-controlled installation, to understand the interaction between the piles of the cluster. A force-controlled installation, more representative of the field installation, was finally simulated. All the numerical models described here were built using the PFC3D code (Itasca Consulting Group, 2019).

2 METHODOLODY

2.1 *Preparation of the DEM sample*

The DEM framework is used to create a virtual centrifuge environment with an enhanced constant gravity field (Ng = 60), in order to simulate small-scale model and compare the results with actual centrifuge tests in the future. The sample is composed of sand particles only and represents a dry sand bed. However, it is possible to calculate an equivalent gravity scaling factor in saturated sand, i.e. the gravity scaling that would create the same initial stress distribution, assuming that the sand behaviour is drained throughout the simulation (Li *et al.*, 2010).

$$G_{sat} = \frac{\rho_d}{\rho'} G_d \qquad (1)$$

where G_d (=Ng = 60 here) is the acceleration applied to a dry sample and G_{sat} (= 100) is the equivalent acceleration that would be applied to a saturated sample. The dry and buoyant densities are ρ_d and ρ' respectively. Therefore, the prototype piles in an offshore environment are scaled to 1:100. Dimensions are given at prototype scale in the following.

The sample was prepared according to the methodology detailed in (Ciantia *et al.*, 2018). A representative elementary volume (REV) is first prepared. It consists of a cylindrical slice, whose diameter is equal to the sample diameter, but whose height is one fourth of the sample height. The sample outer diameter is equal to 40m at prototype scale, and its height is equal to 60m.

A polydisperse assembly of particles is adopted here to realistically represent the HST95 sand, characterised by Lauder, (2010). Representing the sand particles at their true scale would lead to samples composed of millions of particles, which are far too computationally expensive. Therefore, the particle size distribution (PSD) was discretised into ten bins, each representing 10% of the total solid volume. Each marker in Figure 2 represents the particle diameter of one bin. A scaling factor (SF) was applied to the PSD to reduce the number of particles. The same SF is applied to all particles of the PSD, therefore the shape of the PSD is maintained, but it is shifted in size, as shown in Figure 2. The homogeneous upscaling of all particles ensures that the soil mass behaviour is the same whatever the SF, providing there are enough particles to form a REV. Such an upscaling is quite common for DEM simulations (Evans & Valdes, 2011; Ciantia *et al.*, 2019b; Zhang *et al.*, 2019).

The REV is composed of seven zones extending radially (Figure 1) populated with PSD affected by increasing scaling factors, ranging from 18 (centre) to 205, as shown in Figure 2). This gradation in

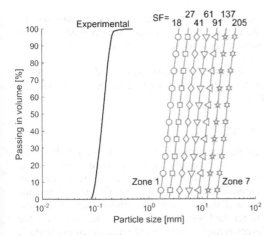

Figure 2. Comparison between experimental (from Lauder (2010) and scaled up (SF = scaling factor) particle size distributions.

Table 1. HST95 sand properties and DEM parameters.

Sand properties [unit]	Symbol	Value
Minimum void ratio [-]	e_{min}	0.467
Maximum void ratio [-]	e_{max}	0.769
Critical state friction angle [°]	ϕ	32
Sand-steel friction coefficient [°]	δ	0.445
Particle dimension [mm]	d_{10}	0.09
Particle dimension [mm]	d_{50}	0.141
Particle dimension [mm]	d_{100}	0.213
Particle density [kg/m³]	ρ_s	2650
Dry density [kg/m³]	ρ_d	1637
Buoyant density [kg/m³]	ρ'	992
Coefficient of earth pressure at rest	K_0	0.47
DEM properties [unit]		
Particle shear modulus [GPa]	G	3
Particle Poison's ratio [-]	ν	0.3
Particle friction coefficient [-]	μ	0.264

particle size is similar to mesh refinement in the finite element method, using smaller particles where soil-structure interaction occurs, while the far field can be modelled by larger particles. For the smallest particles, the shaft diameter (D_c) to average particle (d_{50}) ratio equal to 6. This number is greater than what is usually used in the literature for pile penetration such as CPTs (Ciantia et al., 2019b). The diameter of the central zone (zone 1) is equal to 5.3m.

The particles were randomly generated within each zone of the REV to achieve a target relative density (on average 54% in central core). Such a process creates large contact forces and velocities within the sample. Therefore, a dissipation phase takes place (Khoubani & Evans, 2018), during which the kinetic energy of particles is zeroed every few time steps and large local damping is applied to the particles (0.7).

Once the system has reached a static equilibrium, four REV are stacked to create the final sample. The gravity is set up to 60g and the target stress state ($K_0 = 0.47$) is achieved within the sample. The physical properties of the HST95 are given in Table 1, together with the properties of the Hertz-Mindlin contact model adopted for the particles. The DEM parameters were obtained by back-calculation of triaxial tests (Sharif et al., 2020). The average relative density in the central core of the model is equal to 54%. Particles are free to move in translation, but their rotation was fixed. This is commonly used to simulate the additional restraint of non-spherical particles (Ciantia et al., 2019a) in order to capture the macroscopic response of the soil.

2.2 Pile model and loading

A cluster of four 1.52m (60") core diameter (D_c) piles is simulated in this work. The spacing of the piles is equal to 1.5 pile diameter from centre to

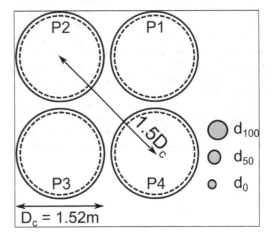

Figure 3. Cluster arrangement and comparison with the particle diameter in the core section.

centre, as shown in Figure 3. The pile wall thickness is equal to (2.5") 63.5mm. Each pile is modelled as a rigid body and split into several parts (base and five shaft segments) to identify the different contribution to the penetration resistance. The steel to particle coefficient of friction is equal to 0.4.

The pile penetration must remain quasi-static during the installation. Therefore, the DEM simulation must limit the inertial effects. The inertial number (I) is typically estimated and must be maintained below a certain threshold to ensure the simulation is quasi-static. The adopted threshold varies between 10^{-3} (Khoubani & Evans, 2018) to 10^{-2} (Janda & Ooi, 2016; Ciantia et al., 2019b), although recent work has shown that lower values could be necessary for polydisperse particle distribution (Shire *et al.*, 2020). In this work, the maximum pile velocity ($v_{z,max}$) when the pile tip is at a certain depth is calculated based on the approach of (Sharif *et al.*, 2020)

$$v_{z,max} = 3D_c \frac{I}{d_{50}} \sqrt{\frac{p_0'}{\rho_s}} \qquad (2)$$

where the inertial number I is equal to 10^{-2}, d50 is the average particle diameter of the central zone, p_0' is the initial confining pressure at the considered depth and ρ_s is the particle unit weight. The penetration rate is increased stepwise as the pile tip reaches greater depths, in order to minimise the CPU time while maintaining quasi-static conditions.

3 DISPLACEMENT-CONTROLLED

The push-in piles concept was first simplified by simulating a fully-displacement controlled installation. First, all piles were installed at the same penetration rate (synchronous phase) until 15m depth. During a second phase (asynchronous)., each pile is moved individually (0.5m stroke) while the other piles remained fixed. The cluster moves downwards by 0.5m every sequence of four steps. During the first step, pile P1 is moved downwards while piles P2, P3 and P4 are fixed. P3 is moved during step 2 (other piles are fixed), then P2 during step 3and finally P4 during step 4. The imposed displacement is represented in Figure 5a as a function of normalised steps,

one normalised step being the time necessary for one pile stroke. Several sequences were applied during the asynchronous phase until the cluster reached 19m depth.

Closed-ended and open-ended piles were tested to evaluate the effect of plugging on the installation force requirements.

3.1 *Macroscopic forces*

The macroscopic force acting on the pile base and shaft were calculated by summing the vertical component of contact forces acting on them. The shaft component include contact on the outside and the inside if the pile was open-ended. All piles of the cluster had similar behaviour, therefore only the forces acting on pile P1 are represented in Figure 4. Results for a single open-ended pile, i.e. not part of a cluster and installed at the same position as P1, were added for comparison. Figure 4c shows that during the synchronous installation phase (0-15m depth), the total force measured is approximately 30% greater for a closed-ended pile in the cluster than for the single pile. It is only 15% greater if the piles were open-ended. The increase in total force is due to the interaction between the piles of the cluster, while the difference between closed- and open-ended is due to the pile partial plugging, which will be discussed below. The shaft contribution is greater for open-ended piles as friction is mobilised on the inside and the outside of the pile. On the contrary, the base contribution is smaller for the open-ended piles.

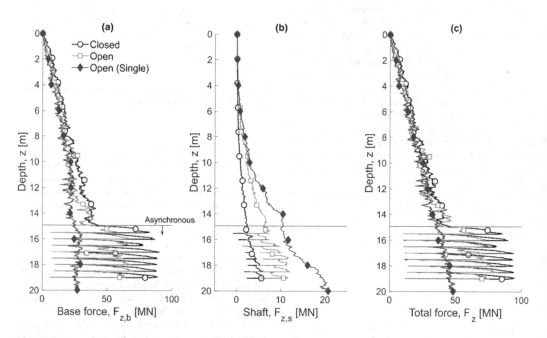

Figure 4. Comparison of the force acting on pile 1 (total, base only or shaft only) in three configurations: closed-ended pile cluster, open-ended pile cluster and single open-ended pile.

The asynchronous installation of the pile starts at a depth of 15m. The measured base and total forces vary cyclically (Figure 4) between a peak and a low value. This is one of the main features of the push-in process. It can be further illustrated by inspecting the evolution of the base force for each pile of the cluster during one asynchronous sequence, i.e. each pile is moved downwards of 0.5m successively (Figure 5a). Figure 5b shows that the base force increases for each pile when it is pushed downwards, then decreases progressively when the other piles are moved. At the end of the sequence, the force acting on each pile is different, although they have all reached the same depth.

It can be observed in both Figure 4 and Figure 5 that the force necessary during a pile stroke is approximately twice the force that was applied to the pile during the synchronous installation (phase 1, 0-15m). However, the total measured force (sum of P1 to P4) is lower than the force necessary for synchronous installation, because the compressive force decrease on all other piles. It should be noted that open-ended piles require a lower force per stroke, indicating that they are not fully plugged.

3.2 Microscopic observations

Figure 6 depicts a cross-section of force chains around the piles before and after pile P1 was moved downwards. At the beginning of the asynchronous sequence (Figure 6a), the force chains are not symmetrical, which is consistent with Figure 5b. At the end of an asynchronous sequence, the force acting on P4 is the greatest, followed by the force on P2, then P3 and P1. When P1 is moved (Figure 6b), the force chains acting on its base increase in magnitude. Their magnitude decreases on P2, but the pile

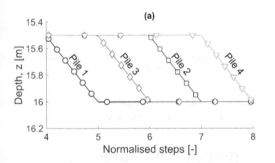 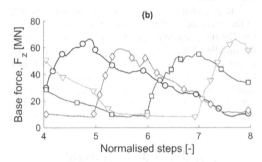

Figure 5. Zoom-in on one asynchronous installation sequence for closed-ended piles, displacement-controlled installation.

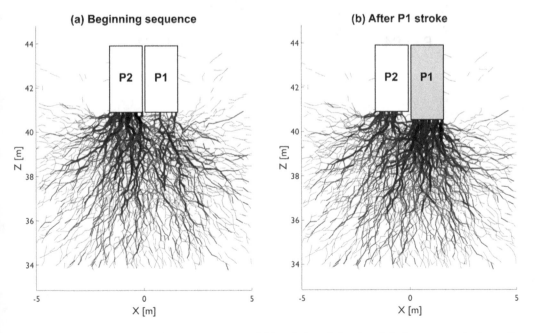

Figure 6. Force chains under the piles during the sequential installation of closed -ended piles, sequence begins at 19m depth.

283

remains in compression. It can be assumed that the fixed-displacement piles adjacent to the pile being moved constrain the pile penetration failure mechanism.

Figure 7 depicts the particle displacement between the beginning (all piles at 19m depth) and the end of the asynchronous sequence (all piles at 19.5m). Each dot in the figure represents a particle, irrespectively of its diameter. The comparison of Figure 7a and Figure 7b shows that the installation of pile P4 mainly induced a displacement of the particles to the right (positive displacement), with only a marginal additional displacement of particles to the left. This also highlights the asymmetry of the installation process. The last pile to be installed (P4) will be more constrained than the first pushed pile (P1). However, it is interesting to note that the displacement field at the end of the sequence (Figure 7b) is still fairly symmetrical. Although this can be due to the impose displacement control.

The plugging of the piles was measured internally in the open-ended piles and in the middle of the cluster. The plugging behaviour is important, especially in the middle of the cluster, because the absence of soil in between the piles will inhibit stress redistribution between the different piles. The evolution of the plug length, measured from the base of the piles/cluster, as a function of the tip depth is depicted in Figure 8. This figure shows distinct behaviour between the

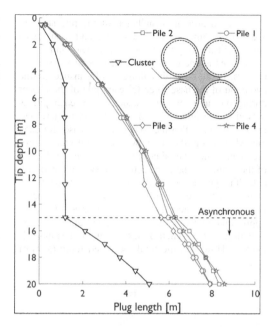

Figure 8. Plugging of the cluster and the individual piles for a displacement-controlled installation.

synchronous and asynchronous installation phases. During the first phase, the inside of the cluster

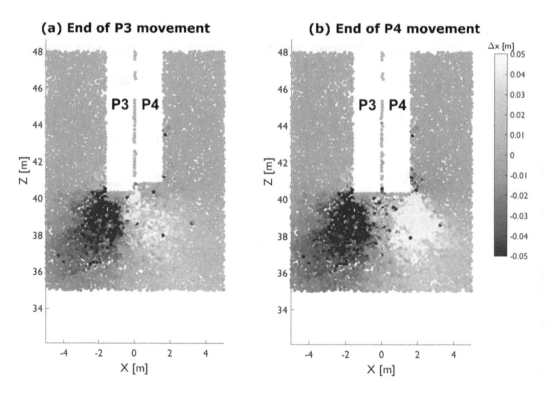

Figure 7. Horizontal displacement at the end of a sequential installation of the close-ended piles, sequence begins at 19m depth, displacement capped at +0.05 or -0.05m.

284

plugged very early and the plug length is equal to 1.2m. This length increases rapidly when the asynchronous installation sequence starts (15m depth). On the contrary, the open-ended piles are only partially plugged. At the end of the simulation, less than half of the pile length was filled with particles.

However, the ratio of the smallest dimension of the cluster inside space ($=0.5D_c$) to the largest particle diameter (d_{100}) is equal to 2. The ratio of the inside pile diameter to d_{100} is equal to 3.8. There is no guidance available in the DEM literature on the minimum value of this ratio to ensure there is no particle scale effect on plugging behaviour. It could be expected that larger particles can create premature plugging of the cluster or the pile, similarly to the effects coarse gravel, cobbles or boulders could have on piles in the field. Further simulations with smaller particles or experiments are necessary to verify the observed cluster plugging behaviour.

4 FORCE-CONTROLLED

The basic mechanisms of the push-in method have been detailed in the previous section. However, the necessary forces measured largely exceed what is typically available in the field. A more realistic installation process was then simulated.

The synchronous phase took place between 0-3m depth. At approximately 3m depth, the force necessary to jack all the piles together (penetration resistance) is equal to the total dead weight of the piles (4 times W_{piles} = 4MN) and the installation tool (W_{tool} = 20MN here).

During the force-controlled asynchronous installation phase, each pile is pushed downwards by a 0.5m stroke in the same sequence as previously. However, while one pile is pushed downwards (the active pile), the other piles are moved upwards together to mobilise some tensile capacity along their shaft. Therefore, throughout the asynchronous phase, the following equilibrium condition must be met

$$\sum_{i=1}^{4} F_{z,i} = W_{tool} + 4W_{pile} = W_{tot} \qquad (3)$$

where $F_{z,i}$ is the total vertical force acting at the top of each pile (positive in compression) and W_{tot} is the total dead weight. In the DEM, this condition is ensured by a servo controlling the vertical uplift displacement of the reaction piles. Each pile is successively active in compression, then becomes a reaction pile. One sequence of cluster installation consists of a series of strokes applied to piles P1, P3, P2 and finally P4.

The base, shaft and total vertical force related to pile P1 are depicted in Figure 9. Figure 9c shows that the total force acting on the cluster is approximately equal to the total weight at a depth equal to 3m. This was selected as the maximum penetration depth during the synchronous phase. Below this depth, the total force

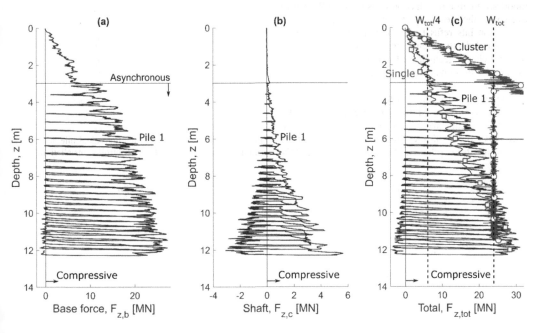

Figure 9. Comparison of the force acting on pile 1 (total, base and shaft) and the total force acting on the cluster during the load-controlled installation with the total force for a single pile.

acting on the cluster is constant and equal to the dead weight.

Figure 9c also shows that the total force necessary to install pile P1 (as part of a cluster) is larger than the total force necessary if a single pile was installed. However, the extra force required is lower than in the previous displacement-controlled case. At greater depths (11-12m), the force necessary to install one pile in the cluster tends to that necessary for a single pile.

In the force-controlled simulations, the reaction piles are moved upwards, rather than being fixed (displacement-controlled). This upwards movement mobilises the tensile capacity of the shaft, but will also reduce the constraining effect of the reaction piles on the active pile penetration failure mechanism. This explains why the vertical total force per pile is not that different from the force necessary to penetrate a single pile.

One of the main mechanisms of the load-displacement installation is the compressive load redistribution between the piles. At the end of the synchronous phase (3m depth), the total weight of the tool and piles is equally distributed between the piles. When the reaction piles are moved upwards, the vertical total force is reduced and the piles can even be loaded in tension along the shaft (Figure 9b). Therefore, the resulting compressive force (W_{tot}) is only balanced by the active pile. This happens between 3-10m depths in Figure 9c. Beyond this depth, the necessary installation force is greater than the dead weight W_{tot}. Tension must be mobilised in the reaction piles to maintain the equilibrium. It can be observed that as the depth increases, the base force in the reaction pile tends to zero (Figure 9a) and the shaft is mobilised in tension (Figure 9b). At some point, the tensile capacity mobilised in all three reaction piles is not sufficient to overcome the necessary penetration force and the installation hits refusal at approximately 12m depth. The refusal depth depends on the soil conditions, the pile and cluster geometries. Further work is necessary to optimise those parameters to maximise the penetration depth.

The capacity of the cluster at a depth of 12m could be approximated from Figure 4. Indeed, the compressive capacity can be assumed as the force necessary to install the cluster in a synchronous manner at 12m depth. This force is equal to 120MN

approximately. Therefore, it can be concluded that a 20MN weight tool can install a cluster whose capacity is 6 times its own weight.

The corollary effect of the reaction pile uplift during the installation, is that the average cluster penetration displacement reached after each sequence of four strokes is lower than the stroke applied to each individual pile (Figure 10a). As the cluster penetrates further down, a greater uplift displacement is necessary to mobilise the full friction along the reaction pile shaft, reducing the efficiency of the installation. Consequently, more strokes will be necessary to achieve the target depth.

Figure 11 shows the displacement over one installation sequence (four strokes) of each pile, as a function of the average cluster depth. In this example, the displacement of piles P3 and P4 is significantly lower than the two other piles, beyond a depth of approximately 9m. Consequently,

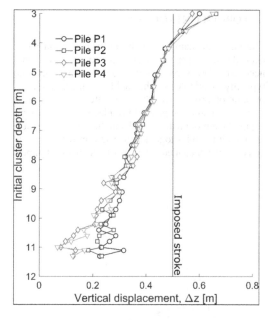

Figure 11. Displacement of each individual pile at the end of one cycle (4 strokes) of load-controlled installation.

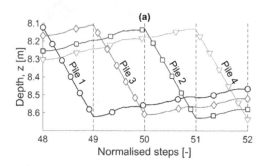

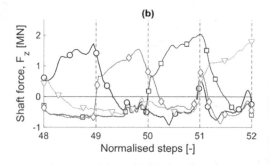

Figure 10. Zoom-in on one asynchronous installation sequence for closed-ended piles, force-controlled installation.

a differential penetration accumulates between the piles. Although this does not pose any problem for the DEM simulation, this differential installation must be continuously corrected in the field, as the installation tool cannot accommodate such a large difference.

5 CONCLUSIONS

This paper presents numerical simulations of the installation of a novel silent piling concept. This concept uses a cluster of four closely spaced piles which are jacked asynchronously. During one sequence of installation, each pile is successively pushed downwards of a stroke equal to 0.5m. The dead weight of the installation tool and the piles, together with the tensile capacity of the three remaining piles are used as a reaction for the necessary jacking force.

The DEM technique enables an insightful investigation of the installation mechanisms, as both macroscopic forces and microscopic observations are possible. It also permits simulation of the large soil displacement inherent in the pile penetration process.

The simulations compared synchronous (all piles together) and asynchronous (successive jacking of individual piles by 0.5m strokes) installations. It was shown that the asynchronous installation under a constant dead weight enables the piles to reach a much greater depth (12m) than the synchronous installation (3m). This is due to the redistribution of the dead weight between the four piles for synchronous loading, to a single pile during the asynchronous installation.

The force necessary to install a pile as a part of a cluster, was shown to be greater than the force necessary to install a single pile. This is due to the restraining effect of adjacent piles on the penetration failure mechanism. The DEM particle movements are laterally limited by the adjacent piles. This constraint is progressively reduced as the adjacent piles are moved upwards to provide tensile reaction for the installation tool.

A vertical upwards displacement is necessary to mobilise the tension along the reaction pile. As a consequence, the net cluster penetration displacement for a cycle of four strokes (one for each pile) is lower than the displacement imposed on each individual pile. This loss of efficiency increases with depth. Beyond a certain depth, the efficiency tends to zero and/or the tensile capacity of the reaction pile is insufficient to push the active pile further.

This work has shown that the group push-in pile concept is worthy of further investigation. The numerical simulations showed that a cluster of piles can be jacked 'silently' and indicated that a capacity equal to six times the tool weight necessary for installation. Further work is necessary to assess the installations requirements as a function of the pile spacing, predict the tensile capacity of

the pile with depth or optimise the pile control to minimise the loss of efficiency for each asynchronous cycle.

REFERENCES

Arroyo, M., Butlanska, J., Gens, A., Calvetti, F. and Jamiolkowski, M. (2011) Cone penetration tests in a virtual calibration chamber. *Geotechnique*, vol. 61, no. 6, pp. 525–531, 2011 doi: 10.1680/geot.9.P.067.

Bailey, H., Senior, B., Simmons, D., Rusin, J., Picken, G. & Thompson, P. M. (2010) 'Assessing underwater noise levels during pile-driving at an offshore windfarm and its potential effects on marine mammals', *Marine Pollution Bulletin*, 60(6), pp. 888–897. doi: 10.1016/j.marpolbul.2010.01.003.

Ciantia, M. O., Arroyo, M., Butlanska, J. & Gens, A. (2016) 'DEM modelling of cone penetration tests in a double-porosity crushable granular material', *Computers and Geotechnics*, 73, pp. 109–127. doi: 10.1016/j.compgeo.2015.12.001.

Ciantia, M. O., Arroyo, M., O'Sullivan, C., Gens, A. & Liu, T. (2019a) 'Grading evolution and critical state in a discrete numerical model of Fontainebleau sand', *Geotechnique*, 69(1), pp. 1–15. doi: 10.1680/jgeot.17.P.023.

Ciantia, M. O., Boschi, K., Shire, T. & Emam, S. (2018) 'Numerical techniques for fast generation of large discrete-element models', *Proceedings of the Institution of Civil Engineers - Engineering and Computational Mechanics*, 171(4), pp. 147–161. doi: 10.1680/jencm.18.00025.

Ciantia, M. O., O'Sullivan, C. & Jardine, R. J. (2019b) 'Pile penetration in crushable soils: Insights from micromechanical modelling', in *Proceedings of the XVII ECSMGE-2019*. Reykjavik, Iceland, Iceland, pp. 298–317. doi: 10.32075/17ECSMGE-2019-1111.

Evans, T. M. & Valdes, J. R. (2011) 'The microstructure of particulate mixtures in one-dimensional compression: numerical studies', pp. 657–669. doi: 10.1007/s10035-011-0278-z.

Huisman, M., Ottolini, M., Brown, M. J., Sharif, Y. and Davidson, C. (2020). Silent deep foundation concepts: push-in and helical piles. *Proceedings of the 4th Int. Symp. on Frontiers in offshore Geotechnics* Austin, Texas.

Itasca Consulting Group (2019) 'PFC3D 6.17'.

Janda, A. & Ooi, J. Y. (2016) 'DEM modeling of cone penetration and unconfined compression in cohesive solids', *Powder Technology*, 293, pp. 60–68. doi: 10.1016/j.powtec.2015.05.034.

Khoubani, A. & Evans, T. M. (2018) 'An efficient flexible membrane boundary condition for DEM simulation of axisymmetric element tests', *International Journal for Numerical and Analytical Methods in Geomechanics*, 42(4), pp. 694–715. doi: 10.1002/nag.2762.

Koschinski, S. & Lüdemann, K. (2013) *Development of Noise Mitigation Measures in Offshore Wind Farm Construction 2013, Report commissioned by the Federal Agency for Nature Conservation (Germany)*.

Lauder, K. (2010) *The performance of pipeline ploughs*. University of Dundee, UK.

Li, Z., Haigh, S. K. & Bolton, M. D. (2010) 'Centrifuge modelling of mono-pile under cyclic lateral loads', *Proceedings of the 7th International Conference on Physical Modelling in Geotechnics*, 2, pp. 965–970. doi: 10.1680/ijpmg.2010.10.2.47.

Liu, J., Ph, D., Duan, N., Ph, D., Cui, L., Ph, D., *et al.* (2019) 'DEM investigation of installation responses of jacked open-ended piles', *Acta Geotechnica*, 14(6), pp. 1805–1819. doi: 10.1007/s11440-019-00817-7.

O'Sullivan, C. (2011) *Particulate Discrete Element Modelling, Particulate Discrete Element Modelling*. Spon Press. doi: 10.1201/9781482266498.

Sharif, Y. U., Brown, M. J., Ciantia, M. O., Cerfontaine, B., Davidson, C., Knappett, J., *et al.* (2020) 'Using DEM to create a CPT based method to estimate the installation requirements of rotaty installed piles in sand', *(in press) Canadian Geotechnical Journal*. doi: 10.1139/cgj-2020-0017.

Shire, T., Hanley, K. J. & Stratford, K. (2020) 'DEM simulations of polydisperse media: efficient contact detection applied to investigate the quasi-static limit', *Computa-tional Particle Mechanics*. doi: 10.1007/s40571-020-00361-2.

White, D., Finlay, T., Bolton, M. & Bearss, G. (2002) 'Press-in piling: Ground vibration and noise during pile installation', in *Deep Foundations 2002: An International Perspective on Theory, Design, Construction, and Performance*, pp. 363–371.

Zhang, N., Arroyo, M., Ciantia, M. O., Gens, A. & Butlanska, J. (2019) 'Standard penetration testing in a virtual calibration chamber', *Computers and Geotechnics*, 111(March), pp. 277–289. doi: 10.1016/j.compgeo.2019.03.021.

Zhang, Z. & Wang, Y. H. (2015) 'Three-dimensional DEM simulations of monotonic jacking in sand', *Granular Matter*, 17(3), pp. 359–376. doi: 10.1007/s10035-015-0562-4.

Proceedings of the Second International Conference on Press-in Engineering 2021, Kochi, Japan – Matsumoto et al (eds)
© 2021 Taylor & Francis Group, London, ISBN 978-1-032-10414-0

Rehabilitation of brownfield sites using the Gyropiler to remove existing bored cast in-situ concrete piles

J.P. Panchal
Keltbray Piling, Esher, London, UK

A.M. McNamara
City, University of London, London, UK

ABSTRACT: Reinforced concrete deep foundations are typically abandoned after a building is demolished. This is a particular problem in highly developed cities, such as London, as they undergo redevelopment cycles. Many London sites are now in their third generation of deep foundations. The previous foundations present significant economic, environmental and safety challenges for subsequent developers, as the ground is increasingly more congested with piles; this is clearly not sustainable. Reuse of existing piled foundations is particularly difficult as insurance cannot be easily obtained. The piling industry has adopted construction methods to core out existing piles, only to replace them with larger diameter or deeper concrete piles. This paper explores the use of the Giken Gyropiler to overcore and remove concrete piles in their entirety. This technique, combined with more sustainable foundations could rehabilitate sites that have been polluted with piles, and has the potential to improve the value of brownfield sites.

1 BACKGROUND

Bored cast in-situ concrete piles have been used widely on commercial buildings in London for over 50 years. Therefore, a large proportion of brownfield sites are now on their third generation of piled foundations, and the ground is becoming increasingly littered with man-made infrastructure, see Figure 1.

RuFUS (Butcher *et al.*, 2006) published guidance and a number of case studies relating to the 'Reuse of Foundations in Urban Sites', however there is still little appetite amongst developers to reuse existing concrete piles as, amongst other factors, it is difficult to assess their condition. Consequently, new piles are often installed in favor of reusing existing piles, which cannot be easily removed, during redevelopment.

A high profile example of a bored piled foundation polluted site is illustrated in Figure 3. The Shard development in Southwark, central London, was previously home to a 26 storey structure which had been founded on underreamed piles. The previous development, Southwark Towers, had limited basement space, therefore the structures had been founded on relatively shallow piles (Moazami & Slade, 2013). Aerial views comparing the Southwark Towers and the Shard development are presented in Figure 2.

Consequently, the piles could not be reused and the presence of the underream bell eliminated the option to remove the piles. The new piles were therefore positioned in between the existing underream piles, as shown in Figure 1.

Loads from the Shard superstructure was vastly different from those that had been imposed on the underream foundations from the Southwark Towers. Therefore, to accommodate these loads, additional, deeper piles were required. Importantly, none of the existing piles were mobilised, and still remain in the ground as obstructions to future developments.

The scheme comprised the installation of over 100no. bearing piles 1.8m in diameter constructed to depths of over 60m deep, through the London Clay and where founded in Thanet Sands (Beadman et al., 2012).

2 INTRODUCTION

If the situation remains where developers do not reuse existing piles, a solution must urgently be sought to enable the efficient and economic removal of existing piled foundations. Failure to do so, would inhibit the future development potential of large cities, ultimately reducing the value of such areas.

Current pile removal techniques rely on a large piling rig, typically weighing over 120 tons, to core out existing piles, which can range in diameter from 450mm up to 1800mm or more. This process involves

DOI: 10.1201/9781003215226-26

Figure 1. Géotechnique artist impression of sub-surface congestion (courtesy of Keller).

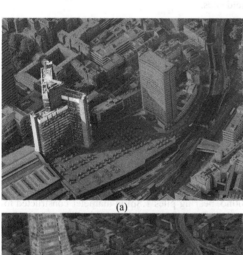

(a)

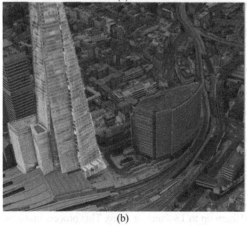

(b)

Figure 2. Aerial views of (a) Southwark Towers and (b) the Shard, post development.

Figure 3. Existing underreamed piles from Southwark Towers and new pile layout (shown in dark grey) adopted at the Shard, shown as dark circles (Moazami & Slade, 2013).

the use of a segmental casing, with 50mm thick walls, with cutting teeth. The piling rig repeatedly advances the casing along the length of the pile to grind through existing obstructions and remove sections of the unwanted pile in short segments, as shown in Figure 4 and Figure 5. This is a laborious, time consuming, high maintenance and expensive process. The environmental concerns arising as a result of these coring operations are also significant; as noise, dust and vibration levels are frequently excessive.

The construction industry is generally keen to consider the use of new technology to establish a competitive advantage over other contractors, and thereby maintaining and expanding on a significant market share.

As a vast majority of brownfield development sites in London have already been littered with pile foundations, there is an urgent requirement for an efficient and reliable procedure for removing existing piles. This paper investigates the suitability of the Gyropiler to develop a bored cast in-situ piled removal process.

Figure 4. Pile removal on 900mm diameter piles.

Figure 5. Pile coring in progress on 1800mm diameter piles.

This project considers the removal of existing concrete bored piles through a series of centrifuge model tests (Gorasia *et al.*, 2013). A bespoke auger was designed to drive through the soil. The purpose of this tool was to remove the soil directly adjacent to the piled foundation, as well as the bored element, in its entirety.

3   GIKEN GYROPILER

Giken (Giken, 2018) recognizes that infrastructure projects heavily rely on concrete construction, which are constructed without regard for future developments and are therefore inherently difficult to remove at demolition stage; this all contributes to a congested subsurface environment.

The Giken Gyropress Method (Giken, 2018) has the potential to resolve the deep foundations industry by reimagining new construction method concepts. The Gyropress, shown in Figure 6, enables the installation of tubular piles with cutting teeth attached to the pile toe. The piles are rotated and jacked simultaneously. This technique respects the "Five Construction Principles" (Giken, 1994).

Expansion for the applicability of the Giken Gyropress into the pile removal market aligns with the "Five Construction Principles". This application would further enhance the Giken offering, as significant improvements would be realized in environmental

Figure 6. Giken Gyropiler (Giken, 2018).

Environmental Protection	Construction work should be environmentally friendly and free from pollution.
Safety	Construction work has to be carried out in safety and comfort with a method implementing the highest safety criteria.
Speed	Construction work should be completed in the shortest possible period of time.
Economy	Construction work must be done rationally with an inventive mind to overcome all constraints at the lowest cost.
Aesthetics	Construction work must proceed smoothly and the finished product should portray cultural and artistic flavour.

Figure 7. General construction principles (IPA, 2016).

protection, safety, speed and economy, as construction practices would move away from inefficient, slow and hazards incremental pile breaking methodologies, to more effective solutions.

4   GYROPILER BESPOKE PILE REMOVAL AUGER

This discussion paper seeks to explore an experimental technique for bored pile removal that has been developed by City, University of London. This technique has been trialed in laboratory model tests and aims to allow for the removal of complete piles by over excavation and remolding of the soil from the pile annulus using a specially developed auger, as shown in Figure 8.

Figure 8. Over coring auger for model tests on bored pile extraction.

In principle, the results from previous tests (Gorasia *et al.*, 2013) showed that the cutters reduced the applied torque force between approximately 20-30%.

The auger was designed to remove a small annulus of soil from around an existing concrete pile. The tool is advanced along the full depth of the pile before being twisted to shear the soil just below the pile toe.

Special consideration was given to developing a means of ensuring that the pile could be removed as a complete element. 'Cheesegrater' cutting teeth (visible in Figure 9) allow for a small annulus of soil to also be removed along the face of the pile. Without this modification, the Gyropiler would easily shear lightly reinforced piles as friction between the inside face of the tube and the pile would exceed the structural capacity of the piled element (McNamara *et al.*, 2013).

If a pile was sheared in this way, the removal process would be incredibly time consuming and laborious, as the Gyropiler would be required to repeatedly lift the cutting auger which would then need to be cleaned, before repeating the activity to the toe of the pile.

Figure 9. Pile over coring auger showing 'cheese grater' type cutters inside a steel tube.

The auger is effective because the cutting teeth scrape some of the soil from around the pile, remolding it and forcing it to the outside of the tube where it can be removed on the flights. This allows the tube to advance over the length of the pile with very little friction acting on the inside of the tube and leaving a concrete column that is no longer in contact with the soil along its shaft and that can be lifted out of the ground using a crane.

If similar principals could be applied to tubular steel piles then it may be possible to remold the soil around the base of an advancing pile with careful design of the tip. If the soil is remolded then structure or fabric may be removed during rotation of the pile and this could contribute to an overall reduction in vertical jacking force required.

5 APPLICABILITY TO BROWNFIELD SITES

The operation of the Gyropiler relies on tubular driven piles on which it is supported. The tubular piles provide reaction to both driving forces and extraction forces.

Brownfield sites are characterized by many unknown and known obstructions in the ground. These include the piles to be removed and also other old shallow foundations, redundant buried structures and both live and disused services.

Whilst the Gyropiler is easily capable of penetrating such ground conditions and obstructions, the removal of existing obstructions would nonetheless be time-consuming as temporary tubular steel piles would be required to provide the necessary reaction forces to implement the Gyropiler in its current configuration. Furthermore, on congested sites in well-developed urban areas, previous pile layouts will have been designed to avoid existing buried infrastructure. Therefore, the installation of temporary tubular piles may impinge on such infrastructure which would make the Gyropiler a non-viable method of extracting existing concrete piled foundations.

The obvious solution to ensure the successful deployment of the Giken Gyropiler for all applications on brownfield sites would be to develop or adapt existing tracked piling rigs to accommodate the Gyropiler system.

This approach to pile removal would complement a new concrete piled foundation that may be adopted on a site remediation development. The bespoke Gyropiler mechanism could be attached to a tracked piling rig and would be capable of removing existing foundations. The same piling rig could either be used to extend the newly bored pile, or continue to drill or extract piles across the site without the need for other pile removal machinery, importantly, negating the need for environmentally damaging percussive methods.

Benefits of this adaptation include limiting the number of items of heavy plant on site, which is particularly important in helping to manage safety and logistics on piling sites.

This process greatly improves on current practice, where piles are cored, backfilled with lightly cemented concrete, only to be drilled through and replaced with a structural concrete pile. The environmental impact of this process is catastrophic and adds considerable time to the construction program. The Gyropiler mechanism, combined with a tracked piling rig, would enable site teams to remove existing concrete piles and construct new piles in continuous single operation.

In addition, a piling mat would be already be required for the installation of concrete bored piles. Consideration would need to be given to the existing piling mat design to ensure sufficient bearing capacity for the Gyropiler attachment when extracting existing foundations.

6 PILE REMOVAL SEQUENCE

Removal of existing piles would be carried out using the following process, as illustrated in Figure 10:

  i. Drill through the centre of pile to the pile toe
 ii. Resin the end of the bar to provide a method of lifting the pile out of the bore; see (a)
iii. Install Gyropiler attachment; see (b)
 iv. Use conventional segmental casing to advance Gyropiler attachment; see (c)
  v. Extend Gyropiler attachment below existing pile toe; see (d)
 vi. Attach pile cropped to top of casing and lift pile head from central bar; see (e)
vii. Continue to lift pile from bore and incrementally crop, resulting in concrete falling away from central bar; see (f)
viii. Extend pile bore if required
 ix. Extract segemental casing and Gyropiler to competent ground conditions; see (g)
  x. Place reinforcement and structural concrete to complete construction of new textured pile; see (h).

7 IMPLICATIONS FOR THE PERFORMANCE OF FUTURE BORED CONCRETE PILES

During a previous experimental trial (McNamara et al., 2013), two piles were cast in a clay sample and over-cored with one straight shafted tubular cutter, and the other with a profiled tubular cutter.

Upon removal the plain pile shaft was especially clean and no clay had adhered to the tubular cutter. The profiled pile cutter plugged at the base and a significant amount of clay had adhered to the surface, as illustrated in Figure 11.

Figure 12 shows the profile of the straight tubular pile cutter, whilst Figure 13 clearly shows a more textured bore profile arising as a result of using the

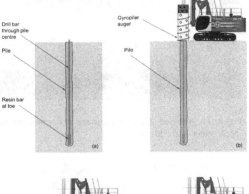

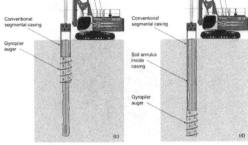

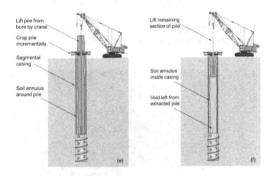

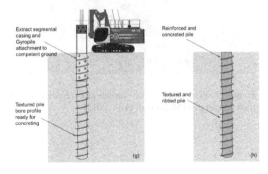

Figure 10. Gyropiler pile extraction sequence.

profiled tubular cutter. The implication of such a pile bore surface is that the pile newly constructed in this bore would possess a much improved adhesion factor, compared with a standard bored pile.

Figure 11. Adhered clay on the surface of the profiled pile (McNamara et al., 2013).

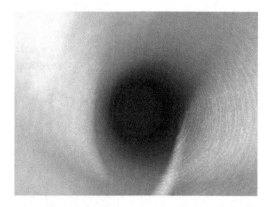

Figure 12. Photograph of clay surface after removal of plain tubular pile (McNamara et al., 2013).

Figure 13. Photograph of clay surface after removal of profiled tubular pile (McNamara et al., 2013).

8 CONCLUSIONS

There is an urgent need to develop new systems and approaches to brownfield site remediation.

A potential method of removal of existing bored concrete cast in-situ piles has been proposed using Gyropile technology.

The method has potential to allow removal of deep piled foundations whilst also creating enhanced shear capacity at the soil-pile interface.

The processes described in this paper extends the application of Giken plant to sites where tracked machinery is required.

A feasible construction sequence has been outlined which has considered the practicalities and typical site limitations likely to be experienced. The process has genuine potential to improve health and safety in construction, reduce the environmental impact of demolition works, and accelerate construction programs.

REFERENCES

Beadman, D., Pennington, M. & Sharratt, M. (2012) Pile test at the Shard London Bridge. *Ground Engineering Magazine*, January 2012.

Butcher, A.P., Powell, J.J.M. & Skinner, H.D. (Eds). 2006. Reuse of Foundations for Urban Sites: *Handbook and Conference Proceedings on the Re-use of foundations*, Building Research Establishment CRC, London.

Giken, 2018. Gyropress Method, Accessed: https://www.giken.com/en/wp-content/uploads/press-in_gyropress.pdf.

Gorasia, R.J., McNamara, A.M. & Rettura, D., 2014. Reducing driving forces for pressed-in piles. *In Proceedings of the Institution of Civil Engineers - Geotechnical Engineering 2014* 167(1), pp. 19–27.

IPA, 2016. Press-in retaining structures: a handbook. Editors, Kusakabe, O., Kikuchi, Y., Matsumoto, T. & Terashi, M.

McNamara, A., Gorasia, R., Halai, H., Philips, N. (2013) Press-in Engineering 2013: *Proceedings of 3rd IPA International Workshop in Singapore*, Singapore.

Moazami, K. & Slade, R. (2013) Engineering Tall in Historic Cities: The Shard. *Proceedings in the Council on Tall Buildings and Urban Habitat Journal*, Issue 2, pp. 44–49.

Session B: Infrastructure development

Proceedings of the Second International Conference on
Press-in Engineering 2021, Kochi, Japan – Matsumoto et al (eds)
© 2021 Taylor & Francis Group, London, ISBN 978-1-032-10414-0

Numerical simulation for centrifuge model tests on cantilever type steel tubular pile retaining wall by rigid plastic FEM

K. Mochizuki, H.H. Tamboura & K. Isobe
Hokkaido University, Sapporo, Japan

J. Takemura
Tokyo Institute of Technology, Tokyo, Japan

K. Toda
Nippon Steel Corporation, Tokyo, Japan

ABSTRACT: Constructions of retaining walls in soft rocks using cantilever type steel tubular pipe are increasing in Japan because the Press-in construction methods using tubular piles with a large-diameter and high flexural rigidity are now available. However, the current design method to determine the necessary embedment length in such a case is empirical. In this paper, the rigid plastic FE analysis was used to simulate centrifugal model tests aimed for developing a new design method. Besides, the stability of the wall in two-layered ground and the influence of seismic inertia forces on the stability of the wall were studied. As the horizontal seismic intensity increases, it affects the earth pressure distribution and the failure mechanism, and it decreases the safety factor. In two-layered ground, the failure mechanism changes according to the thickness of the top layer and the embedment length of the wall in the rock layer.

1 INTRODUCTION

In these days, in Japan, the number of construction cases of earth retaining structures in narrow spaces and firm soil such as soft rock is increasing. This increase is due to the widespread availability of a quite new construction method, called the press-in method. The cantilever type steel tubular pile retaining wall with a large-diameter and high flexural rigidity is more suitable for such a construction site. Hence, the demand for the cantilever type steel tubular pile retaining wall is increasing. However, in the current design of this kind of structure for roads, the requisite embedment length is determined based on the following methods: (a) Limit equilibrium method, (b) Prerequisites for the elastic calculation. In the first method, the necessary embedment length is obtained as the minimum embedment length that satisfies the stability of the structural system. In the second method as described in the design manual of cantilever type steel tubular pile retaining wall (2009), the necessary embedment length is obtained as, 2.5/β - 3.0/β, where β is the characteristic value of the pile obtained by Chang's equation and shown as below;

$$\beta = \sqrt[4]{\frac{k_H D}{4EI}} \quad (m^{-1}) \tag{1}$$

where, k_H = the horizontal subgrade reaction coefficient, D = the pile diameter and EI = the flexural rigidity.

The necessary embedment length for the cantilever type steel tubular pile retaining walls with a large diameter and high flexural rigidity in soft rock is mainly determined based on the second method, according to the past design cases. However, the method of calculation of β for layered grounds is not clearly addressed and there is a high possibility of over-estimation by using the first method. Therefore, it is important to establish theoretical and reasonable design methods determining the necessary embedment length of cantilever type steel tubular pile retaining walls in soft rock.

In the previous studies, (1) the validity of the rigid-plastic finite element was confirmed, and (2) the effect of shortening the embedment length on stability was evaluated using the same method (Kunasegram et al., 2018; Ishihama et al., 2018; Mochizuki et al., 2019).

DOI: 10.1201/9781003215226-27

In this study, the impact of shortening the embedment length, the safety factor, the failure mode and the earth pressure are evaluated (1) under the influence of seismic intensity (evaluation of earth pressure during an earthquake) and (2) under the effect of a two-layered ground by rigid-plastic finite element analysis. (3) The effect of the tensile strength cut-off model in the soft rock material is evaluated as well.

2 OUTLINE OF CENTRIFUGAL NUMERICAL MODEL

2.1 Centrifuge model

The model tests under a centrifugal acceleration of 50 G were conducted as shown in Figure 1 (Kunasegram et al., 2018 and Ishihama et al., 2018). The model-grounds consisted of the two ground layers; the lower layer was a hard ground simulating sand rock or mud rock, and the upper layer was Toyoura sand, modeling the backfill. A rubber bag filled with water was installed at the front of the retaining wall which is made of aluminum plate (the bending rigidity of the plate EI is equivalent to that of the real steel tubular pipe although the yield moment and full plastic moment is not the same as those of the real pipe), and a rubber bag with an open-topped filled with Toyoura sand was installed behind the retaining wall. In this series of experiments, the stored water in the rubber bag in front of the retaining wall was drained to model the excavation of the ground, and the drained water was stored in a water tank. When the horizontal displacement of the retaining wall head was small enough and the failure did not happen, the stored water in the tank was poured into the backfill to apply more horizontal pressure acting on the retaining wall by raising the water level. In the tests, the horizontal displacement and the strain of the retaining wall, the settlement of the backfill, the pore water

pressure and the earth pressure were measured. This process was repeated until the retaining wall falls over by the horizontal load, and the failure mode in the limit state is evaluated.

2.2 Rigid plastic FEM

In this research, the safety factor and displacement rate are calculated using slope stability analysis based on the rigid plastic FEM. (Tamura et al., 1984, Hoshina et al., 2010 and Yagi et al., 2010). The rigid-plastic constitutive equation is derived using the Drucker-Prager type yield function as follows.

$$f(\boldsymbol{\sigma}) = \omega I_1 + \sqrt{J_2} - \psi = 0 \qquad (2)$$

where, I_1 = the first invariant of stress, J_2 = the second invariant of deviatoric stress, ω and ψ = the coefficient associated with the cohesion c and the shear resistance angle φ.

Stress which generates plastic deformations was disassembled to determinate stress $\boldsymbol{\sigma}^{(1)}$ derived from the plastic strain rate, and indeterminate stress $\boldsymbol{\sigma}^{(2)}$ is not derived from the plastic strain rate. Determinate stress $\boldsymbol{\sigma}^{(1)}$ is shown as follows by the associated flow rule.

$$\boldsymbol{\sigma}^{(1)} = \frac{\psi}{\sqrt{3\omega^2 + \frac{1}{2}}} \frac{\dot{\varepsilon}^p}{\dot{e}} \qquad (3)$$

$$\dot{e} = \sqrt{\dot{\varepsilon}^p : \dot{\varepsilon}^p} \qquad (4)$$

where, $\dot{\varepsilon}^p$ = the plastic strain rate and $\dot{e}$ = the equivalent plastic strain rate.

Indeterminate stress $\boldsymbol{\sigma}^{(2)}$ is a stress component along the yield function, so this stress cannot be obtained from the yield function. However, the component of indeterminate stress can be expressed as follows by using the stress on the yield function and considering constraint on volume change like Eq. (6).

$$\boldsymbol{\sigma}^{(2)} = \alpha \frac{\partial h}{\partial \dot{\varepsilon}^p} = \alpha \left\{ \boldsymbol{I} - \frac{3\omega}{\sqrt{3\omega^2 + 1/2}} \frac{\dot{\varepsilon}^p}{\dot{e}} \right\} \qquad (5)$$

$$h(\dot{\varepsilon}^p) = \dot{\varepsilon}_v^p - \frac{3\omega}{\sqrt{3\omega^2 + \frac{1}{2}}} \dot{e} = \dot{\varepsilon}_v^p - \beta \dot{e} = 0 \qquad (6)$$

where, α = an indeterminate multiplier, $\boldsymbol{I}$ = unit tenser, $\dot{\varepsilon}_v^p$ = plastic volume strain rate. The rigid plastic constitutive equation for the Drucker-Prager type yield function is finally expressed as following.

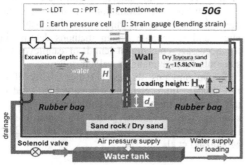

LDT: Laser displacement transducer
PPT: Pore pressure transducer

Figure 1. Schematic view of the centrifuge model tests (Kunasegram et al., 2018).

$$\sigma = \sigma^{(1)} + \sigma^{(2)} = \frac{\psi - 3\omega\alpha}{\sqrt{3\omega^2 + \frac{1}{2}}} \frac{\dot{\varepsilon}^p}{\dot{e}} + \alpha I \qquad (7)$$

This α can be obtained by solving the boundary value problem. Because rigid-plastic constitutive equations are applied to deformed objects, they cannot basically be applied to rigid objects. However, a rigid body is included in the stability analysis. The rigid-plastic constitutive equation should be extended as follows. When the equivalent plastic strain rate $\dot{e}$ falls below a certain threshold $\dot{e}_0$, the equation (8) can be obtained by replacing $\dot{e}_0$ with $\dot{e}$. This corresponds to the operation of reducing the yield function, and a similar structural relationship is established for the stress in the yield function by allowing a small plastic deformation in the rigid part.

$$\sigma = \sigma^{(1)} + \sigma^{(2)} = \frac{\psi - 3\omega\alpha}{\sqrt{3\omega^2 + \frac{1}{2}}} \frac{\dot{\varepsilon}^p}{\dot{e}_0} + \alpha I \qquad (8)$$

For the slope stability analysis, the safety factor is defined by the shear strength reduction factor. Therefore, the yield function and dilatancy characteristics are expressed as follows using the safety factor.

$$f(\sigma, F_s) = \frac{\omega}{F_s} I_1 + \sqrt{J_2} - \frac{\psi}{F_s} = \hat{\omega} I_1 + \sqrt{J_2} - \hat{\psi} = 0 \qquad (9)$$

$$h(\dot{\varepsilon}^u, F_s) = \dot{\varepsilon}_v^p - \frac{3\hat{\omega}}{\sqrt{3\hat{\omega}^2 + \frac{1}{2}}} \dot{e} = \dot{\varepsilon}_v^p - \hat{\beta}\dot{e} = 0 \qquad (10)$$

In this research, in order to accelerate the calculation speed, the penalty method is used as the method to directly consider the constraint condition.

$$\sigma = \frac{\hat{\psi}}{\sqrt{3\hat{\omega}^2 + \frac{1}{2}}} \frac{\dot{\varepsilon}^p}{\dot{e}} + \kappa\left(\dot{\varepsilon}_v - \hat{\beta}\dot{e}\right)\left[I - \frac{3\hat{\omega}}{\sqrt{3\hat{\omega}^2 + \frac{1}{2}}} \frac{\dot{\varepsilon}^p}{\dot{e}}\right] \qquad (11)$$

where, K = penalty constant.

The slope stability analysis is conducted by substituting this rigid-plastic constitutive equation (11) for the virtual work formula. After conducting some expansions, the following equation is obtained.

$$\int_V \left[\frac{\hat{\psi} - 3\hat{\omega}\kappa\left(\dot{\varepsilon}_v^p - \hat{\beta}\dot{e}\right)}{\sqrt{3\hat{\omega}^2 + \frac{1}{2}}} \frac{\dot{\varepsilon}^p}{\dot{e}}\right] : \delta\dot{\varepsilon}^p dV$$

$$+ \int_V \kappa\left(\dot{\varepsilon}_v^p - \hat{\beta}\dot{e}\right) I : \delta\dot{\varepsilon}^p dV$$

$$= \int_V x \bullet \delta\dot{u} dV + \int_{S_\sigma} t \bullet \delta\dot{u} dS \text{ for } \forall\delta\dot{u} \qquad (12)$$

where, x = the body force, t = the surface force, $\dot{u}$ = the displacement rate.

Since a magnitude of displacement rate is indeterminate in the rigid-plastic constitutive equation, the safety factor can be obtained by considering the following constraints;

$$\int_V x \bullet \dot{u} dV + \int_{S_\sigma} t \bullet \dot{u} dS = 1 \qquad (13)$$

In the rigid-plastic constitutive model, the strength parameters and dilatancy characteristics of the ground change depending on the safety factor as shown in equations (9) and (10). So, equation (11) is a non-linear equation on the safety factor. Therefore, equation (12) uses an iterative solution method that calculates and updates the safety factor and the displacement rate by solving the equation assuming the safety factor $\hat{F}_s$ and the initial displacement rate $\hat{\dot{u}}$. When the constraint condition (13) is incorporated into the equation (12) by using the penalty method, the following equation is obtained.

$$\int_V \hat{F}_s\left[\frac{\hat{\psi} - 3\hat{\omega}\kappa\left(\dot{\varepsilon}_v^p - \hat{\beta}\dot{e}\right)}{\sqrt{3\hat{\omega}^2 + \frac{1}{2}}} \frac{\dot{\varepsilon}^p}{\dot{e}}\right] : \delta\dot{\varepsilon}^p dV$$

$$+ \int_V \kappa\left(\dot{\varepsilon}_v^p - \hat{\beta}\dot{e}\right) I : \delta\dot{\varepsilon}^p dV$$

$$= \mu\left(\int_V x \bullet \dot{u} dV + \int_{S_\sigma} t \bullet \dot{u} dS - 1\right)$$

$$\left(\int_V x \bullet \delta\dot{u} dV + \int_{S_\sigma} t \bullet \delta\dot{u} dS\right) \text{ for } \forall\delta\dot{u} \qquad (14)$$

where, μ = the penalty constant.

If the displacement rate $\dot{u}$ is obtained by solving equation (14), the safety factor can be obtained by the following equation.

$$F_s = \mu\left(\int_V x \cdot \dot{u} dV + \int_{S_\sigma} t \cdot \dot{u} dS - 1\right) \qquad (15)$$

2.3 Outline of the numerical analysis

The centrifugal model experiment was reproduced using the analysis model shown in Figure 2. The boundary conditions were set large enough to simulate an infinite soil mass. The density of the mesh elements was refined near the wall to capture the higher expected strain. The sides of the domain were pinned, and the bottom boundary was fixed.

The analysis parameters are shown in Table 1. Figure 2 shows the FE mesh for Cases 1, 6 and 7. d_e = the embedment length (m), H = the height of the wall, D = the pile diameter (m), ϕ = the shear resistance angle (deg.), ψ = dilation angle (deg.), c = cohesion (kPa), γ_t = the unit volume weight (kN/m³). The embedment length of each case set nearly $1/\beta$ considering the stiffness of the soil and the wall in the experiments so as to become much shorter than the design value. The parameters of Toyoura sand and the sand rock were determined based on the uni-axial tests etc. (Kunasegram et al., 2018). The cohesion of the wall was set much larger as rigid material than the soil materials. In the experiment, a membrane was wound around the retaining wall to neglect the skin friction, so in the analysis also the skin friction between the wall and ground is reduced by providing the pseudo joint element with a width of 0.01 m and a very small shear strength.

2.4 Validation of numerical analysis

Comparing the numerical analysis results (Figure 3) considering the hydrostatic pressure in the Toyoura sand of the back ground and the experimental results (Figure 4), the water level at which the horizontal displacement of the retaining wall head rapidly increases and the water level at which the reciprocal

Table 1. The analysis cases and parameters.

		ϕ	ψ	c	γ_t
		deg.	deg.	kPa	kN/m³
Case 1	Sand rock	0	0	700	20.1
(d_e=2.5m)	Dry sand	40	40	0	15.8
(H=12m)	Wall	0	0	50000	26.5
Case 6	Mud rock	0	0	500	18.1
(d_e=2.5m)	Dry sand	40	40	0	15.8
(H=12m)	Wall	0	0	50000	26.5
Case 7	Sand rock	0	0	650	20.1
(d_e=2.5m)	Dry sand	40	40	0	15.8
(H=12m)	Wall	0	0	50000	26.5
Case 2	Sand rock	0	0	550	20.1
(d_e=1.8m)	Dry sand	40	40	0	15.8
(H=9.0m)	Wall	0	0	50000	26.5
(D=0.5m)					
Case 3	Dry sand	40	40	0	15.8
(d_e=9.8m)	Dry sand	40	40	0	15.8
(H=12m)	Wall	0	0	50000	26.5
Case 4	Sand rock	0	0	650	20.1
(d_e=3.0m)	Dry sand	40	40	0	15.8
(H=12m)	Wall	0	0	50000	26.5
Case 5	Mud rock	0	0	500	18.1
(d_e=3.0m)	Dry sand	40	40	0	15.8
(H=12m)	Wall	0	0	50000	26.5

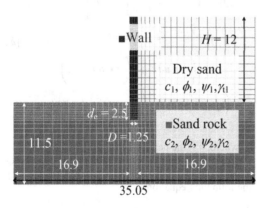

Figure 2. FE mesh for Cases 1, 6 and 7 (unit: m).

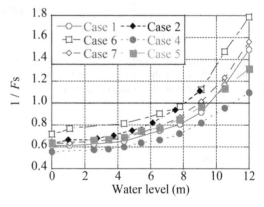

Figure 3. Relationship between the inverse of the safety factor and W.L. obtained in the simulation (See Mochizuki et al., 2019).

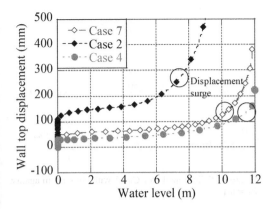

Figure 4. Wall top displacement as increasing the water level (See Kunasegram et al., 2018 and Mochizuki et al., 2019).

of the safety factor F_s is less than 1 are almost the same. From this, it was confirmed that this numerical analysis method can appropriately evaluate the results of the centrifugal model experiment.

3 INFLUENCE OF THE SEISMIC INTENSITY

3.1 The seismic intensity method

The effect of seismic inertial force based on the seismic intensity method is evaluated based on the analysis model of Case 1 with the embedment length of 1.5 m. The analysis was carried out considering the body force obtained by applying the horizontal seismic intensity coefficient (K_h) to its own weight as the seismic inertial force. K_h was varying from 0.1 to 0.5, and the inclination angle and unit volume weight due to the seismic inertial force were corrected as follows. The corrected inclination angle and unit weight are shown in Table 2.

$$\theta = \tan^{-1}(K_h) \qquad (16)$$

$$\gamma = \gamma_0 \times \sqrt{1 + \tan \theta^2} \qquad (17)$$

3.2 Analysis result

Figure 5 shows the relationship between the horizontal seismic intensity coefficient (K_h) and the safety factor obtained by RPFEM (rigid plastic FEM) for the case with the embedment length of 1.5 m and water level of 0 m (dry sand condition). It was confirmed that the safety factor decreased as the seismic inertial force increased according to this figure.

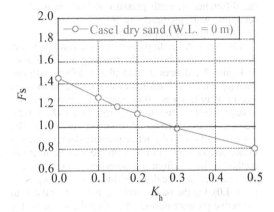

Figure 5. Change of Safety factor due to the increase of K_h.

Table 2. Simulation parameters.

$K_h = 0.1$ ($\theta = 5.71$)				$K_h = 0.15$ ($\theta = 8.53$)		
	γ_0	γ			γ_0	γ
	kN/m³	kN/m³			kN/m³	kN/m³
Sand rock	20.1	20.2		Sand rock	20.1	20.3
Dry Sand	15.8	15.9		Dry Sand	15.8	16.0
Pile	26.5	26.6		Pile	26.5	26.8
$K_h = 0.2$ ($\theta = 11.3$)				$K_h = 0.3$ ($\theta = 16.7$)		
Sand rock	20.1	20.5		Sand rock	20.1	21.0
Dry Sand	15.8	16.1		Dry Sand	15.8	16.5
Pile	26.5	27.0		Pile	26.5	27.6
$K_h = 0.4$ ($\theta = 21.8$)				$K_h = 0.5$ ($\theta = 26.6$)		
Sand rock	20.1	21.6		Sand rock	20.1	22.5
Dry Sand	15.8	17.0		Dry Sand	15.8	17.7
Pile	26.5	28.5		Pile	26.5	29.6

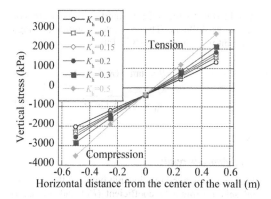

Figure 6. Vertical earth pressure distribution on the bottom of the retaining wall.

From the distribution of vertical earth pressure on the bottom of the retaining wall shown in Figure 6, the difference in earth pressure at both ends of the retaining wall increases as K_h increases, and the over-turning mode is predominant.

The horizontal earth pressure (active earth pressure) distribution acting behind the retaining wall as shown in Figure 7 indicates a triangular distribution (linear change) similar to Rankin's earth pressure when K_h is small (safety factor is 1.0 or more). However, when K_h is large and the safety factor is less than 1 ($K_h = 0.5$), a non-linear earth pressure distribution was shown. The point of load due to the active pressure during an earthquake obtained by RPFEM with the seismic intensity method is 4.07 m from the boundary surface between the sand rock and the dry sand for the case of $K_h = 0.3$ ($F_s = 1.0$). On the other hand, the point of load due to the active pressure during an earthquake based on the seismic intensity method by Mononobe-Okabe is generally assumed to be $(1/3) \times H$ (4.0 m) from the lower end of the retaining wall following Rankin's theory of

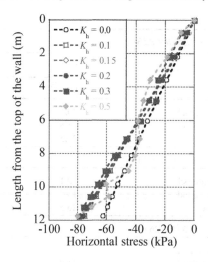

Figure 7. Horizontal earth pressure distribution acting on the behind the retaining wall.

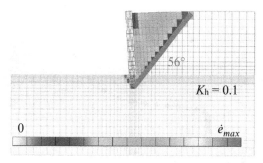

Figure 8. Failure mode for Case1 with $de = 1.5$ m against $K_h = 0.1$.

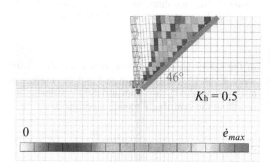

Figure 9. Failure mode for Case1 with $de = 1.5$ m against $K_h = 0.5$.

earth pressure. Thus, it indicates that such a simple assumption for the point of load of the horizontal earth pressure due to seismic inertial force may underestimate the overturning moment.

Figures 8 and 9 show the failure mode of the case for $K_h = 0.1$ and for $K_h = 0.5$, respectively. As K_h increased, the angle of the slip surface in the failure mode became smaller and agreed with the theoretical solution considering the reduction of the strength by F_s, but a failure mode different from that of pure wedge was also confirmed for $K_h = 0.5$. The exact reason is unknown at this point, but it is thought that the behavior of the backfill soil near the retaining wall like a rigid body was linked to the change in active earth pressure distribution acting on the retaining wall.

4   INFLUENCE OF THE TWO-LAYERED GROUND

4.1   Analysis model

In order to establish a more rational method for determining the embedment length, the stability of the cantilever type steel tubular pile retaining wall in a two-layered ground, which consists of the sandy soil and soft rock, was examined. A two-layered ground of hard ground and Toyoura sand for the embedment part was modeled. The length of the dry sand layer at the embedded part is defined as d_{es} as shown in Figure 10. The analysis was performed for the cases with the embedment length d_e from 1.5 to 7.5 m and d_{es} from 0 to $0.8 \times d_e$.

4.2 Analysis result

Figure 11 shows that the safety factor decreases with the increase in sand layer thickness (decrease in the thickness of the hard ground in the embedment length, d_e- d_{es}) regardless of the embedment length d_e. The longer the embedment length, the greater the rate of decrease in the safety factor due to the two-layered ground.

Figures 12 through 15 show the vertical earth pressure distribution on the bottom of the retaining wall and the horizontal earth pressure at the front of and behind the wall, respectively. As the surface thickness of the sand layer increases, drastic changes in the vertical stress distribution are observed. In particular, for the case with the longer embedded length, as the surface thickness of the sand layer decreases, tensile and compressive stress became much smaller and non-linearly distributed. Thus, as

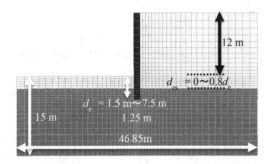

Figure 10. Wall embedded in two-layered ground FE mesh.

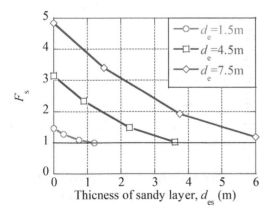

Figure 11. Safety factor when shortening the embedment length.

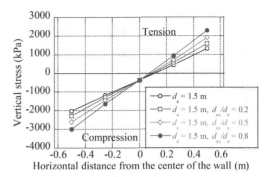

Figure 12. Vertical earth pressure distribution on the bottom of the retaining wall of two-layered ground for d_e = 1.5 m.

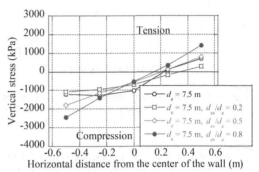

Figure 13. Vertical earth pressure distribution on the bottom of the retaining wall of two-layered ground for the case with d_e = 7.5 m.

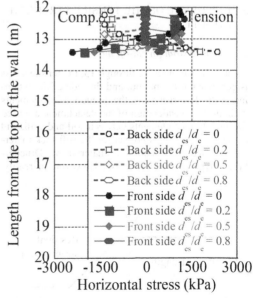

Figure 14. Horizontal earth pressure distribution on the back and front of the retaining wall embedded in two-layered ground for the case with d_e = 1.5 m.

303

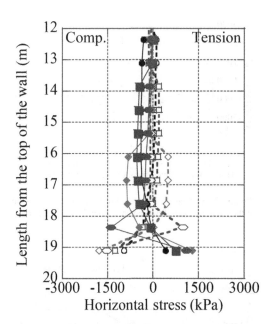

Figure 15. Horizontal earth pressure distribution on the back and front of the retaining wall embedded in layered ground for the case with d_e = 7.5 m.

the embedment length increases, the change of the failure mode from over-turning to sliding was extrapolated. On the other hand, non-negligible tensile stress in the front and behind of the wall embedded in the soft rock was observed.

5 INFLUENCE OF THE TENSILE STRENGTH CUT-OFF

5.1 *Tensile strength cut-off model*

As shown in Figures 12 through 15, the tensile stress is generated at the bottom and the side of the retaining wall. The reason is the pseudo joint elements without the consideration of the detachment are used for the interface elements between soil and wall, and the tensile strength is not set for the soft rock material modeled as c cohesive material by Drucker-Prager type of yield function. However, in reality,

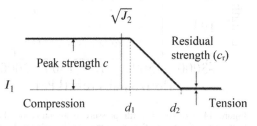

Figure 16. Schematic diagram of tensile strength cut-off model.

Table 3. The analysis parameters for the tensile strength cut-off model (unit: kPa).

Soft rock	c	c_r	d_1	d_2
Value	700	1.0	0	699

tensile resistance is not considered by the design methods, it leads to overestimating the stability.

In order to avoid such a problem and evaluate the effect of the tensile failure of the soft rock, a tensile strength cut-off model (TSC) was used as shown in Figure 16. The tensile strength was set as the same as the shear strength (c = 700 kPa) while considering

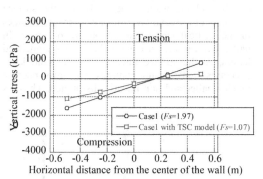

Figure 17. Comparison of the safety factor and the vertical earth pressure distribution on the bottom of the retaining wall with and without tensile strength cut-off model.

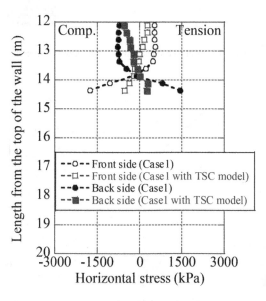

Figure 18. Comparison of the horizontal earth pressure distribution in the front and behind the retaining wall with tensile strength cut-off model.

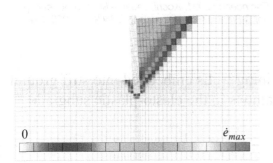

Figure 19. Failure mode of the wall without the tensile strength cut-off model.

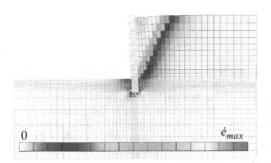

Figure 20. Failure mode of the wall with the tensilestrength cut-off model.

the residual strength 1.0 kPa to avoid the instability of numerical analysis as shown in Table 3.

5.2 Analysis result

Figures 17 and 18 show the safety factor of the analysis results, vertical and horizontal earth pressure distribution around the retaining wall, respectively. It is found that the safety factor was considerably reduced by using the tensile strength cut-off model and the tensile stress around the retaining wall decreased.

Figures 19 and 20 show the comparison of the failure mode with and without the tensile strength cut-off model. Although the different slip surface angle depending on the factor of safety was observed, the predominant failure mode was overturning for both cases.

6 CONCLUSION

(1) The effect of seismic inertial force was evaluated based on the seismic intensity method. The safety factor decreases as the seismic inertial force increased. The increase in the horizontal seismic intensity affects the failure mode and earth pressure distribution during the earthquake, inferring that the assumptions for the point of load by the horizontal earth pressure may be underestimated.

(2) The influence of the two-layer ground on the stability of the cantilever type steel tubular pile retaining wall was evaluated. The longer the embedment length is, the greater the rate of decrease in the safety factor becomes due to the two-layered ground. For the case with the longer embedded length, as the surface thickness of the sand layer decreases, tensile and compressive stress became much smaller and non-linearly distributed. Thus, as the embedment length increased, the change of the failure mode from overs-turning to sliding was extrapolated.

(3) The effect of setting the tensile strength was evaluated by using the tensile strength cut-off model. Although the model drastically decreased the safety factor and the resistance against the lifting of the wall, the predominant failure mode was over-turning as ever.

REFERENCES

Hoshina, T., Takimoto, H., Tanaka, T., Isobe, K. and Ohtsuka, S. 2010. Estimation of Reinforcement Effect on Slope Stability by Nailing Method based on Rigid Plastic Finite Element Analysis, *Journal of applied mechanics*, 13: 379–389. (in Japanese)

Ishihama, Y., Fujiwara, K., Takemura, J. and Kunasegram, V. 2018. Evaluation of Deformation Behavior of Self-Standing Retaining Wall Using Large Diameter Steel Pipe Piles into Hard Ground, *Proc. of the First International Conference on Press-in Engineering*: 153–158.

Kunasegram, V., Takemura, J., Ishihama, Y., and Ishihara, Y. 2018. Stability of Self-standing High Stiffness-Steel Pipe Sheet Pile Walls Embedded in Soft Rocks, *Proc. of the First International Conference on Press-in Engineering*: 143–152.

Mochizuki, K., Isobe, K., Takemura, J. and Ishihama, Y. 2019. Numerical simulation for centrifuge model tests on the stability of self-standing steel pipe pile retaining wall by Rigid Plastic FEM, *Geotechnics for Sustainable Infrastructure Development (Proceedings of the 4th Geotech Hanoi), Hanoi*: 481–488.

Mononobe, N. and Matsuo, H. 1929. On determination of earth pressure during earthquake, *Proc. of World Engineering Congress*, Vol.9, pp.177–185.

Okabe, S. 1926. General theory of earth pressure, *Proc. of Japan Society of Civil Engineers*, 12(1): 123–134.

Tamura, T., Kobayashi, S. and Sumi, T. 1984. Limit analysis of soil structure by rigid-plastic finite element method, *Soil and Foundations*, 24(1): 34–42.

Yagi, K., Nakamura, M., Isobe, K. and Ohtsuka, S. 2010. Prediction Method for Seepage Failure of River Bank based on Rigid Plastic Finite Element Analysis, *Journal of applied mechanics*, 13: 391–400. (in Japanese)

Proceedings of the Second International Conference on
Press-in Engineering 2021, Kochi, Japan – Matsumoto et al (eds)
© 2021 Taylor & Francis Group, London, ISBN 978-1-032-10414-0

Reliability analysis on cantilever retaining walls embedded into stiff ground (Part 1: Contribution of major uncertainties in the elasto-plastic subgrade reaction method)

N. Suzuki
GIKEN LTD., Tokyo, Japan

K. Nagai
The University of Tokyo, Tokyo, Japan

T. Sanagawa
Railway Technical Research Institute, Tokyo, Japan

ABSTRACT: Reliability analysis was performed for cantilever retaining walls embedded in two-layer ground with deep stiff ground. The analysis was conducted by treating soil/rock properties, uniform surcharge, depth of the rock layer surface and yield strength of steel as random variables based on various previous works. We calculated sensitivity factors of each variable for the limit states in the persistent design situation. The contributions of scatters of the depth of the rock layer surface are high in the order of rotational failure, deformation failure, and flexural failure. The results suggest that the scatter of the depth of the rock layer surface should be considered, especially if varied horizontal layer is expected and the deformation failure or the rotational failure are determinants of the design.

1 INTRODUCTION

1.1 Overview

More and more embedded cantilever retaining walls have been applied when retrofitting structures and widening roads in Japan to minimize the impact on daily traffic (Miyanohara et al. 2018 and Suzuki & Kimura 2021). Since the walls became higher and more rigid, the required embedment depth became longer in adapting the semi-infinite embedment condition. That resulted in a longer construction period and a more expensive construction, especially in stiff ground.

Thus, TC1 of International Press-in Association (IPA) has tried to rationalize the design of cantilever steel pipe pile retaining walls in the stiff ground. The authors believe that two issues are important for applying short embedment; the varied horizontal layer (Figure 1) and the strict limits on the displacement of the wall top.

Fine boundaries of strata require high quality investigation and sampling for hard/medium rock (EN 1997-2. 2007). However, detailed geotechnical investigations are rarely conducted for retaining walls with long horizontal extensions, while a varied horizontal layer has been taking on renewed importance by pile foundations (e.g. Zhang & Dasaka 2010 and JGCA 2017).

The failure modes of cantilever retaining walls are flexural and rotational failure in the overall stability as well as piping and heaving. However, since cantilever steel pipe pile retaining road walls are usually constructed close to existing buildings (Suzuki & Kimura 2021), the displacement of the wall top has to be strict, and the wall deformation can be the determining factor of the wall specifications.

1.2 Previous studies

Many studies on the cantilever retaining walls have been conducted. For example, the observational data have been collected (Moormann 2004 and Michael 2001) and an easy method for reliability analysis has been introduced (Bak 2017).

Honjo & Otake (2014) compared the design values based on JRA (1999) with the field data for temporary earth retaining walls (including prop and anchor types walls) in Japan. They reported the statistics of model errors (including transformation

DOI: 10.1201/9781003215226-28

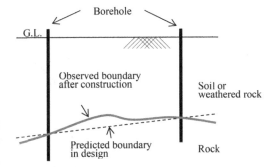

G.L.

Borehole

Observed boundary
after construction

Soil or
weathered rock

Predicted boundary
in design

Rock

Figure 1. Illustrative example of varied horizontal layer.

errors) on the displacement (mean=0.70 and Coefficient of Variation [COV]=0.42) and on the bending moment (mean =0.69 and COV=0.62). The model errors in the design method were also reported by Zhang D. M. et al. 2015. They compared FEM and the Mobilized Strength Design (MSD) proposed by Osman & Bolton (2004), as well as 45 field data, and reported the model error of the displacement (FEM: mean=1.01 and COV=0.21).

Sivakumar & Basha (2008) performed sensitivity analysis with the cantilever steel sheet pile retaining walls against rotation about the base point and flexural failure using the Inverse First-Order Reliability Method (IFORM). They reported that the rotational failure is strongly influenced by the angle of shearing resistance and the yield strength for the flexural failure. In Japan, Shiozaki et al. (2010) conducted a reliability analysis of steel pipe pile walls for flexural failures and drew similar conclusions.

However, few studies consider the deformation failure, and no previous studies consider two-layer ground with the deep stiff ground. It is necessary to grasp the effect of embedment depth on the wall deformation to achieve a rational design by short embedment.

1.3 Objective

This study aims to grasp the effect of the depth of stiff ground on the failure of the cantilever retaining walls embedded in the two-layer ground. The results will provide useful information on which parameters are dominant in the reliability and when the reliability can be improved (i.e. in geotechnical investigation, design, or construction).

Section 2 introduces the methods of reliability analysis and random variables. Section 3 reports the contribution of random variables on each failure: deformation, flexural, and rotational failure.

2 METHODS

2.1 Overview

A calculation model was based on the one which has been commonly used in Japan. Analysis cases were decided with reference to the Japanese case histories of the rotary cutting press-in wall (Suzuki & Kimura 2021). The uncertainties of each random variable were obtained from various previous works, and then the sensitivity factors and the contribution factors of the variables were calculated.

2.2 Models and analysis methods

A cross-sectional view of the calculation model is shown in Figure 2a. The model assumed a cantilever retaining wall with 5 m excavation and two layers of soft soil and sandstone. The wall consisted of continuously pressed-in steel pipe piles with a distance of 180 mm between the pile surfaces. Because the drainage treatment was carried out between the steel piles, water pressure balanced between the excavation side and the backside. The average of the uniform surcharge was 10 kN/m² on the backside.

The deformation of the wall was calculated by the elasto-plastic subgrade reaction method, and the rotational stability and the critical embedment depth were confirmed by the limit equilibrium (Figure 2b). The subgrade reaction method is one of the simplest soil-structure interaction analyses, which models the wall as a beam and the ground as a series of horizontal springs (e.g. Gaba et al. 2017). Rotational resistance due to ground reaction at the pile tip was not taken into account since it was not observed in the FEM analysis for open-ended piles (Ishihama et al. 2019).

The allowable lateral displacement at the wall top was set as 50 mm, which is often used in the persistent design situation for road retaining walls in Tokyo, Japan. The lateral displacement was the alternative to the settlement of backside since the analysis of beam-columns on the elastic foundation could not estimate the settlement directly.

2.3 Analysis case

We analyzed 27 cases, including three cases of the depth of the rock layer surface, N-value of the rock layer, and the stiffness factor (Table 1).

The mean depths of the rock layer surface were 1.0, 3.0, and 5.0 m. About 3.0 m was the dominant depth for pile deformation $(1/\beta_0)$.

The surface soil had an N-value of 10, and the sandstone had three converted N-values of 50, 300, and 1500 (Table 2). The stiffness, E_0, and strength parameters, c and ϕ, were estimated from SPT N-value by the equations in Appendix. This was

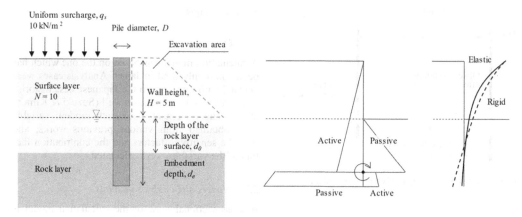

(a) Cross-sectional view of the calculation model   (b) Idealized ultimate stress distribution   (c) Deformation

Figure 2. Illustrative analysis model of cantilever retaining wall.

Table 1. Analysis case.

	Symbol	Unit	Values
Depth of the rock layer surface	d_0	m	1.0, 3.0, 5.0
N-value of the rock layer	N_r	-	50, 300, 1500
Factor of the stiffness of the wall	$\Sigma\beta_i d_i$	-	1.5, 2.0, 3.0

Table 2. Typical parameter of ground.

	Soil	Sandstone (Rock)		
N	10	50	300	1500
E_0 kN/m^2	2.8E+04	1.6E+05	5.5E+05	1.7E+06
ϕ degree	32	38	42	46
c kN/m^2	-	55	98	166
γ kN/m^3	18.0	18.2	21.2	23.9

because it is sometimes difficult to sample rock masses that are heavily weathered or cracked, or due to existing buildings. And since the transformation equations in the range of N-values over 300 does not exist, this paper expanded the N value range of the equations shown in NEXCO RI (2016) to 1500. The typical values of the ground are shown in Table 2.

Stiffness factor, $\Sigma\beta_i d_i$, was 1.5, 2.0, and 3.0, which means rigid short piles, intermediate piles, and elastic long piles respectively (e.g. JRA 1999 used and Michael & John 2007 introduced the inverse of β_i);

$$\beta_i = \sqrt[4]{\frac{k_H B}{4EI}} \qquad (1)$$

where B: Unit wall width, k_H: coefficient of lateral subgrade reaction, and EI: wall stiffness. As the embedment depth decreases, the behavior changes from the elastic to the rigid (Figure 2c), and the rotational failure becomes dominant.

Finally, pile cross-sections and embedment depths were determined by the allowable displacement of the wall top, about 50 mm. Table 3 shows the embedment depth and pile diameter finally used in calculations.

2.4 Random variables

The uncertainty consisted of soil scatter, transformation error, model error, and others (Figure 3). The soil and rock had statistics for each random variable

Table 3. Embedment depth and wall stiffness finally used in each case. The values in the table represent d_e [m] (EI [MNm^2/m]).

| | | $\Sigma\beta_i d_i$, | | |
| | | a) 1.5 | b) 2.0 | c) 3.0 |
d_0	N_r	(short pile)	(intermediate)	(long pile)
1	50	3.48 (208)	4.06 (127)	5.64 (101)
	300	2.69 (169)	3.02 (90)	4.09 (72)
	1500	2.13 (114)	2.42 (71)	3.20 (59)
3	50	5.19 (753)	5.53 (321)	6.83 (165)
	300	4.48 (646)	4.76 (288)	5.82 (166)
	1500	4.08 (596)	4.25 (248)	5.04 (147)
5	50	5.83 (872)	6.06 (331)	7.47 (182)
	300	5.37 (646)	5.66 (294)	6.91 (197)
	1500	5.28 (646)	5.49 (282)	6.40 (186)

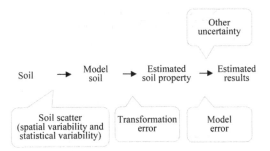

Figure 3. Types of uncertainty in ground property estimates.

with reference to previous works (Table 4). The load and the depth of the rock layer surface were also random variables. The subscripts s and r represent soil and rock, respectively. The transformation error from E_0 to k_H was omitted in this paper, though that COV was assumed 0.25 by Nanazawa et al. (2019). This was because the estimated total uncertainty was almost equal to that of the previous observations as shown later.

Table 4. Statistics of random variables.

	bias	COV	Distribution	Note	References
Soil scatter					
d_0	1.0	-	Normal	SD = 0.5m	Ohki et al. (2005)
N_s	1.0	0.3	Log-normal		Phoon et al. (1995)
γ_s	1.0	0.07	Log-normal	const.	Phoon et al. (1995)
N_r	1.0	0.3	Log-normal		
Transformation error from N-value					
E_{0s}	1.1	0.7	Log-normal	Eq. A2	PARI (2009), Nakatani et al. (2009)
$\tan\phi_s$	1.1	0.1	Log-normal	Eq. A1	Ching et al. (2016)
γ_r	1.0	0.07	Log-normal	Eq. A3	NEXCO RI (2016)
E_{0r}	1.5	1.2	Log-normal	Eq. A6	NEXCO RI (2016)
$\tan\phi_r$	1.2	0.2	Log-normal	Eq. A4	NEXCO RI (2016)
c_r	1.2	0.5	Log-normal	Eq. A5	NEXCO RI (2016)
Other uncertainty (Load and material)					
q_s	1.0	0.2	Normal		Phoon & Kulhawy (1999)
f_y	1.2	0.07	Log-normal		PARI (2009), Shiozaki et al. (2010)

The statistics of the log-normal variables are converted as;

$$\mu_{\ln X} = \ln \mu_X - \frac{1}{2}\sigma_{\ln X}^2 \tag{2}$$

$$\sigma_{\ln X} = \sqrt{\ln\left[1 + \left(\frac{\sigma_X}{\mu_X}\right)^2\right]} \tag{3}$$

where the subscript $\ln X$ was a logarithm of a variable X.

Besides, although EN 1997-1 (2004) stated that the overdig should be considered as 10% of the wall height (limited to a maximum of 0.5 m) for Ultimate Limit State (ULS), it was not considered because the surface level was easy to be controlled in the situation of Figure 2a and Serviceability Limit State (SLS) was the primary target of the paper.

2.5 Reliability analysis

The elasto-plastic subgrade reaction method was difficult to formulate and obtain an exact analytical solution of the performance function at the design point. So, we calculated the difference of the performance function Z for $\pm 0.2\sigma_{Xi}$ and $0.4\sigma_{Xi}$ change in each random variable (Shiozaki et al. 2010) and conducted First-Order Reliability Method (FORM). The sensitivity factors are the normalized partial derivatives of the performance functions when there is no correlation between random variables (Figure 4):

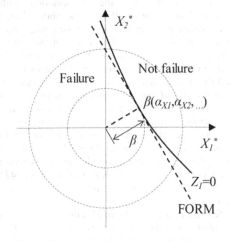

Figure 4. Illustration of FORM.

$$a_{X_i} = \left(\frac{\partial Z}{\partial X_i}\bigg|_{X_f}\right) \cdot \frac{\sigma_{X_i}}{\sigma_Z} = \left(\frac{\partial Z}{\partial X_i^*}\bigg|_{X_f}\right) \cdot \frac{1}{\sigma_Z} \quad (4)$$

$$X_i^* = (X_i - \mu_{X_i})/\sigma_{X_i} \quad (5)$$

where X_f represents the random variable X at the failure, which is defined as the state where the following performance functions Z are less than zero, respectively;

$$Z_{\delta_{top}} = \delta_a - \delta_{top}(X_i) \quad \text{(deformation failure)} \quad (6)$$

$$Z_M = M_y(f_y, S) - M_{max}(X_i) \quad \text{(flexural failure)} \quad (7)$$

$$Z_{d_e} = d_e - d_{eq}(X_i) \quad \text{(rotational failure)} \quad (8)$$

The calculation points were around the deformation failure. Since they were not around the flexural failure, the calculated maximum bending stress was converted to the yield strength of the wall, f_y, for comparison with other random variables.

3 ANALYSIS RESULT AND DISCUSSION

Figures 5, 6, and 7 show the contribution of uncertain sources and COVs to each performance function. The contribution factors were determined to be the squares of the sensitivity factors, and the COVs were the SDs divided by the means.

Firstly, COVs on the deformation failure decreased in the order of (a) rigid short piles, (b) intermediate piles, and (c) elastic long piles. As for rigid short piles (a), the distribution of the contribution factors changed as d_0 did. When the depth of the rock layer surface, d_0, was 1 m, the scatters of d_0 and the transformation error of E_{0r} were dominant, whereas when d_0 was 5 m, each parameter of the surface layer was dominant. An intermediate d_0 of 3 m shows a trend in between. These trends were also true for intermediate piles (b) and elastic long piles (c). Also, COVs were about 35% in the elastic long piles, which were generally consistent with the one reported by Honjo & Otake (2014) above (42%), and the assumption of this study could be reasonable.

Secondly, as for the flexural failure (Figure 6), the distributions of the contribution factors and COVs of (a), (b), and (c) did not differ significantly. Each parameter of the surface layer was dominant because the bending moment was mostly determined by the active earth pressure in the excavation area. The contribution of the scatter of d_0 was large when d_0 was 1 m. When d_0 was 5 m, the sensitivity factors of $\tan\phi_s$ in (c), was about 0.8, which is in good agreement with the report by Shiozaki et al. (2010) if the soil scatters of N_s and N_r are omitted.

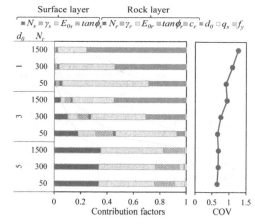

(a) rigid short piles ($\Sigma\beta_i d_i \approx 1.5$)

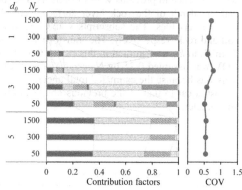

(b) intermediate piles ($\Sigma\beta_i d_i \approx 2.0$)

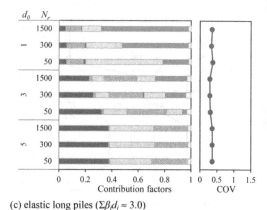

(c) elastic long piles ($\Sigma\beta_i d_i \approx 3.0$)

Figure 5. Contribution factors of uncertain sources and COVs of the performance function on the deformation failure.

Finally, in the case of rotational failure (Figure 7), since the performance function is independent of E_0, the contribution factors of E_0 are naturally zero. As in Figure 4, when d_0 was 1 m, the depth scatters of the rock layer surface and the transformation error of the strength of the rock layer (especially c_r) were

Surface layer   Rock layer

$\blacksquare N_s$ $\boxtimes \gamma_s$ $\square E_{0s}$ $\boxtimes tan\phi_s$ $\blacksquare N_r$ $\boxtimes \gamma_r$ $\square E_{0r}$ $\boxtimes tan\phi_r$ $\blacksquare c_r$ $\blacksquare d_0$ $\square q_s$ $\boxtimes f_y$

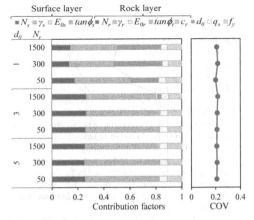

(a) rigid short piles ($\Sigma\beta_i d_i \approx 1.5$)

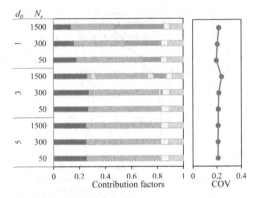

(b) intermediate piles ($\Sigma\beta_i d_i \approx 2.0$)

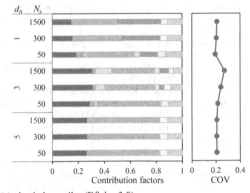

(c) elastic long piles ($\Sigma\beta_i d_i \approx 3.0$)

Figure 6. Contribution factors of uncertain sources and COVs of the performance function on the flexural failure.

dominant, whereas when d_0 was 5 m, each parameter of the surface layer is dominant.

COVs of the deformation failure depended on the stiffness factor (Figure 5), but those of the flexural failure did not (Figure 6). The distributions of the contribution factors depended on the depth and

Surface layer   Rock layer

$\blacksquare N_s$ $\boxtimes \gamma_s$ $\square E_{0s}$ $\boxtimes tan\phi_s$ $\blacksquare N_r$ $\boxtimes \gamma_r$ $\square E_{0r}$ $\boxtimes tan\phi_r$ $\boxtimes c_r$ $\blacksquare d_0$ $\square q_s$ $\boxtimes f_y$

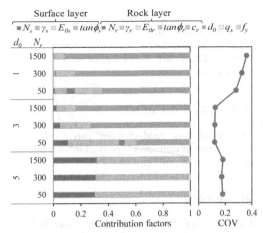

Figure 7. Contribution factors of uncertain sources and COVs of the performance function on the rotational failure.

N-value of the rock, but little on the stiffness factor. They result in that when the depth of the rock layer surface is shallow, the rigid short piles require larger safety margins of all variables especially the depth of the rock layer surface.

4 SUMMARY

In this paper, a reliability analysis was performed for cantilever steel pipe pile retaining walls embedded in the stiff ground, treating the uncertainty of the depth of the rock layer surface as variables. Since the sensitivity factors and COVs with sufficiently deep embedment were in good agreement with previous studies, the results of this study could be reasonable.

The following conclusions were drawn:

– The uncertainty of the depth of the rock layer surface had a significant contribution to each limit state in the order of rotational failure (ULS), deformation failure (SLS), and flexural failure (ULS).
– While COVs of the performance function on the deformation failure of the small stiffness factor were higher than those of the large stiffness factor, those on the flexural failure scarcely depended on the stiffness factor.
– The distributions of the contribution factors depended on the depth and N-value of the rock, but little on the stiffness factor.

When the contribution factor of the depth of the rock layer surface is large, it can be difficult to design cantilever walls rationally because we don't know its scatter in general. In this case, therefore, we believe that construction data that observe the

311

actual geotechnical situation directly improve the reliability of the structure.

Also, the quantitative sensitivity factors and contribution factors are useful for rational design and construction.

ACKNOWLEDGEMENTS

This study was conducted as part of IPA-TC1 (Application of cantilever type steel tubular pile wall embedded to the stiff ground).

REFERENCES

Bak Kong Low. 2017. EXCEL-Based Direct Reliability Analysis and Its Potential Role to Complement Eurocodes, *Joint TC205/TC304 Working Group on "Discussion of statistical/reliability methods for Eurocodes"*. – Final Report (Sep 2017)

Ching, J., Li, D. Q., & Phoon, K. K. 2016. Statistical characterization of multivariate geotechnical data. Chapter 4 in *Reliability of geotechnical structures in ISO2394*. Florida: CRC Press: 89–126.

EN 1997-1. 2004. *Eurocode 7: Geotechnical design. Part 1: General rules*. Belgium: European Committee for Standardization (CEN).

EN 1997-2. 2007. *Eurocode 7: Geotechnical design. Part 2: Ground investigation and testing*. Belgium: European Committee for Standardization (CEN): 24–29.

Gaba, A., Hardy, S., Doughty, L., Powrie, W. & Selemetas, D, 2017. *Guidance on embedded retaining wall design*. C760. London: CIRIA. https://www.ciria. org/ProductExcerpts/c760.aspx

Honjo, Y. & Otake, Y. 2014. Consideration on Major Uncertainty Sources in Geotechnical Design. *Second International Conference on Vulnerability and Risk Analysis and Management (ICVRAM) and the Sixth International Symposium on Uncertainty, Modeling, and Analysis (ISUMA), Liverpool, UK*: 2488–2497.

Ishihama, Y., Takemura, J. & V. Kunasegaram. 2019. Analytical evaluation of deformation behavior of cantilever type retaining wall using large diameter steel tubular piles into stiff ground, *The 4th International conference on geotechnics for sustainable development, GEOTEC Hanoi 2019*.

Japan Geotechnical Consultants Association (JGCA, or ZENCHIREN). 2017. *(Draft) Report of the Study Committee on "Investigating Method for Pile Foundations on Rock Layer"* (In Japanese) https://www.zenchiren.or. jp/geocenter/pdf/201701.pdf

Japan Road Association (JRA). 1999. *Road Earthwork, Guideline for the Construction of Temporary Structures*. Tokyo: JRA. (in Japanese)

Long, M. 2001. Database for retaining wall and ground movements due to deep excavations. *Journal of Geotechnical and Geoenvironmental Engineering*, 127(3): 203–224.

Michael Tomlinson & John Woodward, 2007. *Pile Design and Construction Practice*, Fifth Edition, Florida: CRC Press: p.330

Miyanohara Tomoko, Kurosawa Tatsuaki, Harata Noriyoshi, Kitamura Kazuhiro, Suzuki Naoki & Kajino Koji, 2018. Overview of the Self-standing and High Stiffness Tubular Pile Walls in Japan, *Proceedings*

of the First International Conference on Press-in Engineering 2018, Kochi*. IPA: Tokyo: 167–174.

Moormann, C. 2004. Analysis of wall and ground movements due to deep excavations in soft soil based on a new worldwide database. *Soils and foundations*, 44 (1): 87–98.

Nakatani S., Nanazawa, T., Shirato, M., Nishida, H., Kono, T. & Kimura, S., 2009. Research on Stability Verification Method of Highway Bridge Foundation in Performance Based System, *Public Works Research Institute Report* No.4136 (In Japanese)

Nanazawa, T., Kouno, T., Sakashita, G., & Oshiro, K. 2019. Development of partial factor design method on bearing capacity of pile foundations for Japanese Specifications for Highway Bridges. *Georisk: Assessment and Management of Risk for Engineered Systems and Geohazards*, 13(3): 166–175.

NEXCO RI. 2016. *Design Guide Vol. 2: Bridge Construction*. (In Japanese)

Nishioka Hidetoshi, Anzai Ayako & Koda Masayuki, 2010. Estimation of deformation modulus of ground and coefficient of subgrade reaction depending on ground investigation method. *RTRI report*, 24(7): 11–16 (in Japanese)

Ohki, H., Nagata, M., Saeki, E. & Kuwabara, H., 2005. Fluctuation on bearing strata levels of piles (part 2). *Proceedings of the 40th Technical Report of the Annual Meeting of the Japan Geotechnical Society, Japan*: 1549-1550 (In Japanese)

Osman, A. S., & Bolton, M. D. 2004. A new design method for retaining walls in clay. *Canadian geotechnical journal*, 41(3): 451–466.

Otake Yu, Nanazawa Toshiaki, Honjo Yusuke, Kono Tetsuya & Tanabe Akinori, 2017. Improvement of deformation modulus estimation considering soil investigation types and strain level, *Journal of Japan Society of Civil Engineers*, Ser. C (Geosphere Engineering), Volume 73 Issue 4: 396–411.

Phoon, K., & Kulhawy, F H. 1999. Evaluation of geotechnical property variability, *Canadian Geotechnical Journal*. 36(4): 625–639.

Phoon, K., Kulhawy, F H. & Grigoriu, M D. 1995. *Reliability-based design of foundations for transmission line structures*. Final report, Electric Power Research Inst., Palo Alto, EPRI-TR-105000

The Port and Airport Research Institute (PARI). 2009. *Technical standards and commentaries for port and harbour facilities in japan*. Part III FACILITIES, The Oversies Coastal Area Development Institute Japan, p.464 http://ocdi.or.jp/tec_st/tec_pdf/tech_364_551.pdf

Robert L. Parsons, Matthew C. Pierson, Isaac Willems, Jie Han & James J. Brennan, 2011. Lateral Resistance of Short Rock Sockets in Weak Rock: Case History. *Transportation Research Record Journal of the Transportation Research Board* 2212(-1): 34–41

Shiozaki, Y., Maeno, K., Otsushi, K., Kakimoto, R. & Ohtsuki, M. 2010. Partial Factors for Stress Verification of Self-Supporting Sheet Pile Quay Using Steel Pipe Sheet Pile. *Proceedings of the 65th Japan Society of Civil Engineers Annual Meeting, Japan*, II-094 (In Japanese)

Simpson, B. 2005. Eurocode 7 Workshop–Retaining wall examples 5-7. *In ISSMGE ETC23 workshop, Trinity College, Dublin*.

Sivakumar Babu, G. L. & Basha, B. M. 2008. Optimum design of cantilever retaining walls using target reliability approach. *International journal of geomechanics*, 8 (4): 240–252.

Suzuki N. & Kimura Y. 2021. Summary of case histories of retaining wall installed by rotary cutting press-in method. *Proceedings of the Second International Conference on Press-in Engineering 2021, Kochi* (under review)

Zhang, D. M., Phoon, K. K., Huang, H. W. & Hu, Q. F. 2015. Characterization of model uncertainty for cantilever deflections in undrained clay. *Journal of Geotechnical and Geoenvironmental Engineering*, 141(1).

Zhang, L. M. & Dasaka, S. M. 2010. Uncertainties in geologic profiles versus variability in pile founding depth. *Journal of Geotechnical and Geoenvironmental Engineering*, 136(11): 1475–1488.

APPENDIX 1

The transformation equations for the ground (sandy soil) from an N-value to ϕ and E_c are;

$$\phi = \sqrt{15N} + 20 \qquad (A1)$$

$$E_c = 700N \qquad (A2)$$

The transformation equations from a converted N-value of rock (sandstone) to γ, ϕ, c and E_c are as follows (NEXCO RI 2016), respectively;

$$\gamma = 1.7 \ln(N) + 11.5 \qquad (A3)$$

$$\phi = 5.10 \ln(N)/2.3 + 29.3 \qquad (A4)$$

$$\ln(c) = 0.327 \ln(N) + 2.72 \qquad (A5)$$

$$\ln(E_c) = 0.69 \ln(N) + 7.9 \qquad (A6)$$

The coefficient of the lateral subgrade reaction, k_H, is calculated from the deformation coefficient of the ground E_c, taking into account the size effect of the loading width B_H (= 10 m).

$$k_H = \left(\frac{4E_c}{0.3}\right)\left(\frac{B_H}{0.3}\right)^{-3/4} \qquad (A7)$$

Nishioka et al. (2010) and Otake et al. (2017) are also helpful, which showed the relationship between the deformation coefficients for each geotechnical investigation method. It is also noted that the applicability of the subgrade reaction method for rock is not clear. The socket pile is a well-known structure embedded in bedrock, and the researches on them are informative though there are a few studies on the horizontal deformation (e.g. Robert et al. 2011).

Active and passive earth pressures, p_A and p_{Pu}, are calculated by the angle of shearing resistance, cohesion, and wall friction angle δ_f which is a constant of 15 degrees. Effective unit weight below the groundwater level is uniformly the unit weight minus 9 kN/m³.

$$p_A = \left[K_A\left(\sum \gamma z + q\right)\right] \cos \delta_f$$

$$K_A = \frac{\cos^2 \phi}{\cos \delta_f \left(1 + \sqrt{\frac{\sin(\phi + \delta_f)\sin \phi}{\cos \delta_f}}\right)} \qquad (A8)$$

$$p_{Pu} = \left[K_P\left(\sum \gamma z\right) + 2c\sqrt{K_P}\right] \cos \delta_f$$

$$K_P = \frac{\cos^2 \phi}{\cos \delta_f \left(1 - \sqrt{\frac{\sin(\phi - \delta_f)\sin \phi}{\cos \delta_f}}\right)} \qquad (A9)$$

The differential equation of the lateral displacement, δ, is expressed using the depth z (upward direction is positive, and the depth at the excavation surface is zero);

$$EI\frac{d^4\delta}{dz^4} = -pB$$

$$p = \begin{cases} -p_A & (z > 0) \\ \min(k_H \delta, p_{Pu}) & (z \le 0) \end{cases} \qquad (A10)$$

Where both ends of the wall are free, the equation can be solved under the four boundary conditions;

$$\left[EI\frac{d^3\delta}{dz^3}\right]_{z=H} = 0, \left[EI\frac{d^3\delta}{dz^3}\right]_{z=-d_e} = 0$$

$$\left[EI\frac{d^2\delta}{dz^2}\right]_{z=H} = 0, \left[EI\frac{d^2\delta}{dz^2}\right]_{z=-d_e} = 0 \qquad (A11)$$

Young's modulus of elasticity of steel, E, is 2.0E+05 N/mm². The corrosion of the steel was not considered in this paper. The second moment, I, and the section modulus, S, of the steel pipe pile wall are;

$$I = \pi(D^4 - D_{in}^4)\pi/64 \qquad (A12)$$

$$S = \pi(D^4 - D_{in}^4)/32D \qquad (A13)$$

APPENDIX 2

Symbols and abbreviations are as follows.

B	m:	Unit wall width
B_H	m:	Equivalent Loading Width
D	m:	Outer pile diameter
D_{in}	m:	Inner pile diameter
d	m:	Installed pile depth
d_e	m:	Embedment depth
d_{eq}	m:	Critical embedment depth for extreme equilibrium
d_0	m:	Depth of the rock layer surface, or thickness of the surface layer
E	N/mm^2:	Young's modulus of elasticity of steel
E_c	N/mm^2:	Deformation coefficient of the ground (horizontal loading test in the borehole)
E_0	N/mm^2:	Deformation coefficient of the ground (flat-plate loading test)
f_y	N/mm^2:	Yield strength of steel
H	m:	Height of structures, that is excavation depth
I	m^4:	The second moment of area of the wall section
k_H	kN/m^3:	Coefficient of lateral subgrade reaction
K_A	-:	Active earth pressure coefficient
K_P	-:	Passive earth pressure coefficient
M_y	kNm/m:	Yield bending moment
M_{max}	kNm/m:	Maximum bending moment
N	-:	SPT N-value (blows per 300 mm penetration)
p_A	kN/m^3:	Active earth pressure

(Cont.)

p_{Pu}	kN/m^3:	Ultimate passive earth pressure
q_s	kN/m^3:	Any uniform surcharge at the ground surface
S	m^3:	Section modulus
t	mm:	Steel pile thickness
X_i	-:	Random variable
X^*	-:	Standardized value by the standard deviation
z	m:	Depth
Z	-:	Performance function
$Z_{\delta top}$	-:	Performance function on the deformation failure
Z_M	-:	Performance function on the flexural failure
Z_{de}	-:	Performance function on the rotational failure
α_i	-:	Sensitivity factor for i-th variable
β_i	m^{-1}:	Stiffness factor of i-th ground layer
γ	kN/m^3:	Effective unit weight of the ground
δ	m:	Displacement of wall
δ_a	m:	Allowable displacement
δ_f	degree:	Angle of friction between soil and wall
δ_{top}	m:	Displacement of wall top
μ	-:	Arithmetic mean
σ	-:	Bending stress
σ	-:	Standard deviation
ϕ	degree:	Angle of shearing resistance
COV	:	Coefficient of Variation
FORM	:	First Order Reliability Method
SD	:	Standard Deviation
SLS	:	Serviceability Limit State
ULS	:	Ultimate Limit State

(*Continued*)

314

Proceedings of the Second International Conference on
Press-in Engineering 2021, Kochi, Japan – Matsumoto et al (eds)
© 2021 Taylor & Francis Group, London, ISBN 978-1-032-10414-0

Reliability analysis on cantilever retaining walls embedded into stiff ground (Part 2: Construction management with piling data)

N. Suzuki
GIKEN LTD., Tokyo, Japan

Y. Ishihara
GIKEN LTD., Tokyo, Japan

K. Nagai
The University of Tokyo, Tokyo, Japan

ABSTRACT: Part 1 showed that the scatters of the depth influenced the rotational and deformation failure of the cantilever retaining wall, especially when it is embedded short into stiff ground. This part considers on how to deal with the uncertainty using the piling data. Reliability and cost analyses draw following conclusions. The effect of the piling data on the expected total cost are about 8% for Serviceability Limit State and 27% for Ultimate Limit State at most. The construction management with the piling data has advantages especially when the uncertainty of the depth of the rock layer surface are large and additional geotechnical investigations are conducted, and when embedment depth is short. Furthermore, with the piling data, the expected total cost becomes less susceptible to the scatters of the depth of the rock layer surface, which make the proposed method effective in practice.

1 INSTRUCTIONS

1.1 *Part 1*

The reliability analysis of Part 1 (Suzuki et al. 2021) found that the contribution of the scatters of the depth of the rock layer surface to the rotational and deformation failure of the cantilever walls was large.

In the case of pile foundations, it has been considered important to confirm the rock layer surface during construction, and the quality assessments during piling becomes necessary. We believe that the piling data also improve the reliability of the retaining walls.

1.2 *Proposed method with piling data*

The rock layer with an SPT (Standard Penetration Test) N-value of 50 or more can be confirmed by seismic exploration and boring. However, the accuracy of the seismic exploration is estimated to be around 2 m (e.g. JGCA 2017) and it is difficult to carry out boreholes throughout the construction area. Borehole investigations are generally conducted every 30-300 m for road lines (Tony 2009). Therefore, it is difficult to determine accurately the depth of the rock layer surface all along the structures that have long horizontal sections such as a retaining wall, from the above-mentioned ground investigation alone.

Also, Ishihara et al. (2015) proposed a method of estimating geotechnical information from the piling data of the rotary press-in piling. This technique can estimate the boundaries of the rock layer with all piles.

This paper proposes the construction management system with the piling data (Figure 1). The depth of the rock layer surface is estimated from the piling data and compared with its design value determined from the preliminary ground investigations. If the estimated depth is deeper than the design value, the countermeasure is taken to extend the pile length. Otherwise, the pile length is maintained as designed. Compared with the measure that simply extending all piles, the proposed method is expected not to take unnecessary countermeasures.

In the observational method for the retaining construction, the prediction of the wall deformations and force are updated step by step, and the number of props can be controlled accordingly (e.g. Young & Ho 1994). In contrast, the proposed method is expected to give another approach as an observational method for the cantilever retaining wall.

DOI: 10.1201/9781003215226-29

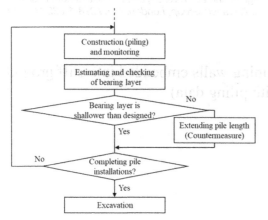

Figure 1. Procedure of construction management with piling data.

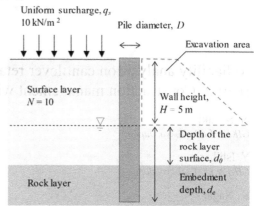

Figure 3. Illustrative analysis model of cantilever retaining wall.

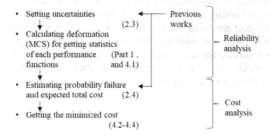

Figure 2. Flow of research steps.

1.3 Objective

This paper aims to study the cost effectiveness of the construction management using piling data for cantilever walls in stiff ground. A reliability analysis is conducted setting geotechnical uncertainties, including the depth of the stiff ground, as the random variable, X_i, to estimate probability failures. Then, a cost analysis is conducted to compare the expected total cost between the cases with and w/o the piling data (Figure 2).

2 METHOD

2.1 Overview

A reliability analysis was performed on a two-layer cantilever retaining wall (Figure 3), as in Part 1. The expected total cost was calculated based on Monte Carlo Simulation (MCS, e.g. Honjo 2008) and determine the optimized embedment depth.

This paper considered the rotational failure (Ultimate Limit State, ULS) and the deformation failure (Serviceability Limit State, SLS). This is because flexural failure seldom happens if the allowable displacement of the wall top, δ_a, is designed to be about 50 mm in the persistent design situation.

$$Z_{d_e} = d_e - d_{eq}(X_i) \quad \text{(rotational failure)} \quad (1)$$

$$Z_{\delta_{top}} = \delta_a - \delta_{top}(X_i) \quad \text{(deformation failure)} \quad (2)$$

where d_e: embedment depth, d_{eq}: critical embedment depth for extreme equilibrium, δ_{top}: displacement of wall top.

2.2 Analysis cases

The cross sections of piles were determined so that the effect of the piling data can be large with reference to Part 1, and they were common in all cases; a pile diameter, D, was 1000 mm and a plate thickness was 10 mm; the N-value of the rock layer, N_r, was 1500; the depth of the rock layer surface, d_0, was 3.0 m; and stiffness factor $\Sigma\beta_i d_i$ was 1.5-5.0 (which corresponds to the embedment depth (d_e) of 4.1-9.7 m). The calculation method and other conditions followed Part 1.

Three cases were carried out: Case A as the standard: Case B with a large standard deviation (SD) of d_0: and Case C with a lower coefficient of variation (COV) of E_0 (Table 1) with reference to Phoon et al. 2016. This was because it is unlikely that d_0 is known in advance, and also because uniaxial compaction tests and in-situ horizontal loading are sometimes conducted in actual projects.

Table 1. Analysis cases.

Case	SD of d_0	COV of E_0	Piling data
A-1	0.5 m	1.2	Not utilized
A-2	0.5 m	1.2	Utilized
B-1	1.0 m	1.2	Not utilized
B-2	1.0 m	1.2	Utilized
C-1	0.5 m	0.4	Not utilized
C-2	0.5 m	0.4	Utilized

The reliability index and the expected total cost were calculated in each case, and the effect of the construction management (i.e. utilization of the piling data) on these values were examined.

The SD of d_0 in Case B, 1.0 m, was judged as realistic based on Ohki et al. (2004). It was for comparison with Case A. In Case C, the in-situ horizontal loading test was assumed to supply the mean value of the deformation coefficient, E_0, which was consistent with the predesigned of Case A.

Since there was no information on the estimation accuracy of the depth layer, it was assumed that the boundary could be estimated by the piling data with no error when the difference of N-value between the boundaries was large.

2.3 Reliability analysis

2.3.1 Method

MCS was performed 5000 times in each case to obtain the reliability index, β. Since it is difficult to obtain the probability of failure, P_f, directly from the MCS when P_f is smaller than 10^{-3} (the general target reliability index for ULS is from 3.1 to 4.3, ISO 2394), the following method was used. First, the mean and SD of each performance function, Z, were calculated by MCS, then the log-normal distribution was assumed. The conformity was confirmed by the Quantile-Quantile (Q-Q) Plot, which will be discussed later in Section 4.1. Finally, the reliability index was obtained by assuming the standard normal distribution.

$$P_f = \text{Prob}\,(Z<0) \tag{3}$$

$$\beta = -\Phi^{-1}\left(P_f\right) \tag{4}$$

where $\Phi(.)$ is the standard normal cumulative distribution function.

Since the reliability index increased monotonically with the embedment depth, it was linearly complemented between MCS results.

2.3.2 Random variables and their correlations

Part 2 assumed the correlation between the random variables for MCS setting 0.3 as poor correlation and 0.7 as high correlation with reference to Kulhawy & Mayne (1990), Ito & Kitahara (1985), Ogawa & Matsumoto (1978) (Table 2). The remaining statistics of the random variables (mean, coefficient of variation, and probability distribution) were the same as in Part 1.

2.4 Cost analysis

2.4.1 Expected total cost

The expected total cost, C_{tot} is sum of the building cost, C_b, and the maintenance cost, C_m, and the expected cost of the failure, C_f, which would be minimized for the optimal design (e.g. ISO 2394):

$$C_{tot} = C_b + C_m + P_f C_f \tag{5}$$

Since the cost for excavation (a part of C_b) and the maintenance cost were independent of the wall specification, these were omitted in this paper, and the simplified expected total cost C_{tot}^{*} was introduced as follows:

$$C_{tot}^{*} = C_p + P_c C_c + P_f C_f \tag{6}$$

where C_p is the piling cost, C_c is the countermeasure cost and P_c is the probability of the countermeasure.

2.4.2 Piling cost

Pile installation cost was estimated based on JPA (2019), which is a cost estimation standard of

Table 2. Correlation coefficients between variables.

X_i	N_s	γ_s	E_{0s}	$\tan\phi_s$	N_r	γ_r	E_{0r}	$\tan\phi_r$	c_r	d_0	q_s
N_s	1.0	0.0	0.0	0.0	0.0	0.0	0.0	0.0	0.0	0.0	0.0
γ_s		1.0	0.3	0.3	0.0	0.0	0.0	0.0	0.0	0.0	0.0
E_{0s}			1.0	0.7	0.0	0.0	0.0	0.0	0.0	0.0	0.0
$\tan\phi_s$	Surface layer			1.0	0.0	0.0	0.0	0.0	0.0	0.0	0.0
N_r					1.0	0.0	0.0	0.0	0.0	0.0	0.0
γ_r						1.0	0.3	0.3	0.0	0.0	0.0
E_{0r}							1.0	0.7	0.0	0.0	0.0
$\tan\phi_r$								1.0	-0.7	0.0	0.0
c_r					Rock layer				1.0	0.0	0.0
d_0		Symmetry								1.0	0.0
q_s											1.0

Note: Subscripts s and r represent surface layer and rock layer, respectively.

Press-in operation in Japan. Its outline in English has been shown in Suzuki & Kimura (2021). Also, in order to focus on the effect of construction data, steel pipe pile welding was not considered.

2.4.3 Failure cost

Failure cost includes function loss and fatalities as well as repair cost of structures (e.g. Kanda & Shaf 1997). It widely varies depending on the structural type, location, evaluation method of human life, etc. In this paper, the failure cost at the SLS was set to be 1.5 million Japanese Yen (JPY) per meter, with the reference to the subsidence repair of small-scale buildings due to liquefaction (such as underpinning or grout injection) (AIJ 2008).

On the other hand, the expected cost of failure at the ULS was assumed to be 30 times the piling cost so that the expected total cost was minimized with β=3.0, although the relationship between failure costs and construction costs varies depending on the structures and surroundings. Also, though failures at the SLS and ULS had a correlation, the probability of failure at these limit states were different by an order of magnitude, so the failure cost was considered separately at these two limit states.

2.4.4 Countermeasure cost

Assuming that the depth of the rock layer surface followed a normal distribution and the designed depth was its mean, the probability of countermeasure (P_c) became 50%. Figure 4 shows the pile extension length and its frequency, which is half-normal distribution. The mean length of the extension became about 0.8 times the SD of d_0.

The cost of countermeasures included the cost of the operations and the materials of the pile (instead of concrete cap), but not the time and cost for decision making on the implementation of the countermeasure.

Besides, Kakurai et al. (2006) stated that more than 2 m difference in pile length took a lot of time to repair a pile foundation. In the case of the cantilever retaining wall, we believe that the difference from the

design can be absorbed by the concrete cap at the top of the pile after excavation, even if it is greater than 2 m.

3 EFFECT OF THE PILE EXTENTION AS COUNTERMEASURE AGAINST FAILURE

This chapter describes the effect of the countermeasures on preventing the deformation failure and rotational failure before the cost analysis in Section 4, and check the applicability of the countermeasure.

3.1 Deformation failure (SLS)

Figure 5 shows the relationship between the wall top displacement and the scatter of the depth of the rock layer surface, d_0, by MCS. When d_0 were less than the design value of 3.0 m, the countermeasures were not conducted, so the plots were omitted in the figure.

Without the piling data, the displacements were weakly correlated with d_0. Naturally, the variations in the same depth were due to variations of other variables such as the deformation coefficient, E_0.

On the other hand, with the piling data, the scatter of the displacement decreased. And the correlation between the displacement and d_0 became weaker.

3.2 Rotational failure (ULS)

Figure 6 shows the relationship between the critical embedment depth and the scatter of d_0. The dot line represents the boundary of the occurrence of the rotational failure.

The critical embedment depth and d_0 had a strong correlation. When the embedment depth was designed

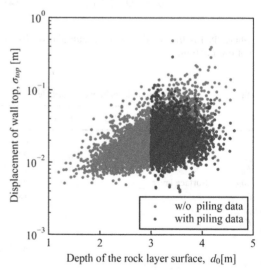

Figure 5. An example of the effect of countermeasures on the displacement of the wall top (Case A, d_e =4.9 m).

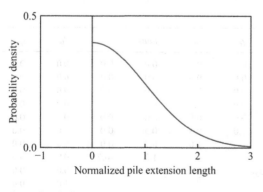

Figure 4. Distribution of the pile extension length normalized by the standard deviation of the depth of the rock layer surface.

318

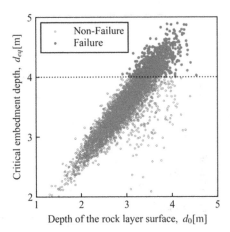

(a) without piling data (Case A-1)

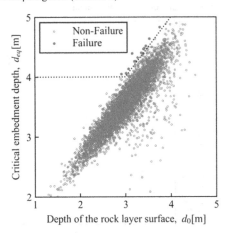

(b) with piling data (Case A-2)

Figure 6. Examples of rotational failure due to different construction managements (d_e =4.0 m).

to be 4.0 m for example, the probability of failure was about 17 % without the piling data (as shown in red dots of Figure 6a), but it became as small as 0.3 % with the piling data (Figure 6b).

The proposed method, i.e. the countermeasure against the scatter of the depth of the rock layer, was valid against the deformation failure and the rotational failure. It was also efficient, comparing with the countermeasure that simply extending the entire pile length.

4 RESULTS AND DISCUSSION ON THE RELIABILITY AND COST ANALYSIS

First, we confirm that the performance function calculated by MCS fit a lognormal probability density function. Then, we show the results of the cost analysis, and discuss the influence of the variation of d_0 and E_0 on the expected total cost.

4.1 Probability density distribution

Figure 7 shows an example of a histogram of the wall displacement and the critical embedment depth, a probability density assuming the lognormal distribution, and a Q-Q plot for the lognormal distribution.

Q-Q Plots showed that the MCS results were in good agreement with the log-normal distribution, although the log-normal distribution slightly overestimated the probability in the upper part. So, the assumption of the log-normal distribution was valid.

4.2 Results of reliability index and expected total cost

Figure 8 shows the relationship between the embedment depth and the reliability index and the expected total cost in Case A. The reliability index increased monotonically with the embedment depth for both construction managements (that is, with and w/o the piling data), and settled to a certain value (Figure 8a and c). This was because embedding more than a sufficient length did not affect the behavior of the cantilever wall retaining wall. The difference between the construction managements was larger at ULS than at SLS. This might be due to both the greater effect of pile embedment on rotational failure and the greater cost at the ultimate failure. In addition, since the reliability index converged at SLS, the piling data was not be useful when the embedment depth was long enough.

The costs were convex parabolic to the pile embedment, and the reliability index for SLS where the cost was minimized was about 1.5 (Figure 8a and b). Since 1.5 is the target value for SLS in ISO 2394, the subsidence repair cost assumed in Section 2.4 was generally reasonable. Also, most of the cost difference between cases with and w/o the piling data was the cost of countermeasures when the embedment depth was large.

Next, Figure 9 shows the reliability index and the expected total cost in Case B, with SD of d_0 of 1.0m. The reliability index increased monotonically and converged at enough embedment (Figure 9a and c), as in Case A, and the differences of the construction managements were larger in Case B than in Case A. And the convergent reliability index at SLS was less than in Case A. The reliability index for ULS also showed similar trend as Case A.

Finally, Figure 10 shows the results of Case C, where the uncertainty of E_0 is updated by the ground investigations. In both construction managements, the reliability indices were larger than that in Case A, and the slope of the reliability index were also

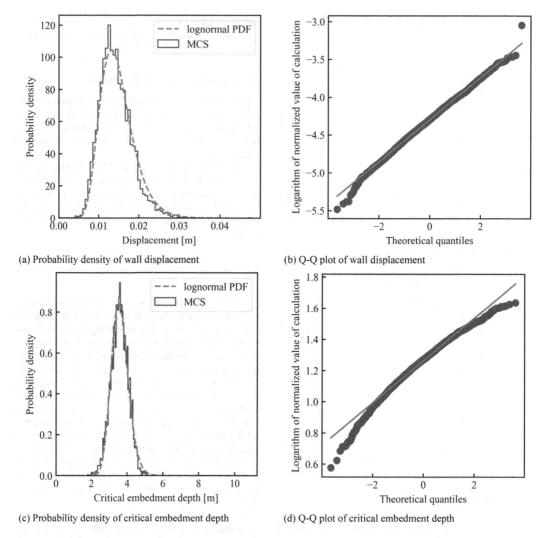

(a) Probability density of wall displacement

(b) Q-Q plot of wall displacement

(c) Probability density of critical embedment depth

(d) Q-Q plot of critical embedment depth

Figure 7. The probability density and lognormal distribution (d_e=4.9, Case A-1, without piling data) and its Q-Q plot.

larger than that in Case A. As a result, the minimum expected costs of both construction managements were smaller than those in Case A (Figure 8b).

4.3 *Influence of the scatters of the depth of the rock layer surface on expected total cost*

Table 3 summarizes the minimum expected total cost for each case normalized by the minimum expected total cost of Case A.

The differences of the expected total costs with and w/o the piling data were maximized in Case B both at SLS and ULS. The difference is about 8% at SLS and about 27% at ULS.

The minimum expected total cost of Case B-1 was 6% larger at SLS and 17% larger at ULS

than those of Case A-1 respectively. On the other hand, those of Case B-2 was only 2% larger at SLS and 3% at ULS than those of Case A-2. So, the piling data enabled the cost to be stable regardless of the scatters of the depth of the rock layer surface, d_0.

In this paper, the scatter of d_0 was assumed to be constant and known before construction. However, it varies greatly in different regions and it is difficult to estimate qualitatively in advance from ground investigation. Therefore, it is not practical to consider the depth variation of the rock layer in the preliminary design. It is advisable to use the piling data, especially when the inhomogeneity of the ground is foreseen but not qualitatively in advance.

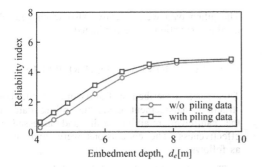

(a) Reliability index at SLS

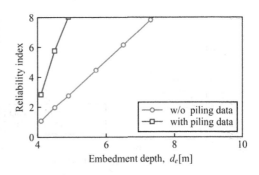

(c) Reliability index at ULS

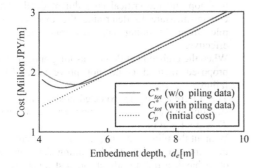

(b) Expected total cost at SLS

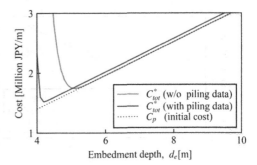

(d) Expected total cost at ULS

Figure 8. Analysis results of Case A.

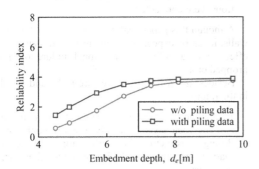

(a) Reliability index at SLS

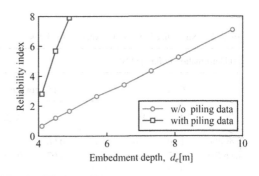

(c) Reliability index at ULS

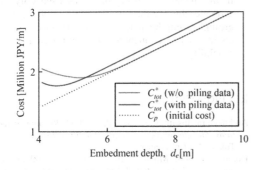

(b) Expected total cost at SLS

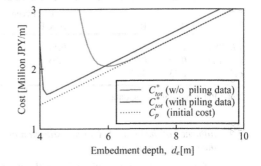

(d) Expected total cost at ULS

Figure 9. Analysis results of Case B.

321

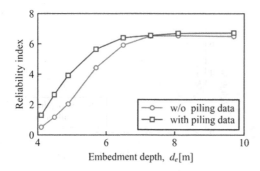

(a) Reliability index

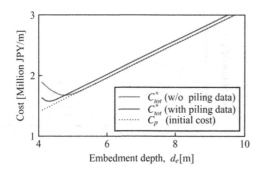

(b) Expected total cost

Figure 10. Analysis results of Case C at SLS.

Table 3. Normalized minimum expected total cost.

(a) Deformation failure (SLS)

	Case			diff. between	
	A	B	C	A & B	A & C
w/o PD.	1.00	1.06	0.94	0.06	0.06
with PD.	0.97	0.99	0.88	0.02	0.09
diff.	0.03	0.08	0.06		

(b) Rotational failure (ULS)

				diff. between
	A	B		A & B
w/o PD.	1.00	1.17	-	0.17
with PD.	0.87	0.90	-	0.03
diff.	0.13	0.27		

Note: PD. represents piling data.

4.4 *Influence of the geotechnical investigation on the expected total cost*

The difference of the costs with and w/o the piling data was larger in Case C than in Case A (Table 3).

The piling data were effective also when geotechnical investigations were conducted.

5 CONCLUSIONS AND FUTURE ISSUES

This paper described a framework of the construction management using the piling data for the cantilever retaining wall, and assessed its effectiveness. The conclusions were summarized as follows;

- The scatter of the depth of the rock layer surface, d_0, was correlated with the displacement of the wall top and the critical embedment depth. The counter-measure to determine the extension of piles from the piling data was appropriate and effective.
- When the embedment depth was long enough, the proposed method could not prevent the SLS failure.
- The difference of the expected cost by the construction management was about 8% at SLS and 27% at ULS at most.
- Without the piling data, higher SD of d_0 increased the expected total cost. With piling data, on the other hand, it did not affect the cost much. Considering that the variation of d_0 is usually unknown, the proposed method can be practical.
- The piling data were effective to reduce the expected cost, even when geotechnical investigations were conducted.

Although this paper dealt with cantilever retaining walls, it can be expected that the proposed method is effective in avoiding the rotational failure of the propped or anchored walls as well.

Besides, the following issues are to be addressed in the future;

- Estimation accuracy In this paper, it was assumed that the rock layers could be reliably confirmed during construction, but verification of the estimation accuracy and its consideration are necessary.
- Use of estimated ground strength/stiffness
- In this paper, the piling data was used only to estimate the depth, but the estimation of the ground strength/stiffness from the piling data could also contribute to improving the reliability.

REFERENCES

Architectural Institute of Japan (AIJ). 2008. *Recommendations for Designing of Small Buildings Foundations*, Tokyo: AIJ (in Japanese).
Honjo, Y. 2008. Monte Carlo simulation in reliability analysis. *Reliability-based design in geotechnical engineering: computations and applications*: 169–191.

International Standard ISO/FDIN 2394. 1998. *General Principles on Reliability for Structures*, Zurich: ISO. Appendix E

Ishihara, Y. Stuart Haigh & Malcolm Bolton. 2015. Estimating base resistance and N value in rotary press-in, *Soils and Foundations*, Volume 55, Issue 4: 788–797.

Ito, H., & Kitahara, Y. 1985. The actual condition and some considerations about the scattering of the mechanical properties of a rock masses, *Report of Central Research Institute of Electric Power Industry (CRIEPI)*, No.384025. Tokyo: CRIEPI (in Japanese).

Japan Geotechnical Consultants Association (JGCA or ZENCHIREN). 2017. *(Draft) Report on the Investigation for Pile Foundations on Rock Layer*. (In Japanese) https://www.zenchiren.or.jp/geocenter/pdf/201701.pdf

Japan Press-in Association (JPA). 2019. *(Draft) Standard Cost Estimation Material: Gyropress Method – Steel tubular pile Press-in Method assisted by rotary cutting* (in Japanese).

Kakurai M., Tsujimoto, K., Kuwabara, F. & Manabe, M. 2009. Relationships soil exploration and pile construction (Part 1), *Summaries of technical papers of annual meeting 2009 of Architectural Institute of Japan (AIJ)*: 595–596. (in Japanese).

Kanda, J. and Shah, H. 1997. Engineering role in failure cost evaluation for buildings, *Structural Safety*, 19(1): 79–90.

Kulhawy, F. H., & Mayne, P. W. 1990. Manual on estimating soil properties for foundation design, *Electric Power Research Inst.*, EPRI-EL-6800, Calif.: Palo Alto.

Ogawa, F. & Matsumoto, K. 1978. The Correlation of the Mechanical and Index Properties of Soils in Harbour Districts, *Report of the Port and Harbour Research Institute (PHRI)*, Vol.17, No.3, Tokyo: PHRI (in Japanese).

Ohki, H., Nagata, M., Saeki, E. & Kuwabara, H., 2005. Fluctuation on bearing strata levels of piles (part 2). *Proceedings of the 40th Technical Report of the Annual Meeting of the Japan Geotechnical Society, Japan*: 1549–1550 (in Japanese).

Phoon, K. K., W. A. Prakoso, Y. Wang, & J. Ching. 2016. Uncertainty Representation of Geotechnical Design Parameters. Chap. 3 in *Reliability of Geotechnical Structures in ISO2394*, Rotterdam: CRC Press: 49–87.

Suzuki N. & Kimura Y. 2021. Summary of case histories of retaining wall installed by rotary cutting press-in method, *Proceedings of the Second International Conference on Press-in Engineering 2021*, Kochi (under review).

Suzuki N., Nagai K. & Sanagawa T. 2021. Reliability analysis on cantilever retaining walls embedded into stiff ground (Part 1: contribution of major uncertainties in the elasto-plastic subgrade reaction method), *Proceedings of the Second International Conference on Press-in Engineering 2021*, Kochi (under review).

Tony Waltham. 2009. *Foundations of Engineering Geology*, Third Edition. Rotterdam: CRC Press: p.47.

Young, D. K., & Ho, E. W. L. 1994. The observational approach to design of a sheet-piled retaining wall. *Géotechnique*, 44(4): 637–654.

APPENDIX

Symbols and abbreviations used in the paper are as follows:

C_{tot}	JPY	:	Expected total cost
C_{tot}^{*}	JPY	:	Simplified expected total cost
C_f	JPY	:	Cost of failure
C_p	JPY	:	Cost of piling operation
C_c	JPY	:	Cost of countermeasure
D	m	:	Outer pile diameter
d_i	m	:	Depth of each layer
d_e	m	:	Embedment depth
d_{eq}	m	:	Critical embedment depth for extreme equilibrium
d_0	m	:	Depth of the stiff ground, or thickness of the surface layer
E_0	N/mm^2	:	Deformation coefficient of the ground (flat-plate loading test)
H	m	:	Height of structures, that is excavation depth
N	-	:	SPT N-value (blows per 300 mm penetration)
P_f	-	:	Probability of failure
P_c	-	:	Probability of countermeasure
q_s	kN/m^3	:	Any uniform surcharge at the ground surface
X_i	-	:	Random variable
Z	-	:	Performance function
$Z_{\delta top}$	-	:	Performance function on the deformation failure
Z_{de}	-	:	Performance function on the de rotational failure
β_i	m^{-1}	:	Stiffness factor of i-th ground layer
β	-	:	Reliability index
δ_{top}	m	:	Displacement of wall top
δ_a	m	:	Allowable displacement of wall

COV	:	Coefficient of Variation
MCS	:	Monte Carlo Simulation
PDF	:	Probability Density Function
SD	:	Standard Deviation
SLS	:	Serviceability Limit State
ULS	:	Ultimate Limit State

Proceedings of the Second International Conference on
Press-in Engineering 2021, Kochi, Japan – Matsumoto et al (eds)
© 2021 Taylor & Francis Group, London, ISBN 978-1-032-10414-0

Dynamic behavior of cantilever tubular steel pile retaining wall socketed in soft rock

S.M. Shafi & J. Takemura
Tokyo Institute of Technology, Tokyo, Japan

V. Kunasegaram
South Eastern University of Sri Lanka, Oluvil, Sri Lanka

Y. Ishihama & K. Toda
Nippon steel corporation, Chiba, Japan

Y. Ishihara
GIKEN LTD., Kochi, Japan

ABSTRACT: Stability against extreme loads, such as earthquakes, water rise behind the wall and its combination, is a major problem in the application of the cantilever steel tubular pipe wall (CSTP). Centrifuge model tests were carried out to study the mechanical behavior of the CSTP wall with a retain height H=12m and a pipe diameter Φ=2m subjected to such extreme loads in 50g for two different wall socket depth (d_e) of 3 m and 2.5 m. Sequential loadings were applied to the wall with cohesionless backfill (dry & wet). Apart from these loadings, white noises were applied before each load to confirm the dynamic characteristics of the wall. The stability of the wall against dynamic and static loads has been significantly increased by 0.5 m increase in socket depth, and the resilience of the wall has been ensured until the end of the sequential loads for the wall with d_e=3.0m.

1 INTRODUCTION

The cantilever retaining wall is one of the old geotechnical structures used to retain earth with moderate height. These structures are interesting as their stability relies on the generation of the earth pressures on either side of the wall which is based on complex soil-structure interaction. An extensive research has been done to investigate this complex soil-structure interaction (Terzaghi 1934a, b; Bica and Clayton 1998; Madabhushi et al 2005). However, those studies work on the static loading condition. So, by following the Rankine or Coulomb earth pressure, adequate design safety can be provided to the wall.

The soil structure becomes more complex when the wall is under dynamic loading. The wall may behave mysteriously when subjected to dynamic loading. The general design practice which is based on stress, for example, the famous pseudo-static approach by Mononobe-Okabe (1929) is adopted by the designer in the early stage. A recent research revealed that the stress-based approach may give an over-conservative design. Steedman (1998) suggested that the stress-based design

approach will underestimate the lateral displacement. So, it is important to understand not only the earth pressure generated behind the wall but also the wall displacement mechanism for the stability of the retaining wall under dynamic loading. Also, the characteristics of the retaining wall (rigid or flexible) govern the wall behavior. Terzaghi (1934 a) explain a generation of pivot point near the base of the relatively stiff sheet pile wall. The passive earth pressure generated below the retaining wall changes its regime below the pivot point. According to the 1g experiment conducted by Bica and Clayton (1998) on cantilever retaining wall, showed that the earth pressure below the pivot point was smaller than the Rankine passive earth pressure as the wall friction acted downward below the pivot point. So, it is essential to provide a sufficient penetration depth to the retaining wall so that the passive earth pressure generated below the retaining wall can provide stability to the wall.

Current design practice in Japan is based on the famous Chang's (1937) equation which limits the minimum embedment depth required by (2.5β~3β) where β is the characteristics value obtained by the equation:

DOI: 10.1201/9781003215226-30

$$\beta = \sqrt[4]{\frac{k_H B}{4EI}} \qquad (1)$$

where,

k_H= Horizontal subgrade modulus
B= Width of sheet pile
E= Young's modulus of steel.

One of the major limitations of this method is that the required embedment depth is based on the flexural rigidity (EI) of the wall without considering the wall height, and the design is based on the elastic linear behavior. It does not consider the nonlinear soil structure relationship observed in the real field. Also, when the wall height increases the lateral earth pressure increases which will yield large deflection. One way to prevent this large deflection is by increasing the flexural rigidity of the wall, for example, circular pipe wall with a large diameter, but the embedment depth will increase as well. This large embedment depth may not be economical considering the construction point of view. Also, if the structure becomes more rigid, then its application becomes limited to a comparatively stiff ground condition like a soft rock which will make the construction process much harder. Through cutting-edge technology like Gyro-press, it is made possible to install circular pipe wall into the stiff ground yet the cost of construction may not be economical every time if the embedment depth and the length of the wall are comparatively large.

Cantilever retaining wall is normally used to retain moderate height Madabhushi et al (2005); B. V. S. Viswanadham (2009) Different researcher has conducted a study on the cantilever retaining wall with a large retain height with cohesionless backfill soil and penetrating dense sand Jo et al. (2014;2017). Also, Madabhushi et al (2006) conduct a comprehensive numerical analysis for dry and saturated cohesionless backfill as saturated backfill soil may cause structural failure due to high stress or may cause large deformation which will lose the serviceability of the wall. This research includes a large diameter cantilever type steel tubular pipe (CSTP) wall with a wall height H =12m and a pipe diameter Φ=2m embedded in artificially prepared soft rock with embedment depths of 1.25Φ & 1.5Φ. The behavior of the retaining wall has been investigated under static and dynamic loading with dry and wet backfills conditions.

The main intention of this model study is to investigate the dynamic stability of a steel tubular pile wall embedded in soft rock with relatively small embedment depth than Chang's proposed minimum depth and observe the deformation and failure behaviors. Two different cohesionless backfill conditions like dry backfill and wet backfill condition were maintained. The undrained condition was maintained in the backfill to utilize the maximum of the pore pressure generated behind the wall so that

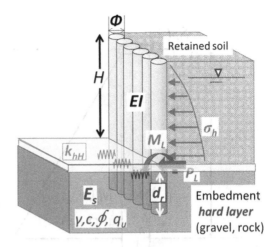

Figure 1. Cantilever retaining wall.

the applied lateral loads on the wall can be increased up to 2.5 times that of the dry backfill condition using the water feeding technique. Figure 1 shows the typical load acting on the wall. Therefore, the stability of the retaining wall was investigated under two extreme loading conditions: large lateral thrust by water feeding and earthquake motion as shown in Figure 1.

2 METHODOLOGY

The description of the model is shown in Figure 2. The whole model was designed for 50g centrifugal acceleration. Different sensors like Leaser Displacement Transducer (LDT), Earth Pressure Cell (EPC), Pore Pressure Transducer (PPT), and Accelerometer, and Strain Gauges (SG) were used to record various responses during the experiment. The wall displacement and force in the forward direction are

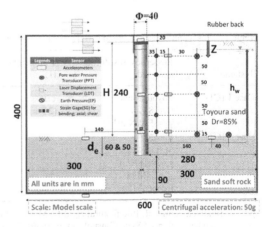

Figure 2. Model setup.

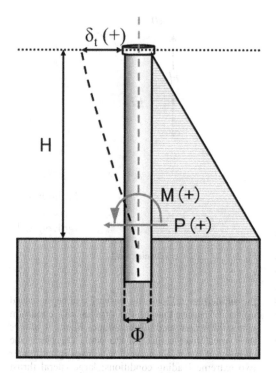

$\delta_t (+)$

H

M (+)

P (+)

Φ

Figure 3. Sign convention used for study.

(max of 5 piles) the breadth of the container was shorten using a lateral acrylic spacer on the inner face of the back-face panel. Before the casting, 0.5 mm thick Teflon sheets were pasted in the front and rear internal container wall faces and lubricated by silicone grease for easy detachment of wall from the hardened soft rock ground.

The artificial soft rock was prepared by mixing cement, sand, clay, and water with a target 14th day unconfined compressive strength of $q_u=1.4$ MPa. Toyoura sand and Sumi clay were mixed with Portland cement and water to prepare the mixture. The mechanical properties of the artificially prepared soft rock were reported by Vijay et al. (2019) and the main mechanical properties are shown in Table 1 along with the other properties. The soft rock ground was then constructed by compacting the mixture layer by layer at every 30 mm thickness, up to the final height of the planned rock layer with the help of a mechanical vibrator. The density of the compacted mixture was carefully controlled by the volume of each compacted layer and the required mass of the mixture for the layer.

During the preparation of the rock ground, 10 samples were prepared for the unconfined compression test on the 3rd, 7th & 14th days. A mold with a diameter of 50 mm and a height of 100mm was used to prepare the sample. After installation of a pile, in the ground, the ground was covered with a wet towel to avoid any moisture loss. Special care was taken to avoid any crack on the rock surface. After one week, both the container and walls were removed and a new Teflon sheet was attached to the rare wall. 10x10 mm mesh was then made on the front of the ground surface to help in the image analysis. Two days before the test, the gap between the wall was closed by silicon paste so that the surface of the wall facing the backfill may become a uniform plane. After that grease was applied on both sides of the wall in the backfill direction. A membrane rubber bag was used to create an undrained backfill condition as shown in Figure 4 (a). The use of Latex rubber (see

considered as positive and the anti-clockwise moment is also considered as positive. Different sign assumed for the analysis is shown in Figure 3.

The model container had the original internal dimensions of 600 mm in length, 250 mm in breadth and 400 mm in depth. The container was made up of a removable rear-side aluminum wall and a front-side transparent thick acrylic wall and an aluminum hollow frame to stiffen the acrylic. Both wall plates were bolted with the main container body to form a rigid box. To secure the plane strain conditions and to model the maximum possible width of the wall

Table 1. Test conditions, and the material properties of wall model.

Test code	Properties of soft rock and sand	Wall height: H_w	Rock socket depth: d_R []: βd_R	Pile Properties Φ, t, EI, M_y
Case 1 (C1)	Toyoura sand (Dr=85%): $\gamma_d=15.8$kN/m^3 $\phi'=42°$	12m (240mm)	3.0m (60 mm) $^$ [1.2]	Φ=2 m (40 mm), t = 25 mm (0.5mm) Spacing: 2.15m (43mm)
Case 3 (C3)	Soft rock: $\gamma_t=20.1$kN/m^3 $q_u=1.4$MPa $E_S=660$MPa		2.5m (50mm) [1.0]	EI= 6.8 GNm^2/m (5.4x10^{-5}GNm^2/m) M_y= 9.0 MNm/m (3.6x10^{-3} MNm/m)

βd_R: normalized depth, EI: Pile flexural rigidity, M_y: Bend moment causing pile yielding, ϕ': friction angle from triaxial compression test with σ_3=98kPa (Tatsuoka et al, 1986) [3)], E_S: Secant modulus of SR (Kunasegaram et al, 2019) [2)] $^$: Model scales are given in parenthesis

Figure 4. (a) Membrane rubber bag (b) Model before experiment.

Figure 4) bag may create interference of the transmission of the shear stress in between the rock surface and the bottom of the rubber bag. Therefore, to effectively transmit the input accelerations, the rubber membrane bag was designed by using a carbon fiber base (PZ-564 real carbon) at the interface between soil and rock which is also a watertight membrane but quite stiffer compare to the latex rubber.

The prepared rubber bag was then placed behind the wall and filled up with Toyoura sand up to the wall top by maintaining a relative density Dr =85%. The air pluviation method was adopted to fill out the backfill soil so that a uniform dense sand layer can be obtained. While backfilling, different sensors were placed in the backfill soil to measure the earth pressure, acceleration, pore water pressure. The front

view of the model before the experiment is shown in Figure 4 (b). Two centrifuge tests were conducted using the Tokyo Tech Mark III centrifuge at a centrifugal acceleration of 50g. Figure 5 shows the detail of the loading sequence along with the typical shape of the input motion. A controlled sinusoidal wave of predominant frequency 1Hz was applied as the input motion for dynamic loading. For this paper, the amplitude of the input motion is defined by the absolute maximum value in the entire time history of the accelerogram and denoted by (a_i). Also, the acceleration in the negative direction is considered as the cause of the forward displacement. The number of the cycle (n_c) referred to the number of effective cycles counted in each shaking. The increment of wall top displacement ($\Delta\delta_t$) for each shaking is also shown in Figure 5. Each case is consisting of three dynamic events along with two static events by water feeding. Case 1 consists of 11 shakings and case 3 consists of 4 shakings. White noise was applied to study the model condition after the application of different loading. The magnitude of the shaking was maintained in such a way that the effect of the loading history could be studied.

3 RESULT AND DISCUSSION

All the results are shown in the prototype scale unless stated otherwise by the author.

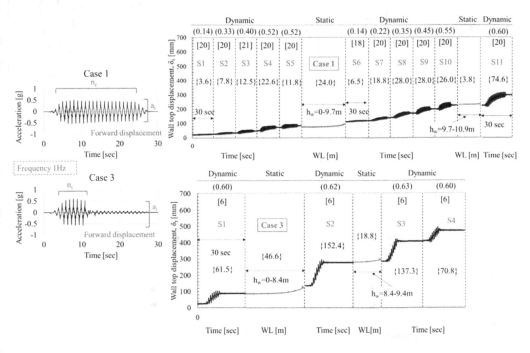

(a_i)= maximum amplitude of input motion (g); [n_c] = no of cycle during the shaking; S1,2,3… = Shake number; { }=$\Delta\delta_t$

Figure 5. Typical shape of input motion, detail of loading sequence and observed wall top displacement.

3.1 Acceleration response

Ten accelerometers were installed at various locations of the wall, backfill, rock surface, and container base to record the acceleration response during the dynamic loading as shown in Figure 2. Figure 6 (a), (b) & (c) shows the time history of wall top acceleration, backfill top acceleration and wall top displacement. From Figure 6 (c) it is seen that the maximum accumulation of wall displacement is taking place within the time frame of 5-10 sec. From Figure 6 (d) it is seen that, the wall displacement in forward direction is the combined effect of wall inertia force (acceleration multiplied by negative mass) and lateral earth pressure. Small time lag between the wall acceleration and the backfill acceleration can also be confirmed from the acceleration time history. The peak of the wall acceleration comes earlier than the peak of the earth pressure meaning the wall will push the soil.

The acceleration of all shakings of case 1 & 3 when the inertia force become maximum in forward direction is plotted against the input acceleration (a_i) as defined in Figure 5 is plotted in Figure 6 (e) & (f). It is observed that the acceleration observed in the top of the wall and backfill are over the reference line with slope 1:1 meaning amplification takes place in both wall top and backfill top. Also, it is

observed that the amplification in the dry condition is smaller than in the wet condition. One of the reasons could be the addition of water in the backfill soil reduce the predominant frequency of the soil which may cause this large amplification. The amplification in the wall top and the backfill top indicates that the wall and backfill soil do not behave as a rigid body as assumed by the Mononobe-Okabe method. The acceleration response in a dry condition increases almost linearly with the input acceleration meaning the amplification ratio remains almost the same in dry conditions. However, the acceleration ratio in wet conditions tends to decrease for high magnitude input motion.

3.2 Dynamic displacement behavior of the wall

To investigate the dynamic behavior of the wall, the displacement, earth pressure, and bending moment time history have been shown in Figure 7. By following the elastic beam theory, bending moment was computed from the bending strain measured from the strain gauge measurement. To show the trends of accumulation of the residual displacement, earth pressure, and bending moment, the moving average of the data recorded during dynamic shaking has been taken and indicated by the thick lines. It is observed that, all the trend lines increase before and

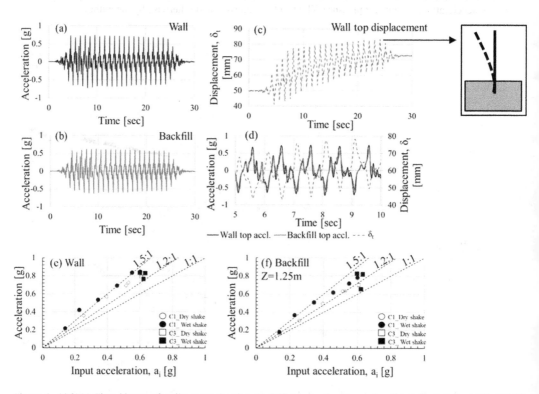

Figure 6. (a)(b)(c) Time history of wall top acceleration, backfill top acceleration and wall top displacement [C1_S4] (d) Relationship between wall and backfill acceleration with wall top displacement [C1_S4] (e)(f) Amplification of wall and backfill top.

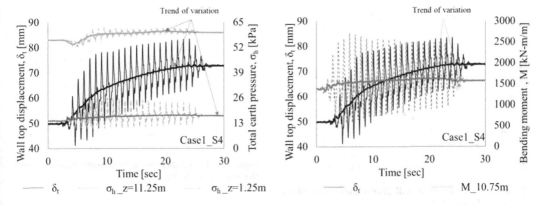

Figure 7. (a) Time history of wall displacement and earth pressure (b) Time history of wall displacement and measured bending moment at 1.25m from rock surface.

after the shaking meaning that accumulation takes place before and after the shaking in displacement, earth pressure, and bending moment measurement.

From the time history of the earth pressure at a shallow and deepest depth, the accumulation of the residual earth pressure is found to be greatest at a deeper depth than at a shallow depth, but the amplification is higher at a shallow depth which makes the shallow depth more critical for the dynamic loading. Small time lag can be seen between the wall displacement and the bending moment response indicating the phase difference occurring between the applied load and the resisting load from the rock surface. In the wall displacement and the earth pressure response, it is seen that the earth pressure increases though the wall displacement is increasing. The mechanism behind this behavior is that due to the wall resilience (property of the wall to move back to its original position) the wall will try to push the soil behind which will create a passive condition thus increase the earth pressure. As the wall requires a large load to push the soil back to its original position that causes a permanent displacement to the wall.

To further investigate the wall displacement behavior, the residual wall displacement increment $(\Delta \delta_t)$ has been plotted against arias intensity (A_i) which is defined by equation 2 as shown in the Figure 8.

$$A_i = \frac{\pi}{2g} \int_0^{Td} a(t)^2 dt \qquad (2)$$

where, g is the acceleration due to gravity and Td is the duration of signal above threshold.

For each case, one shake in dry condition and one shake in wet condition are presented in Figure 8. It is seen that the wall displacement during wet shaking is much larger than the dry shaking. The reason behind this behavior is that due to the deterioration of the rock confinement at the later shaking events which yield

large displacement. Also, the wall displacement experienced by case 3 is much higher than case 1 which indicates that by changing 0.5m rock socketing depth, accumulation behavior of the residual displacement changes significantly. Observing the trends line for shake 4 & 11 of case one in Figure 8 it can be confirmed that the accumulation of residual displacement which takes place in the earlier part is higher than the later part. A steady increase at the earlier part may expand as the rock confinement deteriorate which is observed in the later shaking (shake 11). To see the effect of the rock socketing depth, the accumulated wall displacement of case 1 and case 3 is plotted against cumulative arias intensity which is defined by Equation 2.

In Figure 9, arias intensity is considered as different shaking was applied to the two models. It is seen that the wall displacement largely increases for case 3 than case 1 though the cumulative arias intensity is much less compared to case 1. Comparing the shake 4 and shake 5 of case 1 and shake 3 and shake 4 of case 3 it is seen that the wall displacement reduced for the later shaking if they are under a similar condition which

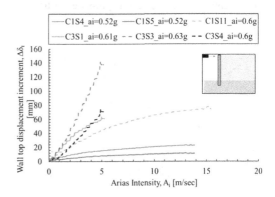

Figure 8. Wall displacement relationship with arias intensity.

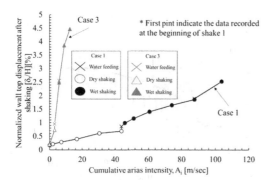

Figure 9. Comparison of wall displacement accumulation behavior of case 1 & 3.

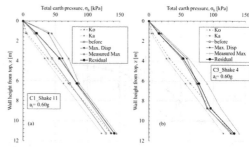

Figure 11. Distribution of earth pressure during dynamic shaking (wet backfill).

confirms the effect of the wall resilience in the determination of the accumulation of the displacement.

3.3 Effect of rock socketing depth on the wall behavior

Comparing Figure 10 (a) & (b), it is seen that the difference in earth pressure before and residual for case 1 is much smaller than case 3. As Figure 10 (a) represents the fourth shaking of case one, due to the previous loading history, the earth pressure doesn't change much which is observed in Figure 10 (b) for the first shake of case 3, where the earth pressure changes from active to at-rest earth pressure. In all the cases, the measured earth pressure at the time of maximum displacement is smaller than the maximum earth pressure recorded by each sensor. However, for the dry backfill condition, the earth pressure distribution is not linear as assumed by the Mononobe-Okabe method.

Figure 11 (a) shows the earth pressure distribution of case 1 for shake 11, and Figure 11 (b) shows the earth pressure distribution for shake 4 of case 3. It is seen that the residual earth pressure experience by case 1 is higher than case 3 which means that the wall of case 1 provides more resistance than case3 which indicates the clear effect of 0.5m rock socketing depth. Also, the earth pressure distribution

significantly increases (more than at-rest earth pressure) in wet shaking compared to dry shaking which justifies the use of a closed membrane rubber box.

To further investigate the effect of the earth pressure and the rock socketing depth, the distribution of earth pressure, wall displacement, and bending moment during the time of first water feeding are shown in Figures 12 & 13. Figure 12 shows the variation of normalized effective earth pressure at the deepest depth with wall displacement at different water levels. The deepest earth pressure cell has been

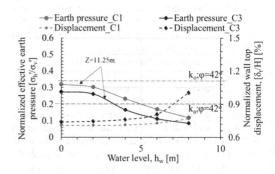

Figure 12. Earth pressure and wall displacement variation during static loading (water rise).

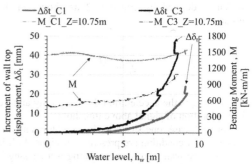

Figure 13. Increment of wall displacement and bending moment variation during static loading (water rise).

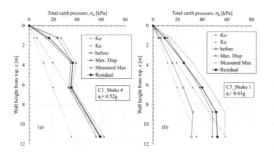

Figure 10. Distribution of earth pressure during dynamic shaking (dry backfill).

considered as the water start to fill from bottom to top. When h_w=0m the normalized effective earth pressure of case 1 is higher than case 3 which indicates the effect of the loading history on the accumulation of the earth pressure along with the effect of 0.5m rock socketing depth. As the water level increases, the effective earth pressure starts to decrease from at-rest to active condition though the wall displacement does not change significantly. The wall displacement starts to increase significantly after a certain water level which can be explained by the increase of bending moment after a certain water level (see Figure 13). At the end of the water feeding, the wall displacement experienced in case 1 is smaller than case 3 which indicates the effect of the 0.5m rock socketing depth.

To further study the effect of the rock socketing depth on the overall behavior of case 1 and case 3, the bending moment and earth pressure relationship to the wall top displacement are shown in Figures 14 & 15. Figure 14 shows the residual bending moment (M_r) relationship to the residual wall top displacement (δ_{tr}). For the comparison purpose the strain gauge which is located in the same distance from the rock surface (1.25m from rock surface) for case 1 and 3 have been considered. It is seen that the wall displacement increases as the bending moment

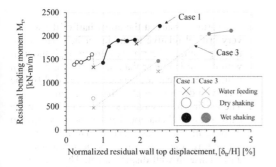

Figure 14. Relationship between residual wall displacement and residual bending moment (at 1.25m from rock surface).

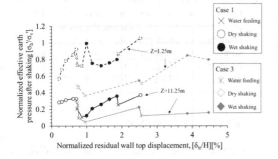

Figure 15. Relationship between residual wall displacement and residual earth pressure at shallow and deepest depth.

increases, which mean the force behind the wall has increased. The increase of the residual bending moment can be explained by the increase of the residual earth pressure shown in Figure 15. As seen in Figures 10 and 11, the earth pressure at the top half is more critical than the bottom half, because the earth pressure at the top part is higher than the reference K_o pressure, where the bottom half is close to K_o pressure. So, to further understand the earth pressure variation at shallow and deepest depth, the residual effective earth pressure at z=1.25 & 11.25 m is plotted against the residual wall top displacement in Figure 15. The earth pressure has been normalized by effective vertical earth pressure (σ_v') to understand the behavior more clearly. It is seen that the earth pressure at shallow depth increases significantly than the deepest depth. In case 1, the earth pressure at shallow and deepest depth increases from the first shaking till the last shaking which causes the gradual increase of the residual bending moment as observed in Figure 14. Similarly, in case 3, the earth pressure at shallow and deepest depth increase from the first shaking till the last shaking in the deepest depth but decreases between the shake 3 & 4 at the shallow depth which causes an overall small increase in the bending moment between shake 3 and 4. Though case 3 has large wall displacement but the bending moment of case 3 is much smaller than case 1, which means that the reaction given by 3m socketing wall is higher than a 2.5m wall. Also, the increase of the bending moment at the final shaking proves that the rock confinement can still provide resistance to the applied load. However, considering the bending moment of the final shaking of case 1 & 3, it is confirmed that the wall with a 3m socketing depth has more resilience left compared to case 3.

According to Wu and Prakash (1996), the failure by horizontal wall displacement will take place at 10%H. Huang et al. (2009) limit the seismic displacement criterion based on the soil strength mobilization and proposed that 5%H can be considered as a significant wall movement to cause failure. According to the IPA (2016) standard, on the top of the sheet pile wall should not exceed 1%H in normal condition and 1.5%H under Level 1 seismic motion. From Figure 14 & 15 it is seen that, case 1 can withstand many shakings before reaching to the allowable limit provided by IPA (2016) standard but in case 3 the wall almost reaches to the allowable limit (about 0.8%H) just after 1 shake confirming the significant effect of change in 0.5m rock socketing depth.

Figure 16 shows the observed crack and deformation of the soft rock after each test. Considering the post softening characteristics in the stress-strain behavior of the soft rock, catastrophic failure was expected during the experiment. Though the wall displacement experience by case 3 is almost close to 5%H yet no catastrophic failure was observed. One of the possible reasons could be the presence of large overburden pressure (about 180kPa for dry and

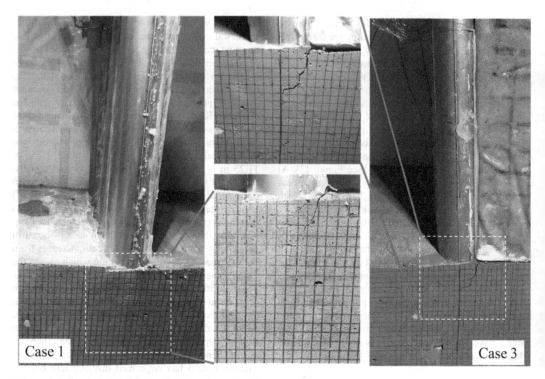

Figure 16. Observed crack and deformation of the soft rock after the test.

240 kPa for wet) due to the large retain height which prevents the wall from undergoing any catastrophic failure.

4 CONCLUSION

The main intention of this model study is to investigate the dynamic stability of a steel tubular pile wall embedded in soft rock with relatively small embedment depth than Chang's proposed minimum depth and observe the deformation and failure behaviors. The d_e=3m and 2.5m were considered for this study and the effect of 0.5m change in the embedment depth are discussed in this paper. Based on the discussion following conclusion can be drawn:

1. Under similar test conditions, when two similar earthquake loads are applied, the wall displacement will be smaller for the second loading than the first loading meaning the wall resilience will increase in the second loading.
2. The performances of the wall are considerably improved by the increase of the depth of the rock socketing by 0.5 m.
3. Although the wall with d_e=2.5m had a displacement of approximately 4.5%H, there were no catastrophic failures, demonstrating that d_e=2.5m can provide an adequate margin of safety to avoid catastrophic failures even under level 2 earthquake.

4. The earth pressure at the upper half of the wall is very critical relative to the lower half. However, the experience of earth pressure through the wall with d_e=3m is higher than the wall with d_e=2.5m which means that the resistance provided by the rock with d_e=3m is higher than d_e=2.5m.
5. The loading history has a significant effect on the earth pressure behavior that has been confirmed in dynamic and static events. During dynamic loading, residual earth pressure accumulates with an increase in wall displacement. The accumulation of the residual earth pressure of the wall with d_e=3m is greater than the wall with d_e=2.5m confirming the effect of 0.5m of change in the depth of the rock socketing.
6. Contrary to the dynamic load, the earth pressure tends to decrease with the increase in the displacement of the walls during the static load. The reduction of earth pressure takes place at a very small change in the displacement. However, the movement will begin to change considerably once the water level reaches a certain level.

ACKNOWLEDGEMENT

The authors gratefully acknowledge the individual advice and guidance provided by the members and advisers of the IPA TC1 (Committee on the application of cantilever type steel tubular pile wall embedded to the stiff ground) in connection with the preparation of this paper.

REFERENCES

Bica, A. V. D. & Clayton, C. R. I. 1998. An experimental study of the behaviour of embedded lengths of cantilever walls. *Geotechnique* 48(6), 731–745.

Chang, Y.L. 1937. Lateral pile loading tests. *Transaction of American Society of Civil Engineering* 102, 273–276.

Huang, C.-C., Wu, S.-H. & Wu, H.J. 2009. Seismic Displacement Criterion for Soil Retaining Walls Based on Soil Strength Mobilization. *Journal of Geotechnical and Geoenvironmental Engineering* 135:74–83.

International Press-in Association (IPA), 2016. Press-in Retaining Structures: a handbook, 1st ed. AD II–76

Jo, S., Ha, J., Yoo, M., Choo, Y.W., & Kim, D. 2014. Seismic behavior of an inverted T-shape flexible retaining wall via dynamic centrifuge tests. *Bull Earthquake Engineering* 12:961–980. DOI 10.1007/s10518-013-9558-9.

Jo, S., Ha, J., Lee, J. & Kim, D. 2017. Evaluation of the seismic earth pressure for inverted T-shape stiff retaining wall in cohesionless soils via dynamic centrifuge. *Soil dynamics and earthquake engineering* 92, 345–357. http://dx.doi.org/10.1016/j.soildyn.2016.10.009.

Kunasegaram, V., Shafi, S.M., Takemura, J. & Ishihama, Y., 2019. Centrifuge model study on cantilever steel tubular pile wall embedded in soft rock. *Geotechnique for sustainable infrastructure development, Lecture Notes in Civil Engineering 62*, https://doi.org/10.1007/978-981-15-2184-3_135

Kunasegaram, V. & Takemura, J. 2019. Deflection and failure of high stiffness cantilever retaining wall embedded in soft rock. *International Journal of Physical Modelling in Geotechnics*, https://doi.org/10.1680/jphmg.19.00008

Madabhushi, S.P.G. & Chandrasekaran, V.S. 2005. Rotation of cantilever sheet pile walls. *Journal of Geotechnical and Geoenvironmental Engineering* 131(2), 202–212.

Madabhushi, S.P.G. & Zeng, X. 2006. Seismic response of flexible cantilever retaining walls with dry backfill. *Geomechanics and Geoengineering: An International Journal* 1(4), 275–289.

Mononobe, N. & Matsuo, M. 1929. On the determination of earth pressures during earthquakes. *Proceedings: World Engineering Conference*, Japan, Vol. 9.

Tatsuoka, F., Goto, S. & Sakamoto, M. 1986. Effects of some factors on strength and deformation characteristics of sand at low pressures. *Soils and Foundations* 26(1): 105–114.

Terzaghi, K. 1934a. Large retaining wall tests. I. Pressure of dry sand. *Engineering News-Rec.* 136–140.

Terzaghi, K. 1934b. Large retaining wall tests. II. Pressure of dry sand. *Engineering News-Rec.* 259–262.

Viswanadham, B., Madabhushi, S. Babu, K. & Chandrasekaran, V. 2009. Modelling the failure of a cantilever sheet pile wall. *International Journal of Geotechnical Engineering* 3:2, 215–231, DOI: 10.3328/IJGE.2009.03.02.215-231

Wu, Y. & Prakash, S. 1996. On seismic displacements of rigid retaining walls. *ASCE Geotechnical Special Publication: Analysis and Design of Retaining Structures Against Earthquakes*, S. Prakash, ed., ASCE, New York, 21–37.

Proceedings of the Second International Conference on
Press-in Engineering 2021, Kochi, Japan – Matsumoto et al (eds)
© 2021 Taylor & Francis Group, London, ISBN 978-1-032-10414-0

A centrifuge model study on laterally loaded large diameter steel tubular piles socketed in soft rock

V. Kunasegaram
South Eastern University of Sri Lanka, Oluvil, Sri Lanka

J. Takemura
Tokyo Institute of Technology, Tokyo, Japan

Y. Ishihama
Nippon Steel and Sumitomo metal Corporation, Chiba, Japan

Y. Ishihara
GIKEN LTD, Tokyo, Japan

ABSTRACT: This paper discusses the influence of rock socket depths on the deformation and failure mechanism of rock socketed piles under a constant vertical eccentricity of 6.5 m. For the centrifuge model study, two types of model soft rock ground were prepared, a single soft rock layer and a soft rock layer with overlying sand. Lateral resistance of piles with three different rock socket depths were investigated in both model grounds at 50g centrifugal acceleration. From the loading tests, two different failure modes were observed, i.e., ground failure and pile structural failure depending on the embedment depth and the ground conditions. For the piles with relatively small socket depth (d_R) in a single rock layer, the increase of d_R can increase the lateral and moment resistance. However, as the d_R increases, the effect of d_R becomes less significant, especially for ultimate resistance due to the pile structural failure.

1 INTRODUCTION

Thanks to the high structural stiffness of large diameter piles and the development of novel installation technique, recently the application of large diameter steel tubular piles in hard ground, such as soft to medium-hard rock has increased in engineering projects around the world (IPA). Large diameter steel pipe piles can be applied for various large geotechnical structures, such as the mono-pile foundation for offshore wind turbine and large height cantilever type retaining wall. Lateral response of mono-pile foundations in sand was quite deeply investigated through 1g laboratory tests and more sophisticated centrifuge models by several researchers. However, as described in Lehane and Guo (2017), the documented literature to illustrate the mechanical behaviour of rock socketed large-diameter steel tubular piles under lateral loading is extremely rare. Perhaps it could be attributed to the difficulties of conducting large scale tests in a hard medium to observe the critical behaviour, which is controlled by several influential factors, such as pile factors (diameter, stiffness), ground factors (strength and stiffness), and loading factors

(loading height, monotonic and cyclic). On the other hand, a few field tests have been conducted on concrete shafts and some centrifuge tests were also conducted for modelling caisson and solid piles in various types of rock as summarized in Kunasegaram and Takemura,(2020). In most of the previous field and physical model studies, the points of loading were almost closer to the ground surface except the field tests done by Digioia and Rojas- Gonzalez (1994). These loading conditions are different from the abovementioned target structures. A typical loading condition for both, the mono-pile foundation and the large height self-standing wall is a relatively large moment load due to one-way cyclic lateral loads induced by wind loads and seismic excitations, respectively. Therefore, as a preliminary study, authors have conducted few single pile lateral loading tests with a constant vertical eccentricity of 6.5 m at 50g centrifugal acceleration to understand the deformation and failure mechanism of rock socketed single piles. This paper reports the influences of rock socket depth on the deformation and failure mode of single piles embedded in stiff grounds. Also it describes the loading history on the behavior of single piles.

DOI: 10.1201/9781003215226-31

2 CENTRIFUGE MODELLING

2.1 Centrifuge models and test procedures

Two different ground conditions were made in the centrifuge models, namely soft sand rock and soft rock overlaid by the medium dense Toyoura sand (Dr =80%). Centrifuge model arrangement for the piles embedded in above grounds are shown in Figures 1(a) and 1(b) respectively. A container with the internal dimensions of 700 mm length, 500 mm depth and 150 mm width was used in both models. Model tubular piles used in this study were thin wall pipes with 40 mm outer diameter and 0.5 mm thickness, made of stainless steel (SUS304) having the young's modulus (E) of 193 GPa and yield stress (σ_y) of 255 MPa. At the pile top, a solid circular pile cap made of aluminum with 30mm socketed depth was tightly fixed to form a solid loading head (Figure 1(c)). Sectional and mechanical properties of steel tubular piles are described together with the other test conditions in Table.1. Detailed mechanical properties of Toyoura sand were reported by Tatsuoka et al, (1986) including the effects of density, and the mechanical properties of model soft rock used was described in Kunasegaram and Takemura, (2020). The model piles were equipped with bending strain gauges on both sides of the pile along the loading direction, and Wheatstone circuits were made with the help of bridge boxes. Bending strains were measured by using full bridge circuits along the pile, while at the pile tip a pair of half bridge circuits were utilized to measure the axial strains at the loading and the opposite sides independently. Centrifuge model arrangement for three single piles embedded (d_e=40 mm, Pile-S; 60 mm, Pile-M; 80 mm, Pile-L) in soft sand rock (Model-4) is described in Figure 1 (a). In the preparation of soft rock model, 300 mm thick acrylic plates stack was tightly placed in the container bottom to reduce the depth to 200 mm. A 190 mm thick layer of soft sand rock was made by compacting sand-clay-cement mixture layer by layer with 30 mm thickness, confirming the target unit

weight of each layer of compacted mixture. Immediately after casting the mixture, the model tubular piles were installed vertically into the unsolidified mixture with a pile guide to the specified depth, and fixed the pile position. The casted mixture was cured for 14 days in order to achieve the targeted strength (q_u) and stiffness of the embedded medium. The detailed preparation procedures and mechanical properties of the soft sand rock material are reported by Kunasegaram et al, (2015). It is important to note that sand and soft rock was filled inside the pile up to the rock surface level with the pile installation process employed in the model preparation stage, which was confirmed by means of physical measurements. In the preparation of Model-5 with two-layers ground (Figure 1(b)), 160 mm thick acrylic plates stack was placed at the container bottom to make 340 mm depth for the sample. Then a 200 mm thick soft rock layer was made and the model piles were installed in the soft rock layer with the rock socked depths (d_R) same as the d_e of Model-4. After 14 days curing for the soft rock, a 130 mm thick top sand layer with Dr=80% was made by air pluviation. It must be emphasized that the pile SP_SR_2* in Model-4 was accidently preloaded about 1 mm prior to the test without instrumentation and the stiffness and resistance could not be obtained in the intact condition for the small displacement range. Therefore, to study the behavior of pile with short socket (d_e=2 m) and to confirm the repeatability, three single piles were tested in Model-8 with the identical embedment depths of model-4 and slightly different imposed displacement cycles as described in Table- 2.

Upon completion of the model, the loading jack and laser displacement sensors (LDTs) were mounted on the container. Thereon the container moved to centrifuge platform and rigidly fixed, then the centrifugal acceleration was increased up to 50g. One-way horizontal load cycles were applied by the jack from small to large pile top displacements as described in Figure 2 at 50g environment. Applied horizontal load at the pile top (P_L) was measured by a load cell and horizontal

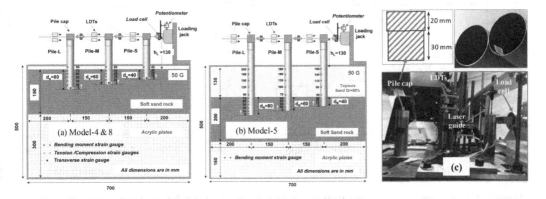

Figure 1. Centrifuge model arrangements.

335

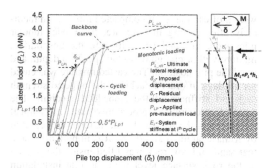

Figure 2. One way cyclic loading and sign conventions defined in the loading sequence.

displacement (δ_t) and rotation (θ_t) at the pile top was obtained by means of LDTs at two elevations as described in Figure 1(a,b). Having completed one loading test, the centrifuge was once stopped and the loaded pile was removed. Resetting the jack and LDTs to the next pile and the same horizontal loading was repeated. The loading was conducted in the sequence of Pile-S, Pile-M and Pile-L. In the following chapter, the test results are shown in prototype scales.

3 RESULTS AND DISCUSSION

3.1 *Observed load-displacement behaviours*

A typical cyclic load (P_L) –displacement (δ_t) behaviours for the piles embedded in SR (SP_SR_4) and MS_SR ground (SP_MS_SR_4) are drawn in Figure 3(a). The Figure 3(a) describes the imposed displacement cycles with increasing mean load and cyclic displacement amplitudes. Corresponding moment load (at the ground level, $M_L = P_L * h_L$) - pile top rotation (M_L-θ_t) relation for the piles is also

illustrated in Figure 3(b). The imposed pre-maximum displacements (δ_{Pi}) and applied pre-maximum loads (P_{LPi}) in each cycle are summarised in Table 2 for all the piles given in Table 1. The loading sequence in this study consists of a limited number of one-way load cycles and a subsequent monotonic loading up to the ultimate failure of the system, determined by either the failure of embedment ground or structural buckling.

In each unloading-reloading cycle, a certain amount of residual displacements can be seen (Figure 3(a)) after the unloading, which pinpoints the plastic deformations of the embedded medium even at relatively small δ_{Pi}'s. Although this residual displacement (δ_{ri}) is eventually accumulated in each cycle, the P_L- δ_t relation in reloading processes returns to a unique curve, which is the envelope (Figure 2) of cyclic load-displacement behaviour. Here onwards the envelope curve will be written as the backbone curve.

Figure 4(a) and Figure 4(b) show the backbone curves of P_L- δ_t and M_L- θ_t relationships obtained in the loading sequences of all the piles, respectively.

The influence of embedded medium and embedment depth (d_e) on the lateral and rotational resistances of large diameter piles under identical loading conditions can be confirmed from Figure 4(a) and Figure 4(b), respectively. From Figure 4(a), the deeper the rock socketing is, the larger the mobilized resistance of pile can be observed in the overall load displacement behaviour. However, based on Figure 4 (a) and Figure 4(b), there is no significant difference in the lateral and rotational resistances between the two piles Model-5 (SP_MS_SR_3, SP_MS_SR_4). This observation implies the insignificance in the increment of socketing over 3 m (1.5Φ) in the underlain rock strata for the case of MS_SR ground. Although increasing the socketing depth from 3 m to 4 m has no significant influence on the lateral resistance of socketed piles in the two-layer profile, the comparison between SP_MS_6.5 and SP_MS_SR_2

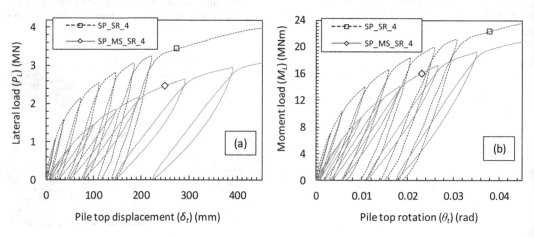

Figure 3. Typical (a) cyclic load – displacement and (b) moment load- rotation behaviour observed at pile top for the piles in single rock layer (SP_SR_4) and two layer (SP_MS_SR_4) profile.

Table 1. Model conditions and properties of embedment medium.

Model condition	Properties of embedment medium	Pile notation	Embedment depth (d_e)	E_e/G^*	Pile properties EI, M_y, M_p
Model-4 Soft rock (SR)	γ_{tr} =20.1 kN/m³ γ_{dr}=16.8 kN/m³ q_u =1.4 Mpa E_r =660 Mpa	SP_SR_2* SP_SR_3 SP_SR_4	2 m [40 mm] 3 m [60 mm] 4 m [80 mm]	60 60 60	Φ=2 m (40 mm)
Model-5 Soft rock (SR) overlain by Toyoura sand (MS) Dr=80%	γ_{tr} =20.1 kN/m³ γ_{dr}=16.8 kN/m³ q_u =1.4 Mpa E_r =660 Mpa γ_{ds}=15.5 kN/m3 $\phi' = 41°$	SP_MS_SR_2 SP_MS_SR_3 SP_MS_SR_4	d_s=6.5 m [130 mm] d_R=2m [40 mm] d_s=6.5 m [130 mm] d_R=3m [60 mm] d_s=6.5 m [130 mm] d_R=4m [80 mm]		t = 25 mm (0.5 mm) EI=14.6 GNm² (2.34 kNm²)
Model-8 Soft rock (SR)	γ_{dr}=16.8 kN/m3 q_u=1.4 Mpa E_r=660 MPa	SP_SR_2× SP_SR_3× SP_SR_4×	2 m [40 mm] 3 m [40 mm] 4 m [40 mm]	60 60 60	M_y=19.3 MNm (154 Nm) M_p=24.8 MNm (198 Nm)

* Preloaded prior to the test without instrumentation
d_s=Embedment depth in sand layer
d_R=Socketing depth in rock
Loading height, h_L= 6.5 m for all the piles
M_y: Theoretical yielding bending moment of the pile
M_p: Theoretical bending moment causing the plastic failure of pile
All dimensions are given in prototype scale, model scales are given in brackets

Table 2. Imposed pre-maximum displacements and pre-maximum loads applied in each cycle.

| Pile notation | Imposed pre-maximum displacement (δ_{Pi} in %ϕ)/Pre-maximum load ($PLPi$ inMN) in each one way loading cycles, N = cycle number | | | | | | | |
	N=1	N=2	N=3	N=4	N=5	N=6	N=7	N=8
SP_SR_2*	2.3/0.56	4.1/0.78	5.8/0.90	7.7/0.98	9.6/1.06	-	-	-
SP_SR_3	0.8/0.84	1.8/1.24	3.7/1.54	5.6/1.74	7.6/1.90	9.4/2.02	-	-
SP_SR_4	1.0/1.08	1.9/1.58	3.7/2.14	5.6/2.56	7.3/2.82	9.3/3.08	11.1/3.26	-
SP_MS_SR_2	1.3/0.48	2.4/0.78	4.9/1.14	7.4/1.40	9.9/1.62	14.7/2.00	19.6/2.34	-
SP_MS_SR_3	1.1/0.44	2.4/0.82	4.7/1.32	7.2/1.66	9.8/1.98	14.3/2.40	19.4/2.78	-
SP_MS_SR_4	1.4/0.52	2.7/0.88	5.0/1.42	7.4/1.86	9.8/2.18	14.6/2.66	19.5/2.96	-
SP_SR_2×	1.5/0.58	2.6/0.72	4.9/0.84	7.4/0.92	10.3/0.96	13.2/0.96	16.0/0.96	19.1/1.00
SP_SR_3×	1.1/1.14	2.3/1.44	4.7/1.78	7.1/1.98	9.3/2.14	11.6/2.28	13.9/2.40	16.3/2.48
SP_SR_4×	1.1/1.04	2.4/1.48	4.6/2.00	6.9/2.40	9.2/2.66	11.7/2.92	14.4/3.14	16.7/3.28

* Preloaded prior to the test without instrumentation
Imposed displacements and Pre-maximum loads are given in prototype scale

indicates the significant contribution of 2 m socketing into relatively hard layers. Lateral resistance of pile SP_MS_SR_2 increased more than twice in the overall response than that of the non-socketed pile SP_MS_6.5. Structural failures of the pile were confirmed with clear local buckling at a point below the ground level as shown in Figure 5(b) for all three socketed piles in the two-layer profile.

However, the buckling point of SP_SR_4 is located at the ground surface level which could be attributed to the large lateral confinement of soft

rocks and the plugged portion of the rock inside the piles. Once the pile failed by the structural buckling, the further increase of socketing depth (d_R >3 m for socketed piles) could have no significant influence on the lateral resistance of piles for the abovementioned loading conditions. The reductions of resistance in Figure 4(a) after the peak load for the piles in MS_SR ground are the indication of clear structural failure, while for non-socketed pile SP_MS_6.5, the resistance increased until large pile top displacement over 50% of pile diameter (Φ)

337

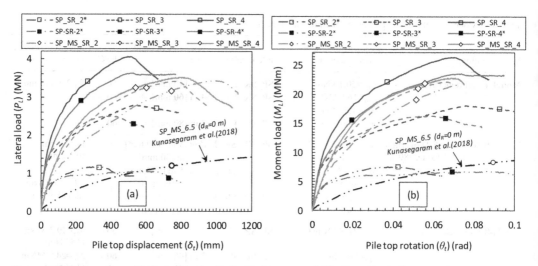

Figure 4. Backbone curves of (a) Lateral load - pile top displacement and (b) moment load-pile top rotation.

without showing peak resistance also no structural failure can be observed from Figure 4(a). For the piles embedded in the soft sand rock, 1 m increment in embedment depth can significantly increase the ultimate lateral resistance up to d_e=4 m. As a clear structural failure with local buckling was observed slightly above the ground level for SP_SR_4 with d_e=4m as shown in Figure 5(a), it can be inferred that further increase of the embedment could not provide substantial contribution to the pile lateral resistance. The depth over which the effect of embedment increment cannot be obtained is considered as an "optimum embedment depth (OED)". The difference of OED in the soft rock and socketed piles can be attributed to their rigidity or confinement. However, the pile with identical embedment (SP_SR_4^x) in Model-8 exhibited the rock splitting as the ultimate failure mode, and no visible local deformations appeared in the pile (Figure 5(a)). The different failure mechanisms of piles with identical d_e could be attributed to two reasons as follows.

The d_e =4 m is closer to the critical d_e for the implemented loading condition, around this d_e the

failure mode can be easily alternated from the ground failure to the pile structural failure and vice versa, by loading cycles or loading sequence. The other reason is the difference in the imposed pre-maximum displacement (δ_{Pi}) sequence of two piles, where the pile SP_SR_4^x experienced larger δ_{Pi}'s compared to SP_SR_4 in each cycle (see Table 2). Besides the ultimate failure mechanism of these two piles, smaller mobilised resistance of pile SP_SR_4^x than the pile SP_SR_4 at small δ_{Pi}'s (δ_t =2%Φ (40mm)) could be resulted by unforeseen irregularities of rock in the toe back regime. It can be confirmed from normalized deflection profiles discussed in the subsequent chapter, where the bottom tip displacements of pile SP_SR_4^x is higher from the early stage of loading. Furthermore, structural failures were not observed for the piles SP_SR_2*, SP_SR_2^x, SP_SR_3 and SP_SR_3^x, but the load displacement curves of these piles also showed a peak resistance and subsequent reduction. These behavior of the piles with no pile failure but ground failure of the soft rock is different from that of the pile (SP_MS_6.5) in sand. This can be attributed to the strain softening of stress-strain relationship of the rock material and smaller d_e of the soft rock model than that of sand model. Although the piles with structural failure and the piles embedded in the soft rock with relatively small embedment depth (SP_SR_2*, SP_SR_2^x, SP_SR_3 and SP_SR_3^x) also showed strain softening behavior in the load displacement curves, the post peak resistance reduction are different between the piles failed by the structural failure and the ground failure. Once the structural buckling initiated a sudden reduction of load against the displacements can be seen. On the other hand, the observed post peak behavior related to ground failures exhibited smaller post peak stiffness and much more ductile especially for the deeper embedment condition (SP_SR_3 and SP_SR_3^x)).

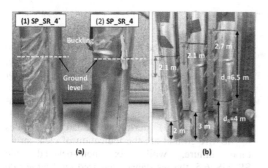

Figure 5. Observed structural deformation of piles in (a) single rock layer and (b) two layer profile.

3.2 Influence of embedment depth

The influence of embedment (d_e) or socketing depth (d_R) on lateral resistance of rock socketed piles from small to large pile top displacements (δ_t) are depicted in Figure 6. It is important to note that the load corresponding to the pile top displacements of δ_t =40% Φ was considered as ultimate resistance (P_{L_ult}) for the pile (SP_MS_6.5) without peak resistance in the load - displacement relation. From Figure 6, a distinctive behavior of piles at different displacement levels depending on the stiffness of embedded medium can be observed. The horizontal resistance of piles socketed in SR ground (d_e=d_R) (Model-4 & 8) increased with d_e, but the trend is highly influenced by the imposed displacement. In Model-4 and 8, significantly large increase of lateral resistance can be attained by a small increment (0.5Φ and1Φ) of d_e from 2 m to 4 m, exhibiting the large lateral confinement of rock type materials even with small socketing (Φ to 2Φ) depths. The variation of lateral resistance is almost linearly increasing with d_e for all displacements, except the piles with d_e over 3 m at small displacements. Lateral resistance of the pile with d_e=3 m (SP_SR_3, SP_SR_3^x) is almost the same as d_e = 4 m (SP_SR_4, SP_SR_4^x) at δ_t = 0.5%Φ and 1%Φ in Model- 4, similar behaviour can be seen up to δ_t = 2%Φ in Model-8. The distinctive behaviour at these small imposed displacements could be attributed to the lateral confinement given by shallower depth of the rock (less than 3m) at their intact condition. As the loading progresses beyond δ_t =1%Φ, the softening of rock by the increase of displacement deteriorates the subgrade reaction of shallow rock layers and cause the difference in resistance between the piles with d_e = 3 m and d_e = 4 m. Although the pile with d_e = 4 m in Model -8 exhibits smaller resistance compared to that of Model-4 from δ_t =2%Φ to P_{L_ult}, the increasing tendency is similar in both models. Furthermore, the smaller ultimate resistances of all three piles in Model-8 could be attributed to the cyclic weakening mechanism due to relatively large imposed displacements (Table-2) than that of Model-4. Influence of socketing depth (d_R) on the lateral resistance of piles embedded in MS_SR ground also could be explained from Figure 6. Comparing the piles with d_R = 0 (SP_MS_6.5) and d_R = 2 m (SP_MS_SR_2), the lateral resistance of piles can be increased about three times from the early stage of loading to the ultimate condition, by a d_R of Φ into the underlain soft rock layer. In Model-5, the influence of d_e (d_e=d_s+d_R) is differently appeared in the mobilised resistance depending on the imposed δ_t. From Figure 6, d_R over 2 m has no significant influence on the lateral resistance up to δ_t = 2%Φ. However, a clear influence of d_R tends to appear from δ_t = 4%Φ to 10%Φ and again the influence of d_R over 2 m becomes insignificant on the P_{L_ult} of piles in MS_SR ground as seen in Figure 6. From the variation of ultimate resistances against d_e in Figure 6 and observed failures (Figure 5) of rock socketed piles in Model-4,5 and 8 it can be confirmed that, if the pile structural failure determines the ultimate failure condition the effect of d_R becomes insignificant, contrarily ultimate resistances of ground failure conditions are highly influenced by d_R. As overall behaviour, it can be concluded that the lateral resistance of pile embedded in sand and soft rock increases with increasing d_e if the ground stiffness determines the ultimate resistance. However, the effect of the embedment over the optimum depth has no significant contribution. On the other hand, due to the softening of ground materials and the change of failure mode from the ground failure to pile structural failure, the optimum embedment depth changes depending on the conditions of resistance, which is shallower for the initial stiffness than the ultimate loading conditions in the soft rock.

3.3 Measured nominal bending moments

Measured bending moments along the pile at different pile top displacements (0.1%, 0.5%, 1%, 2%, 4% & 10% Φ) for the piles in all three centrifuge models are presented in Figure 7. From Figure 7(a) to 7(f), the bending moments of the pile in the overlain sand layer increase almost linearly with the distance from the loading point to the depth of -2 m. The observation clearly indicates that the confining stresses from shallow layers (up to Φ) of sand has less significant influence on the lateral resistance of piles. On the other hand, the piles in the soft rock showing the maximum moment slightly above the ground surface and an abrupt change in the bending moments above and below the ground surface also can be observed. This observed bending behavior is another evidence of very high confinement of shallow layers of soft rock. A high degree of radial and circumferential restraint could be expected closer to the rock surface due to the high confinement of rock type materials. It results in higher circumferential membrane stresses

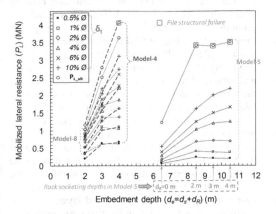

Figure 6. Influence of rock socketing depth on the lateral resistances of the piles single rock layer and two layer profile.

339

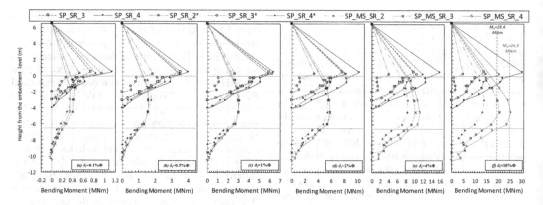

Figure 7. Observed bending profiles at different Imposed displacement amplitudes.

in a pile, and the influence of membrane action could be expected up to a distance of Φ above the rock surface, as described in Kunasegaram and Takemura (2021). As a consequence of membrane stresses, the measured nominal bending strains become larger than that of real bending strains which corresponds to a pure bending behaviour of the pile. Therefore, the observed abrupt changes could be resulted by stress discontinuity at the embedment level and a certain degree of local deformations of tubular pile.

From Figure 7, at small imposed displacements (δ_t <2%Φ), the difference between the bending moments of piles with different d_R in MS_SR ground is insignificant. However, as the displacement increases, pile with larger d_R shows the larger bending moment. Also, the depths of maximum bending moment in the sand layer of MS_SR ground are deeper for the pile with large d_R than small d_R. The observation implies that the effect of the deterioration of soft rock stiffness at the shallow layers affected the lateral resistance over δ_t=2%Φ, it can be confirmed from the variation of lateral resistance illustrated in Figure 6 for the piles in Model-5. Up to the displacement, δ_t=4%Φ (δ_t=80 mm) the bending moments of all the piles are smaller than the yielding moment (M_y). At δ_t =10%Φ (δ_t=200 mm), SP_SR_4, SP_SR_4[x], SP_ MS_SR_3 & SP_ MS_SR_4 exhibit the bending moments more than M_y. In which, the bending moment of SP_SR_4 and SP_MS_SR_4 became more than that of the plastic hinge (M_p). It should be noted that the moments over M_y are not actual mobilized moments in a pile, but they are the nominal ones calculated from the strain measurement. None the less, these results well agree with the structural deformations observed in Figure 5.

3.4 Deformation modes and failure mechanism

As described in Figure 8(a), the PT displacement (δ_t) can be divided into three components; (a) displacement at the RS (δ_{RS}), (b) displacement caused by the rotation at RS ($h_L*\theta_{RS}$) and (c) displacement caused

by the bending of pile above the RS (δ_{t_b}). These three components are considered as indices representing the effects of translation, rotation, and bending of a laterally loaded pile in SR ground, respectively. Similarly, the three components of PT and SS displacements for the piles in MS_SR ground are defined in Figure 8(b) in terms of displacements and rotations at SS and RS levels. The corresponding variation of displacements and rotations against the moment load for the piles in SR (Figure 8(c), (d)) and MS_SR grounds (Figure 8(e), (f)) are illustrated in Figure 8. From Figure 8(c,d), relatively small displacements and rotations at RS level with a linearly increasing trend prior to the yielding deformations and subsequent progression of a nonlinear variation against the M_L could be observed as a typical behaviour of piles in SR ground. However, the piles in MS_SR ground exhibits relatively large displacement and rotations at SS with a nonlinear variation even at small M_L (Figure 8(e,f)). Mechanism behind these distinctive behavior of piles in SR and MS_SR ground can be explained using Figure 9. The percent fractions of δ_t representing the effects of translation, rotation and bending against the δ_t/Φ for the piles in both SR and MS_SR grounds are described in Figure 9. Comparing Figure 9(a) and Figure 9(b) the translational fraction (δ_{RS}) of piles in SR ground are less than 10% at the beginning and much smaller compared to bending and rotational fractions even at large δ_t/Φ. Thanks to the higher initial stiffness of intact rock sockets, which effectively hold the pile prior to the rock yielding. However, the translational fraction (δ_{GS}) of piles in MS_SR ground governs more than 30% of δ_t from the beginning of loading. The above observation could be attributed to relatively small subgrade modulus of overlain sand compared to SR and the higher fixity of rock sockets, which allows significant bending of the pile in the sand layer. It can be confirmed from the bending fraction representing the SS displacements as illustrated in Figure 9(c) and the bending profiles in Figure 10.

340

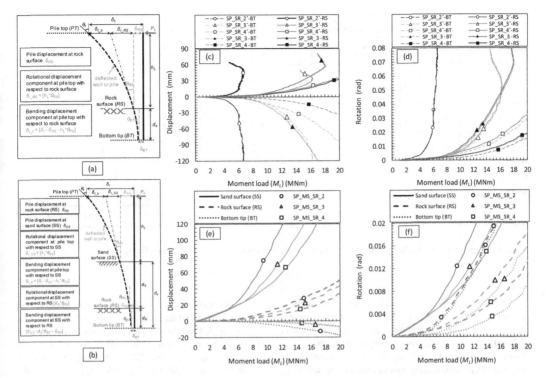

Figure 8. (a,b) Typical deflection profiles and the variation of (c,e) displacements and (e,f) rotations at sand surface, rock surface and bottom tip against the applied moment load ($M_L=P_L*h_L$) for the rock socketed piles.

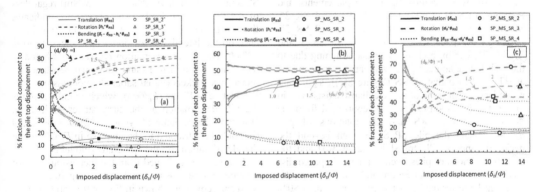

Figure 9. Variation of % fraction displacement components of (a,b) pile top (PT) displacement and (c) sand surface (SS) displacement against the normalized imposed displacements for rock socketed piles.

Deformation modes of rock socketed piles in a single rock layer, and the mode change with the increase of moment load and imposed displacements could be explained from Figure 8(c,d), Figure 9(a) and the variation of normalized deflection profiles illustrated at different δ_t/Φ values in Figure 10. From Figure 8(c,d), Figure 9(a) the pile SP_SR_2$^\times$ with a short socket depth ($d_R/\Phi = 1.0$) exhibits a vertical increase of RS, BT displacements and rotations beyond a moment load of 6 MNm. Furthermore,

identical rotations at RS(θ_{RS}), BT(θ_{BT}) in Figure 8(d) and a larger rotational fraction (about 65%) from the initiation loading with a continuously increasing and decreasing trends of both rotational ($h_L*\theta_{RS}$) and bending ($\delta_t - \delta_{RS} - h_L*\theta_{RS}$) fractions against δ_t/Φ (Figure 9(a)) clearly dictate a brittle failure with an activated rotational mode. The above failure took place due to the extension of toe- back shear deformations (Kunasegaram and Takemura (2021)) with a pivot point of pile SP_SR_2$^\times$ located around

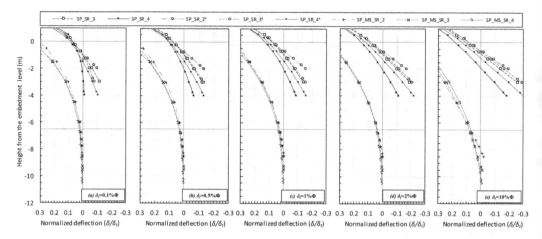

Figure 10. Estimated normalized deflection profiles at different imposed displacement amplitudes.

0.6 m (0.3Φ) below the rock surface as seen in Figure 10(c-e). Thanks to the large lateral confinement of rock type materials even at shallow depths, a small increment of socket depth $\Delta(d_R/\Phi) = 0.5$, from $d_R/\Phi = 1.0$ and $d_R/\Phi = 1.5$ significantly increased the redundancy against the brittle softening and changed the deformation and failure mechanism from brittle to the ductile one.

The above can be confirmed from Figure 8(c,d) and Figure 10, while comparing the displacements and rotations at RS, BT for the piles SP_SR_3, SP_SR_3$^\times$, SP_SR_4, SP_SR_4$^\times$ with those of pile SP_SR_2$^\times$. Although the piles with d_R/Φ =1.5 (SP_SR_3 and SP_SR_3$^\times$) exhibit a clear post-peak softening due to the rock failure and comparatively large displacements at BT level than that of RS (Figure 8(c)), significant redundancy against the ultimate collapse exists even in the post-peak response. It can be confirmed from the difference between RS and BT rotations of pile SP_SR_3$^\times$ and the pile SP_SR_3 even beyond the peak resistance as illustrated in Figure 8(d) and the bending fractions of piles even at a large imposed displacement of δ_t/Φ =6, in Figure 9(a). Furthermore, the observed large residual resistances of more than 80% of M_{L_ult} in Figure 8(c,d) also another evidence of redundancy against brittle failures. As the d_R/Φ increases from 1.5 to 2, the ultimate failure mode has changed from ground failure to the structural buckling (SP_SR_4) as seen in Figure 5(a-2). The observed bending profile (Figure 7(f)), displacements and rotations at RS and BT, even at a large moment load (SP_SR_4) as illustrated in Figure 8(c,d) and the decrease of rotational and increase of bending fractions in Figure 9(a) also support this physical observation. The above observations of piles with d_R/Φ =1,1.5 and 2 indicates the change of load transfer mechanism, i.e., as d_R/Φ increases from 1.5 to 2, the depth-dependent bearing factors provides large rotational resistance due to high confinement at deep locations.

It allows more strains in the pile front shallow rock layers (Figure 10) and causing significant deterioration of soft rock modulus. As a consequence, the deformation at the pile front rock surface is higher for the pile with d_R=4 m (SP_SR_4, SP_SR_4x) than d_R =3 m (SP_SR_3, SP_SR_3x) under same imposed displacement as seen in Figure 10.

From Figure 9(b), all three piles MS_SR ground exhibit almost same displacements and rotations at SS and RS up to δ_t/Φ =2% (M_L =4 MNm) regardless of the rock socket depths. The observation indicates that, socketing over $d_R = \Phi$ has no significant influence on the lateral and rotational resistances up to δ_t =2%Φ. The above observation could be attributed to large lateral confinement and rotational resistance of intact rock sockets (initial subgrade modulus or smaller relative stiffness (E_e/G^*)), which mainly controls the pile deformation at the early stages of loading. However, the deviations beyond δ_t =2%Φ in Figure 9 (b) could be attributed to the increase in E_e/G^* due to the deterioration in the subgrade modulus of soft rock. It can be confirmed from the increasing trend of rotation and decreasing trend of bending fractions of SS displacement as shown in Figure 9(c) and the normalized deflection profiles given in Figure 10. Furthermore, Figure 11(a) summarizes the influence of d_R/Φ on the % fractions of SS displacement. The deformation mechanism of piles below the SS is dominated by the bending fractions in the overlain sand layer up to certain % of δ_t at which the rotational fraction overcomes the bending as observed in Figure 9(c). This δ_t is about 2%, 4% and 8%Φ for the piles with d_R/Φ =1,1.5 and 2, respectively. These values of δ_t tends to increase with d_R and exhibit significant recovery of rotations and increase of bending fractions as d_R/Φ increases in Figure 11(a). Based on above observations, at relatively small δ_t (< 2%Φ) the displacement at overlain SS is dominated by the pile bending in the sand layer due to large lateral

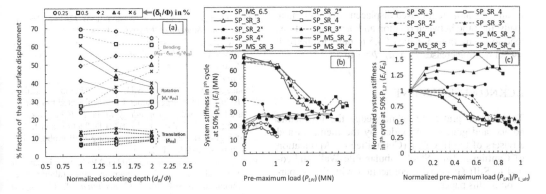

Figure 11. Influence of rock socketing depth on the lateral resistances of the piles single rock layer and two layer profile.

confinement of intact rock sockets. However, as E_e/G^* of the system increases due to the deterioration of soft rock modulus, the displacement caused by the rotation at RS tends to dominate the bending fraction at SS. This ratio between rotation and bending fraction at a δ_t of 10% Φ tend to shrink from 4.5 to a value of unity as the d_R increases from d_R/Φ =1 to 2.

3.5 Change of system stiffness

To investigate the embedment ground, structure conditions and the consequences of cyclic loading on the deterioration of ground stiffness, an initial system stiffness (E_i) is defined at $50\%P_{LP1}$ as described in Figure 2. The variation of E_i values against P_{LPi} and P_{LPi}/P_{L_ult} are plotted in Figure 11(b) and 11(c), respectively. From Figure 11(b), the differences of E_0 values between rock socketed piles with and without the overlain media can be confirmed. Except for the short socketed piles (SP_SR_2* and SP_SR_2ˣ), the E_0 of piles directly socketed in the soft rock (SR) is about 2.5 and 3 times higher than those of embedded in MS_SR ground, at 50% P_{LP1}. The observation implies that the E_0 values highly depends on the stiffness of embedded ground rather than the embedment depth. Comparing the pile SP_SR_2ˣ with SP_SR_3 and SP_SR_3ˣ in Figure 11(b), the E_0 values could be increased about 50% by the increment of rock socketing $\Delta(d_R/\Phi) = 0.5$ from $d_R/\Phi = 1.0$. Meanwhile, a further increment from $d_R/\Phi = 1.5$ (SP_SR_3 and SP_SR_3ˣ) to 2 (SP_SR_4 and SP_SR_4ˣ) has no remarkable contribution on the E_0 values of rock socketed pile as seen in Figure 11(b). Still, the deterioration of the system stiffness gradually occurred for those with a socket depth of $d_R/\Phi = 2$ compared to those with $d_R/\Phi = 1.5$ and the observation is more perceptible from Figure 11(b) than the normalised profiles illustrated in Figure 11 (c), showing the better redundancy for the former socket depth than the latter.

Similarly, in MS_SR ground, a 2 m ($d_R/\Phi = 1.0$) rock socketing in the underlain rock layer has increased the E_0 values by about 4 times compared

to the pile with zero socketing (SP_MS_6.5), it can be confirmed from the comparison of pile SP_MS_6.5 with the pile SP_MS_SR_2 in Figure 11 (b). Furthermore, a clear contribution of additional socketing also visible in the normalised relations illustrated in Figure 11(c), where E_i/E_0 increases with the increase of socket depth even at relatively large P_{LPi} values (up to 60% P_{L_ult}). The observed constant trend of E_i's even at large P_{LPi} values is a clear indication of system redundancy against the softening of underlain rock layer. It is also true from the ultimate failure mode as structural buckling for all three piles shown in Figure 5. Unlike the piles in MS_SR ground, for the piles in SR ground, the E_i values steadily decreased from the second loading cycle, which indicates the softening in the stress-strain behaviour of soft rock and significant deterioration of the subgrade modulus or the foundation stiffness.

4 CONCLUSIONS

In the stiff ground like the soft rock, the lateral resistance changes significantly in a small range of embedment depth. A small increment of embedment depth (0.5Φ to 1Φ) in soft sand rock remarkably increased the redundancy against the brittle softening and changed the deformation mechanism from brittle to the ductile one.

The ultimate failure mechanism of rock socketed piles in a single rock layer is mainly caused by the rotation at RS when the ultimate resistance is determined by the rock failure. However, when the structural buckling determines the ultimate resistance, the mechanism is a combination of rotation at RS, translation at RS and the bending above the RS with the domination of rotational fraction.

In the two-layer profile; displacement at SS is dominated by the pile bending in the sand layer from the initiation of loading to a critical δ_t/Φ, over which the rotational fraction tends to dominate the bending. This critical δ_t/Φ increases with the increase of d_R,

343

meanwhile at large imposed displacements or ultimate loads, the ratio between rotation and bending fractions tend to shrink from a value of 4.5 to unity as the rock socket depth increases from d_R/Φ =1 to 2.

ACKNOWLEDGEMENT

The authors gratefully acknowledge the invaluable advice and guidance provided by the members and advisers of the IPA TC1 (Committee on Application of Cantilever Type Steel Tubular Pile Wall Embedded to Stiff Ground) in connection with the preparation of this paper.

REFERENCES

Digioia AM, Rojas-Gonzalez LF (1994) Rock socket transmission line foundation performance. *IEEE transactions on power delivery* 9(3):1570–1576.

IPA (International Press-in Association) (2016) *Press-in Retaining Structure: A Handbook*, 1st ed.

Kunasegaram V, Akazawa S, Takemura J, Seki S, Fujiwara K, Ishihama Y, and Fujii Y,(2015) "Modeling of soft rock for a centrifuge study", *Proc. 12th Geo-Kanto*, pp. 15–19.

Kunasegaram V, Hsiao W.H, and Takemura J,(2018) "Behavior of a large diameter piles subjected to moment and lateral loads", *Proceedings of the 1st ICPE*, Kochi.

Kunasegaram V, Takemura J (2020) Deflection and failure of high-stiffness cantilever retaining wall embedded in soft rock. *IJPMG*: 1–21.

Kunasegaram V, Takemura J (2021). Mechanical behaviour of laterally loaded large-diameter steel tubular piles embedded in soft rock. *Geotechnical & Geological Engineering*, (*under review*).

Lehane BM, Guo F (2017) Lateral response of piles in cemented sand. *Géotechnique* 67(7):597–607.

Tatsuoka F, Goto S, Sakamoto M (1986) Effects of some factors on strength and deformation characteristics of sand at low pressures. *Soils and foundations* 26(1):105–114.

Proceedings of the Second International Conference on
Press-in Engineering 2021, Kochi, Japan – Matsumoto et al (eds)
© 2021 Taylor & Francis Group, London, ISBN 978-1-032-10414-0

Discussion about design method for embedded length of self-standing steel tubular pile walls pressed into stiff ground

T. Sanagawa

Railway Technical Research Institute, Tokyo, Japan

ABSTRACT: Application of the steel tubular piles as the retaining walls are increasing. Especially, Gyropress method are often used because this method can install the piles penetrating stiff ground and underground obstacles. Design methods of the self-standing retaining walls in Japan vary depending on the objects of structures and assumed actions, and there are mainly three analysis methods. The self-standing retaining walls using steel tubular piles pressed into stiff ground are also designed conformed to each design method, and it is known that the calculated embedded lengths differ depending on the analysis methods. However, there is few case studies about the evaluation results depending on the design method and the ground condition. In this paper, the overview of each analysis method is introduced, and the optimum method is discussed by the comparing the calculated embedded length depending on different analysis methods and the design condition.

1 INTRODUCTION

The press-in method is a piling method which accurately installs pre-formed piles using static loading without noise and vibration. There are several penetration modes available by using driving assistance, therefore the Press-in Method is universally applicable and can be used in soft to extremely hard ground conditions due to these wide varieties of penetration modes.

Gyropress method, one of the press-in method, is a reaction based rotary press-in method to install tubular piles with cutting bits (shown in Figure 1) with self-walking functions. It can penetrate the gravels, hard rocks, and existing reinforced concrete structures (Figure 2, 3) and expand the scope of the application of press-in method. In addition, amount of soil displacement is controlled by rotary cutting mechanism which excises just the obstacle part.

Recently, piles with a large diameter are widely used for road and port structures against the large earthquake motions. Considering design methods, a simple method based on the beam on elastic subgrade and the Winkler's hypothesis is widely used. According to this method, embedded lengths of the piles is determined only by the stiffness of the piles and ground. This method may make the designed embedded length long and lead increment of the cost and construction period depending on ground and structure conditions.

Considering this background, rationalization of the design calculation method is required. This paper describes the outline of the design method and discuss about the effect on the design for the cantilever pile wall embedded into stiff ground by trial calculation.

2 DESIGN METHOD

2.1 Outline of design calculation method of embedded length of self-standing walls

There are three major design calculation methods for embedded length. (a) Limit Equilibrium Analysis Method, (b) prerequisite of Elastic Subgrade Reaction Method, (c) Combined Subgrade Reaction Method. The outline about these design methods are described below.

2.2 Limit equilibrium analysis method

The limit equilibrium analysis can evaluate the member stress and required embedded length of the cantilever wall from passive and active soil pressure at the ultimate condition. Figure 4 shows the outline of this method. The earth pressure from background is usually calculated at the range from top to toe of wall by the Coulomb and Rankine-Resal equation.

Horizontal resistance of piles is calculated from the balance of the subgrade rection at ultimate condition of soil and the external forces. This is to assume empirically the distribution of subgrade reaction, and may include one that assumes quadratic parabolic

DOI: 10.1201/9781003215226-32

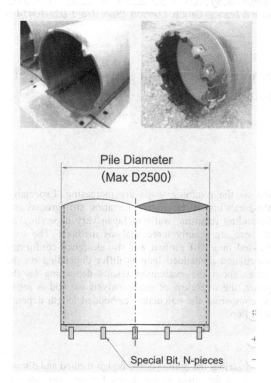

Figure 1. Cutting bits (Giken, 1999).

Figure 2. Coring reinforced concrete (Giken, 1999).

relationship, and one that assumes the subgrade reaction by linear relationship or by arbitrary curves.

2.3   Elastic subgrade reaction method

Figure 5 Shows the outline of the elastic subgrade reaction method. Assuming that the soil and the pile are elastic, the behavior of the pile can be calculated as a beam on an elastic floor, as the following equation.

$$\frac{EI}{D}\frac{d^4y}{dx^4} + p = 0 \qquad (1)$$

Where,
E: Young's modulus of pile (kN/m²)
I: secondary moment of pile (m⁴)
D: pile diameter (m)
p: subgrade reaction (kN/m²)
x: depth from the ground surface (m)
y: deflection at a depth of x (m)
Elastic subgrade reaction method uses Winkler spring model (Winkler, 1867), and p is shown as

$$p = P(x, y) \qquad (2)$$

The elastic subgrade reaction method is divided into three methods by modeling of $P(x,y)$. The linear elastic subgrade reaction method includes Chang's equation, where $p=k_h y$, (k_h: coefficient of subgrade reaction). In this equation, it is assumed that the horizontal subgrade reaction has a linear relationship with displacement. Under this condition, the pile behavior is shown as a simple equation, then widely used.

Generally, it is considered that piles have semi-infinite length when embedded length is more than π/β (usually $3/\beta$ are used in some design standards). β is called as characteristic value of pile and described as:

$$\beta = \sqrt[4]{\frac{k_h D}{4EI}} \qquad (3)$$

Under this condition, the behavior of the piles like displacement and stress, and design embedded length can be easily calculated.

Some design standards describe that the required embedded length is $3/\beta$ because it ensures that sufficient stability, and can simplify pile behavior, such as

346

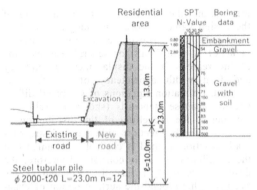

(a) Cross section and ground information

(b) Pressing-in steel tubular pile

Figure 3. Example of construction (Suzuki, 2018).

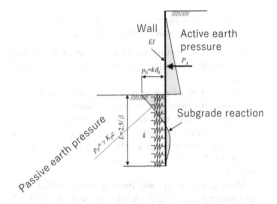

Figure 5. Outline of elastic subgrade reaction method.

deflection and member stress can be simple. However, displacement of wall a should be restricted at small value (e.g. 15mm or 1% of pile diameter) because non-linear behavior of k_h become large and to check that the coefficient of subgrade reaction k_h is within the elastic range.

2.4 Combined subgrade reaction method

In this method, it is assumed that the soil around the surface of the excavation subgrade is perfectly plasticized because the pile displacement is large, and lower subgrade is elastic. The former is called as the plastic region, the latter is the elastic region, and the limit equilibrium analysis method is adopted to the plastic region, and the elastic subgrade reaction method is adopted to the elastic region. The sample of the distribution of the subgrade reaction is shown in Figure 6.

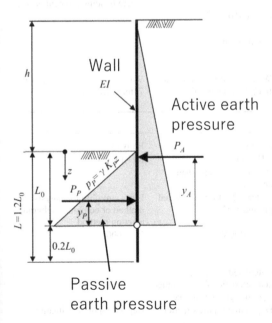

Figure 4. Outline of limit equilibrium analysis method.

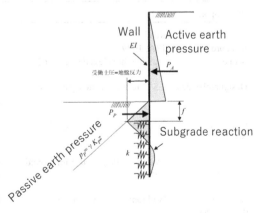

Figure 6. Outline of combined subgrade reaction method.

2.5 Issue of design method of self-standing steel tubular pile walls pressed into stiff ground

Issues of the design method for cantilever steel tubular pile walls embedded into stiff ground is described below.

(1) Selection of a design calculation method

Design methods of embedded length of a self-standing retaining wall used in design standards are shown in Table 1. Many design standards use the elastic subgrade reaction method, but not unified.

(2) Evaluation of characteristic values of stiff ground

Most standard ground investigation methods for determining the characteristic of soil is SPT in Japan. However, SPT cannot evaluate the stiff ground like sandstone, then the engineers must select the other investigation method. However, it is not ruled.

(3) evaluation of design value of k_h

All standards use the elastic subgrade reaction method or the combined subgrade reaction

method, and they set the estimation equations of the design value of horizontal subgrade reaction coefficient k_h. However, evaluation of design value of k_h is difficult for several reasons. One of the reasons is the displacement dependency and load width dependency of k_h. This As a result, estimation equations of k_h are all different in Japanese standards (shown in Table 1). Another is the modelling of the distribution of k_h in the depth direction. When the elastic subgrade reaction method is used, it is assumed that the distribution of k_h is constant. Therefore, the design value of k_h is usually evaluated as the average value around the surface region of excavation subgrade. However, when the self-standing steel tubular pile walls is designed by the elastic subgrade reaction method, the tubular piles have high stiffness because of allowable displacement in seismic, therefore, β becomes smaller, and required embedded length (e.g. π/β or $3/\beta$) become longer.

Table 1. Design calculation method of Japanese standards.

(a) Permanent structures

	Road	River	Port and Harbour
Standard	ACTEC, JASPP, 2007	PIIANDP, 2018	PHJA 2007
Scope	Sheet pile wall whose height is less than about 4.0m	–	–
Design embedded length	$3/\beta$ from excavated ground surface	$3/\beta$ from virtual ground surface	$1.5\,l_{m1}$ from virtual ground surface l_{m1}: depth of first point where bending moment is zero
Estimation formula of k_h	$k_h = k_{h0}\left(\frac{B_H}{0.3}\right)^{-3/4} k_{h0} = \frac{1}{0.3}\alpha E_0$	$k_h = 6910 N^{0.406}$	
Calculation method of k_h	Average in range $1/\beta$ from ground surface	No mention	–
Allowable displacement	Wall top: 1% of wall height (non-seismic) 1.5%of wall height (seismic) Excavated ground level: 15mm	Wall top: 50mm (non-seismic) 75mm (seismic)	Wall top: 100mm (non-seismic) 75mm (seismic)

(b) Temporary structures

	Road	Architecture	Railway
Standard	JRA, 1999	AIJ, 2001	RTRI 2008
Scope	Wall height is less than about 3m	Wall height is less than about 5m	–
Design embedded length	$2.5/\beta$ from excavated ground surface	At least $2/\beta$ from excavated ground surface	Limit equilibrium analysis method or combined subgrade reaction method
Estimation formula of k_h	$k_h = k_{h0}\left(\frac{B_H}{0.3}\right)^{-3/4} k_{h0} = \frac{1}{0.3}\alpha E_0$	$k_h = 1000N$	$k_h = 600N$
Calculation method of k_h	Average in range $1/\beta$ from ground surface	Calculated from average characteristic of soil in range $1/\beta$ from ground surface	–
Allowable displacement	Wall top: 3% of wall height	Depending on the situation	Depending on the situation

3 TRIAL CALCULATION RESULT OF SELF-STANDING STEEL TUBULAR PILE WALLS PRESSED INTO STIFF GROUND

3.1 Overview

In the design of a self-standing wall, the embedded length is usually determined by stability and displacement of the wall. However, the effect of embedded length to the behavior on the wall is not clearified.

In this chapter, this effect is studied by trial calculations using the three methods described in Chapter 2. Specifically, the required embedded length was calculated by the Equilibrium Analysis method and the elastic subgrade reaction method at first. Next, the displacement of the wall is calculated by the combined subgrade reaction method, and the effect of embedded length was explored from the relationship between the embedded length and the wall displacement.

3.2 Calculation conditions

This section describes the calculation condition. Figure 7 shows the outline drawing of the calculation condition. The wall height is 5.0m, background is sand (N-value is 10), and bearing stratum is sand (N-value is 15) or rock. The diameter of the tubular pile has 1.0m and the thickness is 25mm. Table 2 shows the specification of piles, and Table 3 shows the specification of soil. Variation parameters are the kind of bearing stratum and the thickness of soft

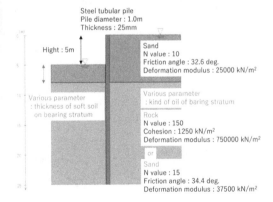

Figure 7. Outline drawing of calculation condition.

Table 2. Condition of piles.

Diameter	1.0 m
Thickness	25 mm
Young modulus	200 GPa
Area of cross section	649 cm²
Secondary moment	772000 cm⁴

Table 3. Characteristic value of soil.

	Soil	Rock
N value	10	150
Friction angle (deg.)	32.6	0
Cohesion (kN/m²)	0	1250
Deformation modulus (kN/m²)	25000	750000
Unit weight (kN/m³)	18	20

ground layer (sand with N-value 10) on the bearing stratum.

The earth pressure and the soil springs (e.g. the coefficient of horizontal subgrade reaction k_h and the limit value of k_h as the passive earth pressure) are modelled from soil characteristic values based on Japanese Railway structure standard (RTRI, 2008).

The active earth pressure is calculated by equation (4), (5).

$$p_a = K_a \left(\sum \gamma_t \cdot z - p_w \right) + p_w + p_s \quad (4)$$

Where
p_a: active earth pressure (kN/m²)
K_a: coefficient of active earth pressure
γ_t: unit weight (kN/m³)
z: depth (m)
p_w: pore water pressure (kN/m²)
p_s: earth pressure from overload (kN/m²)

$$K_a = tan^2 (45° - \varphi/2) \quad (5)$$

Where φ: friction angle of soil (deg.)
The passive earth pressure (used as the limit value of the reaction force of the soil spring) is calculated by equation (6), (7).

$$p_p = K_p \left(\sum \gamma_t \cdot z - p_w \right) + p_w \quad (6)$$

Where
p_p: passive earth pressure (kN/m²)
K_p: coefficient of active earth pressure

$$K_p = \frac{cos^2 \varphi}{\left\{ 1 - \sqrt{\frac{sin(\varphi + \delta) sin\varphi}{cos\delta}} \right\}^2} \quad (7)$$

δ: friction angle between pile and soil (deg.)
In addition, the characteristic value of pile β is calculated as average value of k_h in $1/\beta$ surface

349

region. k_h is estimated from the deformation modulus of soil by following equation (RTRI, 2008).

$$k_h = 0.24E_d \qquad (8)$$

Where, E_d: deformation modulus of soil (kN/m²)

3.3 Calculation results of required embedded length

The required embedded length calculated by the Equilibrium Analysis method and the elastic subgrade reaction method are compared. The results of 4 cases are shown in Table 4.

In all cases, the required embedded lengths of limit equilibrium analysis method are smaller than that of the elastic subgrade reaction method. And especially the gap is larger when the embedded length into rock is large.

This result shows that the embedded length of self-standing steel tubular pile walls pressed into stiff ground can be economically designed by using other method like the limit equilibrium analysis method or the combined subgrade reaction method, not only the elastic subgrade reaction method.

Table 4. Required embedded length of pile.

			Required embedded length (m)	
	Bearing stratum	Thickness of soft ground	Limit Equilibrium Analysis method	Elastic Subgrade reaction method $(3/\beta)$
Case1	Rock	0m	0.87	7.26
Case2		3m	4.55	10.87
Case3		5m	6.75	15.28
Case4	Sand	0m	10.43	15.35

3.4 Calculation results of relationship between embedded length and wall displacement

In usual, the limit value of displacement at the wall top is set as required performance considering the surrounding environment, and the specifications are mainly determined by this factor. Then the displacement at the wall top are calculated at each embedded length by the combined subgrade reaction method.

Figure 8 shows the relationship between displacement at the wall top and embedded for each case.

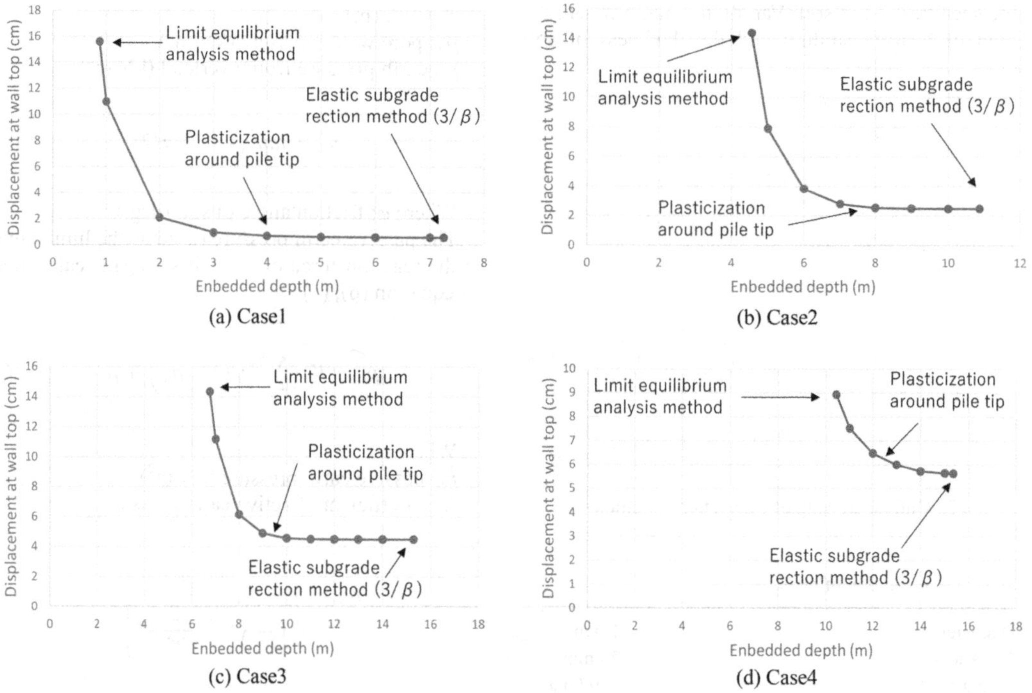

(a) Case1

(b) Case2

(c) Case3

(d) Case4

Figure 8. Relationship between embedded length and wall displacement.

350

The relationship is not linear, and the displacement is rapidly increase after plasticization around pile tip regardless of the thickness of the soft soil when the bearing stratum is rock.

On the other hand, displacement is gradually increasing before the plasticization. This behavior is different from soil type of baring stratum, Especially, the gap between the embedded length with rapid increase of displacement and required embedded length by limit equilibrium analysis method is not large, so it is especially important to check the embedded length in design and construction when the pile is installed into stiff ground like hard rock.

4 CONCLUSION

In this paper, the overview of each analysis method is introduced, and the optimum method for the self-standing retaining walls using the steel tubular piles pressed into the stiff ground is discussed by the comparison with the calculated embedded length depending on the analysis method and the design condition.

Use of the limit equilibrium analysis method or the combined subgrade reaction method may lead the economical results of design when the piles are pressed into stiff ground.

On the other hand, the embedded length influence on the stability of a self-standing wall pressed into the stiff ground, therefore a ground investigation and a check at construction become very important.

REFERENCES

Advanced Construction Technology Center (ACTEC), Japanese Association of for Steel Pipe Piles (JASPP). 2007. Design manual of self-standing steel pipe walls.

Architectural Institute of Japan (AIJ). 2001. Design Recommendations for Architectural Foundation.

Giken Ltd. 1999. Rotary Press-in. https://www.giken.com/en/press-in_method/penetration_tech/rotary_cutting/ (accessed 2020.11.15).

Japan Road Association (JRA). 1999. Guideline of road works and temporary structures.

Port and Harbour Association of Japan (PHJA). 2007. Technical Standards of port and harbour facilities and explanations.

Public Interest Incorporated Association of Nationwide Disaster Prevention (PIIANDP). 2018. Design Guideline for Disaster Recover work.

Railway Technical Research Institute 2008. Design Standards for Railway Structures ana Commentary (Cut and Covered Tunnel).

Suzuki, N. and Kajino, K. 2018. Issues for the Reduction of the Embedded Length of Cantilevered Steel Tubular Retaining Wall Pressed into Stiff Ground, Proceedings of the First International Conference on Press-in Engineering 2018, pp.159–166.

Winkler, E. 1867. Die Lehre Von Elasticitaet Und Festigkeit, 1st Edn.

Proceedings of the Second International Conference on
Press-in Engineering 2021, Kochi, Japan – Matsumoto et al (eds)
© 2021 Taylor & Francis Group, London, ISBN 978-1-032-10414-0

3D fem analysis of partial floating steel sheet piling method on two-layered ground

K. Kasama
Tokyo Institute of Technology, Tokyo, Japan

H. Fujiyama
Oyo Corporation, Saitama, Japan

J. Otani
Kumamoto University, Kumamoto, Japan

ABSTRACT: This paper presents the result of a finite element method to evaluate the effectiveness of the partial floating steel sheet piling method (PFS method) for the stability of embankment on two-layered soft ground. In order to clarify a suitable ground layer structure for the PFS method, ground is assumed to consist of a sand layer as a surface layer and a clay layer as a bottom layer, and the settlement and the lateral displacement of soft ground due to the construction of an embankment are investigated changing the thickness ratios of sand and clay layers. In addition, the effect of the length of a floating sheet pile of the PFS method is investigated on the ground deformation under different ground layer structures. Finally, suitable ground layer structures for the PFS method and the minimum length of floating sheet piles are discussed to reduce the lateral displacement of soft ground.

1 INTRODUCTION

There is one of ground reinforcement technologies, called a steel sheet piling method to reduce the ground settlement and the lateral displacement of soft ground due to the construction of a river embankment. In the steel sheet piling method, there are a conventional method (called "the CS method") penetrating steel sheet piles to the supporting layer, a floating steel sheet piling method (called "the FS method") not penetrating to the supporting layer, and the partial floating steel sheet piling method (called "the PFS method") that is a combination of the FS method and the CS method. A steel sheet piling method including the PFS method has been widely used as a reinforcement technology for river embankments in Kumamoto Prefecture, Japan.

Ochiai et al (1991) verified the effectiveness of CS and FS methods for the settlement reduction countermeasure based on the field test and the finite element analysis. They concluded that the stress cutoff effect of steel sheet piles was effective for reducing settlement and lateral displacement of soft ground due to embankment construction. Kimizu and Otani (2010) compared the reinforcement effect

of the CS and PFS methods on the soft ground deformation due to embankment construction using the 3D finite element analysis. Furuichi et al (2015) and Fujiwara et al (2017) evaluated the seismic behavior of embankment with steel sheet pile based on the result of a model test and a numerical analysis.

The objective of this study is to evaluate the effectiveness of the PFS method for the reduction of deformation of two-layered soft ground due to embankment construction using the 3D finite element method. In order to clarify a suitable ground layer structure for the PFS method, soft ground is assumed to consist of a sand layer as a surface layer and a clay layer as a bottom layer, and the settlement and the lateral displacement of soft ground due to the embankment construction are investigated by changing the thickness ratios of sand and clay layers. In addition, the effect of the length of a floating sheet pile of the PFS method is investigated on the soft ground deformation under different ground layer structures comparing with/without the reinforcement of the CS method. Finally, suitable ground layer structures for the PFS method and the minimum length of a floating sheet pile are discussed to reduce the lateral displacement of soft ground.

DOI: 10.1201/9781003215226-33

2 NUMERICAL PROCEDURE

A 3D soil-water coupled finite element method developed by Nakai (2007) was used in this study. Figure 1 shows the schematic diagram of soft ground consisting of a sand layer as a surface layer and a clay layer as a bottom layer which is a typical ground structure of the Kumamoto Plain in Japan. It is assumed that the depth of the supporting layer is 30 m and the ground water level is at the top of the ground. Clay layer was modeled as Sekiguchi-Ohta model while sand layer was modeled as elastic material. Figure 2 shows finite element mesh used in this study. The bottom boundary in Figure 2 is fixed for x, y and z directions while side boundary is fixed for x and z directions (free for y direction). In addition, bottom and side boundaries are drained condition. The embankment (wet unit soil density = 17.4 kN/m^3) is assumed to be 3.4 m in height, 6.0 m on the top side and 19.6 m on the bottom

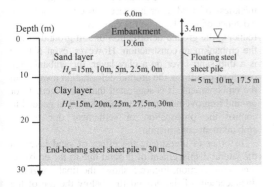

Figure 1. Schematic diagram of model ground.

side. The embankment construction was completed in 76 days.

Firstly, in order to investigate the influence of the sand layer on the clay layer on the deformation of soft ground, the settlement and the lateral displacement of soft ground due to the embankment construction are analyzed by changing the thick-ness ratios of sand and clay layers from 0 (no sand layer) to 1.0 (15 m sand layer and 15 m clay layer). Table 1 summarizes the input parameters for sand and clay layers, which are determined by element tests conducted to soil samples taken from the Kumamoto Plain.

Next, in order to reduce the settlement and the lateral displacement of soft ground, the geometry of the PFS method consisting of floating sheet piles with an end-bearing sheet pile was carefully modelled in the 3D FEM analysis. Namely, the unit of five floating sheet piles with one end-bearing sheet pile for the typical PFS method was modelled as a solid element. The sheet pile was installed at the toe of the embankment. It is assumed that the length and the width of an end-bearing sheet pile is 30 m and 0.9 m respectively. The length of a floating sheet pile of the PFS method is changed from 5 m, 10 m and 17.5 m to propose an effective length of a floating sheet pile to reduce the soft ground deformation. For the comparison, the CS method consisting of only one end-bearing sheet pile was also analyzed. From the practical application of PFS method in Kumamoto Prefecture in Japan (Kasama et al., 2020), the mean length ratio of floating sheet pile and end-bearing sheet pile is 0.66 ranging from 0.27 to 0.9.

Table 2 indicates the material properties of the sheet pile. The modelling of steel sheet pile as a solid element in the FE analysis was referred to Nakai et al (2017).

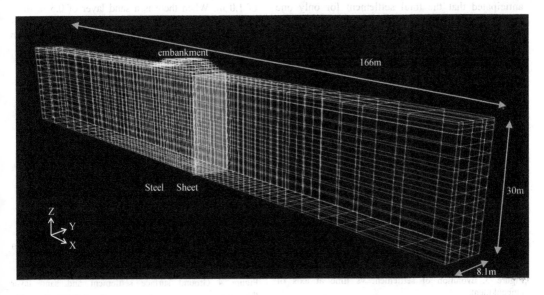

Figure 2. 3D finite element mesh.

Table 1. Input parameters for sand and clay layers.

	Sand layer	Clay layer
Elastic modules E (kPa)	9806	5295
Poisson's ratio v	0.30	0.41
slope of the critical state line in e-ln$p$ space λ	-	0.54
slope of the overconsolidated line in e-ln$p$ space κ	-	0.04
slope of the critical state line in p-q space M	-	1.63 ($\phi = 39.8$ °)
Initial void ratio e_0	0.89	2.59
Wet unit weight γ_t (kN/m^3)	18.24	14.32
Permeability (m/day)	$1.15*10^{-2}$	$5.30*10^{-3}$

Table 2. Input parameters for steel sheet pile.

Elastic modules E (kPa)	$3.2*10^7$
Moment of inertia of area (m^4)	$6.7*10^{-4}$
Cross sectional Area (m^2)	$2.0*10^{-1}$

3 EFFECT OF SAND LAYER

Figure 3 shows the evolution of settlement vs time at the axis of the embankment due to the embankment construction for a given thickness ratio of sand clay layers. It is seen that instantaneous settlement was remarkable irrespective of thickness ratios, which was finished within 76 days (the day of completion for embankment construction). After causing instantaneous settlement, consolidation settlement for all thickness ratios gradually occurred and completed within 912 days. It was anticipated that the total settlement for only one clay layer is 230 cm. It can be characterized that the total settlement greatly reduces due to the

existence of the sand layer. Namely, even if there is a sand layer of 0.5 m thick on the clay layer, the total settlement sharply reduces to 70 % of that for only one clay layer. In addition, the total settlement for the soft ground with a sand layer of 15 m thick on a clay layer of 15 m thick is 11.7 cm, which corresponds to only 5 % of that for only one clay layer.

This is because the embankment load was widely distributed and reduced due to the existence of sand layer, and the heaving deformation of clay layer due to embankment load was restrained due to the self-weight of sand layer.

Figure 4 shows the total ground settlement after consolidation depending on the thickness ratios of sand and clay layers. It is noted that the location of the toe of the embankment is 10 m from the axis of the embankment. It is seen that there is a large settlement at the toe of the embankment while there is also ground heaving behind the toe. The magnitudes of the total settlement and the ground heaving decrease as the thickness of sand layer increases. In other words, the existence of sand layer on the clay layer is effective for reducing the ground deformation of soft ground due to the embankment construction. However, even if there is a thin sand layer on a clay layer, the embankment construction affects the ground deformation away from the embankment. It is suggested that reinforcement or ground improvement for the embankment is needed to control the propagation of settlement due to the embankment construction.

In order to investigate the effect of sand layer on the lateral displacement of ground due to the embankment construction, Figure 5 shows the final horizontal displacement of the ground just below the toe of the embankment against the ground depth. When there is no sand layer on the clay layer, the maximum horizontal displacement of 80 cm is obtained at the depth of 1.0 m. When there is a sand layer of 0.5 m thick on the clay layer, the maximum horizontal displacement reduces 50 % of that for no sand layer. It can

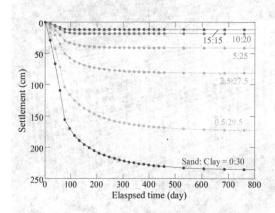

Figure 3. Evolution of settlement vs time at axis of embankment.

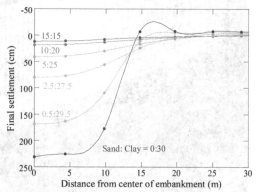

Figure 4. Ground surface settlement and sand layer thickness.

354

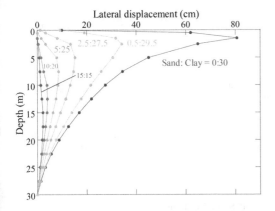

Figure 5. Lateral displacement and sand layer thickness.

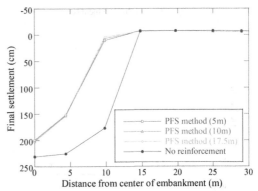

Figure 7. Ground surface settlement with/without PFS method.

be characterized that the maximum horizontal displacement greatly reduces due to the existence of the sand layer similar to the ground settlement. For cases with the sand layer, the maximum horizontal displacement is obtained slightly below the boundary between the sand layer and the clay layer.

4 EFFECT OF PFS METHOD

In order to investigate the effectiveness of the PFS method for the settlement countermeasure, Figure 6 shows the evolution of ground settlement vs time at the axis of the embankment for the soft ground without a sand layer. It is seen that consolidation settlement is reduced by the PFS method while instantaneous settlement due to the embankment construction is similar to that for no reinforcement. The total settlement of ground reinforced by the PFS method is 86 % of that for no reinforcement irrespective of the length of floating piles for the PFS method.

Figure 7 shows the total ground settlement after consolidation for the ground with/without the PFS

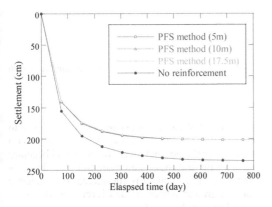

Figure 6. Evolution of settlement vs time with/without PFS method.

method for the ground without a sand layer. Due to the reinforcement of the PFM method, ground settlement and heaving beyond the toe of the embankment is well reduced and the propagation of ground deformation due to the embankment construction is well controlled irrespective of the length of a floating pile for the PFS method. Therefore, it can be suggested that the PFS method is effective for restricting the ground settlement and heaving due to the embankment construction.

In order to evaluate the effective length of a floating pile of the PFS method for the ground without a sand layer, Figure 8 shows a lateral displacement of ground just below the toe of the embankment for the floating pile length of 5 m, 10 m and 17.5 m together with the result for no reinforcement. It is seen that the lateral displacement for the end-bearing pile reduces to 50 % of that for no reinforcement irrespective of the length of the floating pile. Namely, the influence of the length of floating pile for the PFS method is very small on the lateral displacement on the end-bearing pile. On the other hand, the lateral displacement for the floating pile is slightly larger than that of the end-bearing pile. In addition, the tip of the 5 m floating pile shows a large lateral displacement compared to those for other floating piles. Therefore, it is considered that a 5 m floating pile is not enough to restrain the lateral displacement of ground due to the embankment construction.

In order to discuss the effective two-layered ground condition for the PFS method, Figure 9 shows the lateral displacement of ground just below the toe of the embankment for a given thickness ratio of sand and clay layers. In this figure, the lateral displacement at the end-bearing pile is shown as dots and that at the floating pile is shown as lines. The difference between the end-bearing pile and the floating pile becomes large when the length of the floating pile is 5 m and the thickness of the sand layer less than 2.5 m. In addition, the tip of the

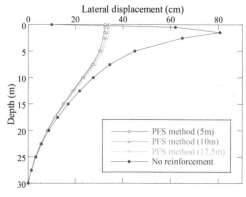

a) End-bearing pile

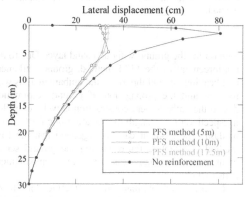

b) Floating pile

Figure 8. Ground surface settlement with/without PFS method.

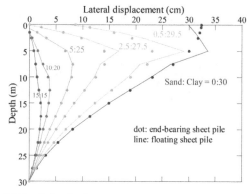

a) PFS with 5 m depth.

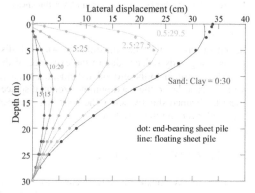

a) PFS with 10 m depth.

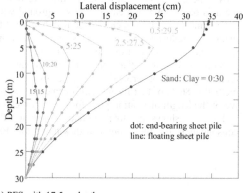

c) PFS with 17.5 m depth.

Figure 9. Lateral displacement of ground reinforced with PFS.

floating pile for 5 m length is not well fixed indicating the maximum horizontal displacement for small thickness ratio. However, the lateral displacement is well restrained with 10 m and 17.5 m floating piles for a given thickness ratio.

In order to evaluate the effectiveness of the PFS method for the settlement countermeasure of two-layered soft ground due to the embankment construction, the reduction ratio R, which is the final lateral displacement divided by that for the ground without a sand layer and reinforcement, was calculated. Figure 10 shows the reduction ratio R against the thickness ratio of sand and clay layers. The reduction ratio R sharply increases as the thickness ratio increases, which suggests that the effect of a sand layer on a clay layer is very large on the restrain of soft ground deformation due to the embankment construction. When the thickness ratio is more than 0.2 (which corresponds to the sand layer of 5 m on the clay layer of 25 m), the lateral displacement reduces 85 % of that for no sand layer and no reinforcement.

Figure 11 shows the reduction ratio R only due to the effect of the PFS method. The reduction ratio R is 0.6 for the no sand layer condition and becomes zero for the thickness ratio more than 0.2. In other words, the PFS method is considered to be suitable for soft ground without a sand layer or with a sand layer less than 5 m thick.

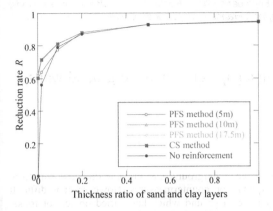

Figure 10. Reduction rate of lateral displacement.

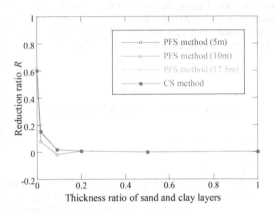

Figure 11. Effect of PFS on lateral displacement reduction.

5 SUMMARY

In order to investigate the effectiveness of the partial floating steel sheet piling method (PFS method) for the settlement countermeasure of two-layered soft ground due to embankment construction, a series of 3D soil-water coupled FE analyses was conducted. The main conclusions are as follows:

1. The total settlement and lateral displacement of soft ground greatly reduces due to the existence of the sand layer. This is because the embankment load was widely distributed and reduced due to the existence of the sand layer and the heaving deformation of the clay layer due to embankment load was restrained due to the self-weight of the sand layer.

2. Due to the reinforcement by the PFM method for embankment, the ground settlement and heaving beyond the toe of the embankment is well reduced and the propagation of ground deformation due to the embankment construction to surroundings is well controlled irrespective of the length of a floating pile for the PFS method.

3. The floating pile of 10 m for the PFS method is enough to restrain the lateral displacement of ground due to the embankment construction irrespective of thickness ratio of sand and clay layers because the tip of the floating pile for 5 m length is not well fixed indicating the maximum horizontal displacement for the thickness ratio less than 0.2.

ACKNOWLEDGEMENT

This study is part of the activity of Technical Committee on Expansion of Applicability and Assessment of Dynamic Behavior of the PFS Method (chairperson: Prof. Otani, Kumamoto University) under the International Press-in Association. The field data of the steel sheet piling method constructed in the Kumamoto Plain in Japan was provided by the Ministry of Land, Infrastructure, Transport and Tourism, Kyushu Regional Development Bureau.

REFERENCES

Fujiwara, K., Kobori, Y., Yashima, A., Sawada, K., Taenaka, S., Otsushi, K. & Toda, K. 2017. Study on reinforcement effect of double sheet-pile wall installed in dyke as the countermeasure against large scale earthquake, *Japanese Geotechnical Journal*, Vol. 12, 1, 109–122 (in Japanese). https://doi.org/10.3208/jgs.12.109

Furuichi, H., Hara, T., Tani, M., Nishi, T., Otsushi, K & Toda, K. 2015. Study on Reinforcement Method of Dykes by Steel Sheet-pile against Earthquake and Tsunami Disasters, *Japanese Geotechnical Journal*, Vol. 10, 4, 583–594 (in Japanese). https://doi.org/10.3208/jgs.10.583

Kasama, K., Yamamoto, S., Ohno, M., Mori, H., Tsukamoto, S. & Tanaka, J. 2020. Seismic damage analysis on the river levees reinforced with steel sheet pile by the 2016 Kumamoto earthquake, *Japanese Geotechnical Journal*, 15, 2, 395–404 (in Japanese). https://doi.org/10.3208/jgs.15.395

Kimzu, M. & Otani, J. 2010. 3D numerical analysis on the settlement reduction of PFS method due to embankment construction, *JSCE*, 479–480 (in Japanese).

Nakai, T. 2007. Modeling of soil behavior based on t_{ij} concept, *13th Asian Regional Conference on Mechanics and Geotechnical Engineering, Keynote Paper*, Vol. 2, 69–89.

Nakai, K., Noda, T. & Kato, K. 2017. Seismic assessment of sheet pile reinforcement effect on river embankments constructed on a soft foundation ground including soft estuarine clay, *Canadian Geotechnical Journal*, 54, 10, 1375–1396. https://doi.org/10.1139/cgj-2016-0019

Nakai, K., Noda, T., Taenaka, S., Ishihara Y. & Ogawa N. 2018. Seismic Assessment of Steel Sheet Pile Reinforcement Effect on River Embankment Constructed on a Soft Clay Ground, *First International Conference on Press-in Engineering*, 221–226.

Ochiai, H., Hayashi, S., Otani, J., Umezaki, T & Tanaka, Y. 1991. Numerical verification of sheet-pile countermeasure in soft ground, *7th International conference on computer methods and advances in geomechanics*, 387–392.

Proceedings of the Second International Conference on
Press-in Engineering 2021, Kochi, Japan – Matsumoto et al (eds)
© 2021 Taylor & Francis Group, London, ISBN 978-1-032-10414-0

Experimental study for liquefied soil in a gap between underground walls

K. Fujiwara & E. Mallyar
Tokai University, Hiratsuka-city, Kanagawa, Japan

ABSTRACT: As the steel or concrete walls constructed underground can prevent liquefied soil from flowing, they have been used as the effective anti-liquefaction countermeasures. However, it is difficult that these countermeasures are constructed perfectly under a ground which has buried objects or rocks, so the walls may have some tiny gaps. As the behaviour of liquefied soil flowed out of a gap is not clarified. Therefore, the author carried out two steps of experiments. Firstly, the author focused on a small gap between the walls and carried out shaking model tests. Secondly, the author investigated the effect of the gaps on a whole construction. Shaking model tests for an embankment reinforced by the sheet-piles which has gaps, or no gaps were performed. Through the second model tests, it was confirmed that the soil was pushed from the gaps between sheet-piles, which induced the settlement of the embankment.

1 INTRODUCTION

The severe seismic damage to constructions in Japan's Tohoku region was caused by the 2011 off the Pacific Coast of Tohoku Earthquake. A large earthquake such as Nankai Trough Earthquake is also concerned to occur in the near future and there is a fear that construction or residential areas will sink by liquefaction. As the steel or concrete walls constructed underground can prevent liquefied soil from flowing, they have been used as the effective anti-liquefaction countermeasures (Fujiwara, Koseki, Otsushi & Nakayama 2013). However, it is difficult that these countermeasures are constructed perfectly under a ground which has buried objects or rocks, so the walls may have some tiny gaps. In addition, there are also some methods originally designed to be with gaps such as the PFS (Partial Floating Sheet-pile) method (Fujiwara, Nakai & Ogawa 2019). As the behaviour of liquefied soils flowed out of a gap is not clarified, the effect of the gaps on a whole construction is also not clarified. Therefore, the author carried out two steps of experiments. Firstly, the author focused on a small gap between the walls and carried out shaking model tests to investigate the liquefied soil behaviour. Secondly, the author investigated the effect of the gaps on a whole construction. Shaking model tests for an embankment reinforced by the sheet-piles which has gaps or not were performed. Through the two tests, a possibility of modelling structures with gaps in the 2D numerical analysis was discussed.

2 SOIL BEHAVIOUR BETWEEN GAPS

2.1 Test condition

2.1.1 Device
A test box shown in Figure 1 was used for "the wall gap test". The rigid box was used in this study. The dimension of box is 353 mm of width, 170 mm of height and 200 mm of depth. These dimensions are approximately 1/25 of actual size. The box was fixed on a table of the shaking device during the motion.

2.1.2 Test cases
Three cases shown Table 1 were carried out to confirm the behavior of liquefied sand in a small gap. Case-A (without walls) was also carried out in order to compare to Case-B and C (with walls).

2.1.3 Procedure
The procedure of the test is explained in numerical order;

(1) Two wooden walls were put in the box with a gap. The bottom of the walls was fixed to the bottom of the test box by adhesives. The walls have enough thickness (13 mm) to be considered as rigid body.
(2) Water and Toyoura sand were put in the box in this order. Toyoura sand with the characters of soil density (ρ_s=2.64 g/cm^3), maximum void ratio (e_{max}=0.927), minimum void ratio (e_{min}=0.635), permeability coefficient (k=1.94×10^{-2} cm/s

DOI: 10.1201/9781003215226-34

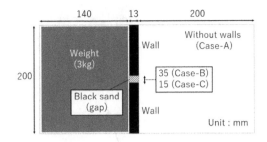

(a) Plain figure

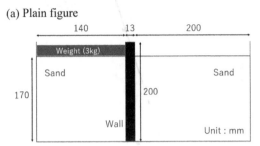

(b) Cross section

Figure 1. Test box.

Table 1. Test cases.

	Type	Gap (mm)	Dr (%)
Case-A	Without walls	-	50
Case-B	Two walls	35	55
Case-C	Two walls	15	59

(D_r=45%)) and equal factor (U_c=1.37), 50% diameter (D_{50}=0.325 mm), was used as liquefiable soil.

(3) Black sand that the same sand as (2) colored with black ink was put between the two walls, which indicated the behaviour of liquefied soil. In advance, the authors put thin plastic sheets to enclose the gap in order to make the black sand form. Then the authors put the black sand in the enclosed area, after that pulled the sheets out softly.

(4) The saturated sand layer which had a thickness of 170 mm and relative density (Dr) of 50 % approximately was made.

(5) A weight which was gravels wrapped with a plastic sheet was put on one side of the soil surface. The mass of the weight was 3 kg which was equivalent to a common embankment for this model size.

(6) After the model was put on the shaking table, a sine wave motion was given to the shaking table. The sine wave had a frequency of 5 Hz, a duration of 5 seconds and a maximum amplitude of 5 m/s^2, shown in Figure 2. This condition can be converted to 0.27Hz, 56 seconds and 5 m/s^2 in actual

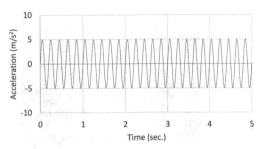

Figure 2. Input motion.

size based on the similarity rule (Iai 1988), which had an extremely large motion.

(7) After shaking, the authors absorbed the extra surface water by a sponge, took off the weight and dug the sand to measure the horizontal displacement of the black sand.

2.2 Test result

Figure 3(a) shows a picture of the surface after shaking for Case-B as an example of test results. Almost whole ground was liquefied by the large motion. The weight was already taken off in Figure 3(a). As the sand was liquefied by the motion, the sand under the weight was settled shown, which moved the black sand laterally. The settlements at the side of the gap under the weight were almost the same as Case-A (40 mm), Case-B (38 mm) and Case-C (35 mm).

The authors dug the sand and took pictures at 10mm, 25mm, 50mm, 75mm and 100mm under from the surface. Figure 3(b) shows a picture of 10 mm under the sand surface for Case-B as a typical picture. The black sand was pushed toward the right side by the weight, which had an arc shape.

Figure 4 shows the residual horizontal displacement of the black sand along the vertical direction. The horizontal displacement of the black sand is defined as the longest distance from the gap to the outside of the black sand arc. As the black sand was mixed with non-colored sand, it was difficult to be observed clearly. Therefore, these values were evaluated in 5 mm unit.

As for Figure 4, the residual horizontal displacement became larger as the width of the gap became larger under the depth of 25 mm (Case-A is considered to have an infinity gap). The horizontal displacement of Case-B was close to that of Case-A. The horizontal displacement of Case-C which had the smallest values was approximately 25 % less than that of Case-A. On the other hand, Case-A had the smallest horizontal displacement over the depth of 25 mm. It is considered that this is because, although the surface sand was pulled back to the weight for Case-A, shown as Figure 3(c), the walls prevented the sand from pulling toward the weight for Case-B and C.

359

(a) Surface ground

(b) 10 mm under the surface

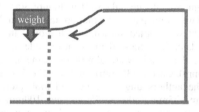

(c) Conceptual figure of soil back for Case-A

Figure 3. Test results.

3 CONSTRUCTION BEHAVIOUR WITH GAPS

The authors focused on an embankment reinforced by the sheet-piles as a construction which has gaps inside it. Sheet-piles are installed into the ground and lined in the depth direction along the toe of an embankment occasionally, which reinforce an embankment against earthquake (Fujiwara, Taenaka, Otsushi, Yashima, Sawada, Hara, Ogawa, & Takeda 2017). As a specific type of the reinforcement, the different lengths of sheet-piles are installed alternately, called the PFS (Partial Floating Sheet-pile) method shown in Figure 5 (Fujiwara, Nakai & Ogawa 2019). In this chapter, the authors focused on the reinforcement with the sheet-pile skipping alternately, which made a gap between two sheet-piles. The behaviour of the embankment with and without sheet-piles gaps is discussed by a series of shaking model tests.

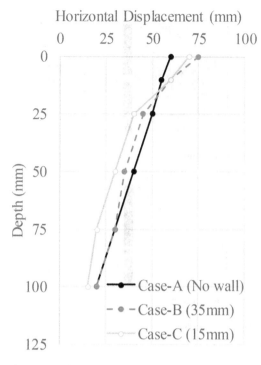

Figure 4. Horizontal displacement.

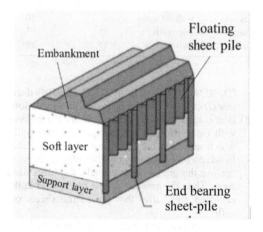

Figure 5. An example of construction with gaps (the PFS method).

3.1 Test condition

3.1.1 Dimension
A rigid container was used for the tests. The dimensions were set as approximately 1/20~1/30 of actual size shown in Figure 6. The tests were conducted under 1 g gravity field. The target embankment was cut in half size considering symmetry for the

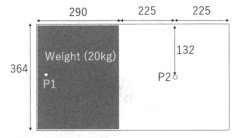

(a) Plain figure for Case-1

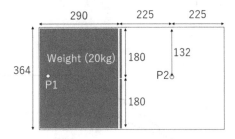

(b) Plain figure for Case-2

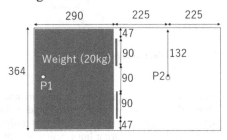

(c) Plain figure for Case-3

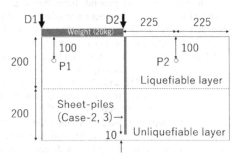

(d) Cross section for all Cases

Figure 6. Dimensions of test cases.

limitation of the container's size. A steel plate was used for the test as an embankment, mentioned later.

3.1.2 Soil character
The test ground was made of Toyoura sand same as used in Chapter 2. The relative density and weight per unit volume for the liquefiable layer are indicated in Table 2.

Table 2. Test cases (liquefiable layer).

	Case-1	Case-2	Case-3
	No sheet-piles	Sheet-piles (without gaps)	Sheet-piles (with gaps)
γt (kN/m^3)	18.7	18.7	18.6
D_r (%)	41	44	38

3.1.3 Procedure
The test ground was made by putting water after soil. Firstly, the unliquefiable layer which had a thickness of 200 mm and relative density (D_r = 95 %) was made by sticking with a wood bar. Secondly, model sheet-piles made from acryl which had a thickness of 5 mm, a length of 400 mm and Young's ratio 3.2 MPa, was installed 190 mm into the unliquefiable layer. The thickness and the material are decided considering the similarity rule against the actual size (Iai 1988). Thirdly, the liquefiable layer which had a thickness of 200 mm and relative density (D_r = 40 %) was made by sprinkling the sand. Finally, a steel plate was used instead of an embankment. The plate which had a size of 360 x 290 x 20 mm and a weight of 20 kg equivalent to the embankment was put on one side of the surface ground. The steel plate and model sheet-piles were put with 1~10 mm clearance from the side of the container not to be attached each other.

3.1.4 Test cases
Three cases were conducted as follows;
   Case-1: Without sheet-piles
   Case-2: Sheet-piles without gaps
   Case-3: Sheet-piles with gaps
   Two sheet-piles whose width was 180 mm were used for Case-2 and 3. There was a gap of 90 mm equivalent to approximately 2 m in actual size for Case-3.

3.1.5 Measurement points
The measurement points are also indicated in Figure 6. The vertical displacements of the steel plate were measured at D1 and D2 for Figure 6(d). The excess pore water pressures were measured at P1 and P2 100 mm under the surface. The bending strain of the sheet-piles were measured at four points; 100 mm, 150 mm, 200 mm and 260 mm from the bottom of the container respectively.

3.1.6 Input motion
A sine wave which has a frequency of 3 Hz, duration of 5 seconds and a maximum acceleration of 3.0 m/s^2 shown in Figure 7, was given to the bottom of the test container.

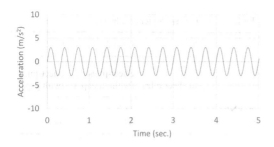

Figure 7. Input motion.

3.2 Test result

3.2.1 Excess pore water pressure ratio

The time histories of excess pore water pressure ratio at P1 and P2 are shown in Figure 8. The excess pore water pressure ratio means the value that the excess pore water pressure is divided by the initial effective stress, calculated from the weight and the unit weight of the soil. The excess pore water pressure ratio at P1 and P2 closed to 1.0 during shaking, which indicated that the ground was liquefied during shaking. The excess pore water pressure ratio was reduced as time passed.

3.2.2 Settlement

The time histories of vertical displacement at D1 and D2 are shown in Figure 9. The weight plate was settled down during liquefaction for all cases. As the settlement at D1 was different from that at D2, the plate was inclined. The residual average values of D1 and D2 were 56mm (Case-1), 34mm (Case-2) and

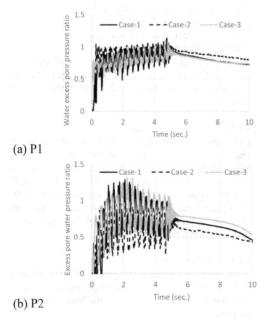

(a) P1

(b) P2

Figure 8. Excess pore water pressure ratio.

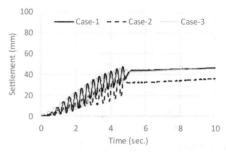

(a) D1

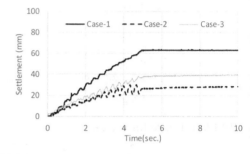

(b) D2

Figure 9. Weight settlement.

45mm (Case-3) respectively, which indicated that sheet-pile with or without gaps could reduce the settlement. The ground was pushed laterally by the weight of the plate during liquefaction, which caused the settlement. The sheet-pile installed next to the weight prevented the ground from moving, which reduced the weight settlement accordingly. Although the settlement at D2 is larger than that at D1 for Case-1, the settlement at D1 is larger than that at D2 for Case-2 and 3. It is assumed that the settlement at D2 was specially reduced by the near sheet-pile. The average values of D1 and D2 for Case-3 was larger than that for Case-2. It is considered that there are two reasons. One reason is that the soil passed through the gap for Case-3. As this width of 90 mm was much larger than 35 mm, the soil passed through the same as without sheet-piles, as discussed in chapter 2. The other reason is that the deformation of the sheet-pile for Case-3 was larger than that for Case-2, to be discussed in the next section.

3.2.3 Bending strain

The residual and maximum bending strain are shown in Figure 10. The sheet-pile had bending strains caused by the ground moving laterally. The bending strain in Case-3 was larger than that in Case-2. It is considered that more comprehensive earth pressure applied to the sheet-pile per width, which made the weight settle more in Case-3. In addition, although the width of the sheet-piles in Case-3 was half of that in Case-2, the bending strain in Case-3 was not

362

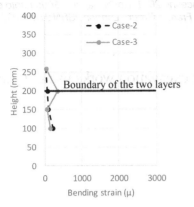

(a) Residual

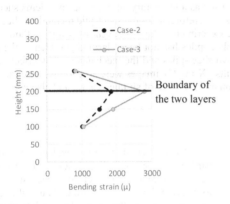

(b) Maximum

Figure 10. Bending strain.

half of that in Case-2 exactly. The relationship of the earth pressure with the width of sheet-piles should be discussed. It is assumed that the bending of the walls makes the lateral displacement of soils large.

4 DISCUSSION FOR NUMERICAL ANALYSIS

Through the two model tests, the way of numerical analysis applied to a construction with gaps is discussed in this chapter. As the structure has gaps, there are multiple sections in the depth direction. Therefore, the 3D numerical analysis is more suitable to analyze than the 2D analysis to gain a precise result (Fujiwara, Nakai & Ogawa 2021). On the other hand, the 2D numerical analysis is more suitable in design work for the easiness and time-shortening. Generally, the rigidity of sheet-piles is adjusted to be averaged value in the depth direction when a 3D model is converted to a 2D model. However, there should be a limitation of this way because soil passes through gaps and the bending deformation of the sheet-piles are different in the depth direction, as indicated in chapter 2 and 3. The limitation for numerical analysis should be discussed by the comparison of 3D and 2D methods in the future.

5 CONCLUSIONS

The authors carried out two model tests for liquefaction; one was the wall gap test which had different width of gaps to investigate the soil behaviour between walls, the other was the embankment test which had sheet-piles with or without gaps at the toe of it with and without gaps to investigate the effect of gaps on a whole construction. From the above, following results were gained.

1) Through the wall gap test, the wider the width of the gaps became, the larger the lateral displacement of the liquefied soil between gaps became. The lateral displacement of soil in the width of 35 mm gap was close to that without wall considered to be the same as the infinity width.

2) Through the embankment test, the soil was pushed laterally by the weight during liquefaction. The effectiveness of sheet-piles against settlement can be confirmed not only for the case without a gap but also the case with a gap. Additionally, it was confirmed that the sheet-piles with a gap had more earth pressure per width than that of without a gap per depth.

3) After the two tests, the application of numerical analysis for a structure with gaps was also discussed. It is considered that there should be a limitation when a 3D numerical model is converted into a 2D numerical model, because soil passes through gaps and the bending deformation of the sheet-piles are different in the depth direction. The limitation for a numerical analysis should be discussed by the comparison of the 3D and the 2D method in the future.

REFERENCES

Fujiwara, K., Koseki, J., Otsushi, K., & Nakayama, H. 2013. Study on reinforcement method of levees using steel sheet-piles, *Foundation and Soft Ground Engineering Challenges in Mekong Delta*, pp.281–289

Fujiwara, K. Taenaka, S. Otsushi, K. Yashima, A. Sawada, K. Hara, T. Ogawa, T. & Takeda, K. 2017. Study on Coastal Levee Reinforce Using Double Sheet-piles with Partition Walls, *International Journal of Offshore and Polar Engineering*, Vol. 27, No. 3, September 2017, 310–317.

Fujiwara, K. Nakai, K. & Ogawa, N. 2019. Quantitative evaluation of PFS (Partial Floating Sheet-pile) Method under liquefaction, *Geotechnics for Sustainable Infrastructure Development*, 467–472.

Fujiwara, K. Nakai, K. & Ogawa, N. 2021. 3-D numerical analysis for partial floating sheet-pile method under liquefaction, *Journal of Japanese Society of Civil Engineering* (Submitted).

Iai, S. 1988. Similitude for Shaking Table Test on Soil structure-Fulid Model in 1 g Gravitation Field, *Report Port Harbor Res Inst*, 27(3), 3–24.

Proceedings of the Second International Conference on
Press-in Engineering 2021, Kochi, Japan – Matsumoto et al (eds)
© 2021 Taylor & Francis Group, London, ISBN 978-1-032-10414-0

Model test on double sheet-pile method for excavation works using X-ray CT

H. Sugimoto, S. Akagi, T. Sato & J. Otani
Kumamoto University, Kumamoto, Japan

H. Nagatani & A. Nasu
Kajima Corporation, Tokyo, Japan

ABSTRACT: Steel sheet-piles are driven into the ground to prevent failure of soils and intrusion of water. The current main function of constructing sheet-piles used as a temporary structure has problems such as low productivity, few adoption opportunities, and high cost. Therefore, a double sheet-pile method has been developed. Also, from centrifugal model tests or full-scale experiments, the effectiveness of this method had been confirmed. However, the soil behavior between two sheet-piles has not been clarified yet. Hence, the purpose of this study is to clarify the soil behavior between two sheet-piles and the mechanism of its effectiveness of the structure using X-ray CT. From experimental results,, X-ray CT images were able to visualize the location where the maximum force is applied during excavation and the form of sliding failure, and also confirmed the importance of the soil between the two sheet-piles.

1 INTRODUCTION

Steel sheet-piles are driven into the ground to prevent failure of soils and intrusion of water. The specific type of sheet-piles makes the construction easier, its period shorter and the cost lower, as well as it is environmentally friendly. Also, sheet-piles are used in temporary structures such as earth retaining and temporary revetments. Furthermore, permanent structures such as revetments and embankments prevent slippage damage and ground settlement.

In this study, sheet-piles used as temporary structures are the subject of this study. There are some construction methods that include the beam type, the retaining pile type, the anchor type, and the free-standing wall type. The most general construction method is the beam type; however, it has a problem of low productivity and a long construction period because of limited space (Sugimoto et al. 1993). In the case of the retaining pile type and the anchor type, the productivity is high, however, it needs sufficient space on the background of the sheet-pile (Tamano. 1983) (Miyoshi et al. 2009). The free-standing wall type costs much money because it requires the use of more substantial materials. (Eto et al. 2002). Thus, there is a demand for the development of a new construction method that can (1) make an open space inside the construction area, (2) reduce the area of the background of sheet-pile (3) minimize costs.

Takahashi et al. (2013) developed the diagonal earth retaining method as a new construction technology. In general, a sheet-pile needs support work to prevent large deformation. Thus, the diagonal earth retaining method is used at a relatively shallow excavation depth. On the other hand, this method has been confirmed to reduce the earth pressure acting on the retaining wall and the horizontal displacement of the sheet-pile, and it is possible to perform deep excavation.

Furthermore, as an alternative diagonal earth retaining construction method, the double sheet-pile method has been developed. In this method, the heads of two sheet-piles are joined together. Thus, it is expected that the two sheet piles and the soil between the two sheet-piles become one rigid material and that the effect as the earth retaining material will become more than two times. The effect of the double sheet-pile method under certain conditions has been confirmed from centrifugal model tests or full-scale experiments. On the other hand, the soil behavior between two sheet-piles and adequate construction conditions have not been clarified yet.

Besides, recently, in the field of geotechnical engineering, an X-ray CT system that enables non-destructive and three-dimensional visualization of geotechnical phenomena has been attracting attention. It is one of the most effective methods to study soil behavior at the micro-level using the X-ray CT system because a mechanical method to directly observe the internal structure of soil has not been

DOI: 10.1201/9781003215226-35

established yet. Thus, actual understanding phenomena, such as soil mechanics, have been improved, and previously unidentified phenomena have been clarified (Otani et al. 2000).

The purpose of this study is to investigate the reinforcement mechanism of double steel sheet piles using model experiments with X-ray CT and to investigate adequate construction conditions. In this study, the experiment equipment used in the X-ray CT system is developed, and the authors conducted model tests with this equipment. Besides, the behavior of the soil is quantitatively evaluated by visualizing the inside of the soil with an X-ray CT scan.

2 EQUIPMENT AND METHOD

2.1 Experimental equipment

In this study, a new experimental equipment was developed to perform experiments in the X-ray CT room. In this chapter, a newly developed experimental equipment is introduced. In addition, the experiment equipment for horizontal loading of the head of the sheet-pile is also introduced.

Figure 1(a) shows a photo of the developed equipment for this study. Furthermore, the side view of the developed experimental equipment in this study is shown in Figure 1(b). The top and the bottom of the soil box are made of aluminum, and the sides are made of polyvinyl chloride. These materials are highly permeable to X-rays, which minimizes the negative effects of using the X-ray CT system. When creating the ground, a shaking table is mounted under the soil box. The lower part of the sheet pile is fixed with acrylic material in order to install the sheet-pile vertically. Polycarbonate was selected to be used for the model sheet pile in this experiment considering the permeability of X-rays. Also, Figure 2(a) and (b) show that the bottom and the top of sheet-piles on this experimental equipment. On the bottom of the sheet-piles, both sides of the sheet-piles are fixed by the fixture to put soil between the two sheet-piles, like Figure 2(a). During the experiment, the two sheet-piles are set by a fixture on the top of the sheet-piles, like Figure 2(b).

In this experiment, a 300 mm-deep excavation was conducted. During the excavation, the depth of the excavation was measured by a ruler. Simultaneously, the horizontal displacement of the head was measured every 50 mm using a laser displacement meter.

Next, sheet-piles were loaded by a horizontal loading device after excavation because the displacement caused by the excavation was minimal. Figure 3 shows the equipment for the horizontal load. This equipment was made with reference to the horizontal load experiments (Matsumura et al. 2017) (Takano et al. 2006).

2.2 Overview of model test

Toyoura sand was used for this model test. Table 1 shows the physical property of Toyoura sand. The model ground was prepared using the vibration method, and the relative density of the model ground was between 80% and 82%.

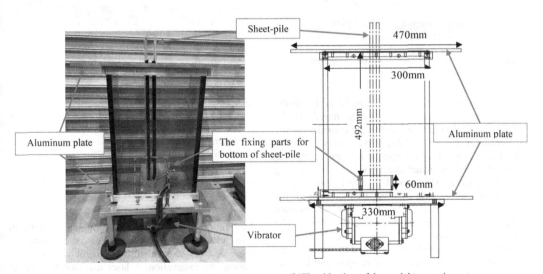

(a) The photo of the model test equipment   (b)The side view of the model test equipment

Figure 1. The photo of developed equipment.

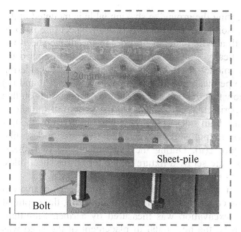

(a) Jig for bottom of the sheet-pile

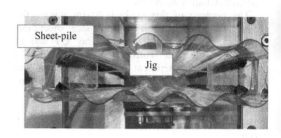

(b) Jig for top of the sheet-pile during making ground

Figure 2. The photo of bottom and top of experiment equipment.

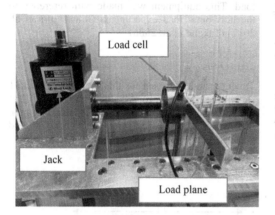

Figure 3. Experiment equipment for the horizontal load.

Figure 4. The model sheet-pile.

Figure 4 shows the model sheet-pile in this study. The model sheet pile was selected based on the results of preliminary experiments, considering the deflection of the material due to its rigidity. From the results of preliminary experiments, polycarbonate was selected. The model sheet-pile was 565mm long and 6mm thick. This model sheet-pile had a wave shape, and in the case of using two sheet-piles, the two waves were put so that they were in opposite phases, and the narrowest point was 20mm. The wavelength of this was 32mm, and the amplitude was 9mm. The sponge was also used to prevent leaking out the sand between the sheet-pile and the soil box wall. The size of the sponge was 8mm thick and 12mm wide.

Figure 5 shows the procedure of this experiment. Initially, the head and the bottom of the sheet-pile were fixed and placed in the soil box center. After

Table 1. The property of Toyoura sand.

Soil particle density	2.65(g/cm³)
Maximum dry density	1.65(g/cm³)
Minimum dry density	1.33(g/cm³)
Average diameter	0.18(mm)
Uniformity coefficient	1.29

that, the ground was created at the same height on both sides of the sheet-piles. Next, the ground on one side was excavated to a depth of 300 mm. During the excavation, a laser displacement meter was used to measure the horizontal displacement of the head per 50 mm mining. Additionally, the top of the sheet-pile was loaded by the horizontal load equipment, and load and horizontal displacement

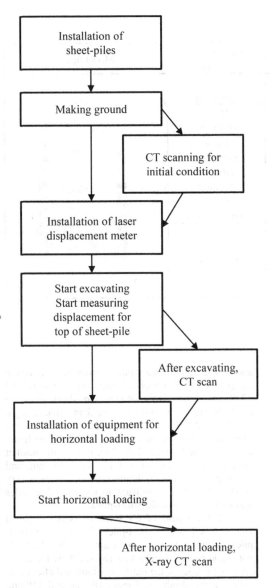

Figure 5. The procedure of this experiments.

were measured. In this model experiment, the horizontal load was measured with the load cell as shown in Figure 3, and the horizontal displacement is measured with a laser displacement meter on the other side.

Table 2 shows the data of this experiment. In this study, a single sheet-pile, a double sheet-pile without being fixed at the top of the sheet-pile, and the double sheet-piles being fixed at the top of the sheet-pile were tested. Additionally, a test in which the soil between the double sheet-pile was excavated to a depth of 140mm was conducted.

2.3 X-ray CT scan

X-ray CT scanners have been used mainly for medical purposes, but nowadays, they are widely used for industrial purposes. Especially in recent years, non-destructive and three-dimensional understanding of phenomena at the microscale has been attracting attention in geotechnical engineering as a new method for elucidating the mechanical behavior of soils. (Otani et al. 2005)

In this study, an industrial X-ray CT scanner owned by Kumamoto University X-Earth Center was used. Figure 6 shows the setup for the experiment during the X-ray CT scanning. When CT imaging is started, an X-ray is first irradiated, and the projection information from one direction is obtained. When the specimen table reaches the endpoint, it rotates and then moves parallel to obtain the projection information from a new direction. By repeating this process, the information is accumulated, and the cross-sectional image is constructed. CT images consist of values called CT values, which are highly correlated with the specimen's density and are calculated by the computer. The images are displayed in 256 gray and white levels. The CT value is calculated by

$$CTvalue = \frac{\mu_t - \mu_w}{\mu_w} K \qquad (1)$$

where μ_t is the coefficient of absorption at the scanning point; μ_w is the coefficient of absorption for water, and K is the material constant. It is noted that this constant is fixed to a value of 1000. Thus, the CT value of air should be -1000, and the coefficient of absorption for air is zero. (Otnai et al. 2005)

The scanning conditions are shown in Table 3. The voltage is 300 kV, the current is 2.00 mA, the filter function is FC1, the scan area is 400 mm diameter, the slice thickness is 1 mm, the scan speed is FINE, the matrix size is 2048×2048, and the scan mode is double full scan.

Furthermore, Table 4 shows the imaging cases and the area to be photographed. In the beginning, three cross-sections with a pitch of 100 mm were scanned for Case 1 (single sheet-pile), Case 2 (double sheet-piles without head fixing), and Case 3 (double sheet-piles with head fixing) in the range from 50mm to 250mm of the initial condition and after excavation. Besides, in Case 3, 28 cross-sections with a pitch of 10 mm were imaged in the range of 50-320 mm at the depth of the soil layer in the initial state, after excavation and horizontal loading of the head. In addition, based on the results of these CT scans, 150 cross-sections at a 0.2 mm pitch were taken at the depth of 210-240 mm, where the slip surface could be identified, and the soil behavior was studied in more detail.

Table 2. Test case.

Case	Condition	300mm Excavation	Horizontal loading	CT scanning (Pitch)	Model figure
1	Single sheet-pile			Done (100mm)	
2	Double sheet-pile without fixture			Done (100mm)	
3	Double sheet-pile with fixture	Done	Done	Done (100mm) (10mm) (0.2mm) Only 210mm~240mm	
4	Double sheet-pile with fixture (140mm excavation between double sheet-pile)			Not-Done	

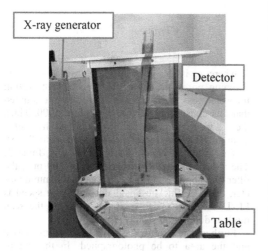

Figure 6. Model test using X-ray CT.

3 TEST RESULTS AND DISCUSSION

3.1 Comparison of double sheet-piles with and without head fixing

In this chapter, the Case1 (single sheet-pile), Case2 (double sheet-piles without head fixing), and Case3 (double sheet-piles with head fixing) are comparing the relationship between the excavation depths and horizontal displacements of the head of sheet-piles. Figure 7 shows the results. From these results, the horizontal displacement of Case1 (single sheet-pile) and Case2 (double sheet-pile without head fixing) at an excavation depth of 300mm was 16.89mm and 12.00mm, respectively. When two sheet piles are used, the rigidity should be more than twice as high, but Case 2 does not have more than twice the reinforcement effect than Case 1. On the other hand, in Case 3, the horizontal displacement of the head at an excavation depth of 300mm was 1.19 mm, and the displacement from the initial condition was very small. Thus, it is significant to use two sheet-piles and to fix the top of the sheet-piles.

Next, the results of CT scanning were compared among Case1 (single sheet-pile), Case2(double sheet-pile without head fixing), and Case3 (double sheet-pile with head fixing). Figure 8(a), (b), (c) show the cross-sectional images at the initial condition and after excavation at the depths of 50mm, 100mm and 150mm. The cross-sectional images were taken at depths of 50, 150, and 250mm. In the case of Case1(single sheet-pile), the sheet-pile deformed significantly from the initial condition, and some slip lines were found in the background of the sheet-pile. Comparing between Case 1 and Case 2, Case 2 did not show significant deformation of the sheet pile on the background side as in Case 1, but the sheet pile on the excavation side was found to be significantly deformed. Although no-slip lines were found on the ground behind the sheet-piles, a few slip lines were found in the soil between two sheet-piles. Thus, the sheet-pile of the excavation side is considered to be pressed by the soil between two sheet-piles.

In Case3, there were few deformations of the sheet-piles, and the slip line has not appeared on the

Table 3. Scanning condition.

Voltage	300kV
Current	2.000mA
Filter function	FC1
Scanning area	φ400mm
Scanning velocity	FINE
Matrix size	2048 × 2048
Scan mode	Double full scan (720 deg)

ground behind the sheet-piles or soil between them. No-slip lines were found on the background side of the excavation or the soil between the two sheet-piles because the two sheet-piles and the soil between them were unified as one rigid body. The X-ray CT scan results show that the head fixture of the two sheet-piles effectively increases the earth-retaining strength.

3.2 In-ground behavior and the importance of the soil between two sheet-piles

3.2.1 Investigating the behavior of the soil between two sheet-piles

From Chapter 3.1, it is clarified that double sheet-piles with head fixing sufficiently increase the strength of the earth retaining. However, it is not clear from the experiments in Chapter 3.1 how the head fixing effect increases the retaining capacity of the sheet piles. Therefore, to understand the reinforcement mechanism of this construction method, in the Case3, 28 cross-sections of the initial condition, after excavation and after horizontal loading of the head were scanned by industrial X-ray CT at a pitch of 10mm for a depth range of 50 to 320mm. Since there was no significant change in the ground behavior after excavation, the behavior of the ground was observed from the X-ray CT images after the top of the sheet-piles was loaded horizontally.

Figure 9 shows the cross-sectional images obtained from the X-ray CT scanning at the depth of 200mm~310mm of the sheet-piles after horizontal

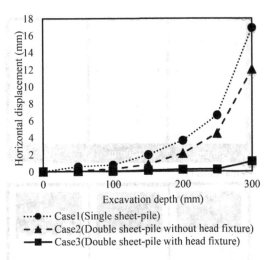

Figure 7. The relationship between horizonal displacement and Excavation depth.

loading. This image confirmed that there were some slip lines in the soil between the two sheet-piles.

Besides, the 210mm to 240mm depth position was scanned in more detail to visualize some slip lines three-dimensionally. Figure 10 is the cross-sectional view of a combination of 150 CT images scanned at a pitch of 0.2mm for the depth range from 210 to 240mm. This figure shows that there were two slip lines between the two sheet-piles. Considering the direction of the slip line, it is assumed that the sheet-pile on the excavation side at the depth from 210 to 240mm moved and caused the soil to slide.

Moreover, the 28 images in Case3 at the after horizontal loading and Figure 10 were cut in half and combined vertically. The combined image is shown in Figure 11. In addition, the average data of the image of a specific section was made so that the positions of continuous slip lines could be identified even in a vertical section (Figure 12). From these results, it is estimated that the slip line between the

Table 4. The list of scanning case.

			Case				Condition		
Scan Range (Depth)	Pitch	The number of cross-section	1	2	3	4	Initial	After Excavation	After Horizontal load
50mm~250mm	100mm	3	Done	Done	Done	Not-Done	Done	Done	Not-Done
50mm~320mm	10mm	28	Not-Done	Not-Done	Done	Not-Done	Done	Done	Done
210mm~240mm	0.2mm	150	Not-Done	Not-Done	Done	Not-Done	Not-Done	Not-Done	Done

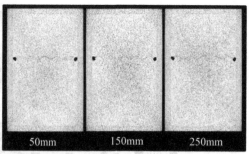

50mm | 150mm | 250mm

Initial Condition

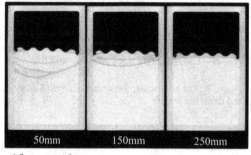

50mm | 150mm | 250mm

After excavation

(a) Single sheet-pile

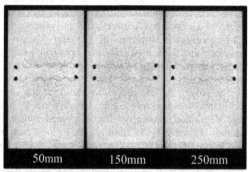

50mm | 150mm | 250mm

Initial Condition

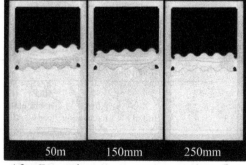

50m | 150mm | 250mm

After Excavation

(b) Double sheet-pile without head fixing

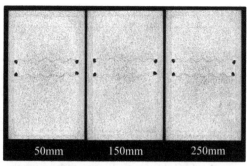

50mm | 150mm | 250mm

Initial Condition

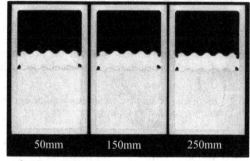

50mm | 150mm | 250mm

After Excavation

(c) Double sheet-pile with head fixing

Figure 8. Cross-section image at each depth.

two sheet-piles was 150mm~230mm on the excavation side and 170mm~260mm on the background side of the excavation. Combining each cross-section into a three-dimensional form, the start and endpoints of the slip line were revealed.

Next, from the CT scan image of the Case3, the coordinate of the sheet-piles in each section was taken, and the displacement of the two sheet-piles was graphed (Figure 13). This Figure shows that the displacement of the sheet-piles on the excavation side is greater than the displacement of the sheet-piles on the background side of the excavation. In other words, when the soil between two sheet-piles slips, the sheet-pile on the excavation side was pushed by the soil, causing the sheet-pile to deflect. Also, it is considered that the slip of the soil between two sheet-pile was dominated by the earth pressure of the backside soil. The shear force on the soil between two sheet-piles was caused by the active failure of the backside soil, and the slip was caused.

Further, the difference in deflection between the two sheet-piles was graphed by correcting the deflection angle (3 to 6 degrees) from the Figure 13. Figure 14 shows the results after correction. This Figure shows that the difference between the two sheet-piles increases to the point where the slip line is found in the soil between the two sheet-piles.

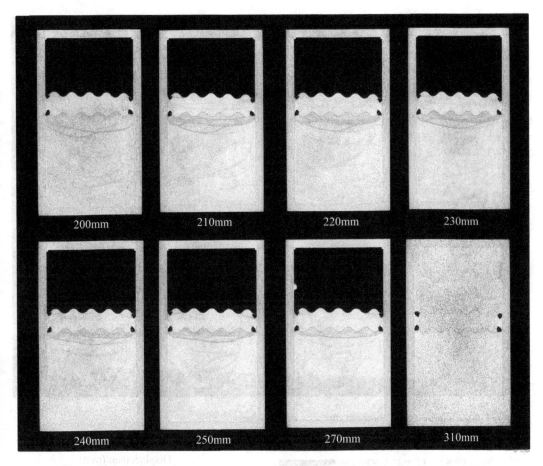

Figure 9. The cross-sectional image of case3(with head fixing) after horizontal loading.

Figure 10. Three-dimensional CT image at the depth of 210~240mm in after excavation of case 3.

In consideration of the above results, it is suggested that the sheet piles on the excavation side were pushed out at the depth of 50-320 mm, where the slip line was found. Also, it is essential to increase the strength of the sheet pile because the rigidity of the sheet pile is low at the location where the slip line was found.

3.2.2 Importance of soil between two sheet-piles

From the X-ray CT scan results, in Case3(double sheet-piles with head fixing), slip lines were found in the soil between two sheet-piles at the depth of 150~260mm, a little above the excavation depth (300mm). However, no-slip lines were found in the upper part of the soil between the two sheet-piles.

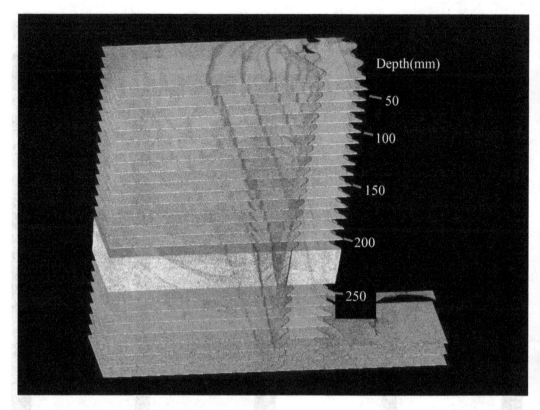

Figure 11. Three-dimensional CT image in after excavation of case3.

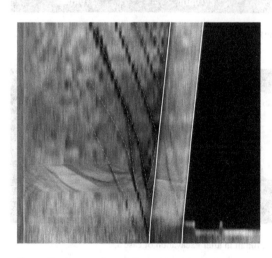

Figure 12. Average data of Figure 11.

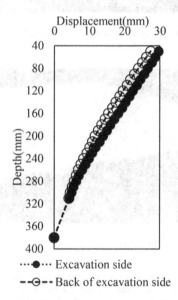

Figure 13. The value of displacement of two sheet-piles.

Therefore, to investigate the importance of the soil between the two sheet-piles in detail, the upper layer of the soil between the two sheet-piles was removed from which no slip line was found and we conducted an excavation experiment. (Case4). Figure 15 shows the result of the horizontal head displacement under

excavation. A comparison of the horizontal head displacement between Case 1 (single sheet pile) and Case 3 (double steel sheet pile with head fixing) was carried out using this experiment as shown in

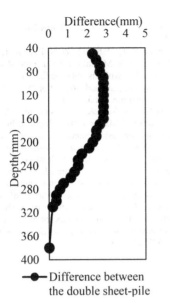

Figure 14. The value of difference between two sheet-piles.

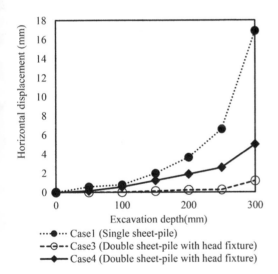

Figure 15. The relationship between excavation depth and horizontal displacement.

Case 4. From these results, the horizontal displacement of the head after the excavation of 300mm was larger in Case4 (140mm of the soil between two sheet-piles) than Case3 (double sheet-piles with head fixing). In other words, it was confirmed that the strength of the sheet piles as an earth retaining structure was lowered in case of the absence of the soil between two sheet-piles in the upper layers. Thus, it is essential not only to fix the head of the sheet pile but also to have the soil between two sheet-piles. It is possible to enhance the reinforcement effect of the sheet pile by increasing the density of the soil between two sheet piles and restraining the movement of the soil between two sheet-piles.

4 CONCLUSION

In this study, a new construction method using two sheet-piles was focused. Model experiments were conducted to evaluate the strength of double sheet-piles. The new experimental equipment that can take a CT scan was developed, and excavation experiments were conducted. Simultaneously, the effectiveness of this construction method was confirmed by measuring the horizontal displacement of the head of the sheet-piles. Besides, the behavior of the ground with the double sheet-piles was investigated by using industrial X-ray CT. Furthermore, this study focused on the soil between the two sheet-piles to effectively increase the reinforcement of the sheet-piles.

The displacement of the head with two sheet piles was smaller than that with only one sheet pile. Besides, the use of two sheet piles and simultaneous fixing of the head of the pile effectively reduced the displacement. The effectiveness of the proposed method was confirmed by the results of this construction method.

From the X-ray CT scan image, in the case where two sheet piles were used and the heads were fixed, multiple slip lines were found in the soil between two sheet-piles. The position of the slip lines was located a little above the excavation surface, and this position is considered to be essential for the effectiveness of this construction method. Also, the coordinates of the two sheet piles were calculated from the CT images, and the deformation mode of the sheet piles was discussed. The results confirmed that the sheet pile on the excavation side was more deflected than the sheet pile on the background side. Furthermore, the direction of the slip line was identified by visualizing the CT image.

Since the reduction of the soil between two sheet-piles decreases the overall stiffness of the sheet pile, the condition of the soil between two sheet-piles might have a significant effect on the behavior of the sheet-pile. This result suggests that increasing the density of the soil between two sheet-piles and constraining the movement of the soil between two sheet-piles may further increase the strength of the sheet-pile.

These findings indicate that the strength of the sheet piles as a retaining wall was increased by using two sheet piles to fix the head. The ground behavior of the two sheet piles used to fix the head in place was also clarified. In addition to the fixation of the heads, the presence of soil between two sheet-piles was also found to be important for increasing the strength of the sheet piles as retaining walls.

373

REFERENCES

D, Takano & K, Dang & J, Otani. (2006). 3-D visualization of soil behavior due to laterally loaded pile using X-ray CT. *Journal of applied mechanics*. 9: 513–520

H, Sugimoto & H, Yamamoto & T, Sasaki & J, Mitsuo. (1993). On design optimization of retaining wall structures by genetic algorithm. *Journal of Japan Society of Civil Engineers*. 474: 105–114

J, Otani & T, Mukunoki & K, Sugawara. (2005). Evaluation of particle crushing in soils using X-ray CT data, *Soils and Foundations* 45.1: 99–108

J, Otani & Y, Obara & K, Sugawara & T, Mukunoki. (2000). Application of Industrial X-ray Computed Tomography Scanner to Geotechnical Engineering. *Soil mechanics and foundation engineering* 48. 2: 17–20

M, Eto & Y, Toyosawa & N, Fujita & M, Sato & M, Eguchi (2002). Effect of counterfort retaining wall in cantilever sheet pile method, *The 37th Japan National Conference on Geotechnical Engineering*: 1561–1562

S,Matsumura & T, Matsubara & N, Fujii & T, Mizutani & Y, Morikawa & M, Sato. (2017). Lateral resistance of coupled piles with its intermediate soil stabilized by cement treating method, *Report of the port and airport research institute*. 56. 3: 3–27

S, Takahashi & S, Sugie& S, Matsumoto & Y, Shimada & Y,Sakahira & T,Maeda. (2013). Inclined-braceless Excavation Support using Sheet Piles, *Report of Obayashi corporation technical research institute*: No.77

T, Miyoshi & M, Yoshida & S, Tashiro & K, Gouda & O, Kiyomiya. (2009). Earthquake resistance of sheet pile quay reinforced by ground anchors, *Journal of Japan Society of Civil Engineers SerA1* 65. 1: 345–353

T, Tamano. (1983). A case study of a sheet-pile wall multi-tied with ground anchors for excavation with field measurements, *Journal of Japan Society of Civil Engineers*. 332: 127–136

Proceedings of the Second International Conference on
Press-in Engineering 2021, Kochi, Japan – Matsumoto et al (eds)
© 2021 Taylor & Francis Group, London, ISBN 978-1-032-10414-0

A study on the effect an earth-retaining wall's rigidity and embedded depth on its behavior

N. Matsumoto & H. Nishioka
Chuo University, Tokyo, Japan

ABSTRACT: In the design of an earth-retaining cantilever, it is necessary to establish the balance between the retaining wall's rigidity and its embedded depth. In this study, an experiment was conducted using an aluminum-layered ground model that could easily simulate ground failure; the main parameters considered were the rigidity and embedded depth (wall length) of a cantilever-type earth-retaining wall. We demonstrated that highly rigid earth-retaining walls built using traditional design standards are susceptible to brittle collapse. Therefore, the safety against collapse may be improved by increasing the embedding length.

1 INTRODUCTION

For ground excavation using the earth-retaining wall in a neighboring construction site, the displacement of the wall should be limited to minimize its influence on the integrity of the surrounding ground. Therefore, in such cases, the bracing method is generally preferred to the cantilever method. However, to ensure workability inside the excavation, the cantilever method may be adopted by increasing the rigidity of the earth-retaining wall and eliminating the need for struts. In the design for this case, a balance between the rigidity of the earth-retaining wall and the embedded depth must be reasonably established. In some design standards, the embedded depth of the earth-retaining wall should be a depth that can be regarded as a semi-infinite length of the pile. According to this principle, the embedded depth must also be increased as the rigidity of the earth retaining wall increases. However, this may not necessarily result in rational design.

In addition, only a few studies have been conducted wherein the behavior leading up to collapse was simulated using both rigidity and embedded depth as parameters.

Therefore, in this study, we experimented using an aluminum-layered ground model, which can easily simulate failure and subsequent collapse. The main parameters considered were the rigidity and embedded depth (wall length) of the cantilever-type earth-retaining wall.

We also investigated the effects of these parameters on the safety of the cantilever-type earth-retaining wall in terms of collapse and allowable deformation.

2 EXPERIMENT OUTLINE

2.1 *Ground model and experimental equipment*

Two types of aluminum rods with diameters of 3.0 mm and 1.6 mm shown in Figure 1 were used in this experiment. The aluminum-layered ground model was made by mixing these rods well and layered them 40% and 60% in weight content, respectively.

The ground model's weight per unit volume γ and angle of internal friction φ was measured using a measuring box (width = 250 mm and depth 50 = mm). Here, the angle at which the measuring box was tilted and the aluminum rod collapsed (Figures 2 and 3) was defined as the angle of internal friction φ. After three iterations, the average values of $\gamma = 21.0$ kN/m^3 and $\varphi = 30$ ° were obtained.

The ground model was created in an experimental container with a width of 500 mm and a depth of 300 mm, as shown in Figure 4.

In addition, to observe the slip surface in the ground model, horizontal lines were drawn on the end face of the aluminum rod at intervals of 20 mm.

2.2 *Earth-retaining-wall model*

The earth-retaining wall was simulated by inserting a 70-mm-wide aluminum plate into the center of the ground model. There were nine experimental cases involving three thicknesses (i.e. $t = 0.5$, 1.0, and 2.0 mm) and three lengths (i.e. $L_0 = 200$, 250, and 300 mm) of the earth-retaining-wall model. L_0 is the embedded depth before excavation, which does not include the length of the protrusion above the ground.

DOI: 10.1201/9781003215226-36

Figure 1. Two types of aluminum rods.

Figure 2. Measurement of internal angle of friction (just before the aluminum rod collapsed).

Figure 3. Measurement of internal angle of friction (just after the aluminum rod collapsed).

The validity of modeling the earth-retaining-wall model was evaluated based on the product of the characteristic values β (Eq. 1) and L_0.

$$\beta = \sqrt[4]{\frac{k_h B}{4EI}} \qquad (1)$$

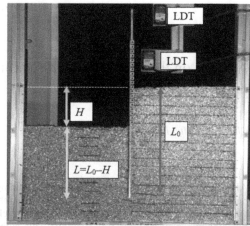

Figure 4. Layout of experimental apparatus.

where EI is the flexural rigidity of the earth-retaining-wall model, k_h is the coefficient of subgrade reaction, and B is the contact width between the ground and the earth-retaining-wall, which in this experiment is the aluminum rod's length (B=50mm).

This βL_0 value is a dimensionless parameter that represents the ratio of the rigidity of the wall relative to the ground. In other words, if modeled relative rigidity (βL_0) corresponds to that of an actual retaining wall, the model is considered valid.

To obtain the characteristic value β, the horizontal-loading test of each earth-retaining-wall model (Figure 5) was conducted separately prior to excavation. The value of β was calculated by applying Chang's equation (Eq. 2).

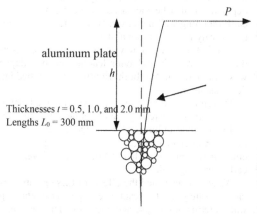

Figure 5. Horizontal loading test of the model earth retaining wall.

376

$$y_0 = \frac{P}{3EI\beta^3}\left\{(1+h\beta)^3 + 1/2\right\} \qquad (2)$$

where y_0 is the pile-head displacement; P is the horizontal load; h is the loading height.

The value of β was calculated backward for each earth-retaining-wall model from measured P and y_0 values shown in Figure 6. Table 1 presents the reciprocal of the measured characteristic value β of the pile and the value βL_0. This range of βL_0 values (i.e., 1.4–5.9) is wider than the general range of an actual cantilever-type earth-retaining wall, and the modeling of the wall in this experiment can be considered valid.

2.3 Excavation

The excavation was simulated step by step, removing and leveling the aluminum rods at a pitch of about 10 mm depth (Figure 7). At each excavation step, the horizontal displacement δ_0 of the earth-retaining-wall at the original ground surface height was measured. The horizontal displacement δ_0 was converted measured values of two laser displacement transducers (LDT, shown in Figure 4) at 40mm and 100mm height from the original ground surface of the protruding part of the earth retaining wall. The change in the unexcavated side ground's surface height due to excavation was not taken into account. However, if the range of LDT was exceeded, δ_0 was measured by tracking the reference point of the picture captured by a digital camera.

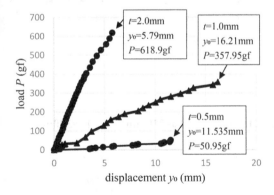

Figure 6. Horizontal loading test result.

Table 1. The characteristic value β of the pile.

Thickness	$t = 0.5$mm	$t = 1.0$mm	$t = 2.0$mm
$1/\beta$	51mm	75mm	134mm
βL_0	3.9–5.9	2.7–4.0	1.4–2.2

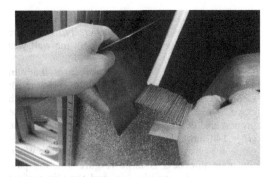

Figure 7. A state of simulating excavation.

3 EXPERIMENTAL RESULT

Sample results are shown for cases 0.5 mm and 1.0 mm thickness, for the same length of the earth-retaining wall (i.e., $L_0 = 200$ mm).

Figure 8 shows the deformation of the ground model and the earth-retaining-wall model one step before collapse (excavation depth $H = 110$ mm for both). The point of collapse is the point at which passive collapse can be confirmed visually. The βL values shown in Figure 8 do not meet the criteria for semi-infinite length (generally $\beta L \geq 2.5$–3.0), which is the principle of some design standards. When the

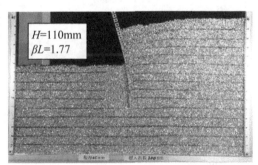

a) $t = 0.5$mm

b) $t = 1.0$mm

Figure 8. One step before collapse ($L_0 = 200$mm).

plate thickness was small (Figure 8a), the wall's deflection was significant. Conversely, when the plate thickness was large (Figure 8b), the wall was rigid (no deflection).

Figure 9 Shows the results for cases 0.5 mm and 1.0 mm for the same length L_0 (250 mm); i.e., only plate thickness was different for both cases. These figures indicate the behavior of the ground and wall models at the time of collapse ($H = 140$ mm). The dotted line in the figure indicates the visually confirmed slip surface. When the plate thickness is large (Figure 9b), the passive collapse is considerable, and the slip surface can be easily observed. In this case, when the plate thickness was small (Figure 9a), passive collapse did not occur, and a small slip surface could be observed near the wall model. Even at $t = 2.0$ mm, a significant slip surface (passive collapse) could be observed. From this, it was found that the ground's behavior on the excavation side is significantly affected by the rigidity of the earth-retaining wall.

Figure 10 shows the relationship between the excavation H_c at collapse and the plate thickness t during the collapse. It can be observed that H_c mainly depends on the retaining wall length L_0 and is hardly affected by rigidity.

Figure 11 shows the relationship between the excavation depth H and the horizontal displacement δ_0 of the earth retaining wall for all cases. Figure 12 shows the relationship between the horizontal displacement δ_0 and plate thickness t when excavation depth H was 50 mm. Figure 13 shows the relationship between the horizontal displacement δ_0 plate thickness t when $H = 100$ mm. Further, the effect of suppressing the horizontal displacement of the earth-retaining wall by increasing the embedded depth could not be verified; this effect was evaluated by increasing the plate thickness (flexural rigidity of the wall).

Therefore, from a perspective of displacement suppression, it was found that increasing the embedded depth had no effect on suppression, which solely depended only on the flexural rigidity of the retaining wall.

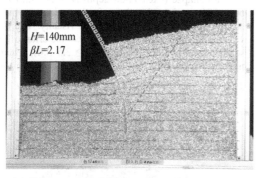 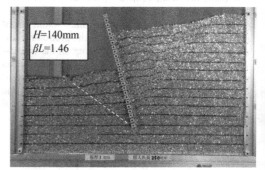

a) t=0.5mm   b) t=1.0mm

Figure 9. Model earth retaining wall at the time of collapse.

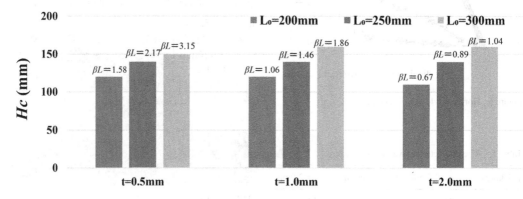

Figure 10. The relationship between the excavation depth H_c and plate thickness t during collapse.

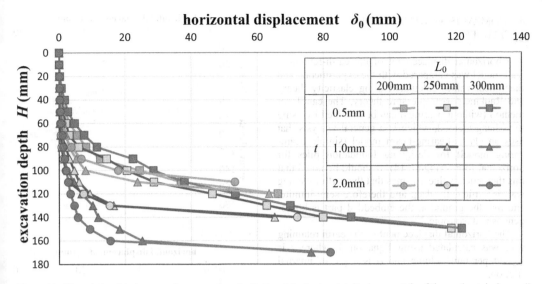

Figure 11. The relationship between the excavation depth H and the horizontal displacement δ_0 of the earth retain-ing wall in all cases.

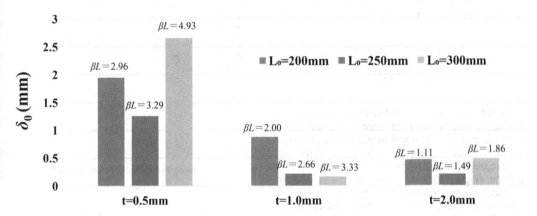

Figure 12. Relationship between horizontal displacement δ_0 and plate thickness t when the excavation depth $H = 50$ mm.

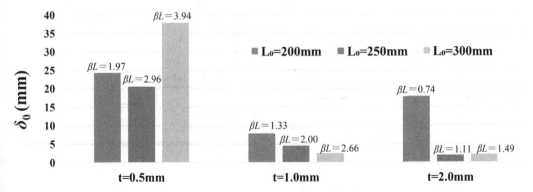

Figure 13. The relationship between the horizontal displacement δ_0 and plate thickness t when the excavation depth $H = 100$ mm.

4 COMPARISON WITH THEORETICAL VALUES

The horizontal displacement of the earth-retaining-wall model was compared with the experimental and theoretical values calculated using elasticity theory and Rankine's earth-pressure theory. The cantilever method (without bearings) can be obtained by using the Chang method, which is a general analysis that uses an elastic-bearing-beam model (simple beam-spring model) for designing foundation piles for horizontal forces (Figure 14). It should be noted that this theoretical value assumes that the earth-retaining wall is sufficiently deep and therefore semi-infinite, and that the ground at the embedded part is homogenous and isotropic.

The horizontal displacement of the earth retaining wall was calculated using Equation 3, where γ is weight per unit volume and φ is angle of internal friction.

$$\delta = \frac{Ph}{EI\beta^2}\left\{\frac{(1+\beta h)^3 1/2}{3\beta} + \frac{(1+\beta h)^2 l}{2}\right\} \quad (3)$$

$$P_h = 1/2 \times \gamma \times H^2 \times \tan^2(45° - \varphi/2)$$

$$h = 1/3H$$

Figure 15 shows the relationship between the horizontal displacement obtained using the above equation and the horizontal displacement measured in the experiment for each plate thickness. A comparison of the horizontal displacement of $t = 0.5$ mm (Figure 15a) and the theoretical value from elasticity theory indicates that they are almost in agreement. However, the horizontal displacement of $t = 2.0$ mm (Figure 15c) deviates significantly from the theoretical value from elasticity theory as the excavation depth increases.

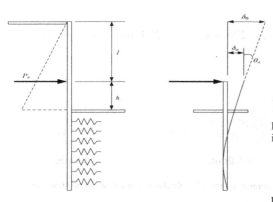

Figure 14. The Chang method.

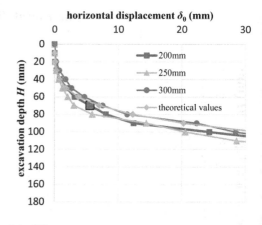

a) $t = 0.5$ mm

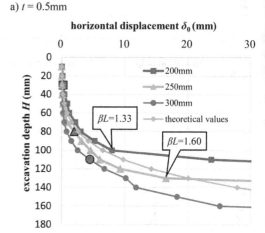

b) $t = 1.0$ mm

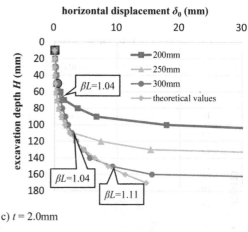

c) $t = 2.0$ mm

Figure 15. Comparison of experimental values and theoretical values.

In Figure 15, the depth ($\beta L < 2.5$), which cannot be regarded as semi-infinite, is plotted in red. As shown in the chart, even if this point is exceeded, the value initially approaches the theoretical value considered from the elasticity theory.

In addition, βL of the point at the excavation depth immediately before the displacement starts to increase; this trend deviates sharply from elasticity theory, as shown in the chart.

Notably, when $\beta L = 1.0$–1.6, the displacement of the elastic theory deviates. Hence, it is inferred from this observation that the displacement can be estimated by the theoretical value of elasticity theory if at least 1.0–1.6 is secured even if $\beta L \geq 2.5$ is not satisfied.

5 CONCLUSION

The results obtained from this study were as follows.

1) The decay depth of the earth-retaining wall does not depend on the wall's flexural rigidity; the flexural rigidity only affects the embedded depth. In addition, if the rigidity of the earth retaining wall is small, the extent of passive collapse (length of the slip face) becomes small as well.
2) The displacement-suppression effect of the earth-retaining wall during excavation was affected by the increase in the wall's flexural rigidity. In contrast, the effect of increasing the embedded depth on the wall's retaining capacity was not verified.
3) Even if the condition $\beta L \geq 2.5$ is not satisfied if $\beta L \geq 1.0$–1.6 is met, the displacement, which can be predicted from the theory of elasticity, does not deviate significantly.

In addition, the design standard $\beta L \geq 2.5$ is considered to cause the brittle failure of a highly rigid retaining wall. Consequently, the safety against collapse may be improved by increasing the embedded depth.

In the future, by clarifying the uncertainties to be considered and their effects, we would like to study a more reasonable and simple design method with an appropriate safety margin in a different way from $\beta L \geq 2.5$.

REFERENCES

Japanese Geotechnical Society: *Method for lateral load test of piles*, pp32 Maruzen Publishing, 2010. (in Japanese)
Japan Road Association: *Guidelines for Temporary structure of Road earth works*, pp153–154 Maruzen Publishing, 1994. (in Japanese)

Proceedings of the Second International Conference on
Press-in Engineering 2021, Kochi, Japan – Matsumoto et al (eds)
© 2021 Taylor & Francis Group, London, ISBN 978-1-032-10414-0

Physical and numerical modeling of self-supporting retaining structure using double sheet pile walls

A. Nasu, T. Kobayashi, H. Nagatani, S. Ohno, N. Inoue, Y. Taira & T. Sakanashi
Kajima Corporation, Tokyo, Japan

Y. Kikuchi
Tokyo University of Science, Tokyo, Japan

ABSTRACT: The main objective of this research was to observe the mechanical behavior of double sheet pile walls with model experiments and FE analysis. The model experiments were conducted in 1/4 scale under Earth's gravity conditions. The experiments were conducted in the following steps: first, excavating the ground, then loading the head of the sheet piles by hydraulic jacks. The physical model experiments showed a much smaller deformation compared to the conventional method and proved the earth-retaining mechanisms. The connection of the heads was thought to be the main reason for minimizing the deformation. Furthermore, the monitored axial forces of double sheet pile walls indicated the frictional resistance between the sheet pile walls and the ground contributes to resisting the load. These findings are supported by a 2D FEM with an elastoplastic constitutive model.

1 INTRODUCTION

1.1 *Background*

In excavation projects, members, such as struts and pin piles, are often used to support temporary earth retaining walls. These support members obstruct the construction work in excavated space and reduce efficiency. The self-supporting method, which does not need support members, is one of the solutions to this issue. Ground anchors and soil mixing walls (SMWs) are typical self-supporting methods. They, however, are not always applicable because of the limitation of space, high costs, and special equipment. This was our motivation to develop a cost-effective, self-supporting retaining structure by double sheet pile walls for urban construction.

Historically, double sheet pile walls have been used for cofferdams and breakwaters. One of the early studies is Sawaguchi (1974), who conducted model experiments and developed analytical solutions for the deflection of double sheet pile walls. In his work, the importance of the interaction between both sheet piles and the filling material was emphasized. Kikuchi et al. (2001) performed centrifuge model experiments and Finite Element analysis to investigate the effect of the filling material by comparing sand and cemented dredged soil. It was concluded that the adhesion with the sheet pile needs to be strong enough for the cemented filling material

to utilize its rigidity. Khan et al. (2006) performed centrifuge model experiments to study the stability of cofferdams against high floodwater. They also compared the stability of cofferdams on a thick clay deposit and a sand deposit. One of their key findings was that the failure mechanism of such cofferdams was dominated by the shear deformation of the filling material. More recently, Fujiwara et al. (2017) studied the reinforcement of bank using double sheet piles with diaphragm walls. They reported that the settlement was reduced by 10% against an earthquake compared to the case without the reinforcement.

1.2 *Concept of double sheet pile walls*

Figure 1 illustrates the concept of double sheet pile walls as a temporary earth retaining structure. In this method, two sheet piles are pressed into the ground in parallel, typically with a distance of 1 m. The heads of both sheet piles are connected by a rigid member. The relatively narrow distance between the sheet piles distinguishes our study from previous studies on cofferdams and breakwaters. This structure utilizes the following effects to resist horizontal loads.

  i. Rigidity enhancement by head fixing
 ii. Frictional resistance between the sheet pile walls and the ground
iii. Shear stiffness of internal soil

DOI: 10.1201/9781003215226-37

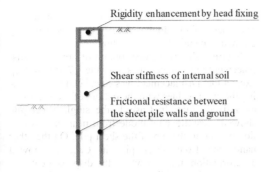

Figure 1. The concept of temporary earth retaining by the double sheet pile walls.

The main objective of this research was to observe the mechanical behavior of the double sheet pile walls through physical model experiments. The model experiments were conducted in 1/4 scale under the Earth's gravity conditions. The experiment procedure consisted of two phases: excavating the ground and loading the head of the structure using hydraulic jacks. In addition, a series of numerical modeling was performed to examine the results of the model experiments.

2 PHYSICAL MODEL EXPERIMENTS

2.1 Overview

Figure 2 shows a schematic view of the model experiments in 1/4 scale under the Earth's gravity. The model ground was 2,500 mm in width, 6,000 mm in length, and 3,700 mm in depth. Two model experiments were conducted in this study. Case 1 was a single sheet pile wall (conventional method), and Case 2 was a double sheet pile wall. For both cases, sheet piles had a length of

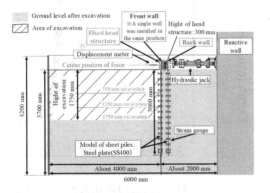

Figure 2. A schematic view of the physical modeling experiments in 1/4 scale under the Earth's gravity conditions.

3,300 mm. A rubber plate was attached to the end of the sheet pile to reduce friction between the sheet pile and the soil tank wall. Parameters for model sheet piles are shown in Table 1. For Case 2 the horizontal distance between the front and the back sheet piles was 250 mm. A total 1,750 mm excavation was performed in a stepwise manner.

2.2 Material and method

Natural sand in Chiba Prefecture was used as the ground material. Figure 3 shows the basic properties and the grain size distribution of the sand. After the steel sheet piles were placed, a certain amount of sand was poured and compacted every 30 cm per layer to achieve the degree of compaction of 85%. Soil samples were acquired after the experiments to determine the density and water content of the model ground. Measured values for each test are listed in Table 2. Because of the natural water content of the soil, apparent cohesion existed due to suction forces between the particles.

2.3 Experiment procedures

The model experiments consisted of two steps. First, the ground was excavated by manpower down to 1,750 mm in depth. The excavation was separated into steps of 750 mm, 1,250 mm, and 1,750 mm to evaluate the gradual deformation behavior of the walls. Second, the head of the walls was loaded statically with 0.2-0.3 mm per minute in horizontal direction by using jacks up to 190 mm. The horizontal earth pressure due to the excavation was not large enough to deform the sheet piles to a demanded

Table 1. Parameters for model sheet piles.

Parameters		Steel plate (SS400)	
Young's modulus	E	1.9×10^8	kN/m^2
Thickness	t	12	mm
Moment of inertia of area	I	14	cm^4/m
Flexural rigidity	EI	2.8×10	$kN \cdot m$

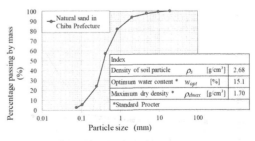

Figure 3. Basic properties and particle size distribution of natural sand in Chiba Prefecture.

Table 2. Measured values of the model grounds.

Parameters		Case 1	Case 2	
Water content	w	15.2	14.5	%
Dry density	ρ_d	1.45	1.48	g/cm^3
Wet density	ρ_t	1.70	1.70	g/cm^3
	D_{50}	0.37	0.37	mm
Coefficient of Uniformity	U_c	3.2	3.2	
Degree of Compaction	D_c	85	87	%

level in the model scale. Thus, the jacks were used to apply further horizontal displacement at the head although the mode of deformation differed from that of the excavation. The deformation behavior of the retaining wall was evaluated using strain gauges attached to the steel sheet piles and a horizontal displacement gauge for the head. Figure 4 shows an image of the excavation and the horizontal loading.

2.4 Results

2.4.1 Excavation
Figure 5 shows the change in head displacement with excavation depth. In Case 1, the single sheet pile wall did not show significant displacement up to an excavation depth of 1,250 mm. Then, the head displacement was increased to 35 mm with the depth

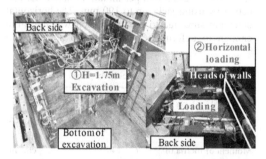

Figure 4. An overview of the model experiments in 1/4 scale under the Earth's gravity.

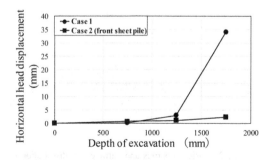

Figure 5. The head displacement with excavation depth.

from 1,250 to 1,750 mm. In Case 2, the head displacement of the double sheet pile walls was only 2 mm at the end of the excavation.

Figure 6 shows the distribution of horizontal displacement in the sheet plies after the 1,750 mm-deep excavation. Displacement was calculated by integrating the strain values with the bottom of the sheet pile as a fixed point. In Case 1, the single sheet pile showed a cantilevered mode from near the excavation bottom to the top of the sheet pile. On the other hand, the front sheet pile in Case 2 showed a deformation mode in which the displacement was largest in the middle. This behavior is similar to those of anchor-supported sheet piles.

Figure 7 shows the distribution of bending moments after the excavation of 1750 mm. The bending moment was calculated from the strain gauges attached to the front and back of the sheet pile. In Case 1, positive bending moment was largest near the bottom of excavation. The negative bending moment above the bottom of excavation is assumed to be caused by frictional resistance between the edge of the sheet pile and the container or the shear resistance between the sheet pile and the soil. In Case 2, the front and back sheet piles showed different distributions of bending moment. In the back sheet pile, only a negative peak appeared near the head indicating the back sheet pile behaved like a raked pile to resist the earth pressure.

2.4.2 Horizontal loading
Figure 8 shows the horizontal displacement of the head during horizontal loading by the jacks. The

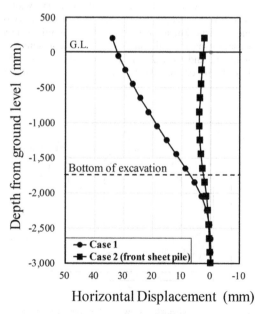

Figure 6. The distribution of horizontal displacement after 1750 mm excavation.

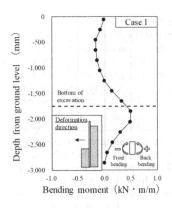

(a) Case 1

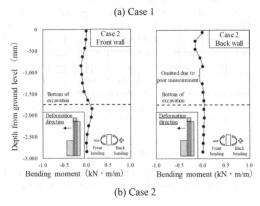

(b) Case 2

Figure 7. The distribution of bending moments after 1750 mm excavation.

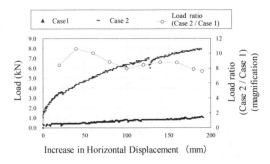

Figure 8. The relationship between the horizontal displacement of the head during loading and the load.

ratio of the load of Case 2 to Case 1 for the same amount of displacement is also shown by hollow circles. Note that the horizontal displacement discussed here is the incremental value after the end of excavation. At the displacement of 190 mm, the load required was 1.1 kN with the single sheet pile wall and 8.0 kN with the double sheet pile walls. Throughout the loading process, the load was about

7.6 to 10.6 times higher in the double sheet pile walls than in the single sheet pile wall.

Figure 9 shows the horizontal displacement distribution when the head displacement was 100 mm. In Case 1, the displacement showed a cantilevered mode. While, in Case 2, the displacement of the back sheet pile was slightly larger than that of the front sheet plie at the ground level of -500 mm. This result may indicate the contraction of internal soil in this particular part.

Figure 10 shows the distribution of the bending moment before applying the horizontal load, at 100 mm displacement, and at 190 mm displacement of the pile head. In Case 2, as the head displacement increased, bending moment distributions of both the front and the back sheet piles showed similar modes. The front sheet pile had a sharper positive peak below the bottom of excavation. Moreover, the back sheet pile showed a larger negative bending moment near the head.

Figure 11 shows the distributions of axial forces in the loading experiments in Case2. Compressive and tensile forces were observed in the front and back sheet plies, respectively. Below the bottom of excavation, axial forces decrease with depth in both sheet piles. This trend indicates the shear resistance of the soil against the push-in and pull-out of each sheet pile.

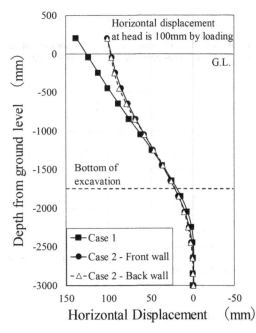

Figure 9. The distribution of horizontal displacement of the compulsory head horizontal displacement for 100 mm by loading.

385

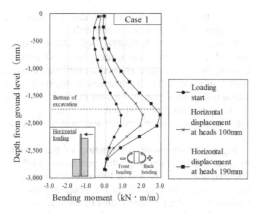

(a) A single sheet pile (conventional method)

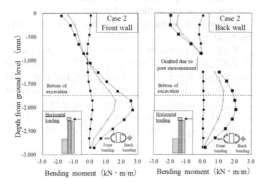

(b) Double sheet pile walls

Figure 10. The distribution of bending moments of the compulsory head horizontal displacement by loading.

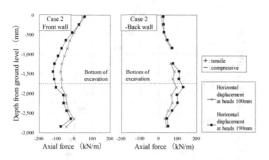

Figure 11. The distribution of axial forces of Case 2 in the loading experiments.

3 NUMERICAL MODELING

3.1 Conditions for numerical analysis

A series of numerical analyses was conducted to study the retaining mechanism of the double sheet piles in greater detail. Our purpose was to get an insight into the behavior of the soil and its interaction with the double sheep piles. Such behavior is difficult to observe directly in physical modeling experiments. The process of the physical modeling experiment was simulated by a 2D Finite Element (FE) program. The model for the FE analysis is shown in Figure 12. For the soil, the Drucker-Prager model whose parameters are given in Table 3 was used. These parameters are derived by laboratory testing of the same soil used in the physical modeling experiments. Young's modulus varies with depth, and it is proportional to the 0.5 power of the mean effective stress of soil σ'_m. Both sheet piles and the head-fixing member are modeled by beam elements. Between the solid and the beam elements, Goodman's joint elements are placed to account for slip and separation. Displacements are fixed at the lower boundary while only horizontal displacement is fixed at the left and right boundaries. The process of the excavation is modeled by removing the corresponding soil elements layer by layer. At the same time, equivalent force is released at the adjacent nodes. For the process of the jack-loading, horizontal displacement is prescribed at the head of the double sheet piles.

3.2 Results and Discussion

First, the results of the FE analysis were compared to the physical modeling experiment to examine its

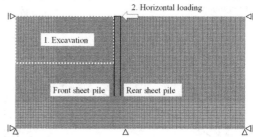

Figure 12. The model for the FE analysis. First, the part of the soil shown by the white dashed line was excavated. Second, horizontal loading was applied at the head of the double sheet piles.

Table 3. Parameters for the Drucker-Prager model used in the FE analysis.

Parameters			
Unit weight	ρ_t	16.5	kN/m^3
Young's modulus	E	15500	kN/m^2 (at σ'_m = 90 kN/m^2)
Poisson's ratio	ν	0.31	
Friction angle	φ	34.5	degrees
Cohesion	c	2.0	kN/m^2
Dilatancy angle	ψ	0.0	degrees

validity. Here, we focused on the experiment conducted with the double sheet piles (Case 2). Figure 13 shows the horizontal displacements of the sheet piles when the excavation was finished and when the head-displacement reached 100 mm. The markers indicate the physical modeling while the solid line indicates the FE analysis. The mode of deformation was similar between the experiment and the FE analysis both at the excavation and the jack loading. Moreover, the bending moments of both sheet piles were compared in the same manner in Figure 14. Both results agreed well except the peaks

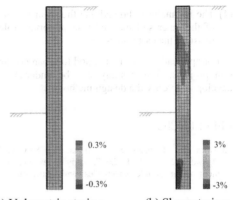

(a) Volumetric strain ε_v (b) Shear strain γ_{xy}
(Compression positive)

Figure 15. Strain distributions of the soil elements in the double sheet piles when the head displacement is 100 mm. The shear is dominant over the volumetric change.

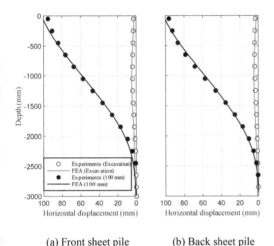

(a) Front sheet pile   (b) Back sheet pile

Figure 13. Horizontal displacements of the (a) front and (b) back sheet piles. The experiment and FEA results are compared at the end of the excavation and at the head displacement of 100 mm.

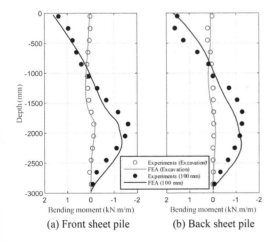

(a) Front sheet pile   (b) Back sheet pile

Figure 14. Bending moments of the (a) front and (b) back sheet piles. Note that the values of the bending moments are shown for a unit width of 1 m.

of the bending moment occurred at slightly deeper positions in the FE analysis. Overall, the behavior of the double sheet piles was well demonstrated by the FE analysis.

Second, the behavior of the soil between the double sheet piles was investigated. Figure 15 illustrates the strain distributions of the inner soil when the head displacement is 100 mm. The volumetric strain is less than 0.3% but the shear strain is as large as 3% at maximum. The positive peak of shear strain is at the middle depth of the excavation. This implies that the shear resistance of the inner soil takes an important role in the retaining mechanism of the structure. If the inner soil is dense enough to dilate, which is not considered in this analysis, the shear resistance is expected to be even larger. The effect of the inner soil could possibly be an advantage of the double sheet piles over the conventional method.

4 CONCLUSIONS

The following conclusions were drawn from the model experiments in 1/4 scale under the Earth's gravity and the FE analysis.

(1) The horizontal displacement at the end of 1,750 mm-deep excavation was 2 mm in double sheet pile walls compared to 35 mm in the single sheet pile wall.
(2) The distribution of bending moment indicated that the back sheet pile behaved as a raked pile to resist the earth pressure.
(3) The distributions of axial forces indicated the frictional resistance of the soil against the push-in and pull-out of both sheet piles.

387

(4) The FE analysis showed that the shear resistance of the inner soil may take an important role in resisting the earth pressure.

For applications of this method to urban construction projects, further study will be undertaken to develop and verify the design method.

REFERENCES

Fujiwara, K., Taenaka, S., Yashima, A., Sawada, K., Ogawa, T. & Takeda, K. 2017. Study on levee reinforcement using double sheet-piles with partition walls. *Japanese Geotechnical Society Special Publication* 5(2): 11–15.

Khan, M.R.A., Takemura, J. & Kusakabe, O. 2006. Centrifuge model tests on behavior of double sheet pile wall cofferdam on clay. *International Journal of Physical Modeling in Geotechnics* 6(3): 01–23.

Kikuchi, Y., Kitazume M., Suzuki M., & Okada, T. 2001. Structural property of double steel sheet pile walls filled with premixed-soil. *Technical Note of the Port and Harbour Research Institute* No. 997 (JUN 2001): (in Japanese)

Sawaguchi, M. 1974. Lateral behavior of a double sheet pile wall structure. *Soils and Foundations* 14(1): 45–59.

Proceedings of the Second International Conference on
Press-in Engineering 2021, Kochi, Japan – Matsumoto et al (eds)
© 2021 Taylor & Francis Group, London, ISBN 978-1-032-10414-0

Study on liquefaction countermeasure method of river embankment using wood and sheet pile

G. Hashimura, K. Okabayashi & D. Yoshikado
National Institute of Technology, Kochi College, Nankoku City, Kochi, Japan

Y. Kajita
Chiyoda Engineering Consultants Co., Ltd, Chiyoda Ward, Tokyo, Japan

ABSTRACT: In this study, the construction method using wooden piles and steel sheet piles was examined as a countermeasure against liquefaction of river embankments. The behavior of river levees was confirmed, and the effect of countermeasures was evaluated in each of the models.

As the research method, liquefaction experiments using a dynamic centrifuge and analysis using effective stress analysis (LIQCA) were performed.

As a result, with no measures taken, the lateral flow due to liquefaction occurred on the horizontal ground on both sides of the river embankment and the embankment subsided.

In the case where countermeasures were taken, displacement of the embankment could be suppressed. By penetrating the wooden piles, the ground density increased, and the effect of liquefaction countermeasures was obtained. Also, it was confirmed that the deformation of the embankment and the lateral flow was suppressed by penetrating the steel sheet piles.

1 INTRODUCTION

The 2011 off the Pacific coast of Tohoku Earthquake caused large-scale levee damage along the river coasts. According to a survey by the Ministry of Land, Infrastructure, Transport, and Tourism (2011), a total of 1195 damages were confirmed in 5 of the 12 water systems managed by the Tohoku Regional Development Bureau on the Pacific Oceanside. Also, the main cause of the levee damage in the upstream area was levee settlement due to liquefaction of the ground.

In the Nankai Trough, located off the coast of Kochi Prefecture, the probability of an earthquake occurring in the next 30 years is 70 to 80% (Japan Meteorological Agency (2020)). According to the assumption of Kochi Prefecture (2013), it is said that land settlement due to liquefaction will occur mainly in Kochi City, and it is predicted that the damage caused by the run-up of the tsunami will increase due to the loss of the levee function.

The purpose of this study is to consider and propose a liquefaction countermeasure construction method for river embankments, the effect of liquefaction countermeasures on river embankments due to the penetration of wooden piles and steel sheet piles will be examined. The wooden piles aim at suppressing liquefaction by increasing the ground density, and the steel sheet piles aim at suppressing the deformation of river embankments. The method of increasing the ground density using wooden piles is expected to have the effect of suppressing liquefaction. The construction method was used when actually constructing the new Kochi City Government Building (Kochi City (2015)). The application of this method to the river embankment is examined. Also, various studies have already been conducted on the construction method using steel sheet piles in the embankment structure. In our laboratory, the effectiveness had been examined from the viewpoint of the penetration depth of steel sheet piles (Nakazawa (2018)). In this study, we focus on the construction position of the steel sheet pile and examine the effect of countermeasures. It is compared to the results of the experiment and analyzed to confirm the suitability. Besides, the effectiveness of each liquefaction countermeasure method is examined.

2 RESEARCH METHOD

2.1 *Research case*

Figure 1 shows the study model. In this study, we examine five models (Case1: Unmeasured model with only embankment constructed, Case2: 4m

DOI: 10.1201/9781003215226-38

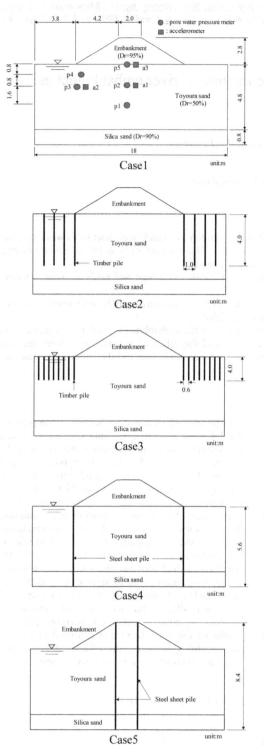

Figure 1. Research model.

wooden piles placed at 1m intervals, Case3: 2mwooden piles placed at 0.6m intervals, Case4: Penetration of the steel sheet pile from the top of the embankment to the base layer, Case5: Penetration of the steel sheet pile from the slope of the embankment to the base layer). Regarding the placement position of wooden piles, considering the workability for existing river embankments, the construction method that penetrates only the horizontal ground around the embankment is considered. The extraction points of the results in all models of displacement, pore water pressure, and horizontal acceleration are shown in Case1.

2.2 Liquefaction experiment by dynamic centrifuge

The liquefaction experiment is performed using the centrifuge force model test equipment owned by our school. Figure 2 shows the centrifuge model test equipment. The experimental soil layer is placed on the shaking table, the centrifugal force is loaded, and then the seismic waves are loaded to the experiment.

Table 1 shows the experimental condition, Table 2 shows physical properties (Hashida (2016)), Figure 3 shows experimental soil layer and Figure 4 shows shaking table seismic wave. In this study, the

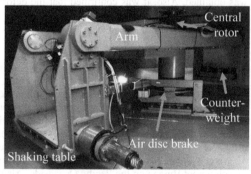

Figure 2. Centrifuge model test equipment.

Table 1. Experiment condition.

Experimental tank	Rigid earthen tank (W:455mm, H:365mm, D:140mm)
Centrifuge force field	40G
Base layer	Silica sand, Dr=95% (Air fall method)
Liquefaction layer	Toyoura standard sand, Dr=50% (Water fall method)
Embankment	Silica sand, Dr=95%
Input seismic wave	sin wave, period 20s, displacement 3mm, frequency 20Hz
Measurement items	Acceleration of in the ground and the quay
	Pore water pressure in the ground Settlement process and cross-sectional displacement process of back ground of sheet pile

Table 2. Physical properties of the sample.

	Unit	Embankment Silica sand (Dr=95%)	Liquefaction layer Toyoura sand (Dr=50%)	Base Layer Silica sand (Dr=90%)
Relative density (Dr)	%	95	50	90
Wet density (ρt)	g/cm³	1.608	1.478	1.65
Maximum density ($\rho tmax$)	g/cm³	1.624	1.644	1.683
Minimum density ($\rho tmin$)	g/cm³	1.357	1.342	1.402
Volume (V)	cm³	1508.2	7506	1251
Weight (m)	g	2425.1	11093.9	2064.2
Sand weight (ms)	g	2146.1		
Water weight (mw)	g	279		

Figure 3. Case1 (before vibration).

experiment was conducted in a centrifugal force field of 40G. Also, according to the similarity rule, an experimental model size of 1/40 and groundwater using a methylcellulose aqueous solution with a viscosity of 40 mPa · s was used. The experimental soil layer was made in an experimental container with a width of 450 mm, a depth of 139mm, and a height of 355mm. The base layer was made of silica sand No. 5 with a relative density of 90%. Afterward, viscous

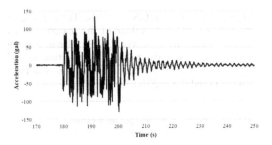

Figure 4. Shaking table seismic wave (Case1).

water was poured and a liquefaction layer was made by the underwater fall method. The liquefaction layer was made using Toyoura sand with a relative density of Dr=50%. The embankment model packed silica sand No. 7 in a dedicated formwork. After that, it is frozen, installed after the liquefaction layer is completed, and melt to make it. A wooden pile with a diameter of 5.0mm and a steel sheet pile with a thickness of 1.0mm was used. Also, the input seismic wave was tested by inputting a sine wave (frequency 20Hz, 20times, displacement 2.0mm) to the shaking table.

2.3 Liquefaction analysis

2.3.1 Analysis procedure
Liquefaction analysis is performed using the liquefaction analysis software SoilWorks for LIQCA. LIQCA is a liquefaction analysis program based on effective stress. Figure 5 shows the analysis procedure.

2.3.2 Material parameter
In this study, the physical property values of the ground material were determined based on the liquefaction analysis conducted by Saito (2016). Also, the liquefaction layer Toyoura sand (Dr=50%) was subjected to element simulation within Soilworks for LIQCA, and the one consistent with the results of the direct shear test conducted at our school was used (Tanimoto, W e t al. (2019)). The physical property values of steel sheet piles and wooden piles were determined from the references (Maruzen Co., Ltd. (1989), NIPPON

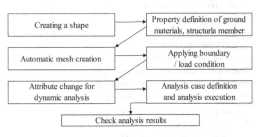

Figure 5. Analysis procedure.

STEEL CORPORATION. (2019)). Table 3 shows the physical property values of the materials used for the countermeasure work, Tables 4 and 5 show the physical property values of the ground, and Figure 6 shows the elemental simulation results of Toyoura sand (Dr = 50%).

2.3.3 *Input seismic wave*

The seismic waves used in the liquefaction analysis used the horizontal acceleration observed on the shaking table of the liquefaction experiment. Figure 7 shows the seismic waves used in the liquefaction analysis.

Table 3. Structural physical property values.

		Unit	Steel sheet pile	Timber pile
Elastic modulus		kN/m^2	205000000	100000
Unit volume weight		kN/m^2	76.44	6.860
Rigidity	Area	m^2	0.392	0.031
	Second moment of area	m^4	0.00005227	7.854
	Effective shear area ratio		0.833	0.900
	Section modulus	m^3	0.003	0.001
	Plastic section modulus	m^3	0.004	0.001
non-linear Consideration	Yield stress	kN/m^2	175000	100
	Bending moment	Mf1	$kN*m$	457.333	
		Mf2	$kN*m$	686	
		a1		0.500	0.500
	Reduction factor	a2		0.100	0.100
		a3		0.500	0.500

Table 4. Parameters for dynamic analysis.

	Unit	Embankment Silica sand (Dr=95%)	Liquefaction layer Toyoura sand (Dr=50%)	Base Layer Silica sand (Dr=90%)
Wet unit volume weight(γ_t)	kN/m^3	15.758	14.484	16.170
Satureted unit volume weight(γ_{sat})	kN/m^3	18.700	18.774	19.845
Element depth	m	1.00	1.00	1.00
Poisson's ratio		0	0	0
Static earth pressure coefficient		0	0	0
Nondimensional initial shear coefficient		873	910	1043
Initial gap ratio (e_0)		0.856	0.791	0.600
Compression exponent (λ)		0.018	0.0039	0.025
Sweling index (K)		0.006	0.00022	0.00020
Pseude-consideration ratio		1	1	1
Dilatancy factor (D_0)		5.000	0.500	0.1000
Dilatancy factor (n)		1.500	5.000	9.000
Water permeability/ Unit volume of water	$m/sec/kN/m^3$	8.67E-05	0.0001	0.0001
water volume modulus of elasticity	kN/m^2	2000000	2000000	2000000
S wave velocity	m/sec	0	0	0
P wave velocity	m/sec	0	0	0
Transformation stress ratio (M_m)		0.909	0.909	0.909
Fracture stress ratio (M_f)		1.122	1.229	1.551
Prameter in hardeining function (B_0)		2200	3500	5000
Prameter in hardeining function (B_1)		30	60	60
Prameter in hardeining function (C_f)		0	0	0
Anistropic loss parameter (C_d)		2000	2000	2000
Plasticity reference strain ($\gamma P * r$)		0.005	0.003	0.010
Elastic reference strain ($\gamma E * r$)		0.010	0.006	0.200

Table 5. Parameters for static analysis.

	Unit	Embankment Silica sand (Dr = 95%)	Liquefaction layer Toyoura sand (Dr = 50%)	Base Layer Silica sand (Dr = 90%)
Wet unit volume weight (γ_t)	kN/m³	15.758	14.484	16.170
Satureted unit volume weight (γ_{sat})	kN/m³	18.700	18.774	19.845
Element depth	m	1.00	1.00	1.00
Poisson's ratio		0.33	0.33	0.33
Effective soil covering pressure	kN/m³	1.00E-08	1.00E-08	1.00E-08
Ststic earth pressure coefficient		1	1	1
Propertionality coefficient of Young's modules (E_0)		3494.07	2775.2	2775.2
Constant (n)		1	1	1
Adhesive force	kN/m³	0	0	0
Internal friction angle (φ)		31.30	37.75	42.00

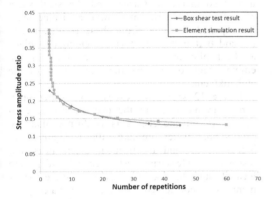

Figure 6. Element simulation.

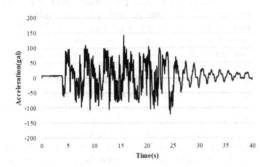

Figure 7. Input seismic wave.

2.3.4 Analysis model

In the liquefaction analysis, a model assuming the actual ground (hereinafter referred to as a real model) and a model assuming a liquefaction experiment (hereinafter referred to as an experimental model), two models are analyzed. The differences between these two models are the model size and boundary conditions. In the real model, the model size is increased, the sides are a cyclic boundary, and the bottom is the viscous boundary. The dimensions of the experimental model are the same as in the liquefaction experiment, and fixed boundaries are applied to the sides and bottom. The results of the real model are mainly used as the analysis results, and the experimental model is used for comparison with the experimental results. Also, the relative density increases to about 70% in the ground around the wooden piles in Case 3. Therefore, the relative density around the wooden stake is changed in the analysis model as shown in Figure 9. Figure 8 shows the schematic diagram of real model, Figure 9 shows the analysis model of Case3 and Figure 10 shows the drainage and boundary condition.

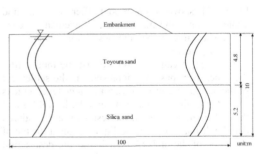

Figure 8. Real model.

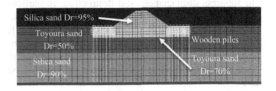

Figure 9. Analysis model (Case3).

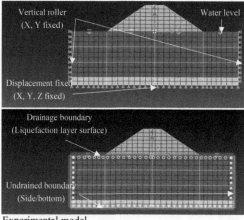

Experimental model

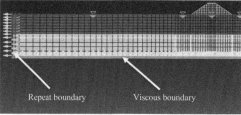

Real model

Figure 10. Drainage and boundary conditions.

3 RESEARCH RESULT

3.1 *Experimental result*

Figure 11 shows each model after shaking, and Table 6 shows the maximum displacement at each point. The values in parentheses in Table 6 show the actual displacement.

In Case1, it can be observed that the model after the experiment has variations such as the settlement of the supporting ground, rise in water level, cracks at the top of the embankment, and the lateral flow of the embankment. From these results, it is considered that the displacement due to liquefaction at the toe of the slope of the embankment led to the destruction of the embankment. It was confirmed that unmeasured river embankments may lose their river embankment function due to liquefaction.

In the experiment in which the wooden pile was penetrated, both Case2 and Case3 showed a tendency to suppress displacement and destruction. In particular, in Case3, the settlement of the top of the embankment was suppressed by 240 mm compared to Case1, and no noticeable evidence of liquefaction was confirmed on the embankment or supporting ground after the experiment. In Case3, the penetration depth of the

wooden pile is shallower than in Case2, but the pile spacing is narrow, so it is considered that the effect of increasing the density of the ground shallow layer is high. Liquefaction is a phenomenon that is particularly likely to occur in shallow areas where effective stress is low. Therefore, it is presumed that Case 3 showed a greater effect of suppressing liquefaction.

In the experiment in which the sheet pile was inserted, the tendency was confirmed to suppress displacement in both Case4 and Case5 models. In Case5, there was no noticeable destruction at the top of the embankment, and it is considered that the shape of the top of the embankment was kept by the steel sheet pile. Even if the slope runs out due to liquefaction, it is unlikely that the embankment function will be lost. On the other hand, in Case4, a larger displacement and destruction of the embankment were confirmed compared to Case5, and the steel sheet pile after the experiment showed a displacement that opened outward. From these results, it can be said that in Case4, it is not enough to place a sheet pile on the toe of the slope of the embankment to suppress the lateral flow of the embankment.

Figure 12 shows the excess pore water pressure ratio measured in the experiments of each model. In all models, it is confirmed that the excess pore water pressure ratio is significantly higher in p3 and p4 than in other measuring instruments. Measuring instrument p3 and p4 are located under the toe of the slope of the embankment. It is considered that the pore water pressure tends to rise in the toe of the slope because the restraint pressure due to the embankment load does not affect the toe of the slope part.

In Case1, the excess pore water pressure ratio reached 1.0 in both p3 and p4, confirming that complete liquefaction occurred. From this, it is important to focus on the toe of the slope of the embankment as a countermeasure against liquefaction in the embankment ground. Also, the excess pore water pressure ratio, which theoretically does not exceed 1.0, greatly exceeds 1.0 at p3 and p4. This may be since the measuring instrument sank due to liquefaction and that the water pressure exceeded the theoretical value due to the occurrence of the lateral flow of the embankment. In the model in which the wooden pile was penetrated, the increase in pore water pressure was suppressed in Case2 compared to Case1, but the excess pore water pressure ratio increased to nearly 1.0 in p4. Therefore, it can be said that the effect of suppressing liquefaction as a countermeasure is imperfect.

On the other hand, in Case3, the increase in excess pore water pressure ratio is suppressed to about 0.4 even at p4. From this, the effect of

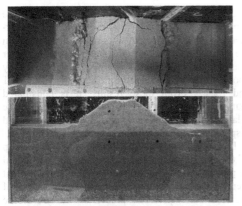

Case1

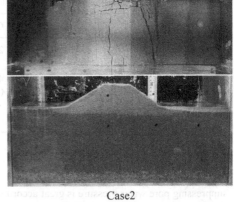

Case2

Case3

Case4

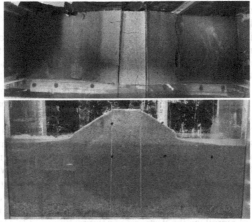

Case5

Figure 11. Observed movement.

Table 6. Maximum deformation (experimental result).

		Case1	Case2	Case3	Case4	Case5
Top of slope	displacement	6 (240)	5 (200)	1 (40)	2 (80)	2 (80)
	settlement	14 (280)	4 (160)	4 (160)	2 (80)	4 (160)
Settlement of crest		7 (280)	5 (200)	1 (40)	2 (80)	4 (160)

unit: mm

suppressing pore water pressure is great according to the dense placement of wooden piles.

In the experiment using steel sheet piles, p4 showed the highest excess pore water pressure ratio in Case4, and the numerical value was about 0.6. In Case5, it was confirmed that the excess pore water pressure ratio increased more than in Case4 in all measuring instruments. In this experiment, the steel sheet pile was inserted to the bottom of the container, the inflow and outflow of interstitial water were completely blocked.

This prevented the divergence of pore water pressure at the toe of the slope, which is also considered to be one of the reasons why the excess pore water pressure ratio exceeded Case4.

3.2 Analysis result

Figure 13 shows the analysis results, Table 7 shows the maximum displacement, and Figure 14 shows the excess pore water pressure ratio. In Figure 13, the red part indicates that the excess pore water pressure ratio is high. In Case1, the settlement of about 1m occurred at the top of the embankment and about 1.2 m at the bottom of the embankment. Also, it is confirmed that the embankment and the area directly below the embankment are not completely liquefied. From these results, the increase in excess pore water pressure was suppressed in the part where the suppressing pressure of the embankment was acting. It is confirmed that complete liquefaction occurs from the toe of the slope of the embankment to the horizontal ground. Therefore, it is considered that the horizontal ground part lost its bearing capacity due to liquefaction and the lateral flow was generated by the embankment load.

In Case2, the penetrated wooden piles were deformed to the extent that it was likely destruction. SoilWorks for LIQCA cannot reproduce the destruction of structural members. Therefore, in reality, the wooden piles of Case2 are expected to break. Also, the excess pore water pressure ratio is not different from Case1 at each point, and no liquefaction suppressing effect is observed.

In Case3, the lateral flow and settlement were suppressed at the toe of the slope of the embankment, so the settlement was also suppressed in the embankment. Although the excess pore water pressure ratio fluctuated sharply, it decreased by about 0.1 to 0.2 at each point compared to Case1 and Case2. From these results, it is considered that the penetration of wooden piles increased the relative density and suppressed liquefaction and displacement of the embankment.

In Case4, the amount of settlement at the top of the embankment and the amount of horizontal displacement at the bottom of the embankment were 1409 mm, which exceeded Case1. Even if the steel sheet pile was placed on the toe of the slope of the embankment, it was not possible to suppress the settlement and the lateral flow of the embankment. From Figure 13, it is confirmed that the excess pore water pressure ratio is smaller on the inside of the steel sheet pile than on the outside. It is considered that the water pressure did not increase because the steel sheet pile blocked the inflow of pore water to the area directly below the embankment.

In Case5, the settlement at the top of the embankment was most suppressed. On the other hand, the horizontal displacement of the embankment was the largest. Therefore, it can be said that the deformation of the top can be suppressed by penetrating the sheet pile into the top of the embankment, but the lateral flow of the toe of the slope cannot be suppressed. In the construction method using sheet piles such as Cases 4 and 5, the sheet pile head is deformed to open. Therefore, it is considered that deformation can be further suppressed by connecting the upper ends of the sheet pile with a tie rod.

3.3 Comparison of analysis and experimental result

Figure 15 shows a comparison diagram of the deformation results of the experiment and analysis. The analysis results here show the results of the experimental model. In this study, the analysis results tended to be more deformed than the experimental results. The maximum difference between the analysis results and the experimental results was about 1300 mm. The following reasons can be considered for this. (1) Suppression of deformation due to wall friction between the experimental soil layer and the container, (2) When making the embankment of the experimental soil layer, the frozen and compacted embankment was thawed and used. Therefore, it is considered to be more solid than only compressing it. (3) Consolidation of experimental soil layer by centrifugal force.

Regarding the excess pore water pressure ratio, in the experiment, there was a difference depending on the countermeasure method. But in the analysis, the same result was obtained for all models. That is

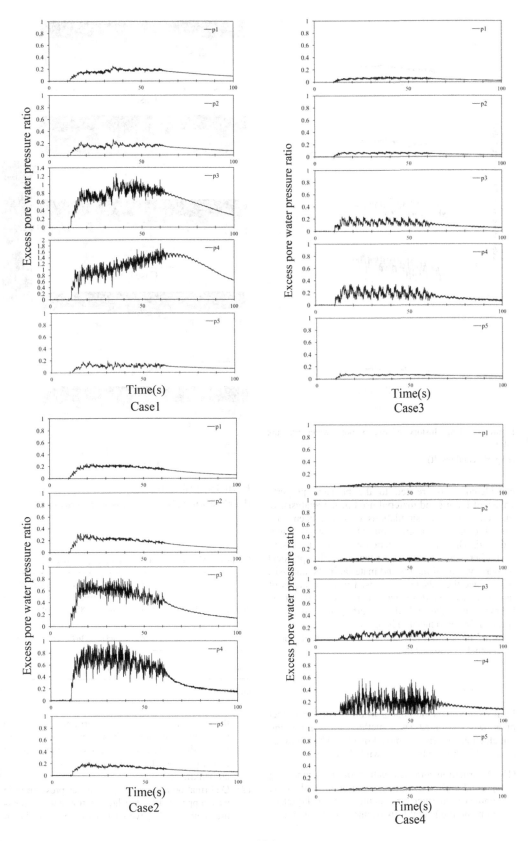

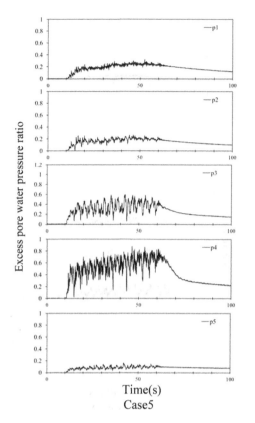

Figure 12. Time history of excess pore water pressure ratio.

(Experimental result)

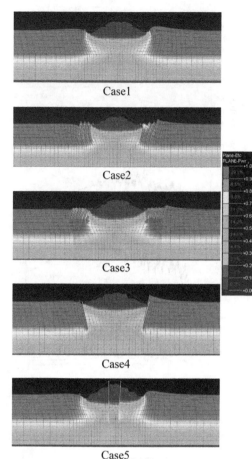

Figure 13. Analysis result.

(Displacement and excess pore water pressure ratio)

considered to be related to the physical property values of the ground material used in the analysis. In this study, element simulations have not been confirmed for ground materials other than Toyoura standard sand, which is a liquefaction layer. It is necessary to further confirm the consistency of the physical property values of the ground material in the analysis. Also, since the seismic wave used in the analysis of this study is about 40 seconds, it is considered that the behavior after the end of the earthquake cannot be sufficiently reproduced. Therefore, it is necessary to perform the analysis in a sufficient time after the principal shock is completed.

Table 7. Maximum deformation (analysis result).

		Case1	Case2	Case3	Case4	Case5
Top of slope	displacement	1279	860	856	1409	1026
	settlement	613	451	367	515	725
Settlement of crest		1045	878	851	1062	811

unit: mm

4 CONCLUSION

In this study, a method using wooden piles and sheet piles was examined as a method for suppressing liquefaction of river embankments. The findings obtained in this study are shown below.

(1) Deformation and destruction of the supporting ground and embankment were confirmed in the unmeasured river embankment. Also, liquefaction and the lateral flow occurred from the toe of the slope to the horizontal ground, leading to subsidence of the embankment. It was confirmed that there is a risk of loss of embankment function due to liquefaction of the river embankment.

(2) Deformation and excess pore water pressure ratio was suppressed by the liquefaction countermeasure method using wooden piles. In particular, it

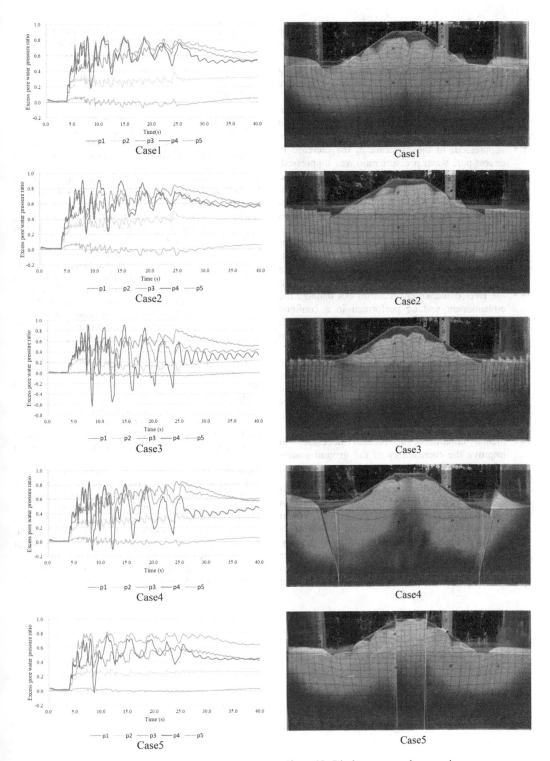

Figure 14. Excess pore water pressure ratio.
(Analysis result)

Figure 15. Displacement result comparison.

399

can be said that the effect was remarkable in Case 3 in which wooden piles were penetrated into the shallow layer. Therefore, it can be said that the method of increasing the ground density using wooden piles is effective as a countermeasure against liquefaction of river embankments.

(3) In Case4, where a steel sheet pile has penetrated the top of the embankment, deformation and settlement of the top of the embankment were suppressed. In the experiment, the increase in excess pore water pressure ratio was suppressed. In Case 5, where the sheet pile has penetrated the toe of the slope of the embankment, deformation on the slope of the embankment was decreased, but it was confirmed the deformation of the steel sheet pile due to the settlement of the embankment. Case4 is effective for reinforcement at the top of the embankment, and Case5 is effective for reinforcement of the slope. Also, the penetration of the sheet pile into the river embankment must do not damage the embankment and be performed in a confined construction space, so the press-in method with a static load with less vibration is considered to be suitable.

(4) Comparing the analysis results and the experimental results, a large difference was found in the deformation of the model. In the liquefaction experiment, deformation is considered to be suppressed depending on the soil layer construction method and experimental conditions. It is also necessary to improve the consistency of the ground material used for the analysis.

From the results, the effect of liquefaction countermeasures was confirmed for the construction method using wooden piles and steel sheet piles in river embankments. It can be said that the construction of wooden piles is more effective in suppressing liquefaction by narrowing the pile spacing rather than increasing the penetration depth. It was confirmed that the construction method using steel sheet piles is effective in suppressing deformation. Also, it is necessary to consider the flow of pore water in the construction method using steel sheet piles.

REFERENCES

Hashida, K. 2016. Development of dynamic centrifugal force model experimental device and research on oscillating earth pressure and embankment experiment. *Kochi National College of Technology Graduation Thesis Collection No.15* . 9-16

Japan Meteorological Agency. 2020. Information related to the Nankai Trough earthquake. https://www.data.jma.go.jp/svd/eew/data/nteq/index.html

Kochi City. 2015. Kochi City New Government Building Implementation Design [Summary Version]. *Kochi City New Government Building Construction Office.* 17

Kochi Prefecture. 2013. [Kochi version] Calculation results of damage estimation due to Nankai Trough giant earthquake (Drawing collection). *Kochi Prefecture Crisis Management Department Nankai Trough Earthquake Countermeasures Division Public materials.* 20-31

Ministry of Land, Infrastructure, Transport and Tourism. 2011. Damage and restoration status of river and coastal facilities in the Great East Japan Earthquake. *Materials provided by the River Department, Tohoku Regional Development Bureau, Ministry of Land, Infrastructure, Transport, and Tourism .* 2-46

Nakazawa, Y. 2018. Seismic response analysis and dynamic centrifuge model test on river embankment reinforcement method. *Kochi National College of Technology Graduation Thesis Collection.* 4-5

NIPPONCORPORATION STEEL. 2019. Steel sheet pile. 7 *https://www.nipponsteel.com/product/catalog_download/pdf/K007.pdf*

Tanimoto, W., Okabayashi, K., Mukaidani, M., and Hama, K. 2019. Comparison of Kochi College type cyclic box shear apparatus with cyclic triaxial apparatus. *Reiwa 1st Japan Society of Civil Engineers National Convention 74th Annual Scientific Lecture .* 2162-311

YONDEN CONSULTANTS co., Inc, Saito. 2016. Simulation analysis of centrifugal force model experiment using liquefaction analysis program LIQCA. 14

Maruzen Co., Ltd. 1989. National Astronomical Observatory of Japan, Science Chronology, 1990, Vol. 63. 444

Session C: Disaster prevention and mitigation

Proceedings of the Second International Conference on
Press-in Engineering 2021, Kochi, Japan – Matsumoto et al (eds)
© *2021 Taylor & Francis Group, London, ISBN 978-1-032-10414-0*

A preliminary numerical model for erosion at the flow-soil interface based on the sediment transport model

Y. Yuan, F. Liang & C. Wang*
Tongji University, Shanghai, China

ABSTRACT: In recent years, using numerical methods to calculate fluid-particle interaction has become a new trend in scour research. To simulate the flow field and soil particle movements during scour, the traditional Euler equation and the sediment transport model were combined under the condition of single-phase flow in this study. Based on the assumptions and theories of Engelund and Bagnold, the whole calculation process from the particle initiation to the final deposition is completed. Through the sediment transport model and particle movement analysis, the topographic parameters of the grid can be updated, and the iterative calculation of the water-soil interface is realized. Using the above numerical model, the erosion process can be calculated under various test conditions, and the results can be validated by the test data. Finally, the parameter sensitivity of the model has also been discussed.

Keywords: Erosion, Numerical model, Sediment transport, Flow-soil interface

1 INTRODUCTION

Local scour under the condition of current is a typical physical phenomenon in water conservancy projects, which is common in projects such as wading bridges and offshore wind turbines. More than half of the bridge water damage problems are related to scour in recent years, which has also caused more and more scholars to pay attention to the scour problem (Søren, 2013; Escobar, 2018; Wang, 2017; Melville, 2000). Many experimental studies on this problem have been carried out (Ma, 2018; Liang, 2020; Lagasse, 2007). The flow structure in the local scour hole has been investigated using laboratory experiments, and the water flow's attenuation law around the foundation with the development of scour was obtained by scholars such as Rajaratnam (1977), Melville (1977), Graf (2002), Karim (2000) and Loópez (2008). To further clarify the scour mechanism under different water flow conditions, Dongol (1994) explored the law of scour evolution under the conditions of live water through flume tests, and the results are consistent with the test results of Chiew (1984) and Baker (1986). While Yang et al. (2010) conducted laboratory experiments on the phenomenon of river bed coarsening under the condition of clear water.

Although laboratory tests can effectively simulate on-site scour conditions to a certain extent, which intuitively shows the scour results under different working conditions, there are still shortcomings, such as time-consumption and high requirements for on-site equipment. It is worth noting that when the particle size of the soil is small, the water flow is prone to turbidity after scour occurs, which brings difficulties to observing the scour results. Therefore, numerical simulation can be a good choice as the research method for multi-condition analysis. The research results can be divided into two aspects: (1) Combining the standard k-ε turbulence model and Navier-Stokes equation to realize the numerical simulation of the three-dimensional complex flow field. The instantaneous change of the riverbed elevation coordinate during the scour process can be obtained (Zhou, 2016; Wang, 2014; Zhang, 2020; Zhu, 2011). (2) Exploring the local sediment transport situation based on the sediment initial motion theory (Einstein, 1942; Engelund, 1976a; Laursen, 1998; Wu, 2002). The above two studies mainly focused on the complex flow fields and sediment transport and carried out scour analysis and prediction based on the instantaneous results. However, these studies did not establish the relationship between flow, particle migration, and scour development, which considers the dynamic evolution of the erosion process. Meanwhile, most of the existing theoretical calculations and numerical analysis methods only consider the erosion effect of the flow, while the particle accumulations after the initial movement are usually ignored.

DOI: 10.1201/9781003215226-39

This research aims to establish the connection between water flow, particle migration, and scour development under a three-dimensional flow field. A preliminary numerical model for erosion at the flow-soil interface based on the sediment transport model is proposed. The effects of water erosion and particle accumulations on the riverbed elevation are comprehensively considered. Finally, taking the simple scour resistance test as an example, the numerical model was validated by previous SSRT results.

2 THEORETICAL MODEL

2.1 Flow field governing equation

The governing equations for fluids can be divided into continuity equations and motion equations. Among them, the continuity equation establishes an equal relationship between the sum of the inflow mass of the control surface and the mass increase in the control body from the perspective of mass conservation. According to Gauss's theorem, it can be written as the following tensor form:

$$\frac{\partial \rho}{\partial t} + \frac{\partial (\rho u_i)}{\partial x_i} = 0 \tag{1}$$

where ρ is the fluid density, u_i is the flow velocity, t is the time.

The conservation of momentum is then used to calculate the fluid motion equation. For the convenience of calculation, it is usually assumed that the fluid is incompressible Newtonian fluid flow, and its motion equation is called Navier-Stokes equations (abbreviated as N-S equation). The ideal fluid N-S equation is used in this study, namely Euler's equation:

$$\begin{cases} f_x - \frac{1}{\rho}\frac{\partial p}{\partial x} = \frac{\partial u}{\partial t} + u\frac{\partial u}{\partial x} + v\frac{\partial u}{\partial y} + w\frac{\partial u}{\partial z} \\ f_y - \frac{1}{\rho}\frac{\partial p}{\partial y} = \frac{\partial v}{\partial t} + u\frac{\partial v}{\partial x} + v\frac{\partial v}{\partial y} + w\frac{\partial v}{\partial z} \\ f_z - \frac{1}{\rho}\frac{\partial p}{\partial z} = \frac{\partial w}{\partial t} + u\frac{\partial w}{\partial x} + v\frac{\partial w}{\partial y} + w\frac{\partial w}{\partial z} \end{cases} \tag{2}$$

where u, v, w are the flow velocity components in the x, y, z-axis directions, respectively.

2.2 Plane sediment transport equation

The first step in the sediment calculation model is to judge the initiation of the surface particles under the calculated flow field conditions. So it is necessary to conduct a force analysis. Affected by water flow, surface particles are mainly subjected by drag force F_D, upward force F_L and gravity W (Einstein, 1942; Laursen, 1998). The initial movement occurs under the combined action of the three forces, and the calculation expressions are as follows:

$$F_D = C_D \cdot \frac{\pi D^2}{4} \cdot \frac{\rho u_0^2}{2} \tag{3}$$

$$F_L = C_L \cdot \frac{\pi D^2}{4} \cdot \frac{\rho u_0^2}{2} \tag{4}$$

$$W = \frac{\pi D^3}{6} \cdot (\gamma_s - \gamma) \tag{5}$$

where C_D is the drag coefficient, C_L is the uplift coefficient, u_0 is the flow velocity acting on the surface of the river bed, D is the particle size, γ_s is the dry unit weight, and γ is the unit weight of soil.

When the instantaneous drag force, F_D, is greater than the particle interaction force, the surface particles will occur initial movement, which can be judged according to the following equation:

$$F_D = f \cdot (W - F_L) = (W - F_L) \cdot \tan\varphi \tag{6}$$

where φ is the underwater repose angle of the surface particles.

After the initiation, the surface particles are continuously eroded by flow and transported downstream. At this moment, the scour begins to occur. To realize the quantitative calculation of the scour development process, the formula of sediment transport rate proposed by Engelund (1976b) is used to calculate the particle loss. Then the relationship between the elevation of the riverbed and the particle migration can be established. When calculating sediment transport, Engelund simplified the sediment particles into spheres. There are $1/D^2$ sediments in the unit area at this time, and it is assumed that $p\%$ of them will undergo initial movement. So, the moving sediment number for per unit area n is:

$$n = \frac{p}{D^2} \tag{7}$$

Eq. (8) shows the formula proposed by Engelund, which can calculate the sediment transport rate. Combining calculation assumptions, n and u_b are critical parameters in theoretical and numerical calculations. Meanwhile, n is used to calculate the deformation of mesh nodes, and u_b is for particle migration. The drag force, F_D, and friction resistance of the particles, F_x, after the initial movement, can be calculated according to Eq. (9) and Eq. (10).

$$g_b = \frac{\pi D^3}{6} \gamma_s \cdot \frac{p}{D^2} \overline{u_b} \tag{8}$$

$$F_D = C_D \cdot \frac{\pi D^2}{4} \cdot \frac{\rho}{2} \cdot (a_0 U_* - \overline{u_b})^2 \tag{9}$$

$$F_x = \frac{\pi D^3}{6}(\gamma_s - \gamma) \cdot \beta \qquad (10)$$

where g_b is the unit sediment transport rate and $\overline{u_b}$ is the initial velocity of particle movement, β is the friction coefficient, and $a_0 U_*$ reflects the flow velocity at which the bed load moves near the bed surface. If the distance from the bed surface when the bed load moves are about twice the particle size, the recommended value range for a_0 is [6,10].

When the scour reaches the equilibrium state, the drag force is equal to the friction force, Eq. (9) and Eq. (10) can be combined:

$$\frac{\overline{u_b}}{U_*} = a_0 \cdot \left(1 - \sqrt{\frac{\Theta_0}{\Theta}}\right) \qquad (11)$$

where, Θ_0 is equivalent to the water flow intensity value corresponding to the condition that the surface particles just no longer move ($\overline{u_b}= 0$), which can be connected with Θ_c; Θ is the flow intensity parameter, which can be calculated as follows (Albertson, 1958):

$$\Theta = \frac{\tau_0}{(\gamma_s - \gamma)D} = \frac{U_*^2 \rho}{(\gamma_s - \gamma)D} \qquad (12)$$

where τ_0 is the shear stress of flow.

In particular, the intensity of the water flow when the surface particles just reach the initial motion conditions is recorded as Θ_c. Then Eq. (12) can be further derived that:

$$\Theta_0 = \frac{4\beta}{3a^2 C_D} \qquad (13)$$

After establishing a connection based on the measured data, Eq. (11) can be written as the following equation:

$$\frac{\overline{u_b}}{U_*} = a_0 \cdot \left(1 - 0.7\sqrt{\frac{\Theta_c}{\Theta}}\right) \qquad (14)$$

On this basis, combined with the concept of Bagnold (from the perspective of energy) (Bagnold, 1966), the shear stress τ_0 of the sand-laden flow is composed of two parts, the particle shear force T and the flow shear force τ' of the static riverbed, and it is considered that the flow shear force τ' is equal to the initial shear force τ_c of the sediment particles. The particle shear force is generated by particle friction, which can be obtained from Eq. (15):

$$T = n\overline{F_x} \qquad (15)$$

From this, the moving sediment number for per unit area n and the local grid elevation change ΔH can be derived:

$$n = \frac{6}{D^2 \pi \beta} \cdot (\Theta - \Theta_c) \qquad (16)$$

$$\Delta H = \frac{6D}{\pi \beta} \cdot (\Theta - \Theta_c) \qquad (17)$$

2.3 Inclined plane sediment transport equation

However, the actual riverbed conditions are not ideally flat. Furthermore, with the occurrence of scour, the riverbed topography will continue to change. The scour calculation method derived in the planar state often cannot meet complex topographic changes. So it is necessary to derive the calculation formula on the slopes. It is worth noting that, unlike the movement in a flat state, particle movement under slope conditions is more complicated (as shown in Figure1). It is no longer a simple linear motion but should be considered in several stages for analysis.

In the derivation of the plane condition, the component force of the particle gravity along the water flow direction is ignored. When the river bed condition is a slope, the existence of the gravity component will promote the initial movement of the particles, which will affect the subsequent development of scour. In order to clarify the particle movement under slope conditions, this study established

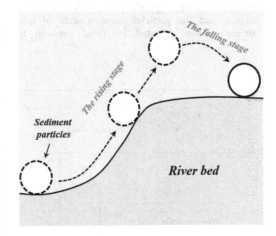

Figure 1. Schematic diagram of particle movement on a slope.

a standard particle slope force model. The force analysis diagram is shown in Figure 2. It is assumed that the slope angle is θ, the particles will initially move along the direction at an angle of α to the horizontal axis of the slope under the action of the water flow.

Affected by the slope angle and the current direction, the drag force and the gravity will form a component force, F, with a certain angle with the horizontal axis. The equation for determining the initial movement of particles at this time is as follows:

$$F_D = (W \cos \theta - F_L) \cdot \tan \varphi \qquad (18)$$

Similarly, the calculation formula of the moving sediment number for per unit area n and the local grid elevation change ΔH also needs to be modified:

$$n = \frac{6}{D^2 \pi \beta \cos \theta} \cdot (\Theta - \Theta_c) \qquad (19)$$

$$\Delta H = \frac{6D}{\pi \beta \cos \theta} \cdot (\Theta - \Theta_c) \qquad (20)$$

Eq. (19) and Eq. (20) can calculate the erosion depth after the surface particle eroded by the flow. However, the sedimentation will also affect the final elevation of each grid node when the scour reaches the equilibrium status. To determine the impact of particle sedimentation on the scour development, it is necessary to analyze the subsequent movement of particles after leaving the soil surface. It can be seen in Figure 1 that when the riverbed is in a slope state, the particles can be roughly divided into two stages after the initiation. In the first stage, called the ascending stage, the particles continue to move with the flow after leaving the soil surface but slowly deposit due to gravity. In the second stage, which is called the falling stage, the particles fall to the soil surface, and the particles show a state of near-surface motion. Limited by force between the

particles, the particles quickly reach a stable equilibrium state. For the convenience of analysis, the water flow velocity and particle movement velocity are decomposed into U_{*x}, U_{*y}, u_x, and u_y along the x-axis and y-axis. When the particle starts on the inclined plane, the component of particle motion velocity along the z-axis is recorded as u_z. When the particle movement is in the first stage, it can be known from the momentum equation:

$$\begin{cases} F_x = m \frac{\partial u_x}{\partial t} \\ F_y = m \frac{\partial u_y}{\partial t} \\ -F_z - W = m \frac{\partial u_z}{\partial t} \end{cases} \qquad (21)$$

where,

$$\begin{cases} F_x = C_D \cdot \frac{\pi D^2}{4} \cdot \frac{\rho}{2} \cdot (U_{*x} - u_x) |U_{*x} - u_x| \\ F_y = C_D \cdot \frac{\pi D^2}{4} \cdot \frac{\rho}{2} \cdot (U_{*y} - u_y) |U_{*y} - u_y| \\ F_z = C_D \cdot \frac{\pi D^2}{4} \cdot \frac{\rho}{2} \cdot u_z^2 \end{cases} \qquad (22)$$

Specially, the transformation of partial differential equations can be considered:

$$\frac{du_z}{dt} = \frac{1}{t} \frac{dz}{dt} - \frac{u_z}{t} \qquad (23)$$

Eq. (23) can be used to transform Eq. (21) into a differential equation related to displacement and time. Through iterative calculation, the movement state of the particles along with the x and y directions when the particles sink to the surface of the soil can be calculated. The movement of the particles enters the second stage, and the momentum equation at this time is established:

$$\begin{cases} F_x - \overline{F_x} = m \frac{\partial u_x}{\partial t} \\ F_y - \overline{F_y} = m \frac{\partial u_y}{\partial t} \end{cases} \qquad (24)$$

where, $\overline{F_x}$ and $\overline{F_y}$ are the component of resistance along the x- and y-axis, which can be calculated as follows:

$$\begin{cases} \overline{F_x} = \frac{\pi D^3}{6} (\gamma_s - \gamma) \cdot \beta \cdot \cos \theta \cdot \sin \alpha \\ \overline{F_y} = \frac{\pi D^3}{6} (\gamma_s - \gamma) \cdot \beta \cdot \cos \theta \cdot \cos \alpha \end{cases} \qquad (25)$$

Combining Eqs. (25) and (23), iteratively solve the displacements in the x and y directions to determine the coordinates of particle accumulation. Then using Eq. (20) to calculate the elevation change, which realizes the dynamic update of the grid node elevation.

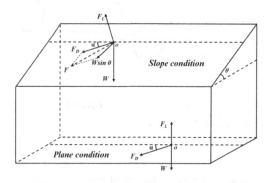

Figure 2. The force analysis diagram.

3 CASE ANALYSIS AND MODEL ESTABLISHMENT

3.1 *Simple Scour Resistance Test (SSRT)*

As a new type of test method for studying meso-scour (Wang, 2018), the simple scour resistance test has dramatically alleviated the drawbacks of the flume experiment due to its convenience and time-saving process. Through the rotation of the blade, different local flow field conditions can be simulated. With the test results for various soil conditions, the performance of different soil properties is obtained. To validate the accuracy and applicability of the theoretical model and numerical calculations, we selected SSRT experiments as a example because it involves the results and processes of both erosion and deposit. The test device comprises the following three parts: power device, transmission device, and sample container. The schematic diagram of the structure is shown in Figure 3. The device effectively simulates the actual flow field conditions during scour, and restores soil erosion and particle migration. The inner diameter of the sample container is 90mm, the total height is 128mm, and the thickness of the soil sample in the container is 40mm. The blades used are 76mm wide and 22mm high. The speed range is 0-150r/min, and the corresponding simulated flow velocity range is 0-0.6m/s.

3.2 *Model establishment and calculation process*

Unlike the flume tests, this method also has apparent mesoscopic particle erosion and accumulation phenomena. In order to simulate the SSRT experiment well through the model, the deposition of particles needs to be considered. which is rarely considered in traditional methods. However, the scour theory proposed in the previous section satisfies the numerical calculation requirements of SSRT well.

The numerical calculation simulated four soil samples with a median diameter of 0.075mm, 0.25mm, 0.5mm, and 2mm under the conditions of the rotation speed of 70r/min and 90r/min. To simplify the description, take the speed condition of 70r/min as an example to introduce the overall modeling calculation

process. This study mainly focuses on calculating local scour using the sediment transport formula, which does not have such high accuracy requirements for complex flow fields. Therefore, in order to improve the calculation efficiency, combined with the theoretical basis of the previous section, numerical calculations are carried out using coupled modeling of COMSOL and MATLAB. The schematic diagram and meshing of the COMSOL model are shown in Figure 4. The *k-ε* model in COMSOL is used to calculate and update the real-time flow

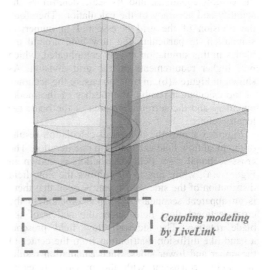

Coupling modeling by LiveLink

(a) Model building

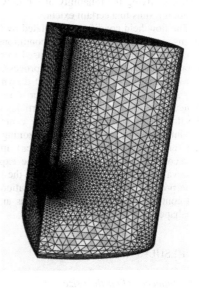

(b) Meshing result

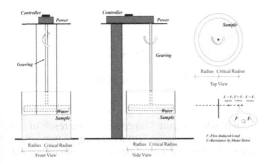

Figure 3. SSRT setup (Wang 2018).

Figure 4. Model building and meshing result.

field. In each iterative calculation, for the flow field calculation results of COMSOL, the movement calculation of sediment particles is realized through MATLAB. After obtaining the instantaneous riverbed elevation changes, the model parameters can be adjusted through the LiveLink module. As shown in Figure 4(a), the red dashed area is the coupling modeling area. The numerical model is established as a 1/4 prototype model, which maximally reduces the calculation time.

A grid is the basis of numerical simulation calculation. Grid division determines the division of the calculation domain and the degree of dispersion of the control equations, and its scale determines the stability and accuracy of the calculation. Therefore, the division of the grid is essential in numerical simulation. In particular, the flow field around the blades in this simulation is very complicated, which puts higher requirements on the grid division. As shown in Figure 4(b), in order to ensure the accuracy of the calculation results, the meshes of the model boundary and the surrounding area of the blade are locally refined.

Figure 5 shows the numerical calculation results of the local flow field calculated in one iteration. The cross-sectional flow field distribution is shown in Figure 5(a), while Figure 5(b) shows the flow field distribution of the sidewall. It can be seen that there is an apparent secondary flow phenomenon in the cross-sectional flow field under the action of the blade rotating. The sidewall flow field presents a band-like diffusion distribution from the center to the upper and lower sides. Such calculation results are highly consistent with the previous research results of many scholars (Zhang, 2020; Einstein, 1926), verifying the reliability of the flow field calculation results to a certain extent.

The flow field conditions calculated by COMSOL are used as the initial flow field conditions, and the scour theory mentioned above is used to realize the iterative calculation of the scour process. Figure 6 shows the shape of the scour result drawn based on the scour calculation results after one iteration. Affected by the secondary flow, particles eroded by the flow continue to gather into the central area of the intermediate device. Apparent zoning phenomena (eroded area and deposition area) appeared in the scour pattern, consistent with the experimental observation results (Wang, 2018). As the scour iteration progresses, erosion and accumulation phenomena continue to develop and appear as an apparent hat shape at last.

4 RESULTS AND ANALYSIS

4.1 *Influence of particle size*

Figure 7 shows the scour development curve of the numerical simulation and the SSRT test of soil with different particle sizes under rotation speed of

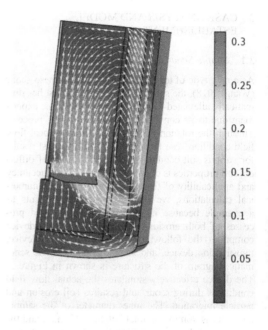

(a) Cross-section

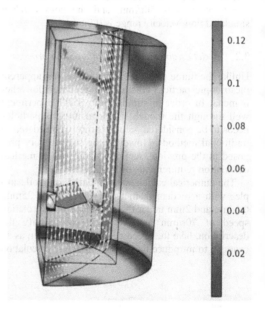

(b) Sidewall

Figure 5. Flow field calculation results.

70rpm, where Figure 7(a) compares the result of erosion depth, and Figure 7(b) for the deposition height. Since the flow field conditions in the numerical calculation are all instantaneous loading, which assumes that the water flow velocity reaches the test

Figure 6. Scour simulation results after one iteration.

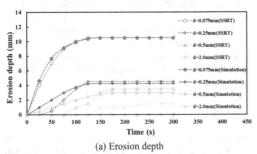

(a) Erosion depth

(b) Deposition height

Figure 7. Scour development curve of numerical simulation and SSRT at 70rpm.

requirement of 70rpm at the beginning of the test, the numerical simulation curve develops rapidly from 0s. However, there are specific differences in actual test conditions. After the blade rotates, the flow field conditions will slowly generate the expected conditions. As the flow velocity increases, only some of the particles have initiated. Therefore, the initial slope of the test curve is delayed than the

numerical results. This phenomenon is more apparent in the case of large particles than small ones. With the continuous development of the scour depth, the intensity of the water flow decays along the depth direction. When the strength caused by the flow is equal to the erosion resistance of the soil particles, the scour reaches an equilibrium status. Therefore, the differences at the beginning will not affect the final scour depth and deposition height. The numerical model can provide reliable equilibrium erosion depth and deposition height.

Comparing the erosion depth and accumulation height after the erosion reaches the equilibrium status, the numerical calculation result of the soil sample with a particle size of 0.075 mm is approximately equal to the test results. When the median particle size is changed to 0.25mm and 0.5mm, the accuracy of the numerical simulation will decrease. In general, the numerical calculation method used in this study has higher accuracy for small-particle soil samples than large ones. This is because the grid node elevation in the calculation model uses the particle size as the minimum standard calculation unit. Calculated results (erosion depth or deposit height) less than the calculation unit would be automatically ignored in every step. Hence, the deviation of simulation results for large particles is larger than small ones. Especially, when the particle size is 2.0mm, it can be known from Eq. (18) that the particles will not initiate at this time, which makes the calculated erosion depth and deposition height 0. After disturbed by the water flow, the particles may roll over under the flow condition. Therefore, there will be a certain erosion depth and deposition height in the experiment but are observed smaller than the size of a single particle.

4.2 Influence of rotating velocity

The comparisons of numerical simulation and test results at the speed of 70rpm and 90rpm are shown in Table 1 and Table 2. After the water flow rate increased, the numerical calculation results for different particle sizes have changed to various degrees. At the speed of 70rpm, for soil samples with particle sizes of 0.25mm and 0.5mm, the calculation deviations of the erosion depth are 5.6% and 14.3%. When the speed is 90rpm, the deviation of the erosion depth under two soil conditions becomes 5.6%

Table 1. Comparison of numerical simulation and test results at 70r/min.

		Erosion depth (mm)			Deposition Height (mm)		
Cases	D_{50} (mm)	SSRT	Simulation	Deviation	SSRT	Simulation	Deviation
1	0.075	10.5	10.5	0%	12.0	12.0	0%
2	0.25	4.5	4.25	5.6%	10.5	10.25	2.4%
3	0.5	3.5	3.0	14.3%	7.5	7.0	6.7%

Table 2. Comparison of numerical simulation and test results at 90r/min.

Cases	D_{50} (mm)	Erosion depth (mm)			Deposition Height (mm)		
		SSRT	Simulation	Deviation	SSRT	Simulation	Deviation
4	0.25	9.0	9.5	5.6%	13.0	13.5	3.8%
5	0.5	5.0	5.5	10.0%	10.0	11.0	10.0%

and 10.0%, and the calculation accuracy is higher than that under the low-velocity condition. However, the simulation results of the deposition depth show an opposite situation. It increased from 2.4% and 6.7% at 70rpm to 3.8% and 10% at 90rpm. Nevertheless, the maximum numerical simulation deviations of both the erosion depth and the deposition height of the two soil samples did not exceed 10%. This verifies the reliability and applicability of the numerical calculation method for different flow rate conditions to a certain extent.

5 CONCLUSIONS

Based on the sediment transport model, a method for calculating the erosion depth of particles under is derived. From the perspective of conservation of momentum, the migration and accumulation process of particles after initiation is discussed. Based on these, a preliminary numerical model for erosion at the flow-soil interface was proposed. The calculation model is then used for the numerical simulation of the SSRT test, which explores the difference between the numerical simulation results and the test results, and draws the following conclusions:

1. The scour numerical model proposed in this study establishes the relationship between water flow, particle migration, and scour development under the conditions of a three-dimensional flow field. It performs well under most scour conditions.
2. The load loading mode of the flow field assumed by the model is different from the practice. The initial erosion development simulation curve is different from the experimental one, but it does not affect the simulation accuracy of the final erosion depth and deposition depth.
3. The numerical simulation accuracy is related to the flow field conditions and particle sizes in the experiments. For the same flow field conditions, the smaller the particle sizes, the higher the model's accuracy. With the increase of the flow rate, the method performs well for the small particle size soil show strong stability. For large particles, the results of numerical calculations are generally reliable, except for those that will not initiate.
4. The model can study the accumulation process of particles, which can simulate the process with evident particle accumulation, e.g., what happens in the SSRT test.

ACKNOWLEDGEMENTS

This work was supported by the National Natural Science Foundation of China (Grant No. 51908421) and "Chen Guang" project supported by Shanghai Education Development Foundation and Shanghai Municipal Education Commission (Grant No. 19CG21). Financial support from these organizations is gratefully acknowledged.

REFERENCES

Albertson, M.L., Simons, D.B., Richardson, E.V. 1958. Discussion of "mechanics of sediment-ripple formation" by H. L. Liu. *Journal of Inorganic & Nuclear Chemistry*, 39(8), 1437–1442.

Bagnold, R.A. 1966. An Approach to the Sediment *Transport Problem From General Physics*. U.S. Geol. Survey, Prof. Paper No.422, 1996, p.37.

Baker, R.E. 1986. Local scour at bridge piers in non-uniform sediment. Rep.No402, School of Eng., University of Auckland, Auckland, New Zealand, 1986.

Chiew, Y.M. 1984. Local scour at bridge piers. Rep.No.355, University of Auckland, Auckland, New Zealand, 1984.

Dongol, D.M.S. 1994. Local scour at bridge abutments. Rep. No.544, School of Eng., University of Auckland, Auckland, New Zealand.

Einstein, A. 1926. Die ursache der manderbildung der flulufe und des sogenannten baerschen gesetzes. *Naturwissenschaften*, 14(11), 223–224.

Einstein, H.A. 1942. Formulas for the transportation of bed load. *Transactions of the American Society of Civil Engineers*, 107, 561–597.

Engelund, F., Fredsee, J. 1976a. A sediment transport model for straight alluvial channels. Iowa Publishing. Sediments. IFCEE 2018.

Engelund, F., Jørgen, F. 1976b. A sediment transport model for straight alluvial channels. *Hydrology Research*. 7(5), 293–306.

Escobar, A., Negro, V., López-Gutiérrez, J.S. 2018, et al. Assessment of the influence of the acceleration field on scour phenomenon in offshore wind farms. *Renewable Energy*. 136, 1036–1043.

Graf, W.H., Istiarto, I. 2002. Flow pattern in the scour hole around a cylinder. *Journal of Hydraulic Research*, 40 (1):13–20.

Karim, O.A., Ali, K.H.M. 2000. Prediction of flow patterns in local scour holes caused by turbulent water jets. *Journal of Hydraulic Research*, 38(4):279–287.

Lagasse, P.F., Clopper, P.E., Zevenbergen, L.W., Girard, L.W. 2007. National cooperative highway research program (NCHRP REPORT 593): Countermeasures to protect bridge piers from scour. Washington D.C.: Transportation Research Board.

Laursen, E.M., Papanicolaou, A.N., Cheng, N.S., et al. 1998. Pickup Probability for sediment entrainment. *Journal of Hydraulic Engineering*, 125(7):789–789.

Liang, F.Y., Zheng, H.B., Zhang, H. 2020. On the pile tension capacity of scoured tripod foundation supporting offshore wind turbines. Applied Ocean Research, 35:295: 306.

Loópez, R., Barragaón. 2008. Equivalent roughness of gravel-bed rivers. *Journal of Hydraulic Engineering*, 134(6):847–851.

Ma, H.W., Yang, J., Chen, L.Z. 2018. Effect of scour on the structural response of an offshore wind turbine supported on tripod foundation. *Applied Ocean Research*, 73: 179–189.

Melville, B.W., Raudkivi, A.J. 1977. Flow characteristics in local scour at bridge piers. *Journal of Hydraulic Research*, 15(4):373–380.

Melville, B.W., Coleman, S.E. 2000. Bridge scour. *Water Resources Publications*, Colorado, USA.

Rajaratnam, N., Berry, B. 1977. Erosion by circular turbulent wall jets. *Journal of Hydraulic Research*, 15(3):277–289.

Søren, P.H.S., Ibsen, L.B. 2013. Assessment of foundation design for offshore monopiles unprotected against scour. *Ocean Engineering*, 63:17–25.

Wang, C., Yu, X., Liang, F.Y. 2017. A review of bridge scour: mechanism, estimation, monitoring and countermeasures. *Natural Hazards*, 87(3):1881–1906.

Wang, C., Yu, X., Liang, F.Y. 2018. A preliminary design of apparatus for scour resistance test in riverbed. International Foundations Congress and Equipment Expo (IFCEE), ASCE, Orlando, U.S.A., 746–757.

Wang, J.J., Ni, F.S. 2014. Numerical simulation of coarse-sandy bed scour by 2D vertical submerged jet. *Science Technology and Engineering*, 14 (3): 108–111. (In Chinese)

Wu, F.C., Lin, Y.C. 2002. Pickup probability of sediment under log-normal velocity Distribution. *Journal of Hydraulic Engineering*, 128(4):438–442.

Yang, F., Liu, X., Cao, S., et al. 2010. Bed load transport rates during scouring and armoring processes. *Journal of Mountain Science*, 7(3):215–225.

Zhang, W., Zapata, M.U., Bai, X., et al. 2020. Three-dimensional simulation of horseshoe vortex and local scour around a vertical cylinder using an unstructured finite-volume technique. *International Journal of Sediment Research*, 35 (2020):295–306.

Zhou, L.D. 2016. Scour mechanism and characteristics of vortex -induced vibrations of a submarine free spanning pipelines. Tianjin: Tianjin University. (In Chinese)

Zhu, Z.W., Liu, Z.Q. 2011. Three-dimensional numerical simulation of local scour around cylindrical bridge piers. *China Journal of Highway and Transport*, 24 (02):42–48. (In Chinese)

Proceedings of the Second International Conference on
Press-in Engineering 2021, Kochi, Japan – Matsumoto et al (eds)
© 2021 Taylor & Francis Group, London, ISBN 978-1-032-10414-0

Design calculation method for sheet pile reinforcement method in liquefiable ground

K. Kasahara, T. Sanagawa & M. Koda
Railway Technical Research Institute, Kokubunji, Tokyo, Japan

ABSTRACT: A reinforcement method utilizing steel sheet piles (sheet pile reinforcement method) is sometimes adopted in Japan for reinforcing existing foundation structures. This is a reinforcement method where steel sheet piles are installed into the ground so as to surround the existing footing and are integrated with the footing. Since the widening width of the footing can be minimized, this reinforcement method is often adopted at locations where land usage is severely restricted. However, it hasn't been clarified whether this reinforcement method can be applied to structures in the liquefiable ground. Therefore, in this study, model vibration experiments are conducted to clarify the effect and the mechanism of this reinforcement method for pile foundation structures in the liquefiable ground. Furthermore, considering of the experimental results, a structural analysis method is proposed to design this reinforcement method, which can correspond to changes in strength and stiffness of the ground due to liquefaction.

1 INTRODUCTION

When the liquefaction occurs during an earthquake, the ground suddenly loses strength and rigidity, causing great damage to the foundation structures of bridge piers and viaducts. As a result, the structural members of the foundation might be damaged or the bridge might collapse. For example, in the 1995 Hyogo-ken Nanbu Earthquake, many pile foundation structures were damaged due to liquefaction. Subsequent investigations and researches clarified that the damage of pile foundation structures was not only caused by the inertial force acting on the superstructure, but also by the increase in the ground displacement due to liquefaction. With the background of such disaster cases, the foundation structures are designed considering the effect of the liquefaction in the current design standards. However, not a few older structures have to implement liquefaction countermeasures because in these structures the effect of the liquefaction was not considered in the design stage.

As conventional liquefaction countermeasures for existing pile foundation structures, there are enumerated ones by additional piles (Kishimoto et al. 1998) and by soil improvement (Kiryu and Sawada. 2005). However, as the additional pile method requires a large expansion of the land usage and the reinforcement work becomes large-scale, resulting in high cost. Furthermore, the construction is often difficult in urban areas where there are many adjacent structures under bridge girders. On the other hand, the soil improvement

may not be applicable to structures in rivers due to environmental considerations. For this reason, it is expected to develop a liquefaction countermeasure method that is excellent in economic efficiency as well as workability for foundation structures in narrow areas and under harsh overhead clearance restrictions.

Incidentally, a reinforcement method is proposed, which utilizes steel sheet piles (hereinafter called "sheet pile reinforcement method") (Nishioka et al. 2010) (Figure 1). This is a reinforcement method where steel sheet piles are installed into the ground so as to surround the existing footing and are integrated with the footing. This reinforcement method mainly targets existing pile foundations of relatively small and medium-sized, where footing width is about 5 to 10 m. Although the specifications such as the steel sheet piles length are decided by design calculations, the embedded depths are relatively short (about the same length as the footing width) under general conditions. Therefore, this reinforcement method is superior in that it doesn't require a large pile driving machine. Since the widening width of the footing can be minimized, this reinforcement method with excellent economic efficiency and workability has been often adopted. However, it hasn't been clarified whether this reinforcement method can be applied to structures in the liquefiable ground. If the effect of this reinforcement method on the liquefiable ground can be confirmed, it can be a reinforcement method with excellent economic efficiency and workability on the liquefiable ground.

DOI: 10.1201/9781003215226-40

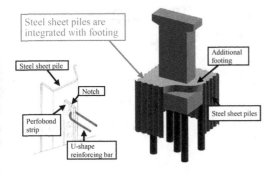

Figure 1. Sheet pile reinforcement method (Sanagawa et al. 2015).

Therefore, in this study, model vibration experiments are conducted to clarify the effect and the mechanism of the sheet pile reinforcement method for pile foundation structures in the liquefiable ground. Furthermore, considering the experimental results, a structural analysis method is proposed to design this reinforcement method, which can correspond to changes in strength and rigidity of the ground due to liquefaction.

2 CONFIRMATION OF REINFORCEMENT EFFECT BY MODEL EXPERIMENT (SANAGAWA ET AL. 2015)

At first, we conduct a model vibration experiment to clarify the reinforcement effect of the sheet pile reinforcement method on the liquefiable ground. Figure 2 shows an outline of the model. The prototype of the target structure is a pile foundation pier, where body height: 6.0 m, yield seismic intensity: 0.586, equivalent natural period: $T_{eq} = 1.34$ sec, ground natural period: $T_g = 1.95$ sec. The foundation

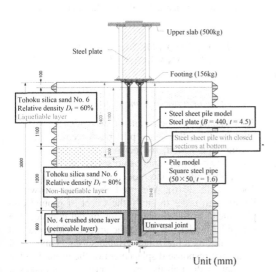

Unit (mm)

Figure 2. Overview of model experiment.

type is a driving pile with a pile length of 15.0 m using PHC piles (the diameter is 500 mm, and the number of piles is 2 × 2). We assume 25H type hat-shaped steel sheet piles for reinforcement, and the pile base is driven into the non-liquefiable layer. We perform the vibration experiments with a 1/6 scale model of the prototype, setting the specifications according to the similarity laws (Kagawa. 1988) (Table 1). The pile model is a square steel pipe with dimensions of 50 mm × 50 mm and a thickness of 1.6 mm, and the sheet piles model is a steel plate with a thickness of 4.5 mm. Regarding the sheet piles model, we confirm by pre-analysis that the axial force of the steel sheet pile is dominant as the mechanism for the sheet pile reinforcement method in the liquefiable ground. Therefore, the sheet pile model simulates the base machined steel sheet pile (Nakayama et al. 2007) with excellent vertical resistance. Specifically, we weld U-shaped steel plates (height: 33 mm, width: 88 mm) to the section with a base of 300 mm (Figure 3). Here, the base machined steel sheet pile is a steel sheet pile in which the closed cross-section is provided by

Table 1. List of similarity laws.

Items	Model (M)	Actual thing (A)	M/A	Target value
Height of column (mm)	1000	6000	0.167	$1/\lambda = 0.167$
Pile length (mm)	2849	15000	0.190	
βL (pile)	5.18	4.86	1.07	1.0
βL (steel sheet pile)	5.49	4.88	1.13	1.0
Natural frequency of the ground f_g (Hz)	7.35	1.95	3.77	$\lambda^{3/4} = 3.83$
Natural frequency of the structure f_s (Hz)	4.50※	1.34	3.35	

※ A case when inputting

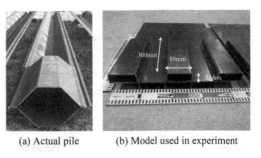

(a) Actual pile   (b) Model used in experiment

Figure 3. Steel sheet piles with closed sections at the bottom (Website of NIPPON STEEL CORPORATION. (date of last access is November 10, 2020).

combination processing to improve the vertical bearing capacity.

The model vibration experiment is carried out using the large vibration test device and the laminated shear soil tank owned by the Railway Technical Research Institute. The internal dimensions of the laminated shear soil tank are 3000 mm in width, 1100 mm in depth, and 3000 mm in height. The part which is 2400 mm deep from the top of the soil tank is composed of 12 stages of shear frames. Each shear frame is supported by a linear guide and can be deformed depending on the movement of the ground. A rubber membrane is installed inside the soil tank, and after installing the model inside of this, the model ground is constructed. The geo material is Tohoku silica sand No. 6, and it is saturated by injecting water from the bottom of the soil tank after construction. The vibration waveform is Level 2 seismic motion (spectrum I that models an earthquake of plate boundary type) at bedrock used in the design of railway structures (Ministry of Land, Infrastructure, Transport and Tourism. 2012. Seismic Design). We use this waveform with the time axis compressed according to the similarity laws (Figure=4). Table 2 shows a list of measurement items. We confirm the effect of the sheet pile reinforcement method for the ground conditions that have become completely liquefiable due to vibration (Figure 5).

Figure 6 shows the outline of the model vibration experiment results. We confirm that the maximum response rotation angle is reduced by about 30% and the maximum horizontal response displacement is reduced by about 5% at the upper slab by the sheet pile reinforcement method. Furthermore, the maximum shear force is suppressed by about 30% and the

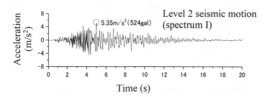

Figure 4. Time history data of input waves (in model scale).

Table 2. List of measurement items.

Items	Positions	Methods
Displacement	Upper slab, Footing, Ground surface, Soil (soil tank)	Displacement sensor
Acceleration	Upper slab, Footing, Ground	Accelerometer
Excess pore water pressure	Ground	Piezometer
Strain	Pile body, sheet pile	Strain gauge

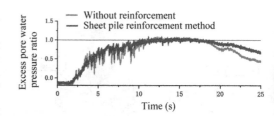

Figure 5. Time history data of excess pore water pressure ratio.

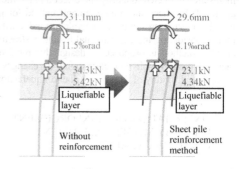

Figure 6. Effect of sheet pile reinforcement method.

maximum axial force is reduced by about 20% at the pile head by the sheet pile reinforcement method. This is because the vertical resistance of the steel sheet pile suppresses the rotational behavior of the superstructure, and the stress generated at the pile head is reduced.

3 VERIFICATION OF NUMERICAL ANALYSIS MODEL (SANAGAWA ET AL., 2015; TODA ET AL., 2016)

3.1 Structural analysis model

To evaluate the behavior of the pile foundations reinforced by the sheet pile reinforcement method during liquefaction, we perform numerical analysis using a two-dimensional beam spring model to verify the applicability of the model (Sanagawa et al. 2015, Toda et al. 2016). We used the model shown in Figure 7 in order to study the effect of ground displacement on the structure. This model connects the soil pillar model simulating the ground with piles and sheet piles by a horizontal ground spring. This model inputs the response displacement of the ground to the foundation through the ground spring. This makes it possible to directly consider the effects of interactions

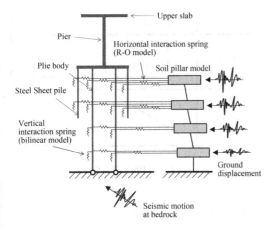

Figure 7. Overview of analytical model.

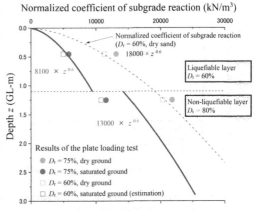

Figure 8. Modeling of normalized coefficient of subgrade reaction.

with structures in a dynamic analysis. In this analysis, we input the time history waveform of the ground displacement measured at each depth in the model vibration experiment into directly the soil pillar model.

3.2 Horizontal interaction between pile and soil

We set the horizontal interaction spring considering the displacement level dependency, the pile width dependency, and the decrease in rigidity due to the increase in excess pore water pressure.

(1) Normalized horizontal coefficient of subgrade reaction

Suzuki et al. (2009) conducted a horizontal flat plate loading experiment (φ= 300 mm) using a screw jack. They examined the subgrade reaction of the pile for dry ground (D_r = 60% & D_r = 75%) and for saturated ground (D_r = 75%). These experimental results show that the normalized horizontal coefficient of subgrade reaction is roughly proportional to the 0.6th power of depth. Therefore, we evaluate the coefficient of subgrade reaction by this power law (Figure 8). In addition, the initial shear stiffness G_0 at D_r = 80% obtained from the triaxial compression test was about 1.6 times the value of D_r = 60%. Thus, we also estimate the normalized coefficient of subgrade reaction by multiplying 0.6th power of the depth. The formula for calculating the normalized horizontal coefficient of subgrade reaction applying to the interaction spring model is shown below.

$$k_{hrB=300}(z) = 8100 \times z^{0.6} (0 \leq z \leq 1.1) \quad (1)$$

$$k_{hrB=300}(z) = 13000 \times z^{0.6} (1.1 \leq z \leq 2.3) \quad (2)$$

where z is the depth (m) from the ground surface, and $k_{hrB=300}$ is the normalized horizontal coefficient of subgrade reaction when the loading displacement is 1% of the plate width.

(2) Size effect on horizontal coefficient of subgrade reaction

The width of the loading plate used in the above study differs from the width of the pile and sheet pile model in this vibration experiment. For this reason, we have to consider the influence of the size effect. Referring to the specifications for highway bridges (Japan Road Association 2012) and the design standard for railway structures (Ministry of Land, Infrastructure, Transport and Tourism 2012. Foundation Structures), we calculate the value of the normalized horizontal coefficient of subgrade reaction of the model pile and model sheet piles. Specifically, we used the relational expression that the normalized horizontal coefficient of subgrade reaction is -3/4 power of the pile width as shown in equation (3).

$$k_{hr}/k_{hrB=300} = (B/0.3)^{-3/4} \quad (3)$$

where B is the loading width (m). In this vibration experiment, the pile width is 0.050 m and the sheet pile width is 0.450 m.

(3) Displacement level dependency

Regarding the displacement level dependency, the R-O model (Jennings 1964) is applied. We calibrate the parameters of the R-O model for the displacement level dependency of the horizontal coefficient of subgrade reaction by using the result of the vibration generator and static horizontal loading test conducted on the dry ground of the same ground material (Figure 9). At this time, the normalized displacement is 1% of the pile width (0.050 m) and steel sheet piles width (0.450 m). In this analysis, the displacement level dependency of the horizontal coefficient of subgrade reaction changes from moment to moment by using the time history of the pile displacement.

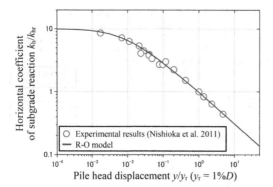

Figure 9. Modeling of displacement level dependency of horizontal coefficient of subgrade reaction.

(4) Reduction of coefficient of subgrade reaction due to liquefaction

During liquefaction, the coefficient of subgrade reaction reduces because the effective confining pressure decreases as the excess pore water pressure rises. Previous studies obtained the effect of the excess pore water pressure ratio on the coefficient of subgrade reaction from experiments (Sawada et al. 1998, Matsumoto et al. 1987, Yoshizawa et al. 2000, Kawai et al. 2001, Igarashi et al. 2003). Many of these studies reported that the coefficient of subgrade reaction decreases in proportion to the power of the water pressure ratio as shown in equation (4).

$$k_h \propto (1 - u)^\alpha \qquad (4)$$

where u is the excess pore water pressure ratio. In this analysis, a low-pass filter ($f_c = 1.0 H_z$) is applied to the time history waveform (Figure 5) of the excess pore water pressure ratio u obtained in the vibration experiment to remove the short-period component. In addition, the reduction rate of the coefficient of subgrade reaction due to the liquefaction is changed with time history by setting $\alpha = 0.5$.

3.3 Vertical interaction between pile and soil

For the vertical interaction spring of the piles, we apply a bilinear model based on the results of the separately conducted steel pipe pile ($\varphi50$ mm) push-in/pull-out test. Regarding the vertical soil spring of the steel sheet piles, we also apply a bilinear model based on the results of the steel sheet pile push-in test (Sanagawa et al. 2010) (Figure 10). The effect of liquefaction is assumed to be proportional to the power of the excess pore water pressure ratio ($\alpha = 0.5$), similar to the horizontal interaction spring.

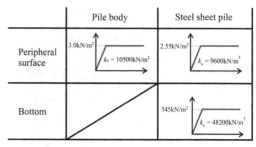

Figure 10. Modeling of vertical interaction springs.

3.4 Modeling of members

The structural members are modeled as a linear model because the stress of the members in the experiment did not exceed the yield point. The results of the bending test and axial compression test are applied to the pile body and the steel sheet pile. For the slab, column, and footing, the cross-sectional rigidity is calculated based on the dimensional specifications as Young's modulus is set to 2.05×10^8 (kN/m²). Table 3 shows a list of cross-section specifications.

3.5 Modeling of damping effect

Typical examples of structural attenuation include structural damping, historical damping, and radiation damping. Since historical damping is considered in the non-linear model of the soil spring set in the previous section, Rayleigh damping is applied as another damping. From the results of a parametric study, we set $\alpha = 0.3$ and $\beta = 0.003$ so that the damping constants don't vary around the natural frequency of the structure to the natural frequency of the ground.

3.6 Reproduction analysis of model experiment

Using the analysis model constructed so far, we performed a reproduction analysis of the model experiment. As a result of the reproduction analysis, Figure 11 shows the time history waveform of the

Table 3. List of member specifications used in the analysis.

Structural member	Cross-sectional area (m²)	Moment of inertia of area (m⁴)
Upper slab	Rigid body	Rigid body
Pier body	Rigid body	Rigid body
Footing	Rigid body	Rigid body
Pile body	3.03×10^{-4}	1.05×10^{-7}
Steel sheet pile	6.55×10^{-4}	3.42×10^{-9}

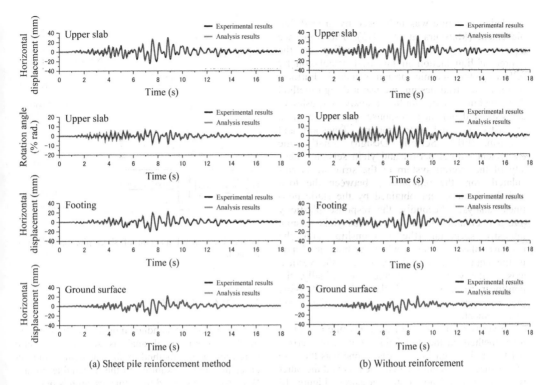

(a) Sheet pile reinforcement method   (b) Without reinforcement

Figure 11. Comparison between experiment and analysis (time history response data).

response displacement, and Figure 12 shows the maximum bending moment distribution diagram. These results show that this analysis model can accurately reproduce the experimental results.

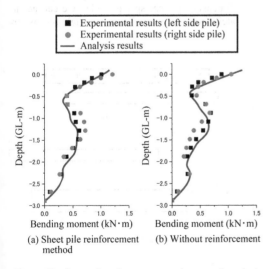

(a) Sheet pile reinforcement method

(b) Without reinforcement

Figure 12. Comparison between experiment and analysis (Maximum bending moment distribution).

Consequently, we confirm that even a two-dimensional beam spring model can evaluate the behavior of the sheet pile reinforcement method for pile foundation structures in the liquefiable ground during an earthquake if several conditions can be evaluated accurately (the dynamic behavior of the liquefiable ground, the non-linear characteristics of the interacting spring, and the reduction in rigidity and strength of the soil spring due to an increase in excess pore water pressure).

4 TRIAL CALCULATION OF THE ACTUAL STRUCTURE (TODA ET AL. 2016)

We confirm the applicability of the two-dimensional beam spring model for the structural analysis of the sheet pile reinforcement method. In this chapter, we show how to consider the effect of liquefaction in the design of this reinforcement method. Additionally, we carried out the trial calculation.

Chapter 2, and 3 show an experimental and numerical analysis of the case of general structures. On the other hand, the previous study confirmed that when there was no superstructure, the response of the structure with the sheet pile reinforcement method may increase because the foundation structure was influenced by ground displacement (Matsuura at el. 2015). Therefore, we consider the ground model depending on the degree of liquefaction in this trial calculation by following the seismic design standard. Furthermore, in this trial design, we use a design method that combines static nonlinear analysis (pushover analysis) and nonlinear response spectrum method (Ministry of Land, Infrastructure, Transport and Tourism. 2012. Seismic Design). Firstly, the equivalent natural period and yield seismic intensity of the overall system of the structure is calculated from the relationship between the load and the displacement obtained by the static nonlinear analysis. Secondly, the response plasticity rate μ is obtained from the required yield seismic intensity spectrum (the relationship with the response plasticity rate μ when the horizontal axis is the equivalent natural period and the vertical axis is the yield seismic intensity). Finally, the nonlinear response is calculated by multiplying the response plasticity rate μ by the yield displacement.

To verifying the effect of the sheet pile reinforcement method on the liquefiable ground, we carried out the trial design for two cases: one was the case of the sheet pile reinforcement method and the other was the case of no countermeasures. Figure 13 shows a general view of the sheet pile reinforcement method and ground conditions. The liquefiable layer is the section of (2) sandy soil (Figure 13).

Figure 14 shows the load-displacement relationship obtained from the trial design results. We confirm that although the maximum response seismic intensity during liquefaction increases, the support yield of the existing pile and the shear failure of the pile head are suppressed because the deformation is suppressed by the sheet pile reinforcement method.

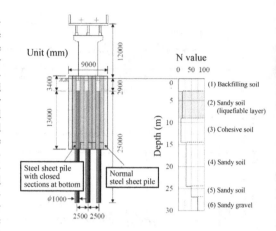

Figure 13. General diagram of sheet pile reinforcement method.

5 CONCLUSIONS

In this paper, we conduct model vibration experiments and numerical analysis so as to develop a reinforcement method with excellent economic efficiency and workability in the liquefiable ground. What we have learned from this research is below.

(1) In the case of structural conditions mainly based on inertial force, by carrying out the sheet pile reinforcement method, the displacement is suppressed and the cross-sectional force of the pile head is also reduced by the vertical resistance of the sheet piles.

(2) As a result of the reproduction analysis of the model vibration experiment by using a two-dimensional beam spring model, we can accurately reproduce the experimental results by

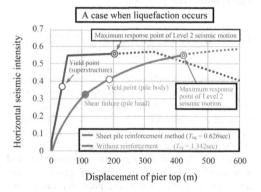

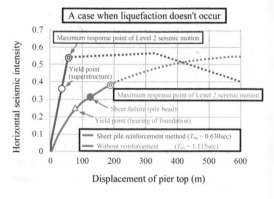

Figure 14. Relationship between load and displacement.

considering the dynamic behavior of the ground and the ground reaction force coefficient during liquefaction. Therefore, we consider that the design calculation can be performed using a general structural analysis model.

REFERENCES

Igarashi, H., Yasui, M., Mori, T., Hirade, T. & Mizuno, N. 2003. Horizontal ground reaction force coefficient during liquefaction of pile foundation -Based on large laminated shear soil tank experiment-, *The 38th Japan National Conference on Geotechnical Engineering* (in Japanese).

Japan Road Association. 2012. *Specifications for highway bridges*, IV Substructure, p.285.

Jennings, P.C. 1964, Periodic Response of a General Yielding Structure, *Proc. ASCE*, EM2, pp.131–163.

Kagawa, T. 1988. On the similitude in model vibration tests of earth-structures, *Proceedings of the Japan Society of Civil Engineers*, Vol. 275, pp.69–77 (in Japanese).

Kawai, E., Tsuchiya, T., Uchida, A., Hamada, J., Takahashi, K., Yamashita, K. & Kakurai, M. 2001. Horizontal loading experiment of model piles on liquefiable ground (Part 2), *The 36th Japan National Conference on Geotechnical Engineering* (in Japanese).

Kishimoto, T., Saito, E., Yamane, T., Ohtani, Y., Miura, F. & Tanifuji, M. 1998. High-strength micropile seismic reinforcement of existing structural foundations, *The Japan Earthquake Engineering Symposium, proceedings*, Vol. 10-1, pp.161–166 (in Japanese).

Kiryu, S. & Sawada, R. 2005. Experimental study on liquefaction countermeasure work by improving only the peripheral surface of pile foundation, *The 28th JSCE Earthquake Engineering Symposium*, Vol. 28 (in Japanese).

Ministry of Land, Infrastructure, Transport and Tourism, Edited by Railway Technical Research Institute, 2012. *The Design Standards for Railway Structures and Commentary (Seismic Design)*, Maruzen Publishing Co. Ltd (in Japanese).

Ministry of Land, Infrastructure, Transport and Tourism, Edited by Railway Technical Research Institute, 2012. *The Design Standards for Railway Structures and Commentary (Foundation Structures)*, Maruzen Publishing Co. Ltd (in Japanese).

Matsuura, K., Nishioka, H., Sanagawa, T., Kita, N., Higuchi, S., Tanaka, R. & Toda, K. 2015. Model experiment on existing pile foundation sheet pile reinforcement method in liquefiable ground, *The 50th Japan National Conference on Geotechnical Engineering* (in Japanese).

Matsumoto, H., Sasaki, Y. & Kondo M. 1987. Ground reaction force coefficient in liquefiable ground, *The 22nd Soil Engineering Research Presentation*, pp. 827–828 (in Japanese).

Nakayama, H., Taenaka, S., Nagatsu, R., Harada, N. & Kato, A. 2007. Experimental study on evaluation of bearing capacity characteristics of steel sheet piles with a closed cross section at the base, *The 10th Symposium on Ductility Design Method for Bridges*. pp.345–350 (in Japanese).

Nishioka, H., Higuchi, T., Nishimura, M., Koda, M., Yamamoto, T. & Hirao, J. 2010. Seismic proving tests on aseismic reinforcement method for existing pile foundations utilizing sheet piles, *Japanese Geotechnical Journal*, Vol. 5, No.2, pp.251–262 (in Japanese).

Nishioka, H., Shinoda, M., Sanagawa, T. & Koda M. 2011. Resonance experiment and static loading experiment on displacement level dependency of horizontal ground reaction force coefficient of pile, *The 46th Japan National Conference on Geotechnical Engineering*. pp.213–218 (in Japanese).

Sanagawa, T., Nishioka, H., Koda, M., Nakayama, H., Harada, N., Kato, A. & Toda, K. 2010. Model loading experiment on horizontal shear resistance characteristics of steel sheet piles, *10th Geotechnical Symposium*.

Sanagawa, T., Nishioka, H., Matsuura, K., Higuchi, T., Toda K. & Taenaka, S., 2015. Experimental and numerical study on seismic behavior of pile foundations utilizing sheet pile reinforcement in liquefiable ground, *Japanese Geotechnical Journal*, Vol. 12, No.2, 197–210 (in Japanese).

Sawada, R. & Nishimura, A. 1998. Study on dynamic behavior of foundation structures in liquefiable ground, *The Japan Earthquake Engineering Symposium* (in Japanese).

Suzuki, S., Ota, T., Koda, M., Nishioka, H. & Kondo, M. 2009. Horizontal flat plate loading experiment in model soil layer focusing on horizontal ground reaction force coefficients with different depths, *the 64th Annual Conference of the Japan Society of Civil Engineers*, pp.III-133-134 (in Japanese).

Toda, K., Sanagawa, T., Nishioka, H., Higuchi, S., Matsuura, K., Taenaka, S. & Otsushi, K. 2016. *RTRI REPORT*, Vol. 30, No.5, May (in Japanese).

Website of NIPPON STEEL CORPORATION. Steel sheet piles with closed sections at the bottom, https://www.nipponsteel.com/tech/nssmc_tech/construction/02.html. Date of last access is November 10, 2020.

Yoshizawa, M., et al. 2000, Ground reaction force coefficient of model steel pipe pile in liquefiable ground in large laminated shear soil tank, *Architectural Institute of Japan* (in Japanese).

Proceedings of the Second International Conference on
Press-in Engineering 2021, Kochi, Japan – Matsumoto et al (eds)
© 2021 Taylor & Francis Group, London, ISBN 978-1-032-10414-0

Anticorrosive effect by inserting sheet piles on the sides of underground tunnel at shallow depth

M. Oka, H. Takeda, Z. Wang & K. Maekawa
Yokohama National University, Yokohama, Kanagawa, Japan

ABSTRACT: In recent years, deterioration due to salt damage in underground transmission tunnels which locate near coastal areas has been reported. In the case of the open-cut construction, earth-retaining walls are adopted at the time of excavating workspace. It is expected that the earth-retaining wall may protect underground transmission structures from the stray current. In this study, the macro-cell corrosion of the underground structures is investigated and the possibility of anti-corrosive effect by inserting sheet piles or utilizing existing steel plates is experimentally evaluated. Specimens were prepared by using the mixture made from starch and water to represent soil and steel plates to stand for structures with sheet piles. A simple systematic corrosion experiment was performed. The external charge was applied and ion profiles are measured, and corrosive states were observed. By electrically jacking sheet piles, the suppressing polarization effect with sheet piles against corrosion was found from the experiment.

1 INTRODUCTION

1.1 Background

The urban tunnels for the underground transmission system to contain cables usually have two types: excavation tunnels and shield ones. The former ones are constructed by producing open spaces with steel sheet piling and constructing the RC ducts thereafter in the workspace followed by backfill. They are electricity facilities of importance to ensure the stable energy supply in urban areas (see **Figure 1**). They were mostly developed for stable power supply during the period of high economic growth in 1960s. Some of the underground tunnels which have passed over 40 years have corrosive deterioration (Enya *et al.* 2011a, b, Liu *et al.* 2018). The deterioration caused by seawater intrusion has been reported for the underground shield tunnels near seashores (see **Figure 2**).

The groundwater including the detrimental ions flows in the tunnels through segment joints, and it remains at the lower part of the underground tunnel, and steel devices and reinforcement corrode accordingly (Aoki *et al.* 2019). For the open-cut construction, sheet piles are driven in the foundation and earth-retaining walls are established during excavation of the workspace (see **Figure 3**). It is expected that these sheet piles may protect the tunnel structures suffering from the toxic substance and corrosion due to the stray current or large-scale electric circuit. Speaking of anti-corrosion methods of reinforced concrete, as commonly used methods, the cathodic protection in use of sacrificial anodic materials and desalination are being applied (Allen & Larry 2001, Cao 2008).

They are somehow local scale means focusing on the anodic and cathodic polarization whose distance is not large (Hsu *et al.* 2000). In reality, considering the gaps of natural potentials (different metals, ion concentrations, etc.) or external electric charge such as the stray current, the macro-cell corrosion of larger sizes took place (Otsuki *et al.* 2007, Chen *et al.* 2017). For underground reinforced concrete, multi-ions' interaction may exist in the pore solution such as Na^+, K^+, Mg^{2+}, SO_4^{2-}, seawater (Cl^-) and cement hydrates of Ca^{2+}, Al^{3+}, OH^-, Si^{2+}. As these ions have much to do with concrete solid, it is required to consider the multi-ion kinetics when we intend to actively control the electric potential which affects structural reinforced concrete (Bazant 1979a, b). This issue has been the engineering problem when railroad tracks are operated with electrical energy (Chen *et al.* 2017, Wang *et al.* 2017).

The objective of this study is to verify the macro-cell corrosion of underground facilities and to discuss the feasibility of using steel sheet piles as an anti-corrosion method. Experiments with specimens made of pseudo-concrete materials were reported (Maekawa *et al.* 2019, 2020, Aoki *et al.* 2020). Here, we may see the locations of polarized anodic-

DOI: 10.1201/9781003215226-41

Figure 1. Interior of utility duct by shield tunneling.

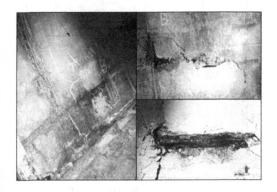

Figure 2. Example of corrosion by salt damage.

cathodic reactions which accompany corrosion products and hydrogen gas. The authors focus on the ionic concentration, since the rising ionic concentration caused by electric protection may deteriorate concrete solid such as ASR (Takahashi *et al.* 2016). Through different cases, we aim to propose an effective anti-corrosion method with steel sheet piles.

1.2 *Scheme of development*

A general flow of the research plan is summarized in **Figure 4** to reach a goal of assessing the macro-cell corrosion with integrating the electric field and ionic

Figure 3. Support system during excavation.

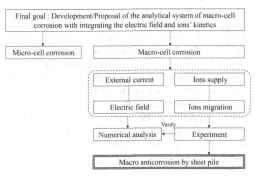

Figure 4. Examination flow.

kinetics. A macro-cell circuit generally brings about severe corrosion which may deteriorate serviceability and ultimate states of underground structural concrete. Electric potential gaps among soil, structure and adjacent facilities will be a source of a large-scale circuit which produces anodic and cathodic polarization. In addition, various ions exist in the soil. Seawater ingredients are included in groundwater near the shore (Otsuki *et al.* 2007). When plural ions exist, the ionic equilibrium may produce self-non-uniform potentials. As the multi-ion kinetics are affected by the electrical field as well and also have an impact on the electrical circuit, which is the coupled effect of electrical field and chemical ion field (Elakneswaran & Ishida 2014).

The electric field and the ion are closely connected to be described by the Nernst-Planck theorem (Na & Xi 2019). Then, the numerical scheme to consider the relation of the electric potential field and the ion concentration gradient is required. In this study, the experiment was carried out to verify the macro-cell corrosion system in association with ion profiles and the locations of anodic-cathodic polarization. The experimental facts serve to examine the qualitative understanding for sheet pile usages as a protection against corrosion of underground facilities and will take part in validation of the numerical analysis.

2 CORROSION EXPERIMENT

2.1 Specimens preparation

Pseudo-concrete materials were used instead of concrete and soil in order to measure the ion concentration easily. The pseudo-concrete materials in the current study were starch mixture, which was different from the past research where transparent polymer was adopted (Maekawa *et al.* 2019, 2020, Aoki 2020). The mix proportion of the specimen is listed in **Table 1**.

NaCl and Ca(OH)$_2$ were mixed with water and starch. The mixture was poured into a plastic container with dimension of 220x150x45 mm^3 and the height of specimen was 15 mm. Steel plates were utilized as electrodes which were inserted vertically between the starch and inner walls of the container on left and right sides. One iron plate was placed in the middle to represent the underground tunnel.

Five specimens were prepared in the current experiment, as shown in **Figure 6**: ① Case 1 was the referenced specimen where the middle steel plate was placed parallelly to the external charge direction; ② Case 2 had two vertical steel plates (to simulate the sheet piles) inserted perpendicular to the charge

Table 1. Mix proportion of the specimens.

starch (g)	H$_2$O (ml)	NaCl (g)	Ca(OH)$_2$ (g)
270	270	5	5

Figure 5. Schematic of non-countermeasure (Case 1).

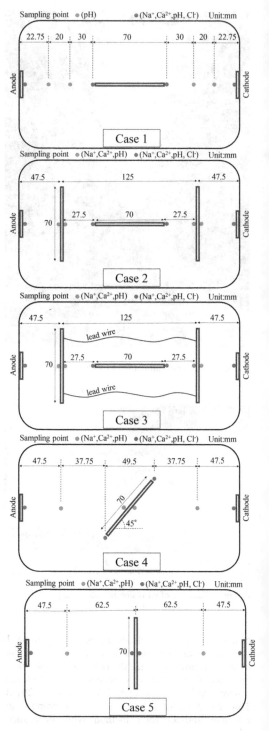

Figure 6. Sampling points.

direction on both right and left sides; ③ Case 3 had a similar arrangement as Case 2 but the difference was that the sheet piles were connected with metal wires;

422

④, ⑤: Case 4 and 5 had similar arrangements with Case 1 but the iron plates had different angles with the external charge direction (45° and 90°). It should be noted that Case 1, the referenced case was simulating the underground tunnels without any protection as shown in **Figure 6**. Case 1, 2 and 3 were performed to confirm the anti-corrosion effect when adopting sheet piles on the sides of the underground tunnel. Case 4 and 5 were adopted to investigate the macro-cell corrosion distribution with different directions of the stray current.

2.2 Electrical charge and ion measurement

The external charge of 6V was applied to the electrode with 4 hours for Case 1, 2 and 3 while 40 hours for Case 4 and 5. The pH and several ions' concentrations were measured at 0h, 2h and 4h after charging. The pH value was measured with a pH sensor. And the chloride testing tubes were used for measurement of chloride ions while the electronic sensors were utilized to measure the sodium and calcium concentrations. The measurement points are shown in **Figure 6** where the green points mean the measured positions of Na^+, Ca^{2+} and pH. The red points mean the measured positions of Na^+, Ca^{2+}, pH and Cl^-. The measurement points of Cl^- become less than other items because the number of Cl^- testing tubes was limited. Samples were taken out and diluted by 100 times to match the limit requirement of the measurement range of measuring sensors for sodium and calcium ions.

3 RESULTS AND DISCUSSIONS

3.1 Visual evaluation of corrosion

The results of macro-cell corrosion were given in **Figure 7**. An obvious polarization was captured where the anodic regions showed oxidation of iron and production of rust while air bubbles (hydrogen) were released at the cathodic regions. Compared with the corrosion products of Case 1, it turned out that the quantity of rust at anode was much less in Case 2 or 3 by naked-eye observation. As a result, it indicated that the underground tunnel (horizontal steel plate) was protected by the sheet piles (vertical steel plates). In this paper, quantitive observation was the focus thus the precise corrosion amount was not measured.

Furthermore, the anti-corrosion effect was more effective if connecting the sheet piles with metal wires (Case 3). Case 4 and 5 showed the symmetric properties of the anodic and cathodic polarization.

3.2 Ion concentration profiles

The description of concentration change with time are shown together with the corrosion conditions in

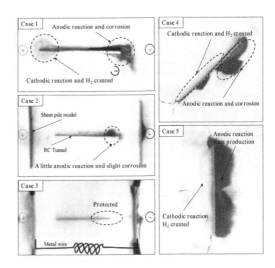

Figure 7. Summaries of the corrosion conditions.

Figure 8. The measured ion concentrations are shown in **Figure 9-12**. The samples were diluted 100 times with water.

In Case 1, the ion concentration changes were found in both anodic and cathodic regions, which could be explained by the polarization reactions:

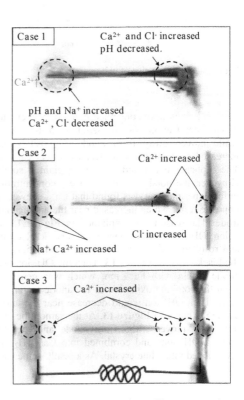

Figure 8. Description of the measured ion concentration.

423

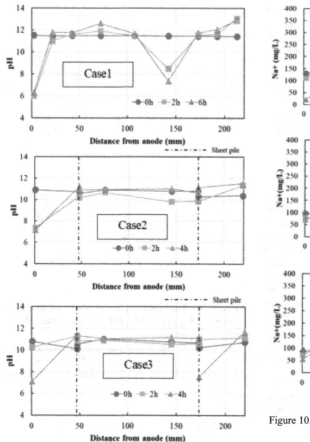

Figure 9. Measured value of pH.

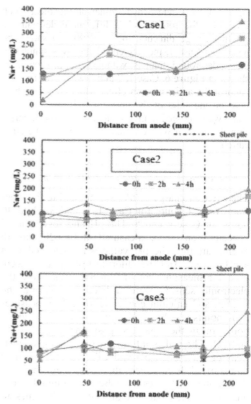

Figure 10. Measured value of sodium ion concentration.

hydroxide ions were consumed in anode and created in cathode; Cation Na^+ gathered towards cathode while moved away from anode; Oppositely, Cl^- increased near anode while decreased near cathode. These phenomena could be well explained so far. However, the notable point was Ca^{2+} concentration in the experiment. It was found that Ca^{2+} showed an opposite result which increased near the anodic electrode. The reason was explained in **Figure 13**: as anion, OH^- ion moved towards the anodic electrode and resulted in the increased ion concentration, see the black curve in **Figure 13**. Then the OH^- ion near anode would with Fe^{2+} ion, which was generated from the combine oxidation of the anode material – steel. Then OH^- ion would decrease near the anode as the red curve in **Figure 13**. At the same time, cations Ca^{2+} moved away from anode and met the coming OH^- ions and combined into $Ca(OH)_2$ and precipitated into white crystal. As a result, some Ca^{2+}

ions were trapped near the anode, which lead to an increasing amount of Ca^{2+} concentration. This phenomenon was also captured by the previous experiment by Maekawa *et al.* (2020), as **Figure 14**.

4 PROPOSAL OF ANTI-CORROSION SYSTEM

From the experiment, the proposal of the anti-corrosion system is the modification of Case 3. In a practical application, it is suggested to apply the electrical charge to the sheet piles as illustrated in **Figure 15**. For example, as corrosion takes place on the left-bottom of the underground facilities because of stagnant salty water, it is possible to apply anodic electrode on the corroded side (left), while the cathodic electrode on the opposite (right). From the knowledge obtained from the experiment, the corrosion part will be protected (cathodic reaction) while a new corrosion would happen on the top-right because of inevitable polarization.

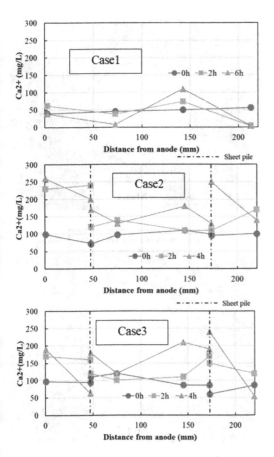

Figure 11. Measured value of calcium ion concentration.

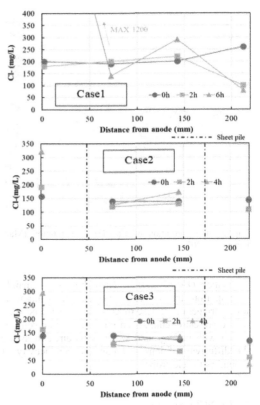

Figure 12. Measured value of chloride ion concentration.

However, it would be acceptable if the corrosion part would be transferred from the mechanically-risky place to non-problematic places (Fan *et al.* 2020). Thus, through this systematic macro-cell analytical method, the corrosion could be controlled not locally but with consideration of the wider domain (underground structures, sheet piles and soil foundation). Although voltage is not directly applied to the sheet pile in Case 3, but it indicates that the underground tunnel is protected. Meanwhile, the protected place would have increased concentration of cation such as K^+ or Na^+ as well as the alkalinity, which may lead to problems of alkali-silica reaction for the concrete. Thus, it is of great significance to have an overall understanding of the coupled chemical and electrical fields as well as the mechanical issue so as to propose the optimized anti-corrosion method.

In current study, the sheet piles are adopted as the anti-corrosion protector. It should be noted that the cost might be quite different in terms of whether the sheet piles are already existing in the ground, or they need additional execution of inserting. Especially, this paper would like to discuss the possibility of using existing sheet piles, which are left in soil after construction to prevent the consolidation of adjacent

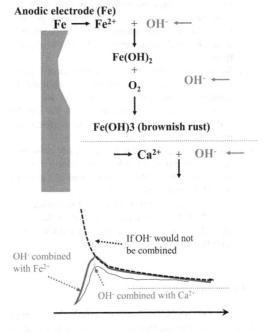

Figure 13. Mechanism explanation with $Ca(OH)_2$ precipitation.

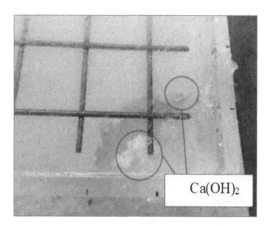

Figure 14. Observation of Ca(OH)₂ precipitation (Maekawa *et al.* 2020).

foundation (JASPP, 2017). Then the cost is mainly coming from the impressed charge, which might be low to trigger the reverse macro-cell current depending on the present circuit. As an important factor, soil saturated with underground water is required to assure the diffusivity of ions and thus the conductivity of ground as the electrolyte.

5 CONCLUSIONS

(1) The experiment was conducted with starch specimens and steel plates under external charge to simulate the underground systems including tunnels, sheet piles and soil foundation. The polarization locations and change of ion concentrations were observed, which obeyed the electrochemical theory.
(2) Besides the common sense where anions gather near anode and cations gather near cathode, the tendency of increasing Ca^{2+} near the anode was observed, which was explained with the ion flux and zero current theories under the coupled electrical and chemical fields.
(3) To apply external charge to the already existing or newly inserted sheet piles was proposed as the anti-corrosion method. But careful attention should be paid on the multi-ion profiles in the whole system with respect to the alkali-silica reaction. And the mechanical behavior of the underground structures should also be clarified along with the anti-corrosion application.

ACKNOWLEDGEMENT

The authors would like to express their sincere gratitude to the financial support by the Grant-in-Aid for JSPS Fellows (No. P20367).

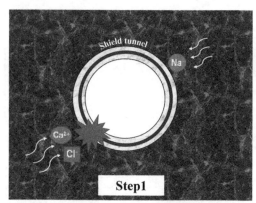

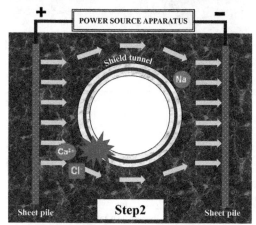

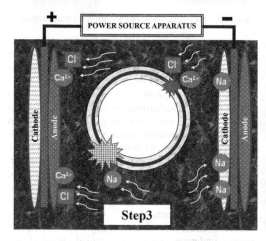

Figure 15. Illustration of the anti-corrosion system.

REFERENCES

Allen, J. B. & Larry, R. F. 2001. *Electrochemical methods fundamentals and applications.* New York: John Wiley & Sons.
Aoki, H., Takahashi, H. & Maekawa, K. 2019. Evaluation study of polarization reaction region by electric

corrosion experiment of reinforcing steel using visualization materials. *Proceedings of Japan Concrete Institute* 42 (1): 761–766.

Bazant, Z. P. 1979a. Physical model for steel corrosion in concrete sea structures – theory. *Journal of the Structural Division* 105 (6): 1137–1153.

Bazant, Z. P. 1979b. Physical model for steel corrosion in concrete sea structures – application. *Journal of the Structural Division* 105 (6): 1155–1166.

Cao, C. 2008. *Principles of electrochemistry of corrosion.* Beijing: Chemical Industry Press.

Chen, Z., Koleva, D. & van Breugel, K. 2017. A review on stray current-induced steel corrosion in infrastructure. *Corrosion Reviews* 35(6): 397–423.

Elakneswaran, Y. & Ishida, T. 2014. Development and verification of an integrated physicochemical and geochemical modelling framework for performance assessment of cement-based materials. *Journal of Advanced Concrete Technology* 12 (4): 111–126.

Enya, Y., Anan, K., Otsuka, M. & Koizumi, A. 2011a. Study on maintenance of the tunnel for the underground transmission. *Proceedings of Japan Concrete Institute* 67(2): 108–125.

Enya, Y., Naito, Y., Anan, K., Otsuka, M. & Koizumi, A. 2011b. Study of reinforcement design for deteriorated shield tunnel. *Journal of Japan Society of Civil Engineers, Ser. F1 (Tunnel Engineering)* 67 (2): 62–78.

Fan, S., Cui, Y. & Maekawa, K. 2020. Failure mode of deteriorated underground RC structures with hollow circular cross-section. *Proceedings of 2020 International Conference on Civil, Architectural and Environmental Engineering, Christchurch, 23-25 November 2020.*

Hsu, K. L., Takeda, H. & Maruya, T. 2000. Numerical simulation on corrosion of steel in concrete structures under chloride attack. *Doboku Gakkai Ronbunshu* 2000 (655): 143–157.

Japan Association for Steel Pipe Piles. 2017. *Kouyaita Q&A (Steel Sheet Piles Q&A). Kawasaki: Hokushin. (in Japanese)*

Liu, S. He, C., Feng, K. & An. Z. 2018. Research on corrosion deterioration and failure process of shield tunnel segments under loads. *China Civil Engineering Journal* 51 (6): 120–128.

Maekawa, K., Okano, Y. & Gong, F. 2019. Space-averaged non-local analysis of electric potential for polarization reactions of reinforcing bars in electrolytes, *Journal of Advanced Concrete Technology* 17(11): 616–627.

Maekawa, K., Takeda, H. & Wang, Z. 2020. Multi-ion equilibrium and migration model in pore solution of sea water concrete. *Proceedings of the Third International Workshop on Seawater Sea-sand Concrete (SSC) Structures Reinforced with FRP Composites, Shenzhen, 11-12 January 2020.*

Na, O. & Xi, Y. 2019. Parallel finite element model for multispecies transport in nonsaturated concrete structures. *Materials* 12(17): 2764.

Otsuki, N., Min, A. K., Madlangbayan, M. & Nishida, T. 2007. A study on corrosion of paint-coated steel with defects in marine environment. *Doboku Gakkai Ronbunshuu* 63 (4): 667–676.

Takahashi, Y., Ogawa, S., Tanaka, Y. & Maekawa, K. 2016. Scale-dependent ASR expansion of concrete and its prediction coupled with silica gel generation and migration. *Journal of Advanced Concrete Technology* 14 (8): 444–463.

Wang, C., Li, W., Wang, Y., Xu, S. & Fan, M. 2018. Stray current distributing model in the subway system: A review and outlook. *International Journal of Electrochemical Science.* 13(2): 1700–1727.

Proceedings of the Second International Conference on
Press-in Engineering 2021, Kochi, Japan – Matsumoto et al (eds)
© 2021 Taylor & Francis Group, London, ISBN 978-1-032-10414-0

Study on countermeasures for liquefaction of individual houses and backfill of quay using SandwaveG

K. Okabayashi, F. Kawatake, Y. Tsuno & G. Hashimura
National Institute of Technology, Kochi College, Nankoku City, Kochi, Japan

ABSTRACT: In recent years, liquefaction damage has been widely reported in Japan due to occurrences of large-scale earthquakes. Even in Kochi Prefecture, it is an urgent task to implement liquefaction countermeasures in individual houses and coastal areas in preparation for the Nankai Trough Giant Earthquake. In this study, regarding the liquefaction countermeasures for individual houses, the effect of liquefaction countermeasures when the ground of the residential land was replaced with SandwaveG was examined. Furthermore, the effect of liquefaction countermeasures when the quay backfill soil was replaced with SandwaveG was also examined. This study was carried out using our university's dynamic centrifuge. As a result, it was confirmed that the countermeasure effect by using SandwaveG as a liquefaction countermeasure work for residential land and quay backfill soil was confirmed.

1 INTRODUCTION

In the 2011 Great East Japan Earthquake, liquefaction occurred in a wide range from the Tohoku region to the Kanto region, and concentrated in the reclaimed land in Tokyo and Chiba Prefecture and the Tone River basin. Most of the damage due to liquefaction were on the foundations of individual houses, and some reinforced concrete low-rise buildings were also damaged (Ministry of Land, Infrastructure, Transport and Tourism. (2011)).

In Japan, the Nankai Trough Giant Earthquake is expected to occur in the near future. Kochi Prefecture is surrounded by steep mountains, so it may not be possible to receive immediate rescue and assistance on land when the disaster happens. Therefore, in Kochi Prefecture, it is necessary to take measures against liquefaction in individual houses and coastal quays (Kochi Prefecture (2013)).

In this study, we will examine the effect of liquefaction countermeasures using SandwaveG. SandwaveG is recycled glass granulation sand that is made by dicing glass into a dice shape by a special crushing method, eliminating sharp edges and allowing safe use. Since it is a recycled material, it contributes to the effective use of resources and is an inexpensive and environmentally friendly material. The material characteristics of SandwaveG are that they have a lower density than sand, high water permeability, a wide range of optimum water content, and

high liquefaction strength. In addition, it has a small swellimg ratio, so if it is applied to the foundations of individual houses sites, it will also prevent reliquefaction.

In this study, we conducted a basic experiment on the liquefaction occurrence condition on horizontal ground using SandwaveG. After implementation, applicability as a liquefaction countermeasure material for individual houses sites and applicability as a drainage material in the backfill part of the sheet pile quay will be examined.

As the research method, the dynamic centrifuge are used. The effect of the proposed method against liquefaction will be examined.

2 FEATURES OF SANDWAVEG

Currently, about 1.24 million tons of glass bottles are supplied to the market as products every year in Japan. As a glass bottle recycling system, there are cases where a glass bottle is used repeatedly and cases where it is crushed and reused as a raw material for a new glass bottle. However, 420,000 tons of glass bottles are discarded annually in landfills. Glass Sourcing Co., Ltd. has established a recycled glass granulation system with a processing capacity of 50 tons per hour. "Sandwave G" is recycled glass granulated sand in which glass bottles are granulated into dice by a special method to eliminate sharp corners (Glass Resourcing Co., Ltd. (2019)).

DOI: 10.1201/9781003215226-42

The grain size distribution curve of SandwaveG is shown in Figure 1. SandwaveG has a particle size of 0.075 to 5 mm.

The compaction curve of SandwaveG is shown in Figure 2. As a compaction characteristic, no clear peak appears and the optimum water content ratio is wide, so construction management is easy and stable quality can be ensured.

Figure 3 shows the liquefaction strength test results at a void ratio of 0.671 (Dr = 35.8%).

Table 1 shows the physical properties of SandwaveG from the maximum density and minimum density test, the maximum density is $\rho_{t\ max}$ = 1.775 g/cm^3, and the minimum density is $\rho_{t\ min}$ = 1.376 g/cm^3.

The coefficient of permeability at a compaction degree of 90% is shown. In the compacted state, the coefficient of permeability is 1.3×10^{-4} m/s, so SandwaveG is as permeable as sand and gravel.

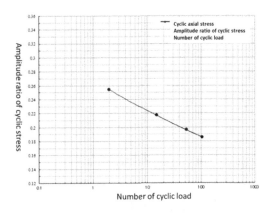

Figure 3. Liquefaction resistance curve of SandwaveG.

Table 1. Physical properties of SandwaveG.

Item	Symbol	Unit	Value
Density	ρs	g/cm^3	2.501
Maximum density	ρtmax	g/cm^3	1.775
Minimum density	ρtmin	g/cm^3	1.376
Maximum dry density		g/cm^3	1.660
Optimum water content	wopt	%	2.6
Coefficient of pemeability (Dr=50%)	k	m/s	2.27×10^{-4}
Swelling ratio		%	0.0013

3 EXPERIMENTAL METHOD

3.1 Experimental case

In this study, four models were used as study cases. Case1 is a horizontal ground model of Toyoura sand, Case2 is a horizontal ground model of SandwaveG, Case3 is a residential land model of Toyoura sand and Case4 is a residential land model of SandwaveG.

Figure 4 shows a horizontal ground full-scale model. The dimensions are full-scale. For the ground layer, Toyoura sand and SandwaveG with Dr = 50% were used. The Underwater drop method was used to prepare the ground.

Figure 5 shows a residential ground full-scale model. The structure was a two-story RC building, and the load on the residential land was 31.4 kN/m^2. The size of the residential land model was 1/40 of the actual size, and the bottom area was 13×12 cm^2. According to the law of similarity, Toyoura sand was placed in the stainless-steel case so that the total weight was 0.994 kg. The ground layer used was Toyoura sand and SandwaveG with Dr = 50%. The preparation used the Underwater drop method.

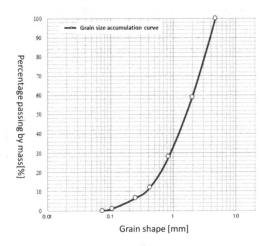

Figure 1. Grain size distribution of SandWaveG.

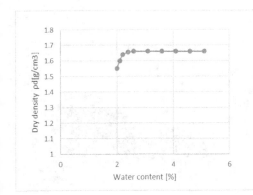

Figure 2. Compaction curve of SandWaveG.

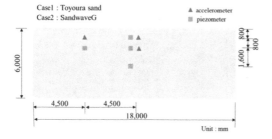

Figure 4. Full-scale model of horizontal ground.

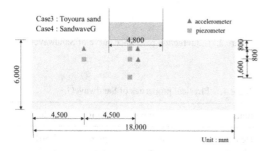

Figure 5. Full-scale model of residential land.

3.2 Seismic wave

The acceleration required to generate liquefaction in the experimental soil layer was calculated by the simple judgment method of the Road Bridge Specification (Japan Road Association (2012)). Figure 6 shows the input seismic wave.

In the centrifugal model test, the acceleration was set to 40 times and the period was set to 1/40 with respect to the actual seismic wave according to the similarity rule.

3.3 Apparatus and model

Liquefaction experiments were carried out using a centrifuge of Kochi National College of Technology. Figure 7 shows the centrifuge. Experiments were carried out in a centrifugal force field of 40 g.

Figure 8 shows an experimental model of the residential land model Case4. The model dimensions

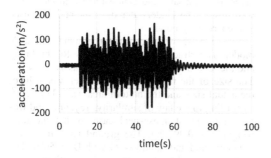

Figure 6. Input seismic wave.

Figure 7. Equipment for dynamic centrifuge.

were scaled to 1/40 of the actual dimensions. Seismic accelerometers and piezometers were installed at 20 mm, 40 mm and 80 mm depth from the ground surface. Targets were placed at 56.25 mm intervals to see the deformation. In the experiment of Case1, the top targets were placed at 20 mm depth.

Figure 9 shows a photograph of a horizontal ground model with SandwaveG before the experiment. The base layer and the liquefaction layer were

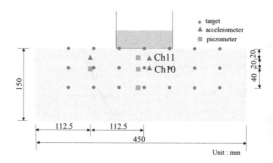

Figure 8. Experimental model.

Figure 9. Model of horizontal ground with SandwaveG (Before experiment).

430

Figure 10. Model of residential land with SandwaveG (Before experiment).

saturated with a methyl cellulose solution having a viscosity 40 times that of water.

Figure 10 shows a photograph of experimental model of the residential land model Case4.

4 EXPERIMENTAL RESULTS

4.1 Settlement of the ground

Figure 11 shows the subsidence diagram of Case1 (Toyoura sand of horizontal ground). The settlement is calculated from the amount of movement of the target before and after the experiment. The maximum amounts of subsidence are about 20 mm at the upper part (20 mm from the ground surface), about 15 mm at the middle part (40 mm from the ground surface), and about 5 mm at the lower part (80 mm from the ground surface). It can be seen that the deeper the depth, the smaller the amount of subsidence.

Figure 12 shows the subsidence diagram of Case2 (horizontal ground with SandwaveG).

It can be seen that the amounts of subsidence are smaller than Case1 and the maximum amounts are about 15 mm at the upper part (ground surface) and about 2 to 5 mm at the middle part and the lower part.

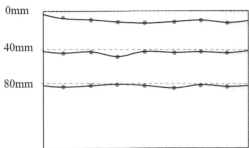

Figure 12. Settlement in Case2 (After experiment).

Figure 13 shows a photograph of the residential land model Case4 after the experiment.

Figure 14 shows the subsidence of Case3 (Toyoura sand of residential ground). The maximum amounts of subsidence are about 15 mm at the upper part (ground surface) and about 2 to 5 mm at the middle part and the lower part.

Figure 15 shows the subsidence of Case4 (SndwaveG of residential ground). The maximum amounts of

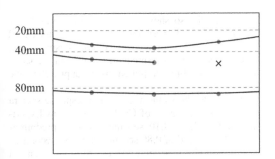

Figure 13. Model of residential ground with SandwaveG (After experiment).

Figure 11. Settlement in Case1 (After experiment).

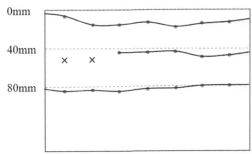

Figure 14. Settlement in Case3 (After experiment).

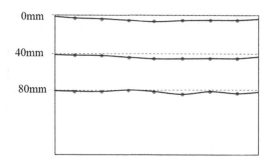

Figure 15. Settlement in Case4 (After experiment).

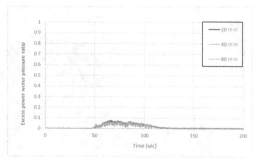

Figure 17. Excess pore water pressure ratio in Case4.

subsidence are about 5 mm at the upper part and about 2 to 4 mm at the middle part and the lower part.

In other words, it can be seen that the amount of subsidence on the ground surface of SandwaveG is smaller than that of Toyoura sand.

4.2 *Excess power water pressure ratio*

Figure 16 shows the measured values of excess pore water pressure ratio in Case3 with Toyoura sand. The excess pore water pressure ratio is almost the same at three locations due to the input of seismic motion. In the upper part (20 mm), the value of excess pore water pressure ratio rises from 0.4 to 0.5, and in the middle and lower parts (40 mm and 80 mm), it rises from 0.4 to 0.6, and then disappears.

Figure 17 shows the measured values of excess pore water pressure ratio in Case4 with SandwaveG. The excess pore water pressure ratio is almost the same at three locations due to the input of seismic motion. The value of excess pore water pressure ratio rises to about 0.08 and then disappears.

Figure 18 shows the measured values of the acceleration in Experimental Case3. The input seismic waves and the measured acceleration in the middle (40 mm) and upper (20 mm) are shown. The measured values are almost the same as the input seismic motion, but due to the rise in pore water pressure, the ground is slightly attenuated due to the liquid state.

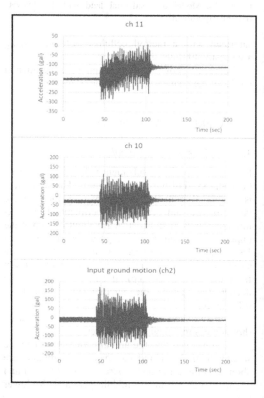

Figure 18. Acceleration in Case3.

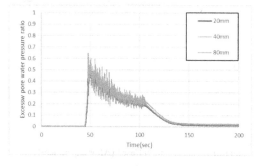

Figure 16. Excess pore water pressure ratio in Case3.

Figure 19 also shows the measured values of the acceleration in Experimental Case4. The measured values are almost the same as the input seismic wave, but the pore water pressure hardly rises, and the acceleration is slightly amplified as it approaches the ground surface.

Figure 20 shows the acceleration response spectra of the seismic waves of Ch10 and Ch11 in Experimental Case3. In Ch10, seismic waves are predominant with a period of 0.80 seconds and 2.39 seconds.

Even in Ch11, seismic waves are predominant at 0.78 seconds and 2.39 seconds, which is almost the same period as Ch10. The maximum acceleration is

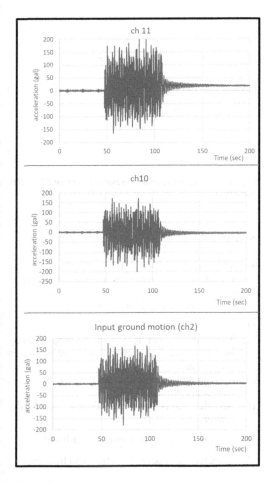

Figure 19. Acceleration in Case4.

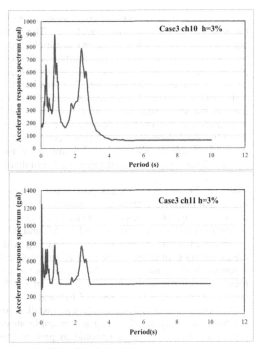

Figure 20. Acceleration response spectrum of Case3.

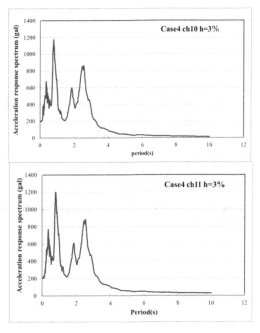

Figure 21. Acceleration response spectrum of Case4.

888 gal for Ch10 and 779 gal for Ch11, which is a slight decrease in Ch11. This is considered to be the effect of liquefaction.

Figure 21 shows the acceleration response spectra of the seismic waves of Ch10 and Ch11 in Experimental Case4. In Ch10, seismic waves are predominant in periods of 0.78 seconds and 12.54 seconds, with maximum accelerations of 1168.2 gal and 864.5 gal, respectively. In Ch11, the seismic wave is 1173.0 gal in 0.80 seconds and 869.9 gal in 2.56 seconds, which is predominant in almost the same period as Ch10. Both the period and the maximum acceleration values have changed little. This is considered to be due to liquefaction has not occurred.

5   APPLICATION TO QUAY BACKFILL SOIL

There was a phenomenon that the excess pore water pressure ratio increases near the wall of the quay compared to the untreated quay (Tokuhisa (2016)). It is necessary to dissipate the excess pore water pressure in the vicinity of the quay wall to prevent liquefaction.

Figure 22 shows the application of SandwaveG to the Sand bag of the full-scale model quay. The Sand bags were piled up with SandwaveG in a highly permeable Non-woven. The permeability steel sheet pile is a steel sheet pile that was made holes at regular intervals so that water is drained from the holes.

In this case, a 1mm thick stainless-steel plate was used as the steel sheet pile. And the stainless-steel plate made 2 mm diameter holes was used as the permeability steel sheet pile. The holes were placed at intervals 25 mm in length and 12.5 mm in width. And the 2 sheet piles were connected by 1 mm diameter wires.

Figure 23 shows the excess pore water pressure ratio when there is no drainage measure in the backfill of the quay. The excess pore water pressure ratio is larger toward the top and rises to 0.7, 0.3, and 0.1.

Figure 24 shows a case where a Sand bag with a SandwaveG is installed on the back of the quay and a permeable sheet pile is used. The excess pore water pressure ratio rose to about 0.4 at the top of Ch15, but was suppressed to 0.2 or less at Ch16 and 17.

A Sand bag with a SandwaveG was installed on the back of the quay. By using a permeable sheet pile as the sheet pile, it is possible to prevent the increase in the excess pore water pressure ratio in the backfill soil. It was confirmed that it has the effect of preventing liquefaction.

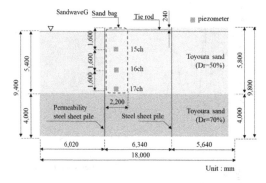

Figure 22. Application of SandwaveG to the Sand bag of the full-scale model quay.

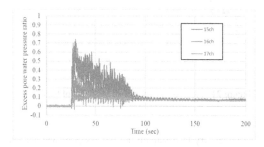

Figure 23. Excess pore water pressure ratio (no drainage measure).

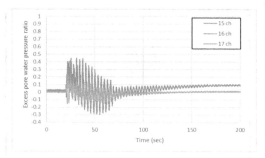

Figure 24. Excess pore water pressure ratio (Sand bag with SandwaveG).

6 CONCLUSION

From this research, we conducted a study to use SandwaveG as a countermeasure method for liquefaction, and the following items were clarified.

By using SandwaveG for residential ground

(1) Resistance to liquefaction increases.
(2) The amount of subsidence can be suppressed by increasing the liquefaction resistance during an earthquake.
(3) It is possible to prevent an increase in excess pore water pressure.
(4) Seismic waves increase slightly but hardly change.

By applying SandwaveG to quay backfill soil

(1) Liquefaction suppression effect can be expected.

REFERENCES

Glass Resourcing Co., Ltd. 2019. *Recycled glass granulated sand "SandwaveG" technical data*: 1–16
Japan Road Association 2012. *Specifications for highway briges PartV: seismic design. (in Japanese)*
Kochi City. 2015. Kochi City New Government Building Construction Office. *Kochi City New Government Building Implementation Design [Summary Version].*: 17
Kochi Prefecture 2013. Kochi Prefecture Crisis Management Department Nankai Trough Giant Earthquake Countermeasures Division Public materials. *[Kochi version] Calculation results of damage estimation due to Nankai Trough Giant Earthquake (Drawing collection)*: 20–31
Ministry of Land, Infrastructure, Transport and Tourism. 2011. Materials provided by the River Department, Tohoku Regional Development Bureau, Ministry of Land, Infrastructure, Transport, and Tourism. *Damage and restoration status of river and coastal facilities in the Great East Japan Earthquake*: 2–46
Tokuhisa, K 2016. Thesis Research Reports, ADVANCE COURSE KOCHI NATIONAL COLLEGE OF TECHNOLOGY, 15th. *A study on the liquefaction countermeasures of wharf considering the Nankai Trough Giant Earthquake*: 12

Proceedings of the Second International Conference on
Press-in Engineering 2021, Kochi, Japan – Matsumoto et al (eds)
© 2021 Taylor & Francis Group, London, ISBN 978-1-032-10414-0

Study on a countermeasure method for liquefaction of fishing ports against the Nankai Trough Earthquake

K. Okabayashi, G. Hashimura, A. Okubo, H. Ogasawara & S. Kadowaki
Kochi National College of Technology, Nankoku city, Kochi, Japan

K. Tokuhisa
Chodai Co., Ltd, Chuo Ward, Tokyo, Japan

ABSTRACT: In the near future, the Nankai Trough earthquake is expected to occur in Japan. It may be difficult to receive the prompt rescue and support in the event of this disaster since the Kochi Prefecture is surrounded by steep mountains. Therefore, Kochi Prefecture has placed earthquakeproof berths, but we propose that fishing ports also be used as disaster prevention bases. In this study, we investigated the liquefaction countermeasures for fishing port quays using sheet piles. As a result, (1) It was found out that the construction method using sheet piles was not enough for liquefaction countermeasures. (2) The method using sandbags and permeable steel sheet piles confirmed the effect of suppressing liquefaction.

1 INTRODUCTION

The 2011 off the Pacific coast of Tohoku Earthquake on March 11, 2011 is the largest earthquake in the observation history of Japan, caused huge damage by tsunami and liquefaction. All the ports on the Pacific side from Hachinohe Port in Aomori Prefecture to Kashima Port in Ibaraki Prefecture were damaged. Several damages such as settlements of structure and large faulting between the quays were seen in the damaged ports. But fishing ports and harbors contributed greatly to recovery after the earthquake because the port function was quickly restored.

The probability of the Nankai Trough Earthquake to occur within the next 30 years is estimated to be between seventy and eighty percent. When the Nankai Trough Earthquake occurs, there is a risk of damage to a wide area from Kyushu region to Tokai region. After the earthquake, Kochi Prefecture may have difficulty receiving help such as transportation of supplies from the land route as it is surrounded by steep mountains.

There are 88 fishing ports in Kochi Prefecture, it is considered that some of the fishing ports will be used as disaster prevention centers at the time of the earthquake, as in the case of the Tohoku Region Pacific Ocean Earthquake. Currently, seismic reinforcement and maintenance of several fishing ports, which are disaster prevention bases, are in progress.

In this study, we investigated an earthquakeproof berth focusing on the drainage near the quay wall.

From the previous study (Kiwa 2016), the following can be seen about the quay wall using the double steel sheet pile method at the time of the earthquake. (1) The quay using the double steel sheet pile method was able to suppress the deformation compared to the untreated quay. However, it was not possible to keep the allowable values for deformation and settlement of the quay wall. (2) There was a phenomenon that the excess pore water pressure ratio increases near the wall of the quay compared to the untreated quay. It is necessary to dissipate the excess pore pressure in the vicinity of the quay wall to prevent liquefaction. Therefore, the purpose of this study is to reduce displacement and reduce excess pore water pressure near the quay wall.

For the research method, the liquefaction analysis by effective stress analysis method (SoilWorks for LIQCA) and the liquefaction experiment by dynamic centrifuge model test were used. We examined the behavior of the quay during an earthquake from the analysis results and the experimental results.

2 RESEARCH CASE

In this study, we examined the four models. No countermeasure (Figure 1), a double steel sheet piling (Figure 2), a combination of double steel sheet piling with sandbags (Figure 3), a double steel sheet piling method of permeable steel sheet piling with sandbags (Figure 4). We call them Case1, Case2, Case3, and Case4 respectively. The

DOI: 10.1201/9781003215226-43

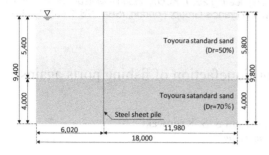

Figure 1. Case1.

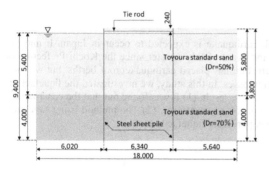

Figure 2. Case2.

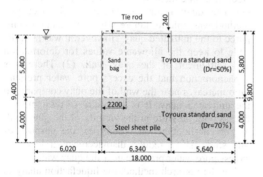

Figure 3. Case3.

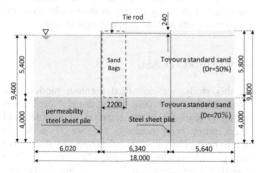

Figure 4. Case4.

Case4 is a model that uses a permeable steel sheet pile for the quay wall of Case3. The Toyoura sand with Dr = 70% is used for the base layer, and that with Dr = 50% is used for the liquefaction layer.

The sandbags used in Case3 and Case4 are made by putting light gravel material (pearl white: $\gamma_t = 16.02 kN/m^2$) in a high permeability bag (non-woven). In the analysis, the integration effect of sandbags is not considered, and it is regarded as an additional effect. Japanese Association for Steel Pipe Piles (2017) proposed permeable steel sheet piles. The steel sheet piles have almost natural water circulation and water permeability by providing small holes for water permeability in the steel sheet piles. A permeability hole is placed in the soil layer portion. The installation location is necessary to consider the future dredging depth, scouring depth, the occurrence of piping, etc.

Regarding the arrangement of permeable holes, we referred to Table 1 Example of the arrangement of permeability holes. In this study, the hole diameter at the effective width of 500 mm was adopted, and the number of holes per unit depth was set to 5.

3 LIQUEFACTION ANALYSIS

3.1 Analysis method

As an analysis method, the liquefaction analysis is performed using SoilWorks for LIQCA. LIQCA (LIQCA Liquefaction Geo-Research Institute 2015) is a liquefaction analysis program based on effective stress, using the u-p formulation with the solid-phase displacement (u) and the pore water pressure (p) as unknowns. Figure 5 shows the analysis procedure.

Table 1. Example of water hole arrangement.

Effective width (mm)	Hole diameter D (mm)	Hole spacing L(mm)	Rate hole ratio
900	80	1000	0.56
600	65	1000	0.55
500	60	1000	0.57

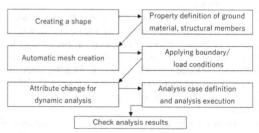

Figure 5. Analysis procedure.

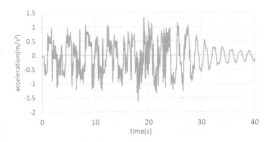

Figure 6. Input seismic wave.

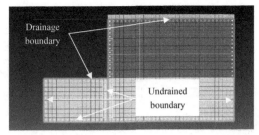

Figure 8. Drainage boundary condition (Experimental model).

3.2 Input seismic wave

The magnitude of the required acceleration was calculated and set based on the simple judgment method in the Specification for Highway Bridges (Japan Road Association 2012). Figure 6 shows the seismic waves used for the analysis. The maximum acceleration is 168.2gal.

3.3 Analysis model

In this study, we analyzed two models, a model based on actual ground (hereinafter, it is called a real model) and a model for liquefaction experiment (hereinafter, it is called an experimental model). In the actual model, the actual fishing port quay was modeled, the analysis area was enlarged, and the side surface was a repeating boundary and the bottom surface was a viscous boundary. The main purpose was to see the effect of the analysis area on the analysis results. The experimental model is used for comparison with the liquefaction experiment.

In SoilWorks for LIQCA2D15(2015), it is necessary to set boundary conditions for soil skeleton and pore water. We defined all boundary and load conditions used for both static and dynamic analyses. Figures 7 and 8 show the boundary conditions of the experimental model, Figures 9 and 10 show the boundary conditions of the real model. In the real model, the width of the analysis model is increased in consideration of the propagation of seismic waves.

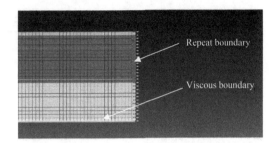

Figure 9. Side and bottom boundaries (Real model).

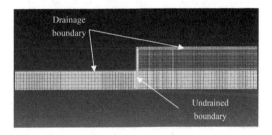

Figure 10. Drainage boundary condition (Real model).

3.4 Materials parameter

Tables 2 and 3 show the physical properties of the ground materials, Table 4 shows the physical properties of the materials used for countermeasures, and Table 5 shows the analysis condition. In this study, ground material properties were set with reference to the parameters used by Saito (2016).

The parameter of Toyoura sand (Dr = 50%) used for the liquefaction layer was determined by performing the element simulations in SoilWorks for LIQCA (2015) to match the results of the box shear test (Tanimoto, 2019) conducted at our university as shown in Figure 11. The physical properties of the steel sheet piles and the permeable steel sheet piles were determined based on the references (NIPPON STEEL CORPORATION 2019), respectively.

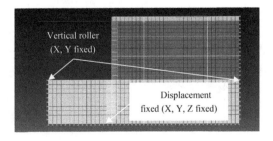

Figure 7. Side and bottom boundary conditions (Experimental model).

Table 2. Parameters for static analysis.

Item	Unit	Toyoura sand (Dr=50%)		Toyoura sand (Dr=70%)	Sand bags	
		Above water	Under water		Above water	Under water
Wet unit volume weight(γt)	kN/m³	14.48	14.48	15.06	17.08	17.08
Satureted unit volume weight(γ_{sat})	kN/m³	18.77	18.77	18.80	19.26	19.26
Element depth	m	1.0	1.0	1.0	1.0	1.0
Poisson's ratio		0.333	0.333	0.333	0.333	0.333
Effective soil covering pressure	kN/m²	1.0E-08	1.0E-08	1.0E-08	1.0E-08	1.0E-08
Static earth pressure coefficient		1.0	1.0	1.0	1.0	1.0
Proportionality coefficient of Young's modulus (E₀)		2775.2	2775.2	2775.2	5098.4	5098.4
Constant(n)		1.0	1.0	1.0	1.0	1.0
Adhesive force	kN/m²	0.0	0.0	0.0	0.0	0.0
Internal friction angle (φ)		37.75	37.75	37.75	37	37

Table 3. Parameter for dynamic analysis.

Item	Unit	Toyoura sand (Dr=50%)		Toyoura sand (Dr=70%)	Sand bags	
		Above water	Under water		Above water	Under water
Wet unit volume weight(γt)	kN/m³	14.48	14.48	15.06	17.08	17.08
Satureted unit volume weight(γ_{sat})	kN/m³	18.77	18.77	18.80	19.26	19.26
Element depth	m	1.0	1.0	1.0	1.0	1.0
Poisson's ratio		0.0	0.0	0.0	15.5	15.5
Static earth pressure coefficient		0.0	0.0	0.0	0.5	0.5
Nondimensional initial shear coefficient		910	910	1040.9	761	761
Initial gap ratio (e₀)		0.791	0.791	0.718	0.8	0.8
Compression exponent (λ)		0.0039	0.0039	0.0039	0.025	0.025
Swelling index (K)		0.00022	0.00022	0.00022	0.0003	0.0003
Pseudo-consideration ratio		1.0	1.0	1.5	1.0	1.0
Dilatancy factor (D₀)		0.0	0.5	0.75	0.0	1.0
Dilatancy factor (n)		0.0	5.0	7.0	0.0	4.0
Water permeability/ Unit volume of water	m/sec/kN/m³	0.0	0.0001	0.0001	0.0	0.00022
Bulk elastic modulus of water	kN/m²	2000000	2000000	2000000	2200000	2200000
S wave velocity	m/sec	0.0	0.0	0.0	100	100
P wave velocity	m/sec	0.0	0.0	0.0	1000	1000
Transformation stress ratio (Mₘ)		0.909	0.909	0.817	0.909	0.909
Fracture stress ratio (Mf)		1.229	1.229	1.245	1.229	1.229
Prameter in hardeinng function (B₀)		3500	3500	5186	2000	2000
Prameter in hardeinng function (B₁)		0.0	60	100	40	40
Prameter in hardeinng function (Cf)		0.0	0.0	0.0	0	0
Anistropic loss parameter (Cₐ)		2000	2000	2000	2000	2000
Plasticity reference strain (γP*r)		1000	0.003	0.005	1000	0.005
Elastic reference strain (γE*r)		1000	0.006	0.02	1000	0.003

Table 4. Structural property value.

		Unit	Iron plate	Steel sheet pile	Wire	Tie rod
Elastic modulus		kN/m²	205000000	210000000	205000000	210000000
Unit volume weight		kN/m²	76.44	73.6	76.44	73.6
Rigidity	Area	m²	0.392	0.0153	0.001256	0.00159
	Second moment of area	m⁴	0.000052267	0.0000874	1.0E-11	1.0E-11
	Effective shear area ratio		0.833			
	Section modulus	m³	0.00261333	0.00087400	——	——
	Plastic section modulus	m³	0.00392			
non-linear Consideration	Yield stress	kN/m²	175000	295000		
	Bending moment	Mf1	kN*m	457.33275	802	——	——
		Mf2	kN*m	686	1.0E+08		
	Reduction factor	a1	0.01	0.01		
		a2	1	1	——	——
		a3	1	1		

Table 5. Analysis condition.

Static analysis data	
Constitutive model	elastic perfectly plastic model
Mesh set	embankment ·liquefaction layer ·base layer
Boundary set	displacement fixed ·drainage boundary
Weight set	Self Weight
Ground acceleration	none
analysis/output control data	
Calculation time increment (s)	0.01
Output time increment (s)	0.1
Time integration method	Newmarkβ method (coefficient:β=0.3025 γ=0.6)
Rayleigh damping	Mass proportional=0 Stiffness proportional=0

Dynamic analysis data	
Constitutive model	cyclic elasto-plasticity model
Mesh set	embankment ·liquefaction layer ·base layer
Boundary set	displacement fixed ·drainage boundary ·attribute change
Weight set	none
Ground acceleration	Input seismic wave
analysis/output control data	
Calculation time increment (s)	0.001
Output time increment (s)	0.1
Time integration method	Newmarkβ method (coefficient:β=0.3025 γ=0.6)
Rayleigh damping	Mass proportional=0 Stiffness proportional=-0.003

Table 6. Experimental condition.

Experimental tank	Rigid earthen tank (W:450mm, H:365mm, D:140mm)
Centrifuge force field	40G
Base layer	Toyoura standard sand, Dr =70%, (Air fall method)
Liquefaction layer	Toyoura standard sand, Dr=50 % (Water fall method)
Input seismic wave	sin wave, period 20s, displacement 3mm, frequency 20Hz
Measurement items	Acceleration of in-ground and quay Pore water pressure in the ground Settlement process and cross-sectional displacement process of back ground of sheet pile

4 LIQUEFACTION EXPERIMENT BY DYNAMIC CENTRIFUGE

4.1 Condition of experiment

The liquefaction experiment was performed using the centrifuge model test equipment by Kochi National College of Technology. In this study, the centrifugal force field was set to 40G. Therefore, the model size used was 1/40 of the actual size. Table 6 shows the experimental conditions (Japan Road Association 2012). Figure 12 shows the input seismic wave in the experiment.

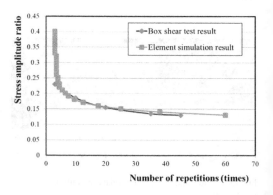

Figure 11. Elemental simulation (liquefaction resistance).

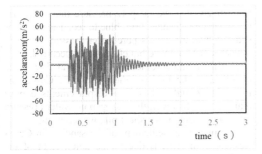

Figure 12 . Input seismic wave.

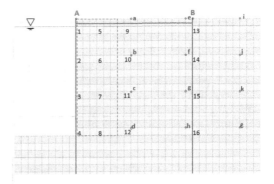

Figure 14. Result extraction point.

water pressure ratio in the specific elements as indicated in Figure 7 are shown in the next paragraph.

5.1 Analysis result

5.1.1 Deformation and settlement
Table 7 shows the final deformation of analysis results. From Table 7, the horizontal displacement is the largest in Case1 (1.57m at the top of the sheet pile) where no countermeasures have been taken, and smaller in Cases2 to 4(0.88m at the top of the sheet pile). In addition, Case3 and Case4 are smaller than Case2 on the back surface of the sandbag (0.46m at point c, about 0.33m at point d). Case1 has the largest backfill on the back of the wall (1.55m at point a), and Cases2 to 4 have the largest backfill on the back of the counterfort (about 1.06m at point i).

5.1.2 Excess pore water pressure ratio
Figure 15 shows the time history of the excess pore water pressure ratio of elements 6 to 8. The excess pore water pressure ratio in Case1 and Case2 reached 1.0 and were completely liquefied. The excess pore water pressure ratios of Case3 and Case4 were small, and it can be seen that the liquefaction countermeasure effect was achieved by increasing the drainage property due to the gravel in the sandbag.

5.1.3 Comparison of each case
Figure 16 shows the final deformation after vibration and the excess pore water pressure ratio distribution,

Figure 13. Experimental soil layer (Case1 before vibration).

4.2 Experiment method

The experimental soil layer is made using Toyoura sand. Base layer Dr = 70%, the liquefiable layer is in Dr = 50%. The sandbags used in Case3 and Case4 were created by putting light gravel material (pearl white: γ_t = 16.02kN/m^2) in a highly permeable bag (non-woven), and stacking it on the back of the quay.

Also, a viscous liquid with viscosity 40 times that of water is used according to the similarity law by mixing water and methylcellulose.

The completed experimental soil layer is placed on the shaking table of the centrifuge model test equipment and the experiment was performed. The centrifugal force up to 40G is applied and the input wave is loaded by the shaking table. After the experiment was completed, the position of the target in the soil layer and the deformation of steel sheet pile was measured. Also measured the deformation of steel sheet pile. Figure 13 shows the completed experimental soil layer (Case1).

5   RESULT AND DISCUSSION

Figure 14 shows the extraction position of the final deformation value and of the time history of the excess pore water pressure ratio. The deformation values at the specific nodes and the excess pore

Table 7. Final deformation (analysis results).

Vertical displacement	a	b	c	d	e	f	g	h	i	j	k	l	Maximum settlement	Maximum uplift
Case1	-1.552	-0.914	-0.484	-0.191	-1.102	-0.780	-0.505	-0.238	-0.670	-0.929	-0.382	-0.254	-1.697	
Case2	-0.340	-0.230	-0.182	-0.139	-0.306	-0.949	-0.785	-0.402	-1.007	-0.610	-0.327	-0.103	-1.069	0.160
Case3	-0.143	-0.069	-0.073	-0.115	-0.511	-0.650	-0.684	-0.399	-1.050	-0.603	-0.322	-0.087	-1.071	0.116
Case4	-0.137	-0.100	-0.076	-0.116	-0.603	-0.696	-0.698	-0.402	-1.003	-0.617	-0.328	-0.100	-1.071	0.121
Horizontal displacement	a	b	c	d	e	f	g	h	i	j	k	l	A	B
Case1	0.032	-0.257	-0.404	-0.326	0.665	0.397	0.101	-0.014	0.894	0.734	0.410	0.194	-1.567	
Case2	-0.859	-0.789	-0.768	-0.466	-0.836	-0.694	-0.543	-0.375	0.331	0.084	-0.177	-0.228	-0.888	-0.901
Case3	-0.925	-0.402	-0.438	-0.330	-0.801	-0.636	-0.509	-0.363	0.337	0.004	-0.206	-0.340	-0.885	-0.900
Case4	-0.923	-0.454	-0.476	-0.339	-0.772	-0.648	-0.510	-0.364	0.353	0.046	-0.201	-0.228	-0.887	-0.901

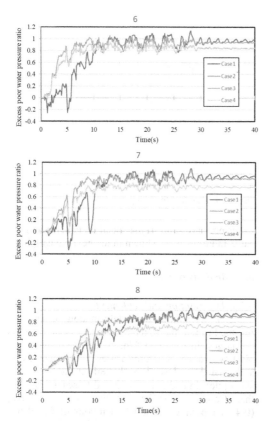

Figure 15. Time history of excess pore water pressure ratio.

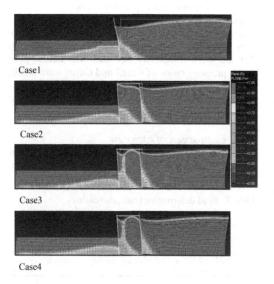

Figure 16. Displacement and excess pore water pressure ratio.

respectively. The excess pore pressure ratio increases as the element changes from blue to red. The red color shows where the excess pore water pressure is 1.0. It indicates that complete liquefaction has occurred.

In Case1, large settlement and lateral flow were observed at the back of the sheet pile. Case2 has lower horizontal and vertical displacements overall than Case1. However, the excess pore water pressure ratio at the back of the sheet pile is increasing.

The maximum settlement of the sandbag in Case3 was 1.071 m. This is due to the unit weight of the sandbag being larger than the Toyoura standard sand, which caused the settlement. There was almost no change in horizontal displacement. The excess pore pressure ratio decreased in the sandbag compared to Case2. In Case4, both the displacement and the excess pore water pressure ratio were not much different from Case3. The excess pore water pressure ratio on the ground surface was smaller than in Case3. But the change was very small.

Figure 17 shows the distribution diagrams of the horizontal stress due to the deformation of Case2 and Case4.

Each figure shows the horizontal stress at 24.4s. The maximum value of the element is 510.93 (kN/m²) for Case2, 560.91 (kN/m²) for Case4, which is larger in Case4 than in Case2. The unit weight of the sandbag is larger than that of Toyoura standard sand, so it is considered that the horizontal stress acting on the steel sheet pile when lateral flow occurred was increased.

5.2 Experiment result

Figure 18 shows the experimental soil layer after vibration. Table 8 shows the displacement at each point, and Figure 19 shows the time history of excess pore water pressure ratio.

5.2.1 Horizontal displacement and settlement
From the Table 8, the destruction occurred in Case1, and the horizontal displacement of the back surface

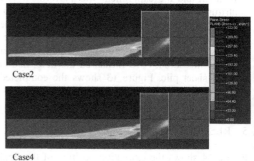

Figure 17. Horizontal stress distribution.

Case1

Case2

Case3

Case4

Figure 18. Experimental result.

of the sandbag was large at all points a to d (Point a is unmeasurable, point b is 1.44 m, point c is 0.92 m). The settlement showed the same tendency as the horizontal displacement.

Vertical displacement	a	b	c	d	e	f	g	h	i	j	k	l
Case1		0.720	0.920	0.160		0.880	0.360	0.080		0.080	0.000	0.000
Case2	0.920	0.080	0.000	0.000	0.920	0.040	0.000	0.000	0.720	0.520	0.200	0.000
Case3	0.280	0.600	0.360	0.080	0.480	0.360	0.240	0.120	0.000	0.480	0.400	0.000
Case4	0.280	0.160	0.000	0.000	0.040	0.120	0.200	0.000	0.520	0.080	0.000	0.000
Horizontal displacement	a	b	c	d	e	f	g	h	i	j	k	l
Case1		1.440	0.920	0.160		0.880	0.360	0.080		0.080	0.000	0.000
Case2	0.800	0.200	0.120	0.080	0.000	0.040	0.040	0.000	0.000	0.000	0.200	0.000
Case3	0.440	0.360	0.400	0.000	0.280	0.240	0.200	0.120	0.000	0.000	0.000	0.000
Case4	0.120	0.000	0.000	0.080	0.040	0.000	0.200	0.000	0.120	0.000	0.000	0.000

Table 8. Final deformation (experimental results)

5.2.2 Excess pore water pressure ratio

From the Figure 19, at the upper part (10, 11) of Case1, complete liquefaction occurred, and settlement and lateral flow at the back of the sheet pile were confirmed. The reason why the pore water pressure ratio exceeds 1.0 at the upper part was considered to be that the pore water pressure gauge sank due to liquefaction and the excess pore water pressure increased.

In Case2, the deformation was suppressed (0.80m at point a), and the excess pore water pressure ratio was also reduced compared to Case1.

In Case 3, the excess pore water pressure ratio was larger in the upper part of the back of the sandbag than in Cace 2. It is considered that this is because the deformation was suppressed (0.44m at point a, 0.36m at point b and 0.40m at point c) and the volume of the void became smaller than that of Case 2.

In Case4, the deformation of the experimental soil layer became very small (0.12m at point a, 0m at points C to d). The excess pore water pressure ratio also became the smallest. This is because the pore water was drained by the permeable steel sheet pile. Therefore, it can be seen that by using the sandbag and the permeable steel sheet pile, the drainage property of the backfill soil and the sheet pile was improved, and the liquefaction countermeasure effect was improved.

6 COMPARISON BETWEEN ANALYSIS RESULT AND EXPERIMENTAL RESULT

Table 9 shows the displacement of the analysis results for the experimental model, and Figure 20 shows a comparison diagram. The analysis results and experimental results were compared for Case1, Case2, and Case4. In Case3, the liquefaction countermeasure effect was not so different from Case2, so it is omitted here.

From the comparison between Case1 and Case2, the deformation was almost the same.

In Case4, the experimental deformation was smaller than the analytical deformation. The reasons for this are as follows: (1) In the experiment, the drainage effect of sandbags and permeable steel sheet

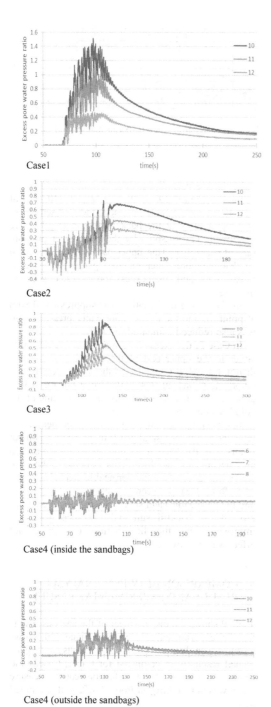

Case1

Case2

Case3

Case4 (inside the sandbags)

Case4 (outside the sandbags)

Figure 19. Experimental results.
(Time history of excess pore water pressure ratio)

Vertical displacement	a	b	c	d	e	f	g	h	i	j	k	l
Case1	-0.897	-0.559	-0.294	-0.126	-0.560	-0.380	-0.217	-0.101	-0.249	-0.083	-0.051	-0.034
Case2	-0.089	0.011	0.029	0.002	-0.621	-0.621	-0.483	-0.158	-0.677	-0.525	-0.370	-0.177
Case3	0.041	0.090	0.101	0.004	-0.269	-0.354	-0.439	-0.203	-0.610	-0.471	-0.330	-0.157
Case4	0.209	0.138	0.110	0.016	-0.245	-0.262	-0.400	-0.198	-0.606	-0.471	-0.330	-0.156
Horizontal displacement	a	b	c	d	e	f	g	h	i	j	k	l
Case1	-0.780	-0.556	-0.373	-0.195	-0.612	-0.170	-0.063	-0.036	-0.513	0.009	0.075	0.033
Case2	-0.887	-0.723	-0.626	-0.374	-0.831	-0.672	-0.500	-0.327	-0.471	-0.176	-0.275	-0.223
Case3	-0.946	-0.369	-0.300	-0.198	-0.766	-0.577	-0.422	-0.272	-0.440	-0.137	-0.250	-0.206
Case4	-1.184	-0.318	-0.238	-0.173	-0.937	-0.570	-0.412	-0.270	-0.432	-0.134	-0.250	-0.206

Table 9. Final deformation (Analysis results of experimental model)

the analysis, the permeability of sandbags and permeable sheet piles was evaluated to be smaller than that of the experimental model.

7 CONCLUSION

In this study, we investigated the liquefaction countermeasures for fishing port quays using double steel sheet piles, sandbags and permeable steel sheet piles. The following conclusion were obtained.

(1) The sandbag had the effect of dissipating the pore water pressure inside the sandbag, and the permeable steel sheet pile had the effect of draining the pore water on the back of the quay. It was confirmed that the effect of liquefaction countermeasures on the quay was improved by using both.

(2) The uncountermeasured model and the double sheet pile model of the centrifuge experiment could be reproduced because the displacement and subsidence were almost the same as the seismic response analysis simulation using LIQCA (2015).

(3) In the actual model with a larger analysis boundary, the amount of settlement was smaller than that of the experimental model, but the horizontal displacement was about the same.

(4) In the Case of the model equipped with sandbags and permeable sheet piles, the centrifugal experiment showed more dissipation of excess pore water pressure than the LIQCA analysis, and the effect of liquefaction countermeasures was higher. The reason may be that the water permeability was evaluated to be smaller than that of the experimental model in the analysis, and the integration effect of sandbags could not be considered.

REFERENCES

Japanese Association for Steel Pipe Piles, 2017. *Steel sheet pile Q&A*: 33–38
Japan Road Association, 2012. *Specifications for highway briges PartV seismic design. (in Japanese)*
LIQCA Liquefaction Geo-Research Institute, 2015. *LIQC A2D15· LIQCA3D15(2015) release version document Part1 Theory*: 1–55

piles was high, and the pore water was dissipated and the liquefaction was suppressed. (2) In the experiment, the deformation of the backfill soil was suppressed by the integration effect of the sandbags, but it could not be considered in the analysis. (3) In

NIPPON STEEL CORPORATION, 2019. *Steel sheet pile, catalog*.7. https://www.nipponsteel.com/product/catalog_download/pdf/K007.pdf

Saito, K. 2016. Simulation analysis of centrifuge model experiment using liquefaction analysis program LIQCA: 14

Tanimoto, W., Okabayashi, K., Mukaidani, M., and Hama, K. 2019. Reiwa 1st Japan Society of Civil Engineers National Convention 74th Annual Scientific Lecture (III-311). *Comparison of Kochi College type cyclic box shear apparatus with cyclic triaxial apparatus.*

Tokuhisa, K. 2016. Thesis Research Reports, ADVANCE COURSE KOCHI NATIONAL COLLEGE OF TECHNOLOGY, 15th. *A study on the liquefaction countermeasures of wharf considering the Nankai Trough earthquake*: 12

*Proceedings of the Second International Conference on
Press-in Engineering 2021, Kochi, Japan – Matsumoto et al (eds)
© 2021 Taylor & Francis Group, London, ISBN 978-1-032-10414-0*

Experimental study on tsunami mitigation effect of pile-type porous tide barrier

K. Toda, A Mori & Y. Ishihara
Giken LTD, Kochi, Japan

N. Suzuki
Giken LTD, Tokyo, Japan

ABSTRACT: Damage to coastal levees and breakwaters due to the tsunami in the Great East Japan Earthquake in 2011 have brought up issues on the necessity of improving the tenacity of these structures. Use of piles in these structures will be effective in enhancing their tenacity. Among several possibilities in structural types of these structures with piles, a pile-type porous tide barrier is expected to be helpful in reducing the tsunami load that passes through it and preventing the existing structures from collapsing. This paper introduces model tests examining the effect of the pile-type porous tide barrier in reducing the surge-type tsunami load. Based on several test results with different porosity ratios and a theoretical approach to interpret the experimental data, it was found that the tsunami mitigation effect of the barrier can be captured by loss factor, which can be defined by combining the porosity ratio and the friction factor.

1 INTRODUCTION

Damage to coastal levees and breakwaters due to the tsunami in the Great East Japan Earthquake in 2011 have brought up issues on the necessity of improving the tenacity of these structures (MLIT, 2013). Use of piles in these structures will be effective in enhancing their tenacity. For example, Kikuchi *et al.* (2015) confirmed the effectiveness of reinforcing the gravity-type breakwater using pipe piles. Suzuki *et al.* (2016a) investigated the effect of the breakwater consisting of arrays of steel tubular piles to mitigate the tsunami load, and proposed its design method through a theoretical approach. In addition to these structures with piles, a pile-type porous tide barrier (hereinafter called as "Implant Barrier" or "Barrier") is expected to serve as one of the countermeasures against tsunami.

The basic structure of the Implant Barrier is shown in Figure 1. It consists of columns, (steel tubular piles), porous sheets (fabric materials) and sheet piles (optionally). The columns are aligned with a certain distance between themselves, and are embedded into the ground to generate enough horizontal resistance to secure the structural stability of the Barrier. The porous sheets are supposed to be made of fabric materials and are fixed to the columns, and have a function of inducing an energy loss of the tsunami when it passes through them. The porosity ratio of the porous sheet is expressed as Equation (1), and the porosity ratio of the Barrier is defined by Equation (2).

$$\lambda = A_P/(L_P H) \qquad (1)$$

$$\lambda_B = L_P \lambda / L_C \qquad (2)$$

where H is the height of the Barrier, D_C is the outer diameter of the column, L_C is the distance between two adjacent columns, L_P is the horizontal length of one porous sheet between two columns and A_P is the area of the porous parts in one porous sheet between two columns, as illustrated in Figure 2. Sheet piles are expected to be optionally used to ensure the stability of the Barrier by avoiding the deformation of the ground surface due to erosion, seepage failure, liquefaction and so on. With the above-mentioned structure, the Barrier is expected to be effective in reducing the load of tsunami that passes through it and preventing the existing structures behind the Barrier from collapsing.

The effect of the Barrier on reducing the tsunami load has partly been investigated so far. Suzuki *et al.* (2016b) provided a theoretical approach to explain the load, velocity and the height of tsunami in front of and behind the Barrier, and confirmed its validity through a series of two-dimensional hydraulic model tests by using a surge-type tsunami. However, in their model tests, the tsunami load was not measured behind the Barrier, and the tsunami mitigation effect of the Barrier was captured by the water depth behind

DOI: 10.1201/9781003215226-44

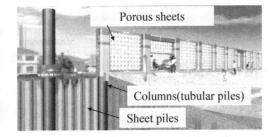

Figure 1. Basic structure of the Barrier.

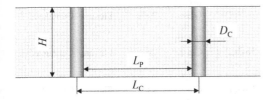

Figure 2. Parameters related to the geometry of the Barrier.

the Barrier. It is necessary to measure the tsunami load behind the Barrier to understand more directly the effect of the Barrier on reducing the tsunami load. Furthermore, the applicability of the theoretical approach to the condition where the height of tsunami exceeds the Barrier has not been investigated, although a tsunami with such a height was dealt with in their model tests. In order to expand the applicability of the Barrier, it is necessary to investigate the applicability of the theoretical approach to such a condition.

This paper reports the results of the two-dimensional hydraulic model tests on the Barrier, where the load of a surge-type tsunami behind the Barrier was measured. In addition, based on the test results, the applicability of the theoretical approach to tsunami higher than the Barrier is studied.

2 MODEL TESTS

2.1 Apparatus

The model tests were conducted using an experimental facility called Tsunami Simulator in Giken (Figure 3). The apparatus consists of a surface tank for storing

water, a gate, a channel and an underground tank for storing water. The channel has a length of 19.5 m, a width of 1.5 m and a depth of 0.8 m. Details of this apparatus can be found in Ishihara et al. (2018).

In this research, a surge-type model tsunami, simulating the front part of tsunamis, were generated by accumulating water in the surface tank and opening the gate instantaneously.

The layout of the test is shown in Figure 4. A slope of 1/7 was placed on the floor of the channel to simulate a geographical feature of coastal areas, and the model of the Barrier was fixed at the end of the slope with the distance from the gate being 10.5 m. The height of the end of the slope was 0.2 m from the floor of the channel, and the initial water level was set as 0.15 m. The height of the Barrier was 0.35 m (i.e. the head of the barrier was 0.55 m above the floor of the channel).

The columns of the Barrier were modelled by polyvinyl chloride pipes with the outer diameter of 48 mm. The scaling law was ignored in terms of their cross-sectional performance (flexural rigidity), as the deformation of the columns were out of scope of this research. The porous sheets were modelled by the glass cloths. The upper and lower edges of the glass cloths were protected by plastic tapes to avoid their fray, as shown in Figure 5.

The layout of measurement is shown in Figure 6. Flow velocimeters and wave gauges were settled at three positions: "Upstream" which was 2000 mm in front of the Barrier, "Front" which was 75 mm in front of the Barrier and "Downstream" which was 490 mm behind the Barrier. The tsunami load on the Barrier was obtained by the readings of the load cells equipped in the lower part of the columns of the Barrier. The load of tsunami that passed through the Barrier was measured by the load measuring device just behind the "Downstream" position as shown in Figure 7, where strain gauges were attached on steel parts which were

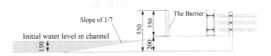

Figure 4. Test layout in "To" test series.

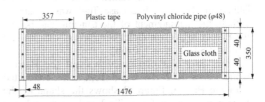

Figure 5. Protection of the upper and lower edges of the model porous sheets.

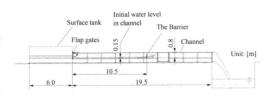

Figure 3. Experimental apparatus (Tsunami Simulator).

445

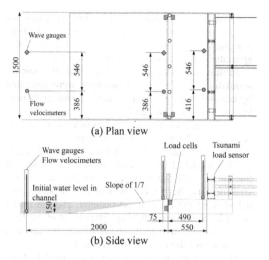

(a) Plan view

(b) Side view

Figure 6. Layout of measurement in "To" test series. (a) Plan view (b) Side view.

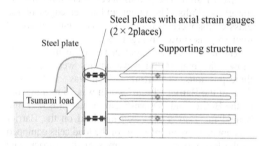

Figure 7. Tsunami load measuring device placed near the "Downstream" position.

fixed to the steel plate facing tsunami and the supporting structure that was fixed to the floor.

2.2 Test cases

A total of 8 tests were newly conducted as shown in Table 1, to compare the effect of the Barrier on reducing the tsunami load. Four different tests were conducted twice. Cases To-1, To-2 and To-3 were conducted with the measurement of the tsunami load behind the Barrier, while Case To-4 was conducted without the measurement of the tsunami load behind the Barrier to allow a direct comparison with the test results of Suzuki et al. (2016b).

Case To-1 was conducted without the Barrier, while Cases To-2, To-3 and To-4 were conducted with the Barrier. Figure 8 shows the shape of the Barrier was flat in Cases To-2 and To-4, and was

Table 1. Test conditions in "To" test series.

No.	Material of porous sheets	Porority ratio	Shape of the Barrier	Measurement of tsunami load behind the Barrier
To-1(1)	-	100%	-	Conducted
To-1(2)	-	100%	-	Conducted
To-2(1)	Glass cloth (#110)	24%	Flat	Conducted
To-2(2)	Glass cloth (#110)	24%	Flat	Conducted
To-3(1)	Glass cloth (#220)	24%	Flexure	Conducted
To-3(2)	Glass cloth (#220)	24%	Flexure	Conducted
To-4(1)	Glass cloth (#110)	24%	Flat	Ommitted
To-4(2)	Glass cloth (#110)	24%	Flat	Ommitted

flexure in Case To-3. The porosity ratio of the porous sheets in Case To-3 was smaller than that in Cases To-2 and To-4, so that the porosity ratio of the Barrier was comparable in these test cases. It should be noted that the tsunami load on the Barrier was not measured in Case To-3, because the bottom of each porous sheet was fixed to the floor with tapes in order to prevent tsunami from leaking and to maintain the flexural shape of the porous sheet.

2.3 Results

Regarding the effect of the Barrier on mitigating the tsunami load, Figure 9 shows the variation of the flow velocity and the water depth at the "Downstream" position with time (t), as well as that of the tsunami load measured just behind the "Downstream" position. The origin of the horizontal axis was taken as the time when the tsunami arrived at the "Upstream" position.

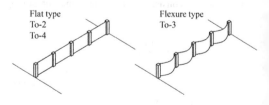

Figure 8. Shape of the Barrier.

446

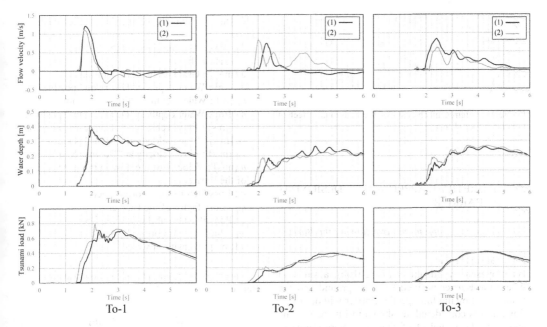

Figure 9. Time-series data of flow velocity and water depth at the "Downstream" position.

The water depth started to increase when t was around 1.5 s and arrived its local peak when 2.0 s $< t < 2.5$ s. After that, it gradually decreased in Case To-1 (without Barrier), while it kept increasing in Cases To-2 and To-3 (with Barrier) and reached its peak when 4.0 s $< t < 5.0$ s. This might be because, as illustrated in Figure 10, the tsunami reflected at the tsunami load measuring device was again reflected at the Barrier in Cases To-2 and To-3, and more amount of water was captured in between the Barrier and the tsunami load measuring device.

The flow velocity started to increase when t was around 1.5 s and arrived its peak by the time when t increased to 2.5 s. After that, it sharply decreased to zero or negative values in Case To-1 (without Barrier) while it gradually decreased in Cases To-2 and To-3 (with Barrier).

Regarding the tsunami load, the trend of variation with time was more similar to that of the water depth rather than that of the flow velocity. It started to increase when t was around 1.5 s, arrived its local peak when 2.0 s $< t < 2.5$ s, and then gradually decreased in Case To-1 (without Barrier) while it kept increasing in Cases To-2 and To-3 (with Barrier) and reached its peak when 4.0 s $< t < 5.0$ s.

The above-mentioned trends of the water depth, flow velocity and tsunami load after the initial local peak will be influenced by the duration of the tsunami, and will better be dealt with by the experiments using a reflux-type model tsunami (i.e. a continuous flow). As the experiment in this paper uses the surge-type tsunami with a limited duration, the discussion hereinafter will be made by focusing on the measured values in $t < 2.5$ s.

Comparing To-1 and To-2, the Barrier with a flat shape reduced the water depth, flow velocity and the tsunami load at or near the "Downstream" position by 33%, 52% and 75% respectively, which demonstrates the effectiveness of the Barrier on reducing the tsunami load. In addition, comparing To-1 and To-3, the Barrier with a flexural shape reduced the water depth, flow velocity and the tsunami load at or near the "Downstream" position by 38%, 57% and 80% respectively. The Barrier with a flexural shape provided additional effect of reducing the tsunami load. It is suggested that this was because the flexural shape of the porous sheets was effective in changing the direction of the flow behind the Barrier, leading to a collision of each flow and the loss of its energy as shown in Figure 11.

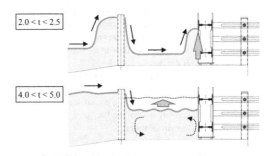

Figure 10. Inferred mechanism of the water depth increase behind the Barrier.

447

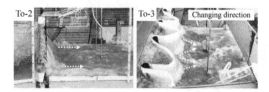

Figure 11. Inferred mechanism of the additional effect of flexural porous sheets.

3 THEORETICAL APPROACH OF DESIGNING THE BARRIER

3.1 *Surge-type tsunami not overflowing the Barrier*

3.1.1 *Basic theory*

As proposed by Suzuki *et al.* (2016b), a situation where the Barrier is fixed to a flat ground and the tsunami is flowing through the Barrier without overflowing it is considered, as shown in Figure 12. The subscripts U, F and D of the parameters represent the positions of "Upstream", "Front" and "Downstream" respectively.

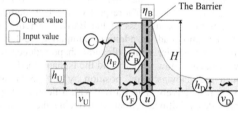

B	[m]	: Width of the channel
H	[m]	: Height of the Barrier
C	[m/s]	: Velocity of the flow reflected by the Barrier
F_B	[kN]	: Tsunami load on the Barrier
f		: Friction factor
g	[m/s²]	: Gravitational acceleration
h_U	[m]	: Water depth at "Upstream" position
h_F	[m]	: Water depth at "Front" position
h_D	[m]	: Water depth at "Downstream" position
v_U	[m/s]	: Flow velocity at "Upstream" position
v_F	[m/s]	: Flow velocity at "Front" position
v_D	[m/s]	: Flow velocity at "Downstream" position
u	[m/s]	: Velocity of the flow in the porous parts of the Barrier
η		: Loss factor of the porous sheet
η_B		: Loss factor of the Barrier
λ		: Porosity ratio of the porous sheet
λ_B		: Porosity ratio of the Barrier
ρ	[t/m³]	: Density of water

Figure 12. Model considered in the theoretical approach.

C [m/s] : Velocity of the flow reflected by the Barrier
F_B [kN] : Tsunami load on the Barrier
f : Friction factor
g [m/s²] : Gravitational acceleration
h_U [m] : Water depth at "Upstream" position
h_F [m] : Water depth at "Front" position
h_D [m] : Water depth at "Downstream" position
v_U [m/s] : Flow velocity at "Upstream" position
v_F [m/s] : Flow velocity at "Front" position
v_D [m/s] : Flow velocity at "Downstream" position
u [m/s] : Velocity of the flow in the porous parts of the Barrier
η : Loss factor of the porous sheet
η_B : Loss factor of the Barrier
λ : Porosity ratio of the porous sheet
λ_B : Porosity ratio of the Barrier
ρ [t/m³] : Density of water

Based on the continuity of the flowrate around the Barrier,

$$Q = Bv_F h_F = Bv_D h_D \quad (3)$$

Considering the law of conservation of momentum around the Barrier, the tsunami load on the Barrier (F_B) is expressed as:

$$F_B = \frac{1}{2}\rho gB\left(h_F{}^2 - h_D{}^2\right) + \rho B\left(h_F v_F{}^2 - h_D v_D{}^2\right) \quad (4)$$

Assuming that the energy loss around the Barrier is caused only by the transmission of the flow in the porous parts, the law of conservation of energy around the Barrier can be written as:

$$f\frac{1}{2g}u^2 = \left(\frac{1}{2g}v_F{}^2 + h_F\right) - \left(\frac{1}{2g}v_D{}^2 + h_D\right) \quad (5)$$

Based on the continuity of the flowrate,

$$u = v_F/\lambda \quad (6)$$

Combining Equations (5) and (6),

$$\frac{1}{2g}\frac{f}{\lambda^2}v_F{}^2 = \frac{1}{2g}\left(v_F{}^2 - v_D{}^2\right) + \left(h_F - h_D\right) \quad (7)$$

On the other hand, from a coordinate system that travels inversely to the tsunami at the velocity of C, the continuity of flowrate and the conservation of momentum at "Downstream" and "Front" positions can be expressed as:

$$B(v_F+C)h_F= B(v_U+C)h_U \qquad (8)$$

$$\frac{1}{2}\rho gBh_F{}^2+\rho Bh_F(v_F+C)^2=\frac{1}{2}\rho gBh_U{}^2+\rho Bh_U(v_U+C)^2 \qquad (9)$$

Based on the experimental results of Suzuki *et al.* (2016b), the flow velocity of the tsunami may be assumed as being independent of the existence of the Barrier.

$$v_D=v_U \qquad (10)$$

By the way, it is complicated from a practical point of view to obtain both the porosity ratio (λ) and the friction factor (f) when designing the Barrier. In addition, it is difficult to obtain an accurate value of λ for a porous sheet with a complicated structure (e.g. a fabric sheet consisting of a lot of strings). To cope with these difficulties, a parameter called "loss factor" (η), as expressed by Equation (11), will be introduced.

$$\eta=\frac{f}{\lambda^2} \qquad (11)$$

As a method to obtain the loss factor (η), the following experimental method will be introduced, by consulting the method of Hasegawa *et al.* (1987). As shown in Figure 13, two PVC (polyvinyl chloride) pipes with flanges are prepared, and the porous sheet is sandwiched between the two flanges. While maintaining a steady flow inside the pipe, the water pressures are measured at both sides of the porous sheet, and the flowrate is measured by an ultrasonic flowmeter in the downstream side. The loss factor is obtained by:

$$\eta=\frac{2\Delta p}{\rho v^2} \qquad (12)$$

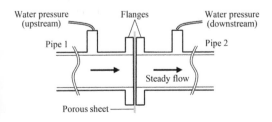

Figure 13. Experimental method to obtain the loss factor.

where Δp is the difference of the water pressure at the two measurement positions and v is the flow velocity calculated from the measured flowrate.

Precisely speaking, the porosity ratio and the loss factor of the Barrier (λ_B and η_B) are different from those of the porous sheets (λ and η) due to the existence of the tubular piles with its porosity ratio being zero. η_B would be assumed to be obtained by:

$$\eta_B=\frac{\lambda^2}{\lambda_B{}^2}\eta \qquad (13)$$

Plugging Equations (11) into Equation (7), and replacing η by η_B, Equation (14) is reduced to:

$$\eta_B\frac{1}{2g}v_F{}^2=\frac{1}{2g}\left(v_F{}^2-v_D{}^2\right)+(h_F-h_D) \qquad (14)$$

Equations (3), (4), (8), (9), (10) and (14), which consist of 9 parameters (C, F_B, h_U, h_F, h_D, v_U, v_F, v_D, η_B), can be solved by considering the input values for h_U, v_U and η_B, to obtain the values of the other 6 parameters.

3.1.2 *Verification of the theoretical approach by comparing with model test results*

The conditions and the results of the model tests conducted by Suzuki *et al.* (2016b) and those introduced in Section 2 are summarized in Table 2. The values of the tsunami load on the Barrier are the initial local peak values, and the values of the water depth at the "Front" position are those recorded at the same timing (when the tsunami load on the Barrier arrived at the initial local peak values). The values of the water depth at the "Downstream" position are the initial local peak values. Note that a slight overflowing was observed in Case Su-6 (with 0% porosity ratio).

The input values for the three parameters (h_U, v_U and η_B) are summarized in Table 3. Regarding h_U and v_U, the values recorded at the "Upstream" position without the Barrier were adopted. The values of η_B were obtained experimentally based on the method introduced in Section 3.1.1. By applying the steady flow with its flow velocity being varied up to the value adopted in the model tests in Suzuki *et al.* (2016b) and in Section 2, η was confirmed to be little influenced by v. In this research, η was taken as the average of the measured values when v was in between 0.6 and 0.8 m/s. The obtained values of η are shown in Table 2, and the values of η_B obtained by Equation (13) are shown in Table 3. Note that the η_B values in the test cases with 100 % porosity (i.e. without the Barrier) and with 0 % porosity were taken as 0.1 and 10000 respectively to be plotted on log-scale graphs, although they are supposed to be 0 and ∞ theoretically.

Table 2. Test conditions and results (surge-type tsunami not overflowing the Barrier).

Case No.	Porous material	Porority ratio λ	λ_B	Loss factor* η	Water depth at "Front" h_F[m]	Tsunami load on the Barrier F_B[kN]	Water depth at "Downstream" h_D[m]	Remarks
Su-1	-	100%	100%	0.1	0.155	0.025	0.118	Suzuki etal.
Su-2	Punching metal	33%	33%	6	0.295	0.386	0.064	(2016b)
Su-3	Punching metal	10	10	147	0.347	0.557	0.021	
Su-4	Grass cloth	40%	40%	14	0.225	0.421	0.031	
Su-5	Grass cloth	10%	10%	353	0.342	0.612	0.037	
Su-6	Steel plate	0%	0%	10000	0.351	0.787	0.027**	
To-4(1)		40%	24%	14	0.346	0.348	0.042	This report
To-4(2)					0.333	0.344	0.046	

* Suzuki etal., 2016b. ** Overflow.

Table 3. Input values for designing the Barrier not overflown by tsunami.

Case No.	Flow velocity v_U [m/s]	Water depth h_U [m]	Loss factor η_B
Su-1	1.28※	0.155※	0.1
Su-2			6
Su-3			147
Su-4			14
Su-5			353
Su-6			10000
To-4(1)	1.27	0.162	39
To-4(2)			

Figure 14 shows the comparison of the water depth at the "Front" position (h_F), tsunami load on the Barrier (F_B) and the water depth at the "Downstream" position (h_D) measured in the model tests and estimated by the theoretical approach. The input values for the estimation are shown in Table 3. Good

agreement between the measured and estimated values is confirmed for the three parameters.

3.2 Surge-type tsunami overflowing the Barrier

3.2.1 Introduction of equivalent porosity ratio and equivalent loss factor of the Barrier

To apply the theoretical approach in Section 3.1 to the cases where the Barrier is overflown by the tsunami, an imaginary Barrier that has the same height with the tsunami will be considered, and the concept of the equivalent porosity ratio of the Barrier (λ_B') and the equivalent loss factor of the Barrier (η_B') will be introduced. As illustrated in Figure 15, λ_B' will be obtained by averaging the porosity ratio of the actual Barrier above its actual height (= 0%) and that below its actual height (= λ_B) over the height of the imaginary Barrier, as expressed by:

$$\lambda_B' = \frac{1 \times (h_F - H) + \lambda_B H}{h_F} = 1 - \frac{H}{h_F}(1 - \lambda_B) \quad (15)$$

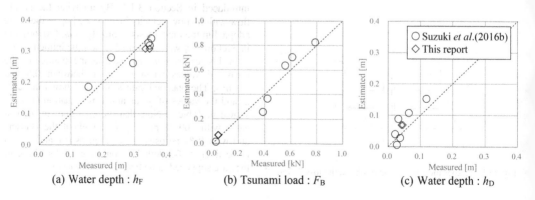

(a) Water depth : h_F (b) Tsunami load : F_B (c) Water depth : h_D

Figure 14. Comparison of measured and estimated results (with tsunami not overflowing the Barrier).

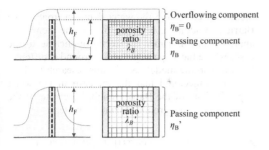

Figure 15. Correction of porosity ratio when tsunami overflows the Barrier.

Table 5. Input values for designing the Barrier overflown by tsunami.

Case No.	Flow velocity v_U [m/s]	Water depth h_U [m]	Loss factor[*] η_B	Corrected Loss factor η_B'
Su-1	1.36※	0.189※	0.1	0
Su-2			6	6
Su-3			147	48
Su-4			14	14
Su-5			353	88
Su-6			10000	76

*Suzuki et al., 2016b

Replacing λ_B and η_B in Equation (13) by λ_B' and η_B', η_B' is expressed as:

$$\eta_B' = \frac{\lambda_B^2}{\lambda_B'^2}\eta_B = \eta_B \frac{\lambda^2}{\left(1 - \frac{H}{h_F}(1-\lambda_B)\right)^2} \quad (16)$$

Adopting λ_B' and η_B' obtained by Equations (15) and (16) instead of λ_B and η_B in Equations (3), (4), (8), (9), (10) and (14), along with the input values for h_U, v_U and η_B', the values of the other 6 parameters (C, F_B, h_F, h_D, v_F, v_D) can be obtained.

3.2.2 Comparison with the model test results

The conditions and the results of the model tests conducted by Suzuki et al. (2016b), where tsunamis overflew the Barrier, are summarized in Table 4. Figure 16 shows the comparison of the water depth at the "Front" position (h_F), tsunami load on the Barrier (F_B) and the water depth at the "Downstream" position (h_D) measured in the model tests and estimated by the theoretical approach based on the input values shown in Table 5. Regarding h_D, a good agreement can be found, with a slight overestimating tendency (which

leads to a slightly conservative design). On the other hand, h_F and F_B were underestimated. This might be partly because the instant increase of the water depth in front of the Barrier just after it was hit by the tsunami is not considered in the theoretical approach.

To cope with the underestimating tendency of h_F and F_B under the condition of tsunami overflowing the Barrier, it was attempted to obtain the corrected water depth (h_{F_o}) based on the correlation between the estimated and measured h_F. Considering the trend line obtained by the least-square method as shown in Figure 16(a), h_{F_o} is expressed as:

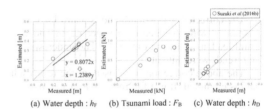

(a) Water depth : h_F (b) Tsunami load : F_B (c) Water depth : h_D

Figure 16. Comparison of measured and estimated results (with tsunami overflowing the Barrier).

Table 4. Test conditions and results (surge-type tsunami overflowing the Barrier).

Case No.	Porous material	Porority ratio λ	λ_B	Loss factor[*] η	Water depth at "Front" h_F[m]	Tsunami load on the Barrier F_B[kN]	Water depth at "Downstream" h_D[m]	Remarks
Su-1	-	100%	100%	0.1	0.189	0.037	0.169	Suzuki
Su-2	Punching metal	33%	33%	6	0.385	0.572	0.089	et al.
Su-3	Punching metal	10	10	147	0.432	0.942	0.085	(2016b)
Su-4	Grass cloth	40%	40%	14	0.412	0.778	0.066	
Su-5	Grass cloth	10%	10%	353	0.436	1.120	0.044	
Su-6	Steel plate	0%	0%	10000	0.518	1.384	0.046	

*Suzuki et al., 2016b.

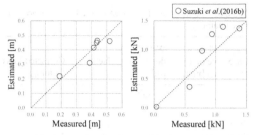

(a) Water depth : h_F (b) Tsunami load : F_B

Figure 17. Comparison of measured and estimated results (with tsunami overflowing the Barrier, with corrected hF in estimation).

$$h_{F\_o}= 1.25h_F \qquad (17)$$

Substituting Equation (17) into Equation (4), the corrected tsunami load on the Barrier (F_{B_o}) is obtained by:

$$F_{B\_o}=\frac{1}{2}\rho g B\left(h_{F\_o}{}^2 - h_D{}^2\right)+\rho B\left(h_{F\_o}v_F{}^2 - h_D v_D{}^2\right) \qquad (18)$$

The corrected water depth and tsunami load were compared with the measured water depth and tsunami load, as shown in Figure 17. The underestimating trends were mitigated.

The above correction was based on the test result and is empirical. It is a future work to generalize the method of correcting the water depth in front of the Barrier. In addition, it is unknown if the theoretical approach in this paper is directly applicable to the cases where the slope angle of the ground is different from that in the model tests in this research (= 1/7), as it is conjectured that the increase in the water depth in front of the Barrier might be greater if the slope angle of the ground becomes smaller.

4 SCALE EFFECT ON THE LOSS FACTOR

In this section, the scale effect on the loss factor is briefly discussed, to confirm the possibility of expanding the findings in this research to actual (full-scale) projects.

According to several researches on the pressure loss when a fluid passes through small apertures (Hamaguchi et al., 1982; Hasegawa et al., 1987), the friction factor (f) becomes constant if the Reynolds number (R_e) obtained by Equation (19) is greater than 500.

$$R_e=\frac{lu}{\nu} \qquad (19)$$

Here, l is the representative length (the size of a small aperture or the porous part of the Barrier), u is the flow velocity of a liquid in the small apertures and v is the coefficient of kinematic viscosity of the liquid.

Table 6 shows the values of l, u and R_e in the model tests introduced in this paper, where l was taken as the aperture of the porous part of the Barrier and u was obtained by Equation (6) using the flow velocity (v_F) recorded when the tsunami load on the

Table 6. Values of Reynolds number in the model tests.

Case No.	Over flow	Porority ratio λ	Aperture width l[mm]	Flow velocity v_F[m/s]	Flow velocity u^*[m/s]	Reynolds number R_e
Su-1	No	100%	-	1.277	-	-
Su-2	No	33%	3.0	1.045	3.167	9502
Su-3	No	10	4.0	0.738	7.377	29507
Su-4	No	40%	0.8	0.013	2.533	2027
Su-5	No	10%	1.2	0.666	6.656	7987
Su-6	No	0%	-	0.611	-	-
To-4(1)		24%	0.8	0.611	2.554	2035
To-4(2)			0.8	0.611	2.554	2043
Su-1	Yes	100%	-	1.368	-	-
Su-2	Yes	33%	3.0	0.765	2.319	6958
Su-3	Yes	10	4.0	0.302	3.023	12092
Su-4	Yes	40%	0.8	0.703	1.758	1407
Su-5	Yes	10%	1.2	0.446	4.463	5356
Su-6	Yes	0%	-	0.330	-	-

* $u = v_F/\lambda$ (Suzuki et al., 2016b)
Dynamic viscosity coefficient of water: $v = 0.000001$ [m²/s]

Barrier (F_B) increased to its initial local peak. The value of v was taken as 1.0×10^{-6} m/s. It can be confirmed that the R_e values in the model tests exceeded 500, which would be suggesting that the scale effect on the loss factor can be ignored in this model test and the findings in this paper can be applied to the full-scale prototype.

When designing the prototype Barrier based on the methods in this paper, it is desirable to obtain the value of the loss factor (η) of the porous sheet by conducting the experiment introduced in Section 3.1.1 under the condition of the R_e value controlled similar to that in the prototype (desirably by using the actual porous sheet). Even if this is difficult, it is necessary to confirm that the R_e values of the prototype and the experiment are greater than 500.

5 CONCLUSIONS

Results of the two-dimensional hydraulic model tests on the Barrier, where the load of a surge-type tsunami behind the Barrier was measured, were reported. The Barrier with the porosity ratio of 24 % was confirmed to reduce the tsunami load behind the Barrier by 75% if the porous sheet was flat and by 80% if the porous sheet was flexural.

A theoretical approach to design the Barrier, based on a model where the Barrier is placed on a flat ground and the tsunami flows without overflowing it, as proposed by Suzuki et al. (2016b), was explained. In this theoretical approach, the loss factor was introduced to simplify the design process. An experimental method to obtain the loss factor was also introduced. This theoretical approach was then expanded to be applied to the condition where the tsunami overflows the Barrier, by introducing the equivalent porosity ratio and the equivalent loss factor.

The validity of the theoretical approaches was confirmed by comparing with the results of the model tests newly introduced in this paper as well as those conducted by Suzuki et al. (2016b). As a result, good agreement between the estimated and measured values of the water depth in front of and behind the Barrier and the tsunami load on the Barrier was confirmed for the cases with the tsunami not overflowing the Barrier. On the other hand, for the cases where the tsunami overflowed the Barrier, good agreement between the estimated and measured values of the water depth behind the Barrier was confirmed, while the water depth in front of the Barrier and the tsunami load on the Barrer were underestimated. It was confirmed that these underestimating trends were mitigated if the water depth in front of the Barrier was corrected based on the model test results.

The scale effect on the loss factor was discussed, and it was suggested that the theoretical approach in this paper would be applied to designing the prototype scale Barrier if the porous sheet was selected so that the Reynolds number become greater than 500.

Further research is necessary to investigate the applicability of the theoretical approach to the cases where the Barrier is placed on the ground with its slope angle being different from that in the model tests in this research (= 1/7). In addition, additional model tests are necessary to investigate the effectiveness of the Barrier under the steady flow (i.e. a tsunami with a long wave length).

REFERENCES

Hamaguchi, K., Takahashi, S. and Miyabe, H. 1982. Flow losses of regenerator matrix: case of packed wire gauges. *Transactions of the Japan Society of Mechanical En-gineers, Series B*, Vol. 48, No. 435: 2207–2216. (in Japanese)

Hasegawa, T., Fukutomi, K., Narumi, T. 1987. A study of a flow through small apertures (1st report, Experiments on the excess pressure drop). *Transactions of the JSME, Series B*, Vol.53, No.496: 3510–3515. (in Japanese)

Ishihara, Y., Okada, K. and Hamada, M. 2018. Comparison of pile-type and gravity-type coastal levees in terms of resilience to tsunami. Proceedings of the First International Conference on Press-in Engineering 2018, Kochi: 251–256.

Kikuchi, Y., Kawabe, S., Taenaka, S. and Moriyasu, S. 2015. Horizontal loading experiments on reinforced gravity type breakwater with steel walls. *The 15th Asian Regional Conference on Soil Mechanics and Geotechnical Engineering, Japanese Geotechnical Society Special Publication*, Vol. 2, Issue 35: 1267–1272.

Ministry of Land, Infrastructure, Transport and Tourism (MLIT). 2013. Guideline for Tsunami-Resistant Design of Seawall (Parapet Wall): 18p.

Suzuki, N., Ishihara, Y. and Isobe, M. 2016a. Experimental study on tsunami mitigation effect of breakwater with arrays of steel tubular piles. *Journal of Social Safety Science*, No. 29, 2016.11: 7–14. (in Japanese)

Suzuki, N., Ishihara, Y., Isobe, M. 2016b. Experimental study on influence of porosity and material of Pile-type porous tide barrier on its tsunami mitigation effect. *Journal of Japan Society of Civil Engineers. Ser. B3 (Ocean Engineering)*: I_491-I_496. (in Japanese)

Session D: Case histories

*Proceedings of the Second International Conference on
Press-in Engineering 2021, Kochi, Japan – Matsumoto et al (eds)
© 2021 Taylor & Francis Group, London, ISBN 978-1-032-10414-0*

Press-in piling applications: Permanent stabilization of an active-landslide-slope

M. Yamaguchi, Y. Kimura & T. Nozaki
GIKEN LTD., Tokyo, Japan

M. Okada
GIKEN LTD., Osaka, Japan

ABSTRACT: The road surface of a national highway failed in Japan, due to a landslide caused by typhoons. For the restoration of the road, the installation of steel tubular piles with ground anchors was selected to protect traffic safety in consideration of the possible occurrence of soil mass erosion and loss at the toe of the slope. In this construction project, the quick restoration of one-way alternating traffic was required by installing the piles without delay. Furthermore, installing the piles into the bedrock (bearing stratum) was a technical challenge. To address the requirement and technical issue, the rotary press-in piling (Gyropress Method) and the Non-staging System were selected. This paper aims to review the applicability of press-in piling technology to the pile installation on an active-landslide-slope by reporting the outlines of the disaster rehabilitation work, design of permanent measures, and the construction plan and implementation.

1 INTRODUCTION

In Japan, natural disasters are increasing and serious damage has been caused by record-breaking heavy rainfall brought by very large typhoons such as the typhoon that hit the Boso Peninsula and another that moved through eastern Japan in 2019. It is an urgent issue to restore damaged urban infrastructure quickly and make reconstructed structures stronger against disasters than before. Walk-on-pile type press-in piling is one of the piling methods that can overcome this issue effectively. In particular, the rotary press-in piling (Gyropress Method), one of the press-in piling methods that rotates a steel tubular pile with pile toe ring bits and press-in into the ground, and the Non-staging System, a system of press-in piling methods that allows all necessary construction machines and equipment to be operated on the previously installed piles so that it is suitable for sites with various constraints, such as on slopes, irregular ground, and water, are increasingly being selected at many construction sites.

In the section of national highway No. 19 reported in this paper (hereinafter referred to as Route 19) in the Minochi district, Shinshushinmachi, a road surface failure occurred because of a landslide caused by Typhoon Nos. 21 (Asian name: *Lan*) and 22 (Asian name: *Saola*) in 2017. The damaged section

of the highway needed to allow at least one-way alternating traffic during rehabilitation work to serve as an emergency transportation route in the post-disaster period. Under such conditions, steel tubular piles with ground anchors were installed using the aforementioned technologies as a permanent measure. The details of the rehabilitation work are described as follows.

2 OUTLINE OF DAMAGE

2.1 Outlines of Route 19 in the Minochi district, Shinshushinmachi

Route 19 runs 272.6 km from Nagoya City, Aichi Prefecture to Nagano City, Nagano Prefecture. It runs through precipitous mountains along the Sai River between Ikusaka Village and Nagano City, Nagano Prefecture. In addition, the ground is not firm enough in this section. Therefore, the section is vulnerable to natural disasters such as landslides, debris flow, and slope toe scouring, and disaster-prevention projects have been conducted by the Ministry of Land, Infrastructure, Transport and Tourism (MLIT).

The Minochi district, Shinshushinmachi, reported in this paper is located at an elbow of the Sai River (a first-class river) passing under Route 19 (Figure 1).

DOI: 10.1201/9781003215226-45

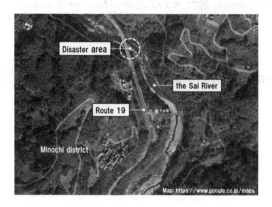

Figure 1. Site location (Map: https://www.google.co.jp/maps).

2.2 Weather conditions brought by Typhoon Nos. 21 and 22

2.2.1 Typhoon No. 21

Typhoon No. 21 (Asian name: *Lan*) made landfall near Omaezaki City, Shizuoka Prefecture, on 23 October 2017, at about 3 a.m. with an extremely large size and strength and proceeded through the Tokai and Kanto regions north-east-ward while maintaining its storm area before reaching off the coast of Fukushima Prefecture on the same day at 9 a.m. It transformed into an extratropical cyclone in the east of Hokkaido. Its central pressure was 950 hPa, and the maximum wind speed was 40m/s at the time of landfall. It was the first time for Japan to be directly hit by such a super-scale typhoon since 1991 (NLMO, 2017).

Because of this typhoon and a front stimulated by it, Nagano Prefecture was hit by intense rainfall, mainly in its north and south areas. A rainfall of 112.0mm/h was recorded in Shinshushinmachi where the site of the road failure was located.

2.2.2 Typhoon No. 22

Typhoon No. 22 (Asian name: *Saola*) which occurred near the Mariana Islands moved northward in the Okinawa region with its storm area on 28 October 2017, and proceeded northeast over the south sea of Honshu island on the 29th while maintaining its power. It became an extratropical cyclone off the Sanriku coast on 30 October 2017, at midnight. A central pressure of 975 hPa and a maximum wind speed of 30m/s were recorded.

Unlike Typhoon No. 21, not many areas were hit by intense rainfall because the path of the typhoon was far off the coast. However, Miyazaki Prefecture received record-breaking heavy rain (NLMO, 2017).

It was unusual for multiple typhoons to make landfall or approach Japan in a short period of time in late October. Because of the weather brought about by Typhoon Nos. 21 and 22, road surface failures occurred in the area in this report due to a landslide.

2.2.3 Circumstances of damage

A landslide with a width of 40m, slope length of 40m, and a slip surface depth of 10m occurred in the Minochi district, Shinshushinmachi. The top of the landslide was located on the road, and a bridge (Mizushino Bridge) was beside it. The end of the landslide was found to have reached the riverbed of the Sai River. Because of this landslide, stepwise cracking occurred in the part of Route 19 from its centre line to a 2m distance toward the mountainside.

For a while after the landslide, the river side slope of the road was moving at a speed of a few centimetres a day. Then, the speed quickly increased to 10-70 cm a day due to the rise of groundwater level and riverbed erosion at the lower end of the slope due to rainfall (Kusatani et al. 2019).

3 DESIGN OF PERMANENT MEASURES

3.1 Design of permanent measures[1)3)]

The following three plans were proposed and considered as permanent measures for the disaster site (Kusatani et al. 2019).

3.1.1 Counterweight fill method

The principle of the counterweight fill method is to increase a resistance against the land sliding force by forming an embankment at the lower end of the sliding soil mass (Figure 2) (JASDM,2020). If there is a potential sliding plane in the slope under the embankment, it may induce the occurrence of the landslide. For this reason, the stability of the foundation of the embankment needs to be examined when considering a counterweight embankment.

If the permeable layer of groundwater lies at a shallow depth in the embankment section, or if the groundwater seeps out at the lower end of the landslide slope, the handling of the groundwater shall be considered because there is a risk of increased instability of the slope due to the blockage of groundwater outlets by the embankment or its load, or the rise of the groundwater level on the back of the slope.

In addition, the toe of the landslide slope was too adjacent to the Sai River, and large-scale earthwork would be required within the river, which would have a large impact on the surrounding environment. As a result, the counterweight fill method was determined not to be suitable for this site.

3.1.2 Installation of a prefabricated platform

An idea to install a prefabricated platform in the sunken part to act as a road was discussed.

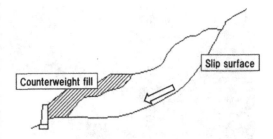

Figure 2. Counterweight fill method.

This idea was determined not to be suitable for this site because continuous movement of the landslide slope was observed, even after the initial landslide, and it was difficult to secure a workspace on the landslide slope for the installation of the prefabricated platform.

3.1.3 Installation of steel tubular piles with ground anchors

Piling work is planned for the purpose of adding shear and bending resistance to the ground by inserting piles into the rigid ground so that those resistance forces will be able to directly bear the sliding force of the sliding soil mass. Steel tubular piles (hereinafter referred to as tubular pile) are typically often used in landslide-prone areas. The use of tubular piles with an outside diameter of over 1000mm has been started recently, which is applicable to civil work sites that require a large resistance force against landslides. In addition, it is possible to integrally stabilize the tubular piles and the ground by installing an anchor on the pile head (Figure 3) so that the load acting on the anchor head will be transmitted to the anchored ground via the tension member (JASDM,2020).

With this structure, the safety of the road can be ensured even if the soil mass on the river side at the lower end of the landslide slope is lost by erosion. Furthermore, it was considered possible to install tubular piles without a temporary platform on the slope, if the rotary press-in piling and the Non-staging System described below, which would enable the installation of tubular piles into the bearing stratum (stratified sandstone, CM class), were used in combination. On the basis of this consideration, the installation of steel tubular piles with ground anchors was selected as a permanent measure.

3.2 Guidelines and manuals for designing tubular piles with ground anchors

The following literatures were consulted for the structural design of a retaining wall composed of tubular piles with ground anchors in the construction reported in this paper.

- Japan Association for Slope Disaster Management (JASDM) 2003. *Design Manual for Landslide Prevention Steel Pile Retaining Wall* (in Japanese)
- Japan Association for Slope Disaster Management (JASDM) 2008. *Design Manual for Landslide Prevention Technologies* (in Japanese)
- Japanese Geotechnical Association (JGA) 2013. *Ground Anchor Design and Construction Standards, Commentary* (in Japanese)
- Japan Road Association (JRA) 2010. *Road Earthwork - Embankment Construction Guide-lines* (in Japanese)
- Japan Road Association (JRA) 2012. *Road Earthwork - Retaining Wall Construction Guidelines* (in Japanese)

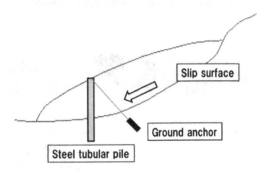

Figure 3. Steel tubular pile with ground anchor.

4 INTRODUCTION OF TUBULAR PILE INSTALLATION METHOD

4.1 Rotary press-in piling (Gyropress Method)

Rotary press-in piling is a piling method to press-in a tubular pile with pile toe ring bits while rotating it. It is applicable not only to granular and cohesive soils but also to hard ground such as rock and obstacles (e.g. reinforced concrete) because the rotated piles cut and penetrate into the ground to the specified pile toe depth (Figures 4 and 5). The applicable range of the outside diameter of a tubular pile is from 600 to 2,500mm.

In addition, equal angle steels or small diameter steel pipes can be installed by rotary press-in piling machinery in the gaps between the installed piles to prevent soil loss, and to ensure the watertightness of the cofferdam (IPA, 2020).

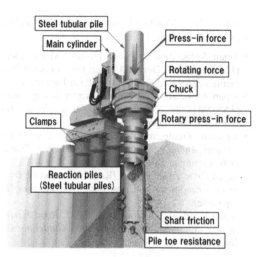

Figure 4. Rotary press-in piling (Gyropress Method).

Figure 5. Cutting the reinforced concrete with the rotary press-in piling.

4.2 Non-staging system

By using the Non-staging System, which enables all the piling processes including the pile transportation, pitching, and press-in piling to be carried out on the installed piles, areas affected by the piling work can be limited to those occupied by the machines operated on the installed piles, and those used for the initial construction base. Furthermore, the machines are self-supporting by gripping the installed pile, and the risk of falling is extremely low (IPA, 2020).

The machine layout of the Non-staging System used at the construction site in this report is shown in Figure 6.

Figure 6. Machine layout of the Non-staging System.

5 DETAILS OF TUBULAR PILE INSTALLATION

5.1 Construction plan

The disaster rehabilitation work for Route 19 in the Minochi district, Shinshushinmachi, conducted in 2018 is reported. In this work, the installation of tubular piles (total 42 of SKK400 made piles with an outside diameter of 800mm and a length of 31.5m, two splices per pile) was planned as a permanent measure because the height difference between the ground surface after the landslide and the newly planned road surface was as large as 12.6m. In addition, removal of existing structures such as the cutting and crushing of road pavement and temporary work of traffic control equipment were also planned. In addition, the construction of a temporary earth retaining wall (500mm wide U-shaped steel sheet piles with a length of 11.5m, type SP-V L, 40 piles) was planned as an additional emergency measure in the other area away from the tubular pile wall.

At the stage of the construction planning, the implementation of the following works in 7 months from the middle of May to the beginning of December in 2018 was planned: detailed planning, structure design review, site reconnaissance, preparatory work such as land surveys, removal of structures, installation of tubular piles, and temporary work. (as an emergency measure, a temporary earth retaining wall was planned to be constructed in October). Estimated work periods were 2 months for the manufacture of tubular piles and 2.5 months for their installation among these works. Note that this paper explains the installation of tubular piles and does not deal with the work for closure piles and ground anchor installations.

5.2 Ground condition

A typical cross-sectional view of the installation of steel tubular piles with ground anchors that was created on the basis of existing geological survey materials and the result of a Standard Penetration Test (Figure 7) is shown in Figure 8.

Pressing in tubular piles into the bearing stratum (stratified sandstone, CM class) to a depth by at least 1 diameter (m), so-called 1D, was a mandatory requirement.

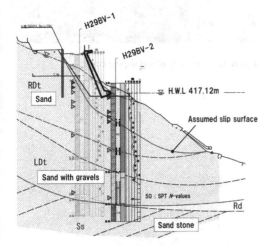

Figure 7. Cross section of soil layers.

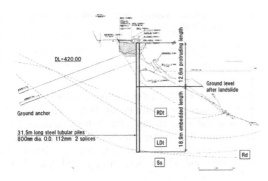

Figure 8. Typical cross section of the installation of tubular piles with ground anchors.

5.3 Pile installation procedures

Tubular piles were installed according to the following procedures:

[1] Secure spaces for a crawler crane and tubular pile storage on the river side of the road by allo-cating traffic control guards and allowing only one-way alternating traffic.

[2] Install a crawler crane (70 ton lifting capacity class) at the secured space, and carry equipment for the installation of reaction piles in the initial piling location.

[3] Press-in steel sheet piles as reaction piles (Figure 9).

[4] Remove equipment used for the installation of reaction piles, and carry equipment for the installation of tubular piles in the initial piling location by a truck and a crane.

[5] Install a Reaction Stand for tubular pile press-in work, and fix it to the reaction piles (Figure 10).

[6] Assemble the rotary press-in piling machine (hereinafter referred to as Gyro Piler), which was carried in the site in three parts, on the Reaction Stand (Figure 11).

[7] Install 15 tubular piles using the Gyro Piler and the crawler crane.

[8] Install the Non-staging System (such as clamping crane, Pile Runner, and transportation trackway) on the installed tubular piles (Figure 12).

[9] Install 37 tubular piles using the Non-staging System (Figure 13).

[10] Have all the equipment used for the installation of the tubular piles self-walk backward to the position within the outreach of the crawler crane after the completion of all piles installation.

[11] Dismantle and remove all equipment.

The construction was completed in 7 months from May 10 to December 5, 2018, according to the original plan.

5.4 Quality control

5.4.1 Steel tubular piles

A quality inspection was performed on prefabricated piles after tubular piles were delivered to the site from a steel manufacturer.

A visual inspection was performed at the site to make sure there were no defects such as deformations. Inspection results of tubular pile dimensions (actual measurements and pass/fail assessments against the tolerances) are shown in Table 1 for pile Nos.1, 10, 20, 30, and 40 as typical examples. On the basis of the inspection results, it was confirmed that the outside diameter, thickness, and length of all tubular piles met the Japanese Industrial Standards.

Figure 9. Installation of reaction piles.

Figure 12. Assembly of a clamping crane.

Figure 10. Installation of a Reaction Stand for tubular piles.

Figure 13. Steel tubular pile installation using the Non-staging System.

Figure 11. Assembly of a Gyro Piler.

5.4.2 *Welded splices of steel tubular piles*

Tubular piles with a length of 31.5m were needed because they had to reach the bearing stratum. For this reason, the tubular piles were transported to the site in three parts, each of them having a length of 9.0, 10.5, and 12.0m, respectively, and spliced by self-shielded arc welding.

Inspection points and the results of welded splices formed along the circumference of the tubular pile are shown in Figure 14 and Table 2 for pile Nos.1, 10, 20, 30, 40, and 41 as typical examples. A visual inspection and a dye penetrant test were performed on every welded splice, and an ultrasonic test was performed on a welded splice randomly selected from every 20 splices. The outlines of the dye penetrant test and ultrasonic test are shown in Tables 3 and 4.

Table 1. Actual values of steel tubular piles dimensions.

Pile No.	Pile Parts	Diameter (mm)	Thickness (mm)	Length (mm)	Total length (mm)	Assesment
0	Bottom	801	12.22	12.029	31.588	Pass
	Middle	800	12.25	9.026		
	Upper	800	12.2	10.533		
1	Bottom	799	12.12	10.535	31.603	Pass
	Middle	802	12.11	9.032		
	Upper	801	12.13	12.036		
10	Bottom	800	12.09	12.03	31.590	Pass
	Middle	800	12.2	9.03		
	Upper	801	12.08	10.53		
20	Bottom	800	12.07	12.035	31.608	Pass
	Middle	800	12.11	9.035		
	Upper	801	12.09	10.538		
30	Bottom	800	12.19	12.04	31.594	Pass
	Middle	800	12.24	9.027		
	Upper	801	12.15	10.527		
40	Bottom	801	12.23	12.035	31.595	Pass
	Middle	800	12.26	9.025		
	Upper	800	12.29	10.535		

Table 2. Inspection results of each welding part.

Pile No.	Splice No.	Welding location	Visual inspection	Dye penetrate test	Ultrasonic test
1	1	Bottom-Middle	Pass	Pass	
	2	Middle-Upper	Pass	Pass	
10	19	Bottom-Middle	Pass	Pass	
	20	Middle-Upper	Pass	Pass	
20	39	Bottom-Middle	Pass	Pass	
	40	Middle-Upper	Pass	Pass	
30	59	Bottom-Middle	Pass	Pass	
	60	Middle-Upper	Pass	Pass	
40	79	Bottom-Middle	Pass	Pass (C)	Pass (C)
	80	Middle-Upper	Pass	Pass (C)	

- Tolerance (Japan Industrial Standards JIS A5330-1994)Outside diameter: ±0.5%, thickness: -0.8 mm (no tolerance specified on the plus side) Length: No tolerance specified on the plus side
- Designed dimensions of the tubular pileOutside diameter: 800 mm, thickness: 12 mmLength: 12.0 m/10.5 m, 9.0 m, and 10.5 m/12.0 m for the bottom, middle, and upper parts, respectively

- Descriptions with the index (C) are results of inspections performed by certified personnel.
- The dye penetrant test was performed on every welded splice and performed by certified personnel on more than 10% of them. As for the ultrasonic test, it was performed by certified personnel on a welded splice randomly selected from every 20 splices.

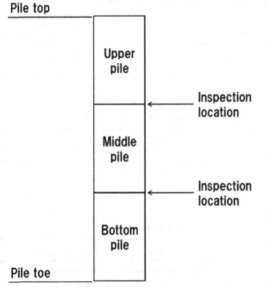

Figure 14. Inspection location of welded splices.

Table 3. Specification of dye penetrant test.

Applicable standard	JIS-Z-2343	Criterion	There is no serious defect occurred by cracks and so forth.
Dye penetrant	Penetrant	R-1AH NT	
	Developer	R-1SH NT	
	Remover	R-1MH NT	
	Manufacture	Eishin Kagaku Co.,Ltd.	
Method & condition	Temperature	30 °C	
	Pre-cleaning	Solvent cleaning with wire brush	
	Penetrant time	10 min or longer	
	Developing time	10 min or longer	

463

Table 4. Specification of ultrasonic test (RSEC, 2020).

● Ultrasonic flow meter

Name	RYOSHO UI-S7	
Serial number	U107AS7164	
Inspection date	2017.11.30	
Amplitude linearity	JIS-Z-2353	+0.5%, -1.1%
time linearity	JIS-Z-2352	
Amplitude linearity line	JIS-Z-2353	+0.5%, -1.1%
DAC No use	Rejection	Off

● **Probe**

Manufacture	PROBE-KGK
Serial number / Name	XA3666· 5Z10x10A70
Inspection date	30 November 2017
STB, Angle of refrection, Probe index	70° 10mm
Dead area	3mm
Resolution, A2sensitivity	4mm, 25.5dB

● **Condition**

Surface state	Texture after removing spatters
Test piece	STB-3A, RB-41
Penetrant range	0~1.0skip
Requirement of sensitivity	RB-41 H1000 80%
Correction of sensitivity	0 dB
Criterion	JIS-Z-3060

5.5 Quality control of installed steel tubular piles

Quality control was conducted on the following construction parameters concerning dimensional accuracy in order to control the compliance of the installed tubular piles with the standards and criteria specified by the client/owner: a) height of the top level, b) embedded depth, c) deviation in plan and d) inclination (KRDB, 2020).

5.5.1 Height of the pile top level

The height of the pile top level was designed for each tubular pile as a specified level (design value) according to the undulation of the surface of the national highway running parallel to the arrangement of the piles. In this project, the height of each tubular pile at the end of the press-in work was controlled using an elevation-measuring instrument (Total Station) to meet the reference levels (control values) within ±50.00 mm tolerance.

The control value and actual measurements (mean value, maximum plus and minus values, and standard deviation) of the installed piles' height are shown in Table 5. The height levels ranged from -18.00 to 12.00mm with a mean of

-4.00mm and a standard deviation of +6.51mm. The data demonstrated that tubular pile installation by rotary press-in piling could achieve high accuracy to the specified levels.

5.5.2 Embedded depth

Inspection results of the embedded depth of the tubular piles are shown in Table 6. It was a mandatory requirement to achieve the specified embedded depth, and all the installed tubular piles were found to have met the control value.

5.5.3 Deviation in plan

Inspection results of the deviation in plan of the tubular piles installed at this site are shown in Table 7.

The required control value was within 100mm. The measurement data showed that the deviations in plan were in a range of 7.00 to 74.00mm with a mean of 30.00mm and a standard deviation of ±19.69, which demonstrated the sufficient piling accuracy.

5.5.4 Inclination

Measurement results of the inclination of the installed tubular piles are shown in Table 8. The control value of inclination was within ±1%.

Table 5. Assessment: Height of pile top levels.

Item	Unit	Control value	Mean value	Max. value	Min. value	Standard deviation	Sample No.	Assessment
Pile top level	mm	±50	-4	12	-18	±6.51	42	Pass

Table 6. Assessment: Embedded depths.

Item	Unit	Control value	Mean value	Max. value	Min. value	Standard deviation	Sample No.	Assesment
Embedded length	mm	18900 or over	19003	19035	18964	±13.06	42	Pass

Table 7. Assessment: Deviations in plan.

Item	Unit	Control value	Mean value	Max. value	Min. value	Standard deviation	Sample No.	Assesment
Deviation in plan	mm	100	30	74	7	±19.69	42	Pass

Table 8. Assessment: Inclinations.

Item	Unit	Control value	Mean value	Max. value	Min. value	Standard deviation	Sample No.	Assesment
Inclination	%	1.00	0.20	0.60	0.00	±0.16	42	Pass

The measurement data showed that the inclination of each of the installed tubular piles was in a range of 0 to 0.6% with a mean of 0.2% and a standard deviation of ±0.16, which demonstrated the successful installation of the tubular piles with sufficient verticality.

5.6 Originality and ingenuity in piling and progress management work

5.6.1 Use of pile roller for tubular piles

It was necessary to crane the tubular piles with care because the work space available right next to the one-way alternative traffic was extremely limited. The way to solve this issue with originality and ingenuity was the use of a Pile Roller for tubular piles (Figure 15).

By using a Pile Roller, the tubular piles were able to be lifted smoothly with reduced friction between metallic objects. In addition, the range of the crane and lifted piles' movement could be minimized so that safety was enhanced.

5.6.2 Use of acrylic board for measuring pile deviation

In order to perform the measurement of the deviation in the plan of the installed tubular piles quickly, an acrylic measuring board was used instead of a traditional measurement using a wooden board (Yamaguchi et al. 2019). On this acrylic measuring board, a large circle (to be matched with the actual pile annulus), its centre and another small circle (with its centre being identical with the centre of the large circle and its radius being the maximum permissible deviation) are indicated (Figure 16).

The specified pile centre was measured using a transit. The two cross points between the pile and a line going through the specified pile centre were marked on the pile top. Then a taut line was fixed on the pile top with magnets by linking the two marks. Doing this in two directions, the specified pile centre was visualized as the cross point of the two taut lines. After that, an acrylic measuring board was placed on the pile top by matching the large circle on it with the annulus of the actual pile. As a result, the positional relationship between the specified pile centre (the cross point of the two taut lines) and the actual pile centre (the centre of the small circle on the acrylic board) was visualized. The visualized distance between the specified pile centre and the actual pile centre was easily judged as being allowable, by confirming that the visualized specified pile centre (the cross point of the two taut lines) was inside the small circle. In addition, the distance between the specified and the actual pile centres was easily measured in X and Y directions on the acrylic board by using a ruler.

In this project, the prime and piling contractors created originality and ingenuity including the aforementioned two examples and made a regional contribution through their collaborative efforts. As a result, this project was commended by the Nagano National Highway Office (of the MLIT Kanto Regional Development Bureau) along with the companies having performed it as an excellent construction that achieved significant outcomes and as excellent construction engineering companies among the constructions completed in fiscal year 2018 (NNHOKRDB, 2018).

Figure 15. Lifting a tubular pile with a Pile Roller.

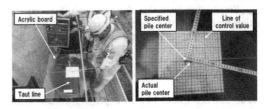

Figure 16. Acrylic measuring board and measurement of pile deviation: Left: Measuring, Right: Example. Gap between specified and actual pile centres.

6 CONCLUDING REMARKS

The following benefits were demonstrated for the installation of tubular piles with ground anchors at landslide disaster sites as a permanent measure and for the combined use of the rotary press-in piling and the Non-staging System that realized the installation of tubular piles at disaster sites:

- A continuous wall constructed with tubular piles with ground anchors is capable to prevent further landslides after the occurrence of the initial landslide even if the soil mass at the edge of the toe of the slope on the river side is lost due to erosion or other reasons.
- By using the rotary press-in piling, it is possible to press-in tubular piles into CM class stratified sandstone (bearing stratum).
- By using the Non-staging System, tubular piles can be installed at active-landslide-slope without additional structures such as a temporary platform.
- By using tubular piles manufactured in plants, it is possible to construct a high-strength and high-quality continuous wall.

- Tubular piles installed by the rotary press-in piling are capable to meet the control values of height of the top levels, embedded depths, deviations in plan and inclinations with a sufficient margin.

In disaster rehabilitation works, it is often the case that a working space is limited and a significant amount of time and costs are spent on the installation of a large-scale temporary platform and embankment. Under such conditions, effective measures need to be devised to bring about safety and security to the neighbourhood and not to disturb their economic and social activities such as traffic when proceeding with the works.

It was demonstrated that the combined use of tubular piles, rotary press-in piling, and the Non-staging System reported in this paper was capable to construct a continuous wall composed of tubular piles on an active-landslide-slope while allowing one-way alternating traffic on the road adjacent to the site. It was also demonstrated that construction cost and duration was reduced because the installation of ancillary facilities such as a temporary platform was not required.

Finally, we hope that the press-in piling application reported in this paper will serve as a reference for similar disaster recovery construction projects.

ACKNOWLEDGEMENTS

The authors would like to express gratitude to the Nagano National Highway Office of MLIT Kanto Regional Development Bureau, Nippon Koei Co., Ltd., and Okayagumi Co., Ltd. for their very helpful support and cooperation in preparing this paper.

REFERENCES

International Press-in Association (IPA) 2020. *Design and Construction Guideline for Press-in Piling*, 540pp. (in Japanese) Tokyo: IPA.

Japan Association for Slope Disaster Management (JASDM) 2003. *Design Manual for Landslide Prevention Steel Pile Retaining Wall.*, 215pp. (in Japanese) Tokyo: JASDM.

Japan Association for Slope Disaster Management (JASDM) 2008. *Design Manual for Landslide Prevention Technologies*, 557pp. (in Japanese) Tokyo: JASDM.

Japan Association for Slope Disaster Management (JASDM) 2020: *Landslide Prevention Construction Method*. Available (online) at: https://www.jasdim.or.jp/gijutsu/jisuberi_gaiyo/taisaku.html (in Japanese) (Retrieved on 15 November, 2020)

Japanese Geotechnical Association (JGA) 2013. *Ground Anchor Design and Construction Standards, Commentary.*, 224pp. (in Japanese) Tokyo: JGA.

Japan Road Association (JRA) 2010. *Guidelines for Road Earthwork, Embankment Construction, 2010 Edition.*, 310pp. (in Japanese) Tokyo: JRA.

Japan Road Association (JRA) 2012. *Guidelines for Road Earthwork, Retaining Wall Construction, 2012 Edition.* 342pp. (In Japanese) Tokyo: JRA.

Kanto Regional Development Bureau, Ministry of Land, Infrastructure, Transport and Tourism (KRDB) 2020. *Standard and Control Values for Civil Work Construction Management*. Available (on line) at: http://www.ktr.mlit.go.jp/ktr_content/content/000743007.pdf (in Japanese) (Retrieved on 15 November, 2020)

Kusatani, Y., Fujiwara, T., Yamashita, T., and Yokoyama, T. 2019. *A construction method applied to road restoration at an active-landslide-slope while allowing one-way alternating traffic*, the 58th Annual Meeting of the Japan Landslide Society, Kumamoto, 2pp. (in Japanese)

Nagano Local Meteorological Office (NLMO) 2017. *Meteorological bulletin for Nagano Prefecture on Typhoon No. 21 in 2017*.Available (online) at https://www.jma-net.go.jp/nagano/topic/topic_2220_1026.pdf (in Japanese) (Retrieved on 15 November, 2020)

Nagano National Highway Office of the Kanto Regional Development Bureau, the Ministry of Land, Infrastructure, Transport and Tourism (NNHOKRDB) 2018. *About the Commendation for Excellent Construction Projects in Fiscal 2018 by the Nagano National Highway Office*. Available (online) at: https://www.ktr.mlit.go.jp/ktr_content/content/000751546.pdf (in Japanese) (Retrieved on 15 November, 2020)

Yamaguchi, M., Kimura, Y., and Morisawa, Y. 2019. *Rehabilitation of the costal road retaining wall using the press-in technology*, Geotechnics for Sustainable Infrastructure Development, Lecture Notes in Civil Engineering (NLCE) volume 62, pp.1323–1330, ISBN978-981-15-2183-6, ISBN978-981-15-2184-3 (eBook), Gateway East: Springer Nature Singapore Pte Ltd.

Ryoden Shonan Electronics Corporation (RSEC) 2020. The operation manual of UI-S7 version2, 80pp. (in Japanese) Available (online) at https://sooki.co.jp/upload/surveying_items/pdf/manual_pdf_051909.pdf (Retrieved on 15 November, 2020)

Proceedings of the Second International Conference on
Press-in Engineering 2021, Kochi, Japan – Matsumoto et al (eds)
© 2021 Taylor & Francis Group, London, ISBN 978-1-032-10414-0

Press-in technology: Advantage of Gyropress method for renovation of the third wharf of Dakar Port in Senegal

Y. Ndoye
GIKEN LTD, International Business Promotion, Tokyo, Japan

Y. Kitano
GIKEN Europe, Amsterdam, The Netherlands

T. Funahara
GIKEN LTD, Manager, International Business Promotion, Tokyo, Japan

ABSTRACT: Dakar Port, located in Senegal, is the second largest port in Western Africa behind Abidjan. It serves as a transit port to inland countries for 14% of its capacity. The 3rd wharf constructed in 1939 was reserved for the Republic of Mali under a bilateral agreement. However, the transit volume of goods had increased 2.5 times from 2010 to 2015 and is expected to continue to increase. The increase in the volume of seaborne goods has caused considerable deterioration of the wharf, which may lead to its collapse. Moreover, large vessels are constrained by draught. This paper describes the construction methodology, named the Gyropress method, used to repair the 350m long quay of the wharf. With the Gyropress method, tubular piles are pressed and rotated with the assistance of cutting ring bits attached to the pile toe. Moreover, berthing and vessel unloading at the wharf were not interrupted during construction.

1 OUTLINE OF THE PROJECT

1.1 Location

Senegal is a small country in West Africa covering a total area of 196,722 km², sharing its borders with Mauritania, Mali, Guinea, Bissau Guinea, and the Atlantic Ocean, in the North, East, South, and West respectively. As of 2013, the population was about 13.5 million and it is expected to reach 25.7 million in 2035 (*National Agency for Statistic and Demography ANSD, Senegal 2013*). Senegal has a semi-arid tropical climate with a rainy season from June to September. Dakar, the capital city, is located at the westernmost point, toward the Atlantic Ocean. Dakar covers only 0.4% of the total area of Senegal, however, 23.2% of the total population is concentrated in this city.

1.2 Project background and objectives

The constant increase in the characteristics of the merchant, cruise and fishing, or even leisure vessels, in particular, the increase in their draught, as well as the demands on port operating equipment, often poses the problem of maintaining port activities, in the oldest harbor basins (*Nicolas R. 2014*). Facing this problem, the port managers have two possible solutions: Building new docks, or strengthening existing ones by increasing the quay depth. The construction of a new wharf is not usually a favorable option, due to the high cost of the project and a lack of space. Thus, port managers often turn to renovation. Quayside renovations are quite complicated due to the location. The quays are the area of loading and unloading of goods, essential operations on which depend several corporations. In addition, the site is often difficult to access so that it requires refurbishment from the seaside. Renovation from the sea is not a preferred option as it would have seriously affected vessel berthing and port operations.

The renovation from the sea could impact on the safety, efficiency, speed, and cost of the project. The structure of quays can be classified into three categories according to their method of construction: Quays on piles, retaining curtains walls quay, and gravity quays. The third wharf of Dakar Port (**Figure 1**) built by a concrete block is a gravity type. The method chosen for the renovation is the Gyropress method. The Gyropress method is an innovative construction method, used in several projects in Japan, particularly for the renovation of the Port of Nagasaki.

DOI: 10.1201/9781003215226-46

Figure 1. Third Wharf of Dakar Port.
Map: https://google.co.jp/maps

1.3 Geotechnical condition

The geotechnical survey conducted onshore and off-shore was allowed to identify three layers from the seabed as shown on **Figure 2**: Plastic and hard marl layer from 1 to 3 meters, a fractured grayish whitish limestone-marl with a moderately low compressive strength from 3 to 12 meters, and fractured grayish

limestone moderately with low compressive strength up to 17 meters. The unconfined compressive strength of the marl and limestone-marl varied from very low to high strength (from 2.6 MPa to 68.8 MPa).

2 STRUCTURE TYPE

Since the quay will be repaired to deepen the depth from -10.0m to -12.0m, for large vessels to berth, a new quay on the seaside was first considered. At the preparatory survey stage (*JICA, 2016*), the con-crete block quay and quay on piles were selected as the structural type for the renovation of the quay based on the "Technical Standards for Japanese Ports and Explanations":

➢ Option 1: Gravity type: L-shaped Cellular block (square block) & Cast-in-place concrete **(Figure 3)**.
➢ Option 2: Tubular pile & self-supporting raker pile **(Figure 4)**.

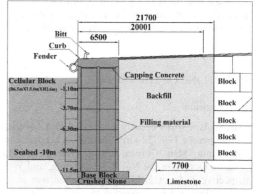

Figure 3. Concrete block quay.

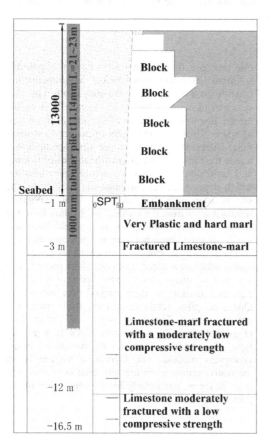

Figure 2. Third Wharf soil condition.

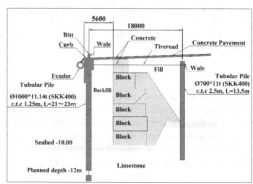

Figure 4. Steel tubular pile quay.

As a result of the comparative examination, the gravity cellular block type quay was initially selected which was similar to the existing quay structure and was planned to be built about 22 m inside the bay from the existing quay. But there was concern that the distance to the opposite quay pier would be narrowed during operation, reducing maritime safety and convenience. In addition, the construction area was divided into two berths, and it was required that the impact on cargo destinated for landlocked countries would be minimized, even when one berth was under construction, by partially using the other berth. Therefore, in the detailed design, the concrete block was changed to a tubular pile type that can reduce the width of the quay inside the bay as shown on **Figure 4**. An additional soil investigation was carried out. The survey revealed that the ground in front of the third wharf was mainly composed of limestone marl and weathered limestone. After the soil survey, the Gyropress method and the All-Casing method were compared.

3 REASON OF SELECTING GYROPRESS METHOD

In the design stage, the large diameter bored pile with temporary steel casing commonly known as the All Casing method was proposed. However, this method would greatly impact the berthing and vessel unloading at the third wharf because it needs to use many working vessels from the seaside. As 95% of the total capacity of the third wharf is destined for Mali and other landlocked countries, stopping or slowing the port operations would have a big impact on those country's economies and food security. In addition, many vessels need to be involved so that, the construction speed and safety would become lower. For all reasons mentioned, the All Casing method and the Gyropress method were compared.

After making a comparative analysis, it was found that renovating the quay with the Gyropress method is more cost effective and needs less time. Moreover, the impact on berthing and vessel unloading could be minimized since it requires less equipment and fewer workers on site. After considering these facts, the Gyropress method was highly evaluated and adopted **(Table 1)**.

4 THE GYROPRESS METHOD

The Gyropress method used in this project is an innovative way for tubular pile installation. In Gyropress Method, a newly developed Silent Piler, called Gyro Piler is used **(Figure 5)**. It installs

Table 1. Comparison between Gyropress method and all casing method.

	Gyropress Method	All Casing Method
Overview	Due to the rotary cutting press-fitting, all processes can be performed on the steel tubular pile. For this reason, adjacent berths can be used even during the construction period, so there is little economic loss. In addition, it does not affect the water area facilities (berths, routes, etc.) in the port.	Using a work boat such as a guide barge or a pile driving ship, a casing is installed first. After placing the steel pipe pile inside, the casing grips it and rotates downward. An excavator is used to remove the crushed stone. This is a general construction method for port piers. Since the work will be carried out offshore by a work boat, the adjacent berths cannot be used during the construction period, so the minimum space is required in advance.
Machines	3 Machines: Gyro Piler, Crawler Crane, power unit.	7 Machines: Casing, guide barge, working barge, crawler crane, hoist ship, tugboat, anchorage ship.
Workability	Since everything is done on land or on piles, the work process and quality are not easily affected by weather and sea conditions. No special skills are required because general construction machines are used.	Work boats need to be circulated and towed. Since it is a work boat, the process and quality are easily affected by weather and sea conditions. It occupies a wide area by using work boats and mooring lines.
Safety	It does not interfere with traffic at sea because no work boat is used.	Due to the large number of work vessels, it is necessary to ensure the safety of vessels moving over the sea.
Environmental	Because it is a vibration-free and noise-free press-fitting method, it has less impact on the surrounding environment.	Since many work boats are used, there is concern about the impact on the natural environment and people.

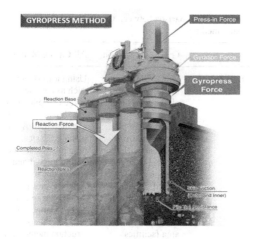

Figure 5 . The Gyropress method.

(a) 318.5 mm tubular pile with cutter bits

(b) 1000 mm tubular pile with cutter bits

(c) Piles Installation Process

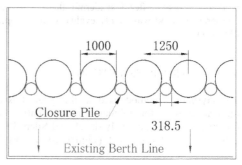

(d) Tubular Piles Alignment

Figure 6. Photos of the Project.

tubular piles with cutting bits attached on pile toe by rotary jack-in system and travels on top of piles which are completely installed (completed piles), (*Giken, 2009*). Thus, piles are installed at the required depth with minimum noise and vibration. The removal of underground obstacles or earthed objects is no longer necessary. In addition, the Gyropress is an efficient construction method for space saving. Since the Gyro Piler has clamps that expand inside the piles, the machine can maintain itself, self-walk on completed piles, and install piles. A temporally working platform is no longer necessary and the total construction yard is reduced. Moreover, the Gyro Piler F401-G1200 used in this project is equipped with the Press-in Piling Total System [PPTSTM] technology that makes it possible to monitor the substrata ground conditions during piling work for scientific analysis. The piling data that gives feedback can be linked to borehole data to optimize the applied press-in force, auger torque, extraction force, and penetration depth for optimum and efficient piling operation work.

4.1 *Structural members*

This project required a high stiffness wall structure because the height of the wall after completion was 13 meters. Robust and stiff tubular piles were selected to ensure enough strength and capacity. The piles of two different diameters were used in this project: 1000 mm and 318.5 mm **(Figure 6)**. The number of bits attached to the 1000 mm pile was 10 units and 6 units for the 318.5 mm pile.

5 CONSTRUCTION SEQUENCE

The construction sequences of this project are described as below:

1000mm and 318.5mm diameter tubular piles diameter were used to construct a 350m long tubular pile wall. The center-to-center distance between each 1000mm diameter tubular pile was 1250 mm. Between the 1000 mm diameter piles, 318.5 mm tubular piles (Closure piles) are installed to serve as particles protrusion protection (**Figure 6) (c)**. The Gyro Piler F401-G1200 used in this project has three clamps of 600mm height each, so they can ensure enough capacity and reaction force during installation. The method uses press-in and rotating force to install the piles, thus, the noise and vibration were kept low. The Gyropress method allows the piles to be installed even there was underground obstacle such as a very hard and plastic limestone marl layer. To prevent soil plugging at the pile toe, a water lubrication system was used. A small water steel pipe is welded inside each 1000 mm tubular pile as shown on **Figure 6 (b)**, then, was connected to the water lubrication system. The required water is taken from water tanks. The pile top is equipped with a swivel to prevent the small steel water pipe from being distorted during piling. At the initial piling, a steel cage grid of few meters long is placed next to the quay on the seaside, on which the reaction is placed **(Figure 8;10)**, then the reaction stand is loaded with counterweights on both sides. Thus, the Gyro Piler is assembled by setting the saddle in the reaction stand first, then the leader mast, and the chuck at the last step. As the diameter is small, the installation of the 318.5 mm tubular piles was facilitated with a water hammer as shown in **Figure 6 (a).** The quay length of 350 meters was

divided into two sections during the construction work. One section was reserved for the berthing and vessel unloading while the other section was being constructed. **Figure 7** shows the plan view during the construction process. After setting the reaction stand at the middle of the section to be constructed, two Gyro Piler F401-G1200 were assembled one by one for the piling work from the left and right of the reaction stand respectively. A 90 tons crawler crane was used to lift and pitch tubular piles to each Gyropiler. **Figure 9** shows the cross section of the construction process. The tubular piles are pitched to the chuck of the Gyropiler one by one. The pitched tubular piles are then pressed and rotated into the ground. When the chuck reaches its full capacity to grip the tubular pile, the driving attachment is fixed to the pile top. Then, the chuck grips on the driving attachment and continues the installation process until the 1000mm diameter tubular pile reaches the required depth. After installing two 1000mm diameter tubular piles subsequently, the closure pile is installed in between (**Figure 11**). To fit inside the chuck, a joint member attachment is inserted into the chuck, then the 318.5mm diameter tubular pile is pressed and rotated to the required depth.

6 PRODUCTIVITY

While the berthing and vessels unloading were continued at the wharf during the piling work **Figure 12**), the tubular pile installation rate was maintained at 2.5 piles per machine per day on average.

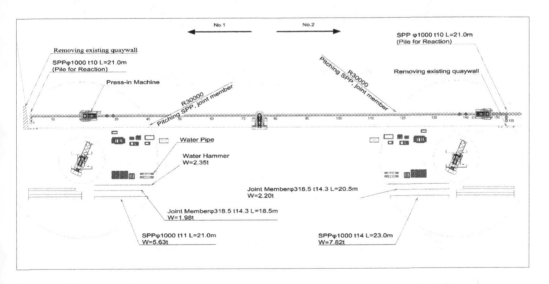

Figure 7 . Piling process: Plan view.

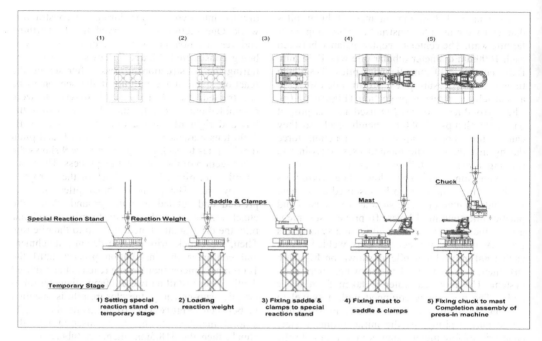

Figure 8 . Gyro piler assembling process (Upper: Plan view, lower: Side view).

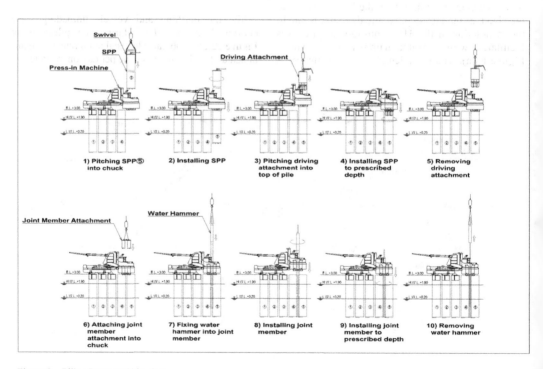

Figure 9 . Piling Process: Side view.

Figure 10 . Initial piling.

Figure 11 . Closure pile installation.

Figure 12 . Berthing during construction.

7 CONCLUSION

This project started in October 2013, in response to a request from the Senegalese government. The government of Japan agreed to provide technical and financial assistance for rehabilitation of the third wharf. The renovation work at the wharf quay at the Port of Dakar was carried out using the Gyropress method with high precision and minimal impact on the wharf and port operations. It was important not to use a method that would further damage the limestone marl layer which is already fractured. This project being a JICA ODA grant, the change of methodology after the preparatory survey stage is a rare case. However, after considering many factors, the initial design was changed from gravity quay type to quay on piles type, thus, the Gyropress method was selected over the All Casing method. That showed the effectiveness of the Gyropress method for a difficult construction site. During all construction periods, the port operation was able to proceed without any major changes.

REFERENCES

ANSD, 2013. National Agency for Statistic and Demography ANSD, Senegal: *General Census of Population and Housing, Agriculture and Livestock (RGPHAE)* (in French), accessed 4 March 2021,
GIKEN. 2009. Gyropress Method Brochure. Self-walk Rotary Press-in Method for Tubular Piles with Tip Bit.
JICA. 2016. Preparatory Study for the Rehabilitation Project of the Third Wharf of the Port of Dakar in the Republic of Senegal (in French).
Kitamura, M. 2018. Construction of Steel Tubular Pile Water Cut-off Wall by the Gyro Press Method and GIKEN Water Tightening System. *Fist International Conference on Press-in Engineering*, Kochi, pp. 445–452.
Kitamura, M. 2018. Cantilevered Road Retaining Wall Constructed of 2,000 mm Diameter Steel Tubular Piles Installed by the Gyro Press Method with the GRB System. *Fist International Conference on Press-in Engineering*, Kochi, pp. 437–444.
Miyanohara, T. 2018. Overview of the Self-standing and High Stiffness Tubular Pile Walls in Japan. *Fist International Conference on Press-in Engineering*, Kochi, pp. 167–174.
Nicolas, R. 2014. Design of Quay Encroachment (in French).
Suzuki, N. 2018. A case study of Design Change in the Press-in Method. *Fist International Conference on Press-in Engineering*, Kochi, pp. 467–474.

Proceedings of the Second International Conference on
Press-in Engineering 2021, Kochi, Japan – Matsumoto et al (eds)
© 2021 Taylor & Francis Group, London, ISBN 978-1-032-10414-0

Construction of retaining wall for river disaster restoration by Gyropress Method

K. Matsuzawa, T. Hayashi & K. Shirasaki
SATO JUKI Corporation, Niigata & Tokyo, Japan

ABSTRACT: A river improvement project has been conducted at the left bank of the Kinugawa River in Chikusei City. The project consisted of the works of levee heightening, construction and renovation of the riverbank and the sluiceway. The repair work of the Ezure water discharge channel was performed using a steel tubular pile wall as a main structure of the retaining wall of the sluiceway. Twenty-five tubular piles were installed at the point where the discharged water flows into the Kinugawa River. The wall was constructed in a hat-shape layout to protect both banks of the mainstream and the sluiceway. The Gyro Piler was set at the pile top level, 3.0m higher than the current bank crest, using a reaction base and reaction sheet piles (type 4). In a construction period of one-month, the tubular pile wall had been constructed steadily. Details of the piling work are reported in this paper.

1 OUTLINE OF THE PROJECT

1.1 *Place*

Chikusei City is located in the northwestern part of Ibaraki Prefecture and about 70km away from the center of Tokyo. Paddy cultivation is prosperous being blessed with the water supply from many rivers for many years. The cultivated area for rice field accounts for about 40 percent of the total area of the city. And the total agricultural field occupies more than half of the city.

1.2 *Background and objectives of the project*

The Kinugawa River is the first-grade river which flows through the eastern Kanto Plain from north to south, and confluents the Tonegawa. It is the longest tributary of the Tonegawa with the total length of 176.7km.

The Kinugawa has been called *the Abaregawa* in Japanese, which means a rampaging river, and the residents have suffered from the damage of the flooding frequently. By heavy rain which occurred in September, 2015, a flooding disaster occurred in the lower stream and the middle of the Kinugawa at 97 places including one bank collapse and seven floods.

The bureau, *Ministry of Land, Infrastructure, Transport and Tourism, Kanto Regional Develop-*ment Bureau, has promoted a large-scale countermeasure project for preventing a flooding disaster which will occur again from now on. Raising and extension of the riverbank was one of the major constructions for the project.

At the meeting point of Ezure discharge channel and the mainstream, a construction of a retaining wall which would supply a sluiceway for preventing flowback of the mainstream to the inland was done. Steel tubular piles were used as a main structure of this retaining wall. They were installed by means of the press-in method with gyration.

2 STRUCTURAL TYPE AND PILING METHOD

2.1 *Site condition*

Photo 1 shows the preparation of the construction yard for the piling. The piling work was performed on the left bank of the Kinugawa River. The high-water floor was used for the construction road, the working space for the crane and stock yard for the tubular pile materials. Because the pile top level was designed at 3.0m higher than the current levee crest, the reaction base for the Gyro Piler had to be set at that level using reaction sheet piles. Eighteen reaction piles,

DOI: 10.1201/9781003215226-47

Figure 1. Perspective view of the pile wall and raised levee.

Photo 1. Preparing construction yard.

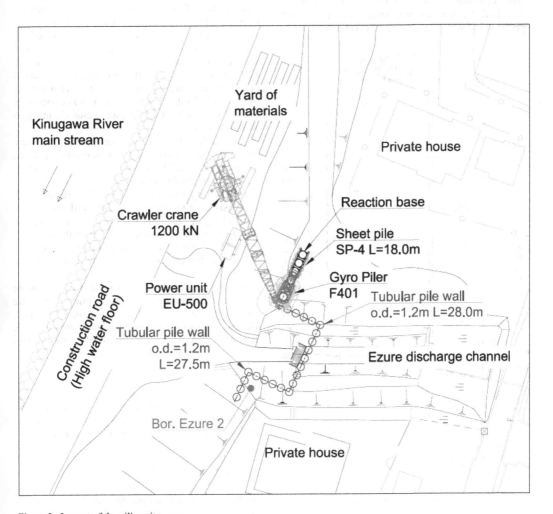

Figure 2. Layout of the piling site.

having length of 18m, were installed by a vibratory hammer before the procedure of press-in work.

2.2 Ground condition

Figure 3 shows soil profiles and SPT N-value at the nearest point of the pile wall. The condition of the ground is as follows.

Near the ground surface there was a refilled material with a thickness of 1m, overlying 7m thickness of clay and a silt layer with SPT N-values of 2 or 3. A fine sand layer having thickness of 8m and SPT N-value over 50 was laid at the depth of 10m to 17m from the ground level. Below the 17m was the silt layer with N-value of 20 to 30. At the depth of 20.8m, fine sand containing gravel, with N-value over 50, was laid for the bearing layer. The piles were penetrated 1.4m deep into the bearing layer.

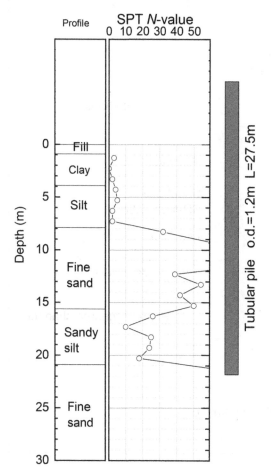

Figure 3. Soil profile with SPT N-value.

2.3 Structural type

Figure 4 shows a structural drawing of the retaining wall. The twelve piles installed to the upstream side had a diameter of 1200mm, a thickness of 12mm and a pile length of 28.0m. The rest, downstream side piles, had the same diameter, a thickness of 14mm and a length of 27.5m. They were installed with a spacing of 1,448mm from center to center. The datum line of piles formed a hat shape, or a crankshaft shape, in the plan view. The upper part of the central three piles were cut and removed at the bottom level of the discharge channel for the construction of the culvert and the flap gate.

2.4 Piling method

In order to install steel tube piles into a fine sand layer, with SPT N-value more than 50, the press-in method with gyration was adopted. It was also easy to cut the concrete blocks which covered the surface of both slopes of the discharge channel without pre-removal. As shown in Figure 4, 1200mm diameter steel tubular piles were used as the main structure of the wall. The pile materials were divided into two upper and lower members and spliced with circumference welds during gyration press-in.

As mentioned above, eighteen sheet piles were driven by the vibratory hammer as reaction piles as shown in Photo 2. The piling machine was supported steadily in the initial stage of piling.

Lubricating systems were used to reduce friction resistance by the surrounding soil and to prevent plugging of internal soil during gyration. A pair of pipelines supplied water at the pile toe. Pipes were protected by the covering angle, protector steel bar and internal cutting bit as shown in Photo 3.

A pair of equal angle steel were used as closure piles between two adjacent tubular piles. They were also pressed-in after each tube had been installed. The enclosed space by two piles and a pair of angle steel was filled with mortar afterward for the purpose of sealing. (see Figure 4)

2.5 Piler machine

The Gyro Piler, F401-G1200, was used for the installation of tubular piles in this site. Table 1 shows the specifications of the piler machine.

3 PRESS-IN PILING

In this section, the author reports the data of the press-in, the cycle time record, recorded by the G-pad.

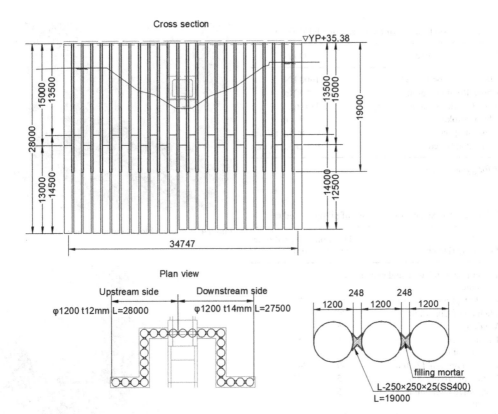

Figure 4. Structural drawing of the pile wall.

Photo 2. Reaction base supported by reaction sheet piles.

Photo 3. Internal pipeline for water supply.

The G-pad is a tablet PC which provides with the Press-in Assistance software. The operator of the piler can keep working while checking data such as press-in force, extraction force, gyration torque, and depth at the pile toe.

3.1 Productivity

Table 2 shows the typical duration of working procedure of piling. In this site, depth of press-in varied because the pile wall passed over the discharge channel as shown in Figure 4. The pile *Upstream No. 3* was the pile installed at the current levee surface, having the longest embedded length of 25m.

As shown in Table 2, it required about six hours of piling duration. In addition, extension of the scaffold was necessary according to the progress of each pile completion. In this site, we performed one pile per day steadily.

Table 1. Specifications of the Gyro Piler, F401-G1200.

Max. Press-in Force*	1500 kN
Max. Extraction Force*	1600 kN
Chuck Rotation Torque	900 kNm
Max. Chuck Rotation Velocity	11.0 rpm
Chuck Stroke	1000 mm
Press-in Speed	0.7 to 4.9 m/min
Extraction Speed	0.7 to 3.5 m/min

* for gyration use

Table 2. An example of durations of piling procedures.

Procedure (Depth m)	Duration (min.)	Elapsed time (min.)
Lower pile installation and center adjustment	27	27
Gyration press-in (6.0 m)	53	80
Pitching of Upper pile	20	100
Welding	35	135
Nondestructive inspection of welded joint	15	150
Gyration press-in (23.0m)	90	240
Setting of driving equipment	15	255
Successive Gyration Press-in (25.0m)	45	300
Angle steel press-in (both riverside and landside)	50	350
Total Gyropress duration	350	

3.2 Encountered difficulties

The concrete blocks for shore protection had been laid over the surface of the slope of the existing riverbank and the slopes of the discharge channel. At the initial gyration, we needed to take care that a tubular pile did not incline. It was better to reinstall the pile after the pile toe had passed over the thickness of concrete blocks. There was a tendency for the pile toe to be pushed out toward the downside of the slope. Piles were extracted and reinstalled after removal of concrete blocks with checking verticality and centering. Photo 4 shows the condition of the initial gyration.

After the pile installation, a pair of equal-angle steel was penetrated by pressing-in, in order to shut-up the interspace between the two piles. Because the length of angle steel was 19m, it seemed difficult to install with no twisting and detaching from the surface of the pile. In order to overcome this problem, a few guide pieces were attached to the pile surface as shown in Photo 5. These were effective to support the angles during installation, which ensured the required quality for following sealing mortar injection.

During the piling procedure, the data concerning press-in were monitored and recorded by the G-pad.

Photo 4. Concrete blocks cut and removed.

Photo 5. Guide pieces attached on the pile surface to prevent deviation of angle steel.

The data sheet from Upstream No. 3 is shown in Figure 5. Press-in force, extraction force and gyration torque were recorded with respect to the level of the pile toe. YP(m) in this graph means Yedogawa Peil and it is 0.84m higher than T.P. (Tokyo Peil) which is used widely in the Greater Tokyo.

The graph shows that large press-in force is not necessary to penetrate tubular piles when using gyration. The maximum press-in force was about 300kN. In the graph, press-in force of 900kN were often recorded. But it was considered that it happened only when gyration stopped for release and re-chucking the pile tube. In fact, the weight of chucking equipment and the tubular pile itself should be considered for the correction of the values of press-in and extract force. About 200kN should be added to the press-in force and should be deducted from the extraction force.

It was also obvious from the graph that gyration torque in sandy soil layer, both at the middle depth and the bearing layer, seems greater than those in silt and clay layers. The graph shows consistency with respect to the soil profile.

478

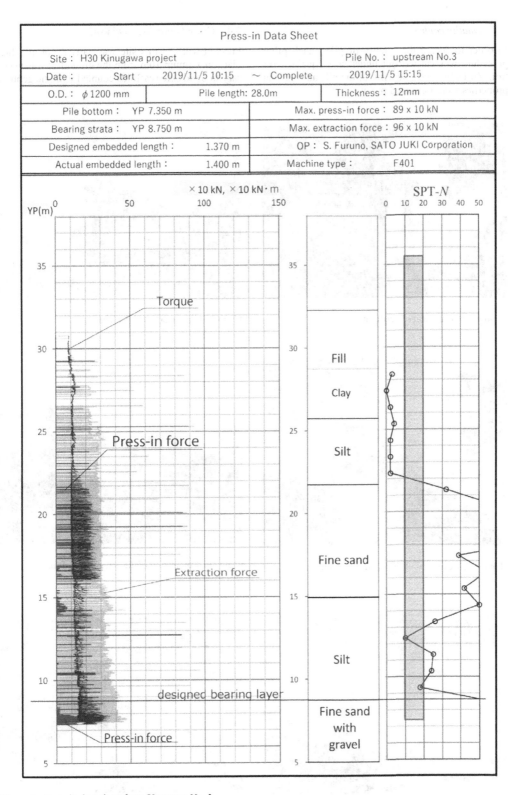

Press-in Data Sheet		
Site : H30 Kinugawa project		Pile No. : upstream No.3
Date : Start 2019/11/5 10:15 ～ Complete		2019/11/5 15:15
O.D. : φ1200 mm	Pile length: 28.0m	Thickness : 12mm
Pile bottom : YP 7.350 m	Max. press-in force : 89 x 10 kN	
Bearing strata : YP 8.750 m	Max. extraction force : 96 x 10 kN	
Designed embedded length : 1.370 m	OP : S. Furuno, SATO JUKI Corporation	
Actual embedded length : 1.400 m	Machine type : F401	

Figure 5. Press-in data sheet from *Upstream No. 3*.

479

3.3 *Quality control*

In order to secure the quality of piles, each piling procedure was carried out under strict quality control. Verticality and centricity of the pile and welding properties were inspected precisely.

3.4 *Contribution to safety work*

Press-in work in this site was carried out at a high place, maximum 10m from the ground. It was necessary to build safe scaffolds coupled with the genuine

Photo 6. Measurement of verticality.

Photo 9. Piling work at high place.

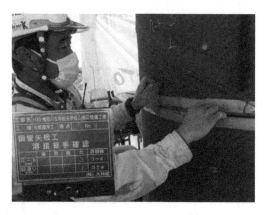

Photo 7. Checking root length and misalignment.

Photo 10. Completion of the construction of the pile wall.

Photo 8. Verticality and centricity were surveyed during the piling procedure.

Photo 11. The Zero Piler used for sheet piling adjacent to the abutment structure.

Photo 12. Successive construction work for the sluiceway.

stages which were attached to the piler machine. These scaffolds have been used in successive works of the project after the series of pile work.

4 CONCLUDING REMARKS

Twenty-five steel tubular piles were installed by Gyropress method to construct the retaining wall for the Ezure discharge channel and the Kinugawa River.

The pile wall is still 3m higher than the current height of the riverbank, which would be raised in the near future.

Not only this pile wall, but many of sheet pile works and other tubular pile works were also used in this project as shown in Photo 11.

Press-in technology has been able to contribute widely to the countermeasure project and helped improve safety of the surrounding residents and their lives.

ACKNOWLEDGEMENTS

The authors express gratitude to people engaged in the Kinugawa left-bank countermeasure project, who recognized this paper contribution.

We also express gratitude to the persons concerned who cooperated with this construction work.

REFERENCES

K.S.S. Kinugawa Sagan Development Office, *http://kinu gawa-sagan.com* (In Japanese)
Ministry of Land, Infrastructure, Transport and Tourism. Kanto Regional Development Bureau (M.L.I.T.), Kanto Regional Development Office. 2015. *About the flood damage in Kinugawa on H27.9 Kanto Tohoku torrential rain.* (In Japanese)
M.L.I.T., Shimodate River Office. 2019. *Aiming at reconstruction of consciousness for flood-defense.* (In Japanese)

Photo 13. Completion after providing with the flap gate at the center of the pile wall.

Proceedings of the Second International Conference on
Press-in Engineering 2021, Kochi, Japan – Matsumoto et al (eds)
© 2021 Taylor & Francis Group, London, ISBN 978-1-032-10414-0

Steel tubular piling by the Gyropress Method in proximity to obstructive existing H-shaped piles

N. Yamazaki

Kajikawa Corporation, Hekinan City, Japan

ABSTRACT: This construction project was aimed at reinforcing the seismic resistance of an existing river dike. Among the 42 steel tubular piles installed, 11 were installed by press-in piling with a GRB System, and 31 were installed by a normal press-in piling. There were existing H-shaped piles in the northern end river section (Benten Bridge) where four new piles were to be installed, and these H-shaped piles were planned to be removed beforehand. An attempt to pull out and remove the existing piles using a vibratory hammer failed, as the piles broke at a depth of about 3m from the riverbed. Therefore, steel tubular piles were installed at deviated locations by rotary-pressing with a Reaction Stand in proximity to the existing H-shaped piles.

1 SUMMARY OF THE PROJECT

1.1 *Outline of the project*

This construction project was aimed at reinforcing the seismic resistance of an existing river wall. It involved block-based soil improvement for the liquefaction prevention by high-pressure jet agitation immediately below an existing river dike; construction of a steel tubular pile earth retaining wall by the Gyropress Method (GIKEN, 2018) on the river side of the existing dike, and soil improvement by high-pressure jet agitation within the river.

The construction site is located on the left bank of the Ebitori River, a national Class A river, adjacent to the area of Haneda 5-chome, Ota-ku, Tokyo. The site is in the uppermost section of the Ebitori River, which is a branch of the Tama River, another Class A river. Haneda Airport is located to the east of the construction site.

1.2 *Objective of the project*

In light of the Great East Japan Earthquake, the Tokyo Metropolitan Government has been working to maintain various facility functions and to prevent flooding caused by tsunamis or other factors, even in the event of the largest possible earthquake in the future.

To reinforce the river dike, an continuous steel tubular pile earth retaining wall was to be constructed by the Gyropress Method on the river side of the dike, and soil improvement was to be carried out by high-pressure jet agitation on the riverbed. The main body of the river dike was designed to be

reinforced by block-based soil improvement, which is as liquefaction prevention, immediately below the dike using high-pressure jet agitation.

2 CONDITIONS AND STRUCTURAL TYPE

2.1 *Site condition*

The site conditions for the pile installation were below:

- Limit noise and vibration in consideration of neighboring residential areas.
- Limit working hours in consideration of neighboring residential areas.
- Do not obstruct the navigation of vessels (fishing boats, houseboats, etc.) and keep a river width of 6.0m during the pile installation.
- The JR Tokaido Freight Line subway with a tunnel diameter of 7,000mm is located on the south (upstream) side of the river section at a depth of A.P. -8.54m or deeper.
- Keep a working distance in the north (downstream) side river section due to proximity to the Benten Bridge.

2.2 *Ground condition*

Figure 1 shows a borehole log obtained on the land (west) side of the embankment at the construction site. The river area where the steel tubular piles (hereinafter called tubular piles) were installed had a riverbed height of A.P. -3.0m, corresponding to a level stuff reading of 7.34m. The soil beneath the

DOI: 10.1201/9781003215226-48

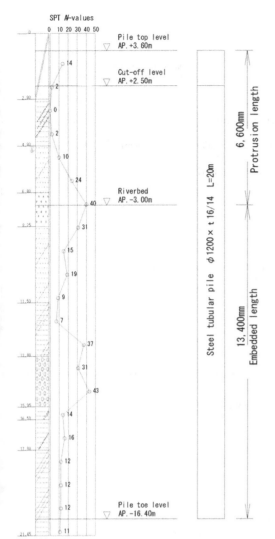

SPT *N*-values

Pile top level
▽ AP. +3.60m

Cut-off level
▽ AP. +2.50m

Riverbed
▽ AP. -3.00m

Protrusion length 6,600mm

Steel tubular pile φ1200 × t 16/14 L=20m

Embedded length 13,400mm

Pile toe level
▽ AP. -16.40m

Figure 1. Borehole log with a tubular pile at the design stage.

riverbed consists of silty fine sand, sandy silt, gravel, gravelly clay, sandy silt, and silty clay. The gravel layer at a level stuff reading of 15m has the highest SPT *N*-value of 43. Below the gravel layer, there is a cohesive soil having SPT-*N* values of around 12 prevails. The pile toe of the tubular piles is at A.P. -16.4m, corresponding to a level stuff reading of 20.74m.

2.3 *Structural type*

The structural type of the wall is a steel tubular pile earth retaining wall. Figure 2 shows an overall plan view, and Figure 3 shows a typical cross-sectional view.

3 STEEL TUBULAR PILE PILING

3.1 *Piling method*

The piles were installed by the Gyropress Method which is also called "Rotary press-in piling". Within the construction site, having a total linear length of 60.4m, the first 11 tubular piles in the north-side river section (closer to the Benten Bridge) were installed by press-in piling with a GRB System (GIKEN, 2018), and the 12th to 42nd tubular piles (up to the southern end of the river section) were installed by a normal press-in piling (without the GRB System). In order to keep a working distance from the Benten Bridge in the north (downstream) side river section and from the JR Tokaido Freight Line subway on the south (upstream) side of the river section, the first Reaction Stand location was set to face the 13th tubular pile counted from the Benten Bridge side. 18 steel sheet piles (hereinafter called sheet piles) were installed by a vibratory hammer as reaction piles, setting up the Reaction Stand.

The reasons for adopting the Gyropress Method are as follows.

[1] It is a press-in piling method suited for reducing noise and vibration to neighboring residential areas.
[2] The press-in piling with a GRB System does not obstruct the navigation of general vessels.
[3] The press-in piling method does not affect the JR Tokaido Freight Line subway on the south (upstream) side of the river section.
[4] The piling method takes into account the neighboring work to the Benten Bridge on the north (downstream) side of the river section.

3.2 *Press-in piling data*

The length of the tubular pile is 20m, and its embedded length is 13.4m. As mentioned in the section 2-2., since the borehole log was obtained on the land side of the embankment, the planned riverbed height of A.P. -3.0m corresponds to a level stuff reading of 7.34m. The acquisition of press-in piling data (Figure 4) started at around 8.5m, and a graph showing a fairly good correlation between the rotation torque values and the SPT-*N* values was obtained. It was possible to press-in a pile by low force from 8.5m (after reaching the riverbed) to 16.0m. At depths of 17.0m and more, although the SPT *N*-values in the borehole log were low, press-in force increased gradually until reaching the highest value at the designed final depth. The pile was equipped with a toe ring with four outer bits and one water pipe allowing a water injection in the pile base.

3.3 *Productivity*

– Construction Period:
From February 12 to April 1, 2020 (8 nonworking days in 4 weeks; 36 days of actual working)

483

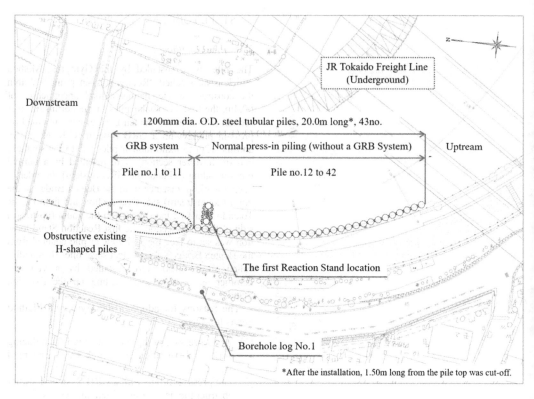

Figure 2. Overall plan view of the steel tubular pile earth retaining wall.

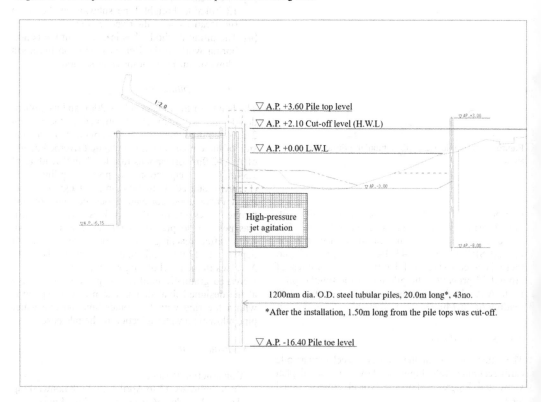

Figure 3. Typical cross-sectional view.

484

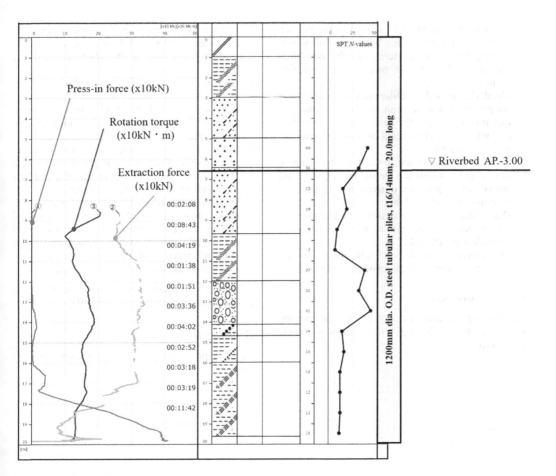

Figure 4. Press-in piling data.

- No. of pressed-in piles per day:
  5 piles/day for the normal press-in piling;
  2 piles/day for one with the GRB System

Among the 42 tubular piles installed at the site, 11 were installed by press-in piling with the GRB System, and 31 were installed by a normal press-in piling method. While the original design of construction was to carry out press-in piling from the Reaction Stand at one location, an additional round of press-in piling from the Reaction Stand and two more rounds of Gyro Piler assembling and disassembling were required for piling in the area of obstructive existing H-shaped piles (hereinafter called H piles). The required days of actual working increased due to the additional assembling and disassembling of the Clamp Crane.

4 STEEL TUBULAR PILING AT THE DEVIATED LOCATION

4.1 Problem at the site

Because of the knowledge of existing H piles in the GRB System section close to the Benten Bridge, an attempt was made to remove the existing piles by using a submersible vibratory hammer. It was possible to extract and remove the 300mm wide H piles, whereas it was not possible to extract and remove the 400mm wide ones as they broke at the joints situated at a depth of about 3m from the riverbed.

- Existing 300mm wide H piles, 8 no., 5.50 m long: The removal was completed
- Existing 400mm wide H piles, 8 no., 27.1 m long: They were unable to be extracted, meaning they became obstructions on the specified tubular pile line.

It was difficult to press-in tubular piles because the existing H piles were situated along the specified line of the tubular piles. After consultation with the client, it was decided to install the tubular piles deviated from their specified locations. Relevant concerns are described below.
[Concerns]

- There are three obstructive existing H piles (H400, L=27.1 m) on the specified installation locations of tubular piles.

485

- Although the height (A.P. -4.0 m) of the pile top of the existing H piles has been revealed by surveying, they may interfere with the new piles because of the unknown precision of verticality of the existing piles.
- The deviation of the tubular piles toward the river should be kept to a minimum because the pile tops might encroach the new structure formed by extending the embankment bottom slab and the width of the river.
- There is a limit to the amount of deviation of tubular piles toward the embankment (land) side because of the presence of sheet piles beneath the embankment bottom slab.
- Depending on the installation locations, the arrangement of tubular piles may not allow the Gyro Piler to self-walk.
- The construction plans should minimize any obstruction to the navigation of general vessels.

4.2 Countermeasures

Overcoming the problem occurred by the existing H piles, the following countermeasures were considered:

[1] Arrange the tubular piles to assure a verticality allowance of 1/100 (=200 mm), regardless of the verticality precision of the obstructive existing H piles.
[2] Arrange the tubular piles on condition that the embankment bottom slab is partially removed.
[3] Keep the distance between piles to allow installation of the Gyro Piler, and determine a piling procedure.
[4] Install some sheet piles, which were used as reaction piles for the initial piling, at the gap between tubular piles where the distance has been widened from the designed plan.
[5] Strengthen the operation and communication systems so as not to obstruct the navigation of general vessels during the installation and extraction of reaction piles using the vibratory hammer, installation and removal of the Reaction Stand, and assembly and disassembly of the Gyro Piler.

Figure 5 shows the arrangement of tubular piles in the area of obstructive existing H piles.

4.3 Piling procedure

The piling procedure of the tubular piles is described as below:

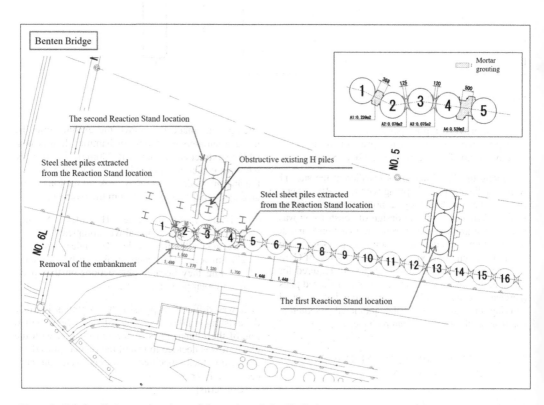

Figure 5. Tubular pile arrangement (area of obstructive existing H piles).

486

Step.1 From the first Reaction Stand location, Pile 12 to 26 are installed by the normal press-in piling.

Step.2 The Gyro Piler turns around and returns to the position of Pile 13 to 15.

Step.3 The Clamp Crane is assembled onto the position of Pile 17 to 19.

Step.4 Pile 6 to 11 are installed using the GRB system.

Step.5 The Reaction Stand is relocated from the position facing Pile 13 to the position facing Pile 3.

Step.6 The Gyro Piler is disassembled and then assembled at the second Reaction Stand location.

Step.7 From the second Reaction Stand location, Pile 2 to 4 are installed.

Step.8 From the second Reaction Stand location, the Gyro Piler self-walks to the position of Pile 2 and 3.

Step.9 Pile No.1 is installed.

Step.10 The Gyro Piler is disassembled and then assembled, and set up onto the position of Pile 6 to 8.

Step.11 Pile 5 is installed.

Step.12 The Clamp Crane is disassembled and removed.

Step.13 The Gyro Piler turns around and presses in Pile 27 to the 42 by the ordinary method until completion.

Note: Regarding the arrangement/numbers of tubular pile, refer to Figure 2 and 5.

Figure 6 to 10 below show the piling states.

Figure 6 shows the state of pile installation in which Step 1 to 6 have been completed.

Figure 7 shows the state of pile installation in which Step 7 is in progress. Pile 3 is being installed from the second Reaction Stand location.

Figure 8 shows the state of pile installation during Step 8 and 9. The Gyro Piler self-walks from the second Reaction Stand location onto already installed steel tubular piles to install the Pile 1.

Figure 6. Piling state after Step 1 to 6.

Figure 7. Piling state at Step 7.

Figure 8. Piling state during Step 8 and 9.

Figure 9. Piling state during Step 11 and 12.

Because the amount of pile deviation due to the obstructive existing H piles exceeds the clamp movable range of the Gyro Piler, Pile 5, which was supposed to be gripped by the third clamp, was pressed in at the final stage.

Figure 10. Completion of pile installation in the area of obstructive existing H piles.

Figure 9 shows the state of pile installation during Step 11 and 12. Because the tubular pile arrangement did not allow the third clamp of the Gyro Piler to grip a pile during installation of Pile 1, press-in piling was conducted after the installation of Pile 1 to 4.

Figure 10 shows the completed state of pile installation of Pile 1 to 5 in the area of obstructive existing H piles. Part of the embankment has been removed because of the deviation of Pile 2 toward the embankment side. Some sheet piles, which were used as reaction piles, were re-installed at the locations where the distance between tubular piles had been widened from the original plan.

5 CONCLUSION

It was possible to successfully complete pile installation in proximity to obstructive existing H piles. Although there was some metallic sound indicating contact with the obstructive piles during the pile installation, tubular piles managed to be pressed in without any problems. Thus, the project was completed without delay from the original plan.

ACKNOWLEDGMENT

We would like to express our gratitude to the staff of the Office of Seismic Retrofitting of the Ebitori River Seawall (No. 203) of Wakachiku Construction Co., Ltd. for kindly supplying data useful for this article.

REFERENCES

GIKEN LTD (GIKEN) 2018. *Gyropress Method*, 9pp. Kochi: GIKEN.
GIKEN LTD (GIKEN) 2018. *Non-Staging Method (GRB system)*, 9pp. Kochi: GIKEN.

Proceedings of the Second International Conference on
Press-in Engineering 2021, Kochi, Japan – Matsumoto et al (eds)
© 2021 Taylor & Francis Group, London, ISBN 978-1-032-10414-0

Case study of oval shaped foundation using the Gyropress Method under overhead restrictions

K. Takeda
KAKUTO CORPORATION, Tokyo, Japan

ABSTRACT: This paper describes the press-in piling method of steel pipe piles to form steel pipe foundations for the expansion of bridge piers to widen an existing bridge on the expressway. In this project, it was necessary to install steel pipes under girders/beams of a road bridge while keeping the bridge in service. It was also necessary to embed piles in a gravel layer having a maximum extrapolated SPT N-value of more than 70. The Gyropress Method using a type of machine capable of low headroom operation was adopted. In this method, steel pipe piles with pile toe ring bits are rotationally pressed-in to cope with hard ground and steel pipe piles can be pitched using a hoisting system for low headroom. By welding short segmental piles together in vertical position under overhead restrictions, it was possible to press in steel pipe piles until reaching the supporting layer.

1 OVERVIEW OF CONSTRUCTION

The construction site is located beneath the Oppe River Bridge, which spans the Oppe River, on the East Japan Expressway of East Nippon Expressway Company Limited. The bridge connects Tsurugashima City, Saitama Prefecture and Higashi-Matsuyama City, Saitama Prefecture (see Figure 1). The purpose of this project is to widen the lanes in the section between the Tsurugashima Junction and the Higashi Matsuyama Interchange on the Kan-etsu Expressway. As regards the press-in piling to form foundations in the bridge substructure for road widening, this article reports the selection of specifications, the procedure of press-in piling, and the outcome of construction.

It was decided to install additional foundation piles near the existing piers in the substructure.

The project involved construction of 12 additional piers: the piers P1 to P6 for both up and down bound lanes.

2 CONSTRUCTION AND SOIL CONDITIONS

2.1 *Construction conditions*

It was necessary to meet the following conditions in this project.

– Maintain safe overhead clearances from the in-service road bridge. (See Figure 2)
– Complete piling during drought seasons.

– Restore the revetment to its original state by each flood season.
– Minimize noise and vibration in consideration of neighboring residents.
– Install piles into a gravel layer that serves as a support ground.

In installing additional piles for the existing piers, two methods were investigated and compared: installation of additional cast-in-place piles, and installation of steel pipe sheet piles. As a result, steel pipe sheet pile foundations were adopted because of their advantage in terms of the ease of construction and short construction period.

Up to three Gyro Pilers were used simultaneously to shorten the construction period and complete work within drought seasons. Moreover, some sections were arranged for continuous work during the day and night (See Figure 3), and a maximum of six teams were employed for the piling operation.

2.2 *Soil conditions*

Among the 12 piers worked on, this paper describes the pier P2.

Figure 4 shows a borehole log obtained on the site. Below the surface layer of the Kanto Loam, the section from the top of an installed pile to a depth of 12.0 m consists of a gravel layer with weathered conglomeric clay, with SPT N-values ranging from 2 to 47. Below an intermediate clayey layer, there is a very dense layer of clayey gravel, with a maximum extrapolated SPT N-value exceeding 70.

DOI: 10.1201/9781003215226-49

Figure 1. Location map.

Figure 2. Overhead clearance under the Oppe River Bridge.

Figure 3. Piling at night.

In selecting a press-in piling method, it was found difficult to apply a standard press-in or a press-in piling with water jetting to a very dense

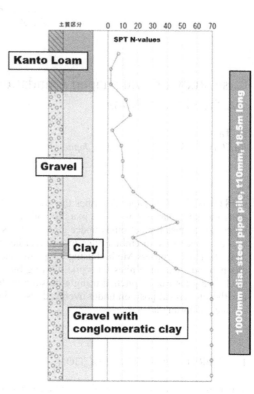

Figure 4. Borehole log.

layer of clayey gravel having a maximum extrapolated SPT N-value of more than 70. One method contemplated was to conduct pre-excavation using a full-slewing excavator or the like and then press steel pipe sheet piles into the ground after replacing the existing soil with fine sand, etc. However, this method was judged to be inapplicable because of the lack of certainty in replacing very hard soil. The Gyropress Method (Figure 5) was used in this project, in which steel pipe piles, instead of steel pipe sheet piles, with pile toe ring bits (Figure 6) were rotary pressed-in into the hard soil by rotary press-in piling.

2.3 Overhead restrictions

There were overhead restrictions because of the site being under the existing bridge beams and girders. At the pier P2, it was necessary to keep overhead clearances under the road bridge in service: about 7.5 m below a girder, and about 5.3 m below a beam. The piling procedure was therefore restricted to attaching a hoisting system for low headroom to a press-in machine, pitching short segmental piles without using a service crane, and then welding the piles together in vertical position. Therefore, the Gyropress Method using a type of machine capable of low headroom operation was adopted.

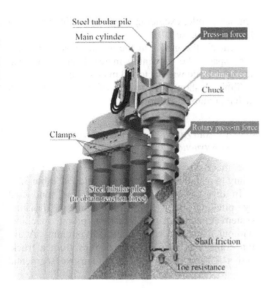

Figure 5. The Gyropress method.

Figure 6. Steel pipe pile with pile toe ring bits.

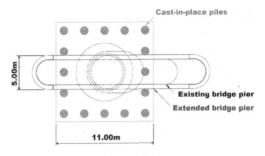

Figure 7. Cast-in-place pile foundation.

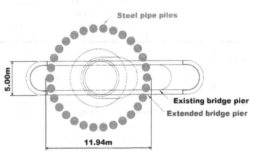

Figure 8. Steel pipe foundation (without interlocks).

3 STRUCTURAL TYPES OF FOUNDATIONS

The structural types of foundations considered in this project are described in Figure 7 and 8. The Gyropress Method, which is applicable to steel pipe foundations (without interlocks), was adopted in this project.

The shape of a foundation is either circular or oval, and the foundation shape and the number of steel pipe piles were decided according to the shape of the existing pier. For each pier, the foundation type was selected as either the cantilever or supporting type of the earth retaining wall in consideration of penetration into the support layer and stiffness.

For the pier P2 of the down bound lanes, 30 steel pipe piles with pile toe ring bits were rotary-pressed in to form an oval shaped foundation. For the 12 piers worked on in this project, a total of 400 steel pipe piles (See Table 1) were installed by the Gyropress Method.

4 PROCEDURE OF PRESS-IN PILING

4.1 Selection of applicable machine

A type of machine that can install steel pipe piles of 1000mm in outer diameter by rotary press-in piling and to which a hoisting system for low headroom (Figure 9) can be attached was selected. The GRAL1015 of GIKEN LTD. was used for press-in piling in this project.

4.2 Selection of segmental pile length

The pile length that allows steel pipe piles to be pitched without getting in contact with the existing bridge in a space that satisfies the overhead clearances was investigated. The overhead clearance is most restrictive beneath the beam of the pier P2,

Table 1. Steel pipe piles.

Pier	P1	P2	P3
Foundation shape	Oval	Circle	Circle
Structural type of retaining wall	Supporting	Cantilever	Cantilever
	Up line	**Up line**	**Up line**
	SKK490	SKK400	SKK400
	1000mm O.D.	1000mm O.D.	1000mm O.D.
	t12mm	t10mm	t10mm
	16m long	18.5m long	16.5m long
Piles	40 no.	30 no.	30 no.
	Down line	**Down line**	**Down line**
	SKK490	SKK400	SKK400
	1000mm O.D.	1000mm O.D.	1000mm O.D.
	t12mm	t10mm	t10mm
	16m long	18.5m long	16.5m long
	40 no.	30 no.	30 no.

Pier	P4	P5	P6
Foundation shape	Oval	Circle	Circle
Structural type of retaining wall	Supporting	Cantilever	Cantilever
	Up line	**Up line**	**Up line**
	SKK490	SKK400	SKK490
	1000mm O.D.	1000mm O.D.	1000mm O.D.
	t16mm	t10mm	t15mm
	16m long	17.5m long	18.5m long
Piles	40 no.	30 no.	30 no.
	Down line	**Down line**	**Down line**
	SKK490	SKK400	SKK400
	1000mm O.D.	1000mm O.D.	1000mm O.D.
	t16mm	t10mm	t10mm
	16m long	17.5m long	12.0m long
	40 no.	30 no.	30 no.

Figure 9. Hoisting system for low headroom in operation.

with the separation between the base surface and the existing bridge being 5.3 m. To join steel pipe piles together safely in this narrow overhead clearance space, the maximum material length of each segmental pile, excluding the bottom pile, was set to 1.5m (see Figure 10).

Beneath the girder in the 7.5 m overhead clearance space, the maximum material length of each segmental pile was set to 4.0 m (see Figure 11).

Table 2 shows the results of consideration into different numbers of joints.

4.3 Consideration of equipment layout during construction

The main equipment used for the rotary press-in piling to install steel pipe piles is listed in Table 3.

In this project, press-in piling was conducted by pitching steel pipe piles using a hoisting system attached to a press-in piling machine, without using a service crane for pitching. A 50-ton crane was used to unload piles from a truck, and a 4.9-ton crane was used to move segmental

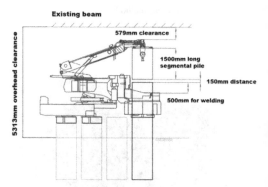

Figure 10. Allowable pile length beneath the beam at pier P2 (5.3m clearance).

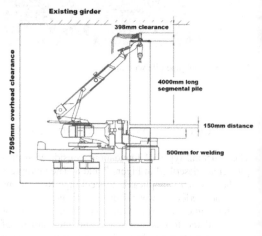

Figure 11. Allowable pile length beneath the girder at pier P2 (7.0m clearance).

Table 2. Specification of segmental piles at the pier P2.

[Under the beam] 5.3m overhead clearance
SKK400, 1000mm O.D., t10mm, 18.5m long
Number of joints: 11 joints
Type-A: 2.5m bottom pile + 10 no. 1.5m piles + 1.0m upper pile
Type-B: 2.0m bottom pile + 10 no. 1.5m piles + 1.5m upper pile

[Under the girder] 7.9m overhead clearance
SKK400, 1000mm O.D., t10mm, 18.5m long
Number of joints: 4 joints
Type-A: 4.5m bottom pile + 4.0m pile + 2 no. 3.5m piles + 3.0m upper pile
Type-B: 4.0m bottom pile + 4.0m pile + 2 no. 3.5m piles + 3.5m upper pile

Table 3. Main equipment.

Machine type	Specification	Notes
Gyro Piler	Press-in force 1500kN, Extraction force 1600kN	Rotary-press-in piling
Power Unit	221kW (300p s)/ 1800min-1	Power source for the Gyro Piler
Telesco-Crane	50 ton class	for loading/unloading
Water lubrication system	60 ℓ/min, 2no.	Reduction of inner friction of steel pipe piles
Welder	600A, 2no.	Splicing segmental piles
Generator	150kVA	Power source for the welder

piles within the space of overhead restrictions. (See Figure 12, 13 and 14).

4.4 Confirmation of reaching the support layer

The Gyro Piler is equipped with a system that acquires press-in data; this system measures and records the press-in force and rotation torque applied to each steel pipe pile during piling. Press-in data was obtained for all the piles on the site, and the horizon at which soil characteristics change was estimated from measurement data to estimate whether the pile has reached the supporting ground.

4.5 Loading test (IPA, 2014)

To check the vertical reaction force of the pressed-in piles, a vertical loading test was carried out by using the Gyro Piler.

At the end of piling after making sure that a pile reached the support ground, the pile was pressed into the supporting ground by a distance of 1D or more without toe lubrication using the water lubrication system. As the outer diameter of the steel pipe pile

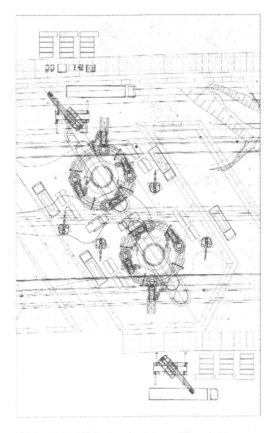

Figure 12. Machine layout for the down bound lane.

Figure 13. Piling work at the open place.

was 1000 mm, a spliced steel pipe pile was pressed in without water by 1.0m or more.

The load applied was set to 1.5 times the maximum normal vertical reaction force stipulated in the design, and the loading test was conducted by applying the press-in force to the pile top by the Gyro Piler for low headroom. For some piers, it was necessary to apply a test load exceeding the upper limit of the press-in force of the Gyro Piler for low headroom. Therefore, the loading

493

Figure 14. Piling work beneath a girder.

Table 5. Measurements.

Planned ground level	21.41 m
Planned pile top level	22.41 m
Planned pile toe level	3.91 m
Planned level of the support layer	8.56 m
Estimated level of the support layer	10.68 m
Actual ground level	21.68 m
Actual pile top level	22.42 m
Actual pile toe level	3.92 m
Specified embedded length to the support layer	4.65 m
Estimated embedded length to the support layer	6.76 m
Confirmation of the embedded length	OK

test was carried out using a dedicated press-in machine compatible with the test load.

The allowable pile top displacement was set to 1% or less of the outer diameter of the steel pipe pile. As the outer diameter of the steel pipe pile was 1000 mm in this project, it was intended to verify that the pile top displacement was within 10mm.

5 RESULTS

5.1 Construction process

The entire construction project was performed during three drought seasons in the years from 2018 to 2020.

The press-in piling of steel pipe piles was completed in a total of nine months over two drought seasons, with the first phase taking approximately five months from late November 2018 to late April 2019 and the second phase taking approximately four months from late December 2019 to late April 2020.

Table 4 shows the cycle time of press-in piling of steel pipe piles at the pier P2 of the down bound lanes. The cycle time was about 300 minutes for

Table 4. Cycle time of piling at the pier P2 of the down bound lanes.

[Under the beam] 5.3m overhead clearance
SKK400, 1000mm O.D., t10mm, 18.5m long
Number of joints: 11 welded joints
Average cycle time: 540 min per pile

[Under the girder] 7.9m overhead clearance
SKK400, 1000mm O.D., t10mm, 18.5m long
Number of joints: 4 welded joints
Average cycle time: 300 min per pile

*Including welding and inspection time

each pile beneath a bridge girder and about 540 minutes beneath a bridge beam.

5.2 Confirmation of reaching the support layer

The press-in data was used to estimate whether the steel pipe pile toe has reached the support layer. Figure 15 shows an example of investigation at the pier P2 of the down bound lanes and Table 5 shows that results of verifying that the piles have reached the support layer.

At the pier P2 of the down bound lanes, the design length of the pile section penetrated into the support layer was 4.65m. According to the press-in data, the rotation torque started to increase at around 11.5m from the ground.

This implies that the toe of the steel pipe pile reached the dense support layer of clayey gravel and that the press-in force applied to the pile toe ring bits caused a greater resistance against ground cutting in this layer than in the overlying gravel layer.

5.3 Loading test

The maximum normal vertical reaction force specified in the design for the pier P2 was 1135 kN. Therefore, the load specified for the loading test using the Gyro Piler was first set to 1702.5 kN (1.5 times 1135 kN), which was then rounded up to 1750 kN. The loading test was conducted by setting the pile top displacement allowance to no more than 10mm, which corresponds to 1% of the steel pipe pile outer diameter of 1000mm. As the upper limit of press-in force was 1500 kN for the Gyro Piler for low headroom, a dedicated press-in piling machine with an upper-limit press-in force of 2600 kN was used in the loading test.

The loading test (See Figure 16) on piles at the pier P2 resulted in a pile head displacement of 9.22 mm, which was within the allowable displacement of 10 mm.

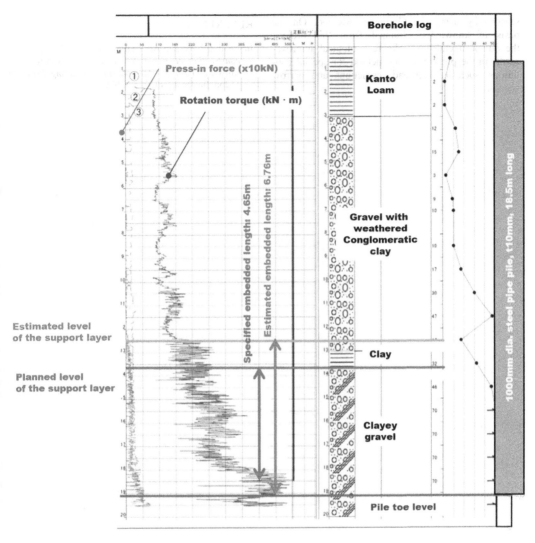

Figure 15. Press-in data.

Figure 16. Vertical loading test.

6 SUMMARY

This paper describes an example of using the press-in pilling method to install steel pipe piles into the hard and supporting ground under overhead restrictions. The applicability of the Gyropress Method has been demonstrated under soil conditions in which the applicability of the standard press-in or the press-in piling with water jetting after replacing existing soil with fine sand by a full-slewing excavator is less certain.

Although many issues arose from the planning to the piling stages under the construction conditions, it was possible to complete the piling for the oval shaped foundation within limited drought seasons by using the Gyropress Method. On the basis of knowledge obtained from this project, the author intends to make technological improvements in the press-in piling method for special conditions, such as overhead restrictions and narrow spaces.

REFERENCES

International Press-in Association (IPA) 2014. *Design and construction manual of steel tubular pile earth retaining walls by Gyropress Method (Rotary cutting Press-in).* 152pp. (in Japanese) Tokyo: IPA

Proceedings of the Second International Conference on
Press-in Engineering 2021, Kochi, Japan – Matsumoto et al (eds)
© 2021 Taylor & Francis Group, London, ISBN 978-1-032-10414-0

Construction of anchor piles for mooring bank by Skip Lock Method

Y. Tada, M. Kitamura, S. Kamimura & Y. Sawada
Gikenseko Co. Ltd., Minato, Urayasu, Chiba, Japan

ABSTRACT: In the Shizuoka Prefectural Port Project, a wharf was necessary to be improved as a cargo passengers quay which can cope with simultaneous port calls with 2 large ships of 150 thousand GT class and cargo ships of 30 thousand DWT class. For this improvement, a retaining wall of steel tubular piles with anchors was adopted under following site conditions: 1) Noise and vibration should be minimized for the existing buildings. 2) The working space was limited. 3) The piles were required to be installed into stiff ground though a rubble mound. In order to overcome these conditions, the "Gyropress Method" in which a steel tubular pile with cutting ring bits is rotated to penetrate underground obstruction/stiff ground, and the "Skip Lock Method" which can install piles with intervals were selected. This paper introduces a case study of installing steel tubular piles as anchor piles for rehabilitated wharf.

1 GENERAL INSTRUCTIONS

1.1 Place

Shimizu Port's Hinode Wharf (hereinafter, Shimizu Port) is a port located in Shimizu Ward, Shizuoka City, capital city of Shizuoka Prefecture. This port is planned to be used as an international logistics base for export-oriented companies such as automobiles, motorcycles and musical instruments, as well as for advanced technology companies such as semiconductors. It has also been attracting attention as a port inviting international cruise ships, triggered by the World Cultural Heritage Registration of Mt. Fuji. (Figure 1).

On the wharf, there are existing buildings and a quay wall (Figure 2), and a construction method was required not to impact them during the construction.

1.2 Background and objectives of the project

The Shimizu Port is divided into the north and south sides (Figure 3). The north quay, both a 30,000 DWT class cargo ships and a 150,000 GT large international passenger ship can berth together. On the other hand, the south quay was not possible to call 2 large passenger ships and cargo ships at the port. In recent years, the number of calls by international passenger ships has been increasing, and further increases are expected in the future. In order to improve the shipping acceptance and deterioration of mooring bank, -this construction order has been placed.

2 STRUCTURAL TYPE AND PILLING METHOD

2.1 Site condition

Since the available space on the existing wharf is limited (see Figure 4 and 5), it was necessary to save space and to shorten the construction time as much as possible in order not to disturb other loading and unloading works. In addition, the construction method was required not to loose the existing ground close to the buildings and quay. Furthermore, it was a condition that steel tubular piles could be installed even if the spacing of the piles was not constant.

2.2 Ground condition

Figure 6 shows the seabed of the location of steel tubular piles. The layer from the seabed to a depth of 9.0m is consisting of approximately 400mm wide diameter rubbles, as shown in the photograph.

From around 9.0m to 18.5m in depth, a silt layer with the SPT N value of about 4 and a gravel layer thereafter (max. SPT N-value is 47) were confirmed from the borehole log (Figure 7).

It was required that the use of Gyro Piler which enables steel tubular pile to penetrate into hard ground by rotary cutting press-in, because it is difficult to carry out the pile installation by other methods such as the press-in piling with water jetting or the vibratory hammer method.

DOI: 10.1201/9781003215226-50

Figure 1. Shimizu Port Hinode Wharf (Map: https://google.co.jp/maps).

Figure 2. Existing buildings and a quay.

Figure 3. Call port status before the construction Hinode Wharf (Map: https://google.co.jp/maps).

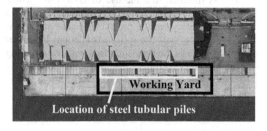

Figure 4. Call port status before the construction Hinode Wharf, (Map: https://google.co.jp/maps).

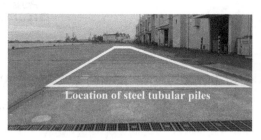

Figure 5. Working Yard before removal of pavement for the installation.

Figure 6. Rubble stones on seabed and installed piles by Gyro Piler in this project.

2.3 Structural type

The mooring quay constructed in this project consists of an interlocking tubular pile wall of 1000mm in diameter and 25.0 to 28.0m in length, and steel tubular piles of 1200mm in diameter and 21.5 to 23.5m in length, which were cast into fillers (sand, stone).

2.4 Piling selection

As the rubble mound does not exist on the sea side, the steel tubular piles with interlocks were planned to be installed by the vibro hammer method using water jets. On the other hand, the anchor piles through the rubble mound were planned to be installed by the Gyropress Method. The Gyro Piler is a piling machine which can install a steel tubular pile with cutting ring bits into the ground through underground obstruction (*i.e.* reinforced concrete) by static load while rotating, without noise and vibration (Figure 9). The pile installation by the Gyro Piler that can realize all of "pile installation to rubble mound", "piling work in narrow space", "Noise and vibration-free piling method" and "pile installation

498

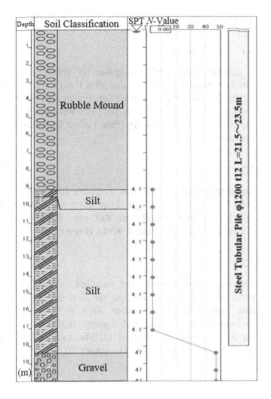

Figure 7. Borehole log.

in a short period", which are problems in this field, were adopted. The internal drilling method was also considered when selecting the piling method, but it was not adopted due to the impact on the existing buildings and quay wall, and the heavy machine size.

2.5 Skip Lock Method

The steel tubular piles as anchor piles should be installed with 3,000mm intervals from center to center, which cannot be applied for the standard Gyro Piler. Therefore, the Gyro Piler with the Skip Lock Method was adopted. Before the development of the Skip Lock Method, the Gyro Piler had difficulty installing steel tubular piles with over 250mm intervals. Therefore, dummy piles between permanent piles were essential for self-walking on the installed tubular piles, however, extracting the installed dummy piles was a problem of this method. The Skip Lock Method was developed to solve this problem.

There are two ways to perform this method: a method of using a Gyro Piler together with a set of Skip Lock attachments, and a method using a "modified Gyro Piler for Skip Lock Method" (refer to Figure 10). When a set of Skip Lock attachments is used, the interval between the pile centers is unchangeable [Pile diameter (mm) × 2.5]. Since it was necessary to handle 2,176mm and 2,496mm

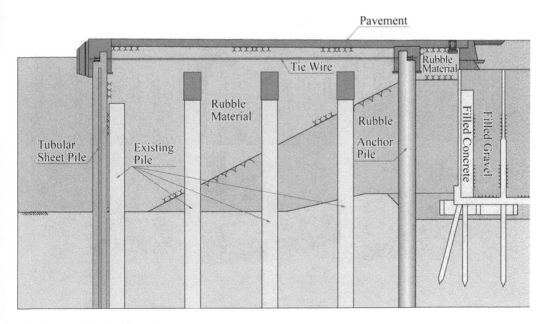

Figure 8. Cross section of mooring bank.

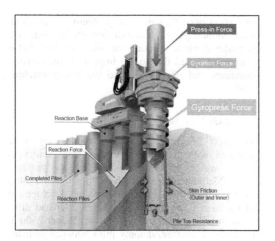

Figure 9. Gyropress method.

intervals between the two pile centers in this site, "modified Gyro Piler for Skip Lock method" was used.

Figure 11 shows the procedure of tubular pile installation.

3 PRESS IN PILING

3.1 *Machine layout*

A service crane works alongside the specified pile line for unloading steel tubular piles from lorries and pitching a pile into the Gyro Piler. For improving the workability and productivity, all piles were placed at the liftable location by the crane before the press-in operation. (Figure 12).

3.2 *Cross section*

Since only one crawler was available for the Gyro Piler in the yard (the Gyro Piler is on the installed piles), the pile installation did not disturb any accesses of lorries for other works (Figure 13).

3.3 *Productivity*

The steel tubular piles were installed by the Gyropress Method in two sections: the first section took five months from December 2019 to May 2020 including the winter and spring holidays, and the second one is expected to take two and half months from September to December 2020.

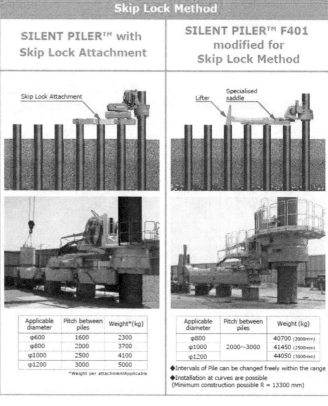

Applicable diameter	Pitch between piles	Weight*(kg)
φ600	1600	2300
φ800	2000	3700
φ1000	2500	4100
φ1200	3000	5000

*Weight per attachmentApplicable

Applicable diameter	Pitch between piles	Weight (kg)
φ800		40700 (2000mm)
φ1000	2000~3000	41450 (2500mm)
φ1200		44050 (3000mm)

◆Intervals of Pile can be changed freely within the range
◆Installation at curves are possible
(Minimum construction possible R = 13300 mm)

Figure 10. Skip Lock Method.

Figure 11. Procedure of the piling operation.

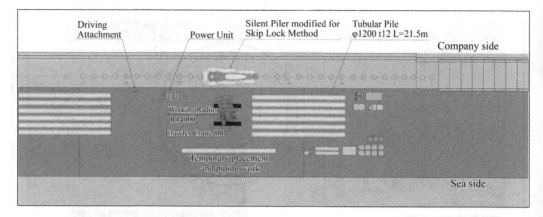

Figure 12. Plan view to show system layout.

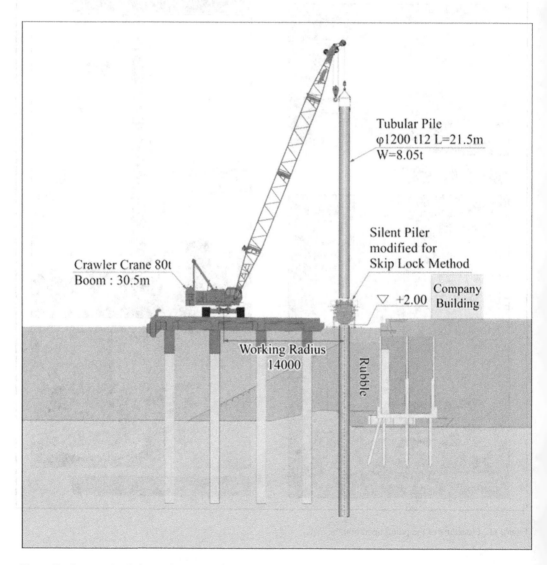

Figure 13. Cross-sectional view to show system layout.

Table 1. Productivity of installing a steel pile.

Material	Specification	Number of Piles	Average Cycle Time per Tubular Pile
Tubular Piler	φ1200, t12mm, L=21.5~23.5m, Joint less	34piles	294min

3.4 Press in data monitoring system

The Gyro Piler can collect press-in data (press-in force, rotational torque, press-in time, *etc.*) in which the piling situation is recorded for each steel tubular pile, and the operator is able to confirm the press-in data, utilize for the analysis of ground condition and so forth, while operating the press-in piling machine. From the data collected at this site (Figure 14), it can be checked that the torque rotational force decreases from around 9.0m depth. It is assumed from the data that the pile toe of the steel tubular pile reached the silt layer through the rubble mound.

In addition, since the torque-rotation force increased from 17.0m depth, it can be anticipated that the pile toe reached the stiff ground layer. In the soil investigation, the borehole log is the result of measuring the "point" in the field. On the other hand, the press-in data, in which steel tubular piles are installed continuously, can confirm the soil condition as a "line". The use of press-in data is an effective tool that can confirm whether the required penetration depth has been ensured for the supporting layer assumed at the design stage.

One cycle of the pile installation including pitching a pile and self-walking took approximately 294 minutes, and the penetration time assumed on the rubble mound was 180 to 240 minutes.

Comparing it with using a set of Skip Lock attachment, the relocation time of a piece of Skip Lock attachment could be reduced about 30min in one cycle. It means that one cycle time per pile was dramatically reduced.

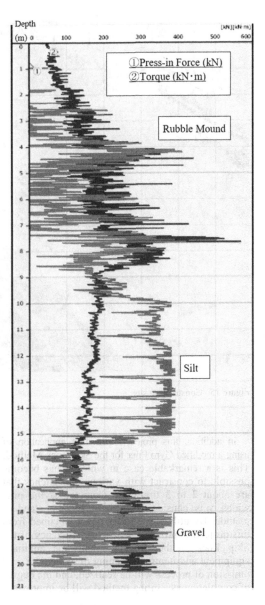

Figure 14. Press in data.

4 CONCLUSIONS

This paper shows the piling application of installing anchor piles by the Gyropress Method and the Skip Lock Method, for steel tubular pile quay wall (Figure 15).

In the result of comparing the Gyropress Method with the Skip Lock Method with the vibro hammer method using water jets after the removal of rubble mound, the Gyropress method was adopted because it can keep a good balance of the "five construction principals" which consists of environment protection, safety, speed, economy and aesthetics. In particular, a large-scale plant is not necessary and steel tubular piles are directly installed without removing the rubble mound, which contributes a great deal toward reducing the total cost and shortening the construction period without disturbing other works.

Figure 15. Construction site.

In addition, this project is the first application of using a modified Gyro Piler for the Skip Lock Method. This is a remarkable case in which it has become possible to construct with various pile pitches that are about 2 to 3 times the diameter of the pile, which was considered difficult by the standard method. By utilizing the experience obtained from this project, improvement of work efficiency of the Skip Lock Method together with the auxiliary equipment and items, shortening of cycle time, and omission of process will be realized, and the degree of completion as a piling method will be improved.

REFERENCES

GIKEN LTD. 2015 *Silent Piling Technologies*, Technical brochure.
GIKEN LTD. 2018. *F401-G100*, Product brochure.
Google Earth, https://google.co.jp/maps
Kitamura, A. 2017 *Construction Revolution*, Diamond. (in Japanese).
Tsukamoto, H. 2013 *To be state of construction five principles of construction scientific assessment model for construction solution section*. Proceedings of 4th International Workshop in Singapore, Press-in Engineering 2013, pp.130–141.

Proceedings of the Second International Conference on
Press-in Engineering 2021, Kochi, Japan – Matsumoto et al (eds)
© 2021 Taylor & Francis Group, London, ISBN 978-1-032-10414-0

Press-in piling applications: Breast walls composed of steel tubular piles and combined wall

M. Yamaguchi & Y. Kimura
GIKEN LTD., Tokyo, Japan

H. Takahagi & M. Okada
GIKEN LTD., Osaka, Japan

ABSTRACT: The disaster rehabilitation work for the Nakanoshima district damaged by the 2011 off the Pacific coast of Tohoku Earthquake was a project carried out for the purpose of constructing the breast wall, quay and related facilities. The most challenging task was the reconstruction of the 542.2m breast wall. Key requirements were [1] build the breast wall in a narrow space, and [2] minimize noise and vibrations during construction not to impact the adjacent buildings and existing piers. To meet these requirements, about 80% of the target length was a structure based on the foundation using a combined wall composed of steel tubular piles and sheet piles, while about 20% of the length formed a structure of cantilever-type steel tubular pile retaining wall. The construction used the rotary press-in piling, Combi-Gyro Method, and Non-staging System. By sharing the project experience, this paper illustrates the advantages of the press-in piling method.

1 INTRODUCTION

In Japan, the 2011 off the Pacific coast of Tohoku Earthquake in March 2011 and the accompanying tsunami damaged many breast walls (seawalls) (NILIM, 2014). For disaster prevention and mitigation in case of assumed big earthquakes such as Nankai Trough Earthquake (JMA, 2020), which is predicted to occur with a probability of 70 to 80% within 30 years, it is urgent to reconstruct the damaged walls as soon as possible and to strengthen the reconstructed walls for disaster prevention. One of the effective piling technologies to overcome that issue is the walk-on-pile type press-in piling.

The Nakanoshima (C) Disaster Rehabilitation Work for the Breast Wall and Other Facilities (hereinafter referred to as the Nakanoshima Disaster Rehabilitation Work) reported in this paper mainly aimed to reconstruct the damaged breast wall and to enhance their functions. For the structure of the reconstructed breast wall, the project selected [1] steel tubular pile retaining wall due to the ground and construction conditions, and [2] a structure that has a combined wall for foundation made up of

Hat-shaped steel sheet piles (hereinafter referred to as Hat sheet piles) for cut-off wall and steel tubular piles (hereinafter referred to as tubular piles) for horizontal resistance to raise the wall height by widening the existing breast wall. This paper presents a case study of the rotary press-in piling, the Combi-Gyro Method, and the Non-staging System, selected in the project as an installation method of tubular piles and sheet piles for constructing breast wall.

2 OVERVIEW OF THE PROJECT

2.1 Location and purpose of the project

The 2011 off Pacific coast of Tohoku Earthquake and subsequent tsunami damaged the breast wall of the Teizan Canal, in the Nakanoshima district managed by Miyagi Prefecture (Figure 1). For this reason, the disaster rehabilitation work was projected to reconstruct the existing 542.2-meterlong breast wall and to construct the small vessel quay, pier, and related facilities.

DOI: 10.1201/9781003215226-51

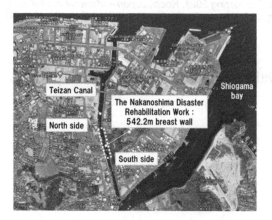

Figure 1. Location Map: Nakanoshima Disaster Rehabilitation Work (Map: https://www.google.co.jp/maps).

2.2 Construction conditions

Private land on the north side or a road on the south side runs side by side behind the existing breast wall, and in addition a pier (Figure 2) to moor small leisure vessels was provided in front of the existing breast wall. As a side note, this pier and its facility had been observed to suffer from subsidence of about 50cm due to the massive earthquake in 2011, for which an application for the disaster rehabilitation work had been submitted as a separate project. This would cause a bottleneck to early reconstruction of the pier because adjustment is required between two works if reconstruction of breast wall impacts the pier. For this reason, the project needed to select a structure form and a construction method that would not impact the reconstruction of the pier in the design phase of reconstructed breast wall.

2.3 Ground condition

Figure 3 shows the soil profile (Soil Section C-1) of the district.

Figure 2. Existing breast wall (seawall) and pier.

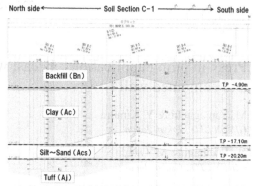

Figure 3. Soil profile: Soil Section C-1.

The geological survey found that the geology is characterized by the distribution of backfill layer (about 4.90m thick) in the surface layer, thick soft clay (about 4.90m thick) with a SPT N-value of 0 to 4 in the next layer, and a deep distribution of the bearing stratum (tuff of T.P. 17.10m to T.P. 20.20m or deeper).

The lack of a bearing stratum in the shallow depth limited the structural form of breast wall to pile foundation or cantilever-type continuous wall, excluding spread footing as it was not suitable.

3 STRUCTURAL REVIEW

3.1 Structural form

As shown in 2.3, under the ground conditions of thickly deposited cohesive soil, it was necessary to apply a foundation form such as a pile foundation or a cantilever-type continuous wall. As the construction section had an existing pier on the canal side (2.2), it was difficult to establish a temporary platform for installing a leader rig mounted press-in system assisted by augering (hereinafter referred to as conventional pile driver, Figure 4). Even on the land side, it was difficult to secure about 10m working width for the conventional pile driver in the hinterland of the existing breast wall (hereinafter referred to as hinterland).

Due to these constraints, it was necessary to select a cantilever-type structure using steel sheet piles (hereinafter referred to as sheet piles) and/or tubular piles, and a structural form that allows "piling by self-walking on the previously installed piles" for installing sheet piles and/or tubular piles. Given the above, three types of forms were reviewed: [1] Tubular pile foundation and cast-in-situ concrete superstructure after removing the existing breast wall, [2] Tubular pile retaining wall without removing the existing breast wall before the pile installation, and [3] Raised structure formed by widening the existing breast wall.

Figure 4. Leader rig mounted press-in system assisted by augering (Yamashita et al. 2010).

3.1.1 Proposal 1: Tubular pile foundation and cast-in-situ concrete superstructure after removal of existing breast wall

This form is a proposal to remove the existing breast wall and to newly construct the breast wall at the same position (Figure 5). As it was difficult to extract the existing H-shaped steel piles, the installation position of the new breast wall had to be moved to the landward side to avoid overlap with the installed tubular piles.

Removal of the existing breast wall was required and there were concerns of a longer construction period due to the larger number of work processes compared to the other two proposals. In addition, moving the piling location of tubular piles to the landward side will exercise the largest impact on the roads and private land in the hinterland compared to the other two proposals, which deemed this inapplicable for the project.

3.1.2 Proposal 2: Tubular pile retaining wall without removing the existing breast wall before the pile installation

This form is a proposal to leave the existing breast wall, install the tubular piles on its front side, and build the tubular pile retaining wall by installing concrete coping on the pile heads (Figure 6). The revetment in front of the existing wall had to be removed to avoid interfering with the tubular pile installation.

There will be no impact on the roads and private land in the hinterland since the tubular pile retaining wall will be built in front of the existing breast wall. However, there was concern that tubular piles would interfere with the floating part of the existing pier. In contrast, this proposal features a shorter construction period than Proposal 1 since the breast wall is built with tubular piles and concrete coping.

The 88-meter-long construction section in the north side in the construction area had an existing pier in front and adjacent private land in the hinterland. For this reason, the raised structure, by widening the existing breast wall, (Proposal 3) was determined to be inapplicable; instead, this structural form was selected for the construction section as it can suppress interference with the pier by installing a tubular pile in the 1.2m space between the existing breast wall and the pier.

3.1.3 Proposal 3: Raised structure formed by widening the existing breast wall

This form is a proposal to raise and widen the existing breast wall since the existing wall suffered minor damage and to install tubular piles as the pile foundation (Figure 7) for enhanced lateral bearing capacity, and reducing the hinterland by 35cm, which is the widen width of the breast wall.

The structural form enables tubular pile installation by installing cut-off sheet piles (Hat sheet piles) behind the breast wall in advance and then having the complete set of piling machinery to self-walk on the pre-installed piles. In addition, the construction cost was estimated to be reduced to about 40% compared with the other two proposals.

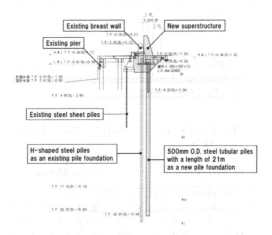

Figure 5. Proposal 1: Tubular pile foundation and cast-in-situ concrete superstructure.

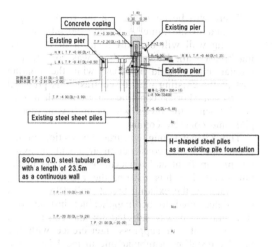

Figure 6. Proposal 2: Tubular pile retaining wall.

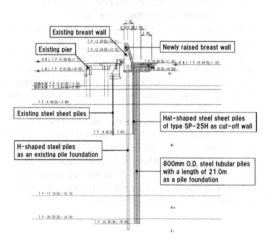

Figure 7. Proposal 3: Raised structure by widening existing breast wall.

Based on the above, this structural form was selected for the 454.2-meter-loncg construction section in the construction area.

4 INTRODUCTION OF PILING METHOD AND SYSTEM

4.1 *Rotary press-in piling (Gyropress Method)*

Rotary press-in piling is a technology to press-in a tubular pile with pile toe ring bits while rotating it. It is applicable not only to cohesive, granular or gravelly soils, but also to harder grounds consisting of bedrocks or containing underground obstacles (such as reinforced concrete structures) since the tubular piles cut and

penetrate into those materials to the planned installation depth (Figures 8 and 9). It is applicable for tubular piles whose outside diameter is from 600mm to 2,500mm (IPA, 2016).

4.2 *Combi-Gyro Method*

The Combi-Gyro Method is a technology of press-in piling for walls that uses a combination of tubular piles and Hat sheet piles (hereinafter referred to as combined wall). The first step of the piling procedure is to press-in a Hat sheet pile. The next step in this method is, by replacing the dedicated chuck for sheet piles with one for tubular piles, to grip a tubular pile, to install it by the rotary press-in piling on the pre-installed sheet piles to gain reaction force. This piling procedure will install a combined wall (IPA, 2016) (GIKEN & NIPPON STEEL, 2017).

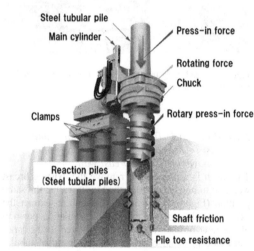

Figure 8. Rotary press-in piling (Gyropress Method).

Figure 9. Reinforce concrete cutting.

Figure 10 Shows the combined wall after the installation in the site.

4.3 *Non-staging System*

The Non-staging System, which enables all the continuous operation including transportation, pitching and press-in pilling of tubular piles and sheet piles to be carried out on the pre-installed piles, can limit the areas affected by the piling work to those occupied by the machinery on pre-installed piles and those used for the working base. Also, the machines are self-supporting by gripping the installed pile, and the risk of falling is extremely low (IPA, 2016).

Machine layout of the Non-staging System in rotary press-in piling is shown in Figure 11.

Figure 10. Combined wall composed of Hat sheet piles and tubular piles.

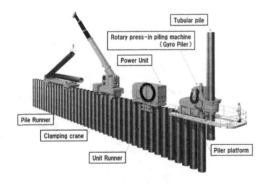

Figure 11. Machine layout of non-staging system in rotary press-in piling.

5 INSTALLATION OF TUBULAR PILES AND COMBINED WALL

5.1 *Construction plan*

This construction section illustrates the construction of the continuous tubular pile wall and the combined wall included in the Nakanoshima Disaster Rehabilitation Work carried out from 2015 to 2016 as a press-in piling application.

This project used tubular piles (SKK400, 800mm O.D., a length of 23.5m, 1 splice, 95 piles) for continuous tubular pile wall, and tubular piles (SKK490, 800mm O.D., a length of 5.0 to 21.0m, non or 1 splice, 159 piles) and Hat sheet piles (25H type, a length of 8.5 to 11.0m, 471 sheets) for the combined wall.

Figure 12 shows the target length and construction sections of each structure while Table 1 shows the quantity of tubular piles and Hat sheet piles required for this project.

5.2 *Procedure for tubular pile installation (Continuous wall)*

In Construction Section 1, the continuous wall of tubular piles was installed by means of rotary press-in piling and the Non-staging System according to the following procedure:

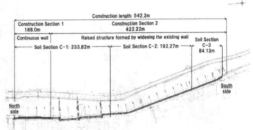

Figure 12. Target length and construction sections.

Table 1. Quantity of tubular piles and Hat sheet piles.

Construction Section	Specification	Remarks
1	Tubular pile - 800mm O.D., a length of 23.5m, 12mm thick, 1 splice, 95 piles	Steel tubular pile retaining wall
2	Tubular pile - 800mm O.D., a length of 5.0m to 21.00m, 16mm thick, non or 1 splice, 159 piles (pile foundation)	Combined wall
	Hat sheet pile SP-25H, a length of 8.5 to 11.00m long 471 sheet(cut-off-wall)	

1) Transport the piles and equipment to the working base and install the service crane.
2) Press-in sheet piles as reaction piles. (Figure 13)

3) Fix a Reaction Stand for tubular piles with the reaction piles. (Figure 14)
4) On the Reaction Stand, assemble the rotary press-in piling machine (hereinafter referred to as Gyro Piler), which was brought into the site in three parts. (Figure 15)
5) Rotary press-in piling of tubular piles with the Gyro Piler and crane. (Figure 16)
6) Install the Non-staging System, including a clamping crane, a Power Unit with Unit Runner, etc., on pre-installed tubular piles. (Figure 17)
7) Lay the pile transportation trackway and the Pile Runner on the pier. (Figure 18)
8) Install the tubular piles using the Non-staging System. (Figure 19)
9) After completion of all tubular pile installation, remove the pile transportation trackway, and let the Gyro Piler and clamping cranes self-walk backward to the working base.

Figure 15. Assembly of Gyro Piler.

Figure 13. Press-in sheet piles as reaction piles.

Figure 16. Rotary press-in piling of tubular piles.

Figure 14. Installation of reaction stand for tubular piles.

Figure 17. Installation of a clamping crane.

Figure 18. Pile runner and transportation trackway on pier.

Figure 19. Non-staging system (tubular piles/continuous wall).

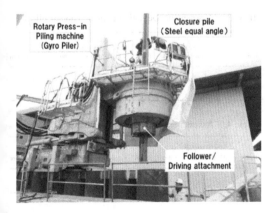

Figure 20. Installation of closure piles.

Figure 21. Steel tubular pile retaining wall.

10) Disassemble and remove the Gyro Piler and clamping crane.
11) Dismantlement and removal of equipment.

Spacing between tubular piles was ensured by press-in piling of closure piles with the Gyro Piler equipped with a follower/driving attachment. (Figure 20)

The steel tubular pile installation for the continuous wall with the rotary press-in piling was completed in about 2 months from March to May in 2016. (Figure 21)

5.3 Procedure for installing combined wall (cut-off sheet piles and pile foundation)

In Construction Section 2, the combined wall was installed using the Combi-Gyro Method and Non-staging System according to the following procedure:

1) Transport the piles/sheet piles and equipment to the working base and install a crane.
2) Standard press-in or press-in assisted with augering of Hat sheet piles using the Combi-Gyro Piler attached to a dedicated chuck for Hat sheet piles and the crane. (Figure 22)
3) Install the clamping crane and Power Unit with Unit Runner on the pre-installed piles.
4) Lay the pile transportation trackway and Pile Runner on the pre-installed piles. (Figure 23)
5) Install Hat sheet piles with the Non-staging System.
6) After completion of all of the Hat sheet pile installation, replace the dedicated chuck for the Hat sheet piles of Combi-Gyro Piler with the dedicated one for tubular piles.
7) Start backward self-walking and install the tubular piles with the Non-staging System. (Figure 24)

8) After completion of all tubular pile installation, the Non-staging System arrives at the working base.
9) Disassemble and remove the Combi-Gyro Piler and clamping crane.
10) Dismantle and remove all equipment.

Figure 22. Press-in assisted with augering of Hat sheet piles.

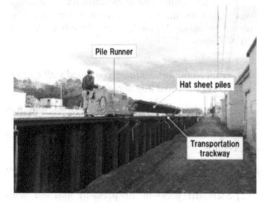

Figure 23. Pile Runner and pile transportation trackway on installed sheet piles.

Figure 24. Non-staging System (Tubular piles as a primary structure of combined wall).

Hat sheet piles were installed in about 2 months from November 2015 to January 2016. Then tubular piles were installed in about 3 months until April of the same year to build the combined wall in a total of 5 months. (Figure 10)

5.4 Quality control of installed piles/sheet piles

Quality control was carried out on required parameters concerning dimensional accuracy, to ensure the installation of Hat sheet piles and tubular piles comprised with the standards and criteria specified by the client/owner (TRDB, 2020) (JRA, 2007). The results are as follows.

5.4.1 Tubular piles

Tables 2 to 5 show the results of measuring [1] height of pile top level, [2] deviation in plan, [3] inclination, and [4] embedded length of installed tubular piles. It was confirmed that sufficient piling accuracy was ensured for each measurement item.

5.4.2 Hat sheet piles

Tables 6 to 11 show the results of measuring [1] height of pile top level, [2] deviation from the planned line, [3] inclination (rightward/leftward and piling direction), [4] embedded length, and [5] wall length of Hat sheet piles. It was confirmed that sufficient piling accuracy was ensured for each measurement item.

Table 2. Assessment: Height of pile top level (Tubular piles).

Item	Unit	Control value	Mean value	Max. value	Min. value	Standrad deviation	Sample no.	Assessment
Pile top level	mm	±50.00	-1.02	15.00	-33.00	7.58	254	pass

Table 3. Assessment: Deviation in plan (Tubular piles).

Item	Unit	Control value	Mean value	Max. value	Min. value	Standrad deviation	Sample no.	Assesment
Deviation in plan	mm	100.00	17.75	55.00	00.00	11.36	254	pass

Table 4. Assessment: Inclination (Tubular piles).

Item	Unit	Control value	Mean value	Max. value	Min. value	Standrad deviation	Sample no.	Assesment
Inclinatio	%	2.00	0.14	1.00	00.00	0.35	254	pass

Table 5. Assessment: Embedded length (Tubular piles).

Item	Unit	Control value	Mean value	Max. value	Min. value	Standrad deviation	Sample no.	Assesment
Embedded length	mm	0 or over	39.59	73.00	14.00	7.80	254	pass

Table 6. Assessment: Height of pile top level (Hat Sheet Piles).

Item	Unit	Control value	Mean value	Max. value	Min. value	Standrad deviation	Sample no.	Assesment
Pile top level	mm	±50.00	-13.30	35.00	-10.00	12.30	24	pass

Table 7. Assessment: Deviation from the planned line (Hat sheet piles).

Item	Unit	Control value	Mean value	Max. value	Min. value	Standrad deviation	Sample no.	Assesment
Deviation from the the planned line	mm	±100.00	-20.50	17.00	-65.00	18	24	pass

Table 8. Assessment: Rightward/Leftward inclination (Hat sheet piles).

Item	Unit	Control value	Mean value	Max. value	Min. value	Standrad deviation	Sample no.	Assesment
Inclination	degree	±10/1000	1.62	7.00	-6.00	3.00	24	pass

Table 9. Assessment: Inclination in the piling direction (Hat sheet piles).

Item	Unit	Control value	Mean value	Max. value	Min. value	Standrad deviation	Sample no.	Assesment
Inclination	degre	±10/1000	1.70	7.00	-2.00	2.60	14	pass

Table 10. Assessment: Embedded length (Hat Sheet Piles).

Item	Unit	Control value	Mean value	Max. value	Min. value	Standrad deviation	Sample no.	Assesment
Embedded length	mm	0 or over	56.70	80.00	35.00	12.3	24	pass

Table 11. Assessment: Wall length (Hat sheet piles).

Item	Unit	Control value	Mean value	Max. value	Min. value	Standrad deviation	Sample no.	Assesment
wall length	mm	0 or over	54.30	300.00	0.00	101.30	7	pass

6 SUPERSTRUCTURE WORK

After the completion of the steel tubular pile retaining wall (5.2), the combined wall composed of cut-off wall and pile foundation (5.3), the concrete coping work with cast-in-situ concrete and the breast wall superstructure work were carried out from March 2016 to March 2017 to complete the project (Figures 25 and 26).

7 SUMMARY

The case study of breast wall reconstruction in the disaster rehabilitation work reported in this paper demonstrates the following advantages of the rotary press-in piling, Combi-Gyro Method, and Non-staging System:

- With the rotary press-in piling, it is possible to install tubular piles as bearing piles, by cutting and penetrating the tuff (bearing stratum) without removing the existing structure (breast walls) in advance.
- The Non-staging System allows tubular piles and sheet piles to be installed in, even under the construction conditions deemed difficult with the other piling method (e.g. conventional pile driver). This provides good grounds for determining the structural form based on the construction conditions.
- By using pre-fabricated tubular piles and sheet piles, high-quality and high-strength tubular pile retaining wall, cut-off wall and tubular pile foundation can be constructed.
- Tubular pile retaining wall, cut-off wall (Hat sheet pile) and pile foundation (mono tubular piles) constructed with the rotary press-in piling or Combi-Gyro Method meet the control values at a high level in pile top level, deviation in plan/ deviation from the planned line, inclination, embedded length and other dimensional factors.

As is often the case with disaster rehabilitation work to reconstruct facilities adjacent to canals such as breast walls (seawalls), it is difficult to secure sufficient work space in narrow areas surrounded by adjacent buildings with many restrictions on construction conditions. Nevertheless, the project must keep the facilities around the construction site functional, bring about safety and security during construction, and be economically efficient to complete the construction in a short time. For addressing these issues, the advantages of the rotary press-in piling, Combi-Gyro Method and Non-staging System were demonstrated at the site.

Finally, we hope that the press-in piling application reported in this paper will serve as a reference for similar breast wall construction projects.

ACKNOWLEDGMENTS

The authors would like to express our gratitude to the Sendai Shiogama Port and Harbour Office and Central Consultant Inc., for their very helpful support and cooperation in preparing this paper.

Figure 25. Newly constructed breast wall as seawall (North side).

Figure 26. Newly constructed breast wall as seawall (South Side).

REFERENCES

GIKEN LTD. and NIPPON STEEL CORPORATION (GIKEN & NIPPON STEEL) 2017. *Combi-Gyro Method Ver 2*, 4pp. Kochi: GIKEN.

International Press-in Association (IPA) 2016. *Design and Construction Guideline for Press-in Piling 2016*, 540pp. Tokyo: IPA.

Japan Road Association (JRA) 2007. Pile Foundation Construction Handbook, 400pp. (in Japanese), Tokyo: JRA.

National Institute for Land and Infrastructure Management Ministry of Land, Infrastructure, Transport and Tourism, Japan (NILIM) 2014. *Field Survey of the 2011 off the Pacific coast of Tohoku Earthquake and Tsunami on Shore Protection Facilities in Ports (II)*, ISSN1346-7328, 56pp. (in Japanese), Tokyo: NILIM.

The Japan Meteorological Agency of the Ministry of Land, Infrastructure, Transport and Tourism (JMA) 2020. *Estimated Intensity of Seismic Waves and Height of Tsunamis Caused by Nankai Trough Earthquake*. Available (online) at: https://www.data.jma.go.jp/svd/eqev/data/nteq/assumption.html (in Japanese) (Retrieved on November 15, 2020)

Tohoku Regional Development Bureau, Ministry of Land, Infrastructure, Transport and Tourism (TRDB) 2020. *Standard and Control Values for Civil Work Construction Management*. Available (online) at: http://www.thr.mlit.go.jp/bumon/b00097/k00910/kyokyou/H27siyousho/H27kyoutuusekoukanri/h27.6sekoukanri.pdf (in Japanese) (Retrieved on 15 November, 2020)

Yamashita, H., Hirata, H. & Kinoshita, M. 2010. Challenges in the past and for the future of design and installation technologies on steel pipe piles in Japan, *Proceeding of Japan society of civil engineering, vol.66 No.3*, pp. 319–336 (in Japanese), Tokyo: Japan society of civil engineering.

Proceedings of the Second International Conference on
Press-in Engineering 2021, Kochi, Japan – Matsumoto et al (eds)
© 2021 Taylor & Francis Group, London, ISBN 978-1-032-10414-0

Press-in piling applications: Seawall pile foundation work

M. Yamaguchi & Y. Kimura
GIKEN LTD, Tokyo, Japan

H. Takahagi & M. Okada
GIKEN LTD, Osaka, Japan

ABSTRACT: In the design phase of the seawall reconstruction project reported in this paper, there was concern that the new seawall was too close to the port facilities. Because of this, it was difficult to secure space to reconstruct the seawall. In addition, vibrations created during construction had to be minimized to prevent interference with an automatic tide-gauge station. Considering these conditions, the following decisions were made: [1] Steel tubular piles shall be used as pile foundations for the seawalls; [2] Rotary press-in piling shall be selected due to the pile installation through existing structures; and [3] A combination of the Skip Lock System, which enables mono-pile installation, and the Non-staging System, which enables piling work on the installed piles shall be used. This paper aims to review the advantages of press-in piling, by reporting the outlines of the project, the structure design and the plan and implementation of the construction.

1 INTRODUCTION

The occurrence of the Great East Japan Earthquake in March 2011, which caused the devastating damage and left more than 20,000 people dead and missing in 12 prefectures, still remains vivid in the memory of many people. The Central Disaster Prevention Council of Japan has estimated the damage by an earthquake of maximum magnitude in the Nankai Trough. According to the estimate, some areas along the pacific coast from Shizuoka Prefecture to Miyazaki Prefecture will possibly be hit by massive seismic waves with an intensity of seven, and wider areas adjacent to those hardest hit areas will suffer strong seismic waves with an intensity of upper six to lower six. In addition, wide areas along the pacific coast from the Kanto to Kyushu regions will be hit by massive tsunamis with a height exceeding 10 meters (JMA, 2020).

Under such circumstances, it is a pressing issue to restore damaged port facilities quickly and make new structures more resilient to disasters than before. Walk-on-pile type press-in piling is one of the piling technologies that can address this issue effectively. The following are increasingly being selected at many construction sites: [1] Rotary press-in piling (Gyropress Method), a pile installation method that rotates a steel tubular pile (hereinafter referred to as tubular pile) with pile toe ring bits, and presses it into the ground and through any existing structures. [2] The Skip Lock System, which enables

the installation of tubular piles at certain intervals. [3] The Non-staging System, which enables piling work in a limited space by a travel crane that can move on the installed piles.

This paper reports a reconstruction project on the damage to Kamaishi's fishing port from the Great East Japan Earthquake. It was difficult to secure space for the dismantling and removal of existing structures and the construction of a new seawall, because part of the planned new seawall was too close to the port facilities. In addition, vibration during the construction had to be minimized to avoid interfering with the automatic tide-gauge station near the construction site. Under these conditions, tubular piles were installed to form the foundations of the new seawall using [1] Rotary press-in piling, [2] the Skip Lock System, and [3] the Non-staging System. Details of the construction work are described as follows.

2 OUTLINES OF THE PROJECT

2.1 *Objectives of the disaster rehabilitation project in Kamaishi's fishing port*

The Great East Japan Earthquake in 2011 and the following tsunami with a maximum height of 6.7 meters in Kamaishi, Iwate Prefecture, seriously damaged Kamaishi's fishing port (Figure 1) and its facilities (managed by the Iwate Prefectural. Government) as

DOI: 10.1201/9781003215226-52

Figure 1. Kamaishi's fishing port in Iwate Prefecture, Japan (Map: https://www.google.co.jp/maps).

well as the surrounding area. To restore the functioning of the damaged Kamaishi port, a disaster rehabilitation project for the fishing port was planned with the aim of reconstructing a 1.9 linear km long seawall and related facilities. The project commenced in the fiscal year of 2011, and the construction work started in the fiscal year of 2013, with the aim of completing it in September 2020.

The disaster rehabilitation project reported in this paper aimed at constructing a new seawall, a land lock gate, a sluice gate, and other facilities that would replace the damaged structures.

2.2 Construction conditions

Since the Kamaishi Port Joint Government Office building occupied the land next to the damaged seawall, there was no space available for the installation and operation of heavy machinery. In addition, there was an automatic tide-gauge station that occupied part of the construction site facing the sea, so it was difficult to secure workspace with access on water by a temporary platform and a rubble mound, or preparing a barge.

Furthermore, it was considered difficult to secure enough working space for the installation of equipment and materials, needed for the dismantling and removal of the damaged structures, and the construction of the new seawall. This is because part of the new seawall was too close to the Kamaishi Port Joint Government Office facility.

Moreover, vibration in construction had to be minimized to avoid interfering with the constant tide level monitoring at the automatic tide-gauge station.

2.3 Ground conditions

The locations of collected borehole logs are shown in Figure 2 (B-1, B-29, and B-30), and the geological cross section at the construction site is shown in Figure 3.

An investigation of the soil revealed the following ground conditions: relatively soft alluvial deposits (composed of silty sand, fine sand, organic soil, and gravel with silt) lie in a range of 30.13 meters (No. B-1) to 16.04 meters (No. B-29) above mean sea

level, and a layer of slate (extrapolated SPT N-values of 100 to 500) lies under the alluvial deposits. Based on this information, a pile foundation was selected as an effective foundation for the new seawall.

3 SEAWALL DESIGN

3.1 Structure type

This project is a disaster rehabilitation work for the damaged seawall. For the superstructure of the new T-shaped sea wall, prefabricated reinforced concrete unit and/or cast-in-situ reinforced concrete were adopted. At the same time, a pile foundation composed of two-row tubular mono piles was adopted as a foundation type (Figure 4).

Three different tubular piles with outside diameters of 800, 1,000, and 1,200mm were examined in the feasibility study for the construction of the pile foundation (Table 1). Installing 1,000 mm diameter tubular piles at a 2.5 meter spacing from centre to centre (C/C) was concluded to be most effective after comparing them in terms of the width occupied by the superstructure, and construction costs and duration.

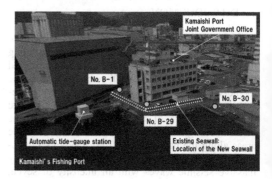

Figure 2. The site before the disaster rehabilitation, including numbers of borehole logs.

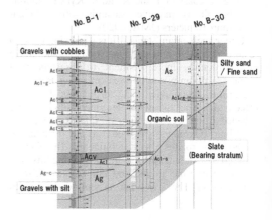

Figure 3. A geological cross section at the site.

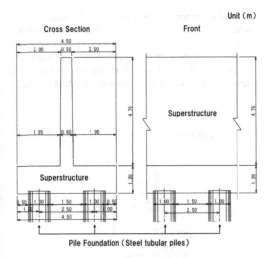

Figure 4. The design of the new seawall (Tubular pile diameter: 1,000 mm, Spacing from C/C: 2.5 m).

Table 1. A comparison of tubular pile diameters on a 10 m long section.

	Option #1	Option #2	Option #3
Spec. of steel tubular piles			
Diameter (mm)	800	1000	1200
Thickness (mm)	19	10	12
No. of lows	2	2	2
Spacing from C/C (m)	2.00	2.50	3.00
Displacement at the wall top			
Due to tsunami (cm)	1.425≤1.500	1.425≤1.500	1.425≤1.500
Due to earthquake (cm)	0.465≤1.500	0.465≤1.500	0.465≤1.500
Superstructure width (m)	3.60	4.50	5.40
Construction cost	1.15	1.00 (base)	1.04
Construction period	#3 plus 21 days	#3 plus 10 days	Shortest
Assesment	Acceptable	Good	Acceptable

3.2 Guidelines and conditions for the pile foundation design

The following guidelines were used for the structural design of the new seawall.

- River Improvement and Management Division, Water and Disaster Management Bureau of the Ministry of Land, Infrastructure, Transport and Tourism. 2012. *Guidelines for the Examination of the Seismic Capacity of River Structures*. (in Japanese)

Table 2. Design conditions for the pile foundation.

Bearing method	End-supported pile
Pile head type	Rigid connection type (either rigid connection or hinge connection type with greater crosssectional force to be selected at the design phase)
Pile toe type	Hinge
Safety factor for allowable bearing capacity (End-supported pile)	3.0 for normal conditions, 2.0 for L1 earthquakes, and 2.0 for L1 tsunamis
Pile layout	Minimum pile edge distance: 1.00 x pile diameter Minimum distance between the centers: 2.50 × pile diameter Maximum distance between the centers: 10.0 × pile diameter, or 4 m (whichever is smaller)
Allowable corrosion depth of steel	1mm (for sections submerged under water or the ground, all of the time)

- Fishing Port and Village Management Department, Agriculture, Forestry, and Fishery Division, Iwate Prefecture. 2012. *Guidance for the structural design of coast protection facilities (measures against tsunamis) for fishing port coasts (draft)*. (in Japanese)
- Japan Road Association. 2012. *Guidelines for Road Earthwork, Retaining Wall Construction, 2012 Edition*. (in Japanese)

The design conditions for the pile foundation are shown in Table 2. The values in the table were based on Part I Common Specifications and Part IV Specifications for Base Structures of the Specifications for Highway Bridges (JRA, 2012) and the Pile Foundation Design Handbook (JRA, 2015).

4 TUBULAR PILE INSTALLATION METHOD

4.1 Rotary press-in piling (Gyropress Method)

Rotary press-in piling is a technology to press-in a tubular pile with pile toe ring bits while rotating it. It is applicable not only to cohesive, granular or gravelly ground, but also to hard ground such as rock mass and underground obstacles (such as reinforced concrete structures) since the tubular piles cut and penetrate into those types of ground to the specified pile installation depth (Figures 5 and 6). It is applicable to a tubular pile with an outside diameter of 600 to 2,500mm (IPA, 2020).

When it is desirable to install tubular piles at a specified interval, not in a continuous wall arrangement, the spacing from C/C can be extended up to a distance 2.5 times larger than the diameter of the tubular pile (D) by using the Skip Lock System (Figure 7).

4.2 The Non-staging System

By using the Non-staging System, which enables all the piling processes including the transport, pitch, and pressing-in of piles to be carried out on the previously installed piles, areas affected by the piling work space can be limited to those occupied by the width of the machinery operated on the pre-installed piles, and those used for the construction base. In addition, the machinery, by gripping the pre-installed piles, is self-supporting, and the risk of them falling over is extremely low (IPA, 2020).

The machine layout of the Non-staging System used at the site in this report is shown in Figure 8.

4.3 Reasons for the piling method selection

For the selection of a tubular pile installation method, the combination of the aforementioned technology and systems were compared with the combination of pre-boring by the all-casing method and the impact driving of tubular piles (hereinafter referred to as the conventional piling method).

Considering the construction conditions in this project described in 2.2, a temporary platform or a rubble mound needed to be installed for the workspace if a conventional piling method was used. After a comparative analysis in terms of construction time and costs, the rotary press-in piling of steel tubular piles combined with the Skip Lock System and the Non-staging System was selected.

5 PILE FOUNDATION WORK

5.1 Construction plan

The pile foundation work carried out in the disaster rehabilitation project at Kamaishi's fishing port from 2017 to 2020 is reported in the following section as a press-in piling application.

The construction plan of this project was to install tubular piles (a total 58 of SKK400 piles with an outside diameter of 1,000mm and a length of 19.5 to 39.0m and each having from 0 to 12 splices) as pile foundations after dismantling and removing the damaged seawall. After that, T-shaped wall by cast-in-situ reinforced concrete was built as a superstructure, completing the new seawall.

The steel tubular pile layout and a standard cross section are shown in Figures 9 and 10.

5.2 Underground obstacles

Before installing the tubular piles, a temporary earth retaining wall was constructed with steel sheet piles, and existing structures were dismantled and removed. Figure 11 shows the dismantling and removal work of the existing seawall.

Aged revetments, stacked blocks, and cast-in-situ concrete under the ground were difficult to completely remove, so they were left as they were. After carrying out pre-cutting by a rotary press-in piling machine (hereinafter referred to as Gyro Piler), the tubular piles were pressed-in until they reached the bearing stratum (slate).

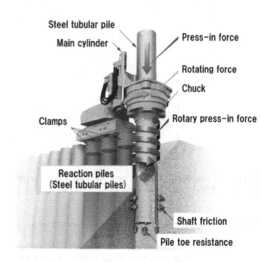

Figure 5. Rotary press-in piling (Gyropress Method).

Figure 6. Reinforced concrete cutting.

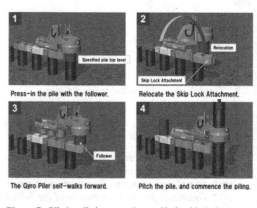

Figure 7. Pile installation procedures with the skip lock system.

Figure 8. The machine layout of the non-staging system.

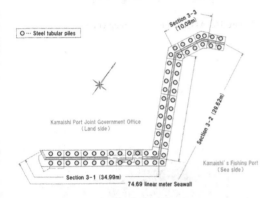

Figure 9. The layout of the steel tubular piles.

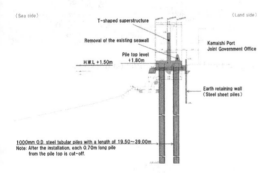

Figure 10. Standard cross section: Section 3-1.

Figure 11. Removing the existing seawall and afterwards.

5.3 Piling procedure

The installation of tubular piles by the rotary press-in piling was carried out according to the following procedure.

1) Equipment and piles were transported to the site, and a crane barge was installed above water adjacent to the site.
2) Install steel sheet piles as reaction piles (Figure 12).
3) Install a Reaction Stand for tubular piles and fix it to the reaction piles (Figure 13).
4) Assemble the Gyro Piler, which was brought to the site in three parts, on the Reaction Stand.
5) Install the tubular piles using the Gyro Piler, the Skip Lock System and the crane barge (Figures 14 and 15).
6) Install the Non-staging System on the pre-installed tubular piles (Figure 16).
7) Install a transportation trackway on land along the pile line in section 3-2 (Figure 17).
8) Install the tubular piles using the Non-staging System (Figure 18).
9) After installing the piles remove the transportation trackway, and move the Gyro Piler and the clamping crane backward by self-walking to the location where they can be disassembled and removed, after the completion of all pile installation.
10) Disassemble and remove the Gyro Piler and the clamping crane.
11) Remove all materials and equipment.

The tubular pile installation by rotary press-in piling was completed in three months from May to August in 2018.

Figure 12. Pressing-in steel sheet piles as reaction piles.

Figure 13. Installation of a reaction stand for tubular piles.

Figure 16. Assembling a clamping crane.

Figure 14. Steel tubular pile installation: Section 3-2.

Figure 17. The pile transportation trackway.

Figure 15. Steel tubular pile installation at the corner.

Figure 18. Tubular pile installation using the non-staging system.

5.4 Quality control of installed tubular piles

Quality control was carried out on the following construction parameters concerning dimensional accuracy, to ensure the installation of tubular piles complied with the standards and criteria specified by the client/owner (TRDB, 2020) (JRA, 2015): 1) height of the top level, 2) deviation in plan, and 3) the inclination.

5.4.1 Height of the top levels

In this project, the top level of each tubular pile was set at T.P. 1.80 meters (design value) to avoid the equipment and materials submerged during the pile installation. The top level of the tubular piles was controlled to meet the control value with a reference level of ±50.00mm tolerance using an elevation measuring instrument (Total Station) at the end of each press-in piling.

The control value and actual measurements (mean value, maximum plus and minus values, and standard deviation) of the installed piles' top level are shown in Table 3. The top levels ranged from 33.00 to 12.00mm with a mean of 17.00mm and standard deviation of +11.61mm, which also means that the pile toes could reach the specified level of bearing stratum.

The data demonstrated that the tubular mono piles installed by the rotary press-in piling with the Skip Lock System could achieve a high level of compliance with the reference level.

5.4.2 Deviations in plan

Inspections for deviations in plan of the tubular piles were conducted for all installed tubular piles at this site using a measuring instrument (Total Station). The inspection results are shown in Table 4.

The required control value was within 100mm. The measurement data showed that deviations in plan of the tubular piles were in the range of 1.00 to 57.00mm, with a mean of 19.00mm per pile and a standard deviation of ±13.04, which demonstrated sufficient piling accuracy.

5.4.3 Inclinations

The inclination of the installed tubular piles was measured by a measuring instrument (Total Station) and a spirit level. The results are shown in Table 5. The control value for the inclination was within 2 degrees.

The measurement data showed that the inclination of the installed tubular piles was in a range of 0.10 to 0.7 degrees with a mean of 0.41 degrees per pile and a standard deviation of ±0.14, which demonstrated the successful installation of the tubular piles.

Table 3. Assessment: Height of the top levels.

Item	Unit	Control value	Mean value	Max. value	Min. value	Standard deviation	Sample No.	Assesment
Pile top level	mm	±50.00	-17.00	12.00	-33.00	±11.68	58	Pass

Table 4. Assessment: Deviations in plan.

Item	Unit	Control value	Mean value	Max. value	Min. value	Standard deviation	Sample No.	Assesment
Deviation in plan	mm	100.00	19.00	57.00	1.00	±13.04	58	Pass

Table 5. Assessment: Inclinations.

Item	Unit	Control value	Mean value	Max. value	Min. value	Standard deviation	Sample No.	Assesment
Inclination	degree	1.00	0.20	0.60	0.00	±0.14	58	Pass

6 SUPERSTRUCTURE WORK

After completion of the pile foundation work, the construction of the superstructure, made of reinforced concrete, was carried out.

It was preferable to use precast reinforced concrete members instead of cast-in-situ reinforced concrete, from the overall consideration of workability at the site and construction time and cost. However, a precast concrete seawall was concluded not to be suitable for these sections because the installation of a temporary platform and a rubble mound for heavy machines, which was essential for the handling of precast concrete members, was difficult along the piling location. As a result, a T-shaped seawall made of cast-in-situ reinforced concrete was adopted (Figure 19).

The in-situ concrete work for the superstructure was carried out for each section from May 2019 to January 2020. After completion of the superstructure, wave-dissipating concrete blocks and steps were installed. This project was completed in February 2020 (Figures 20 and 21).

7 CONCLUDING REMARKS

In this case study, the following advantages were demonstrated for the combination of the rotary press-in piling, the Skip Lock System, and the Non-staging System, in the pile foundation work for the construction of a seawall in a disaster rehabilitation project:

• By combining the rotary press-in piling with the Skip Lock System, tubular mono piles can be installed, which had been considered to be difficult with press-in piling.
• It is possible to press-in tubular piles into slate (bearing stratum) and install end-supported piles without removing underground obstacles (e.g. aged revetments, stacked blocks, and reinforced concrete) beforehand.

Figure 19. Construction of the superstructure on installed piles.

Figure 20. The newly constructed seawall: Section 3-1.

Figure 21. An aerial view of the newly constructed seawall.

- Tubular pile foundations can be constructed in ports and harbours without requiring temporary platforms and rubble mounds in cases where the Non-staging System is used.
- It is possible to construct a high-strength and high-quality pile foundation by using prefabricated tubular piles.
- Mono tubular piles installed using the rotary press-in piling in combination with the Skip Lock

System are capable of meeting the control values for top levels, deviations in plan, and inclinations with sufficient margin.

In disaster rehabilitation projects for port and harbour facilities such as seawalls, it is often the case that space provided for the construction work is limited due to the presence of structures adjacent to the construction site, or the narrowness of the site itself. And a lot of time and cost have to be spent on the installation of large-scale temporary platforms and/or rubble mounds. In such situations effective measures need to be devised to keep the facilities around construction sites functional, ensure safety and security in the construction process, and to avoid interfering with the movement of ships. The advantage of rotary press-in piling, the Skip Lock System, and the Non-staging System, in addressing the aforementioned issues has been proven through this project.

Finally, we hope that the press-in piling application reported in this paper will serve as a reference for similar disaster rehabilitation projects.

ACKNOWLEDGMENTS

The authors would like to express our gratitude to the Fisheries Department of the Iwate Prefecture Coast Area Development Bureau, Sanyo Consultants Co., Ltd., and Yamamoto Constax for their very helpful support and cooperation in preparing this paper.

REFERENCES

International Press-in Association (IPA) 2020. Design and Construction Guideline for Press-in Piling. 540pp. (in Japanese), Tokyo: IPA.

Japan Road Association (JRA) 2012. *Guidelines for Road Earthwork, Retaining Wall Construction, 2012 Edition.* 342 pp. (In Japanese), Tokyo: JRA.

Japan Road Association (JRA) 2012. *Part I Common Specifications and Part IV Base Structure Specifications of the Specifications for Highway Bridges*, 586pp. Tokyo: JRA.

Japan Road Association (JRA) 2015. *Pile Foundation Construction Handbook.* 400pp. (in Japanese), Tokyo: JRA.

Japan Road Association (JRA) 2015. Pile Foundation Design Handbook. 510pp. (in Japanese), Tokyo: JRA.

River Improvement and Management Division of the MLIT Water and Disaster Management Bureau (RIMD) 2012. *Guidelines for the Examination of the Seismic Capacity of River Structures, Commentary.* Available (online) at: https://www.mlit.go.jp/river/shishin_guideline/bousai/wf_environment/structure/index3.html (in Japanese) (Retrieved on 15 November, 2020)

The Japan Meteorological Agency of the Ministry of Land, Infrastructure, Transport and Tourism (JMA) 2020. *Estimated Intensity of Seismic Waves and Height of*

Tsunamis Caused by South Sea Trough Earthquake. Available (online) on at: https://www.data.jma.go.jp/svd/eqev/data/nteq/assumption.html (in Japanese) (Retrieved on 15 November 2020)

Tohoku Regional Development Bureau, Ministry of Land, Infrastructure, Transport and Tourism (TRDB) 2020. *Standard and Control Values for Civil Work Construction Management.* Available (online) at: http://www.thr.mlit.go.jp/bumon/b00097/k00910/kyou/H27siyousho/H27kyoutuusekoukanri/h27.6sekoukanri.pdf (in Japanese) (Retrieved on 15 November, 2020)

*Proceedings of the Second International Conference on
Press-in Engineering 2021, Kochi, Japan – Matsumoto et al (eds)
© 2021 Taylor & Francis Group, London, ISBN 978-1-032-10414-0*

Upgrading earthen levees with press-in piling and the GRB System

T. Takuma
Giken Ltd., c/o Giken America Corp., Orlando, FL, USA

S. Kambe
Blue Iron Foundations and Shoring LLC, Casselberry, FL, USA

M. Nagano
Giken America Corp., New York, NY, USA

ABSTRACT: Sheet piles have been utilized to achieve existing earthen levee upgrades on many projects because of their strength, durability, waterproofness, flexibility in choice of length and rigidity, and relative ease of construction. However, installation of sheet piles into existing levees faces a challenge if the project is very close to homes or in tight working space. This paper will discuss a couple of levee repair and upgrade projects in the U.S. that utilized press-in piling combined with the GRB system to overcome this type of logistical difficulty.

1 PRESS-IN PILING AND GRB SYSTEM

1.1 Press-in piling on earthen levee upgrade

Pressed-in sheet and pipe piles have been increasingly utilized for the upgrades of existing earthen levees for coastal as well as riverine use. The authors have been personally involved in some of the levee and seawall damage recovery projects built after the 2011 Great East Japan Earthquake/Tsunami and existing levees' upgrade projects in California, Florida, and Louisiana in the U.S. The primary reasons for the selection of the press-in piling over other methods are summarized as follows:

- Noise and/or vibration associated with piling need to be mitigated due to the proximity to nearby homes and sensitive structures.
- Project sites are physically confined.
- Piles have to be installed into hard soil, debris, or through existing concrete structures. This can be accomplished with press-in piling assisted with an auger attachment for sheet piles or the rotary press-in piling (Gyropress) method in the case of pipe pile installation.

1.2 GRB system and its advantages

The GRB (the initialism for Giken Reaction Base) System is a combination of proprietary piling equipment which transports, pitches, and presses in piles: all on the top of already installed piles. Because it

eliminates the need for a construction access passage on the ground or above water, depending on the project's conditions, it enables pile installation where the access to the piling location is limited. See Figure 1 for its components and their functions.

In the case of levee upgrades, the access to the pile lines is often limited on a narrow strip of land with a body of water on one side and residential units, businesses, schools, or busy roadways on the other side. Two case studies on levee repair and upgrade that used the GRB System in the U.S. will be discussed hereinafter.

2 EAST GARDEN GROVE - WINTERSBURG CHANNEL NORTH LEVEE EMERGENCY PROJECT (HUNTINGTON BEACH, CALIFORNIA)

The East Garden Grove - Wintersburg Channel is a drainage waterway located in Huntington Beach, California maintained by the Orange County Public Works Department. It was originally constructed back in the Orange County, California, which is operated and 1960's with its levees built primarily of native soil. It has a trapezoidal shape section with a 4.9-to-5.5-meter-wide unlined bottom and 3.0-to-3.5-meter-high embankments on both sides. The side slope is 1.5 to 1.0 (horizontal to vertical). In response to the damage caused by the 2005 storms, the Orange County Board of Supervisors declared

DOI: 10.1201/9781003215226-53

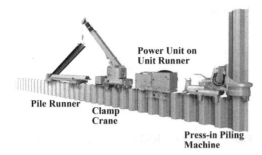

Power Unit on Unit Runner

Pile Runner

Clamp Crane

Press-in Piling Machine

Figure 1. Concept of the GRB system.

a local emergency and its public works department decided to fix the levees with sheet piles to be installed with the press-in piling method. See Figure 2 for the conditions of the levee prior to the emergency repair.

The area behind the levee repair was designated as wetlands to be protected from human activities including construction related trucking or equipment

Figure 2. Channel 05 levee prior to the emergency repair.

placement thereon. Due to this logistical constraint, the GRB System or an equivalent was so specified for the sheet pile installation for the levee repair work. See Figure 3 for the project's specifications on the use of the GRB System or Engineer's approved equal.

The project repaired approximately 1,140 meters of erosion on the north levee with 989 pairs of 13.7-meter-long PZ35 type sheet piles (575mm wide and 378mm deep each). See Figure 4 for the typical soil conditions and the corresponding location of the sheet pile. It was designed to have the pile top elevation to be at 4.5 meters above the channel's invert after planned improvement. The embedment depth was approximately13.2m. Figure 5 is the aerial view of the Channel 05 project and the surrounding area with the emergency repair alignment shown in a yellow line on the north side of the channel, which was worked on from January to February of 2008. Phase 2 (in blue lines) was later upgrade work completed in 2014 with partly single but mostly double sheet pile walls and deep soil mix columns between the double walls (Fayad et al. 2015). Phase 3 (in light green) are currently (2020 to 2021) being worked on.

The GRB System that was deployed on the emergency repair project comprised of a press-in piling machine for Z-shaped sheet piles (Silent Piler SCZ675WM model), a Unit Runner for the power pack for the piling machine, a Clamp Crane (CB3-3 model), and a Pile Runner. See Figure 6 for the profile of the 10-ton-capacity Clamp Crane which was used on the project. It has the maximum lifting height of 32.1 meters above the ground and the maximum operating radius of 30.2 meters.

The sheet pile installation was conducted on a 24-7 basis to complete the repair before the next rainy season. This was achievable only because of the low noise and low vibration nature of press-in piling (White et al. 2002). Actually, there were no noise or vibration related complaints made by the local residents: some of whom were as close to the piling operation as 50 meters. Residents who lived on the

Install Sheet Piles shall be accomplished using the GRBs (Giken Reaction Base System) or ENGINEER'S approved equal. The GRBs shall have all equipment for the piling operation supplied and operated from the top of the pile without the need for external staging beyond the initial Start and End of Project. The:

    1.   Silent Press-In Piler
    2.   Engine Unit
    3.   Clamp Crane
    4.   and Pile runner

or approved equal and shall all work atop of the sheet pile.

Figure 3. Project's specifications (excerpt) on the use of the GRB System or Engineer's approved equal (Pages F-14 and 15, Orange County, California, 2007).

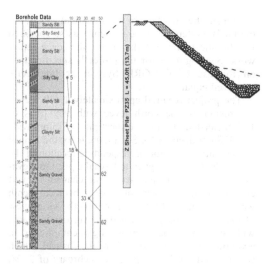

Figure 4. Typical soil conditions and sheet pile location.

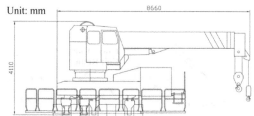

Figure 6. Profile of CB3-3 clamp crane.

Figure 7. Installed line of sheet piles with press-in piling and the GRB system.

opposite side of the channel even voiced their desire to have their side of the levee to be reinforced immediately in a similar manner. See Figure 7 for the installed line of sheet piles with the piling operations in the distance and the wetlands on the left of the levee.

It had been anticipated that the sandy gravel layer at the depth of 11 to 13 meters with N-value higher than 50 would require the use of an auger attachment for press-in piling so the piling subcontractor (Giken America Corp.) brought the attachment to the jobsite to be ready. However, the piling work was able to be completed without using it. The average production

rate was as high as 25 pairs per 12-hour shift and the maximum was 45 pairs in a 12-hour period. Press-in piling impressed the state regulatory agencies (California Fish & Wildlife and California Coastal

Figure 5. Aerial photograph of the levee repair and the surrounding area, base map via Google Map).

Commission) so they recommended the use of the method for future projects by the county (Fayad et al. 2015). The county has been utilizing pressed-in sheet piles for their subsequent levee upgrade projects including the aforementioned Phases 2 and 3 work on the same levee. Additionally, J.F. Shea Construction, Inc. based in Southern California, who had undertaken the emergency levee repair as a general contractor, later purchased a new unit of Silent Piler and the associated auger attachment for their future self-performing work after having seen the advantages of press-in piling on this project. And, they are currently working on the Phase 3 as the general contractor.

3 SANDALWOOD CANAL IMPROVEMENTS (JACKSONVILLE, FLORIDA)

The City of Jacksonville, Florida is in the northeast corner of the Florida peninsula and at the mouth of Florida's longest river, the St. Johns River. The city is integrated with Duval County and has the largest area as a city in the contiguous United States (Wikipedia on Jacksonville, Florida). Its eastern suburb toward the beach is very flat and low-lying where the Sandalwood Canal is located and discharges to Hogpen Creek. The canal is approximately 10 kilometers long, running through densely built single-family residential neighborhoods. See Figure 8 for the project location and the surrounding environment.

The subject project, "Sandalwood Canal In-channel Improvements Project (Hodges Bl. from Beach Bl. to Atlantic Bl., Project No. P-80-01)", was let by the city to repair the damaged earthen levees from an earlier flooding as well as to increase the drainage and retention capacity of the existing canal by widening and deepening with 9-meter-long steel sheet piles to be embedded in a fine sand layer.

Because all of the piling work would have to be done just behind many homes, the project specified the use of press-in piling in order to minimize the piling related noise and vibration impact. See Figure 9 for the corresponding part of the specifications. The Z-shaped PZC18 sheet piles (635mm wide and 387mm deep each) were used. The levee's right of way was 12.2 meters wide with access easement on both shores which would have a minimum of 3.1-meter wide flat shoulders. These shoulders were wide enough for small flatbed trucks to haul in sheet piles but not wide enough for a truck-mounted crane to operate for lifting and pitching sheet piles from a flatbed truck to the piling machines. See Figure 10 for a typical cross section and soil conditions with the vertical locations of the sheet piles.

The piling subcontractor, Giken America Corp., utilized two units of press-in piling machines (Silent Piler SCZ675WM model) with one 2.9-ton-capacity Clamp Crane (CB2-8 model) which had the maximum lifting height of 25.0 meters above ground and the maximum operating radius of 22.6 meters. See Figure 11 for its profile.

Figure 8. Aerial photograph of the project and the surrounding area (sheet pile lines in red and a truck access road in light green with staging areas as light green rectangles on both ends, base map via Google Earth).

Figure 9. Project specifications (excerpt) on the use of press-inpiling.

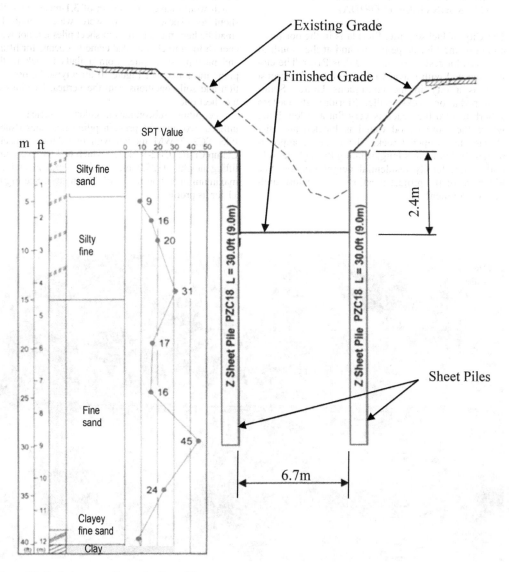

Figure 10. Typical cross section and soil conditions.

528

In order to reduce in-stream exposure of equipment to flood water during construction, the sheet piles were installed during the dry winter season between November 2007 and February 2008.

After clearing vegetation and initial grading including excavation of the levee shoulders for piling work, the press-in piling work was conducted on both levees simultaneously side by side as shown in Figure 12 so the Clamp Crane (painted in yellow) was able to pitch the sheets to both piling machines. Although it was not the full GRB System, the combination of two press-in piling machines and the Clamp Crane successfully accomplished the sheet pile installation in a very narrow but long strip of construction zone. Approximately 950 pairs of sheet piles were installed without causing damage to the nearby homes. The average production rate was 25 pairs per 10-hour shift with the maximum of 35 pairs in a 10-hour period. Figure 13 shows the section with installed sheet piles

Figure 13. Competed sheet pile installation with backfilling and shaping of the levees in progress.

in both levees while backfilling and shaping of the levees were in progress following pile installation.

4 CONCLUSION

- Pressed-in sheet piles are widely used for the repair and retrofit/upgrades of existing earthen levees in the U.S.
- Press-in piling has low-noise and low-vibration advantages over other methods. It enables pile installation in densely populated residential areas without disturbing residents.
- Press-in piling with the GRB System achieves pile installation in a very narrow and long work zone as exemplified with the case study projects.
- Some local government agencies in the U.S. have been specifying press-in piling as well as that in combination with the GRB System for their levee projects.

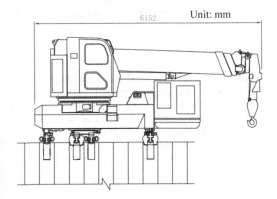

Figure 11. Clamp crane CB2-8.

ACKNOWLEDGEMENT

Authors appreciate the assistance provided by Ian Vaz of Giken America Corp.

REFERENCES

Fayad, A.S., Takuma, T., & Kambe, S. (2015). Levee Repair and Upgrade by Dual Sheet Pile Walls and Accordion Sheet Pile Wall, *Proceedings of 2015 Deep Foundations Annual Conference.*

White, D., Finlay, T., Bolton, M., and Bearss, G. (2002). Press-in Piling: Ground Vibration and Noise During Piling Installation, *Proceedings of the International Foundation Congress, ASCE Special Publication 116.*

Figure 12. Clamp crane pitching sheet piles to press-in piling machines.

Proceedings of the Second International Conference on
Press-in Engineering 2021, Kochi, Japan – Matsumoto et al (eds)
© 2021 Taylor & Francis Group, London, ISBN 978-1-032-10414-0

Repair of flood-damaged New York subway station with pressed-in sheet piles

T. Takuma
Giken Ltd., c/o Giken America Corp., Orlando, FL, USA

S. Nagarajan, M. Nagano & I. Vaz
Giken America Corp., New York, NY, USA

ABSTRACT: The New York subway's Canarsie Line tunnel was flooded by Hurricane Sandy in 2012. Although the emergency repair was completed soon after the storm damage, the tunnel's permanent repair is currently ongoing near the 1st Avenue Station in Manhattan. During the earlier part of the repair work, sheet pile walls were constructed for temporary earth retaining in a busy and relatively narrow street in the densely populated district. The sheet piles were pressed-in day and night to expedite construction without disturbing the area's residents or business owners, achieving substantial cost saving compared to the originally designed earth retaining with secant pile walls.

1 INTRODUCTION

Most of the current BMT Canarsie Line (also called L or 14th Street Line) of the New York subway system was completed back in the 1920's. It has been seeing increased ridership in recent years because of accelerating gentrification along the eastern side of the line in Brooklyn. In 2012 Hurricane Sandy flooded the Canarsie line's tunnel section from the East River west to the terminus in Manhattan. Although the line's train service was restored with temporary repairs soon after the storm, the agency fully suspended the train service in the flooded tunnel section for 18 months as part of the 46-month permanent repair work on its damaged tunnel, rail, and electrical systems on a design-build basis. Support of excavation for the tunnel repair and construction of a new entrance to the 1st Avenue Station was one of the early phases of the project. The support of excavation of a large area was built first to expedite the removal of the corroded and destroyed electric and steel components inside the Canarsie Tunnel before using the same excavated area to create a new entrance to the substation. See Figure 1 for the project's approximate location in New York City. With the site situated in a busy and relatively narrow street (East 14th Street) with a high-rise apartment complex on one side and a row of low-rise stores on the other side, construction

noise and vibration needed to be significantly subdued. Further, the project's ground was composed of hard boulders mixed in sand underlain by a solid rock layer which limited the choice of earth retaining systems and usable construction equipment. See Figure 2 for general soil conditions with N-values along with the vertical location of the sheet pile wall installed.

2 ORIGINAL AND ALTERNATE DESIGNS OF EARTH RETAINIG STRUCTURES

A new entrance structure, an underground substation, and the associated underground utility rearrangement required extensive support of excavation. Although the selection of the shoring method was up to the contractor's preference, the contract documents suggested use of secant pile walls and/or soldier piles and lagging with jet grouting for water cut-off and closure. See Figure 3 for the suggested shoring plan for one of the structures (a new underground substation) to be constructed in the contract document. The lines of small overlapping circles indicate the locations of the secant pile walls with the hatched area to be jet grouted. Mueser Rutledge Consulting Engineers based in New York City was retained by the contractor (a joint venture between Judlau Contracting and TC Electric) for a more

DOI: 10.1201/9781003215226-54

Figure 1. Project location in New York City. (Base map from Wikipedia on "Boroughs of New York City").

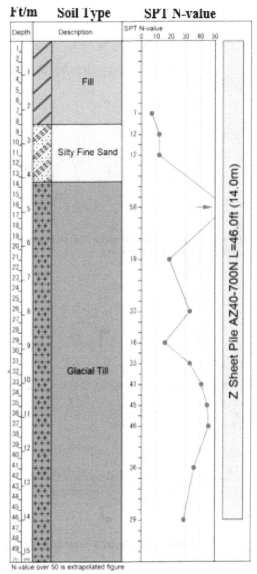

Figure 2. Typical soil conditions.

economical shoring plan. Figures 4 and 5 show the alternate shoring plan prepared by the consultants for the underground substation with the sheet pile walls that would be built in lieu of the combination of secant pile walls, soldier piles and lagging, and jet grouting.

One of the key factors for selecting sheet piles as shoring members was to find a viable way to mitigate noise and vibration during their installation, considering the project location being so close to apartments and operating stores plus full of underground utilities. The consulting engineer chose the press-in piling method to satisfy this requirement (White et al. 2002). Figure 6 shows the part of the shoring specifications on the use of hydraulic pile jacking equipment ("Press-in Method") for the alternate design.

3   SHEET PILE INSTALLATION

Press-in piling started off in the fall of 2017 near the 1st Avenue Station on East 14th Street in Manhattan close to the East River. Figure 7 shows the F401-1400 model Silent Piler which installs a pair of Z-shaped sheet piles at a time. The contractor leased the piling machine from Giken America Corp. for the project. The piling work was safely conducted even at locations where the sheet piles were as close as about 1.5 meters to existing buildings and extremely close to sidewalks open to the pedestrian traffic (Figure 8). 14.0 meters long Z-shaped AZ40-

700N sheets were used for shoring the aforementioned substation construction while 10.7 meters long Z-shaped NZ20 sheets were used for other locations.

An auger attachment was used to install sheet piles through hard soil and obstructions. Some of the obstructions encountered consisted of concrete and some steel materials from foundations of old buildings that stood in the area from early on in the 20th

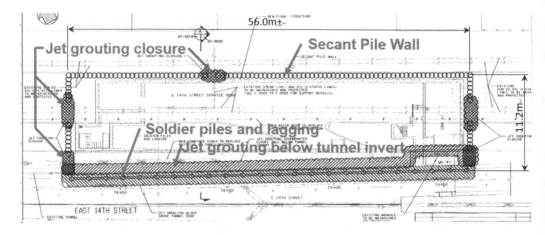

Figure 3. Original shoring design for Avenue B Substation (secant pile and soldier pile and lagging walls plus jet grouting, Courtesy: Judlau – TC Electric JV).

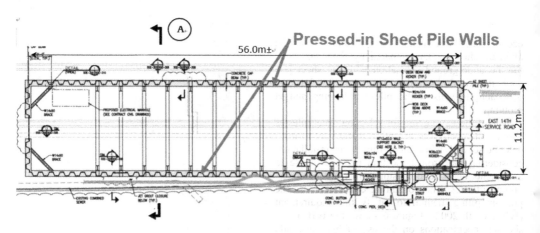

Figure 4. Alternate shoring design (Pressed-in sheet pile walls, Courtesy: Judlau – TC Electric JV).

Century before these buildings were torn down to widen East 14th Street. See Figure 9 for the major components of the auger attachment.

The auger attachment was able to overcome such obstructions and allow for successful pile installation (Takuma et al. 2018). Predrilling with another machine prior to pile installation would have been necessary without the use of the auger attachment, which would have taken much longer time in completing the piling operations.

Piling work in front of some of the stores such as a popular coffee shop could be only done while the stores were closed late at night. See Figure 10. The press-in piling enabled piling installation in the middle of the night without disturbing local residents at rest. The pile installation work was conducted between December 2017 and April 2018.

Additionally, an on-site presentation on the press-in piling was well attended with the representatives of local agencies and consultants in spite of chilly winter weather as shown in Figure 11.

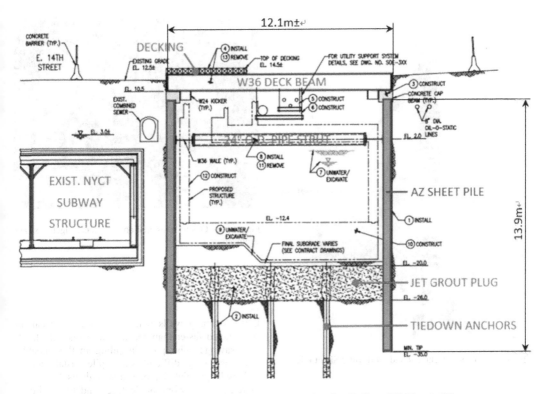

Figure 5. A-A Section of the alternate shoring design for Avenue B Substation (Judlau – TC Electric JV).

PILE INSTALLATION EQUIPMENT: USE HYDRAULIC PILE JACKING EQUIPMENT TO INSTALL SHEET PILES ("PRESS-IN METHOD"). EQUIPMENT SHALL BE SUITABLE FOR THE TOTAL WEIGHT OF THE PILE AND THE CHARACTER OF SUBSURFACE MATERIAL TO BE ENCOUNTERED. OPERATE PILING EQUIPMENT AT THE RATE(S) RECOMMENDED BY THE MANUFACTURER THROUGHOUT THE ENTIRE INSTALLATION.

Figure 6. Part of shoring specifications for sheet pile installation.

Figure 7. Starting press-in piling.   Figure 8. Press-in piling next to sidewalk.

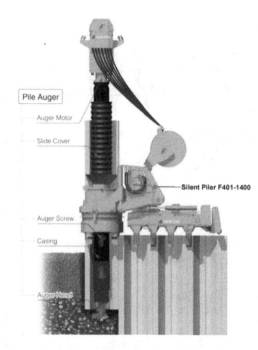

Figure 9. Auger Attachment and F401-1400 Silent Piler.

Figure 11. Onsite presentation for local agencies and engineers.

4   CONCLUSION

Earth retaining walls with steel sheet piles can be constructed faster than secant pile walls with likely cost savings. The press-in piling method enables otherwise very difficult sheet pile installation in a densely populated urban area with its low noise/ extremely low vibration features and the ability to efficiently press sheet piles into hard soil. New York and other metropolitan areas around the world will be further benefitted with the advantages of the press-in piling method.

REFERENCES

Takuma, T., DellAringa, C., and Nagano, M. (2018). Retro-fitting Drainage Systems With Pressed-In Sheet Piles In Very Hard Soil In Southern California, *Proceedings of 2018 Deep Foundations Institute Annual Conference.*
White, D., Finlay, T., Bolton, M., and Bearss, G. (2002). Press-in Piling: Ground Vibration and Noise during Piling Installation, *Proceedings of the International Deep Foundation Congress, ASCE Special Publication 116.*

Figure 10. Press-in piling conducted at night while stores are closed.

Proceedings of the Second International Conference on
Press-in Engineering 2021, Kochi, Japan – Matsumoto et al (eds)
© 2021 Taylor & Francis Group, London, ISBN 978-1-032-10414-0

Flood protection through dyke reinforcement at the river "Elbe" in Germany

F. Geppert

IPP Hydro Consult GmbH, Cottbus, Germany

ABSTRACT: Heavy and prolonged rain caused considerable damage to the infrastructure in June 2013 along the river Elbe region in Germany. Several flood protection dykes were damaged and caused over 12 billion Euros of economic losses throughout Germany. The purpose of dykes is to protect the towns/localities on the land side from being flooded. In order to maintain flood protection, the dykes must be renovated and reinforced. In many cases, the space on the Dam is narrow and there are buildings, monuments and protected trees along the dyke axis. Therefore, the renovation takes place often through a statically fully effective core seal (steel sheet pile wall) directly in the middle of dyke. In Germany, the remediation of the dykes through reinforcement with sheet pile wall is a reliable, efficient, permanent and therefore economical solution and, above all, a method with significantly lower environmental impact.

1 INTRODUCTION

1.1 General preliminary remark

River dykes serve to people to protect all over the world from floods. Damage caused by floods, climate change or simply aging can cause dykes to lose their protective function for human life, the environment, cultural assets and economic activity. Their maintenance is therefore a permanent challenge for competent authorities.

The measure described here serves to repair flood damage to a dyke in the middle reaches of the river Elbe. The Elbe is a European river. Coming from the Czech Republic, it flows through Germany and after about 1095 km (680 miles) it flows into the North Sea. It has a catchment area of approximately 150,000 km² (57,000 sq mi). This corresponds to about 40% of the Japanese territory.

The study described here does not represent a general solution for reinforcing a dyke. It must always be considered in the context of the local boundary conditions. However, it can be used as a method to find economical and sustainable dyke reinforcement.

1.2 Normative references

In Germany the standards of the German Institute for Standardisation (DIN) are used as the main basis for assessing river dykes. The technical standards and regulations of DIN 19712:2013-01 (Flood protection works) apply to the construction and repair of dykes.

Precisely, DIN 19712 contains specifications on:

- Planning criteria,
- Hydrological and hydraulic design bases,
- Requirements for flood protection installations (dykes, walls, mobile systems),
- Required evidence,
- Construction materials, construction execution, and quality assurance,
- Third-party construction,
- Construction work on existing flood protection works
- and operation and maintenance of flood protection works.

1.3 Planning boundary conditions

For the planning and redevelopment of dyke systems, a large number of boundary conditions and specifications must be taken into account. Last but not least, regional requirements of the responsible authorities must be observed. The boundary conditions described here can therefore only provide a small insight into the planning process.

The design of flood protection works is based on the design flood. This is determined by hydrological data and hydraulic models. For the example given here, a flood with a 100-year probability of recurrence is to be used.

DOI: 10.1201/9781003215226-55

To this resulting design-basis water level (HQ100 water level), a freeboard of 1 m is to be added in order to make the dyke overflow-proof even in the event of wave run-up or extreme floods.

Another point is the dyke construction. In addition to the height of the dyke, the width of the crown and the slope inclination, this also includes the soil used to create the cubature. Due to the potential difference of the water head between the land and water side in the event of flooding, water may seep through the dike. Among other things, this geohydraulic load case must be taken into account when planning dyke systems: The cubature and the materials to be used must be coordinated.

Often, the inhomogeneous structure and the unqualified soil placement (insufficient compaction) especially in the case of existing dykes can lead to harmful rearrangements of fine grains in the dyke (inner soil erosion). In extreme cases, these can lead to the failure of flood protection works. Often existing plants have deficits in dyke height with regard to design water levels. The resulting reinforcement of the dykes is accompanied by an increase in the dyke contact area. As a rule, the dyke contact area is extended towards the land side.

This leads to another essential aspect of planning - the availability of land. Thus, the possible construction time and permanent land use is influenced by, among other things, adjacent buildings and infrastructure facilities, agricultural uses and nature conservation law issues, and thus also the technical solution for reinforcing the dyke. These are often technical solutions, in the form of a structurally effective core seal (sheet piling), by which interventions in protected biotopes, existing tree populations, listed buildings or parallel roads (Figure 1) or railway lines can be reduced or avoided. Press-in methods are frequently used as insertion methods because of their reliability and the avoidance of vibration.

Figure 1. Dyke heightening parallel to a federal road.

2 COMPARISON OF VARIANTS FOR THE REINFORCEMENT OF A POLDER DYKE

2.1 Causes

The evaluation of flood damage over the past 15 years clearly shows the need for action. The weak-point analysis carried out on the dykes shows that the existing protective dykes do not meet today's requirements in terms of dyke geometry and construction, and therefore their stability is not guaranteed.

The following deficits were identified in the planning phase:

Although the existing dyke height is above the HQ100 water level in the entire planning section, the required freeboard of 1.00 m is not achieved along the entire section. As a result, there is a risk of overflowing during extreme floods.

The recommended dyke crest width according to DIN 19712 of at least 3.0 m with a dyke height $\geq$ of 2.0 m is largely not achieved. This endangers the stability in the event of flooding.

A land side surcharge filter which fulfils the function of a DIN-compliant filter berm is not existent. Due to permeable layers in the supporting body, it cannot be guaranteed that seepage water will leak out at the foot of the dyke in the event of flooding. If the seepage line above the foot emerges from the embankment on the land side, this endangers the stability of the embankment.

A paved path for dyke defence is only located on the dyke crest. This makes it difficult to access the foot of the dyke on the land side and to defend the dyke in the event of flooding.

The supporting body of the dyke has a highly inhomogeneous structure made of mainly locally extracted materials (alluvial loam, sand and gravel) with sometimes very low layer thickness. This can cause material to be discharged in the event of a flood, confirmed by observations of seepage points during the June 2013 flood. In extreme cases, it may lead to the failure of the flood protection system.

The landward slopes of the dyke are partly steeper than 1:3. The slope is 1:1.8 at some points. This endangers the stability in case of flooding and makes management more difficult.

The measure serves to eliminate flood damage and includes the repair of the existing dyke system in its current position. In addition, the dyke is to be prepared for polder use (damming on both sides).

2.2 Presentation of the variants of the planned measure

2.2.1 Homogeneous dyke of impermeable soil (variant 1)

For the construction of a homogeneous dyke (1-zone dyke), the existing dyke body is to be removed down to the top edge of the terrain. The supporting shell section of the new dyke is to be made of cohesive soil (has to be delivered).

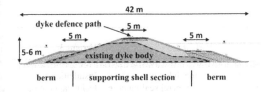

Figure 3. Variant 1.

From the non-cohesive existing material (approximately 50 %) a berm is to be profiled at the foot of the dyke on land and water side. This has two positive effects. The reduction of mass transports reduces the construction costs and furthermore the waterlogging of the actual supporting shell section will be counteracted by raising the terrain dyke (cf. Figure 3).

2.3 2-zone dyke with mineral sealing core and supporting shell section on both sides (variant 2)

As in variant 1, the existing dyke body is completely removed to the top edge of the terrain. Then a watertight core of cohesive material is built up in the area of the dyke top (cf. Figure 4).

A supporting shell section is built on this watertight body on the land and water side. It is estimated that about 50 % of the existing dyke material can be reused.

2.4 Installation of a technical core seal (variant 3)

For the installation of a technical core seal, a sealing element is installed in the existing dyke body. The raising of the dyke to the design height starts in the area of the water-side embankment shoulder towards the dyke land side made of non-cohesive material. Four sub-variants are considered for the creation of a core seal. The embedment depth of 7 m is the same for all sub-variants.

a) Sheet piling (variant 3a)
For this type of core sealing, a steel sheet pile wall is installed in the area of the planned water-side banisters, outside the dyke defence path (cf. Figure 5). In this way, subsidence-related damage to the dyke defence path caused by the sheet pile wall can be avoided.

Due to the more favourable load distribution (larger area), a flexible arrangement in the dike body is possible for variants 3b to 3d. From the point of view of construction technology, the construction on

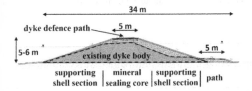

Figure 4. Variant 2.

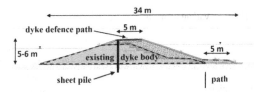

Figure 5. Variant 3a.

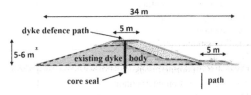

Figure 6. Variant 3b-3d.

the dyke axis (middle of the dyke crest) is advantageous (cf. Figure 6).

b) Slurry wall (variant 3b)
In the slurry wall method, a clamshell is used to excavate the existing soil along the planned dyke axis and replace it with sealing material (usually in-situ concrete).

c) Soil stabilisation (variant 3c)
For the production of the core seal, two methods are used for soil stabilisation, which essentially differ in the production process.

In the Mixed in Place process (MIP), a defined soil volume is prepared along the planned dyke axis with the aid of a single or triple screw with a predetermined quantity of binder suspension (cement) to form a homogeneous self-hardening mass in the wet mixing process and pressed back into the area from the bottom upwards. With the deep soil mixing process (DSM), the water-blocking soil-cement wall is produced by mixing soil and an injected cement suspension in one step.

d) Thin diaphragm wall (variant 3d)
For a thin diaphragm wall, a steel profile (vibrating beam) is vibrated into the substrate. The profile dis-places the existing soil and thus compacts the sur-rounding material. When the beams are pulled, a hollow space is created which is filled with sealing material. This procedure is continued one after the other along the axis of the dyke with an overlap, thus creating the sealing wall.

3 DERIVATION OF THE PREFERRED SOLUTION

3.1 Variant analysis

The first step was to derive the preferred solutions from the three basic variants. For this purpose, an evaluation matrix was created that allows a comparison (cf. Table 1). The consideration includes, for example,

Table 1. Variant analysis.

Evaluation Criteria	Variant 1 Homogeneous dyke of impermeable soil	Variant 2 2-zone dyke with mineral sealing core	Variant 3 Installation technical core seal
Permanent land use	identical, due to uniform cubature specifications		
Building time land use	1	1	3
Constructional expenditure	3	2	1-2
Availability technology	3	3	1-2
Construction time	1	1	3
Quality assurance (effort)	2	1	1-3
Durability	2	2	2-3
Control effort during floods	1	1	2
Nature conservation intervention	1	1	3
Flood protection during construction	1	1	3
Applicability in changing soil conditions	3	3	1-3
Construction costs (net)	2175 €/running meter 1	1345 €/running meter 2	875 - 1330 €/running meter 2-3
Σ	18	17	22-30

Table 2. Variant analysis.

Evaluation Criteria	Variant 3a sheet piling	Variant 3b Slurry wall	Variant 3c Soil stabilisation	Variant 3d Thin diaphragm wall
Technology/Brief description	Installation of steel sheet piles in the area of the dyke crown	Trench construction and placement of hydraulically bound sealing material	Column-shaped loosening by means of a drill or cutter, addition of slurry to produce the sealing wall (MIP, DSM)	Cavity production with injection beam + filling during drawing with self-hardening suspension, application range sandy and gravelly subsoil
Standards	ZTV-W LB 214, DIN EN 10204, DIN EN 12063, DWA 512-1	DIN 4126, DIN EN 1538, EAB, DWA 512-1	DIN EN 14679, DWA 512-1, MIP: general technical approval (Z-34.26-200)	ZTV-W LB209, DIN EN 1538, DWA 512-1
Construction cost (net)	1330 €/running meter 1	1215 €/running meter 1	970 €/running meter 3	875 €/running meter 3
Quality assurance	3	1	1	1
Construction time	2	2	3	2
Durability	3	2	2	1
Building time land use	3	2	2	2
Waste disposal	3	1	2	2
Nature conservation intervention	3	1	1	1
Emissions	2	2	2	1
Flood protection during construction	given for all variants, as the execution is carried out from the existing dyke crown			
Suitability of the existing dyke body	3	3	1	1
Market availability/wide range of suppliers	3	2	1	1
Σ	26	17	18	15

the construction method in terms of its availability and the associated costs (availability technology), the intervention in terms of the nature conservation agreement (nature conservation expert evaluation - impact), and the feasibility of implementation over the entire dyke section (applicability to changing ground conditions).

The criteria were drawn up depending on the local conditions and the requirements of the client and vary from measure to measure. The matrix serves as a summary of the boundary conditions and should allow an objective and comprehensible derivation of the preferred solution.

A scale of points from 1 (negative/disadvantages) to 5 (positive/advantages) has proved useful for the evaluation. In Tables 1 and 2, a scale of 1 to 3 points was used. Points were awarded for the individual criteria according to the scale.

Depending on the process used to manufacture the core seal, there is a range of variation for this variant. Irrespective of the type of core seal, variant 3 with the highest number of points is the preferred solution from the evaluation. For this reason, variants 1 & 2 are not considered in the further planning steps.

3.2 *Variant analysis core seal*

In the second step, the types of technical core seal described in Section 2.4 were considered in more detail. For this purpose, an evaluation matrix was drawn up analogous to the basic variants. The points awarded in Table 2 refer to the evaluation of the individual core sealing types among each other and are to be considered independently of Table 1.

Despite the highest price for this example, the production of a core sealing by means of a sheet pile wall (variant 3a) is the preferred solution with the highest target achievement from the evaluation.

4 DISCUSSION

In summary, it can be said that, in addition to the normative specifications, dyke repair depends on a large number of boundary conditions. Therefore, each planning is an individual process of consideration, and the overall view should result in the most economical method of rehabilitation.

The influence of the solution approach on the feasibility of dyke rehabilitation must already be taken into account in the planning stage. A complicated dyke construction with many cross-sectional elements makes the realization and quality control of the measure more difficult.

The local space conditions must also be taken into account. Often, only by reducing the number of interventions in protected biotopes can the nature conservation agreement be established (cf. Figure 8).

Even in urban areas with adjacent buildings, often only solutions with a statically effective core seal (sheet piling) can be considered. In this case, special solutions must be developed in accordance with the applicable standards (cf. Figure 7).

Changing subsoil conditions and conditions of the dykes make the rehabilitation as pure soil construction more difficult, because they have high demands on the soil to be used. A qualified soil that meets the geotechnical and geohydraulic requirements is often only available locally to a limited extent. Furthermore, it is associated with high costs.

In case of changing subsoil conditions, the installation of a sheet pile wall can be supported by

Figure 7. Confined space conditions during the reinforcement of a dyke.

Figure 8. Freeboard secured by sheet piling.

a combined drill-press system. In addition, that can be installed vibration-free and low-noise by pressing them into the ground, thus avoiding damage to adjacent installations.

Pure earth structures are more sensitive to weather conditions when they are constructed. Excessive moisture, prolonged periods of precipitation, dryness or frost can make it difficult to meet the installation criteria or lead to an interruption of construction activities. In sensitive areas (crossroads, areas with adjacent buildings and areas parallel to roads/railroads) it is important to reliably forecast the construction time, as interruptions due to construction-related impairments often have to be planned for the long term. Due to the high insensitivity to weather influences, the advantages here also lie in the restoration of existing dykes using sheet piling.

5 CONCLUSION

This paper introduced a case history of applying sheet pile walls to the reinforcement of the river dykes in Germany.

In conclusion, it can be said that dyke reinforcement by means of sheet piling is an established part of flood protection measures in Germany. The reasons for this are, among others, the flexible applicability, the reduction of environmental impacts, the constant high quality and reliable and proven technology. Furthermore, the method is an integral part of German standards and technical literature.

REFERENCES

DIN19712:2013-01 2013, *Flood Protection Works*, Berlin, Beuth Verlag GmbH.

Guideline DWA-M 507-1E 2011, *Levees Built Along Watercourses - Part 1: Planning, Construction And Operation*, Hennef, DWA Deutsche Vereinigung für Wasserwirtschaft, Abwasser und Abfall e. V.

Guideline DWA-M 512-1E 2012, *Sealing Systems in Hydraulic Engineering, Part 1: Earthwork Structures*, Hennef, DWA Deutsche Vereinigung für Wasserwirtschaft, Abwasser und Abfall e. V.

Patt, Heinz; Jüpner, Robert (Hrsg.) 2013, *Hochwasserhandbuch: Auswirkungen Und Schutz*, Berlin, Heidelberg, Springer Vieweg.

Proceedings of the Second International Conference on Press-in Engineering 2021, Kochi, Japan – Matsumoto et al (eds)
© 2021 Taylor & Francis Group, London, ISBN 978-1-032-10414-0

Silent Piler in Bangkok MRT Orange Line Project

P. Kitiyodom & I. Boonsiri
Geotechnical & Foundation Engineering Co., Ltd, Bangkok, Thailand

ABSTRACT: Thailand's capital city, Bangkok has been planning to construct many underground infrastructure development projects. Recently, MRT Orange line is considered to be one of the most difficult projects in Bangkok because the tunnel alignment passes through the congested urban areas in the city. There are inevitable interactions be-tween existing structures and new construction, such as tunnels, intervention shafts and stations. Many under-pinning works are required along the route. The low-headroom, limited working space and vibration have become concerned issues to be considered during construction. Silent Piler is used in those constraint areas. The result of underpinning works with Silent Piler is shown in this paper.

1 INTRODUCTION

MRT Orange Line Project is implemented by The Mass Rapid Transit Authority of Thailand (MRTA) to develop a train system network in Bangkok Metropolitan Region. MRT Orange Line is divided into two sections, East Section (Thailand Cultural Centre – Min Buri) and West Section (Taling Chan – Thai-land Cultural Centre). Ch. Karnchang PLC. And Si-no Thai Engineering and Construction PLC. (CKST Joint Venture) have been awarded for MRT Orange Line (East Section) Project, Contract E1 and E2. The Contract E1 and E2 involve the construction of about 9.73km, comprises of 7 underground stations, ventilation and intervention shafts, cut-and-cover tunnels, depot access, stack and parallel twin tunnels. Diaphragm wall is used as a retaining wall with an aid of base slab and bored piles or barrettes for deep ex-cavation. One of the difficult issues for the project is constraints of the working area such as low headroom and insufficient distance between the machine and Right of Way.

2 SOIL CONDITION

The soil condition in Bangkok consists of a thick soft to very soft clay layer on the top deposit, namely Bangkok clay, and encountered with stiff to very stiff clay before reaching the first dense silty sand layer. The hard clay is encountered below the first silty sand and underlain by the second very dense silty sand layer. The subsoil profile of MRT Orange Line Contract E1 and E2 are presented in Figure 1 and Figure 2 respectively. Bangkok clay is very sensitive to deformations and has low shear strength. The foundation in Bangkok is the pile foundation. For low rise buildings, the piling work consists of driven pile and jack in pile. For high-rise buildings and highways, BTS sky train and MRT subway, bored pile and barrette pile with the tip penetrated in the second dense silty sand are utilized. The tunnel alignment needs to be adjusted to evade the existing buildings and elevated the expressway as much as possible. However, some part of the project has to conduct underpinning work to mitigate settlement of existing buildings in congested space. The working space, vibration and pollution, especially PM 2.5, have become concerned issues to be considered during construction.

3 CRITERIA FOR UNDERPINNING

The tunnel alignment compels the underpinning works required for the project. As a result, attention has been paid to the alignment design such that minimal clashes occur between the tunnel alignment and the existing structures. In addition, if such a clash is unavoidable, the underpinning work needs to be conducted. Underpinning for the existing structures may be required if the existing structures either clash with the tunnel alignment or are in its close proximity.

For example, if a clearance between the edge of existing pile and exterior surface of tunnel is less than 900mm, this location requires underpinning. This clearance is inclusive of a tunnel driving tolerance of plus or minus 100mm and a construction tolerance of 1 in 100 for the verticality of the existing pile.

The following information will be taken as bases for the design development of the underpinning

DOI: 10.1201/9781003215226-56

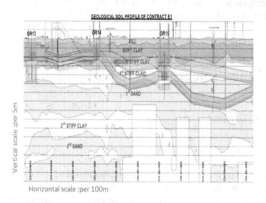

Figure 1. Geological soil profile of contract E1.

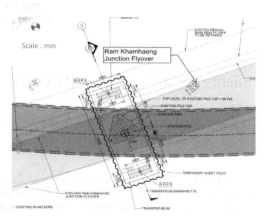

Figure 3. Site location plan of Ram Khamhaeng Junction Flyover to be underpinned at CH.28+200.

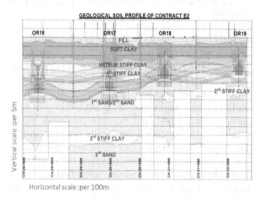

Figure 2. Geological soil profile of contract E2.

work on the project: Tunnel alignment, Station layout, Topographic survey of the site, As-built information and condition survey of the existing structures, Right of Way (ROW) of the project.

4 UNDERPINNING AT RAM KHAMHAENG JUNCTION FLYOVER (CONTRACT E1)

The location of Ram Khamhaeng Junction Flyover has to be underpinned (approximate at Chainage 28+200) as shown in Figure 3. The proposed bored tunnels will pass underneath the existing flyover and will cause an adverse effect on the existing foundation system. Underpinning work is required in order to reconstruct a new supporting system with minimal impact to the operation of the viaduct.

At the beginning, the associated barrettes and temporary steel portal frame will be constructed as the supporting structure for the existing flyover structures. Sheet piles will be installed around the pile cap by Silent Piler as presented in Figure 4 and 5. Where it is acted as permanent structures. The conventional method of pile installation could not be applied due to insufficient headroom under flyover. Temporary

Figure 4 & 5. Site location of Ram Khamhaeng Junction Flyover to be underpinned (CH. 28+200).

vertical jacks between portal frame and existing girder will be placed. Subsequently, preloading will commence to alter the load path from existing flyover piled foundation to the new barrette foundation through the vertical jacks and portal frame.

After the preloading, the existing steel pier will be disconnected, followed by demolishing the existing pile cap and piles. Thereafter remaining portions of the transfer beam will be constructed. A steel pier will be reinstated and connected to the transfer beam in order to remove the temporary jack to complete the load transfer. Finally, TBM works will be carried out underneath the underpinned foundation structure after the completion of load transfer.

5 UNDERPINNING AT RAM KHAMHAENG JUNCTION FLYOVER (CONTRACT E2)

The location of Ram Khamhaeng Junction Flyover has to be underpinned (approximate at Chainage 29+300.) As the proposed bored tunnels will pass underneath the existing flyover and will be partly in conflict with the existing foundation system, underpinning work is required in order to reconstruct a new supporting system with minimal impact to the operation viaduct as presented in Figure 6. Headroom under flyover is less than 5.0m. Therefore, Silent Piler needs to be used for sheet pile installation as shown in Figure 7. The sheet pile needs to be divided into short length and reconnected by welding during the sheet pile installation as presented in Figure 8.

The sequence of underpinning is similar to the underpinning at CH. 28+200. However, after the preloading, a gap between the existing deck soffit and the portal beam will be grouted in order to remove the temporary jack. The second cast of pile cap for connecting to the existing cap will be constructed. TBM works will be constructed underneath the underpinned foundation structure after the completion of load transfer.

6 DESIGN ANALYSIS APPROACH

For underpinning works, internal forces of the structural members have been carried out by two computer programs – PLAXIS 2D and Sap2000. PLAXIS 2D is used to analyze the effects due to TBM tunneling. However, the load transfer and the

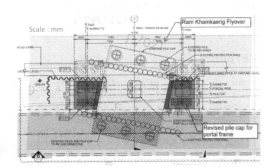

Figure 6. Schematic plan of Ram Khamhaeng Junction Flyover to be underpinned at CH.29+300.

Figure 7 & 8. Installation of sheet pile under Ram Khamhaeng Junction Flyover by Silent Piler (CH. 29+300).

loads from flyover will be analyzed by Sap2000. Figure 9 shows the section of the load transfer system in Sap2000. Jack loads were installed on between the existing deck and the newly constructed portal beam. Installation of a sheet pile by Silent Piler will be applied around the newly piled cap and the existing pier to act as a retaining wall. The load transfer due to volume loss during tunneling will be acquired by PLAXIS 2D as shown in Figure 10.

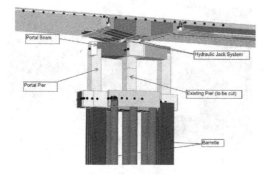

Figure 9. Example section of load transfer system at CH.29+300 by Sap2000.

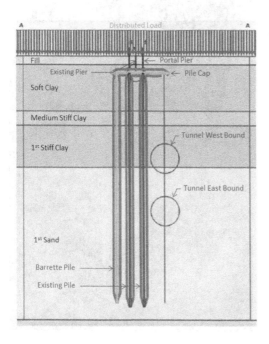

Figure 10. Example of PLAXIS 2D model for underpinning at CH.29+300 with TBM.

The flyover structures are supported on piled foundations where it is affected by sub-surface soil movement due to TBM tunnelling works. The vertical soil movement induces negative skin friction on the piles. The lateral soil movement also induced additional bending moments on the piles.

These additional stresses on the piles and consequences of underpinning works are analyzed on PLAXIS 2D. The results will be checked against the structural and geotechnical capacity of piles to assess the impact on the piles due to ground movement by tunnelling works. The methods of assessment on the flyover are summarized by the following:

First, preliminary assessment on the existing pile according to the free-field ground movements induced by tunnelling using PRAB computer program which is developed by Kitiyodom et al, 2005.

Second, for such pile foundations which are susceptible to damage by tunnelling as determined from preliminary assessment, further assessment is performed by using Sap2000.

7 CONCLUSION

Silent Piler is very useful to be applied in the constraint condition such as low head room, limited working area and nearby operating transportation. For MRT Orange Line Project, it is compulsory that the traffic needs to be operated during the underpinning process. In addition, the conventional sheet pile installation using a vibro hammer causes excessive noise and ground vibration. These results exceed the limit tolerable by human and the vibration may generate adverse effects on existing buildings.

ACKNOWLEDGEMENTS

The authors would like to appreciate to Mass Rapid Transit Authority (MRTA) and CKST Joint Venture for their kind support.

REFERENCES

Basile, F. (2014). Effects of Tunnelling on Pile Foundations. *Soils and Foundations*, 54(3), 280–295.
Jongpradist, P., Kaewsri, T., Sawatparnich, A., Suwansawat, S., Youwai, S., Kongkitkul, W. and Sunitsakul, J. (2013). Development of Tunneling Influence Zones for Adjacent Pile Foundations by Numerical Analysis. *Tunnelling and Underground Space Technology*, 34, 96–109.
Kitiyodom, P. and Matsumoto, T. (2002). A simplified analysis method for piled raft and pile group foundations with batter piles, *International Journal for*

Numerical and Analytical Methods in Geomechanics, 26, 1349–1369.

Kitiyodom, P. and Matsumoto, T. (2003). A simplified analysis method for piled raft foundations in non-homogeneous soils, *International Journal for Numerical and Analytical Methods in Geomechanics*, 27, 85–109.

Loganathan, N. and Poulos. (1998). Analytical Prediction for Tunneling-Induced Ground Movements in Clays. *Journal of Geotechnical and Geoenvironmental Engineering (ASCE)*, 124(9), 846–856.

Pang, C.H. (2006). The Effect of Tunnel Construction on Nearby Pile Foundations. *Ph.D. thesis, National University of Singapore*, 362.

Session E: Miscellaneous

Proceedings of the Second International Conference on
Press-in Engineering 2021, Kochi, Japan – Matsumoto et al (eds)
© 2021 Taylor & Francis Group, London, ISBN 978-1-032-10414-0

Vertical and diagonal pull-out experiments of flip-type ground anchors embedded in dry sand in plane-strain condition

S. Yoshida, X. Xiong & T. Matsumoto
Graduate School of Natural Science and Technology, Kanazawa University, Kanazawa, Japan

M. Yoshida
Daisho Co., Ltd., Shiga, Japan

ABSTRACT: The flip anchor is pressed or driven into the ground. The anchor head rotates to open when pull-out force acts on it. In this study, vertical and diagonal pull-out experiments of flip anchors in plane strain condition were carried out to investigate the pull-out mechanism of flip anchors. Rectangular model anchors with wire rope, which had a width of 98 mm and a breadth of 48 or 80 mm, were used. Experimental parameters were embedment depth of the anchor h, pull-out angle α and initial embedment angle of the anchor plate β. As h became deeper, the maximum pull-out force F_{max} increased. Other parameters had only a minor influence on F_{max}. Soil above the anchor head was lifted in inverted trapezoidal shape when the anchor was pulled vertically. Therefore, calculated F_{max} estimated from the vertical pull-out model of opened anchor could be applied to predict F_{max} at any α.

1 INTRODUCTION

1.1 Flip-type earth anchors

Flip-type earth (ground) anchors (hereinafter referred to as "flip anchor") (Figure 1) are effective means for reinforcing slopes against slope failures. A decisive difference from soil nails and ground anchor, which are also used for slope reinforcement, is that flip anchors do not require grouting for the installation. The flip anchors have been applied also for supporting tower structures against strong winds and can be applied even under water to support floating structures (Anchoring Rope and Rigging Pty Ltd. 2020b). The flip anchors are versatile anchors because the equipment for construction is small-scale and do not require drilling, cement and curing period.

Figure 2 Shows an installation procedure of the flip anchors. The flip anchors are driven or pressed into the ground with its anchor head closed (Figure 2a). When pull-out force acts on the anchor, the anchor head rotates to open so that soil pressure sufficiently acts on it (Figure 2b). Although the installation procedure of flip anchors is simple and quick, the mechanism of their pull-out resistance has not been completely understood yet.

1.2 Review of related researches

The behaviors of embedded plate anchors, such as square anchors, rectangular anchors, or circular

anchors in sand, have been investigated in previous studies. Through mainly laboratory experiments, ground failure patterns were observed and some theoretical approaches to estimate pull-out capacity of those plate anchors have been proposed so far.

Majer (1955) proposed the frictional cylinder model (Figure 3a). The model assumes that the ground is failed in a cylindrical shape with the anchor plate as the bottom surface. The pull-out resistance is calculated from the sum of the weight of the cylindrical soil above the anchor plate and the frictional resistance of the peripheral surface of the soil cylinder.

Mors (1959) proposed the cone model (Figure 3b). The model assumes that truncated cone-shaped model consisting of failure lines extending to the ground surface at $90 + \phi$ from both edge of the anchor plate. In this model, only the weight of the soil in the truncated cone is counted to obtain the pull-out capacity.

Balla (1961) observed a failure pattern consisting of circular failure line from the edge of the anchor (Figure 3c). The circular line meets the ground surface at an angle of approximately $45° − \phi/2$. The pull-out resistance is calculated from the weight of the soil and the friction along the circular failure lines based on the Kötter's equation.

The accuracy of the theoretical formulas derived from these three models depends largely on the ratio of anchor embedment depth h to anchor diameter d (or width B) h/d. For example, Mors' model (Figure 3b) tends to overestimate the pull-out

DOI: 10.1201/9781003215226-57

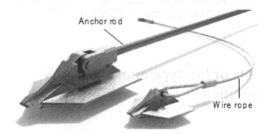

Figure 1. Examples of flip anchor (Anchoring Rope and Rigging Pty Ltd. 2020a).

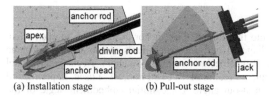

(a) Installation stage (b) Pull-out stage

Figure 2. Installation procedure of flip anchor.

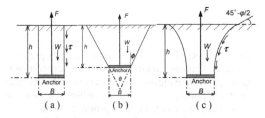

(a)   (b)   (c)

Figure 3. Typical ground failure patterns.

resistance of deep anchors where the slip line does not reach the ground surface. Balla's method (Figure 3c) also overestimates pullout capacity of deep anchor for a similar reason.

Baker & Konder (1966) pointed out the necessity of applying different formulas depending on the ratio h/d from their results of model experiment and field experiment. h/d of 6 was proposed as the border to distinguish between a shallow anchor and a deep anchor. For shallow anchors, Meyerhof & Adams (1968) proposed a practical theory to estimate pull-out capacity. For deep anchors, Vesić (1971, 1975) did the same. However, it is still difficult to determine h/d that distinguishes between shallow and deep anchors depending on various conditions. Tagaya et al. (1988) dealt with the depth that distinguishes between the shallow and the deep anchor as critical depth D_{fcr}. Tagaya et al. (1988) organized and introduced the theoretical formulas and conducted centrifugal experiment and finite element analysis on pull-out capacity of circular and rectangular plate anchors.

Many other researches have been conducted in consideration of other parameters, such as ground density or anchor shape (Das & Seeley 1975, Murray & Geddes 1987, Sutherland 1988, Ilamparuthi et al. 2002). Particle image velocimetry (PIV) or digital image correlation (DIC) methods, which analyze the photographs taken during the experiments of push-up test on a trap door (Tanaka & Sakai 1987) or pull-out test of anchors (Liu et al. 2012), were applied in order to observe ground failure patterns.

A lot of studies have been conducted also using the numerical analysis. Sakai & Tanaka (2007) conducted the FEM analysis on development of shear band for a circulate anchor pulled out in sand ground. Furthermore, some numerical analyses compared the results obtained from past empirical formulas and experiments with the results of numerical analysis (Merifield & Sloan 2006).

It has been reported based on small-scale model experiments that the scale effect exists (Ovensen 1979, Neely et al. 1973, Tanaka & Sakai 1993). It is known that this effect can be avoided by centrifugal experiments (Dickin 1988).

However, those studies focused on embedded horizontal or vertical plate anchors. In other words, the pull-out mechanism of the flip anchor, which is pressed or driven into the ground and rotates and resists in the soil when pull-out force acts, has not been treated in the past studies. Niroumand & Kassim (2013) conducted a pull-out experiment using an actual flip anchor. However, like other studies on plate anchors, the flip anchor was embedded horizontally before being pulled out. It is because that the pull-out test focused on the effects of its irregular shape on the pull-out resistance.

1.3 Objectives of this research

The behaviors of flip anchors have not been investigated sufficiently, as mentioned above. Therefore, in this study, a series of two-dimensional (plane strain condition) pull-out experiments of model flip anchors in dry sand are carried out to obtain the pull-out resistance and to observe the corresponding ground failure patterns. Based on the experimental results, a simplified ground failure model and a simple calculation method for the pull-out resistance of flip anchor under plane strain condition are proposed.

2 OUTLINE OF THE EXPERIMENTS

2.1 Model ground

A transparent acrylic box having a length of 800 mm, a height of 500 mm and a width of 98 mm was used as a soil box for the model ground (Photo 1).

Dry silica sand #3 was used for the model ground. Physical properties of the sand are listed in Table 1. Model ground was prepared in 10 layers of 50 mm thick. Relative density D_r of the model ground was adjusted to be about 80 % (dry density ρ_d = 1.512 ton/m³) by tapping the sand of each layer.

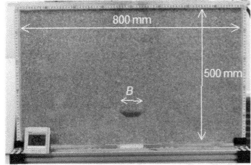

Photo 1. Model ground in a transparent soil box.

Table 1. Physical properties of silica sand #3.

Property	Value
Density of soil particles, ρ_s (ton/m³)	2.632
Max. dry density, $\rho_{d\,max}$ (ton/m³)	1.567
Min. dry density, $\rho_{d\,min}$ (ton/m³)	1.325
Max. void ratio, e_{max}	0.987
Min. void ratio, e_{min}	0.679

Internal friction angle $\phi' = 42°$ was obtained from direct shear tests of the sand with $D_r = 80\%$.

2.2 Model anchors

The model anchor used for the experiment is shown in Figure 4. A steel plate having a breadth B of 48 or 80 mm, a width of 98 mm and a thickness of 5 mm. Wires are attached to eye bolts on the anchor for pulling the anchor. The wires can move on the ring of the eye bolt freely, so that the anchor can rotate to open when pulled.

2.3 Experimental cases and procedure

Figure 5 shows a set-up for the pull-out experiments of anchors. A model anchor with a pulling wire is embedded at a given embedment depth during the preparation of the model ground. A winch for pulling the wire was set on a loading frame. A load cell (LC) to measure pull-out force F was set between the winch and the pulling wire. Pull-out displacement w of the anchor was measured with an encoder (ENC). Five earth pressure gages (EP) were set to a side wall of the box to measure lateral earth pressures at different depths.

The transparent soil box makes particle image velocimetry (PIV) analysis possible. Photos were taken from the front side at an interval of 2 seconds to observe the behavior of the sand particles. The photos were processed using a software named Trackpy (Trackpy Contributors, 2019) to obtain the traces of the soil particles during the pull-out experiment.

Table 2 lists all the cases of pull-out experiments. Figures 6 and 7 show conditions of pull-out experiments. The model anchor was embedded in the model ground at a given embedment depth h. The embedment depth h was varied as 100, 200, 300 and 400 mm, and the breadth of anchor B was 48 or 80 mm. The pull-out angles α was set at 45 or 90 degrees. Under the condition of $\alpha = 90°$, the anchor

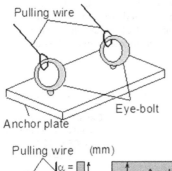

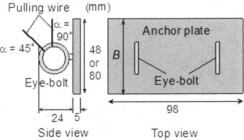

Figure 4. Dimensions of model flip anchor.

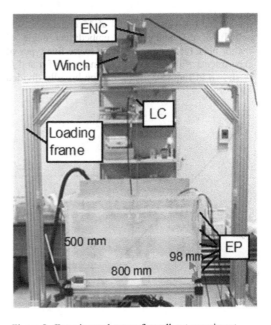

Figure 5. Experimental set-up for pull-out experiment.

Table 2. Experimental cases.

Case	B (mm)	Opened or Closed	h (mm)	α(deg)	β
01	48	Opened	400	90	0
02	48	Closed	400	90	-90
03	48	Opened	400	90	0
04	48	Opened	300	90	0
05	48	Opened	200	90	0
06	48	Opened	100	90	0
13	80	Opened	100	90	0
14	80	Opened	200	90	0
15	80	Opened	300	90	0
16	80	Opened	400	90	0
17	48	Closed	400	90	-90
18	48	Closed	300	90	-90
19	48	Closed	200	90	-90
20	48	Closed	100	90	-90
21	48	Closed	100	90	-90
22	48	Closed	200	90	-90
23	48	Closed	300	90	-90
24	48	Closed	400	90	-90
25	48	Opened	100	45	45
26	48	Opened	200	45	45
27	48	Opened	300	45	45
28	48	Opened	400	45	45
29	48	Closed	100	45	-45
30	48	Closed	200	45	-45
31	48	Closed	300	45	-45
32	48	Closed	400	45	-45
33	48	Closed	100	45	90
34	48	Closed	200	45	90
35	48	Closed	300	45	90
36	48	Closed	400	45	90

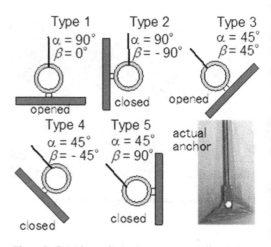

Figure 7. Experimental types.

into Opened or Closed conditions to the pull-out direction.

A total of 30 cases of experiments were conducted as listed in Table 2.

3 RESULTS OF THE EXPERIMENTS

3.1 Pull-out force vs. pull-out displacement

(a) Influence of h

Figures 8 and 9 show comparisons of relationships between pull-out force F and pull-out displacement w of Opened ($\beta = 0°$) anchors under different embedment depths h. As expected, F became larger when the anchors were embedded deeper in the ground. F of the larger anchor ($B = 80$ mm) was larger than F of the smaller anchor ($B = 48$ mm) at the same h. The projected area A of the larger anchor was about 1.7 times A of the smaller anchor. However, F of the larger anchor at every h is about 1.3

was pulled out vertically, and at 45 degrees, it was pulled out diagonally. β is embedment angle of anchor plate (see Figure 6). There were five types of experiments with different combinations of α and β, as shown in Figure 7. They can be broadly divided

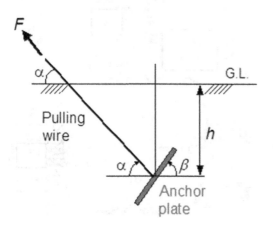

Figure 6. Definitions of experimental parameters.

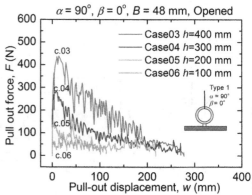

Figure 8. F vs. w of Type 1 with different h ($B = 48$ mm).

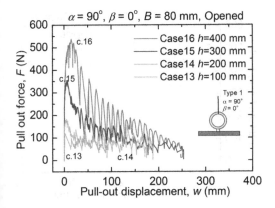

Figure 9. F vs. w of Type 1 with different h ($B = 80$ mm).

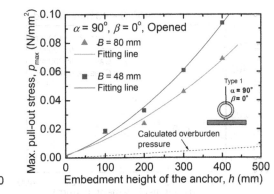

Figure 11. P_{max} vs. h of Type 1.

times larger than F of the smaller anchor That is, F did not increase as much as the ratio of A.

Figure 10 is the comparison of maximum pull-out force F_{max} of each Opened anchor when pulled vertically ($\alpha = 90°$). F_{max} of the both anchors increased with h. F_{max} of the larger anchor was larger than F_{max} of the smaller anchor at any h. However, as mentioned above, F_{max} did not increase as much as the ratio of the projected area A of each anchor.

Figure 11 is the comparison of maximum pull-out stress p_{max} (= F_{max}/A) of each anchor. Contrary to F_{max}, p_{max} became larger as A became smaller. From the results of Figures 10 and 11, it can be seen that p_{max} is affected by other factors besides overburden pressure.

(b) Influences of α and β

Figures 12-15 show comparisons of F vs. w of Opened ($\beta = 0°$) and Closed ($\beta = -90°$) anchors under different h. Regardless of pull-out angle α and embedment angle of anchor plate β, F became larger when the anchors were embedded deeper in the ground. F vs. w of anchors (Types 2 to 5) showed almost similar tendency to that of Type 1 anchor (Figure 8).

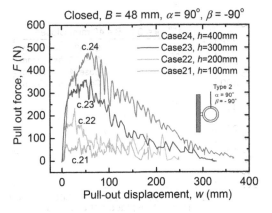

Figure 12. F vs. w of Type 2 with different h.

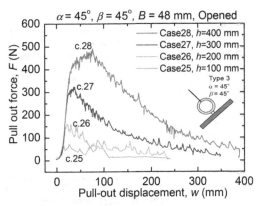

Figure 13. F vs. w of Type 3 with different h.

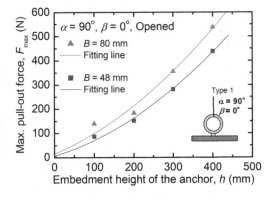

Figure 10. F_{max} vs. h of Type 1.

Figure 16 shows the relationships between F and w of Opened and Closed anchors embedded at the same h were pulled out in the vertical direction ($\alpha= 90°$). F_{max} of Closed anchor (Type 2) was larger than that of Opened anchor (Type 1). However, larger w was necessary to attain F_{max} in Closed anchor.

553

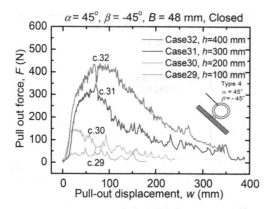

Figure 14. F vs. w of Type 4 with different h.

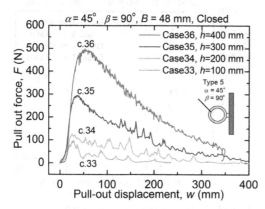

Figure 15. F vs. w of Type 5 with different h.

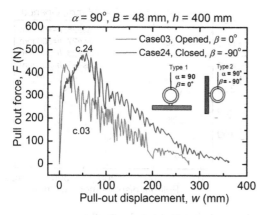

Figure 16. F vs. w of Type 1 and Type 2 with $\alpha = 90°$.

Figure 17 is F vs. w of the anchors with different β and the same h, when they were pulled out in the diagonal direction ($\alpha = 45°$). F_{max} of Type 5 was almost equal to that of Type 3. F_{max} of Type 4 was about 80% of that of Type 3. There were some differences in the amount of w at F_{max}. For example,

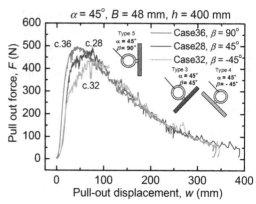

Figure 17. F vs. w of Types 3, 4 and 5 with $\alpha = 45°$.

w of Types 3 and 4 at F_{max} were larger than that of Type 5. In general, β had little effect on F vs. w.

Figure 18 shows F vs. w of Opened anchors under different pull-out angles ($\alpha = 90°$ or 45°). It was found again that α as well as β did not significantly affect F_{max}.

Figure 19 is a similar comparison to Figure 18 for Closed anchors. Same as Opened anchors (Figure 18), α as well as β did not significantly affect F_{max}.

It may be concluded from the results of Figures 16-19 that F_{max} of various anchor types are comparable to F_{max} of Type 1 (Opened anchor pulled out in the vertical direction). Hence, the ground failure in Type 1 is discussed and a simple method of estimating F_{max} is proposed next.

3.2 Modelling of ground failure

In this research, the PIV analysis was also carried out. Figure 20 shows (a) photo of ground deformation with displacement vectors and (b) zoom-up of the traces of the sand particles until pull-out displacement of the anchor w reached 15.1 mm at F_{max}

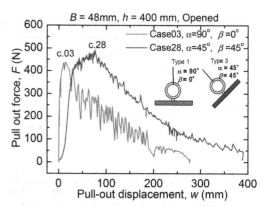

Figure 18. F vs. w of Type 1 and Type 3 with different combinations of α and β.

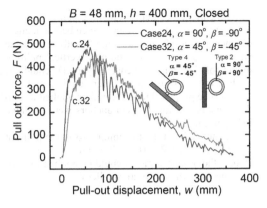

Figure 19. F vs. w of Type 2 and Type 4 with different combinations of α and β.

(a) Photo of ground deformation with displacement vectors

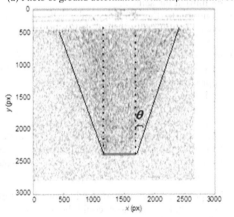

(b) Displacement vectors of ground (Enlarged)

Figure 20. Photo of ground deformation and displacement vectors at F_{max} (Case 16: $B = 80$ mm, $h/B=5$).

= 538 N in Case 16. An inverted trapezoidal failure pattern was clearly observed.

Figure 21 Shows similar results of the PIV analysis of Case 1 ($w = 13.0$ mm at $F_{max} = 438$ N). An inverted trapezoidal failure pattern was observed clearly again.

It is seen from Figures 20 and 21 that the ground failure patterns were similar regardless of anchor width B. The angle of the slip lines from the vertical

(a) Photo of ground deformation with displacement vectors

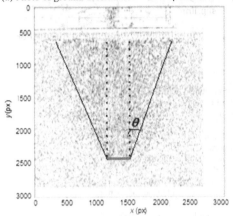

(b) Displacement vectors of ground (Enlarged)

Figure 21. Photo of ground deformation and displacement vectors at F_{max} (Case 1: $B = 48$ mm, $h/B=8.3$).

direction θ was 21° in both cases. And the θ observed in all the vertical pull-out experiments ranged from 20° to 22°. The average was approximately $\theta = 21°$. It is close to an angle of $45° - \phi'/2(\phi' = 42°)$.

The ground failure patterns observed in this study could be simply modelled as Figure 22. It is similar to Mors' model (Figure 3b), although θ is assumed to be equal to ϕ' in Mors' model. In Mors' model (Figure 3b), the pull-out resistance F_{max} is calculated from only the weight of the soil W of the inverted truncated cone above the anchor plate.

On the other hand, in the two-dimensional model proposed in this study (Figure 22), the pull-out resistance F_{max} is calculated from the weight W of the inverted trapezoidal soil wedge and the vertical component of the shear resistance T acting along the slip lines, as Eq. (1).

$$F_{max} = 2(W_1 + W_2 + T \cos \theta) \qquad (1)$$

3.3 Estimate of pull-out resistance of plate anchors

Figure 23 shows measured F_{max} vs. calculated F_{max} for each size of the anchors when pulled-out vertically. In the calculations, $\phi' = 42°$ obtained from the direct shear tests of the sand was used. As mentioned

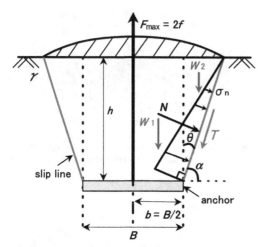

$F_{max} = 2f$

W_2

σ_n

h

N

W_1

θ

T

α

slip line

anchor

$b = B/2$

B

h: embedment depth of anchor
γ: unit weight of soil
W_1: $W_1 = \gamma \times b \times h \times D$ (D: depth of anchor)
W_2: $W_2 = \gamma \times h \times h \tan\theta \times 1/2 \times D$
θ: angle of slip line
N: normal force acting on slip line
T: shear force acting along slip line, $T = N \tan\phi'$
B: width of anchor
b: $b = B/2$
f: force acting on the half of anchor
$f = W_1 + W_2 + T \cos\theta$
F_{max}: total force acting on anchor, $F = 2f$
σ_n: $(1+K_0)\sigma_v'/2 + (1-K_0)(\sigma_v'/2)\cos2\alpha$
σ_v': effective vertical stress
K_0: coefficient of earth pressure at rest

Figure 22. Two-dimensional model of ground failure.

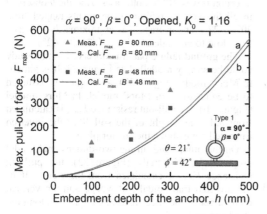

Figure 23. Relationship of measured and calculated F_{max}.

in Figure 5, five earth pressure gauges were set on the side wall of the soil box. After the preparation of the model ground, an average value of $K_0 = 1.16$ was obtained. This relatively high K_0 value may be caused by a tapping method used for the ground preparation. The calculated F_{max} qualitatively agreed well with the measured results.

As mentioned earlier, through the pull-out experiments, the pull-out angle α and the embedment angle of anchor plate β did not significantly affect the value of F_{max}. Hence, the simplest vertical pull-out model of horizontal anchor (Type 1) could be applied to the estimation of other conditions (Types 2 to 5). The proposed model could be applied also to flip anchors, because Closed anchors were included in Types 2 to 5.

It is noticed that the maximum h/B in the pull-out experiments was 8.3 ($h = 400$ mm/$B = 48$ mm). Hence, the proposed model could be applied to shallow anchors, whose h/B are smaller than 8.

4 CONCLUSIONS

In this study, pull-out experiments of model flip-type anchors in dry sand ground under plane strain (two-dimensional) condition were carried out. Main experimental parameters of the pull-out experiments were embedment depth h, breadth of the anchors B, pull-out angle α and embedment angle of anchor plate β.

Main findings from the experiments are summarized as:

(1) The maximum pull-out force F_{max} increased as h increased.
(2) F_{max} of the larger anchor ($B = 80$ mm) was larger than F_{max} of the smaller anchor ($B = 48$ mm) at the same h.
(3) Contrary to the tendency of F_{max}, maximum stress p_{max} (= F_{max}/A) became larger as B became smaller.
(4) α, aswell as β did not significantly affect F_{max}. That is, F_{max} of anchors with various α and β is approximated by F_{max} of Opened plate anchor under the vertical pull-out condition.
(5) The ground failure pattern in the pull-out experiments was simply modelled, based on the observations of ground deformation. F_{max} calculated from the proposed model qualitatively agreed well with measured F_{max} of each pull-out condition.

The proposed two-dimensional ground failure model for shallow anchors could be extended to three-dimensional conditions.

REFERENCES

Anchoring Rope and Rigging Pty Ltd. 2020. https://hulk earthanchors.com, 2020a.
Anchoring Rope and Rigging Pty Ltd. 2020. https://www. arandr.com.au/articles/projects, 2020b.
Balla, A. 1961. The resistance to breaking-out of mushroom foundations for pylons. *Proceedings of the 5th international conference on Soil Mechanics and Foundation Engineering*: 569–576.
Baker, W. H. & Konder, R. L. 1966. Pullout load capacity of a circular earth anchor buried in sand. *Highway Research Record* 108: 1–10.

Das, B. M. & Seeley, G. R. 1975. Breakout resistance of shallow horizontal anchors. *Journal of the Geotechnical Engineering Division* 101 (9): 999–1003.

Dickin, E.A. 1988. Uplift behavior of horizontal anchor plates in sand. *Journal of Geotechnical Engineering* 114 (11): 1300–1317.

Ilamparuthi, K., Dickin, E.A. and Muthukrisnalah, K. 2002. Experimental investigation of the uplift behaviour of circular plate anchors embedded in sand. *Canadian geotechnical journal* 39 (31): 648–664.

Liu, J., Liu M. & Zhu, Z. 2012. Sand Deformation around an Uplift Plate Anchor. *Journal of Geotechnical and Geoenvironmental Engineering* 138 (6): 728–737.

Majer, J. 1955. Zur berechnung von zugfundamenten. *Osterreichister, Bauzeitschift* 10 (5): 85–90.

Merifield, R.S. & Sloan S.W. 2006. The ultimate pullout capacity of anchors in frictional soils. *Canadian Geotechnical Journal* 43: 852–868.

Meyerhof, G. G. & Adams, J. I. 1968. The ultimate uplift capacity of foundations. *Canadian Geotechnical Journal*, 5 (41): 225–244.

Mors, H. 1959. Das Verhalten von Mastgruendungen bei Zugbeanspruchung, *Bautechnik* 39 (10): 367–378.

Murray, E.J. & Geddes, J.D. 1987. Uplift of anchor plates in sand. *Journal of Geotechnical Engineering* 113 (3): 202–215.

Neely, W. J., Stuart, J.C. and Graham, J. 1973. Failure loads of vertical anchor plates in sand. *Journal of Soil Mechanics and Foundations Div., ASCE* 99 (9):669–685.

Niroumand, H. & Kassim, K.A. 2013. Pullout capacity of irregular shape anchor in sand. *Measurement* 46 (10): 3872–3882.

Ovensen, N. K. 1979. The use of physical models in design: the scaling law relationship. *Proceeding of 7th European Conference on Soil Mechanics and Foundation Engineering* 4: 318–323.

Sakai, T. & Tanaka, T. 2007. Experimental and Numerical Study of Uplift Behavior of Shallow Circular Anchor in Two-Layered Sand. *Journal of Geotechnical and Geoenvironmental Engineering* 133 (4): 469–477.

Sutherland, H.B. 1988. Uplift resistance of soils. *Géotechnique*, 38 (4): 493–516.

Tagaya, K., Scott, R.F. and Aboshi, H. 1988. Pullout resistance of buried anchor in sand. *Soils and foundations* 28 (3): 114–30.

Tanaka, T. & Sakai, T. 1987. A trap-door problem in granular materials: Model test and FEA, *Journal of Irrigation Engineering and Rural Planning* 11: 8–23.

Tanaka, T. & Sakai, T. 1993. Progressive failure and scale effect of trap-door problems with granular materials. *Soils and Foundations* 33 (1): 11–22.

Trackpy Contributors, 2019. https://soft-matter.github.io/trackpy/v0.3.2/

Vesić, A. S. 1971. Breakout Resistance of Objects Embedded in Ocean Bottom. *Journal of the Soil Mechanics and Foundations Division* 97 (9): 1183–1205.

Vesić, A. S. 1975. Principles of Pile Foundation Design, *Soil mechanics series* 38.

Proceedings of the Second International Conference on
Press-in Engineering 2021, Kochi, Japan – Matsumoto et al (eds)
© 2021 Taylor & Francis Group, London, ISBN 978-1-032-10414-0

Preliminary results of questionnaire survey on field performance of press-in machine

T. Takeuchi & S. Sato
GIKEN LTD., Tokyo, Japan

T. Takehira, M. Kitamura & H. Murashima
GIKEN SEKO CO., LTD., Tokyo, Japan

ABSTRACT: Field performance of press-in piling greatly depends both on performance of machine and on operators' experiences and skills. A questionnaire survey was conducted on the field performance of press-in piling machine, with the special attention to Gyro Piler. The paper firstly describes key maintenance items of the piling machine for effective piling operation provided by the manufacture. The items listed are a good indication of the parts of machine that may experience malfunction when operators use the machine in a way that mechanical design engineers do not expect. The paper secondly describes the objectives, the methods and results of the questionnaire survey. The survey concludes that operator's experience and skill play an important role for effective press-in piling with a minimum risk for damaging the machine.

1 INTRODUCTION

Modern construction project is performed by a collective and integrated effort of design engineers, construction machinery and its operators on site. It is particularly true for successful press-in piling. Design engineers must select an appropriate piling machine based on the information of soil profiles, construction environments and required performance of piles to be installed. Performance of the final product of piles installed largely depends on the skill of the operator with a proper handling of the machine (Bolton et al., 2020).

It is also of vital importance that the machine used on site must be well maintained for achieving the required performance of piles installed.

Unproper selection of machine and misuse of the machine may lead to disruption or suspension of operations, resulting in delay of the piling project and may even result in damage of the piling machine.

Effective piling operation can only be achieved both by a good combination of a well-maintained machine and a skillful operator.

This paper tries to find out the role of skill and experience of the operator on effective piling with less damage of piling machine by a questionnaire survey.

2 MACHINE MAINTENANCE

Every machine needs a regular maintenance to make a full use of its capability. It is a common practice that manufacturers provide a maintenance service program as well as a user's manual for customers. The manufacturer of press-in piling machine also follows the same practice mentioned above, offering the user's manual and a maintenance system covering over the period of their intended service life.

Fundamental maintenance items are a good indication of the essential parts of the machine necessary for normal effective operation, which in turn infers the parts that may experience malfunction when operators misuse the machine in a way that mechanical design engineers do not expect.

As was pointed out earlier, any machine breakdown may lead to suspension of the piling operation, resulting in delay of the project. Misuse of the machine by the unexperienced operator may lead to failure of the machine.

Key maintenance items for ordinary press-in machine that the manufacture lists up include (1) proper clearance (adjustment): clump, chuck, leader mast, (2) teeth (replacement when excessive wear): at chuck, (3) replacement of packing: main cylinders, hydraulic hoses, (4) replace hanging wires: wire for main body, for power unit (GIKEN, 2019).

DOI: 10.1201/9781003215226-58

Figure 1 Illustrates the parts for the key maintenance items.

Validity of the disclaimer states, in most cases, that the maintenance insurance coverage excludes when operators use the machine, violating the user's manual that the manufacturer provides. It is, therefore, the key for effective piling to foster skillful operators with adequate knowledge of press-in piling and experiences on site.

3 QUESTIONNAIRE SURVEY

3.1 Objectives

This survey aimed at identifying how experienced operators select various driving setting values, and which press-in indices they consider important, depending on a given soil profile and a given piling project. This survey also tried to find out the differences of these setting values between experienced and less experienced operators.

3.2 Method of survey

A questionnaire survey was adopted in this investigation. In this survey Gyro Piler (Gyropress Method) was selected as a target machine among a family of press-in piling machines.

The first-round survey was conducted during the period of April, 2020 to May, 2020 to examine the feasibility of the questionnaire items. Based on the results of the first-round survey, the items of questionnaire were reviewed and modified. The second-round survey was then conducted during the period of May,2020 to June, 2020. Supplementally, some respondents were interviewed to clarify their answers.

3.3 Gyropress method (Rotary Press-in Piling)

As shown in Figure 2, the rotary press-in piling is a technology that installs steel tubular piles by rotating a pile with pile toe ring bits. To reduce rotary press-in resistance, the rotary press-in piling operation uses additional driving assistances such as water supply with a water lubrication system (LS) and air supply with an air system.

The water lubrication system discharges a small water with a discharge rate of 10 to 60 ℓ/min to reduce frictional resistance between a steel tubular pile and ground. Figure 3 shows the configuration of equipment for the rotary press-in piling, while Figures 4 and 5 show water discharge and the specifications of the water lubrication system, respectively. The number of water supply pipes and the water flow rate per pipe (also, the total water flow rate) are to be selected by each operator, considering soil profile and types of pile to be installed.

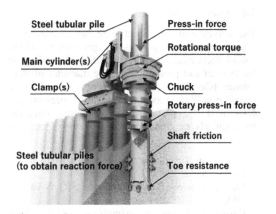

Figure 2. Conceptual diagram of rotary press-in piling.

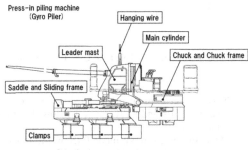

Examples of maintenance items for each component:
1) Clearance (adjustment) : Clumps, Chuck, Leader mast, etc.
2) Inspection, Exchange : Chuck teeth, etc.
3) Replacement of packing : Main cylinders, hydraulic hoses, etc.
4) Replacement of hanging wires

Figure 1. Key parts for regular maintenance.

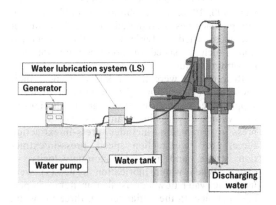

Figure 3. Configuration of equipment for rotary press-in piling.

Figure 4. Water discharge using the water lubrication system.

Water Lubrication System (LS)	Specification
Water pump discharge rate	Max. 60 ℓ /min
Water pump discharge pressure	Max. 6 MPa
Water tank capacity	300 ℓ

Figure 5. Specification of the water lubrication system.

3.4 Questionnaire items

The survey aimed at identifying how experienced operators select key press-in parameters which would affect effective piling operation, depending on the type of soil profile and on the diameter of steel tubular pile. 21m long piles were assumed to be installed at 20m embedment. Figure 6 presents the referenced soil profiles for this survey, covering from soft clayey ground to stiff mudstone ground as below:

- Case 1. Silt/Loam, The SPT N-values: 10 to 30
- Case 2. Sand, The SPT N-values: 10 to 40
- Case 3. Gravels, The extrapolated SPT N-values: 50 to 150
- Case 4. Mudstone, The extrapolated SPT N-values: 50 to 150

The selected outside diameters of the pile were 800, 1000, and 1500mm. Available press-in piling machines designated for each diameter pile were listed in Table 1 with their specifications, such as speed of press-in/extraction and maximum torque.

Respondents were then asked to answer their choices of (1) machine, (2) the number of water supply pipes and their locations, and (3) initial setting values of press-in parameters. The value of press-in parameters includes press-in/extraction force, press-in/extraction speed, press-in/extraction stroke, and rotational torque/velocity of chuck. When water lubrication system(s) were used, the water flow rate is also an item of operator's choice. In the initial setting, three items are selected among given modes as below:

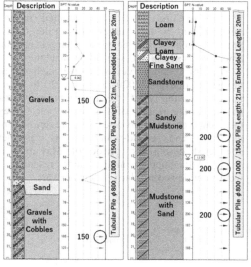

Figure 6. Referenced borehole logs: Case 1, 2, 3 and 4.

- Press-in Speed (m/s): mode 1 (Slow) to mode 5 (Fast) for F301 and F401, mode 1 (Slow) to mode 4 (Fast) for SP8
- Extraction Speed (m/s): mode 0 (Slow) to mode 5 (Fast) for F301 and F401, mode 0 (Slow) to mode 4 (Fast) for SP8
- Chuck Rotational Velocity (min⁻¹): mode 1 (Slow) to mode 5 (Fast) for F301, mode 0 (Slow) to mode 5 (Fast) for F401, mode 1 (Slow) to mode 6 (Fast) for SP8

Table 1. Available machines and their specifications.

Gyro-Piler	F301	F401	SP8
Applicable Diameters of Steel Tubular Piles	φ600-1000 mm	φ800-1200 mm	φ1200-1500 mm
Max. Press-in Force	700 kN	1500 kN	2000 kN
Max. Extraction Force	800 kN	1600 kN	2100 kN
Press-in Speed	0.005-4.3 m/min	0.002-4.9 m/min	0.002-2.0 m/min
Extraction Speed	1.4-8.7 m/min	0.7-3.5 m/min	0.4-3.4 m/min
Chuck Rotational Torque	600 kN·m	900 kN·m	1300 kN·m
Chuck Rotational Velocity	1.0 - 10.0 min⁻¹	1.5 - 11.0 min⁻¹	1.5 - 8.0 min⁻¹
Mass	17,800 kg (φ800)	32,600 kg (φ1000)	41,650 kg (φ1500)

Table 2. Example of a sheet of questionnaire.

Case X (1~4)		A	B	C	D		O
Machine/Material	Machine Model			F301/F401/SP8			
	Pile Diameter	mm		φ800/1000/1500			
	Pile Thickness	mm		9/12/16			
	Embedded Length	m		20			
	Lubrication System	unit					
	Water Supply Pipe	–					
	Pile Toe Ring Bits	–					
Initial Setting	Press-in Force	×10kN				•••	
	Press-in Speed	mode					
	Extraction Speed	mode					
	Press-in Stroke	mm					
	Extraction Stroke	mm					
	Water Flow Rate per Pipe	ℓ/min					
	Total Water Flow Rate	ℓ/min					
	Chuck Rotational Torque	kN·m					
	Chuck Rotational Velocity	mode					
	Remark	–					

3.5 Respondents

15 operators were selected from a piling company and were asked to fill in their answers in the sheet of questionnaire. Table 3 lists their years of experience of press-in machines and those of Gyro-piler.

4 RESULTS

4.1 Survey results

Table 4 shows a selected summary of the survey results for the soil profile, Case 1.

The results were examined from three aspects, namely influence of type of soil profile, influence of pile diameter and influence of operator's experience. The mean value and the standard deviation, the coefficient of variation (COV) were calculated for all the cases.

Table 3. Respondents and their experience.

													Unit : year		
Operator	A	B	C	D	E	F	G	H	I	J	K	L	M	N	O
Experience of press-in piling machinery	22	10	20	27	26	6	20	25	29	28	8	5	4	9	4
Experience of Gyro Piler	12	5	5	9	5	4	4	10	7	7	3	2	2	4	2

Table 4. Summary of the survey (Soil Profile: Case1).

Case 1 : Silt / Loam, The SPT N-values are 0-30			A	G	I	M
Machine / Material	Machine Model	–	F301 / F401	F401	F401	F401
	Pile Diameter	mm	φ1000	φ1000	φ1000	φ1000
	Pile Thickness	mm	12	12	–	12
	Embedded Length	m	20	20	–	20
	Lubrication System	unit	1	1	2	2
	Water Supply Pipe	–	1	1	2	2
	Pile Toe Ring Bits	–	Standard	Standard	Standard	Standard
Initial Setting	Press-in Force	×10kN	20~30	40	25	20
	Press-in Speed	mode	Max.5	2	4	1
	Extraction Speed	mode	6	4	6	2
	Press-in Stroke	mm	800	300	140	300
	Extraction Stroke	mm	200	80	70	150
	Water Flow Rate per Pipe	ℓ/min	40	15	40	each 20
	Total Water Flow Rate	ℓ/min	40	15	80	40
	Chuck Rotational Torque	kN·m	250-280	250	200	200
	Chuck Rotational Velocity	mode	5	6	3~5	4
	Remark	–	A	–	–	–

Table 5 presents the summary of the statistical indices, including the maximum/minimum value, the mean (m), the standard deviation (σ), and the coefficient of variation (COV). From the table, it is noticed that most operators tend to select much less press-in force than the allowable maximum press-in force of the machine regardless of the ground stiffness and steel tubular pile diameter. This is one of noteworthy operator's tendencies observed in this study.

The following points are noticed from the preliminary results.

4.1.1 Influence of type of soil profile

Figure 7 shows the mean values of the selected press-in stroke plotted against the maximum SPT N-values for four different soil profiles for the case of the pile dimeter of 1,000mm. It is seen that there is a variation of the values of operator's selection. However, when excluding a few points far away from the other data, it appears that there is a tendency that the mean values of selected press-in stroke decrease with an increase of maximum SPT N-value. In the figure, three regression lines are drawn for reference. The solid line is all the selected data, the broken line is obtained from the data of the Group A, and the dotted line is obtained from the data of the Group B. These Groups were categorized by operator's experience. The definition of grouping will be described later in 4.1.3.

Figure 8 is the plots of the values of the mean value of selected total water flow rate versus the maximum SPT N-values for four different soil profiles, showing the trend that the total water flow rate is increasing as the stiffness of ground increases.

Some other tendencies can be pointed out.

– The mean values of selected press-in speed are lower, as the ground becomes stiffer.
– The mean value of the selected press-in stroke is decreasing as the ground stiffness increases, accordingly the mean value of extraction stroke follows the same trend. This means that the number of repetitive upward and downward motion (surging) is increasing as the ground stiffness increases.

Table 5. The summary of the statistical indices, including the maximum/minimum, the mean (m), the standard deviation (σ), and the coefficient of variation (COV).

Case1. Silt / Loam — The SPT N-values: 10-30

		Pile Diameter φ800					Pile Diameter φ1000					Pile Diameter φ1500				
		Min.	Max.	Mean	Standard Deviation	Coefficient of Variation	Min.	Max.	Mean	Standard Deviation	Coefficient of Variation	Min.	Max.	Mean	Standard Deviation	Coefficient of Variation
		—	—	(m)	(σ)	(COV)	—	—	(m)	(σ)	(COV)	—	—	(m)	(σ)	(COV)
Machine Model	-	F301/F401	F301/F401	—	—	—	F401	F401	—	—	—	SP8	SP8	—	—	—
Pile Diameter	mm	800	800	—	—	—	1000	1000	—	—	—	1500	1500	—	—	—
Pile Thickness	mm	9	9	—	—	—	12	12	—	—	—	16	16	—	—	—
Embedded Length	m	20	20	—	—	—	20	20	—	—	—	20	20	—	—	—
Lubrication System	unit	1	1.5	1.05	0.15	0.14	1	2	2	0.36	0.18	1.5	3	2.17	0.44	0.2
Water Supply Pipe	-	1	2.5	1.14	0.45	0.39	1	2.5	2	0.39	0.20	2	3	2.25	0.4	0.18
Pile Toe Ring Bits	-	—	—	—	—	—	—	—	—	—	—	—	—	—	—	—
Press-in Force	×10kN	11	40	22	8.13	0.37	11	40	20	7.91	0.40	13.5	40	23	7.71	0.34
Press-in Speed	mode	1	5	2.8	1.38	0.49	1	5	3	1.31	0.44	1	5	2.5	1.43	0.57
Extraction Speed	mode	2	6	4.65	1.42	0.31	1.5	6	4	1.60	0.40	1.5	6	4.25	1.62	0.38
Press-in Stroke	mm	140	1000	402	288.8	0.72	100	1000	406	288.8	0.71	140	1000	379.1	284.3	0.75
Extraction Stroke	mm	50	200	104.5	53.6	0.51	50	200	110	48.6	0.44	50	200	104.5	50.9	0.49
Water Flow Rate per Pipe	ℓ/min	10	40	22	8.57	0.39	10	40	24	9.13	0.38	10	45	24.4	11.19	0.46
Total Water Flow Rate	ℓ/min	10	40	22.7	8.25	0.36	15	80	44	18.2	0.41	20	112.5	47.9	26.7	0.56
Chuck Rotational Torque	kN·m	115	265	210.5	46.5	0.22	115	265	208	46.9	0.23	140	500	248.8	102.5	0.41
Chuck Rotational Velocity	mode	3.5	6	4.5	0.71	0.16	3	6	4	0.74	0.19	3.5	6	4.8	0.65	0.14
Remark	-	—	—	—	—	—	—	—	—	—	—	—	—	—	—	—

Case2. Sand — The SPT N-values: 10-40

		Pile Diameter φ800					Pile Diameter φ1000					Pile Diameter φ1500				
		Min.	Max.	Mean	Standard Deviation	Coefficient of Variation	Min.	Max.	Mean	Standard Deviation	Coefficient of Variation	Min.	Max.	Mean	Standard Deviation	Coefficient of Variation
		—	—	(m)	(σ)	(COV)	—	—	(m)	(σ)	(COV)	—	—	(m)	(σ)	(COV)
Machine Model	-	F301/F401	F301/F401	—	—	—	F401	F401	—	—	—	SP8	SP8	—	—	—
Pile Diameter	mm	800	800	—	—	—	1000	1000	—	—	—	1500	1500	—	—	—
Pile Thickness	mm	9	9	—	—	—	12	12	—	—	—	16	16	—	—	—
Embedded Length	m	20	20	—	—	—	20	20	—	—	—	20	20	—	—	—
Lubrication System	unit	1	3	1.68	0.64	0.38	1.5	3	2.05	0.35	0.17	2	3	2.4	0.52	0.22
Water Supply Pipe	-	1	3.5	2.09	0.86	0.41	2	3.5	2.36	0.55	0.23	2	3.5	2.75	0.59	0.21
Pile Toe Ring Bits	-	—	—	—	—	—	—	—	—	—	—	—	—	—	—	—
Press-in Force	×10kN	13.5	40	22	7.78	0.35	13.5	40	23	7.66	0.33	15	40	24	7	0.29
Press-in Speed	mode	1	4	2.2	1.12	0.51	1	4	2.2	1.12	0.51	1	4	2.2	1.12	0.51
Extraction Speed	mode	1.5	6	4.14	1.53	0.37	1.5	6	4.14	1.53	0.37	1.5	6	4.14	1.53	0.37
Press-in Stroke	mm	110	1000	314	256.2	0.82	110	1000	314	256.2	0.82	110	1000	314	256.2	0.82
Extraction Stroke	mm	7	100	55.8	25.8	0.46	40	105	67.8	23.3	0.34	40	125	72.5	28.6	0.39
Water Flow Rate per Pipe	ℓ/min	10	50	24.5	10.65	0.43	12	50	26.3	10.28	0.39	15	50	29.2	11.86	0.41
Total Water Flow Rate	ℓ/min	20	100	42	23.74	0.57	24	100	53.3	19.8	0.37	30	150	69.3	36.0	0.52
Chuck Rotational Torque	kN·m	115	300	214.1	51.08	0.24	115	300	225	57.97	0.26	140	400	242.7	78.9	0.33
Chuck Rotational Velocity	mode	3	5.5	4.4	0.74	0.17	3	5.5	4.4	0.74	0.17	3	6	4.5	0.88	0.2
Remark	-	—	—	—	—	—	—	—	—	—	—	—	—	—	—	—

Case3. Gravels — The extrapolated SPT N-values: 50-150

		Pile Diameter φ800					Pile Diameter φ1000					Pile Diameter φ1500				
		Min.	Max.	Mean	Standard Deviation	Coefficient of Variation	Min.	Max.	Mean	Standard Deviation	Coefficient of Variation	Min.	Max.	Mean	Standard Deviation	Coefficient of Variation
		—	—	(m)	(σ)	(COV)	—	—	(m)	(σ)	(COV)	—	—	(m)	(σ)	(COV)
Machine Model	-	F301/F401	F301/F401	—	—	—	F401	F401	—	—	—	SP8	SP8	—	—	—
Pile Diameter	mm	800	800	—	—	—	1000	1000	—	—	—	1500	1500	—	—	—
Pile Thickness	mm	9	9	—	—	—	12	12	—	—	—	16	16	—	—	—
Embedded Length	m	20	20	—	—	—	20	20	—	—	—	20	20	—	—	—
Lubrication System	unit	1	3	2	0.65	0.33	1.5	3	2	0.45	0.23	2	4	3	0.62	0.21
Water Supply Pipe	-	1	3.5	2	0.74	0.37	2	3.5	2	0.52	0.26	2	4	3	0.55	0.18
Pile Toe Ring Bits	-	—	—	—	—	—	—	—	—	—	—	—	—	—	—	—
Press-in Force	×10kN	10	35	19	6.51	0.34	10	35	20	6.38	0.32	10	35	20	6.46	0.32
Press-in Speed	mode	1	4	2	1.08	0.54	1	4	2	1.08	0.54	1	4	2	0.98	0.49
Extraction Speed	mode	1.5	6	4	1.53	0.38	1.5	6	4	1.58	0.40	1.5	6	4	1.53	0.38
Press-in Stroke	mm	10	1000	255	253.1	0.99	105	1000	265	245.3	0.93	105	1000	265	245.3	0.93
Extraction Stroke	mm	40	130	69	25.0	0.36	40	130	73	28.3	0.39	40	130	69	24.0	0.35
Water Flow Rate per Pipe	ℓ/min	15	50	27	8.78	0.33	12	50	28	8.79	0.31	15	50	30	9.85	0.33
Total Water Flow Rate	ℓ/min	20	80	51	18.46	0.36	35	100	58	17.3	0.3	40	150	77	29.7	0.39
Chuck Rotational Torque	kN·m	115	275	210	46.62	0.22	115	300	219	53.39	0.24	140	450	231	80.7	0.35
Chuck Rotational Velocity	mode	2	5	4	0.92	0.23	2	5	4	0.92	0.23	2	5	4	0.88	0.22
Remark	-	—	—	—	—	—	—	—	—	—	—	—	—	—	—	—

Case4. Mudstone — The extrapolated SPT N-values: 150-200

		Pile Diameter φ800					Pile Diameter φ1000					Pile Diameter φ1500				
		Min.	Max.	Mean	Standard Deviation	Coefficient of Variation	Min.	Max.	Mean	Standard Deviation	Coefficient of Variation	Min.	Max.	Mean	Standard Deviation	Coefficient of Variation
		—	—	(m)	(σ)	(COV)	—	—	(m)	(σ)	(COV)	—	—	(m)	(σ)	(COV)
Machine Model	-	F301/F401	F301/F401	—	—	—	F401	F401	—	—	—	SP8	SP8	—	—	—
Pile Diameter	mm	800	800	—	—	—	1000	1000	—	—	—	1500	1500	—	—	—
Pile Thickness	mm	9	9	—	—	—	12	12	—	—	—	16	16	—	—	—
Embedded Length	m	20	20	—	—	—	20	20	—	—	—	20	20	—	—	—
Lubrication System	unit	1	4	1.82	0.96	0.53	1.5	4	2	0.75	0.38	2	4	2.77	0.82	0.3
Water Supply Pipe	-	1	4	2.18	1.06	0.49	2	4	3	0.76	0.25	2	4	3.25	0.79	0.24
Pile Toe Ring Bits	-	—	—	—	—	—	—	—	—	—	—	—	—	—	—	—
Press-in Force	×10kN	11	30	20	5.46	0.27	12.5	30	20	5.33	0.27	15	30	22	4.51	0.21
Press-in Speed	mode	0	4	1.7	1.15	0.68	0	4	2	1.13	0.57	0	4	1.6	1.19	0.74
Extraction Speed	mode	1.5	6	3.91	1.46	0.37	1.5	6	4	1.48	0.37	1.5	6	3.91	1.46	0.37
Press-in Stroke	mm	40	300	150.5	81.1	0.54	40	300	157	69.9	0.45	40	300	143.5	79.3	0.55
Extraction Stroke	mm	40	100	73	19.2	0.26	40	100	79	19.8	0.25	40	100	71	17.0	0.24
Water Flow Rate per Pipe	ℓ/min	15	40	25.2	7.94	0.32	12	65	32	13.23	0.41	15	50	29.3	12.25	0.42
Total Water Flow Rate	ℓ/min	20	70	43.6	17.33	0.4	36	100	61	20.0	0.33	40	150	77.7	33.6	0.43
Chuck Rotational Torque	kN·m	115	300	210.5	51.65	0.25	115	300	209	46.67	0.22	140	450	241.8	89.0	0.37
Chuck Rotational Velocity	mode	2	5	3.8	1.06	0.28	2	5	4	1.08	0.27	2	5	4	1.16	0.29
Remark	-	—	—	—	—	—	—	—	—	—	—	—	—	—	—	—

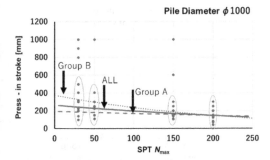

Figure 7. Relationship between press-in strokes and SPT *N*-values (1000mm diameter steel tubular piles).

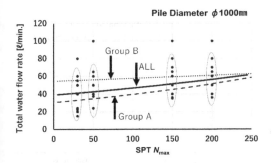

Figure 8. Relationship between total water flow rate and SPT *N*-values (1000mm diameter steel tubular piles).

– The selected number of water supply pipes is increasing as the ground stiffness increases. It is particularly so for the case of experienced operators.
– The values of COV for the press-in force are relatively small and are decreasing with the ground stiffness, while the values of COV for the press-in stroke are larger.

4.1.2 *Influence of pile diameter*

Influence of pile diameter is generally small, compared to the influence of soil profile. A general trend is that the values of COV do not vary. However, it is noticed that the values of COV of the press-in force are small, while the values of COV of press-in stroke are comparatively large. Additionally, the following points are noticed.

- For a given soil profile, the operators tend to select larger values of rotational torque and rotational velocity of chuck when the larger diameter pile is to be installed.
- For a given soil profile, the operators select a larger number of water supply pipes and a larger water flow rate when the larger diameter pile is to be installed.

4.1.3 *Influence of operator*

One of the main interests of this survey is to find out the differences of these setting values between experienced and less experienced operators. The responders were categorized into two groups: the Group A (rotary press-in piling experience equal to or more than five years) and the Group B (equal to or less than four years-experience).

It is seen that experienced operators tend to select a small number of water supply pipes, compared to those of less experienced operators for the case of clayey ground (Case 1). This tendency can be confirmed from the regression curves in Figure 8.

The vertical axis is the mean number of selected lubrication unit in Figure 9, comparing change of the number with soil profile. The figure also compares the selected numbers of lubrication unit by Group A with those by Group B.

It is clear from the figure that Group B selects the same number, regardless of soil profile. In contrast, Group A selects a smaller value for soft ground and a larger value for stiff ground, increasing as the ground stiffness increases. This suggests that the experienced operators consider the role of lubrication important and change their selection according to the stiffness of ground.

Figure 10 shows the arrangements of the water supply pipes and the direction of water flow by an arrow for the three respondents as an example. It is noticed that they arrange the pipe in such a way to discharge the water towards inside the pile for clayey ground and towards outside the pile when the ground becomes stiffer, in addition to an increase in the number of piles. This confirms the importance of a proper selection of the number of water supply pipes and a proper arrangement of the pipes for effective press-in piling.

Similar to Figure 9, the vertical axis of Figure 11 is the mean value of speed mode (The large value means that press-in speed is fast.) for four soil profiles, comparing Group A and Group B. Group

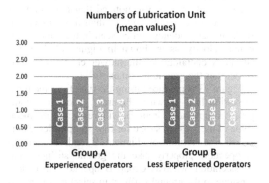

Figure 9. Comparison of experienced/less experienced operators (Unit number of water lubrication system).

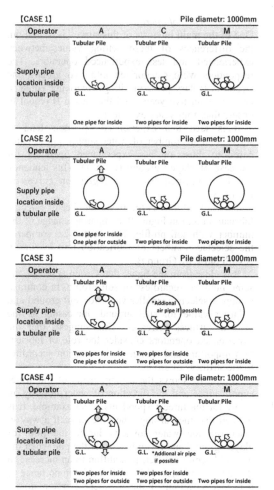

Figure 10. Examples of water supply pipe locations on each case (1000mm diameter steel tubular piles).

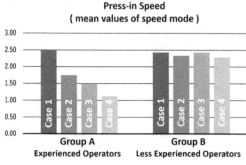

Figure 11. Comparison of experienced/less experienced operators (Press-in speed mode).

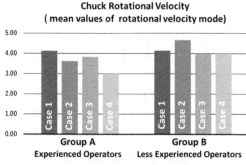

Figure 12. Comparison of experienced/less experienced operators (Chuck rotational velocity).

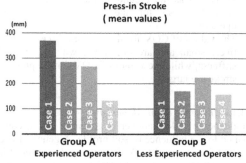

Figure 13. Comparison of experienced/less experienced operators (Press-in stroke).

B operators select almost the same value regardless of the ground stiffness, while Group A operators set the press-in speed gradually lower as the ground becomes stiffer. It is also noticed that the values of Group A are generally smaller than those of Group B. A similar tendency can be seen in the chuck rotational velocity, as is shown in Figure 12. Namely, Group A operators set the slower chuck rotational velocity as the ground becomes stiffer.

Figure 13 is produced using the data from the three regression lines shown in Figure 7. The figure compares Group A and Group B, with respect to the mean value of the press-in stroke for four soil profiles. Similar to the results of Figures 9 and 11, Group A operators properly take the ground stiffness into consideration, whereas Group B operators seem less sensitive to the ground stiffness in selecting the press-in stroke.

This also suggests the importance of operator's skill and experience.

4.2 Summary discussions

The conducted questionnaire survey reveals the interesting findings related to a picture of how operators use Gyro piler on site in different soil profiles. The followings summarize the major findings.

[1] The mean value of the selected press-in force is 200kN regardless of the ground stiffness and steel tubular pile diameter, which is considerably smaller than the specification of the allowable maximum press-in force of the machine. It does not necessarily mean that the machine has unnecessary maximum capacity.

[2] More than 70% operators select the same number of water lubrication units and water supply pipes, regardless of soil profile and pile diameter. A close look at the data, there exists a difference between experienced and less experienced operators.

[3] All the operators tend to select a slower press-in speed as the ground becomes stiffer. Some operators select the same value of the press-in speed even the pile diameter becomes larger.

[4] There is a tendency that shorter strokes both in press-in and extraction are selected. This means that the number of repetitive upward and downward motion (surging) is increasing with an increasing of ground stiffness, in an attempt to avoid the formation of soil plugging at the toe of the pile and to reduce the shaft friction. For a given soil profile, the selected press-in stroke and extraction stroke are unchanged. This implies that the number of repetitive upward and downward motion (surging) is not greatly affected by the pile diameter.

[5] For a given soil profile, there is a tendency that the larger value of chuck rotational torque is selected for a larger pile diameter.

[6] Due to the prevention not to wear out cutting bits at the pile toe, experienced operators select a slower chuck rotational velocity for stiffer ground. As a result of it, machine damages might be reduced. In contrast, less experience operators select the same value of the speed. This finding implies that some causes of machine damages may stem from operator's skill and experience.

[7] The value of COV of the operator's responses becomes larger as the ground becomes stiffer.

5 CONCLUDING REMARKS

With the view that performance of the final product of piles installed largely depends on the operator's skill with a proper handling of the machine, this study began. This paper firstly pointed out that effective piling operation can only be achieved by a good combination of a well-maintained machine and a skillful operator.

The questionnaire survey thus conducted revealed the interesting findings related to a picture of how operators use Gyro Piler on site in different soil profiles. From the above, it is clear that there is a tendency that the experienced and skillful operators carefully chose the values of initial setting of the machine operation and the number and arrangement of water lubrication system in order for smooth piling operation and for avoiding a possible risk of damaging the machine, taking into account the soil profile and the pile diameter. It is also noticed that the less experienced operators tend to select the similar initial setting values regardless of soil profiles.

The survey confirms that the operator's experience and skill play an important role for effective press-in piling with a minimum risk for damaging the machine. The information summarized in this paper may be regarded as valuable rules of thumb from know-how that experienced operators gain on site.

This survey was limited to cases of Gyro Piler. Further survey is planned to carry out for cases of other press-in machines with another group of operators. Further study will be of use for the future development of piling machine for machine designers. The accumulated know-how will become an essential database for developing an automatically operating system as a deep learning database based on AI technology.

ACKNOWLEDGEMENTS

This paper is part of the research efforts conducted by International Press-in Association, Technical Committee 5 on "Influence of operator's skill and experiences on field performance of Press-in Piling" (Chair Kusakabe, O. and Co-Chair Minami K.).

The authors express our gratitude to all the respondents for their cooperation. Supports from Mr. Yamaguchi, M. and Ms. Ogawa, N. during the preparation of the paper are also acknowledged.

REFERENCES

Bolton, M., Kitamura, A., Kusakabe, O. and Terashi, M. 2020. *New Horizons in Piling -Development and Application of Press-in Piling*. Leiden, CRC Press/Balkema.

GIKEN LTD. 2019. *GIKEN maintenance system (in Japanese)*. Retrieved from https://www.giken.com/ja/wp-content/uploads/GMS_ver020ja02.pdf.

Proceedings of the Second International Conference on Press-in Engineering
2021, Kochi, Japan – Matsumoto et al (eds)
© 2021 Taylor & Francis Group, London, ISBN 978-1-032-10414-0

2D/3D FEM embedded beam models for soil-nail reinforced slope analysis

X.C. Lin
Virtuosity, Bentley System(s), Singapore city, Singapore

ABSTRACT: Conventional design of soil-nail structure based on Limit Equilibrium Method (LEM) is adequate, but the efficiency is questionable due to the lack of knowledge in soil-structure interaction. As the computation power nowadays is high and attainable, it becomes increasingly common to use Finite Element Method (FEM) for analyzing complex geotechnical problems. This paper demonstrates the use of 2D *Embedded Beam Row* element for soil-nail group modelling in PLAXIS 2D, comparison of response is drawn against that of 3D *Embedded Beam* element in PLAXIS 3D, as well as the field data from CLOUTERRE-1, the French National Research Project, followed by a discussion on the factor of safety obtained from FEM and LEM. It has affirmed that, the 2D Embedded Beam Row model can effectively handle groups of soil nails in the plane strain condition and produce both quantitative and qualitative predictions of deformation and structural response that concerns practitioners.

1 INTRODUCTION

Soil nailing is an in-situ soil reinforcement technique extensively applied in the slope stabilization problem due to its relatively low cost and easy installation process. Conventional design methods are generally based on limit equilibrium theories, which are adequate, but the efficiency is questionable due to the lack of knowledge on the complex soil behaviors. Finite Element Method (FEM), on the other hand, accounts for a wide range of phenomenon as observed in the real-world, typically the soil-structure interaction; that complements the conventional design method with the possibilities to achieve an economical and reliable design. As the use of Finite Element Method (FEM) prevails, and the required computational power nowadays is attainable at reasonable cost, there is growing popularity in using FEM software package to verify and optimize the design. 2D FEM package has the merits of relatively low cost, easy model execution and result interpretation over 3D FEM package. Soil nail behaviors are however three-dimensional (3D) phenomenon, and technically it is inappropriate to model the application using a 2D program. Prior studies demonstrated that, by modelling the soil nail group as a surface structural element in 2D plane strain condition, comparable results can be found to the field response; however, the numerical artefacts arising from this simplified approach remain unaddressed. The FEM package used in this study is PLAXIS 2D and PLAXIS 3D, with the focus on the structural element namely *Embedded Beam Row*

(2D) and *Embedded Beam* (3D). *Embedded Beam Row* is a line element developed to model a row of piles under 2D plane strain condition, which can be extended to model soil nails in group. The performance of 2D *Embedded Beam Row* is investigated through this study, by referencing to the French National Research Project, e.g. CLOUTERRE-1. The response of 2D *Embedded Beam row* is checked against that of 3D *Embedded Beam*, as well as the available field data from the project, followed by a discussion on the factor of safety obtained from FEM and LEM, respectively.

2 MODELLING OF SOIL-NAIL IN PLAXIS 2D

Before the introduction of *Embedded Beam Row* element, modelling of pile rows or soil nails in group in 2D is nearly impossible. The closest choices of structure elements for the application are *Plate* and *Node-to-Node Anchor (N2N Anchor)*, both however have their own limitations.

Plate element, with both its axial and bending stiffness, can capture the certain structural response of the soil nail; and with the interface element around, the soil-structure interaction, in particular the skin friction, to certain extent, can be accounted for. The limitation is however that, *Plate* as a continuous element, the presence of which causes a discontinued mesh that prevents the soil from flowing through, giving rise to too rigid a lateral response. The practice of using smeared properties to account for the out-of-plane spacing in the soil-nail group, gives

DOI: 10.1201/9781003215226-59

close prediction of the gross structural response; but meanwhile, the approach presents certain numerical artefacts, for instance the unrealistic shear plane, that is less relevant as observed in the field for soil nails in group. Moreover, as the soil-structure interactions take place on both sides of the *Plate* in the 2D plane strain model, the interface properties require some rigorous calibration against the 3D response. The process is cumbersome and discouraging.

N2N Anchor element has axial but no bending stiffness, its interaction with the mesh is realized through the two end points, which means lateral response due to bending, as well as, soil-structure interaction along its length cannot be considered.

The 2D *Embedded Beam Row* element (introduced in 2012) can be used to model a row of piles at a certain spacing perpendicular to the model area. *Embedded Beam Row* is implemented in such a way that, it is not directly coupled with the mesh, but through the special line interfaces (consisting of spring elements and sliders). The program asks for input properties per pile, and calculates the smeared properties based on the out-of-plane spacing defined.

3 THE FRENCH NATIONAL RESEARCH PROJECT, CLOUTERRE-1

3.1 *Project information*

In France, a National Research Project called CLOUTERRE has been conducted from 1986 to 1991 with the objective of better understanding the behaviors of soil-nailed walls during construction, in service and at failure. Within the framework of the first National Research Project CLOUTERRE, three full-scale experimental soil nailed walls were constructed and then pushed to failure by three different modes. The analyses of these three different modes of failures have been performed using limit equilibrium methods and the multi-criteria approach (Plumelle and Schlosser, 1991; Schlosser et al., 1992).

The soil-nailed wall was constructed in an experimental backfill, 7 m high, built with special care at the CEBTP site (Centre Expérimental de Recherches et d'Etudes du Bâtiment et des Travaux Publics) near Paris on a dense sand formation as shown in Figure 1. In particular, the homogeneity and density of the backfill were controlled at each phase of its construction. ontainebleau sand, as the backfill material is in a medium-to-dense state (relative density, D_r =0.6) and has an average unit weight of 16.1 kN/m³.

The underlying foundation sand is of similar properties but much stiffer. The groundwater table is well below the foundation sand, throughout the construction of the nailed wall. Effect of groundwater can thus be ignored since it is below the reinforced zone.

3.2 *FEM modelling*

15 noded triangular element (4th order shape function) was used in the 2D analysis, while 10 noded tetrahedral element (2nd order shape function) was used in the 3D analysis. The compromise in accuracy, as a result of lower order element used in 3D, was compensated by denser mesh refinement. Soil nails were modelled using *Embedded Beam Row* element, and *Embedded Beam* element in 2D and 3D, respectively.

The 2D and 3D numerical model boundary and mesh discretization are shown in Figure 2 and Figure 3, respectively.

The mechanical properties of the backfill and foundation soils were evaluated based on published data on

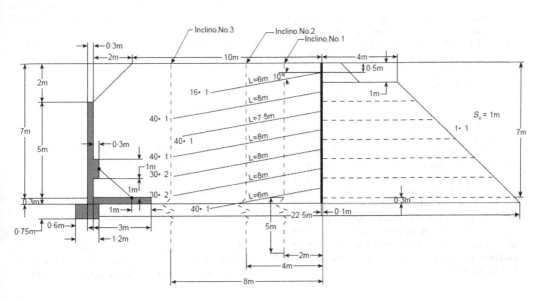

Figure 1. Cross-section of the full-scale experimental nailed wall.

567

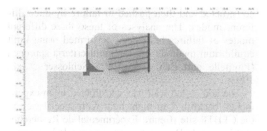

Figure 2. 2D model boundaries and mesh discretization.

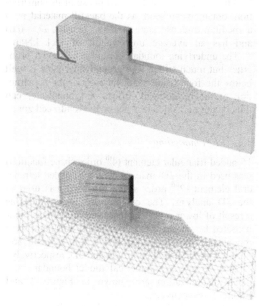

Figure 3. 3D model boundaries and mesh discretization.

Fontainebleau sands, pressuremeter tests conducted at the site, and triaxial tests performed at the CERMES-ENPC=LCPC (Plumelle and Schlosser, 1991).

Two pressuremeter tests were performed within the backfill before construction of the soil nailed wall (Plumelle and Schlosser, 1991). The average Ménard pressuremeter modulus over the 7 m height was 10 MPa.

Two other pressuremeter tests were carried out down to 6 and 20 m in the foundation soil as shown in Figure 4. The results of these deep pressuremeter tests showed the existence of a very hard soil layer (Ménard modulus, E_M=50 MPa and limit pressure, p_l =5 MPa) between 7 and 11 m beneath the backfill in which the soil-nailed wall was constructed. The mesh of the numerical model, therefore, ends at the top of this hard layer. The average pressuremeter modulus in the foundation soil, between 0 and 7 m is 35 MPa.

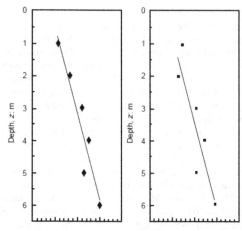

Figure 4. Pressuremeter test result of Fontainebleau sand.

The above mentioned triaxial tests carried out at CERMES (Plumelle and Schlosser, 1991) give an angle of friction in compression varying between 36° and 40° depending on the confining pressure.

The apparent cohesion was found to depend upon the strain level due to the rearrangement of the grains. For the considered sand, with an average water content of 5-8%, the cohesion varies between 2 and 6 kPa (Luo, 2001).

The numerical analysis was carried out using advanced soil model, e.g. Hardening Soil (HS) model, which requires (3) stiffness input of E_{50}^{ref}, E_{oed}^{ref} and E_{ur}^{ref}, that governs the deformation behaviors under deviatoric loading, volumetric loading and unloading, respectively.

For sandy material, the E_{50}^{ref} and E_{oed}^{ref} is in the same order according to Benz (2007), the quantity at reference stress level of 100 kPa can be deduced based on the stress-dependency formula, together with the aforementioned stiffness derived from pressuremeter test.

$$E_{50} = E_{50}^{ref} \left(\frac{c \cos \varphi - \sigma_3' \sin \varphi}{c \cos \varphi + p^{ref} \sin \varphi} \right)^m \quad (1)$$

For (quartz) sand, stiffness is supposed to vary linearly with relative density, RD, according to Brinkgreve et al. (2010).

$$E_{50}^{ref} = 60000 \, RD/100 \, [kN/m^3] \quad (2)$$

The $E_{ur}^{ref}/E_{oed}^{ref}$ ratio falls in the range of 3 to 5 for the material, in this study a lower bound of 3 is considered, which is in line with Brinkgreve et al. (2010),

Table 1. Input parameters for hardening soil model.

Description.	E_{50}^{ref},kPa	E_{oed}^{ref},kPa	E_{ur}^{ref},kPa	m -	c', kPa	φ' °	ψ °
Fontainebleau Sand	36000*/30000	36000*/30000	108000*/90000	0.5	3#/8	38#/35*	25#/10/5.5*
Foundation Sand	100000	100000	300000	0.5	3#/8	38#/35*	20#/10/5.5*

- Unterreiner et.al. (1997) * - Brinkgreve et al. (2010)

$$E_{ur}^{ref} = 180000 \ RD/100 \ \left[kN/m^3 \right] \quad (3)$$

The quantity m measures the rate of dependency that is similar to Janbu's number, except in PLAXIS the range is from 0.5 to 1. For sandy material, m is close to 0.5, correlations to relative density is observed according to Brinkgreve et al. (2010).

$$m = 0.7 - RD/320 \quad (4)$$

Besides the stiffness parameter, Brinkgreve et al. (2010) has the recommendations on strength parameters,

$$\varphi' = 28 + 12.5RD/100 \ (deg.) \quad (5)$$

$$\psi = -2 + 12.5RD/100 \ (deg.) \quad (6)$$

The soil parameters suggested by different literatures are summarized in Table 1.

The nails are driven hollow aluminium tubes grouted in the 63 mm borehole. Structure element, e.g. *Embedded Beam Row* was used to model the soil nail. The structural response, in particular the mobilization of skin friction, is dependent on the complex soil-structure interaction, which is a result of the constitutive response of the selected soil model to the in-equilibrium. The key to a realistic numerical structure response greatly depends on the accurate input skin resistance profile besides the appropriate parameters used in the constitutive soil model.

Data have been compiled on more than 450 pull-out tests carried out by contractors in this field (Plumelle and Schlosser, 1991). This data bank makes it possible to estimate for the preliminary design of the soil/nail interaction parameters.

As the geomaterial considered in this study is sandy material, we hereof refer to the corresponding chart from the data bank, as shown in Figure 5.

Based on the pressuremeter result of a limiting pressure of 1.5 MPa, the corresponding ultimate skin friction for driven nail grouted in sand, is approximately 80 kPa. FEM pullout test was carried out to investigate the numerical response. The mobilization of shear stress, and the corresponding failure mechanism

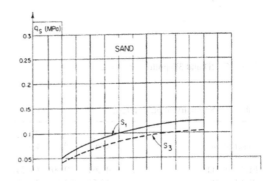

Figure 5. CLOUTERRE pullout test database for sand.

developed, during the numerical pull-out test is shown in Figure 6.

The load-displacement curves from 2D and 3D pull-out test are shown in Figure 7.

A summary is given in Table 2 for the numerical pull-out test.

The input parameters for soil nail as *Embedded Beam Row* are summarized in Table 3. The skin resistance profile is set to *Layer dependent*, whereby the mobilized skin resistance of the soil nail is governed by the failure criterion of the surrounding soil.

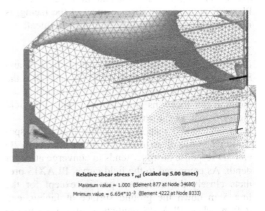

Relative shear stress τ_{rel} (scaled up 5.00 times)
Maximum value = 1.000 (Element 877 at Node 34680)
Minimum value = 6.654*10^{-3} (Element 4222 at Node 8333)

Figure 6. Mobilization of shear stress in FEM pullout simulation.

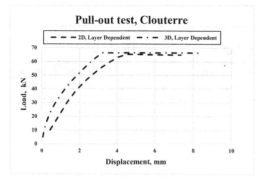

Figure 7. 2D and 3D pullout response.

Table 2. Summary of pullout test.

Description.	τ_{max}, kPa	τ_{min}, kPa	T_{max}, kN
PLAXIS 2D	45.61	39.60	64.745
PLAXIS 3D	47.57	34.59	66.25
CLOUTERRE Recommendation	80	80	–

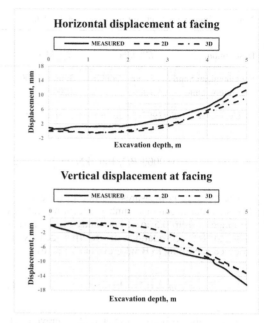

Figure 8. Horizontal and vertical displacement at facing.

Table 3. Input parameters for soil nail.

Description.	E,kPa	γ,kN/m^3	d,mm	$L_{spacing}$, m	M_p, kNm	N_p,kN	Skin Resistance
Soil Nail	30000000	8	63	1.15	1	55	Layer Dependent

3.3 FEM result, CLOUTERRE-1

3.3.1 Displacement at facing
The horizontal and vertical displacement over the top point of the facing, are monitored at different excavation depth and presented in Figure 8. The PLAXIS prediction of displacement at the facing, is in general slightly stiffer than field measurement.

3.3.2 Horizontal deflection
Inclinometers are placed at distance of 0, 2, 4 and 8 m from the facing. In PLAXIS 2D and 3D, the horizontal displacement profiles are extracted at the inclinometer locations and charted against the field measurement, as shown in Figure 9.

In general 3D response is stiffer than 2D, especially when it is close to the ground surface, both 2D and 3D response however tends to converge at larger depth. As compared to the field curve, PLAXIS predicts closely the deflection profile, except for the first 5 m down the ground surface at facing, e.g. d=0 m, where PLAXIS predicts a softer response in horizontal deformation; nevertheless, it still captures the deflection well at the ground surface.

3.3.3 Axial force of soil-nail
The axial force distribution of the four soil nails are extracted from PLAXIS 2D and 3D, and plotted against the field measurement, as shown in Figure 10. It is observed that PLAXIS predicts higher axial force at the nail head, e.g. T_o/T_{max} ratio up to 0.9, which deviates from CLOUTERRE Recommendation that the T_o/T_{max} ratio, close to 1 at the beginning of tension, reaches progressively much smaller values, as a function of the level of the layer considered, going from 0.3 to 0.7 in the case of the soil nailed wall CETBP No.1.

3.4 Factor of safety

Limit equilibrium method considers the strength limit state, by ensuring that the design combined strengths of the nails and the soils, exceed the applied load with a factor of safety appropriate to the level of uncertainty associated with the loads due to the excavation.

Gässler (1988) studied the behaviors of steep slope and had the conclusion that if not a planar, a bi-linear failure surface is critical,

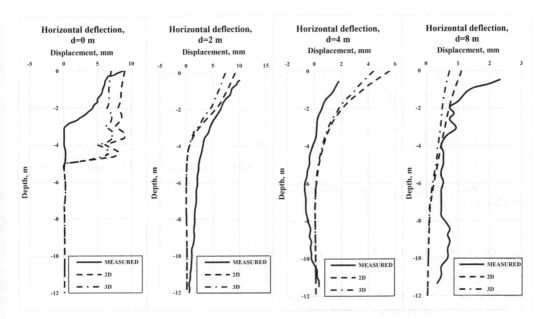

Figure 9. Horizontal deflection profile at 0, 2, 4, 8 m from facing.

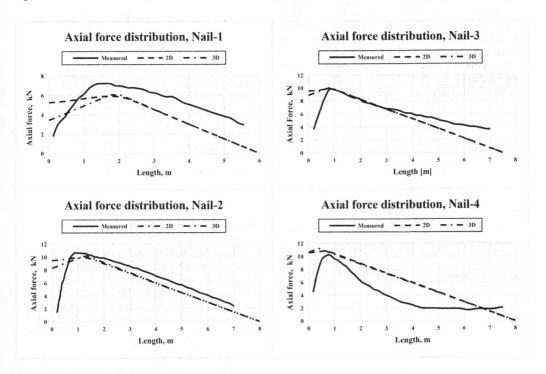

Figure 10. Axial force, Nail-1, 2, 3 & 4.

whereas, for gentler slope, a circular failure surface is critical. This is further affirmed based on the predicted planar failure surface by PLAXIS 2D and 3D.

According to Luo (2001), Global factor of safety (LEM) for both internal and mixed failure modes is checked by Eq. (7) for potential failure surface as shown in Figure 11.

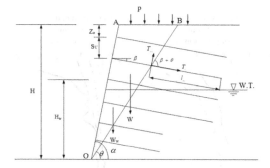

Figure 11. Potential failure surface of a reinforced slope, (after Luo 2001).

$$FS = \frac{\frac{c'H}{\sin\theta} + [(W - W_w)\cos\theta + T_c\cos(\beta + \theta) + T\sin(\beta + \theta)]\tan\varphi'}{W\sin\theta - T_c\sin(\beta + \theta) - T\cos(\beta + \theta)} \quad (7)$$

In PLAXIS, the factor of safety is evaluated using the strength reduction method, e.g. phi-c reduction, where the strength of the soil is reduced in a bid to trigger the ultimate failure, and the factor of safety is calculated in correspondence to the failure mechanism.

The factor of safety and the corresponding ultimate failure predicted by PLAXIS 2D and 3D are presented in Figure 12 & Figure 13, respectively.

For comparison against LEM, the failure mechanism predicted by PLAXIS, e.g. in this case the wedge mechanism, is visually measured for its

Figure 12. Failure mechanisms & FOS by PLAXIS 2D.

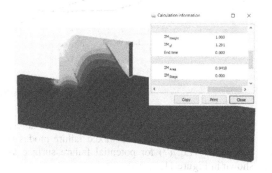

Figure 13. Failure mechanisms & FOS by PLAXIS 3D.

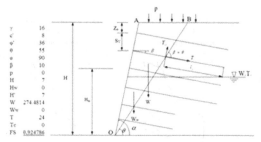

Figure 14. LEM prediction of FOS based on 2D wedge.

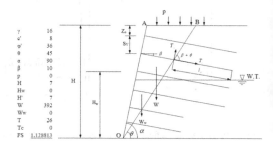

Figure 15. LEM prediction of FOS based on 3D wedge.

Table 4. Summary of FOS by FEM and LEM.

Description.	θ	FOS
PLAXIS 2D	55°	1.18
Limit Equilibrium	55°	0.92
PLAXIS 3D	45°	1.29
Limit Equilibrium	45°	1.13

dimension, for the input to Eq. (7) to evaluate the factor of safety in the LEM way.

The LEM predictions are presented in Figure 14 & Figure 15, which adapt to the wedge dimension from PLAXIS 2D and 3D, respectively. As it turns out, 2D and 3D predicts slight difference in the factor of safety, due to the different size of wedge mechanisms and the nail length in the passive zone. Comparison in factor of safety by PLAXIS and Limit Equilibrium Method is summarized in Table 4.

PLAXIS 3D predicts slightly higher factor of safety than 2D due to a smaller wedge mechanism considered. Limit Equilibrium predicted FOS is in general slightly lower, but sufficiently close to PLAXIS predictions, considering the tolerance effects going into a naked-eye inspection of the wedge size from PLAXIS, given the fact that there is no functionality in PLAXIS to output the dimension of the failure wedge in exact number; after all, the FOS by PLAXIS is based on the amount of soil strength that can be reduced, while maintaining the global equilibrium, rather than the geometric details of the failure mechanism develops as a result of the strength reduction during the safety calculation.

4 CONCLUSIONS

With *Embedded Beam Row* element for soil nail, PLAXIS 2D analysis closely yields the deformation of the reinforced slope, as well as the structural response of the soil nails, that is comparable to the result of PLAXIS 3D analysis and field measurement.

In addition, the 2D and 3D FEM prediction of FOS fluctuates mildly around the LEM based FOS of unity, that the experimental project was based off. It is further affirmed that, in the absence of a 3D program, 2D *Embedded Beam Row* model can handle a group of soil nails or piles in the planestrain condition. Owing to the merit of 2D *Embedded Beam Row*, the analysis of soil-nails or piles in group, is free from numerical artefacts as found in the cases of using alternative structural elements.

It is worth to note that the bearing capacity is an input for the *Embedded Beam Row* element, thus it is essential to validate the numerical pull-out response against field test data, to ensure that the available skin resistance is not exceeded in the analysis at all time; the mobilization of skin resistance is however the result of the constitutive model response, as well as the soil-structure interaction realized through the special line interface; the latter and its Interface Stiffness Factor (ISF), has subtle influence on the load-displacement behavior of geo-structure modelled, and requires calibration on a case-by-case basis just to be rigorous. The default set of ISF, based on the load-displacement behavior, of bored piles founded in sand according to the Dutch Annex of Eurocode, is found to be appropriate for the case of CLOUTERRE-1, due to the similarity in the sub-soil characteristics.

The discrepancies in the prediction of axial force of soil nail close to the facing, could be due to the complex interaction or local failure at the connection, for which an elastic *Plate* element used in the analysis, cannot capture the behavior properly.

Partially the mismatch of numerical result and field measurement, can be attributed to the tolerance effects arising from the experimental setup, e.g. the soil-nailed structure was built without overdesign factor; and in one of the experiments, it failed purely due to water infiltration made on purpose, for the understanding of failure mechanisms; that said, the load-displacement behavior could be erratic when the problem is on the verge of failure, and that is beyond what a FEM numerical analysis at its tolerance could capture.

REFERENCES

Benz, T. 2007. Small-strain stiffness of soils and its numerical consequences. *phD thesis, University of Stuttgart*

Brinkgreve, R.B.J. & Engin, E. & Engin, H.K. 2010. Validation of empirical formulas to derive model parameters for sands. *7th European Conference on Numerical Methods in Geotechnical Engineering, NUMGE (2010)*: 137–142

Gässler, G. 1988. Soil Nailing Theoretical Bars and Practical Design. *International Geotechnical Symposium on Theory and Practice of Earth Reinforcement*, Fukuoka, Japan, 5-7 October 1988: 283–288

Luo. 2001. Soil Nail Behaviors. *phD thesis, National University of Singapore*

Plumelle, C. & Schlosser, F. 1991. Three Full-Scale Experiments of French Project on Soil Nailing: CLOUTERRE. *TRANSPORTATION RESEARCH RECORD 1330*: 80–86

Schlosser, F. Unterreiner, P., & Plumelle, C. 1992. French Research Program CLOUTERRE on Soil Nailing. *Grouting Soil Improvement and Geosynthetics, Geotechnical Special Publication No. 30, ASCE*: 739-749

Unterreiner, P. & Benhamida, B. & Schlosser, F. 1997. Finite element modelling of the construction of a full-scale experimental soil-nailed wall, French National Research Project CLOUTERRE. *Ground Improvement (1997) 1*: 1–8

Proceedings of the Second International Conference on
Press-in Engineering 2021, Kochi, Japan – Matsumoto et al (eds)
© 2021 Taylor & Francis Group, London, ISBN 978-1-032-10414-0

Development of small-sized splice plates applied to steel sheet pile longitudinal joints

H. Nakayama & T. Momiyama
Nippon Steel Corporation, Tokyo, Japan

ABSTRACT: Steel sheet piles are longitudinally combined to form a longer pile in the course of driving them into the ground in places such as under bridge girders, where overhead clearance is limited. To connect piles, sheet piles are welded at each corresponding end, and splice plates are welded to fill a section shortage of the interlock where welding can't be applied. The splice plates tend to be larger and heavier according to the increase of size of sheet piles, and hence welding work becomes laborious. To reduce such burden, a small-sized splice plate that can keep the original splice plate in a like diamond shape was developed. The validity and effectiveness of the newly developed splice plate was confirmed through experiments and construction practices on site.

1 GENERAL INSTRUCTIONS

1.1 Longitudinal joints for steel sheet piles

Steel sheet piles are used as one of the main construction materials for many purposes, such as in retaining walls and in the reinforcement of ports, harbors, embankments, and piers, etc.

According to situations at construction sites or in transportation infrastructure where steel sheet piles are used or carried, the length of each steel sheet pile must be less than a certain level. For example, when it is necessary to set steel sheet piles to strengthen piers under bridge girders, the length of the steel sheet pile should be at least less than the distance between the ground surface and the underside of the bridge girder. Especially, the length needs to be much less at the construction site, where it is necessary to secure space for the driving machine used for installing the steel sheet pile. However, the steel sheet pile length as required by the construction design thus tends to be shorter than the necessary design length—which would be indispensable toward forming the structure. To cope with this kind of situation, the short steel sheet piles are longitudinally combined to make one long pile while these piles are driven into ground.

To form a longer pile consisting of shorter piles, a couple of steel sheet piles are respectively welded at each corresponding end. Because welding along the interlock should be avoided so as not to reduce the internal space of the interlock and so as not to cause deformation of the interlock, in order to secure smooth interlocking during driving, the total section where welding can be applied is less than the total cross-section of the steel sheet pile. The shortage of the cross-section area means a shortage in the strength and stiffness necessary for the steel sheet pile as a construction member. To fill the shortage of welding area, splice plates are commonly used (JASPP 2014). These are attached to the surfaces of the steel sheet piles in the form of bridging over two corresponding piles. Splice plates perform the role of reinforcing longitudinal joints. The reinforcement method using splice plates is very useful because of the simplicity of shape, and hence its handling is very easy when welding the plates onto the steel sheet pile.

In terms of industrial trends at construction sites, labor-saving has recently been required. In the case of steel sheet piles, welding longitudinal joints occupies a large amount of time throughout the driving period. When a larger size of steel sheet piles is used, the size of the splice plates also increases, and this makes the splice plates heavier, leading to longer welding time and then to longer construction time, making the welding task harder. Therefore, the present splice plate form was improved so as to alleviate welding work for longitudinal joints.

1.2 Specifications of splice plates

From a practical point of view, the positions of longitudinal joints are arranged alternately lengthwise along the wall direction so as to avoid stress concentration around the longitudinal joint area as a wall. The location of each longitudinal joint for the respective neighboring steel sheet piles is situated in a staggered form, as shown in Figure 1.

DOI: 10.1201/9781003215226-60

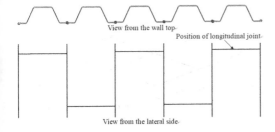

View from the wall top

Position of longitudinal joint

View from the lateral side

Figure 1. Position of longitudinal joints.

A longitudinal joint consists of two kinds of splice plates. One of them is attached to the surface of the web of the steel sheet pile, with the other attached to that of the arm. The latter parts are welded on both arms of the steel sheet pile. Figure 2 illustrates the location of each splice plate. By utilizing both web and arm surfaces where the distance from the neutral axis is almost the same for each, the position of the neutral axis of the steel sheet pile with a longitudinal joint is set to be similar to that of the original steel sheet pile.

The size of the splice plates is determined so as to secure a sufficient section area and capacity of bending moment.

With regard to the section area, the total cross-section area of the splice plates along the longitudinal joints needs to correspond to that of the interlocks where welding should be avoided. When a steel sheet pile is driven into the ground, tension force is repeatedly loaded along the pile in the course of the vertical motion applied from the driving machine.

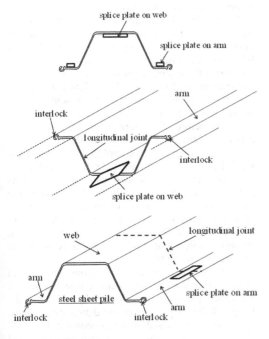

Figure 2. Splice plates.

This motion is necessary so as to realize smooth driving in order to avoid soil consolidation around the bottom of the pile and to reduce friction force along the side surfaces along the pile. Because the interlock can't be welded along the cross-sectional direction of the steel sheet pile, the total section area of the splice plates is designed to be larger than that of the interlocks positioned at the widthwise ends of the steel sheet pile. The sizes of the splice plates are determined by adjusting the thickness and width to match the required section area.

Concerning the capacity of bending moment, the sizes of the splice plates are set to secure the moment of inertia corresponding to the original steel sheet pile. Because longitudinal joints are arranged in a staggered form, the amount of moment of inertia is calculated as a combination of a pair of two steel sheet piles. One of the steel sheet piles is the complete form of the cross-section, which is the same as the original steel sheet pile, and the other is set with the longitudinal joint. The bending stress induced in the splice plate at the edge from the neutral axis is limited to less than the allowable stress.

The outline of splice plates attached to the web of steel sheet piles consists of a diamond-like shape. Because the splice plates are welded to the surface of the steel sheet piles, the amount of welding parts is arranged in order to make the stress level at the welding parts caused by both tension and bending moment less than the allowable stress. The size of throat depth along the welding lines is adjusted to satisfy this amount. On the other hand, it can be possible to reduce the weight of the splice plates, keeping the same amount of welding parts. Because the sizes of the splice plates tend to be larger according to the increase of the size of the steel sheet piles, the welding work would become more difficult if the weight of the splice plates is larger than the weight that one worker can carry/support during welding. Therefore, it is useful to use lighter splice plates so as to mitigate extra welding work. In terms of the shape of splice plates with a certain amount of welding parts, a diamond-like shape is more efficient than a rectangular shape with regard to the weight of the splice plate. As shown in Figure 3, when the perimeters of the diamond-like plate and rectangular plate are set to be the same, keeping the width at the widest part of the splice plate the same, which is noted as "b" in the figure, the area of the diamond-like shape becomes smaller than that of a rectangular plate, and this means that the diamond-shaped plate is lighter than the rectangular plate. Because the widest part of the splice plates is determined so as to secure a minimum area to cope with the tension force, the width cannot be reduced once the thickness of the splice plate is set. From the viewpoint of labor-saving and cost effectiveness, diamond-shaped splice plates are usually adopted for the splice plates attached to the web of steel sheet piles. Because the size of the splice plate attached to an arm is small enough for a single worker to handle, rectangular

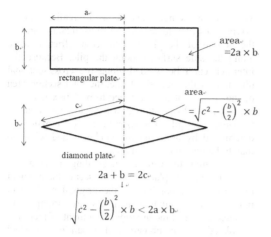

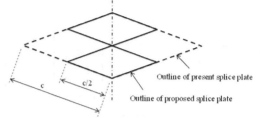

Figure 4. Small-sized splice plate.

Figure 3. Comparison of the area of splice plates.

splice plates are sometimes used because the trimming of plates into rectangular form is easier than diamond-like form.

2 NEW TYPES OF SPLICE PLATES

2.1 Half-size splice plates

Even if diamond-shaped plates are used for the splice plates attached to the web of steel sheet piles, the weight is sometimes still too heavy to support considering workability on site. While welding work is carried out by one welder, the splice plate is usually supported by one hand of the worker, and the welding torch is handled by the other hand. It is normally regarded that one hand can support around 13 kg without worsening the conditions at the construction site. Therefore, it is effective to make the splice plate as light as possible so as to realize less load for welding work. Although the width of the splice plate positioned along the longitudinal line needs to be maintained, it is possible to lighten the splice plate itself by changing other parts. As the outline of the splice plate, the shape is freely decided as long as the perimeter is larger than the required length so as to cover the welding volume necessary for stress transmission between the splice plate and the steel sheet pile.

The present splice plate model consists of only one plate. However, the total weight of the splice plate can be reduced, keeping the total length of the perimeter, utilizing the characteristics of a diamond-like shape. As shown in Figure 4, one diamond-like shape can be divided into two parts, maintaining the total length of the perimeter. The total perimeter length of the present model is calculated as c × 4. On the other hand, the proposed model is c/2 × 8. These two equations produce the same result. The divided two small plates have a form similar to the original splice plate. The total weight of the divided two

plates becomes half that of the original plate, and this leads to 1/4th the weight of the original plate for each splice plate. This small-size type of splice plate is very effective toward realizing smooth welding work, especially when large-size steel sheet piles are used.

2.2 Splice plates with one welding pass

There is another way to reduce the load of welding work in using small-size splice plates. The purpose of the newly proposed model mentioned above is mainly to lighten the weight of the splice plate. On the other hand, the number of welds along the perimeter of the splice plate is also related to workability in terms of welding. Because the welding occupies almost all of the longitudinal joint work time, it is useful to save welding time for one steel sheet pile by reducing the weld amount for each splice plate. The reduction of weld time can lead to an increase in the number of steel sheet piles that can be driven into the ground in one day, and this would result in a decrease of construction cost.

The number of welds along the same line tends to increase according to the increase of the designed throat depth. Because the splice plates are welded by fillet welding along the perimeter of the splice plate, the size of welding volume necessary for stress transmission is decided by throat depth. The thickness of the welded part after one welding pass is limited due to the limitation of welding torch capacity. Therefore, the number of welding passes depends on the designed throat depth. The welding work needs to be repeated until the thickness of the welded part satisfies the designed throat depth. The total volume of the welded part sometimes gets larger than the designed volume. To minimize the welding time, it is effective to minimize the number of welding passes, preferably restraining the number to only one, though it is necessary to keep the total minimum volume of welding corresponding to the designed throat depth.

To reduce the throat depth to a point where it can be completed via only one welding pass, the size of the splice plate can be adjusted by widening the length of the longer side of the diamond-like shape from the first proposed model as shown in Figure 5. The shape of the splice plate, in other words the total

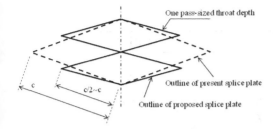

One pass-sized throat depth

Outline of present splice plate

Outline of proposed splice plate

Figure 5. Small-sized splice plate with one welding pass.

length of the perimeter, with one welding pass is determined to secure the amount of welding parts necessary for stress transmission between the splice plate and the steel sheet pile. Even though the total weight of the splice plate for one longitudinal joint becomes heavier than the original splice plate, that of each splice plate can be reduced below a certain level compared with the original. This reduced weight leads to labor-savings, especially when a larger size of steel sheet pile is used.

2.3 New splice plate for hat-type steel sheet piles

Table 1 shows an example of our newly developed splice plate. There are mainly two types of steel sheet piles used in Japan: hat-type and U-type. Hat-type steel sheet piles are more cost-effective than U-type because of the location of the interlocks. In the case of U-type steel sheet piles, in which the interlocks are situated on the neutral axis, the whole shear force cannot be transmitted along the interlocks, and this requires a reduction in sectional properties, such as moment of inertia. On the other hand, no such reduction is required for hat-type steel sheet piles, as the interlocks are positioned at the outmost side of the section.

Therefore, hat-type steel sheet piles are mostly used where cost-effectiveness is demanded, so as to restrain construction costs. Being representative of steel sheet piles on which our new splice plate is attached, hat-type steel sheet piles are being focused on in this development. Because the new splice plate can be used effectively, especially for large-size steel sheet piles with high section properties, the two largest hat-type steel sheet piles, known as "45H" and "50H", are being focused on.

For each steel sheet pile, three types of splice plates are as illustrated in Table 1, that is the present one, a half-size one, and a one-welding-pass model. The latter two models consist of our newly developed splice plate. For each splice plate, two pieces of information, the weight of each and the total set, and the total weld line length with a number of weld passes are added. Instead of throat depth, which can't be directly measured, the leg length is alternatively checked after welding has been completed. The leg length for the respective model is also shown in the table.

When half-sized splice plates are used, the weight of each plate becomes 1/4th the present model and

total weight becomes half. In the case of 50H, the weight of each plate is 5.2 kg and that of the total set is 10.4 kg, which is half of 20.7 kg for the present model. Because the splice plate length is adjusted in 10 mm units, the rounding error is neglected. It is quite hard to continue to carry and support a 20.7 kg plate by one hand all through the construction period during which a lot of longitudinal joint work is included. However, the labor burden can be allayed once the weight of each splice plate is reduced to 5.2 kg. The amount of the weld line length for 50H is totally the same between the present model and the proposed model, which is 6 to 8 m. Welding work of 3 to 4 passes is normally necessary to fill the 12 mm leg length.

To reduce the number of welding passes, the second model, the splice plate with a one-welding-pass model, is effective. In the case of 50H, the total weld line length is 3.4 m, which is a 43–57% reduction from the present and half-sized splice plate model. Although the weight of each splice plate of this model is larger than that of the half-sized splice plate model, the weight is 9.1 kg, which is less than half that of the present model. This weight reduction has great effect for labor-savings. From the viewpoint of total weight of the splice plate set, this model enables a 12% reduction from the present model, and this leads to cost reduction for material expenditures. As for one-pass welding, a 7 mm leg length is adopted, and hence the total length of the longer side of the diamond-like shape becomes larger compared with the half-sized splice plate model.

3 EVALUATION OF BENDING CAPACITY

3.1 Test conditions

The capacity against bending moment for the newly developed splice plate was checked via a bending test (Momiyama 2019). It was expected that the bending stiffness and capacity of the new plate is larger than the designed values.

Two proposed models were investigated, comparing them with the present model. The splice plate positioned on the arm was set to be the same for all three models. The difference among the three models is the specification of the splice plate attached to the web. The details of each splice plate are shown in Table 1. For all three types, each splice plate was designed so as to make the designed bending capacity almost the same. As for the steel sheet pile, the largest hat-type, 50H, was used. The grade of steel of 50H was SWY295 and that of the splice plates was SM490.

Four-point bending tests were carried out, as shown in Figure 6. The longitudinal joint was located at the center of the test specimen, where only bending moment was loaded. In terms of the sectional direction of the steel sheet piles against the direction of force, denoted as "P" in Figure 6, two

Table 1. New splice plates for hat-type steel sheet piles.

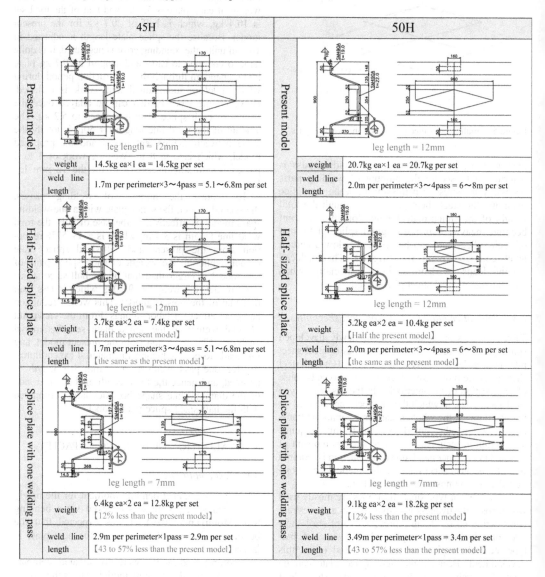

	45H		50H
Present model	leg length = 12mm	**Present model**	leg length = 12mm
	weight: 14.5kg ea×1 ea = 14.5kg per set		weight: 20.7kg ea×1 ea = 20.7kg per set
	weld line length: 1.7m per perimeter×3~4pass = 5.1~6.8m per set		weld line length: 2.0m per perimeter×3~4pass = 6~8m per set
Half-sized splice plate	leg length = 12mm	**Half-sized splice plate**	leg length = 12mm
	weight: 3.7kg ea×2 ea = 7.4kg per set 【Half the present model】		weight: 5.2kg ea×2 ea = 10.4kg per set 【Half the present model】
	weld line length: 1.7m per perimeter×3~4pass = 5.1~6.8m per set 【the same as the present model】		weld line length: 2.0m per perimeter×3~4pass = 6~8m per set 【the same as the present model】
Splice plate with one welding pass	leg length = 7mm	**Splice plate with one welding pass**	leg length = 7mm
	weight: 6.4kg ea×2 ea = 12.8kg per set 【12% less than the present model】		weight: 9.1kg ea×2 ea = 18.2kg per set 【12% less than the present model】
	weld line length: 2.9m per perimeter×1pass = 2.9m per set 【43 to 57% less than the present model】		weld line length: 3.49m per perimeter×1pass = 3.4m per set 【43 to 57% less than the present model】

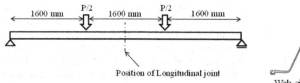

Figure 6. Side view of bending test.

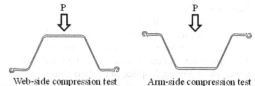

Web-side compression test   Arm-side compression test

Figure 7. Sectional direction view of bending test.

series were investigated, respectively. One of them was a web-side compression test and the other was an arm-side compression test, as shown in Figure 7. Because the interlocks are not welded between corresponding two pile ends, different deformation and hence bending capacity were expected according to the position of the splice plates against the force direction.

3.2 Test results

The relationship between curvature and bending moment for all cases is shown in Figure 8. The designed values of yield moment and maximum moment calculated from both nominal stress and material test are also shown in the figure. The designed maximum moment is defined as a fully plastic moment. The respective values of bending stiffness, yield moment, and maximum moment obtained from the experiment are shown in Table 2. In this table, a comparison between the experimental results and the designed values is shown as well. In the case of the designed values, stress based on the material tests is used.

With regard to bending stiffness, the experimental results correspond to the designed values for all cases. Elastic behavior was kept until the force loaded on the test specimen reached yield moment based on nominal stress.

The capacity of bending moment obtained from the experiments surpassed the designed values for all splice plate models, for both yield moment and maximum moment. In any case, the value of the bending capacity for the arm-side compression tests outnumbered that for the web-side compression tests. The cause is considered to be derived from the compression effect for the interlock section in the arm-side compression tests. The surfaces of the intersection of the respective steel sheet piles touched each other, apparently making the entire cross-section area of the steel sheet piles resist the bending moment, which cannot happen in the web-side compression tests, where the interlock of the arm side is tensioned.

As for the pattern of deformation in the test specimens, a typical tendency was commonly observed for all test cases. At the final stage of the web-side compression tests, tension cracks occurred along the arm of the steel sheet pile and on the splice plate on the arm. As the origin of the breakage point, this crack emerged from the intersection without being welded. The deformation pattern is as shown in Figure 9. On the contrary, the local buckling of the arm of the steel sheet pile around the longitudinal joint line was the main reason to lower the bending capacity at the final stage in the arm-side compression tests, as shown in Figure 10. For all cases, there was no damage for the splice plates attached to the web of the steel sheet piles. Therefore, it can be concluded that the proposed splice plates do not affect the characteristics of the bending behavior of steel sheet piles.

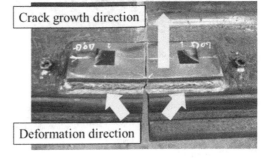

Figure 9. Deformation of splice plate welded on arm in web-side compression test (case1).

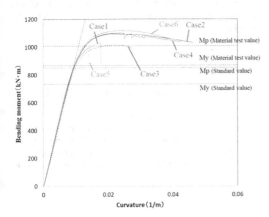

Figure 8. Curvature and bending moment relationship.

Table 2. List of bending test results.

No.	Splice plate model	Compression side	Experimental results			[Experimental result] [designed value]		
			Bending stiffness (kN·m²)	Yield moment (kN·m)	Maximum moment (kN·m)	Bending stiffness	Yield moment	Maximum moment
Case1	Present	web	97093	948	1030	1.03	1.09	1.02
Case2		arm	95860	988	1098	1.01	1.14	1.09
Case3	Half-sized	web	92870	941	1016	0.98	1.08	1.01
Case4		arm	95941	1004	1102	1.01	1.16	1.09
Case5	One welding pass	web	92971	949	1008	0.98	1.09	1.00
Case6		arm	93637	1002	1120	0.99	1.15	1.11

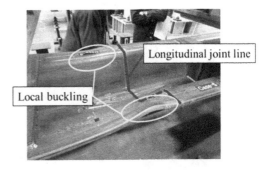

Figure 10. Deformation of arm in arm-side compression test (case4).

4 FIELD TESTS

To check the validity of the new splice plate from a practical point of view, actual welding tests were carried out on site. Figure 11 shows the situation of the field tests. The time of welding is summarized in Table 3. The welding time consumed for the splice plates on the web of the steel sheet piles is separately shown in this table. The number of welding passes was differently set for each model. Three passes were carried out for the present model and for the half-size model. On the other hand, only one-pass welding was carried out for one welding pass model.

Because the total weld line length is the same for the present model and the half-size model, there is no difference in terms of welding time for both the splice plate and the entire section. However, it is confirmed that a one welding-pass model can reduce welding time. Concerning the welding time for the splice plate on the web, a 37% reduction from the present model was observed, leading to the entire welding time reduction by 25% compared with the present model.

Figure 11. Situation of field test.

Table 3. List of field test results.

Splice plate model (Number of welding pass)	Welding time	
	Time for splice plate on web (minute)	Whole welding time (minute)
Present (3)	40	85
Half-Sized (3)	40 [the same as the present model]	85 [the same as the present model]
One welding pass (1)	25 [37% less than the present model]	64 [25% less than the present model]

5 CONCLUSIONS

The validity and effectiveness of the newly developed splice plate for steel sheet piles was confirmed through experiments and field tests. The most-noticeable issue of the new splice plate is the reduction of the weight of each splice plate type and the welding time for longitudinal joints, as this can contribute to labor-saving. A cost reduction for material is also expected. It can be confirmed that the proposed splice plate can maintain a necessary bending capacity similar to the steel sheet piles used on site.

REFERENCES

Japanese Technical Association for Steel Pipe Piles and Sheet Piles (JASPP). 2014. *Design and construction method for steel sheet piles.*
Momiyama, T., Nakayama, H., Kitahama, M., Takeno, M., Nishibe, K., Matsubara, H. 2019. Bending capacity evaluation of steel sheet piles with improved splice plates at longitudinal joint, *Japan society of civil engineers 2019 annual meeting.*

Proceedings of the Second International Conference on
Press-in Engineering 2021, Kochi, Japan – Matsumoto et al (eds)
© 2021 Taylor & Francis Group, London, ISBN 978-1-032-10414-0

Summary of case histories of retaining wall installed by rotary cutting press-in method

N. Suzuki & Y. Kimura

GIKEN LTD., Tokyo, Japan

ABSTRACT: This paper summaries Japanese case histories of retaining walls of the rotary cutting press-in piles in terms of the application, pile materials, project scale, spatial restrictions for working, and ground conditions. The purpose is to grasp the characteristic of the rotary press-in method, and grasp the requirement for the renewal of civil infrastructures under working restrictions. The rotary press-in piles were often used when the pile diameter was 1000 mm, the pile length was 18–20 m, and the total length per project was 20–40 m. About 70% of the projects for reconstruction/rehabilitation of the retaining structures had spatial restrictions, such as insufficient headroom, side space and working space, and unstable work at high places. And over 60% of the projects had problems of hard grounds (SPT N>75) or any underground obstacles.

1 INTRODUCTION

Cantilever steel pile retaining walls are increasing under restrictions of working sites (Miyanohara et al. 2018). Although they have been mainly used as temporary earth retaining walls in the past, they also have recently been used as permanent structures in Japan. IPA-TC1 has been investigating their rational design methods in relatively hard ground and discussing the safety of the walls in short embedded depth (e.g. Kunasegaram et al. 2018).

Types of the earth retaining walls are determined based on many criteria, such as cost, ground conditions, water cut-off, speed of construction, the presence of obstructions, environmental issues, and others (Gaba et al. 2017). Types of wall construction methods include king post walls (as known as soldier piles), sheet pile walls, steel pipe pile walls, cast in-situ piles (contiguous/secant), and diaphragm walls (D-wall) (e.g. Long 2001, JRA 1999). Generally, sheet piles are used for temporary or small-scale excavation, and D-walls are used for large-scale excavation. The steel pipe pile walls are used for the in-between scale, and one of its advantages are it has little danger of soil contamination and it does not require large working area for cage, pumping equipment and concrete plant.

The rotary press-in piling method, which we introduce in this paper, is one of the representative installation methods of steel pipe pile retaining walls (Figure 1). The rotary press-in enables pile rotation by omitting joints (i.e. it use steel pipe piles instead of steel pipe sheet piles). It greatly expands the application range of hard ground, which is generally a problem in steel pipe pile installation methods. And environmental issues such as noise and vibration are less than those of driven pile or other piling methods.

This paper will summarize Japanese case histories of the rotary press-in piling method. The original purpose is to grasp the characteristics of the rotary press-in method. A secondary aim is to grasp the requirements for the retrofits of civil infrastructures under working restrictions.

2 JAPANESE CASE HISTORIES OF ROTARY CUTTING PRESS-IN METHOD

The rotary cutting press-in method was developed by Nippon Steel and GIKEN LTD. It presses and rotates a pile by grasping the previously pressed steel pipe piles for the reaction force. The machine moves over the steel pipe pile heads and constructs the wall in a row (Figure 1). It was first applied to the construction project in 2004, Japan, and now the accumulated number of the projects becomes 442 (Figure 2). Of these, 25 used Clamping Crane, which load piles with self-walking system, and 24 used Clear Piler, which is one of the pilers for operations under low headroom.

Compared with other installation methods of steel pipe piles, the rotary press-in method has following advantages and disadvantages (e.g. Hirata and Matsui 2016);

DOI: 10.1201/9781003215226-61

Figure 1. Rotary cutting press-in method (IPA 2016).

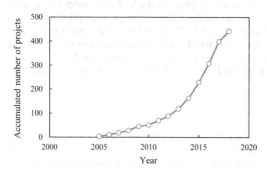

Figure 2. Accumulated number of the application of the rotary press-in method.

Advantages:

- Low noise, low vibration and a small amount of water discharges.
- Short construction period with no temporary structures and little disposal of soil
- Used in a wide variety of soils including underground obstacles such as concrete structures.
- Construction under the girder of the existing bridge or close to a road or railway, is also applicable.

Disadvantages:

- High cost.
- Issues to cut off water.
- Long embedment depth is needed for reaction force if the ground is very soft

2.1 *Application of rotary press-in method*

A total of 442 projects of the rotary press-in piles have been completed as of the end of September 2019, which GIKEN identified. About 40% of the rotary press-in piles has been applied for river structures (Figure 3).

The distribution of the structure types of the rotary press-in piles is shown in Figure 4. The classification of the structure is based on IPA (2016). The rotary

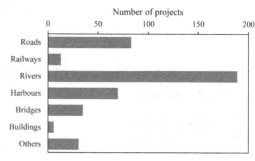

Figure 3. Application of the rotary press-in method.

press-in method was applied to various types of constructions. In road construction, it was mainly used for widening roads and for reinforcing slopes against disasters (a and b). Although fill slopes with retaining walls (a-1) of rotary press-in were less common than cut slopes (a-2), they may be used due to restrictions in the delivery route to the construction site. Constructions close to existing buildings behind the retaining walls made it difficult to use cranes and vibratory equipment (a-2 and a-3). More than half of the projects have been adopted in rivers and harbours (c-e). Since river/sea walls require to be reconstructed or rehabilitated while leaving their functions, discarded rubbles can make it difficult without the piling method. When an urban highway passes over a river revetment, equipment for low headroom would be required (c-2). Constructions under existing piers and abutments in service demand little noise and vibration in order not to interfere with railroad operations (f-1). It was also used for the foundations of seawalls because press-in method has advantage to install in a row (f-2). A few of temporary constructions adopted it especially in hard ground and under narrow working spaces, although steel pipe piles are relatively expensive and difficult to be extracted (g).

To focus on retaining structures, the following will consider 345 cases except for bridge foundations (f), temporary works (g), and others/unknown (h).

2.2 *Pipe pile material*

The most common pile diameter, D, was 1000 mm, which accounted for 40% of the totals (Figure 5). The number of projects whose pile diameters were larger than 1000 mm was increasing and accounted for about 30% of the totals because of the development of new equipment for large diameter piles.

Figure 6 shows ratio of the pile thickness. The ratio was the number of projects divided by the sum of those with the same pile diameter. The lower/upper pile thickness were also drawn. The lower pile thickness is generally 1% D (limited to a minimum of 9 mm) in Japan for workability (e.g. JRA 2017). The minimum thickness was the most commonly used for almost all pile diameters.

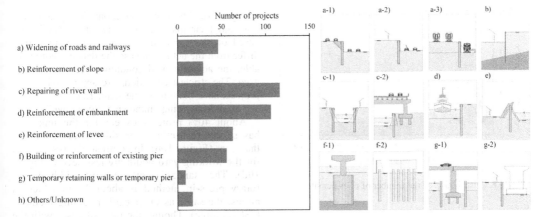

Figure 4. Types of constructions with rotary press-in piles (after IPA 2016).

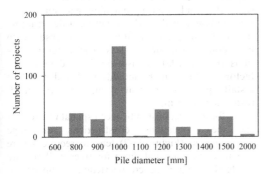

Figure 5. Distribution of pile diameters. Inch size pile diameter was truncated by 100mm. With using multiple piles in the same project, the maximum pile diameter and the maximum pile length were shown.

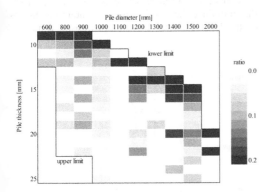

Figure 6. Heatmap of the project ratio of pile thickness.

Figure 7 shows the distribution of the pile length and the number of joints. The most common pile length was 18–20 m, and there were some examples of piles with a length of over 30 m (constructed maximum length is 53.0 m). In Japan, the length of piles is

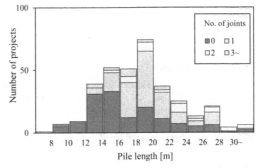

Figure 7. Distribution of pile length and the number of joints.

restricted for the safety of road transportation, and piles exceeding some length are required to have a permit for transportation on the road. Thus, construction work with a pile length of over 12 m was likely to have joints.

One or more vertical joints are required for about half of the total projects. Some restrictions such as low headroom increase joints, which take time to join and inspect the joints.

Though we haven't grasped the height of structures or excavation depth of all cases, 13 m (D: 2000 mm) and 10.5 m (D: 1500 mm) of wall heights are one of the highest cantilever walls observed (Miyanohara et al. 2018).

2.3 Construction scale

The scale of construction per project was mostly 20–40 piles (Figure 8) and 20–40 m in total length (Figure 9), and projects over 40 piles and 40 m length accounted for more than half. Large scale construction can spend more time and cost on geotechnical investigation and design for rationalization.

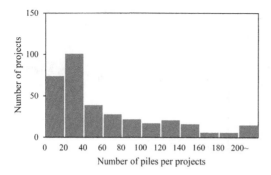

Figure 8. Distribution of the number of piles per site.

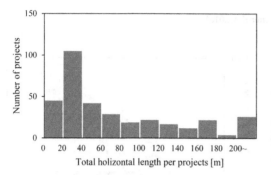

Figure 9. Distribution of total length per project.

2.4 Spatial restrictions for construction operation

The rotary press-in method can be subject to the following spatial restrictions depending on the site conditions: narrow space for installation disturbs the construction, high-level work makes the machine unstable, and insufficient space for machine movement, temporary storage, and crane installation reduces efficiency.

Figure 10 shows the restrictions during construction and their distribution. Since these results are based on the work for a cost estimate, they may differ from the actual. In case the data are not available, the adjacent field conditions are alternatively used. The strictest conditions are applied where the projects have more than one standard cross-sections. We exclude anything unknown as "unknown".

About 40% projects except for the unknown has severe restrictions for side space, with less than 5 m (Figure 10b). In contrast, the restrictions for the working yard width are less severe (Figure 10d). The standard working yard width of the rotary press-in method is about 12 m which is almost the same as other earth retaining methods (JSCE online). Though the ratio of cases with the high-level installation is also not so large (Figure 10c), other common earth retaining methods require the construction on a leveled condition (e.g. all-round rotating machine and machinery for diaphragm wall), which can be a determining factor. Finally, though the ratio of cases with headroom restrictions is less than other restrictions (Figure 10a), it also can be a determining factor, because other methods find it difficult to install piles over water and under head-restrictions of 5 m.

As for the relationship between spatial restrictions and the applications, constructions for slope reinforcement (Figure 4a-2) tend to require severe lateral restrictions, and those for river wall (Figure 4c) tend to require severe restrictions of upper and side space. Also, about 70% of the projects except for "unknown" have one of the spatial restrictions.

2.5 Geotechnical restrictions

A total of 128 projects are used from the projects that introduce the pile layout and ground conditions on the Japan Press-in Association website. Since the

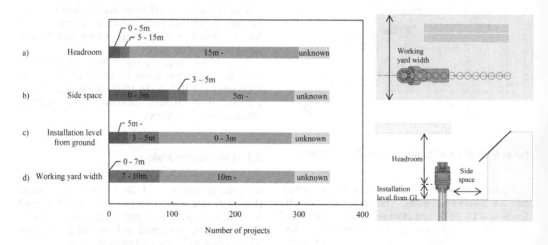

Figure 10. Spatial restrictions for working (based on works shown in Appendix).

information is not exhaustive and is collected from constructors intending to advertise the rotary press-in method unlike the previous, it should be used with caution. However, we believe it contains enough examples to grasp the characteristics of the rotary press-in method.

Figure 11 shows the relationship between the maximum converted SPT-N value and the ground type. Impenetrable layer was represented by N value of 300 or more. If fields had existing structures, obstacles and rubbles which piles had to be penetrated, they set the N value to the maximum N value of the ground other than those and sets the ground type to "other". Otherwise, the ground type represents the one with the maximum N-value. Rock includes consolidated silt. The maximum constructed unconfined compressive strength (UCS) of rocks was 98 MPa (IPA 2016). The "other" (existing structures, obstacles and rubbles) applies for about 20%, and grav-

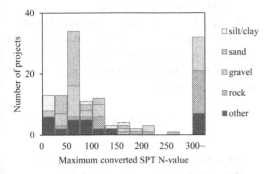

Figure 11. Maximum converted SPT N-value and ground type in the rotary press-in method.

els and rock with an N-value of 300 or higher account for another 20%.

About 40% of the ground had an N-value of 75 or lower and no existing structures, where press-in method associated with water jetting can also install piles (IPA 2016). Also, most of the remaining cases which had no problems of hard ground and obstacles, had the spatial restrictions.

2.6 Environmental restrictions

Low noise and low vibration were also required in most cases, though the qualitative number was not grasped. The power-unit used in the rotary press-in method has been specified as low-noise and low-vibration construction equipment by MLIT (2020).

The environmental restrictions also include the requirement of short construction period. For example, where the construction work must be carried out during the drought season and there is little time to build temporary stages, the press-in method has advantages.

3 SUMMARY

This paper over-viewed Japanese case histories of the rotary press-in piles in terms of the application, pile materials, spatial restrictions, and ground conditions. The case histories supported the characteristics of the rotary press-in method which Hirata and Matsui (2016) mentioned. About 70% of the projects for retaining structures with the rotary press-in method had one of the spatial restrictions; a: headroom restrictions, b: side-space restrictions, c: high-level working, and d: insufficient working space. Besides, over 60% of the projects have problems of hard ground (N>75) and obstacles. Figure 12, an illustrative diagram, summarizes the restrictions of the projects which adopted the rotary press-in method for retaining structures.

Besides, Jitsuhiro (2004) investigated case studies on road business risks in Japan and reported which accidents increased construction costs. The report has listed the following accidents with high costs and a high probability of occurrence: i) difficult land negotiations, discussions on environmental measures, and discussions on routes and structures (at design and planning), and ii) coping with unexpected geological conditions and with underground buried objects (during construction). Since these unfavorable accidents will increase in the future and the rotary press-in method has the potential to avoid them as shown in this paper, we expect more adoptions to occur.

This paper did not intend to introduce the rotary press-in method completely, since the reason for the adoption differed from site to site. A combination of this paper and other case histories (e.g. IPA 2019) will be helpful to grasp the features of the rotary press-in method.

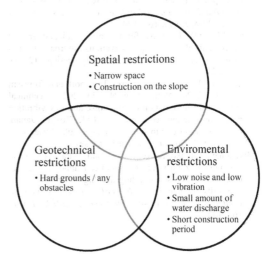

Figure 12. Illustrative diagram of restrictions of construction which adopted the rotary press-in method for building retaining structures.

ACKNOWLEDGEMENTS

This survey has been partly conducted in IPA-TC1, which has worked on the application of cantilevered steel pipe pile retaining walls to hard ground.

REFERENCES

Construction Technology Research Committee of Japan Society of Civil Engineers (JSCE). 11th Recent construction method and key points for selecting construction method-earth retaining construction method-, Retrieved Oct. 2, 2020, from http://committees.jsce.or.jp/sekou05/node/23 (in Japanese)

Gaba, A. R., Hardy, S., Doughty, L., Powrie, W., & Selemetas, D. 2017. *Guidance on embedded retaining wall design*, C760, Ciria.

Hirata Hisashi & Matsui Nobuyuki. 2016. Expanding applications of the Gyro-press Method™, *Nippon Steel & Sumitomo Metal Technical Report*, No.113, 42–48.

International Press-in Association (IPA). 2016. *Press-in retaining structures: a handbook (First edition 2016)*, Tokyo: IPA

International Press-in Association (IPA). 2019. *Press-in Piling Case History (Volume 1, 2019)*, Tokyo: IPA

Japan Road Association (JRA). 1999. *Road earthwork, Guideline for the construction of temporary structures*, 17–24. (in Japanese)

Japan Road Association (JRA). 2017. *Specifications of Highway Bridges, Part IV: Substructures*, 302–303. (in Japanese)

Japan Press-in Association (JPA). 2019. *Standard cost estimation manual (draft): Gyro-press Method – steel tubular pile Press-in Method assisted by rotary cutting*, Tokyo: JPA (in Japanese)

Jitsuhiro Takushi. 2004. Survey and analysis on road business risks, *Japan Institute of Country-ology and Engineering (JICE) report*, No.6, 28–33. (in Japanese)

Kunasegaram, V., Takemura, J., Ishihama, Y. & Ishihara, Y. 2018. Stability of self-standing high stiffness-steel pipe sheet pile walls embedded in soft rocks, *Proceedings of the First International Conference on Press-in Engineering 2018, Kochi*, 167–174.

Long, M. 2001. Database for retaining wall and ground movements due to deep excavations. Journal of Geotechnical and Geoenvironmental Engineering, 127(3), 203–224.

Ministry of Land, Infrastructure, Transport and Tourism (MLIT). 2020. Notification No.1536-2020, Technical standard for designation of low noise and low vibration type construction equipment, MLIT, Tokyo, Japan. https://www.mlit.go.jp/sogoseisaku/constplan/sosei_-constplan_tk_000003.html (in Japanese)

Miyanohara Tomoko, Kurosawa Tatsuaki, Harata Noriyoshi, Kitamura Kazuhiro, Suzuki Naoki & Kajino Koji, 2018. Overview of the self-standing and high stiffness tubular pile walls in Japan, *Proceedings of the First International Conference on Press-in Engineering 2018, Kochi*, 167–174.

APPENDIX

An outline of the cost estimation of the rotary press-in method in Japan is described (JPA 2019). It is usually used by clients for assuming the reasonable construction method and estimating the appropriate cost, before placing an order for Japanese public construction.

a) Construction time

The construction time, T, is the sum of the installation time of the pile itself and the joint member therebetween, T_c and T_h. T_c is calculated as;

$$T_c = (T_S + T_B)/F + T_w \qquad (A1)$$

where T_S: preparation time, T_B: press-in piling time, T_w: welding time, F: work factor.

T_S includes setting piles, centering adjustment, installation/removal of driving attachment and swivel for water lubricating systems, self-walking, preparation for welding, welding of steel sheets on pile heads, and others.

$$T_S = 1.15L + 23.8 + 20n \text{ min} \qquad (A2)$$

where L: pile length [m] and n: number of joints.

Press-in piling time, T_b, is calculated as;

$$T_B = \sum \gamma_i l_i \qquad (A3)$$

where γ_i: unit press-in time per meter for each soil layer, l_i: press-in length for each layer.

γ_i is calculated by the average SPT-N value for each soil layer, N_{avg}.

$$\gamma_i = 0.054N_{avg} + 1.32 \text{ min/m} \qquad (A4)$$

T_w depends on the pile diameter, thickness and the number of joints.

b) Operation cost

The pile installation costs consist of the operation of the press-in machine, rent of the water lubricating

Table A1.   Correction factor for work coefficient.

Correction factor		unit	−0.10	−0.05	0.00	+0.05
f_1	Headroom	m	Under 10	10–15	15 or more	—
f_2	Side space	m	Under 3.0	3.0–5.0	5.0 or more	—
f_3	Installation level from ground	m	5.0	3.0–5.0	Under 3.0	—
f_4	Working yard width	m	-	7.0–10.0	10.0 or more	—
f_5	The number of piles per block	-	30–50	50–100	100–200	200 or more

Note: If there are special working conditions/restrictions, an estimation has to be carried out at each site.

system, labour charge, sundry expenses, and others. Each unit price depends on the country or region.

c) Work factor

Work factor, F, is based on site conditions; headroom, side space, installation level from the ground, working yard width, and the number of piles at the project (Table A1). These conditions effect the efficient work.

$$F = F_0 + \sum f_i \qquad (A5)$$

where F_0: base factor (=1.0), f_i: correction factor with each condition.

587

Proceedings of the Second International Conference on
Press-in Engineering 2021, Kochi, Japan – Matsumoto et al (eds)
© 2021 Taylor & Francis Group, London, ISBN 978-1-032-10414-0

A decade of R&D in press-in technology: Bridging the gap between academia-industry in Malaysia

N.A. Yusoff
Head of Centre, Research Centre for Soft Soil (RECESS), Universiti Tun Hussein Onn Malaysia (UTHM), Batu Pahat, Johore, Malaysia

T.N. Tuan Chik
Lecturer, Faculty of Civil and Built Environment (FKAAB), Universiti Tun Hussein Onn Malaysia (UTHM), Batu Pahat, Johore, Malaysia

M.K. Ghani
Manager, Sustainable Development Division, Construction Research Institute of Malaysia (CREAM), Kuala Lumpur, Malaysia

K.W. Chung
Koye (M) Pte. Ltd., Shah Alam, Selangor, Malaysia

ABSTRACT: The vibrant relationship between academia and industry are needed in fostering an applied research especially for a developing countries, such as Malaysia. This paper will highlight a decade of joint activities between Universiti Tun Hussein Onn Malaysia (UTHM) and several industrial stakeholders in adapting sustainable technology such as Press-in Technology in Malaysia. The technological transfer had been carried out since the establishment of a press-in pile research in 2007 at The University of Sheffield, United Kingdom. In general, conventional teaching and learning sessions, undergraduate and post graduate research and development activities and industrial attachments are the common activities. In addition, industrial sharing sessions and technology showcases are also the common activities initiated to bridge the gap between academia and industry. The activities allow dissemination of latest Press-in technology to both academia-industry by establishing R&D collaborations with local and international stakeholders in the Malaysia construction industry.

1 GENERAL INTRODUCTION

1.1 Preface

The vibrant relationship between academia and industry needs to be strengthened to meet the strategic needs of research and development in the developing countries, such as Malaysia.

The relationship may allow both the academician and the industry to capitalize on the strength of the other. Progressive construction activities in Malaysia worth some RM150 billion last year demand sustainable solution in minimizing the negative impact on the environment. When it comes to substructure construction, conventional dynamic piling methods are ill-suited to urban development because of the emissions of deafening noise and vibrations. These methods may no longer be the best options in certain conditions especially when subjected to the construction of a structure in a busy city such as Kuala Lumpur.

The silent piling technology offers an alternative sustainable solution by using a relatively small press-in machine, so called the "Silent Piler". The technology was developed to satisfy five construction principles consisting of Environmental Protection, Safety, Speed, Economy and Aesthetics. Based on the reaction principle of press-in piling method, pre-fabricated piles are hydraulically jacked-in with minimum noise and vibration. However, the technology can be considered as new and not commonly used in a country such as Malaysia.

1.2 Objectives of review

This paper will highlight a decade of research and development activities between Universiti Tun Hussein Onn Malaysia (UTHM) and several industrial

DOI: 10.1201/9781003215226-62

stakeholders in adapting a sustainable technology such as Press-in Technology in Malaysia. In general, a conventional teaching and learning sessions, industrial sharing sessions, industrial attachments, technology showcases, undergraduate and post graduate research and development activities are among the common activities initiated to bridge the gap between academia and industry.

2 EARLY RESEARCH

2.1 PhD study at the University of Sheffield

The PhD research journey started in 2007 under The International Press-in Association International Grant awarded to Professor Adrian Hyde. During that time, The University of Sheffield received the grant in exploring the potential rate effect of press-in pile.

Press-in pile installation techniques have highlighted the need for a better understanding of rate effects in clays. In order to gain an insight to the problem, a new Rowe Cell-Vane Shear Test (RCVST) apparatus has been developed for this project with a single phase electric servo motor and torque transducer to drive the shear vanes over a wide range of angular velocities. The apparatus as shown in Figure 1 allowed four vanes of different dimensions to be consolidated in a single clay bed.

2.2 Development of Rowe Cell-Vane Shear Test (RCVST) apparatus

The main reason which lead to the development of this apparatus is the existing laboratory apparatus (i.e.: triaxial and direct shear apparatus) have a limited range of strain rate and often cover at a slower rate of loading (strain rate is usually 0.5 % to 1 % per minute). For this study, it was justified that the research interest is to investigate the rate effect at a faster speed as compared to the ordinary laboratory application.

It is hoped that the research finding will be reflected to the similar industrial application in a point of rate of strain. For example, in-situ vane shear tests are normally conducted around 0.03 to 0.13 mm/s, a press-in piling operation is approximately about 25 to 592 mm/s, in-situ cone penetration tests is about 10 to 20 mm/s and direct shear tests are conducted at about 0.02 mm/s.

The apparatus consisted of a modified triaxial test frame, a double Rowe cell consolidometer connected to a pressure unit, a laboratory vanes with servo motor and a gear head connected to a motion controller and a data logging device. (Figure 2).

The test rig is very similar to that used by Srishaktivel (2003) and Hird & Srishaktivel (2005) but with a cross head designed to accommodate a servo motor with a gear head. The test cell, similar to the one used by Srishaktivel (2003) and Emmett (2007) was formed by bolting together with two Rowe consolidation cell bodies, 254 mm in diameter and 126 mm high, while the aluminium cell base was modified in order to accommodate four vanes during consolidation.

In total, it took slightly over two years to complete the development, fabrication and commissioning of the apparatus at the Department of Civil and Structural Engineering laboratory.

2.3 Rate effects from RCVST apparatus

The developed apparatus is capable in conducting both rate effect studies for undisturbed and disturbed soil sample from slurry consolidation. Figure 3 and 4 show the results for undisturbed and disturbed sandy clay specimens using undrained shear strength tests. The vane length ranges from

Figure 1. Early setting of Rowe Cell-Vane Shear Test apparatus at the University of Sheffield.

Figure 2. Rowe Cell Vane Shear Apparatus.

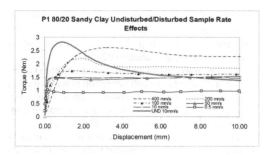

Figure 3. Peak and residual torque for sandy clays by using 80/20 vanes (Plasticity index, PI=22).

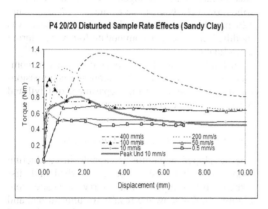

Figure 4. Influence of peripheral velocity on peak and residual torque for sandy clays by using 20/20 vanes (Plasticity index, PI=22).

20mm and 80mm long with 20 mm diameter. The available rate of shearing for this apparatus is in the range from 0.5 m/s to 400 mm/s. Some results are shown in figure below.

3 RESEARCH ACTIVITIES AT UNIVERSITI TUN HUSSEIN ONN MALAYSIA

3.1 Method 1: Statistical studies

Currently, there is still very limited application of press-in technology in Malaysia. This indicates a need to understand the various perceptions of the practitioners in accepting the technology. A qualitative review has been conducted in order to investigate the acceptance of silent piling technology among designers (civil engineering consultants) in Malaysia.

For this reason, sets of questionnaires were distributed to 43 companies from civil engineering consultants registered with Association Consultancy Engineering Malaysia (ACEM). These respondents were given 4 weeks to return the questionnaires. In theory, the response rate from respondents may be

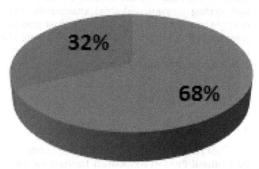

Figure 5. Gender of the respondents (Ibrahim, *et. al*, 2015).

vary. 20% response can be considered adequate and 80% can be considered to be high. For this study, the sample size for data analysis was 52%. Therefore, the data set was adequate and can be considered as valid.

Figure 5 Shows that 68% are male respondents and 32% are female. In addition, in terms of the age of the respondents, 36% of the respondents are in the range of 31 to 40 years old. On the other hand, 32% respondents in the range of 20 to 30 years old and 40 years old and above respectively.

In terms of design experience obtained from the respondents, 50% of the respondents have less than 5 year experience in this sector. The respondents with 5 to 10 year experience represent 14% of the respondents & those with more than 10 year experience are about 36%. The distribution is as shown in Figure 6.

In general, respondents were given several scenarios such as conducting piling work in a sensitive area, which method produce the lowest noise and

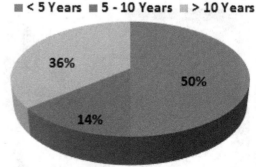

Figure 6. Respondent's experience in particular engineering design (Ibrahim, *et. al*, 2015).

vibration, extension of a railway track without halting the train operation, narrow access site and development above water. The findings show that they are in favor with press-in piling method as compared to other approaches. These show that the respondents agreed that this technology may have a great potential in dealing with a related problems on site in Malaysia.

Figure 7 shows that even though it was observed that most of the respondents were able to recognize the benefits of applying this technology, it was found that 82% of respondents were not familiar and had never used the press-in piling method before. In addition, only 5% were very familiar and often used it. Therefore, it can be concluded that the press-in piling method provides a very positive future for foundation engineering construction but this new technology has not yet widely used in Malaysia.

In order to accelerate the application of this sustainable technology, the respondents advised that some technical support should be provided to the designers. Their recommendation is as shown in Figure 8. In summary, they believed that the designer

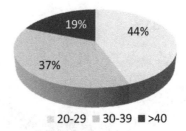

Local Authority

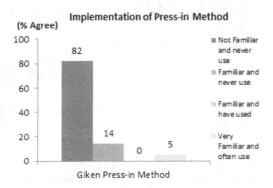

Figure 7. Implementation of press-in method in Malaysia (Ibrahim, et. al, 2015).

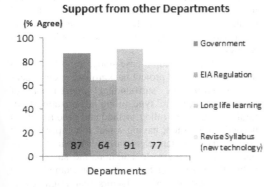

Figure 8. Implementation of press-in method in Malaysia (Ibrahim, et. al, 2015).

Figure 9. Age of respondent.

should emphasize more on their time towards lifelong learning. Majority of them also believed that the government may play a very important role in promoting this kind of technology. At the university level, it was recommended that the current engineering syllabus should be revisited in order to make room for this latest technology (Ibrahim, et. al, 2015).

Based on the finding of the survey, a similar survey had been conducted to the group of engineers working with the local authorities or the government bodies (Azhar, 2018). The findings are as shown in Figure 9 to 11. 43.6% of the respondents are in the range of 20 and 29 years old, 41% in the 30-39 years age group and the rest are 40 years old and above. In respect to gender distribution, 69.2% are male respondents and the rest 30.8% are female. Majority of the respondents are having more than 5 years working experience.

Most of the respondents believe that inadequate skill among the local authorities are the main factors why press-in technology is still not commonly used in Malaysia. The second factor is related to the financial consideration. The result also suggested that incentives should be imply to promote the application of sustainable technology. In addition, the finding also suggesting that mandatory sustainable regulations should be enforced in accelerating the application of this technology. Therefore, further research and development are needed in addressing the issues.

3.2 Method 2: Case study

The application of statistical and survey methods will be helpful in recognizing the general needs of the country in understanding the beauty of press-in technology. However, most of the construction companies in Malaysia do not familiar with the existence of silent piling method as an alternative technique for the pile installation. At present, the availability of press-in machine is also very limited. In order to bridge the needs and current limitations, a case study method is one of the possible options. By having a case study method, deeper understanding of the actual application

591

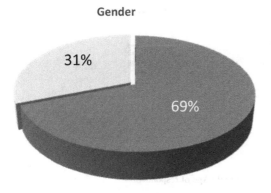

Figure 10. Gender of respondent.

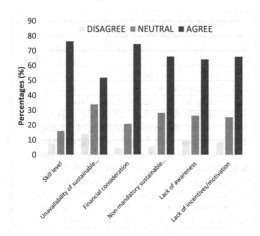

Figure 11. Limitation for the sustainable construction practices with piling construction.

Figure 12. One of the construction site for mass railway Transit in Kuala Lumpur (Yun, 2020).

Figure 13. Student spending their time at the actual construction site (Yun, 2020).

Figure 14. Overall view of one of the construction site in Kuala Lumpur.

on the ground could be written and shared. Hopefully it will be able to facilitate the lifelong learning activities among engineers in order to enhance their awareness such as shown in Figure 12 to Figure 14.

In order to conduct the case study method, several students had been attached at certain period of time at the relevant companies. For example, several engagements with the local contractors such as KOYE (Malaysia) Pte. Ltd. had been initiated in developing the existing case study for press-in application in Malaysia.

In order to develop a case study, important information such as the project background, ground geological properties, surrounding environment of project site, structural type, piling information, structural and pile type, piling method and productivity is needed. In addition, having a real job experience is also a very important element in a student experiential learning process. In addition, the student also indirectly learn other things too. For example, they were exposed to various factors that might cause delays on construction works throughout the project life cycle due to several reasons including the weather, communication and coordination, planning, construction materials, project finance, construction equipment, experience and qualification, construction labour and site management (Durdyev & Hosseini, 2019). However, due to a commercial restriction of the project information, only general information is shareable for this paper.

3.2.1 Case Study 1: KLCC East Station, MRT 2 (SSP Line), Kuala Lumpur

The construction project was located at Mass Rapid Transit (MRT) Line 2 - Sungai Buloh - Serdang - Putrajaya (SSP Line), KLCC East station (SSP21), Kuala Lumpur. The purpose of the project was to conduct the temporary sheet piling works by using the Super Crush Piler for cooling tower at KLCC East station.

According to the published geological maps of Kuala Lumpur area, Kuala Lumpur Limestone Formation underlies in most of the area in Kuala Lumpur. Besides that, KLCC and Bukit Bintang areas are occupied with the Kenny Hill Formation (Tan, 2005). Kenny Hill Formation occupied the depth up to 10 meters below the existing ground level in Kuala Lumpur area with the SPT value greater than 50 which is considered as hard ground. Therefore, hard ground press-in method which known as Super Crush Piler SCU-400M was adopted for this project.

Figure 15 to Figure 17 show the operation of the super crush piler at the construction site. Other equipment and apparatus which assisted the crush piler such as power unit, reaction stand, counter weight, pile auger, crane machine and radio controller were introduced (Yun, 2020).

Figure 16. Lifting and transferring sheet pile to the crush piler by using crane machine (Yun, 2020).

Figure 17. Photo taken with Koye (M) Sdn. Bhd. Project Manager, Mr. Alvin Low (Yun, 2020).

Figure 15. Super crush piler are working on site (Yun, 2020).

Based on the case study, the silent piling technology was recommended to be adopted and applied in the urban and densely populated area. This noise and ground vibration free technology is a more suitable sustainable construction practice as compared to the conventional piling installation method. Therefore, silent piling method can minimize and avoid the risk of destroying the massive underground systems in the urban city such as underground cable system, underground drainage and sewer system and tunnel (Yun, 2020).

3.3 Method 3: Laboratory study

During this period, several laboratory studied related to the press-in technology had been initiated. One of the studies that is worth to share here is the study related to time dependent effect to certain selected soil. This simple undergraduate laboratory study was conducted in order to recognize the understanding related to recovery of pile skin friction concept (Aliff, 2014). Based on Nozaki (2013) site experience as shown in Figure 18 and Figure 19, it was observed that the extraction static force significantly increased after 30 days for sandy gravel. The scenario may trigger a need to establish further understanding especially to estimate the extraction force and the appropriate machinery for pile extraction procedure.

For this study, a model pile with a 450mm length and 25.4mm in diameter had been considered. The study was conducted on a test bed by using a sandy material at the Research Centre for Soft Soil (RECESS). The extraction force was monitored in a sequence of 1, 5, 7, 10 and 25 days after pile installation. Due to a very limited budget, a self-fabricated pile extraction mechanism and a portable scale had been utilized to monitor the extraction force.

Due to the limited depth of test bed and capability of the portable scale, the model pile could only be installed up to 200 mm depth. For reference, a theoretical surface force was estimated based on Meyerhof's Method (1976). In general, the average extraction force after one day installation was approximately double than the theoretical force. It is apparent from this figure that the average extraction force were steadily increased as the numbers of day increased. After 7 days of installation, more than four times force was needed to extract the pile as compared to

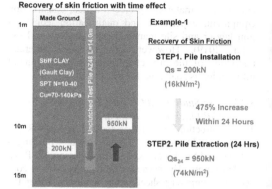

Figure 19. Recovery of skin friction after 24 hours after installation (Nozaki, 2013).

Figure 20. Circular model pipe piles.

Mobile Skin Friction During Installation (kN/m²)	Static Skin Friction after 30 days (kN/m²)	Design Skin Friction (kN/m²)
300 – 800%		260 – 700%
	SILT/ CLAY 21-57	SILT/ CLAY 8
7		
4500%	Sandy GRAVEL 320	Sandy GRAVEL 50
	640%	

Figure 18. Recovery of skin friction for some selected soil samples (Nozaki, 2013).

Figure 21. A self-fabricated pile extraction mechanism and test bed.

Figure 22. A self-fabricated pile extraction mechanism.

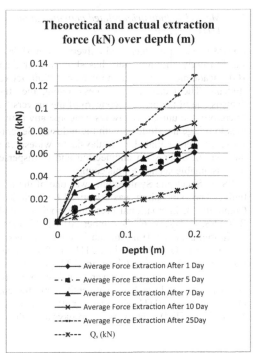

Figure 24. Theoretical and actual extraction force over depth after 1, 5, 7, 10 and 25 days.

topic. However, a better data and results could be established if a proper monitoring device and more appropriate testing programme are established in the near future.

In respect to the economic perspective of the study, this could be among the cheapest option to facilitate the engineering students in appreciating the technology and at the same time to be realistic with the financial readiness of the project.

4 INDUSTRY RELATED INITIATIVES

In general, several industrial sharing sessions, industrial attachment, seminar, conference, technology showcase and development activities are among the common activities initiated to bridge the gap between academia and industry. The activities allow dissemination of the latest Press-in technology to both academia-industry by establishing R&D collaborations with local and international stakeholders in Malaysia construction industry. In addition, the activities triggered more industrial adaptation of this technology in this country.

Figure 23. Pile installation force monitoring setup.

the theoretical force. Some of these laboratory setup has been shown in Figure 20, 21, 22 and 23.

In summary, this simple research shows a similarity with the general trend of the recovery of skin friction due to time effect by Nozaki (2013). It is also useful to help the young engineering students to appreciate their knowledge on this particular

4.1 Method 4: Industrial related activities at the UTHM campus

UTHM is one of the technical universities in Malaysia. Currently, the university has slightly over than 1000 academic staffs and more than 17000 active students. The industrial awareness among the students is one of the primary concern of the university. Therefore, the university always welcome any activities to bridge the students with the industry. Several activities were initiated in Malaysia to widespread the concept of Press-in engineering and to disperse this sustainable agenda to this country.

"Steel Sheet-pile Symposium,, was one of the initiative. The symposium was organized by Technical Committee 3, International Press in Association and RECESS, UTHM on 6th December 2018 at Al-Jazari Auditorium, Tunku Tun Aminah Library, Universiti Tun Hussein Onn Malaysia (UTHM), Parit Raja, Batu Pahat, Malaysia. Close to 100 participants joined this symposium. They were consisted of academicians and students from Universiti Tun Hussein Onn Malaysia (UTHM), Universiti Teknologi Malaysia (UTM), Universiti Malaysia Pahang

Figure 27. A group photo of the participants after the symposium.

Figure 28. Mr. Heng Li presented press-in technology to the honorable Deputy Minister and UTHM Vice Chancellor.

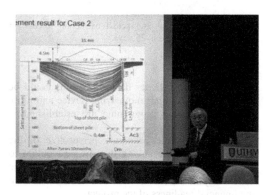

Figure 25. Prof. Jun Otani introducing partially floating sheet-pile concept.

Figure 29. Mr. Syahri Fuddin explaining the concept of tsunami protection.

Figure 26. Dr. Nor Azizi wrap up the symposium with his closing speech.

Figure 30. Undergraduate student, Nur Liyana bte Mohd Nor with Mr. Alvin at the KOYE office.

Figure 31. Nur Liyana operating the Standard Press-In machine for the first time.

Figure 32. A site visit at MRT construction in early 2013 utilizing a crush piler machine for the first time.

Figure 33. Conduction a presentation for a group of civil engineering consultant.

(UMP), Kolej Komuniti Batu Pahat, Kolej Kemahiran Tinggi MARA Sri Gading and RECESS. Several local companies also participated in this symposium. The speaker for the symposium involved Prof. Jun Otani, Mr. Yukihiro Ishihara and Prof. Katsutoshi Ueno from IPA. In addition, the local geotechnical expert, Prof. Ramli Nazir from Universiti Teknologi Malaysia and Dr. Nor Azizi Yusoff are also presented their related topics.

Figure 34. Mr. Takata from Giken Asia (Singapore) presented a Press-in technology to CREAM CEO.

In 2019, the Eid Gathering was initiated on the 13th June 2019 by Research Centre for Soft Soils (RECESS), Universiti Tun Hussein Onn Malaysia. The gathering were attended by our guest of honour YB. Datuk Dr. Shahruddin bin Mohd Salleh (The Deputy Minister of Federal Territories of Malaysia), Datuk Ir. Hj. Abdullah Isnin (Director General, Department of Irrigation and Drainage Malaysia) and Universiti Tun Hussein Onn Malaysia (UTHM) Vice Chancellor, Professor Datuk Ts. Dr. Wahid Razzaly. RECESS fellow researchers and students, UTHM top management and staff also attended the gathering. During the gathering, Mr. Heng Li and Mr. Takata from Giken Sesisakusho Asia (Singapore) Pte. Ltd. delivered a simple explanation of this technology to the guest of honour.

The concept of Press-in technology are also disseminated thru an exhibition. For example, the International Language Exhibition was organized to expose the UTHM students with the Japanese language, culture and some associated technology related to this country. In that exhibition, 'Silent Piling' applications were introduced by GIKEN to the visitors and fellow students.

Each group was assigned and given an ample time to create two replicas of tidal and tsunami defense system using straws and polystyrenes. Then, the presentation were made in Japanese language. Based on the activity, the students found that this technology is interesting and may potentially applicable for Malaysia scenario.

4.2 Method 5: Other industrial related activities outside UTHM campus

Several activities outside the UTHM campus have also been initiated to bridge the gap between the industry and the academia.

For example, a number of students had been attached to the real construction site for their final year project. The activity is needed in order to enhance their real experience on the technology. For that reason, a smart partnership with Press-in piling contractor (Koye (M) Pte. Ltd.) and many more industrial players were initiated several years ago. The partnership may potentially benefit both parties. For students, they could learn and experience themselves the technology. On the other hand, the industry will be able to get a helping hand to highlight their achievement on the ground.

In addition, many site visits have also been established since 2012 especially when there are new established sites using the technology. For now, most of the projects using this technology are more concentrated at Kuala Lumpur area. Normally the site visits are restricted only for a small groups. It is also bound with the regulations by the project owners. The site visit ends in approximately less than two hours. However, during this short period of time, the attendees are able to observe the real application of this technology in Malaysia. In addition, it is also useful as a networking sessions for all the parties.

In addition, many industrial sharing sessions and technology showcases were also initiated. Most of the time, the activity will involve UTHM, Giken Asia and the local press-in piling contractor. The idea is to make sure that the explanation will be able to accommodate the appropriate theoretical, practical and technological needs of the customer. For example, a technical talk was delivered at The Institution of Engineers Malaysia (IEM) Terengganu branch. The very similar presentation was also conducted at Construction Research Institute of Malaysia (CREAM). The event demonstrated RECESS commitment in facilitating CREAM with the cutting edge technology such as Press-in piling technology implementation in Malaysia.

5 CONCLUSIONS

In summary, the activities demonstrated a vibrant relationship and a decade of efforts between academia and industry in Malaysia in order to promote this sustainable technology. The relationship may allow both the academicians and the industry to capitalize on the strength of all. Since the establishment of a research related to the rate effect of press-in pile in 2007 at The University of Sheffield, United Kingdom, many technological transfer practices have been carried out until today. The activities allow dissemination of latest Press-in technology to both academia-industry by establishing R&D collaborations with local and international stakeholders in the Malaysia construction industry.

ACKNOWLEDGEMENTS

The authors are grateful to the International Press-In Association for the long-term support and assistance in the Press-In related issues, in particular it is worth to note the 5th IPA Research Grant Award. The authors appreciate fruitful cooperation with colleagues from Universiti Tun Hussein Onn Malaysia (UTHM) and Research Center for Soft Soil (RECESS). In addition, we would like to express our gratitude to our industrial collaborator, Giken Seisakusho Asia Pte. Ltd., KOYE (Malaysia) Sdn. Bhd., Construction Research Institute of Malaysia (CREAM), The Malaysian Public Works Department (JKR) and Department of Irrigation and Drainage Malaysia (JPS) for their contributions and assistance in preparing this research paper.

REFERENCES

Azhar, Nur Abidah (2018). Acceptance of Silent Piling Technology Among Local Authorities in Malaysia. Masters thesis, Universiti Tun Hussein Onn Malaysia.

Durdyev, S., & Hosseini, M. R. (2019). Causes of delays on construction projects: a comprehensive list. *International Journal of Managing Projects in Business.* 13(1): 20–46.

Emmett, K. (2007). Movement of Soil and Groundwater Around Piles in Layered Ground. The University of Sheffield: Ph.D thesis.

Hird, C.C. & Srishaktivel, S. (2005). Lasboratory Investigation of Permeability Measurement in Clay Using Outflow from Unsupported Cavities. Geotechnique, Vol. 55, No. 5, pp. 393–402.

Ibrahim, Z, Yusoff, N.A., Sufahani, S.F. and Azhar, n.a. (2015). A Quantitative Review On Acceptance Of Silent Piling Technology Among Designers In Malaysia. Applied Mechanics and Materials. ISSN: 1662-7482, Vols. 773-774, pp 1471–1475.

Meyerhof, G.G., (1976). Bearing Capacity and Settlement of Piles Foundations. Journal of the Geotechnical Engineering Division, ASCE, Vol. 102, No. GT, Mar., 1976, pp.197–288.

Mohd Aliff bin Mohd (2014). Recovery of Skin Friction with Time Effect in Sand. Universiti Tun Hussein Onn Malaysia (UTHM). Final year dissertation.

Nozaki (2013). Proposed Research for Each Country or Institutions. The 1st International Research Seminar for Research Works in ASEAN Countries. 10th July 2013. Kochi, Japan.

Srishaktivel, S. (2003). Laboratory Measurement of the Permeability of Clay Soils Assisted by a Self-Boring Device. The University of Sheffield: Ph.D thesis.

Tan, S. M. (2005). *Karstic Features of Kuala Lumpur Limestone.* Retrieved on April 20, 2020, from https://nrmt.files.wordpress.com/2011/04/kl-limestone-paper.pdf

Yun, Lew Wei(2020). The Press-in Method Assisted with Augering for Urban Transportation System Work in Malaysia. Case study: Mass Rapid Transit 2 (SSP Line). Universiti Tun Hussein Onn Malaysia (UTHM). Final year dissertation.

Yusoff, N.A. (2011). Effect of Rate of Shearing on Resistance in Fine Grained Soil. The University of Sheffield: Ph.D thesis.

*Proceedings of the Second International Conference on
Press-in Engineering 2021, Kochi, Japan – Matsumoto et al (eds)
© 2021 Taylor & Francis Group, London, ISBN 978-1-032-10414-0*

Author Index

9781032104164